Flavor Mixing in Weak Interactions

W0036863

ETTORE MAJORANA INTERNATIONAL SCIENCE SERIES

Series Editor:

Antonino Zichichi

European Physical Society
Geneva, Switzerland

(PHYSICAL SCIENCES)

Recent volumes in the series:

A Continuation Order Plan is available for this series. A continuation order will bring delivery of
each new volume immediately upon publication. Volumes are billed only upon actual shipment
For further information please contact the publisher.

Flavor Mixing in Weak Interactions

Edited by

Ling-Lie Chau

Brookhaven National Laboratory
Upton, New York

Plenum Press • New York and London

Library of Congress Cataloging in Publication Data

Europhysics Topical Conference on Flavor Mixing in Weak Interactions (1984: Ettore
    Majorana Center for Scientific Culture)
    Flavor mixing in weak interactions.

    (Ettore Majorana international science series. Physical sciences; v. 20)
    "Proceedings of the Europhysics Topical Conference on Flavor Mixing in Weak Inter-
actions, held March 5–10, 1984, at the Ettore Majorana Center for Scientific Culture,
Erice, Sicily, Italy"—T.p. verso.
    Includes bibliographical references and index.
    1. Weak interactions (Nuclear physics)—Congresses. 2. Quantum flavor dynamics—
Congresses. I. Chau, Ling-Lie. II. Title. III. Series.
QC794.8.W4E97   1984   539.7'54   84-26199
ISBN-13: 978-1-4612-9483-2   e-ISBN-13: 978-1-4613-2439-3
DOI: 10.1007/978-1-4613-2439-3

Proceedings of the Europhysics Topical Conference on Flavor Mixing in
Weak Interactions, held March 5–10, 1984, at the Ettore Majorana Center for
Scientific Culture, Erice, Sicily, Italy

©1984 Plenum Press, New York
Softcover reprint of the hardcover 1st edition 1984

A Division of Plenum Publishing Corporation
233 Spring Street, New York, N.Y. 10013

All rights reserved

No part of this book may be reproduced, stored in a retrieval system, or transmitted
in any form or by any means, electronic, mechanical, photocopying, microfilming,
recording, or otherwise, without written permission from the Publisher

ADVISORY COMMITTEE

K. C. CHOU, Academia Sinica
S. D. DRELL, SLAC
K. KIKUCHI, KEK
L. LEDERMAN, FERMILAB
S. Q. QIAN, Academia Sinica
C. RUBBIA, Harvard–CERN
J. A. RUBIO, University of Madrid
N. P. SAMIOS, BNL
H. SCHOPPER, CERN
V. SOERGEL, DESY
S. C. C. TING, DESY–MIT
S. WEINBERG, University of Texas/Austin
T. Y. WU, Academia Sinica
C. N. YANG, ITP, SUNY/Stony Brook

ORGANIZING COMMITTEE

N. CABIBBO, University of Rome
M. K. GAILLARD, University of California/Berkeley
M. JACOB, CERN
C. JARLSKOG, University of Bergen
M. KOBAYASHI, KEK
J. LEE-FRANZINI, SUNY/Stony Brook
M. POHL, ETH/Zurich
H. SUGAWARA, KEK
D. TADIC, University of Zagreb
B. H. WIIK, Universityof Hamburg
F. WILCZEK, ITP, University of California/Santa Barbara
F. YNDURAIN, University of Madrid

Conference Chairman: LING-LIE CHAU, BNL

Sponsored by Brookhaven National Laboratory, the European Physical Society, General Dynamics, the Italian Ministry of Education, the Italian Ministry of Scientific and Technological Research, and the Sicilian Regional Government

.

PREFACE

The 50-year history of weak interaction since Fermi's proposal of this coupling has been marked with striking direct interplays between experimental results and theoretical understanding, e.g. the discoveries of neutrinos, parity violation, and CP violation. The recent discoveries of the quark hierarchy, the charm and the beauty, and the intermediate vector bosons $W^{\pm}$ and Z^0 have truly made a splendid page in the history of particle physics. It is the purpose of this conference to discuss the questions of quark and lepton generations and mixing, their relations to CP violation, and to ask the questions about what are inside the quarks and the leptons in view of the present and future experimental situation.

Dr. Ling-Lie Chau
Brookhaven National
Laboratory
Upton, New York

ACKNOWLEDGMENTS

I would like to thank all the Advisory and Organizing Committee members for their advice and suggestions during the organization of the conference. The running of the conference could not have gone so smoothly without the help of many participants, I sincerely thank: L. Becker, F.J. Botella, S. Gentile, P. Le Comte, M.E. Machacek, L. Lanceri, W.M. Morse, F.J. Olness, Y.-X. Pham, G. Poulard, K.J. Sliwa, and J.N. Webb. It goes without saying that it would not have been possible for me to organize the conference and to finish the Proceedings without the work of the conference secretary, Mrs. Isabell Harrity. I am especially happy that she could participate in the conference at Erice. My thanks also go to Mrs. Barbara Kponou, Mrs. Patricia Lebitski, and Mrs. Kathleen Touhy, who gave help and support during a critical period of organizing the conference. The excellent work by the Technical Photography and Graphic Arts Division and the efficient services provided by the various Brookhaven staff are greatly appreciated.

It is my great pleasure to thank Dr. S.A. Gabriele and Signora Pinola, and the staff, who have made the Centre an excellent place to have the conference.

Finally, I would like to thank Prof. A. Zichichi and the Europhysics Society for inviting me to organize the conference, and to prepare the Proceedings. They are very trying yet satisfying tasks.

CONTENTS

CONFERENCE PROGRAM

	Chairman	Speaker
Monday March 5	L.-L. Chau	N. Cabibbo P.H. Hansen H.W. Siebert N. Cabibbo
	P. Franzini	G.H. Trilling W.T. Ford P.R. Avery
Tuesday March 6	M. Goldhaber	F.H. Boehm L. Lanceri T.H. Ho L. Wolfenstein M.A.B. Bég
	G. Trilling	J. Lee-Franzini M. Pohl A. Barbaro-Galtieri
Wednesday March 7	R. Cester	G. Bellini K. Niu D. Hitlin S. Ferrara
Thursday March 8	M. Serdaroglu	G.L. Kane R.N. Mohapatra K.C. Wali K. Kleinknecht
	L.-L. Chau	M. Goldhaber P. Franzini G.H. Trilling M. Kobayashi V. Soergel
Friday March 9	V. Soergel	L.S. Littenberg B. Heckel J.S. Hagelin W.M. Morse
	G. Rajasakaran	P. Pavlopoulos K. Kleinknecht K.C. Chou
Saturday March 10	K.-C. Chou	G.C. Segre D.V. Nanopoulos H. Terazawa R. Rückl
	M. Kobayashi	H. Fritzsch B.W. Stech L.L. Chau

OPENING REMARKS

Ling-Lie Chau

Department of Physics
Brookhaven National Laboratory
Upton, New York

With so many friends here, this is clearly a conference on the family problem. We certainly have a complete mixing of generations here. I am so happy to see you all.

I consider that we are fortunate to live in this very interesting period of time. Looking back, the concept of generation mixing was introduced in '63 by Nicola, passing through many rugged tests, as you shall hear from Siebert's talk, the Cabibbo theory, like an aged wine, is better than ever. In '73, a gap of ten years, Kobayashi-Maskawa introduced the beauty and truth quark pair, long before even the charm was found, purely based upon a theoretical requirement that CP violation comes from a simple origin. The discovery of the beauty is a triumph of this clear logical reasoning. There is no better place that we discuss these questions than here in Sicily where scholarly achievements have been treasured ever since its ancient days. Another ten years have passed since the proposal of Kobayashi-Maskawa, the milestone has been the much better understood mixing matrix through the measurements of the b life-time. We may well be at the threshold of better understanding of the question of CP violation, and the dynamics of weak decays.

A fundamental question we must face is, besides the truth quark, are there higher generations? Does this proliferation of quarks really indicate that after all, like protons and neutrons, the quarks are not as fundamental as we thought. All these questions we shall examine during the course of this conference. With so many hardworking and joyous participants, I am sure that the conference will not only be productive but also will be fun.

PRODUCTION OF HIGH MASS eν AND e+e- PAIRS IN

THE UA2 EXPERIMENT AT THE CERN p̄p COLLIDER

The UA2 Collaboration

Presented by Peter Hansen

Cern, Geneva, Switzerland

ABSTRACT

We present new results on Intermediate Vector Boson production at the CERN p̄p collider. A comparison is made with the predictions of the standard model. The observation of high p_T electrons produced in association with hard jets and large missing transverse momentum is presented and discussed.

1. INTRODUCTION

We report here the results from a search for electrons with $p_T > 15$ GeV/c produced at the CERN $\bar{p}p$ collider ($\sqrt{s} = 540$ GeV) during its 1982 and 1983 periods of operation.

Following a general discussion of the topology of the events containing an electron candidate, we shall compare the data with expectations in the framework of the standard model [1] for the reactions:

$$\bar{p} + p \rightarrow W^{\pm} + anything \tag{1}$$
$$\rightarrow e^{\pm} + \nu \, (\bar{\nu})$$

$$\bar{p} + p \rightarrow Z^0 + anything \tag{2}$$
$$\rightarrow e^+ + e^- \quad or \quad e^+ + e^- + \gamma$$

We also report on the observation of high-p_T electrons produced in association with hard jets and large missing transverse momentum. These events are found difficult to understand in terms of standard QCD production processes.

Preliminary results from the study reported here have already been presented elsewhere [2] and a more complete discussion can be found in recent publications [3].

2. THE DETECTOR

The experimental apparatus, shown in Fig.1, has been described in detail elsewhere [4]. At the centre of the apparatus a system of cylindrical chambers (the vertex detector [5]) measures charged particle trajectories in a region without magnetic field. The vertex detector consists of : a) four multi-wire proportional chambers, (C1 to C4), having cathode strips with pulse height read-out at ±45° to the wires ; b) two drift chambers with measurement of the charge division on a total of 12 wires per track. These chambers are used to obtain both tracking information and to evaluate the most likely ionisation I_0 associated with each track. From the reconstructed tracks the position of the event vertex is determined with a precision of ±1 mm in all directions.

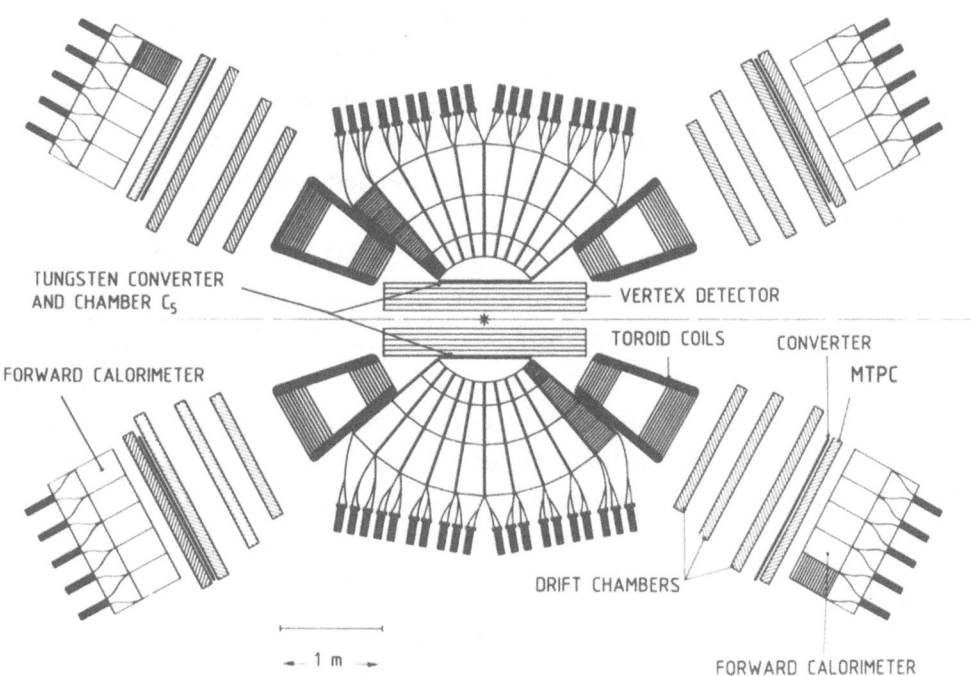

Fig.1 A view of the UA2 detector in a plane containing the beam line.

The vertex detector is sourrounded by an electromagnetic and hadronic calorimeter (central calorimeter [6]), which covers the full azimuth and a polar angle interval $40° < \theta < 140°$. The calorimeter is segmented into 240 independent cells, each covering $10°$ in θ and $15°$ in ϕ and built in a tower structure pointing to the centre of the interaction region. The cells are segmented longitudinally into a 17 radiation length thick electromagnetic compartment (lead-scintillator) followed by two hadronic compartments (iron-scintillator) of ~ 2 absorption lengths each.

In the angular region covered by the central calorimeter a cylindrical tungsten converter, 1.5 radiation lengths thick, followed by a cylindrical proportional chamber (C5), is located just after the vertex detector. This device localises electromagnetic showers initiated in the tungsten with a precision of ±3 mm, as verified using test-beam electrons.

For the first 15 nb^{-1} of integrated luminosity, collected during the Autumn of 1982, the azimuthal coverage of the central calorimeter was only 300°. The remaining interval (±30° around the horizontal plane) was covered by a magnetic spectrometer which included a lead-glass array to measure charged and neutral particle production [7].

The two forward regions (20° < θ < 37.5° and 142.5° < θ < 160°) are each equipped with twelve toroidal magnet sectors with an average bending power of .38 Tm. Each sector is instrumented with :

a) three drift chambers [8] located after the magnetic field region. Each chamber contains three planes, with wires at -7°, 0° and +7° with respect to the magnetic field direction.

b) a 1.4 radiation lengths thick lead-iron converter, followed by a preshower counter which consists of two pairs of layers of 20 mm diameter proportional tubes (MTPC), staggered by a tube radius and equipped with pulse height measurement [9]. This device localises electromagnetic showers initiated in the converter with a precision of ±6 mm.

c) an electromagnetic calorimeter consisting of lead-scintillator counters assembled in ten independent cells, each covering 15° in φ and 3.5° in θ. Each cell is subdivided into two independent longitudinal sections, 24 and 6 radiation lengths thick, respectively, the latter providing rejection against hadrons.

The systematic uncertainty in the energy calibration of the electromagnetic calorimeters for the data presented here amounts to an average value of ±1.5%. The cell-to-cell calibration uncertainty has a distribution with a r.m.s. of 2.2%. The energy resolution for electrons is measured to be $\sigma_E/E = 0.14/\sqrt{E}$ [6] in the central calorimeter and $0.17/\sqrt{E}$ in the forward ones (E in GeV).

3. DATA TAKING AND DATA REDUCTION

In order to implement a trigger sensitive to electrons of high transverse momentum, the photomultiplier gains in all calorimeters were adjusted so that their signals were proportional to the transverse energy.

Because of the cell dimensions, electromagnetic showers initiated by electrons may be shared among adjacent cells. Trigger thresholds were applied, therefore, to linear sums of signals from matrices of 2 × 2 cells, rather than to individual cells.

A signal was generated whenever the linear sum from at least one such matrix exceeded a threshold which was typically set at 8 GeV. To suppress background from sources other than $\bar{p}p$ collisions, we required a coincidence with two signals obtained from scintillator hodoscopes covering the polar angle interval 0.47° - 2.84° with respect to the beams on both sides of the collision region. These hodoscopes, which were part of an experiment to measure the $\bar{p}p$ total cross-section [10], gave a coincidence signal in more than 98% of all non-diffractive $\bar{p}p$ collisions.

Approximately 7×10^5 triggers were recorded during the 1982 and 1983 runs, corresponding to an integrated luminosity $\mathscr{L}$ = 131 nb^{-1}.

A first data reduction is made by requiring the presence of an energy cluster with a transverse energy greater than 15 GeV. In the central, calorimeter clusters are obtained by joining all electromagnetic cells wich share a side and contain at least 0.5 GeV. The forward calorimeter clusters consist of at most two adjacent cells having the same azimuth (here the cell is far from the interaction point and much larger than the lateral extension of an electromagnetic shower, but the dead region between cells at different azimuth does not allow clustering across it).

In the surviving events,a search is made for configurations consistent with the presence of an electron among the collision products.An electron is identified from the observation of:

a) the presence of a cluster of energy deposition in the first compartment of the calorimeters with a small lateral size and a small energy leakage in the hadronic compartment.

b) the presence of a reconstructed charged particle track which points to the energy cluster. The pattern of energy deposition must agree with that expected from an isolated electron incident along the track direction.

c) the presence of a hit in the preshower counter, with an associated pulse height larger than that of a minimum ionising particle (m.i.p.). The distance of the hit from the track must be consistent with the space resolution of the counter itself.

A set of cuts has been defined to satisfy these requirements. A detailed description of these cuts can be found in ref. 3 .

The efficiencies of the simultaneous application of these cuts are 76% and 80% in the central and forward regions respectively.

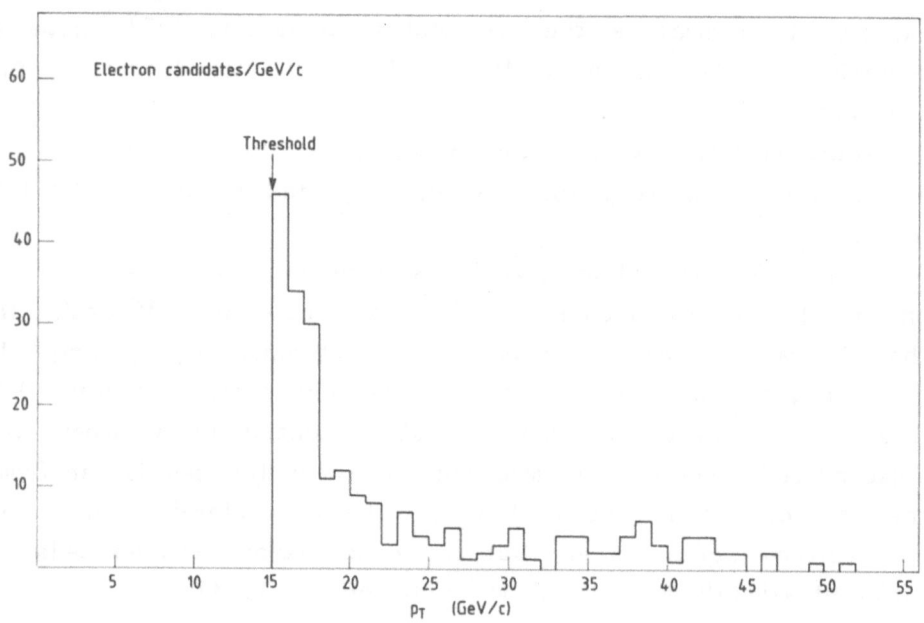

Fig.2 Transverse momentum distribution of the 225 electron candidates satisfying the electron cuts.

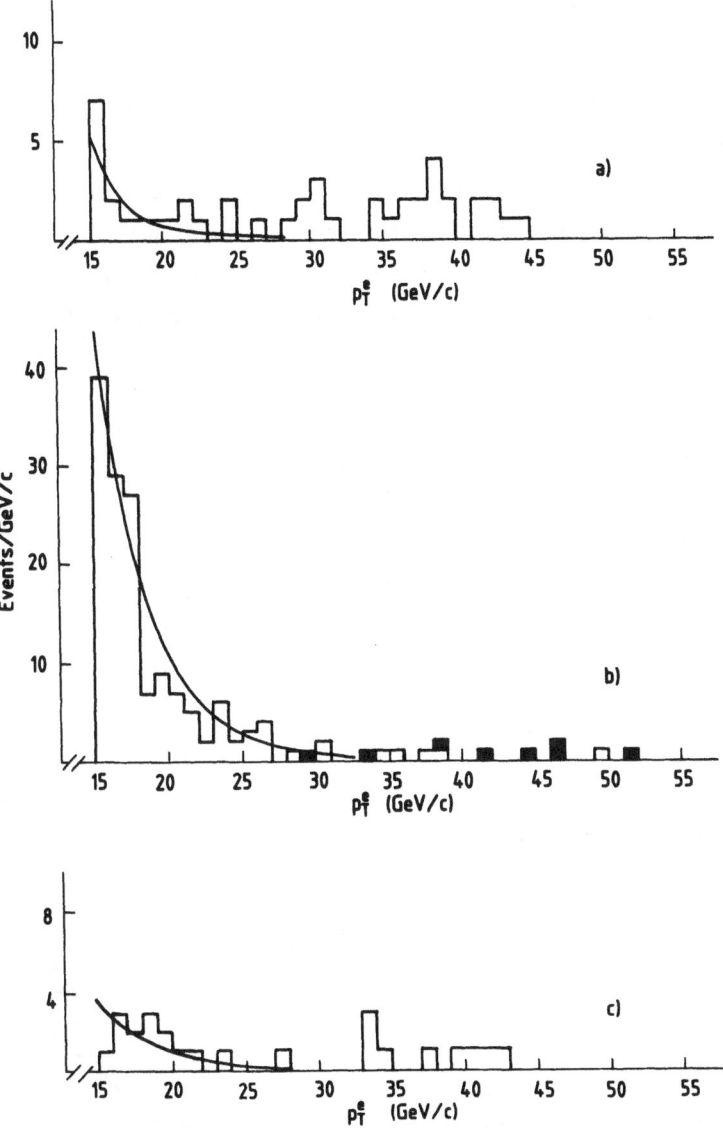

Fig.3 a) Transverse momentum distribution of the electron candidates in events with no additional jet. b) events having $\rho_{opp} > 0.2$. Dark points corresponds to electrons from Z^0 decays. c) events with additional jets having $\rho_{opp} < 0.2$.

4. TOPOLOGY OF EVENTS CONTAINING AN ELECTRON CANDIDATE.

After application of the electron cuts the sample consists of 225 events. Their p_T distribution is shown in Fig.2 .

Some of these contain a genuine electron and some contain a fake one resulting from misidentification of a hadron jet.

The background of fake electrons can be shown to fall mainly into two categories. In approximately 70% of the cases we are

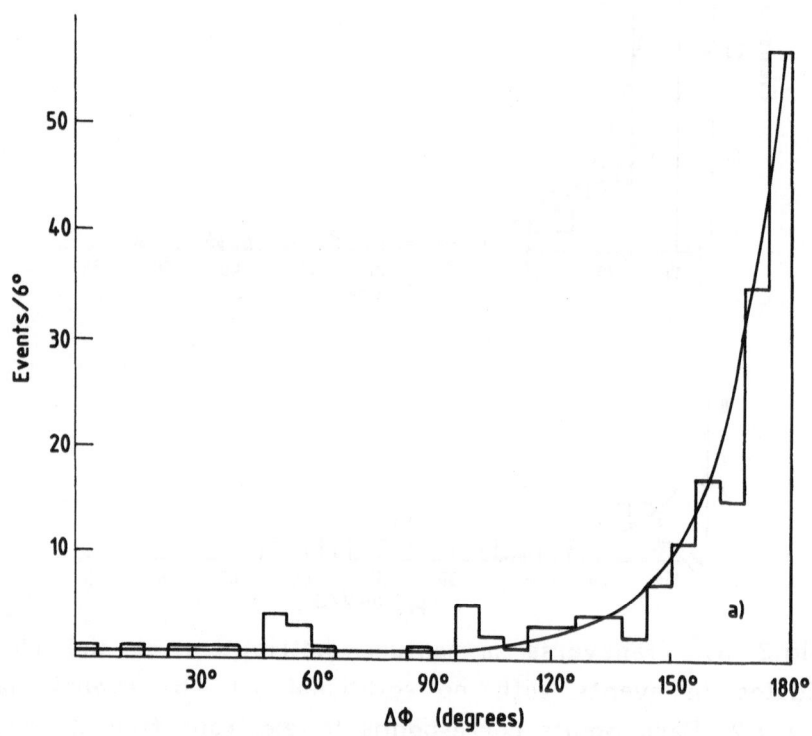

Fig.4 Distribution of the azimuthal separation $\Delta\phi$ between the electron candidate and the associated jet having the highest p_T. Curve : estimated background.

dealing with "overlaps", i.e. jets fragmenting into a hard π^0 with a charged pion nearby in angle. The rest of the background results from π^0s undergoing Dalitz decays or conversions in the beam-pipe.

From studies of hadron jets [11] we expect the fake electrons to be accompagnied by other high-p_T jets at approximately opposite azimuth. We shall therefore search for high-p_T jets by grouping together adjacent cells with energy into clusters using an algorithm described elsewhere [11]. Clusters with more than 3 GeV of transverse energy are retained and called jets. For minimum bias triggers such clusters occur in only 15% of the events.

We find that 45 events contain no jet, the electron candidate being the only high-p_T particle observed. The p_T distribution of these is shown in Fig.3a . Such events contain either a neutrino, as in the case of $W \rightarrow e\nu$ decay, or other high-p_T particles having escaped detection. In the latter case we would expect a rapidly falling p_T-spectrum typical of the jet p_T distribution [11].

The remaining 180 events contain at least one jet (according to our definition), and Fig.4 shows the azimuthal separation of the highest p_T jet from the electron candidate. The events in the peak at $\Delta\phi = 180°$ are likely to be background. Hence we sum all transverse momenta of clusters having $|\Delta\phi| > 120°$ and define

$$p_{opp} = -\bar{p}_T^{\,e} \cdot \Sigma \bar{p}_{T,jet} \Big/ |\bar{p}_T^{\,e}|^2 \qquad (3)$$

The distribution of p_{opp} is shown in Fig.5 . We reject the 156 events having $p_{opp} > 0.2$. Their $p_T^{\,e}$ distribution is shown in Fig.3b . This sample contains the eight $Z^0 \rightarrow e^+e^-$ (or $e^+e^-\gamma$) events recently reported [12].

The $p_T^{\,e}$ distribution of the remaining 24 events is shown in Fig.3c . These events do contain , in addition to the electron candidate, jets which do not carry a significant transverse momentum in the direction opposite to the electron ($p_{opp} < 0.2$).

5.   BACKGROUND TO THE ELECTRON SPECTRA

For the present we shall regard the 148 events containing a jet back to back with the electron candidate as a background sample of fake electrons (after removal of the eight Z^0 events).

These events occur at a rate lower than that of jet production by a factor of 3.6×10^{-4} for the central detector and 2.6×10^{-5} for the forward ones.

The backgrounds to the two event samples of Fig.3a and 3c are now estimated as follows. We consider a sample of events containing a manifestly fake electron not surviving the cuts. These events are now divided into three classes of different topologies, called samples A, B and C, corresponding to the three classes of electron candidates of Fig.3 .

The ratio between the p_T distributions of samples B and A gives directly the probability to miss a jet because of the incomplete coverage of the detector. We find that this probability decreases from 10% to 2% when p_T increases from 15 GeV/c to 25 GeV/c. The p_T^e distribution of the 148 events in the background sample multiplied by this probability now gives the background under Fig.3a . Likewise we estimate the background under Fig.3c by using the ratio between samples C and A. These background estimates are shown as smooth curves in the figures.

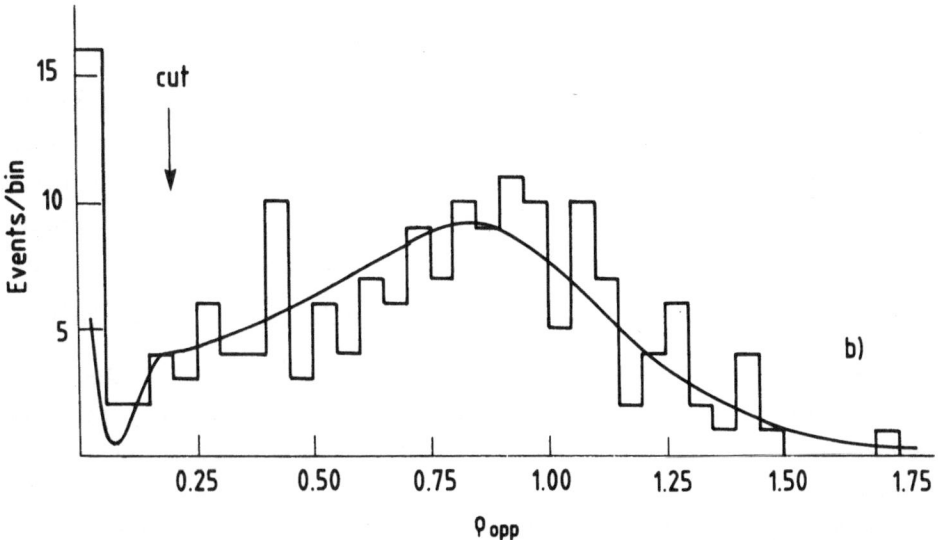

Fig.5 Distribution of ρ_{opp} for events containing an electron candidate and at least one associated jet. Curve : background estimate.

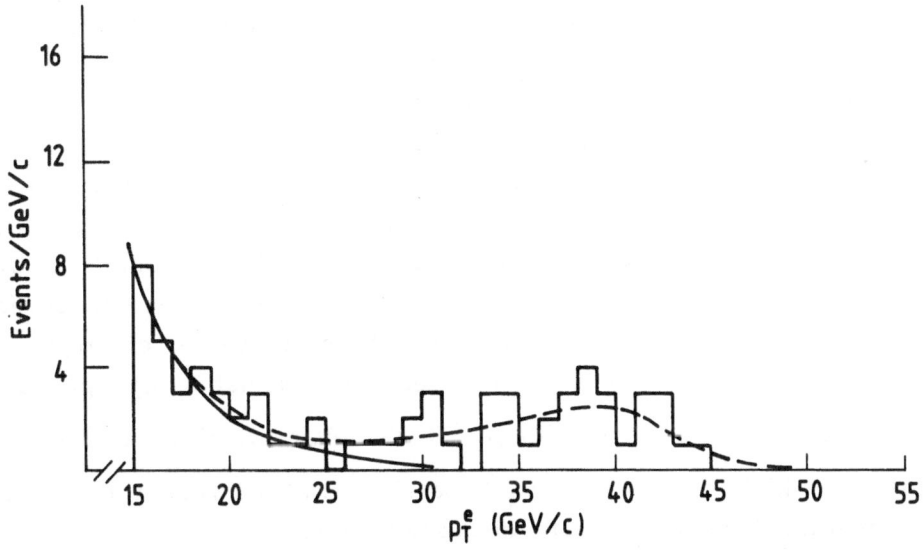

Fig.6 Transverse momentum distribution of the electron candidates in the W → ev event sample. Full curve : background estimate. Dashed curve : sum of all the contributions as discussed in sec.7.

6. THE W → Eν EVENT SAMPLE

The combined p_T^e distribution of the 69 events from Fig.3a and Fig.3c is shown in Fig.6 together with the background estimate. The presence of a significant signal above the background is taken as evidence that most of the electron candidates here are indeed electrons and associated with a high-p_T neutrino. There is a clear accumulation of events near p_T^e = 40 GeV/c, which is distinctive of the Jacobian peak expected for W → eν decay. For p_T^e > 25 GeV/c the distribution contains 37 events with an estimated background of 1.5 ± 0.1 events that we shall consider in the following.

We estimate the transverse momentum $\bar{p}_T^W$ of the W from

$$\bar{p}_T^W = - (\bar{p}_{T,jets} + \xi\, \bar{p}_T^{sp}) \tag{4}$$

where p_T^{sp} is the total p_T of all observed particles not belonging to jets. In an ideal detector ξ would be one. For an incomplete coverage some fraction of the particles in the event is lost (typically among the low transverse momentum particles). We estimate, using the eight Z^0 events, that ξ should be 2.2 ± 0.5 in order to satisfy Eq.(4) on the average.

The distribution of p_T^W is shown in Fig.7a and 7b using ξ = 1 and ξ = 2.2 respectively. The mean value (for ξ = 2.2) is $\langle p_T^W \rangle$ = 6.9 ± 1.0 GeV/c. QCD predictions [13], illustrated by the curve on the figure, are consistent with the observed distribution. The event with the highest value of p_T^W, 29.6 GeV/c, is interpreted as a W recoiling against a high-p_T jet. Among the W candidates with p_T^e < 25 GeV/c we find two events with even higher values for p_T^W clearly separated from the rest of the events. These events are discussed later.

We note that the level of activity in the W events is about twice as high as that in minimum bias events. This feature is reflected in the observed sum of transverse energies $\langle \Sigma E_T \rangle$ = 14.1 ± 1.7 GeV and the average value of the charged

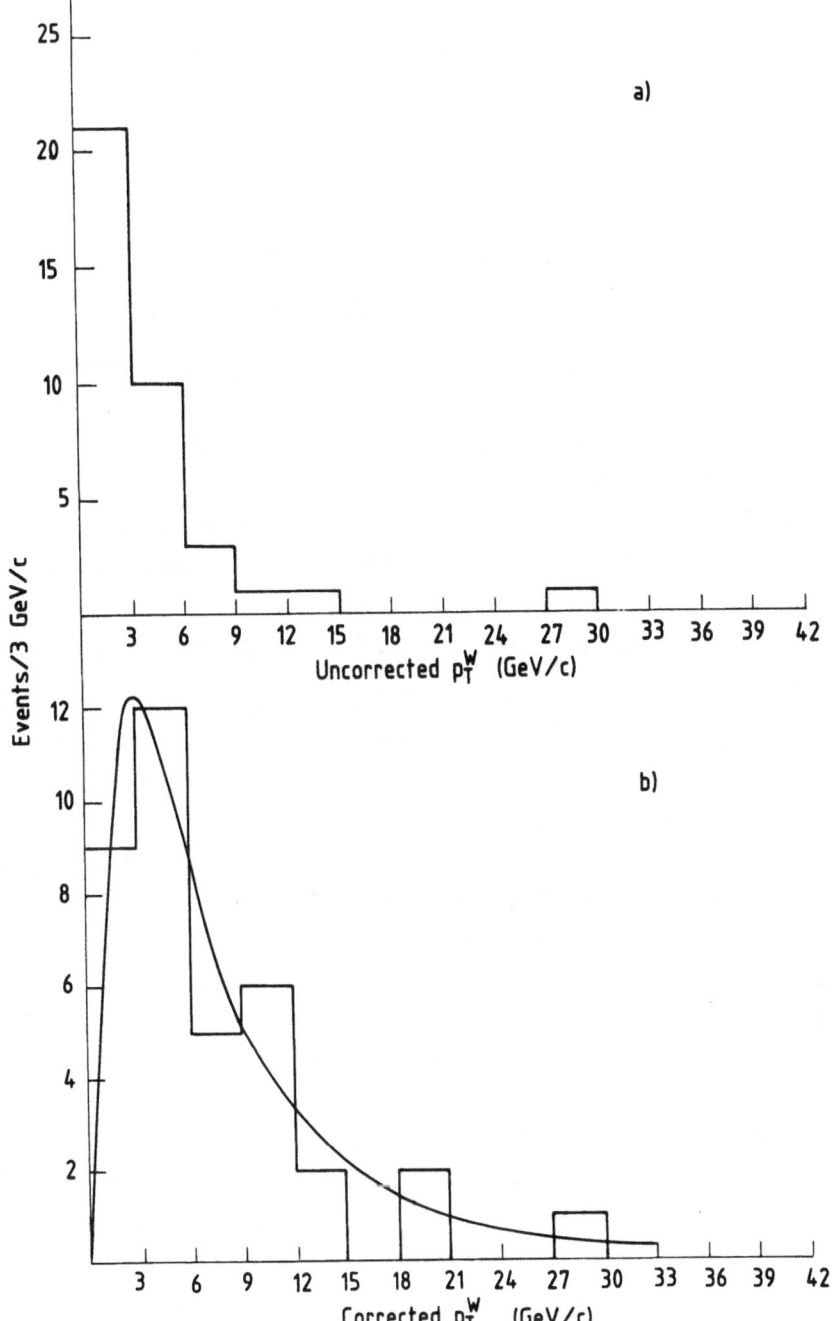

Fig.7 a) Uncorrected distribution of the W transverse momentum (from Eq.(4) with ξ = 1)

b) Corrected distribution (from Eq.(4) with ξ = 2.2). The curve represent a QCD prediction [13].

multiplicity $\langle n_{ch} \rangle$ = 19.3 ± 1.9 in the underlying event to the Ws. The corresponding values for minimum bias events are $\langle \Sigma E_T \rangle$ = 8.8 GeV and $\langle n_{ch} \rangle$ = 14.7.

7.   DETERMINATION OF THE W MASS

The mass M_W of the W is determined from the W → ev candidates with p_T^e > 25 Gev/c by performing a maximum likelihood fit to their two-dimensional distribution $d^2n/dp_T^e d\theta_e$, where θ_e is the measured electron polar angle. In the Monte Carlo program used to generate the distribution $d^2n/dp_T^e d\theta_e$ for different values of M_W, we make the following assumptions:

a) the W longitudinal momentum distribution is obtained using the quark structure functions of the proton (antiproton) with scaling violation, as given by Glück et al. [14]

b) the W transverse momentum p_T^W distribution is taken from ref. [13] allowing for variations of $\langle p_T^W \rangle$ in order to take into account uncertainties of QCD predictions

c) the W → ev decay angular distribution is described by the standard V-A coupling

d) the M_W distribution is generated according to a Breit-Wigner curve with a fixed value of the W width, Γ_W = 2.7 GeV/c² .
The detector response is taken into account.

The full sample of 37 events is contaminated by three background sources:

a) misidentified electrons from two jet background: 1.5 ± .1 events

b) electrons from Z^0 → e^+e^- decay with one electron undetected: 2.5 ± .9 events. This contribution is evaluated by Monte Carlo technique normalizing the result to the total number of Z^0 → e^+e^- detected in this experiment (8 events , ref. 12)

c) electrons from W → τν, τ → $e\bar{\nu}_e \nu_\tau$ decay chain : .9 ± .1 . The detector acceptance for electrons from W → τ → e decay chain has been evaluated by Monte Carlo technique, assuming a branching ratio B_τ = 0.17 for the decay τ → $e\bar{\nu}_e \nu_\tau$ [15]

The sum of all contributions is shown as a dashed curve in Fig.6 . After subtraction of the background events , we are left with a sample of 32.1 ± 6.0 W → ev decays. The best fit to the experimental $d^2n/dp_T^e d\theta^e$ distribution , where the mentioned background contributions are included , gives:

$$M_W = 83.1 \pm 1.9 \text{ (stat.)} \pm 1.3 \text{ (syst.) GeV/c}^2$$

An uncertainty of ±1 GeV/c^2 , which results from the effect of varying $\langle p_T^W \rangle$ between 4 and 10 Gev/c in the fit , is added in quadrature to the statistical error . The systematic error reflects the uncertainty in the overall mass scale arising from the absolute calibration of the calorimeter (±1.5%) and from small differences in the relative calibration of various cells of the calorimeter.

8.   CROSS SECTION FOR INCLUSIVE W PRODUCTION

The cross section σ_W^e for the inclusive process $\bar{p}p \to W^{\pm} + x$, followed by the decay W → ev, at √s=540 GeV is obtained from the relation:

$$N_W^e = \mathscr{L} \, \sigma_W^e \, \varepsilon \, \eta$$

where N_W^e = 32.1 ± 6.0 is the number of W → ev decays obtained by subtracting from the electron sample the background events as described in the previous section, $\mathscr{L}$ = 131 nb^{-1} is the integrated luminosity, ε = .64 ± .01 is the detector acceptance wich includes the effect of the p_T^e threshold, and η = .77 ± .05 is the overall efficiency of the electron identification criteria averaged over the central and forward detectors. The effect of the cut $p_{opp} < .2$ previously described is taken into account in the acceptance evaluation.   Finally we obtain

$$\sigma_W^e = .53 \pm .10 \text{ (stat.)} \pm .10 \text{ (syst.) nb}^{-1}.$$

where the systematic error reflects a ±20% uncertainty in the knowledge of $\mathscr{L}$. This value is in agreement with QCD predictions [13,16] and with the results of the UA1 experiment [17] .

9.   CHARGE ASYMMETRY

It is known that, as a consequence of the V-A coupling, the W is always produced with full polarisation along the direction of the incident $\bar{p}$ beam, and a distinctive charge asymmetry can be observed in the decay W → ev. In the W rest frame the angular distribution has the form $(1 + \cos\theta^*)^2$ for electrons and $(1 - \cos\theta^*)^2$ for positrons, where θ^* is the angle between the charged lepton and the direction of the incident protons. However, precisely the same configuration would result from V+A coupling because in this case all helicities change sign. In order to maintain full generality and allow for different amounts of V and A couplings we write the angular distribution in the form

$$dn/d(\cos\theta^*) \propto (1-q \cos\theta^*)^2 + 2 q \alpha \cos\theta^* \tag{5}$$

where q is -1 for electrons and +1 for positrons, and $0 \leq \alpha \leq 2$ depending on the ratio x between the A and V couplings (time reversal invariance requires x to be real). Under the assumption that x is the same for both Wqq and Wev couplings, α is given by

$$\alpha = [(1-x^2)/(1+x^2)]^2 \tag{6}$$

which gives $\alpha = 0$ for $|x| = 1$.

In the UA2 detector a determination of the charge sign is only possible in the forward detectors where a magnetic field is present.

We consider the sample of 8 events which have an electron with $p_T^e > 20$ Gev/c in the forward detectors.   The estimated background is 0.2 events. A comparison between the electron momentum p and energy E, as measured in the calorimeter, is made

in Fig.8 , which shows the position of these events in the plane
(p^{-1}, E^{-1}), where p is the electron momentum with the sign of the
product q $\cos\theta_e$ (θ_e is the laboratory angle of the electron with
respect to the proton direction). The horizontal error bars in
Fig.8 represent the uncertainty on the measurement of p^{-1}, which
is 0.01 $(GeV/c)^{-1}$.

A clear asymmetry is visible in Fig.8 , with all the events
lying on one side of the plot. In order to extract a value of x
from these data we compute two-dimensional distributions
$f^{\pm}(p_T^e, \theta_e)$ for positrons and electrons separately, using Eqs. (5)
and (6) and taking into account the W longitudinal motion.

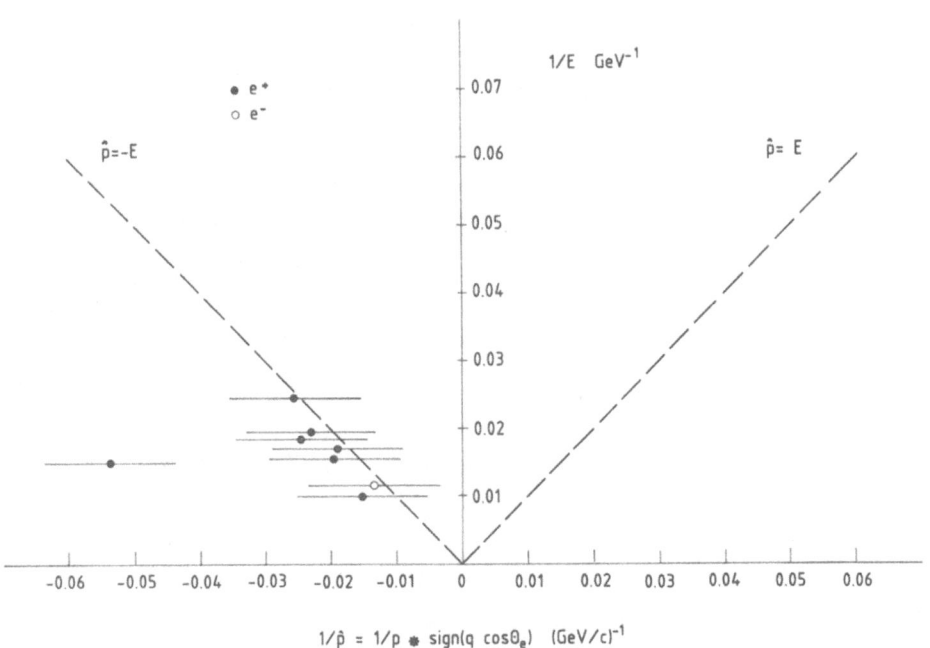

Fig.8 Plot of 1/E vs 1/p for the eight W $\to$ ev candidates with
p_T > 20 GeV/c detected in the forward regions. The quantity p is
the electron momentum with the sign of the product qcosθ where
q=+1(-1) for e⁺(e⁻).

To each event we assign a likelihood $Q_i = f^+\eta^+ + f^-\eta^-$, where $\eta^+(\eta^-)$ is the probability density that the observed particle was a positron (an electron) of momentum p, and the functions $f^{\pm}$ are calculated at the observed values of p_T^e and θ_e. The probability densities $\eta^{\pm}$ reflect the uncertainty in the determination of the charge sign resulting from the error in the momentum measurement. Maximizing the likelihood $\Pi_i Q_i$ we obtain $x = 1.0$ for the ratio between the strengths of the A and V couplings.

We remark that our event sample consists of seven positrons detected in one hemisphere and one electron detected in the opposite one. The probability to observe no more than one electron in one of the two hemispheres is ~ 7%.

10.   ELECTRONS ACCOMPAGNIED BY HARD JETS AND LARGE MISSING P_T

In the analysis described in sec.4 we have rejected events in which a significant amount of transverse energy was detected at opposite azimuth to the electron candidate. This procedure was very powerful to reject background coming from two-jet events in which one of the jets was misidentified as an electron.   Other events, containing a genuine electron, may have also been rejected by this cut, in particular Ws produced at very high p_T^W.

We then reanalyze the sample described in sec.4 releasing any constraint on the event topology.

To compensate for the loss of rejection power against background we ask that both the electron and the "neutrino" have large transverse momenta.

In the following the word "neutrino" must be understood in a broad sense, the detector being unable to distinguish between a neutrino and other possible non-interacting particles, such as the photino postulated by supersymmetric theories [18].

The initial sample of 225 events is then reduced discarding events collected during the 1982 period, when the angular coverage of the central calorimeter was incomplete. We also discard events in which the electron candidate is observed near the interface between the central region and one of the forward ones. In this case our electron identification power is reduced by the presence of dead regions.

The previous sample is then reduced to 190 events and the corresponding integrated luminosity to 116 nb⁻¹.

We apply our clustering algorithm [3] to the events and then calculate the momentum $\bar{p}$ and energy E of the electron candidate (e) and of each individual jet (j) assumed to be massless. We also evaluate the momentum $\bar{p}$, energy E and mass m of various sets of particles such as the system of all jets (J) or the system of electron and jets (Je). The quantity p_T(Je) measures the missing transverse momentum in the set of particles resulting in clusters having $E_T > 3$ GeV.

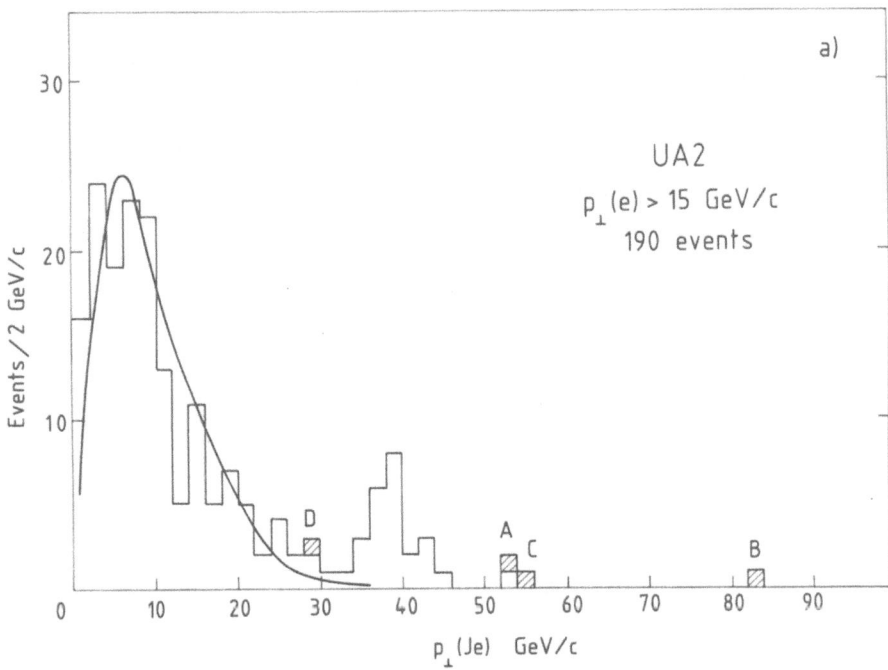

Fig.9 Transverse momentum distribution of the system of electron and jets in the initial sample of 190 events. Curve : background estimate

Given that soft particles carry only a small transverse momentum, if no high p_T particle or system of particles has escaped detection, the presence of a large $p_T(Je)$ reveals the presence of a neutrino (v) with $p_T(v) \approx p_T(Je)$. The distribution of $p_T(Je)$ is shown in Fig.9. The solid curve represents the background evaluation obtained in the same way as described in sec.5. In the following we restrict the analysis to the sample of 35 events with $p_T(Je) > 25$ GeV/c which are candidate for containing a high p_T neutrino in the final state. The background under these events is estimated to be 3.4 ± 0.3 events.

The distribution of the sum of the jet transverse energies is shown in Fig.10. The 31 events with $E_T(J) < 30$ GeV all belong to the samples of Fig.3a and 3c and were already interpreted in terms of W → ev decays.

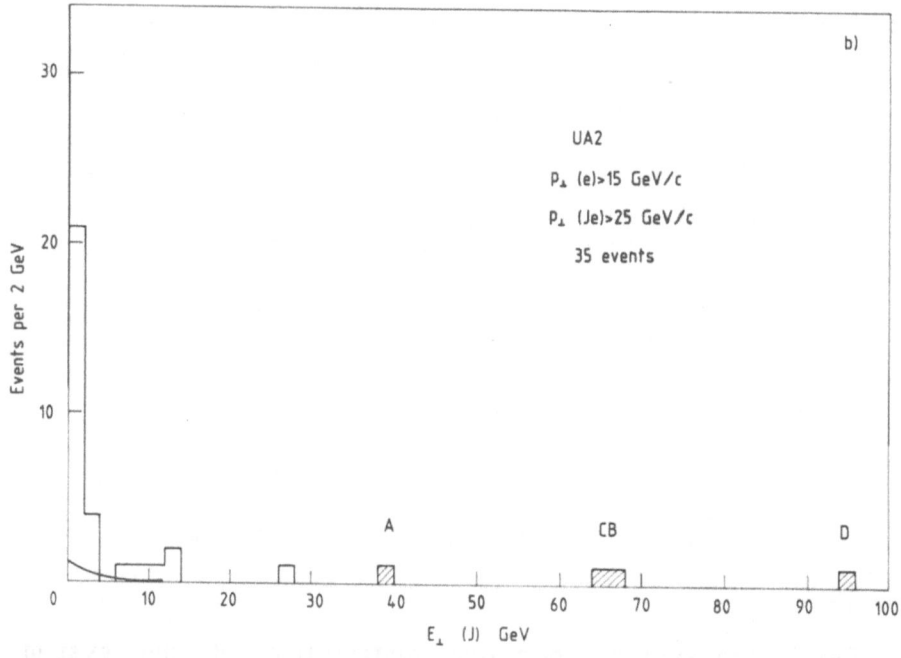

Fig.10 Transverse energy distribution of the system of jets in the sample of 35 events having $p_T(Je) > 25$ GeV/c. Curve : background estimate.

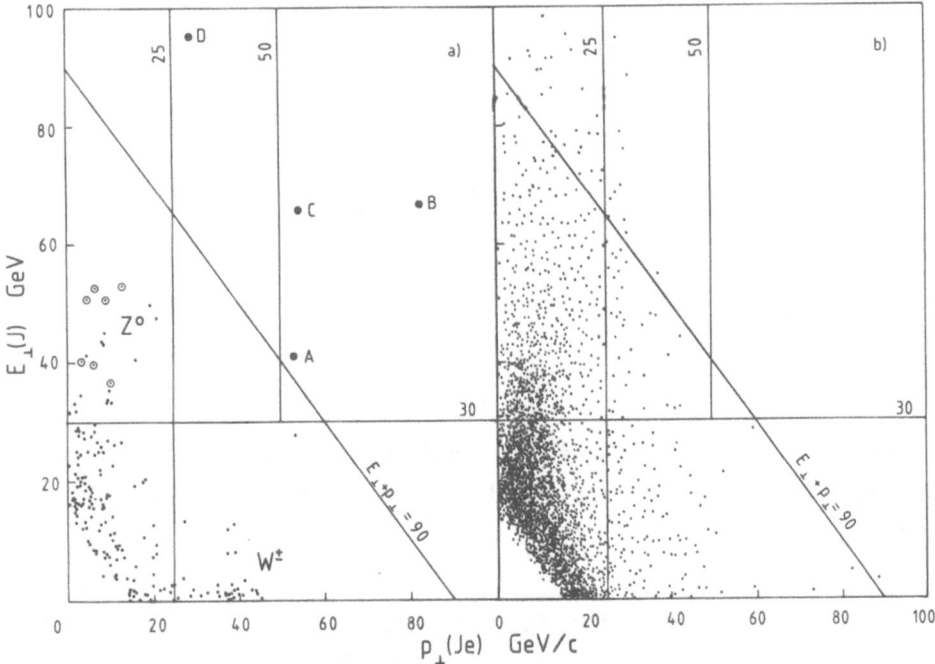

Fig.11 a) Distribution of the 190 events of the initial sample in the $p_T(Je)$, $E_T(J)$ plane. Z^0 events are circled.
b) Distribution of the background sample in the $p_T(Je)$, $E_T(J)$ plane. A reduction factor of 141 must be applied to infer from this sample the background contamination to the sample of Fig.11a).

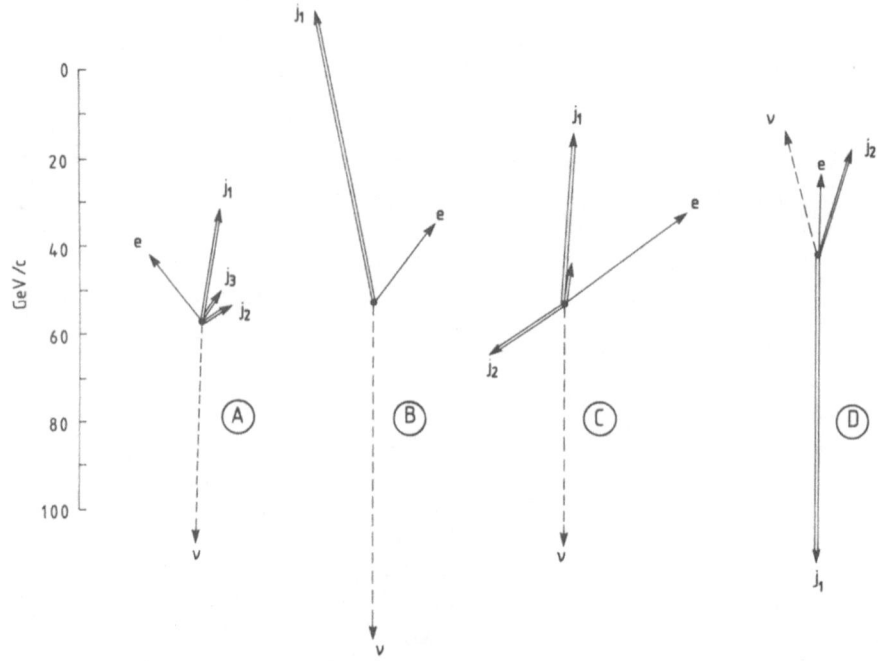

Fig. 12 Transverse momentum configurations of the four events having $p_T(Je) > 25$ GeV/c and $E_T(J) > 30$ GeV. Their relative orientation is arbitrary.

Let's now consider the four events with $E_T > 30$ GeV. The distribution of these events in the plane $p_T(Je), E_T(J)$ is shown in Fig. 11 for each of the signal and the background samples. The background contamination expected in the region $p_T(Je) > 25$ GeV/c, $E_T(J) > 30$ GeV is $\approx 0.45 \pm 0.4$ events.

The transverse plane configuration of the four events is shown in Fig. 12.

For each event, in which $p_T(Je)$ and $E_T(J)$ take values p_o and E_o respectively, we evaluate the expected background contaminations B_1 and B_2 corresponding to the following configurations:

$$B_1 : p_T(Je) > p_o , E_T(J) > 30 \text{ GeV}$$
$$B_2 : p_T(Je) > 25 \text{ GeV/c} , E_T(J) > E_o$$

In each event either B_1 or B_2 is always less than 0.02.

24

No background event is present in the region $p_T(Je) > 50$ GeV/c, $E_T(J) > 30$ GeV, which contains events A to C (background ≤ 0.02 events at 90% c.l.).

To ensure the neutrino identification we have checked that there is no evidence for high transverse momentum particles pointing to passive parts of the detector in the azimuthal region were the neutrino is detected. However in the case of event D, which has in the transverse plane a configuration similar to that of a two jet event, such an interpretation cannot completely be excluded.

Furthermore, in the case of events A to C, the presence of an undetected jet at $\theta = 15°$ (or $165°$), having the same transverse momentum as the neutrino candidate, leads to impossible or very unlikely kinematical configurations.

11.   INTERPRETATION OF THE EVENTS

Here we discuss possible sources of events A to D under the assumption that they contain an ev pair.

Event D separates from the other events by having a small azimuthal angle between the electron and neutrino ($\Delta\phi \approx 17°$) . It contains a jet with $p_T(j_1) = 70$ GeV/c emitted at opposite azimuth to the ev pair and a second jet with $p_T(j_2) = 25$ GeV/c. The mass of the (evj$_2$) system takes its minimum value of 23 ± 6 GeV/c^2, when assigning to the neutrino the same rapidity as the (ej$_2$) system.   In   this   case $m(evj_1j_2) = 145 \pm 15$ GeV/c.   Such   a configuration suggests an interpretation in terms of a quark-antiquark pair, one member of which decays semileptonically. However, because of the absence of other events similar to this , and because its configuration ressembles that of two-jet events , such an interpretation would be premature.

The three events A to C all have a large opening angle in the ev pairs that have the following transverse masses:

$$M_T(ev) = 56 \pm 2 \ , \ 81 \pm 3 \text{ and } 82 \pm 4 \text{ GeV/c}^2$$

This suggests an interpretation in terms of W → ev decay. It is indeed possible to adjust the longitudinal momentum of the neutrino so as to obtain $m(ev) = m_W$. Among the two solutions we choose the one giving the smallest longitudinal momentum fraction $x_F(WJ)$ to the total system:

$$x_F(WJ) = 0.45 \pm 0.04 , -0.01 \pm 0.06 \text{ and } -0.04 \pm 0.05$$

The distribution of m(WJ) resulting from this procedure for all events having $P_T(Je) > 25$ GeV/c is shown in Fig.13. While the 31 events having $E_T(J) < 30$ GeV populate the peak close to m_W, the three events A to C have the following values

$$m(WJ) = 179 \pm 7 , 176 \pm 9 \text{ and } 162 \pm 8 \text{ GeV/c}^2$$

These values are close enough to suggest a decay of some heavy object. However, also the m(WJ) distribution of the background sample peak in this region (see Fig 13) ,simply because of the kinematical constraints. Another obstacle for the decay interpretation is the fact that event C has a high mass jet pair, $m(J) = 63 \pm 5$ GeV/c², whereas in the two other events J consists essentially of a single jet.

We may put an upper limit on the number of W+J events expected from known processes by writing:

$$\sigma (W+J+...) / \sigma (W+...) \leq \sigma (j_1 j_2 +J+...)/\sigma (j_1 j_2 +...)$$

where $j_1 j_2$ is a pair of jets selected to have the same configuration as the ev pair of a W decay, and J a jet (or jet pair) with a transverse momentum (or invariant mass) at least as high as the one observed. Thus the following upper limits are obtained for the expected number of events of type A, B and C:

$$n_{QCD} < 0.4 , 1.2 \times 10^{-2} , 7 \times 10^{-3}$$

These limits make it difficult to understand B and C in terms of known W production processes. Event A, for its part, has an unlikely high longitudinal momentum of 150 GeV/c for the W assigned to it.

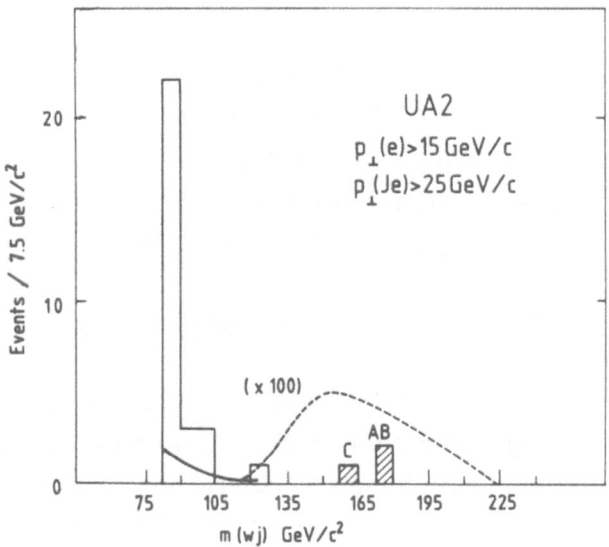

Fig. 13 Distribution of m(WJ) for the 35 events having p_T(Je) > 25 GeV/c. Events having in addition E_T(J) > 30 GeV are cross-hatched. Event D is omitted from the figure because its narrow ev pair does not suggest an interpretation in terms of W → ev decay. Full curve : estimated background. Dashed curve : background (multiplied by 100) for events having E_T(J) > 30 GeV.

It may be remarked that the measured jet energies, because of fragmentation effects, are somewhat smaller than the energies of the parent partons. For W decays into two light quarks we would measure a two-jet mass about 15 GeV/c² below the W mass. Therefore the mass of the jet pair in event C is not inconsistent with the hypothesis that it comes from a W decay.

12.   THE DECAY $Z^0 \rightarrow E^+E^-$

The observation in this experiment of seven $Z^0 \rightarrow e^+e^-$ decays and one $Z^0 \rightarrow e^+e^-\gamma$ decay has already been reported [12]. Following a recent recalibration of the calorimeters, the invariant mass values of these events and their errors have been slightly modified. The updated value of the Z^0 mass is:

$$M_Z = 92.7 \pm 1.7(\text{stat.}) \pm 1.4(\text{syst.}) \text{ GeV/c}^2 \qquad (7)$$

We recall that this value is obtained using only the four events for which the energy of both electrons (and that of the photon in the $e^+e^-\gamma$ event) is unambiguously determined. Within errors, this result agrees with the Z^0 mass value determined in the UA1 experiment:

$$M_Z = 95.6 \pm 1.4(\text{stat.}) \pm 2.9(\text{syst.}) \text{ GeV/c}^2 \text{ [17]}.$$

In order to extract an estimate of the Z^0 width, Γ_Z, from these four events, we first note that the r.m.s. deviation of the four mass values from the value of M_Z given by Eq. (7) is 2.0 GeV/c², which is almost the same as the weighted average of the errors $\sigma = 2.1$ GeV/c². To obtain an upper limit to Γ_Z, we use a Monte Carlo program which generates a large number of event samples, each consisting of four $Z^0 \rightarrow e^+e^-$ decays, according to a Breit-Wigner shape and taking into account the energy resolution of the detector. As an estimate of the upper limit to Γ_Z at the 90% (95%) confidence level, we use the value which gives an r.m.s. of less than 2.0 GeV/c² in 10% (5%) of the event samples. This

28

value is $\Gamma_Z < 6.5$ GeV/c² ($\Gamma_Z < 9.9$ GeV/c²) at the 90% (95%) confidence level.

Within the standard model , this upper limit can be related to the number of additional light neutrinos ΔN_v. We find $\Delta N_v < 22$ ($\Delta N_v < 41$)at the 90% (95%) confidence level, assuming $\sin^2\theta_W = 0.22$ and a value of the t-quark mass $m_t > M_Z/2$.

An independent estimate of Γ_Z can be obtained within the standard model by measuring the ratio $R = \sigma_Z^e/\sigma_W^e$ where σ_Z^e is the cross-section for inclusive Z^0 production followed by the decay $Z^0 \to e^+e^-$ [19]. We obtain σ_Z^e after corrections which take into account the detector acceptance and the efficiency of the electron identification criteria. In this case we use all eight events, for which at least one electron passes all cuts and we find

$$\sigma_Z^e = 0.11 \pm 0.04(\text{stat.}) \pm 0.02(\text{syst.}) \text{ nb} \qquad (8)$$

which is approximately twice as large as the value predicted by QCD [13]. For comparison, the UA1 result [17] is:

$$\sigma_Z^e = 0.050 \pm 0.020(\text{stat.}) \pm 0.009(\text{syst.}) \text{ nb}.$$

The error on R is dominated by statistics, because the value of the total integrated luminosity cancels out. We find $R = 0.21 \pm 0.08$ and $R > 0.116$ at the 90% confidence level. QCD estimates of the ratio between Z^0 and W production cross-sections [19] provide a relation between R and the ratio Γ_W/Γ_Z :

$$\Gamma_W/\Gamma_Z = (9.3 \pm 0.9) \, R \qquad (9)$$

where the error reflects the uncertainty of the QCD calculations. Using the standard model value of Γ_W , $\Gamma_W = 2.77$ GeV/c² (which corresponds to a t-quark mass $m_t = M_Z/2$) we find $\Gamma_Z < 2.6$ GeV/c² ($\Gamma_Z < 3.1$ GeV/c²) at the 90% (95%) confidence level.

As before, we can extract upper limits to the number of additional light neutrinos ΔN_v. We find $\Delta N_v \leq 0$ ($\Delta N_v < 2$) at the 90% (95%) confidence level.

13. THE DECAY $Z^0 \rightarrow E^+E^-\gamma$

We have reported [12] a $Z^0 \rightarrow e^+e^-\gamma$ event containing a photon with an energy $k = 24$ GeV and an 11 GeV electron clearly separated by an angle $\omega_{lab} = 31°$, excluding, therefore, external bremsstrahlung. In Ref. 12 we estimated that the probability that in a $Z^0 \rightarrow e^+e^-$ decay a photon at least as hard as the observed one is emitted as a result of radiative corrections [20], and the e^+e^- opening angle is equal to, or smaller than the measured one, is $\sim 5 \times 10^{-3}$ per event. This calculation, which was performed in the Z^0 rest frame, should not be considered as an estimate of the probability of such a $Z^0 \rightarrow e^+e^-\gamma$ decay because it does not take into consideration all configurations which are less likely than the observed one (for example, the case $k = 23.5$ GeV, $\omega = 90°$).

There are several possible ways to define the relative likelihood of $Z^0 \rightarrow e^+e^-\gamma$ configurations. For cases of non-collinear $e\gamma$ pairs, the event distribution in the Z^0 rest frame is given by the differential cross-section

$$------ = \sigma_0 \ -- \ ------------- \tag{10}$$

where σ_0 is the total cross-section for $Z^0 \rightarrow e^+e^-$ without radiative corrections, and $x_i = 2E_i/M_Z$, E_i being the electron energies. We say that a configuration is less likely than the observed one if its differential cross-section is smaller than that calculated at the point corresponding to the observed $Z^0 \rightarrow e^+e^-\gamma$ event. Integrating Eq. (10) over all configurations which are less likely than the observed one we find a probability of 1.4% per event, or 11% to observe at least one such event in a sample of eight. It should be noted that detectable $Z^0 \rightarrow e^+e^-\gamma$ decays can be divided into two classes of configurations, the first consisting of three clearly resolved energy clusters, and the second containing unresolved $e\gamma$ pairs which result in energy clusters inconsistent with an isolated electron. The corresponding probabilities are $\approx 1.0\%$ and $\approx 0.4\%$ per event, respectively. The remaining $Z^0 \rightarrow e^+e^-\gamma$

decays correspond to configurations which are not detectable in the UA2 apparatus.

Further possibilities to calculate a probability for observing such an event are given in Ref. 3.

14.   SEARCH FOR THE DECAY W → Eνγ

Given the interest in unexpected decay modes of the Z^0 we have also looked for events compatible with the decay W → eνγ in the full data sample. In the central detector, we search for events containing an electron candidate with p_T^e > 8 GeV/c which passes the electron cuts and an additional photon with a momentum k in excess of 8 GeV/c. A photon is defined as an energy cluster which satisfies the same criteria on size and hadronic leakage as an electron, but has no charged particle track pointing to it. If also the photon cluster is seen in the central detector, we require an angular separation ω > 30° between the cluster centroids in order to resolve the eγ pair.

Six events satisfy these conditions. However, none of them survives the additional requirement of a transverse momentum imbalance compatible with the presence of a neutrino having $p_T^ν$ > 10 GeV/c. A Monte Carlo estimate of the number of W → eνγ decays which are expected to satisfy all of these requirements as a result of radiative corrections gives 0.1 event.

In the forward detectors we can identify eγ pairs with very small opening angles by releasing the condition that the electron momentum p and the energy E agree within the measuring errors. In this case, however, we limit our search to electron transverse momenta in excess of 20 GeV/c (as measured in the calorimeter), for which we expect a background contribution of 0.2 events. We find one event which contains a cluster of transverse energy E_T = 41 GeV (E = 86.2 GeV) and a track of 3.6 GeV/c momentum pointing to the cluster and satisfying all other electron identification criteria. Large missing p_T is detected in this event as expected in the case of an associated neutrino. The estimated background for $p_T^e > 35$ GeV/c is 0.01 events. Since the opening angle of this eγ pair is compatible with zero, we also consider the effect of external bremsstrahlung and we obtain a probability of 0.5% per W → ev decay, or 4.5% to observe one such event in the sample of 9 W → ev and evγ candidates with $p_T^e > 20$ GeV/c detected in the forward detectors.

15. COMPARISON WITH THE SU(2)⊗U(1) MODEL

The IVB mass values predicted in the framework of the standard model taking into account radiative corrections are [21] $M_W = 83.0$ GeV/c² and $M_Z = 93.8$ GeV/c². These values are in excellent agreement with our experimental results.

We can extract a value of $\sin^2\theta_W$ from the definition $\sin^2\theta_W = 1-M_W^2/M_Z^2$, where the systematic errors on the mass scale resulting from the uncertainty in the calorimeter calibration cancel out. We find

$$\sin^2\theta_W = 0.196 \pm 0.035 \tag{11}$$

in good agreement with the world average result of deep-inelastic neutrino experiments (including radiative corrections), $\sin^2\theta_W = 0.217 \pm 0.014$ [22].

We can extract a more precise value of $\sin^2\theta_W$ from the relation $M_W = A/\sin\theta_W$, where the numerical value A = 38.65 ± 0.04 GeV/c² is obtained taking into account radiative corrections [21]. In this case we find

$$\sin^2\theta_W = 0.216 \pm 0.010(\text{stat.}) \pm 0.007(\text{syst.}) \tag{12}$$

A test of the standard model is provided by the relationship $\rho = M_W^2/[M_Z^2(1-A^2/M_W^2)]$ which should be equal to 1 for the minimal Higgs structure. We find

$$\rho = 1.02 \pm 0.06 \tag{13}$$

32

in good agreement with the minimal $SU(2) \otimes U(1)$ model.

16.   CONCLUSIONS

We have studied the production of electrons with very high transverse momentum at the CERN $\bar{p}p$ collider. From a sample of events containing an electron candidate with $p_T > 15$ Gev/c, we have extracted a clear signal resulting from the production of the charged intermediate vector boson $W^{\pm}$, which subsequently decays into an electron and a neutrino.

We have also given new and more refined results on the production and decay of the neutral vector boson Z^0.

Our experimental results show good agreement with the predictions of the standard model of the unified electro-weak theory.

In a search for events containing an electron-neutrino pair having $p_T(e) > 15$ GeV/c and $p_T(v) > 25$ GeV/c, four events have been singled out in which the ev pair is produced in association with a jet, or a system of jets, having very large transverse energies. Three of these events contain a large transverse mass ev pair and have been interpreted in terms of W-jet(s) associated production. However, their configurations are such that their production via known processes is very unlikely for at least two of them. In each of the three events the invariant mass of the W-jet(s) system is measured to be in the vicinity of 170 GeV/c², but the significance of this observation is weakened by the fact that this mass region is kinematically favoured by the selection criteria.

REFERENCES

1) For a review see J. Ellis et al., Ann. Rev. Nucl. Part. Science 32 (1982) 443.

2) The UA2 Collaboration, Latest results from the UA2 experiment at the CERN p̄p Collider, Proc. of the Int. Europhysics Conf. on High Energy Physics, Brighton, July 1983, p. 472.
   A.G. Clark, Results from the UA2 experiment at the CERN p̄p Collider, Proc. of the 1983 Int. Symp. on Lepton and Photon Interactions at High Energies, Cornell, August 1983, p. 53.

3) P. Bagnaia et al., A study of high transverse momentum electrons produced in p̄p collisions at 540 Gev, CERN-EP/84/39, submitted to Zeitschrift für Physik C.
   P. Bagnaia et al., Observation of electrons produced in association with hard jets and large missing tansverse momentum in pp̄ collisions at √s = 540 GeV, CERN-EP/84-40, submitted to Phys. Lett. B.

4) B. Mansoulié, The UA2 apparatus at the CERN p̄p Collider, Proc. of the 3rd Moriond Workshop on p̄p physics (1983) p. 609 (éditions Frontières).

5) M. Dialinas et al., The vertex detector of the UA2 experiments, LAL-RT/83-14 (1983).

6) A. Beer et al., The central calorimeter of the UA2 experiment at the CERN p̄p Collider, CERN-EP/83-175 (1983) to be published in Nucl. Instr. Meth.

7) M. Banner et al., Phys. Lett. 115B (1982) 59.
   M. Banner et al., Phys. Lett. 122B (1983) 322.

8) C. Conta et al., The system of forward-backward drift chambers in the UA2 detector, CERN-EP/83-176 (1983) to be published in Nucl. Instr. Meth.

9) K. Borer et al., Multitube proportional chambers for the localization of electromagnetic showers in the UA2 detector, CERN-EP/83-177 (1983) to be published in Nucl. Instr. Meth.

10) R. Battiston et al., Phys. Lett. 117B (1982) 126.

11) P. Bagnaia et al., Z. Phys. C20 (1983) 117.

P. Bagnaia et al., Measurement of very large transverse momentum jet production at the CERN $\bar{p}p$ Collider, CERN-EP/84-12 (1984) to be published in Phys. Lett. B.

12) P. Bagnaia et al., Phys. Lett. 129B (1983) 130.

13) G. Altarelli, R.K. Ellis, M. Greco and G. Martinelli, Vector boson production at Colliders : a theoretical reappraisal, Ref. TH. 3851-CERN (1984).

14) M. Glück et al., Z. Phys. C13 (1982) 119.

15) Review of Particle Properties, Phys. Lett. 111B (1982).

16) F.E. Paige and S.D. Protopopescu, BNL report 31987 (1981).

17) G. Arnison et al., Phys. Lett. 129B (1983) 273.

18) P. Fayet and S. Ferrara, Phys. Rep. 32C (1977) 249, and references therein.
H.E. Haber and G.L. Kane , Signatures and possible evidence for supersymmetry at the CERN collider, preprint UCSC-TH-169-84
March 1984.

19) K. Hicasa, Counting neutrino species at high-energy proton-antiproton collisions, University of Wisconsin preprint MAD/PH/144 (Nov. 1983)

20) D. Albert, W.J. Marciano, D. Wyler and Z. Parsa, Nucl. Phys. B166 (1980) 460.
F.A. Berends and R. Kleiss, Hard Photon Effects in $W^{\pm}$ and Z^0 decay, University of Leiden, The Netherlands, Nov. 1983.

21) For a review see W.J. Marciano, Electroweak interactions, Proc. of the 1983 Int. Symp. on Lepton and Photon Interaction at High Energies, Cornell, August 1983, p. 80.

22) A. Sirlin and W.J. Marciano, Nucl. Phys. B189 (1981) 442.
C. Llewellyn Smith and J. Wheater, Phys. Lett. 105B (1981) 486.

DISCUSSION

TERAZAWA:

What are the invariant masses of e^+e^-, $e^+\gamma$, and $e^-\gamma$ in the anomalous $e^+e^-\gamma$ event?

HANSEN:

The mass of the gamma and the nearest electron is 8.6 GeV and the mass of the two electrons is 49 GeV. The third mass I don't remember – it must be large.

KANE:

Why is the event on the right likely to be background?

HANSEN:

The background estimated from a sample of fake electron events, depends on the amount of missing p_T. This event has relatively low missing p_T.

POHL:

1) In the events $e\nu$ + n jets (n = 1 ~ 3), are the jets single particles or high multiplicity objects?
2) Does UA1 find such events?

HANSEN:

1) These jets are not single particles. Although this has not been studied in detail, I should say that they look like pretty normal jets.
2) Not to my knowledge.

CHAU:

Just a simple comment: You did not really see parity violation. It is more interesting to say that through V–A of W interaction, you have seen quark and antiquark. Then from the e^+e^- distribution you measure V, A structure of the Z^0 interaction.

HANSEN:

That's right. We don't claim to see the V–A coupling to quark and antiquark, but only that the V and A components have equal numerical strength.

QUARK MIXING FROM HYPERON DECAYS

Hans-Wolfgang Siebert

Physikalisches Institut
Universität Heidelberg
D-6900 HEIDELBERG

1. INTRODUCTION

Hyperon and also kaon decays have provided experimental information on quark mixing since more than 20 years ago, when nobody yet
thought of "quark mixing" as such. In 1963 Cabibbo published his
paper on Unitary Symmetry and Leptonic Decays[1], outlining what is
generally known as the Cabibbo model, and especially in the following
decade a lot of experimental effort was devoted to experiments on
semileptonic hyperon decay to provide data for tests of this model.

About 10 years ago, the experimental situation had stabilized
at the level illustrated below in Fig. 1. Bubble chamber experiments
using stopped $\overline{K}$ to produce Λ and Σ hyperons, had already reached
their limits with experiments involving millions of pictures. Further
improvement had come from counter experiments in pion and kaon beams,
and already at the then available beam energies of 20-25 GeV, hyperon
beams at the Brookhaven AGS and the CERN PS had shown the potential
of this new technique. The experimental data agreed quite well with
the Cabibbo model, and the fit result for the Cabibbo angle was
$\sin \theta_c = 0.232 \pm 0.003$, a remarkable accuracy[2].

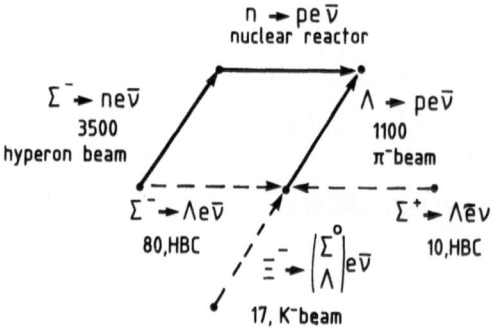

Fig. 1　Experimentally known semileptonic decays in the baryon octet around 1974.　For the experiments with the highest statistics, the number of events and the baryon source are indicated.　HBC: stopped K^- in hydrogen bubble chambers.

Today, a new level of experimental accuracy has been reached in the WA2 experiment at the CERN SPS hyperon beam, which has measured five different semileptonic hyperon decays with high statistics[3]. For the first time, it has been possible to perform a comprehensive test of the Cabibbo model without the inconsistencies which will inevitably occur when results from different experiments have to be combined.　The decay $\Lambda \rightarrow pe\bar{\nu}$ has also been measured at the BNL hyperon beam, with potentially very high statistics[4].

2.　THEORETICAL FRAMEWORK

The formalism

Assuming the usual V,A structure, the matrix element M for the decay

$$B \to B' + \ell + \bar{\nu}_\ell$$

is given by the product of the matrix elements of the baryon and lepton weak currents:

$$M = \frac{G}{\sqrt{2}} \left\langle B' | J_h^\mu | B \right\rangle \bar{u}_\ell(p_\ell) \gamma_\mu (1+\gamma_5) u_\nu(p_\nu) \ ,$$

where G_F is the universal weak coupling constant. The weak hadronic current is expressed in terms of form factors which take into account the effects of the strong interaction:

$$\left\langle B' | J_h^\mu | B \right\rangle = C \bar{u}_{B'}(p') \left\{ f_1(q^2)\gamma^\mu + i \frac{f_2(q^2)}{M} \sigma^{\mu\nu} q_\nu + \frac{f_3(q^2)}{M} q^\mu \right. $$
$$\left. + \left[g_1(q^2)\gamma^\mu + i \frac{g_2(q^2)}{M} \sigma^{\mu\nu} q_\nu + \frac{g_3(q^2)}{M} q^\mu \right] \gamma_5 \right\} u_B(p) \ ,$$

where

$$C = \begin{bmatrix} \cos \theta_C \\ \sin \theta_C \end{bmatrix} \qquad \text{for} \qquad \Delta S = \begin{bmatrix} 0 \\ 1 \end{bmatrix} \text{ transitions } .$$

The θ_C is the Cabibbo angle, p,M and p',M' are the four momenta and masses of the initial and final baryon and $q = p-p'$. In the framework of the Kobayashi-Maskawa (KM) six-quark mixing scheme[5], we have to replace $\cos \theta_C$ by $U_{ud} = \cos \theta_1$ and $\sin \theta_C$ by $U_{us} = \sin \theta_1 \cdot \cos \theta_3$. We have chosen the sign convention that g_1/f_1 is positive for the decay $n \to pe\bar{\nu}$.

For a meaningful experimental analysis, the number of parameters must be reduced drastically. Kinematics allow us to drop the terms with f_3 or g_3, because they are multiplied with a factor $(m_\ell/M)^2$, and are therefore completely negligible in electronic decays. T-invariance implies the f and g are real.

In the Cabibbo model, matrix element terms with the same form factor, but from different decays within the baryon octet, are determined by two parameters F and D. Thus, in principle, if we know two decays, we know all of them. The CVC hypothesis relates the vector form factors to the electric charges and the anomalous magnetic moments of protons and neutrons. For the form factor f_1, we have $F = 1$, $D = 0$, and for f_2 we have $F = \mu_p + \mu_n/2$, $D = -\frac{3}{2} \mu_n$. CVC also predicts $f_3 = 0$, but this term is negligible anyhow. In the limit of exact SU(3), the absence of "second class currents" implies $g_2 = 0$ and $f_3 = 0$.

Thus, in the Cabibbo model, all semileptonic decays in the baryon octet should be described by three parameters only: the parameters F and D determining the form factors g_1 in the different decays, and the Cabibbo angle θ_C.

What can one measure?

Experiments provide three types of information: decay rates, Dalitz plot distributions and asymmetries.

In good approximation, the decay rates are

$$\Gamma = \text{const.} \left\{ \begin{matrix} \cos \theta_C \\ \sin \theta_C \end{matrix} \right\}^2 (1 + 3\lambda^2) \, ,$$

where $\lambda = g_1/f_1$, so they provide information on θ_C and on the absolute value of λ.

The Dalitz plot distribution (see Fig. 2 below) is another source of information: along the $T_{B'}$ axis, the distribution is very sensitive to the absolute value of λ. It is important, however, to specify the assumptions on the q^2 dependence of the form factors made in the analysis, because effects of $|\lambda|$ and of the q^2 dependence cannot be separated. Along the T_e axis, the distribution is weakly sensitive to λ, so that the electron spectrum can be used to determine the sign of λ. The electron-neutrino asymmetry is correlated with the Dalitz plot density, as the electron-neutrino angle $\theta_{e\nu}$ is determined by T_e and $T_{B'}$, thus this asymmetry, which is sensitive

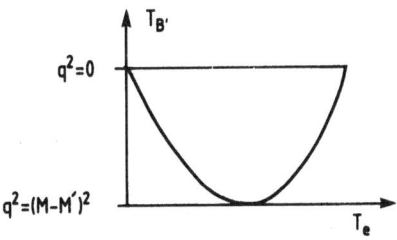

Fig. 2   Typical Dalitz plot contour for semileptonic hyperon decay.
T_e: electron kinetic energy, $T_{B'}$: baryon recoil kinetic
energy, q^2: four-momentum transfer to the $e\bar{\nu}$ pair.

to $|\lambda|$, does not provide any information independent of the Dalitz
plot analysis.

Finally, the lepton asymmetries with respect to the polarization
of either the mother or the daughter baryon are sensitive to both the
absolute value and the sign of λ. Experimentally, these two cases
are very different. Polarized hyperons can be found at large production angles in proton-nucleus collisions[6], but in both experiments
discussed here, the production angles were too small. In the WA2
experiment, however, Ξ^- decays provided a source of Λ with known
longitudinal polarization $P_\Lambda = \alpha_{\Xi^-}$, where the Λ could be reconstructed kinematically from the measured Ξ^- and π^- momenta. On the other
hand, if the daughter baryon is a Λ, then its polarization can be
analysed through the distributions of the decay p and π^-. This is
the case in $\Sigma^- \rightarrow \Lambda e\bar{\nu}$ and $\Xi^- \rightarrow \Lambda e\bar{\nu}$ decays.

Complete expressions for the Dalitz plot distributions and the
decay asymmetries have been given, for example, by Bender, Linke and
Rothe[7]. All these distributions are also sensitive to the form factors f_2 and g_2, but to a much lesser extent.

3. EXPERIMENTAL RESULTS

The CERN experiment

In this section we briefly describe the CERN SPS charged hyperon beam and the apparatus used to measure the semileptonic hyperon decays $\Sigma^- \to n e \bar\nu$, $\Sigma^- \to \Lambda e \bar\nu$, $\Xi^- \to \Lambda e \bar\nu$, $\Xi^- \to \Sigma^0 e \bar\nu$ and $\Lambda \to p e \bar\nu$.

The hyperon beam provided a source of hyperons with intensities 1-2 orders of magnitude larger than at the previous generation of hyperon beams at the CERN PS and the Brookhaven AGS. The beam is shown in Fig. 3. Protons of 200 GeV/c were directed on a production target, and a magnetic channel of 11 m length selected charged secondary particles in a momentum interval $\Delta p/p = 10\%$ (FWHM). Downstream of the magnets a DISC Cherenkov counter was used to select particles in a very narrow velocity range. At the chosen beam momentum of 100 GeV/c, the counter optics could be adjusted so that Σ^- and Ξ^- concurrently triggered different photomultiplier combinations, so that the measurements on Σ^- and Ξ^- could be performed simultaneously. The trigger rates were 3000 Σ^- and 400 Ξ^- per SPS pulse at negative beam polarity and 40 Σ^+ at positive polarity. These rates were limited by the maximum total beam flux the apparatus could tolerate, which was 10^6 π^- or π^+ and p, respectively, per SPS pulse of one second duration. At the chosen production angle of 2 mrad, the hyperons were unpolarized. The momenta of the beam particles could be measured with the aid of MWPCs, with an accuracy better than 1%.

The experimental apparatus is shown in Fig. 4. The principle parts were a magnetic spectrometer and several electron and gamma detectors. The magnetic spectrometer comprised a magnet and drift chambers and allowed the measurement of momenta of charged particles with an accuracy of 1.5% (r.m.s.) at 100 GeV/c. The length of the decay zone between the DISC and the spectrometer was 10 m, which has to be compared to the mean decay length of 3.75 m for Σ^- and Ξ^- at

100 GeV/c. Details of the hyperon beam and the spectrometer can be found in an earlier publication[8].

The electron detectors had to combine good electron efficiency with high π^- rejection, in view of the small semileptonic branching ratios. For momenta above 5 GeV/c, where the bulk of the data is found, a lead glass array and two transition radiation (XTR) detectors were used. The lead glass array consisted of 17 blocks horizontally × 2 blocks vertically × 3 blocks along the beam, with a total thickness of 19 radiation lengths in beam direction. The XTR detectors comprised each a radiator of 650 Li foils and a MWPC filled with a mixture of 80% xenon and 20% carbon dioxide. For one of the detectors, radiator and chamber were separated by the spectrometer magnet, so that the signal from the transition radiation and the signal from the electron track itself appeared on different wires. For electrons at low momenta, a gas-filled Cherenkov counter with a pion threshold at 8.5 GeV/c and lead scintillator sandwiches were used. The redundancy of information provided by these detectors allowed the study of their performance with the semileptonic data themselves. Pion rejection factors of 40,000:1 or greater have been obtained with electron efficiencies of between 70% and 80%, in the presence of other particles. The γ-rays from the Σ^0 produced in $\Xi^- \rightarrow \Sigma^0 e \bar{\nu}$ decay were measured in the lead glass array, and in addition, their conversion points were determined in two layers of 7 mm lead and a MWPC each, placed directly in front of the lead glass array. Details on the electron detectors can be found in Ref. 9.

Of the five decays studied, four have a Λ as final state baryon, which can be precisely measured in the magnetic spectrometer through the decay $\Lambda \rightarrow p\pi^-$. For the measurement of the decay $\Sigma^- \rightarrow n e \bar{\nu}$, an additional neutron counter was placed on the beam axis in front of the lead glass array during special runs. This counter consisted of a sandwich of aluminium plates and MWPCs with a diameter of

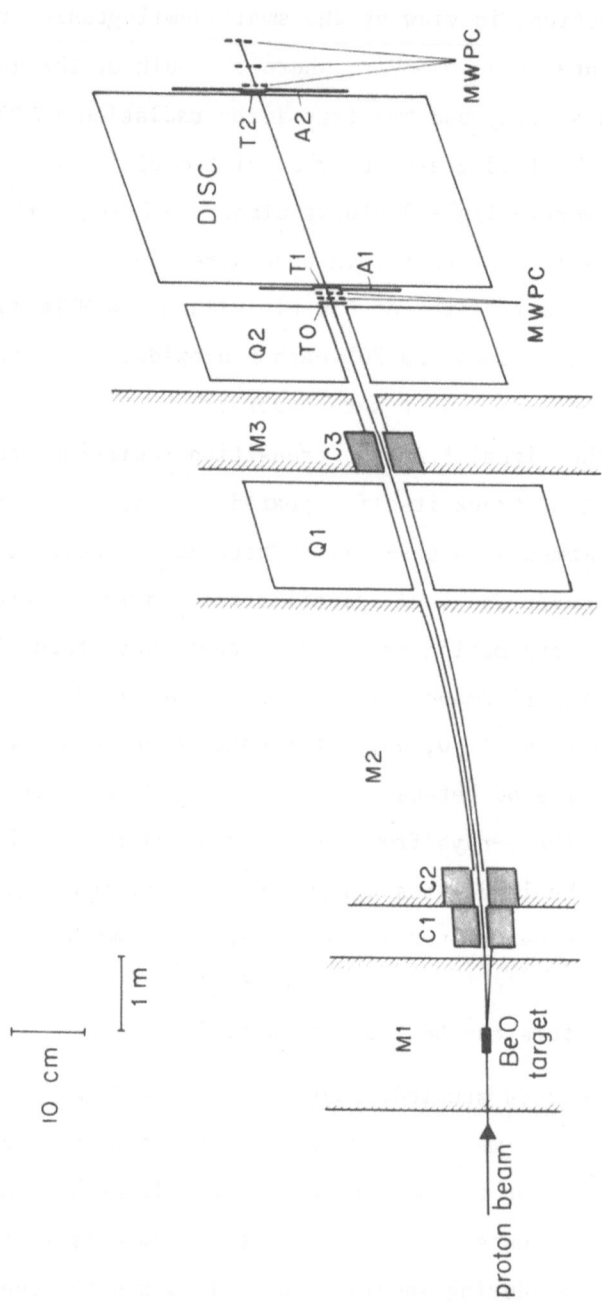

10 cm

1 m

Fig. 3 CERN SPS charged hyperon beam. M_{1-3}: bending magnets; Q_{1-2}: superconducting quadrupoles;
C_{1-3}: collimators; T_{0-2}: beam defining counters; A_{1-2}: beam halo veto counters.

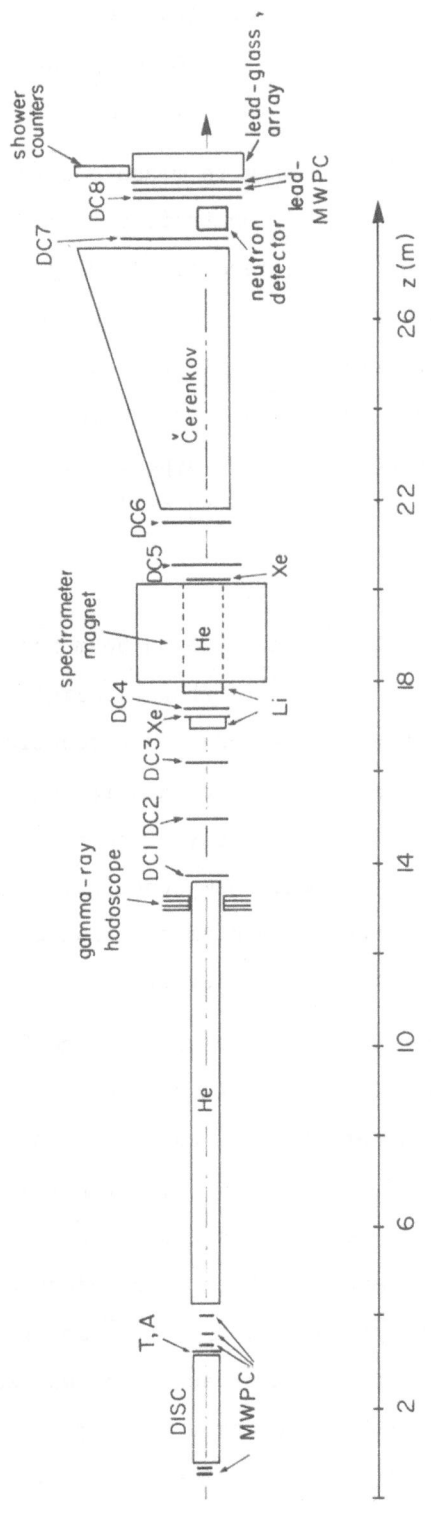

Fig. 4 WA2 apparatus. He: Helium bags; Li: lithium radiators; Xe: xenon proportional chambers; DC: drift chambers.

250 mm, large enough to intercept all neutrons from Σ^- decays. The transverse coordinates of the neutron impact point could be measured with an accuracy of about 6 mm (r.m.s.).

It would have been interesting to measure also the muonic decay modes, by identifying the muons from their passage through a layer of iron placed downstream of the lead glass array. However, 2.5% of the π^- from the nonleptonic decay modes decayed in flight inside the apparatus, and it proved impossible to find the small number of muons from genuine muonic decays in this enormous background from pion decays (the most frequent muonic hyperon decay is $\Sigma^- \to n\mu\bar{\nu}$, which has a branching ratio of 4.5×10^{-4}) [10].

The BNL experiment

The layout of the BNL neutral hyperon beam and of the apparatus used to measure the decay $\Lambda \to pe\bar{\nu}$ is shown in Fig. 5. A 29 GeV/c proton beam from the AGS was directed on a production target, and a collimator positioned in the field of a sweeping magnet selected neutral secondary particles. Gamma rays were removed by a lead converter. The remaining neutral beam contained 500 unpolarized Λ per machine pulse, which had a momentum spectrum extending from 4 to 24 GeV/c, with a maximum at 12 GeV/c.

A magnetic spectrometer with two magnets and 28 automatic spark chambers served to measure the momenta of the charged decay particles. Decay electrons were identified in two gas-filled Cherenkov counters with pion thresholds at 5 GeV/c. Events were accepted only, if the momentum of the electron candidate was below 4 GeV/c, thus good electron efficiency and high pion rejection were ensured. A possible source of background are $K_L^0 \to \pi^+ e\bar{\nu}$ decays, which have the same topology as $\Lambda \to pe\bar{\nu}$. This background was suppressed by vetoing the π^+ in two further Cherenkov counters with pion thresholds at 3 GeV/c. For this purpose a momentum above 5 GeV/c was

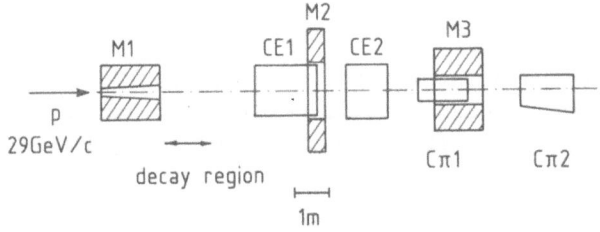

Fig. 5   Layout of the BNL hyperon beam and $\Lambda \rightarrow pe\bar{\nu}$ experiment.
M_{1-3}: bending magnets; C_{E1},C_{E2}: Cherenkov counters for
electron identification; $C_{\pi1},C_{\pi2}$: Cherenkov counters for
pion identification.   Spark chambers are not shown.

required for the proton candidate. The counters from each pair were
calibrated against each other from the data.

Branching ratio results

The branching ratio results are listed in Table 1, together
with the number of observed events. It is important to note that
the numbers quoted are the experimentally observed branching ratios
R_{obs}. For comparison with the Cabibbo model, one has to use the
"bare weak" branching ratios R_{weak}, which differ from R_{obs} because
of radiative corrections: $R_{obs} = R_{weak} (1+\delta)$. The effect of the
radiative corrections on the Dalitz plot density and hence on the
apparatus acceptance has already been taken into account in the
determination of R_{obs}.

Table 1

Measured branching ratios in units 10^{-4}.
Column 1: CERN and BNL results. Column 2:
Previous results. Numbers of events are
also given.

$\Sigma^- \to \Lambda e \bar{\nu}$	0.561 ± 0.031 1650	0.63 ± 0.10 [13] 115
$\Sigma^- \to n e \bar{\nu}$	9.6 ± 0.5	10.5 ± 0.7 [11] 450 10.9 ± 0.6 [12] 600
$\Xi^- \to \Lambda e \bar{\nu}$	5.64 ± 0.31 2600	3.0 ± 1.4 [13] 11
$\Xi^- \to \Sigma^0 e \bar{\nu}$	0.87 ± 0.17 150	< 1.4 [13] No event seen
$\Xi^- \to \binom{\Sigma^0}{\Lambda} e \bar{\nu}$	6.51 ± 0.31 3000	6.8 ± 2.2 [14] 17
$\Lambda \to p e \bar{\nu}$	8.57 ± 0.36 CERN 7100 8.47 ± 0.17 BNL 10000	8.4 ± 0.4 [15] 1100

In column 2 of Table 1 we list the results of the previous experiments with the best statistics. We wish to emphasize that a
direct comparison of the new with the previous results can be quite
misleading because of the different assumptions made in the analysis.
For example, the earlier results on the $\Sigma^- \to n e \bar{\nu}$ branching ratio[11,12]
come from bubble chamber experiments with stopped K^-, where the
upper part of the electron centre-of-mass momentum spectrum had to
be rejected because of the potential background from $\Sigma^- \to n \pi^-$ decays. The fraction of electrons below this cut, and hence the
measured branching ratio, depends critically on the values assumed
for the weak form factors and on the radiative corrections. If one
tries to re-estimate these results with the new values for the

form factors and with radiative corrections, one obtains branching ratios which are lower by 6% or 1.5%, respectively. We conclude that there is no disagreement between the new and the previous results. Also, the two new results on the $\Lambda \to pe\bar{\nu}$ branching ratio from BNL and CERN are in excellent agreement. In the following, we will use only the new results from column 1 for tests of the Cabibbo model.

Form factor results

The results of the form factor analysis are listed in Table 2, again together with the results of the previous experiments with the best statistics. In the analysis, we assumed $g_2 = 0$ and for f_2 the values given by CVC. The q^2 dependence of the form factors was parametrized in the following way:

$$f_1(q^2) = f_1(0)\left[1 + 2q^2/M_V^2\right]$$

$$g_1(q^2) = g_1(0)\left[1 + 2q^2/M_A^2\right]$$

with $M_V = 0.84$ (0.97) GeV/c^2 and $M_A = 1.08$ (1.25) GeV/c^2 for $\Delta S = 0$ ($\Delta S = 1$) decays. The choice of these masses is discussed in Ref. 3a.

These assumptions are not critical, as discussed in Ref. 3d. Only if one assumes constant form factors, which has been done in earlier experiments, the values of g_1/f_1 are shifted significantly, for example by +0.06 in the case of $\Sigma^- \to ne\bar{\nu}$. This again illustrates the difficulties encountered in comparisons of different experiments. For tests of the Cabibbo model, we will use only the new values from column 1 of Table 2.

We have entered in Table 2 the sign of g_1/f_1, or f_1/g_1, respectively, wherever it was determined experimentally. For $\Sigma^- \to ne\bar{\nu}$ and, in the BNL experiment, $\Lambda \to pe\bar{\nu}$, the sign of g_1/f_1 had to be extracted from the electron spectra, which favoured the signs indicated. There exists, however, a measurement performed at Argonne National Laboratory, of the electron asymmetry α_e in the decay

$\Sigma^-_{pol} \to n e \bar{\nu}$, using polarized Σ^- produced by a K^- beam[22]. The result, $\alpha_e = +0.35 \pm 0.29$, favours a positive value of g_1/f_1, which is contrary to the expectation from the Cabibbo model. This experiment is very difficult, because the Σ^- polarization has to be determined through the small decay asymmetry in $\Sigma^- \to n\pi^-$. Clearly, another measurement would be very useful and indeed, a new experiment on $\Sigma^-_{pol} \to n e \bar{\nu}$ is under way at the Fermilab charged hyperon beam.

Table 2

Measured values of g_1/f_1. Column 1: CERN
and BNL results. Column 2: previous results.
Numbers of events are also given.

	f_1/g_1	
$\Sigma^- \to \Lambda e \bar{\nu}$	+0.03 ± 0.08 1650	−0.37 ± 0.20 [16] 190 +0.32 ± 0.30 [13] 120
	g_1/f_1	
$\Sigma^- \to n e \bar{\nu}$	−0.34 ± 0.05 4450	0.435 ± 0.035 [17] 3500 $0.17^{+0.07}_{-0.09}$ [18] 500
$\Xi^- \to \Lambda e \bar{\nu}$	+0.25 ± 0.05 2000	
$\Lambda \to p e \bar{\nu}$	+0.70 ± 0.03 CERN 7100 +0.715 ± 0.026 BNL 10000	+0.63 ± 0.06 [19] 1080 +0.53 ± 0.09 [20] 540

4. TESTING THE CABIBBO MODEL

Hyperon decays

We now have the branching ratios of five different semileptonic hyperon decays and the g_1/f_1 ratios of four of them available for tests of the Cabibbo model. The test can be visualized by plotting the results on F and D extracted from the data. Each g_1/f_1 value with its error defines a road in the F,D plane, according to the parametrization of g_1/f_1 in terms of F and D given in Table 3. In the case of $\Sigma^- \to \Lambda e \bar{\nu}$, however, the form factor result cannot be used in this way, because here $f_1 = 0$ is predicted. Furthermore, each branching ratio can be used to derive a value for g_1/f_1 (or g_1 in the case of $\Sigma^- \to \Lambda e \bar{\nu}$), once the Cabibbo angle is determined from an overall fit. Here one has to take into account the radiative corrections to the branching ratios discussed above. The values of these corrections for the decays involved are also given in Table 3, we have used the most recent calculations of Tóth, Szegö and Margaritisz[23].

The results of that procedure, using the CERN data are shown in Fig. 6. All roads overlap in one region, indicating very good agreement with the Cabibbo model. The results of fits using the CERN data

Table 3

Column 1: g_1 form factors as parametrized in the Cabibbo model. Column 2: radiative corrections δ to the branching ratios, $R_{obs} = R_{weak} (1+\delta)$.

	g_1/f_1	δ (%)
$n \to p e \bar{\nu}$	F + D	+7.3
$\Lambda \to p e \bar{\nu}$	F + D/3	+3.9
$\Sigma^- \to n e \bar{\nu}$	F - D	+1.1
$\Xi^- \to \Lambda e \bar{\nu}$	F - D/3	+1.3
$\Xi^- \to \Sigma^0 e \bar{\nu}$	F + D	
$\Sigma^- \to \Lambda e \bar{\nu}$	$g_1 = D\sqrt{2/3}$	+1.6

alone or averaging the CERN and BNL data on $\Lambda \to pe\bar{\nu}$ are given in columns 1 and 2 of Table 4. Our expectations from the plot are fulfilled, the χ^2 values of 4.5/5 or 6.3/5, respectively, confirm that even at the present level of experiments, no deviation from the standard Cabibbo model can be seen in the hyperon data.

Neutron decay

We have also plotted in Fig. 6 the very narrow roads derived from the neutron lifetime and neutron decay correlations. On the scale of the drawing, one hardly notices that we have a problem here, so we have plotted in Fig. 7 the experimental situation in neutron decay in more detail. There exist two groups of measurements. One can determine g_1/f_1 directly from the decay asymmetries of polarized neutrons[24-26] or from the proton recoil spectrum[27], which reflects the $e\bar{\nu}$ correlation. These correlation experiments yield consistent results with a mean value $(g_1/f_1)_{corr} = 1.263 \pm 0.006$. One can also determine g_1/f_1 from the neutron decay rate, here one is helped by the fact that this rate depends on the cosine of the Cabibbo angle, which has a very small error. Of the three existing results[28-30] one is in violent disagreement with the two others[29], though their mean value happens to coincide with the mean of the correlation data.

If we take only the consistent results of Ref. 28 and Ref. 30, their mean is $(g_1/f_1)_\tau = 1.238 \pm 0.009$. If we now compare these numbers with our expectations from the hyperon decay measurements, keeping in mind that $(g_1/f_1)_{n \to pe\bar{\nu}} = F+D$, we see that there is a large discrepancy between the neutron correlation data and the hyperon data, and even the neutron lifetime data and the hyperon data do not agree too well. The fit results in columns 3 and 4 of Table 4 confirm this impression. So if we believe the correlation measurements, which give a consistent picture, and disregard the confusing lifetime measurements, the Cabibbo model might be in trouble.

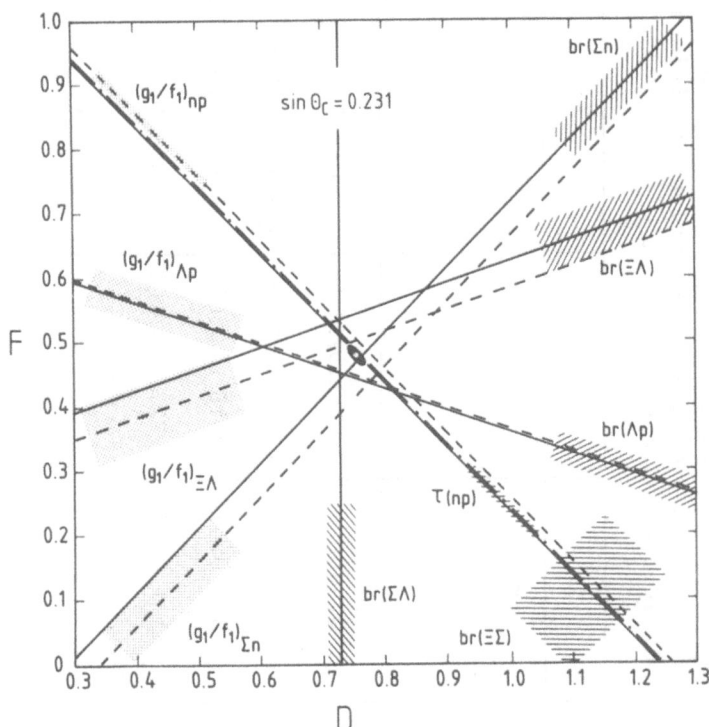

Fig. 6 F and D values from branching ratio measurements (solid lines) and g_1/f_1 measurements (dashed lines). Shaded bands indicate the experimental errors. The result from the $\Xi^- \to \Sigma^0 e\bar{\nu}$ branching ratio is plotted as a dashed-dotted line to distinguish it from the nearby neutron lifetime line. The black ellipse is the one standard deviation contour of the fit to all hyperon decay data and to the neutron lifetime.

Table 4

Results of Cabibbo fits. The basic Cabibbo fit using only the data from the CERN experiment is shown in column 1. The other fits include additional data, as indicated by an appropriate x.

	1	2	3	4	5
BNL data		x	x	x	x
n lifetime			x		
n correlations				x	x
χ^2/NDF	4.5/5	6.3/5	11.5/5	19.0/5	10.9/4
F+D	1.182 ± 0.022	1.185 ± 0.021	1.230 ± 0.008	1.257 ± 0.006	1.256 ± 0.006
α = D/(F+D)	0.617 ± 0.009	0.618 ± 0.018	0.618 ± 0.008	0.618 ± 0.008	0.605 ± 0.009
$\sin \theta_C$	0.235 ± 0.004	0.231 ± 0.003	0.227 ± 0.002	0.224 ± 0.002	$\sin \theta_C^V$ = 0.232 ± 0.004 $\sin \theta_C^A$ = 0.215 ± 0.004

54

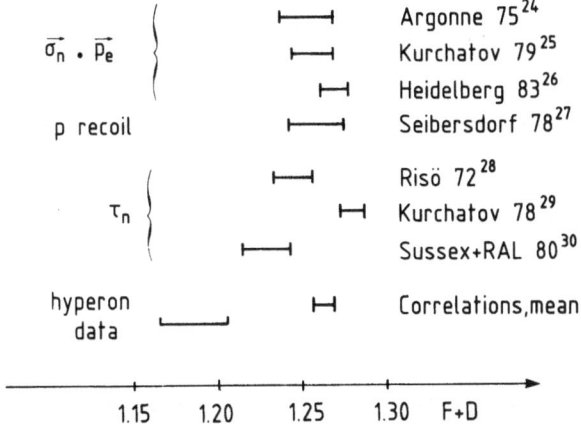

Fig. 7   (F+D) values from neutron decay experiments and from a fit
to all hyperon data (column 2 of Table 4).

The situation can be improved if we perform a fit with different
angles θ_C^V and θ_C^A for vector and axial vector couplings. The fit result is shown in column 5 of Table 4. We now obtain $\chi^2/\text{NDF} = 10.9/4$
compared to 19.0/5 for the one angle fit, certainly some improvement,
even if the χ^2 still is not satisfactory. The difference between
$\theta_C^V = 0.232 \pm 0.004$ and $\theta_C^A = 0.215 \pm 0.004$ does not necessarily mean
different quark mixing for vector and axial vector couplings. As
the difference between $\cos \theta_C^V$ and $\cos \theta_C^A$ is negligible on the level
of the experiments, we can interpret the difference between $\sin \theta_C^V$
and $\sin \theta_C^A$ as a manifestation of different renormalizations for the
vector and axial vector couplings in $\Delta S = 1$ decays. Donoghue and

Holstein[21] have calculated such renormalizations from the s-quark wave function in the bag model, and have obtained renormalization factors of 0.97 for f_1 and 1.08 for g_1. This would be equivalent to $\sin \theta_C^A/\sin \theta_C^V = 1.11$, but the data prefer $\sin \theta_C^A/\sin \theta_C^V = 0.93$. Obviously, we need better understanding of such renormalization effects before drawing further conclusions from the data, and equally obviously, the discrepancies between the neutron decay experiments should be resolved.

5. SIX QUARK MIXING

So far, we have analysed the data in the context of the Cabibbo model. We will now interpret the results in terms of six quark mixing, using the parametrization of Kobayashi and Maskawa[5]. To this end, we have to substitute $\cos \theta_C$ by $U_{ud} = \cos \theta_1$ and $\sin \theta_C$ by $U_{us} = \sin \theta_1 \cdot \cos \theta_3$.

Just the fact that the standard Cabibbo fit works at all, indicates already that $\cos \theta_3$ is close to 1. But if we fit U_{ud} and U_{us} as independent quantities to the hyperon and neutron data, we obtain meaningless results, because the precision on $U_{ud} = \cos \theta_1$ is not sufficient to obtain a sufficiently precise value of $\sin \theta_1$ from it.

A much more accurate information on $\cos \theta_1$ is available from superallowed nuclear β decays, which have been reviewed recently by Wilkinson[31]. From the fit values of these decays one can obtain the vector coupling constant for β decay, $G_\beta^V = G_F \cdot \cos \theta_1$. The universal Fermi coupling constant G_F, on the other hand, can be determined from the muon lifetime. However, to obtain $\cos \theta_1$ from G_β^V and G_F, one has again to take into account radiative corrections. The result is

$\cos \theta_1 = 0.9737 \pm 0.0015$, and hence

$\sin \theta_1 = 0.228 \pm 0.006$.

The error on that number is dominated by the theoretical uncertainty on the radiative corrections[23,32]. Somewhat arbitrarily, we have assumed this uncertainty to be ±0.3%, compared to the total effect of 2.1%.

With the value $\sin \theta_C = \sin \theta_1 \cdot \cos \theta_1 = 0.231 \pm 0.003$ from the hyperon decays, we obtain

$$|\sin \theta_3| < 0.24 \ (90\% \ CL) \ .$$

As discussed in these Proceedings, more stringent limits on $\sin \theta_3$ are now available especially from the B lifetime.

What is the value of U_{us} alias $\sin \theta_C$ really?

We have seen that the renormalization effects suggested by the hyperon and neutron data are just the opposite of theoretical expectations, and we cannot exclude that U_{us} and the effective value of the Cabibbo angle extracted from the hyperon and neutron semileptonic decays differ by several percent.

Another source of information on U_{us} are the K_{e3} decays, to which only the vector current contributes. The average value from K_{e3}^+ and K_{e3}^0, of the effective Cabibbo angle is $\sin \theta_C^V = 0.214 \pm 0.003$. After correction for SU(3) symmetry breaking, one obtains $U_{us} = 0.219 \pm 0.003$ [33].

All this suggests that the "bare weak" value of U_{us} can be somewhere in the region

$$0.215 < U_{us} < 0.235 \ .$$

We conclude that the theoretical understanding of symmetry breaking in hyperon and kaon decays is not sufficient to make full use of the experimental precision in extracting the mixing angle information. On the experimental side, two problems have to be resolved: the neutron lifetime and the electron asymmetry in the decay of polarized Σ^-. Experiments on both problems are under way and we look forward to the results.

REFERENCES

1. N. Cabibbo, Phys. Rev. Lett. 10, 531 (1963).
2. M. Roos, Nucl. Phys. B77, 420 (1974).
3. M. Bourquin et al.,
   a) Z. Phys. C 12, 307 (1982).
   b) Z. Phys. C 21, 1 (1983).
   c) Z. Phys. C 21, 17 (1983).
   d) Z. Phys. C 21, 27 (1983).
4. J. Wise et al., Phys. Lett. 91B, 165 (1980); Phys. Lett. 98B, 123 (1981).
   D. Jensen et al., Proc. Int. Europhysics Conf. on High-Energy Physics, Brighton 1983, p. 255.
5. M. Kobayashi and T. Maskawa, Progr. Theor. Phys. 49, 652 (1973).
6. O.E. Overseth, Proc. Int. Symp. on High-Energy Physics with Polarized Beams and Targets, Lausanne 1980, p. 114.
7. I. Bender, V. Linke and H.J. Rothe, Z. Phys. 212, 190 (1968).
   V. Linke, Nucl. Phys. B12, 669 (1969).
8. M. Bourquin et al., Nucl. Phys. B153, 13 (1979).
9. M. Bourquin et al., Nucl. Instrum. Methods 204, 311 (1983).
10. G. Ang et al., Z. Phys. 223, 103 (1969).
11. B. Sechi-Zorn and G. Snow, Phys. Rev. D 8, 12 (1973).
12. H. Ebenhöh et al., Z. Phys. 266 367 (1974).
13. J.A. Thompson et al., Phys. Rev. D 21, 25 (1980).
14. J. Duclos et al., Nucl. Phys. B32, 493 (1971).
15. K.H. Althoff et al., Phys. Lett. 37B, 531 (1971).
16. P. Franzini et al., Phys. Rev. D 6, 2417 (1972).
17. W. Tanenbaum et al., Phys. Rev. D 12, 1871 (1975).
18. D. Decamp et al., Phys. Lett. 66B, 295 (1977).
19. K.H. Althoff et al., Phys. Lett. 43B, 237 (1973).
20. J. Lindquist et al., Phys. Rev. D 16, 2104 (1977).
21. J.F. Donoghue and B.R. Holstein, Phys. Rev. D 25, 206 (1982).
22. P. Keller et al., Phys. Rev. Lett. 48, 971 (1982).
23. K. Tóth, T. Margaritisz and K. Szegö, CERN preprint TH.3169 (1981)
    K. Tóth, T. Margaritisz and K. Szegö, Private communication and to be published.
24. V.E. Krohn and G.R. Ringo, Phys. Lett. 55B, 175 (1975).
25. B.F. Erozolimskii et al., Soviet. J. Nucl. Phys. 30, 356 (1979).
26. P. Bopp et al., Proc. ILL Workshop on Reactor-Based Fundamental Physics, Grenoble 1983, to be published in J. de Physique.
27. C. Stratowa et al., Phys. Rev. D 18, 3970 (1978).
28. C.J. Christensen et al., Phys. Rev. D 5, 1628 (1972).
29. L.N. Bondarenko et al., JETP Lett. 28, 303 (1978).

30. J. Byrne et al., Phys. Lett. $\underline{92B}$, 274 (1980).
31. D.H. Wilkinson, Nucl. Phys. $\underline{A377}$, 474 (1982).
32. A. Sirlin, Rev. Mod. Phys. $\underline{50}$, 573 (1978).
    A. Sirlin, Nucl. Phys. $\underline{B71}$, 29 (1974).
    A. Sirlin, Phys. Rev. $\underline{164}$, 1767 (1967) and earlier references therein.
33. R.E. Shrock and L.-L. Wang, Phys. Rev. Lett. $\underline{41}$, 1692 (1978).

DISCUSSION

WOLFENSTEIN:

The determination of $\sin \theta_A$ from hyperon decays depends essentially on the $\Delta S = 0$ decay $\Sigma^- \to \Lambda e^- \bar{\nu}$. Thus the conclusion is that $\Sigma^- \to \Lambda e^- \bar{\nu}$ disagrees with $n \to p e^- \bar{\nu}$. Is this correct?

SIEBERT:

The $\Sigma^- \to \Lambda e^- \bar{\nu}$ branching ratio fixes a value for $D \cdot \cos \theta_A$, which has little impact on $\sin \theta_A$. The $\Delta S = 1$ branching ratios are proportional to $\sin^2 \theta_V + 3 \cdot \lambda^2 \cdot \sin^2 \theta_A$ which is equal to $\sin^2 \theta_V + 1.5 \sin^2 \theta_A$ in the $\Lambda \to p e^- \bar{\nu}$ decay. The hyperon-neutron disagreement cannot be blamed on $\Sigma^- \to \Lambda e^- \bar{\nu}$ alone.

WOLFENSTEIN:

The assumption of SU(3) invariance is known to be imperfect. We have in the case of the magnetic moment operator independent measurements of a quantity for half a dozen cases and find significant but not simple deviations from the SU(3) prediction. Cabibbo theory works amazingly well, but one cannot push it too far.

CHAU:

You know that from the B decay, τ_b and $\Gamma_{ub}/\Gamma_{cb} < 0.05$, a more stringent bound on s_3 was obtained. Have you used this number as an input for your analyses?

SIEBERT:

Yes. Once you assume $\sin \theta_3 < 0.1$, you are back at a simple one-angle Cabibbo fit. If we then include the nuclear β-decay in the fits, nothing changes from the quoted results of the simple fits without these data.

LITTENBERG:

What, if anything, could be learned from measuring $K^0_L \to K^\pm e\nu$?

SIEBERT:

I don't know if the $K^0 \to K^+ e^- \bar{\nu}$ matrix element can be calculated with much more reliability than the $K^0 \to \pi^+$ matrix elements in K_{e3} decay. Only an extremely precise measurement of the process $K^0 \to K^+ e^- \bar{\nu}$ would improve the value of θ in K-decays since it is a $\Delta S = 0$ transition.

STECH:

What is known about Ω^- semileptonic decay?

SIEBERT:
The branching ratio for $\Omega \to \Xi^0 e^- \bar{\nu}$ is $(0.48 \pm 0.23)\%$ from a sample of 30 events, half of which is background. This value agrees with simple phase-space extrapolations from octet hyperon semileptonic decays and with an extrapolation from Δ production by neutrinos.

RECENT RESULTS ON WEAK DECAYS FROM THE MARK II EXPERIMENT

George H. Trilling

Lawrence Berkeley Laboratory and Department of Physics
University of California
Berkeley, California 94720

INTRODUCTION

In this paper I shall bring you up to date with respect to
the MARK II Collaboration's work[1] on several aspects of weak decays.
In particular, I shall be discussing our results on the ν_τ mass,
and on τ, D^o and B lifetimes. I begin with a brief description of
the MARK II detector.

The MARK II detector, operating at the electron-positron storage
ring PEP at SLAC, consists principally of (1) a charged-particle
tracking system immersed in a solenoidal magnetic field of 2.3
Kgauss, surrounded by (2) a lead-liquid argon electromagnetic calorimeter which in turn is surrounded by (3) a steel-proportional-tube
muon identifier. The fractions of 4π solid angle covered by the
various systems are 80% for the charged-particle tracker, 64% for
the calorimeter and 45% for the full steel thickness (1 meter) of
the muon identifier. The charged-particle tracking system consists
of two separate parts both concentric with the beam line- (i) a
conventional drift chamber with sixteen layers of sense wires
between radii of 40 cm and 140 cm, and (ii) a high resolution
cylindrical drift chamber (called the vertex detector) located between the beam pipe and the inner wall of the main drift chamber.

63

This vertex detector contains seven layers of sense wires, the inner four of which are closely spaced near a radius of 10 cm and the outer three of which are also closely spaced near a radius of 30 cm. The typical sense wire resolution is 100μ. The vacuum beam pipe is a thin beryllium tube which also serves as the inner wall of the vertex detector to minimize degradation of resolution by multiple scattering. The overall charged-particle momentum resolution is typically $1.0 \times p$ GeV/c% with an additional 2% contribution from multiple scattering.

A NEW UPPER LIMIT ON THE ν_τ MASS

Present upper limits on the ν_τ mass have been set in two independent experiments. Bacino et al[2] have based their limit on the observed electron spectrum in τ decay, whereas Blocker et al[3], have determined a limit from the pion spectrum in the $\tau \to \pi + \nu_\tau$ decay mode. Both limits are 250 MeV/c^2 at the 95% confidence level.

We have used the following new approach to obtain an improved limit. We have identified the decay mode

$$\tau^\pm \to \pi^\pm + \pi^+ + \pi^- + \pi^0 + \nu_\tau$$

and studied the 4π invariant mass spectrum. The upper limit of that spectrum will, in the absence of resolution effects, differ from the known τ mass by just the neutrino mass, and the high multiplicity involved will populate the mass spectrum near the upper end just where the information content is maximal.

The analysis is based on an integrated luminosity of 158 pb^{-1} corresponding, at our c.m. energy of 29 GeV, to about 16,000 produced τ pairs. Candidate events have four charged prongs of total charge zero and at least two detected photons of energy above 250 MeV. The plane perpendicular to the event thrust is used to divide space into two hemispheres and three of the tracks are required to lie in one hemisphere with a total momentum vector acollinear with the fourth track by less than 50°. Cuts are made to ensure good measurements of the tracks, and events in which track pairs are consistent with being Dalitz decays are rejected.

Special care is also exercised to ensure that the photons used in the analysis are well measured, and unconfused with the charged particles. The $\gamma - \gamma$ mass spectrum, shown in Fig. 1, shows a clean π^0 peak, and only events in which there is one such pair with confidence level greater than 10%, and no additional pair with a χ^2 less than 10 (for one degree of freedom) are accepted. Finally, to minimize hadronic background the total energy of the $3\pi \pm \pi^0$ state

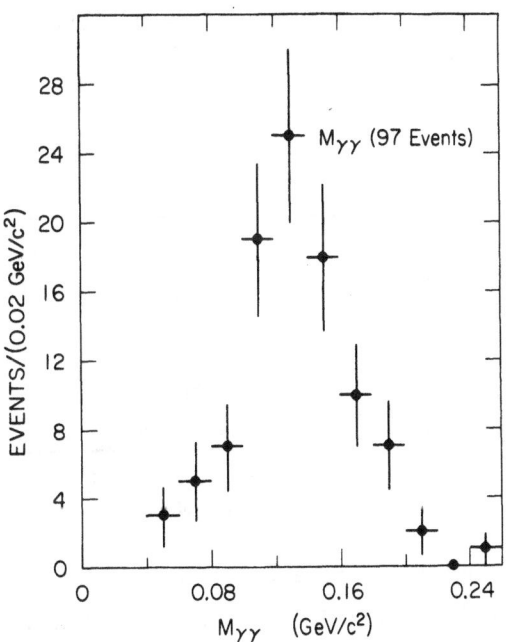

Fig. 1. $\gamma - \gamma$ invarient mass distribution for $3\pi^{\pm}\pi^0$ candidates before requiring the confidence level for the fit to the π^0 hypothesis be greater than 10%.

is required to be greater than 8 GeV for $M_{4\pi} < 1.5$ GeV/c^2 and greater than 10 GeV for $M_{4\pi} > 1.5$ GeV/c^2. The residual hadronic background is estimated to be 3% for the full sample and (10 ± 10)% in the region $M_{4\pi} > 1.5$ GeV/c^2. The 4π invariant mass resolution has an average value of (53 ± 5) MeV/c^2 for $M_{4\pi} > 1.5$ GeV/c^2.

Fig. 2 shows the 4π mass spectrum for the 60 events which survive all the cuts. The curves shown are based on dominance of the 4π state by a ρ' of mass 1570 MeV/c^2 and width 510 MeV/c^2, with ν_τ masses of both zero (solid) and 250 MeV/c^2 (dashed). The fit for zero ν_τ mass is good, and we derive from the data above 1.5 GeV/c^2 a 95% C.L. upper limit of 155 MeV/c^2. If the decay is assumed to follow pure phase space rather than the ρ' shape, the ν_τ limit is unchanged, because the spectrum shape above 1.5 GeV/c^2 is almost completely phase-space dominated. After taking due account of uncertainties in resolution and hadronic background, we obtain as our final result a 95% confidence level upper limit for the ν_τ mass of 164 MeV/c^2.

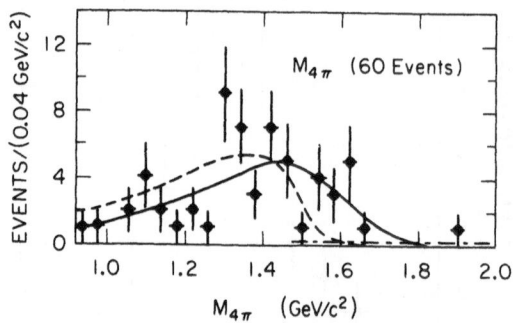

Fig. 2. $3\pi \pm \pi^0$ invariant mass distribution for the selected sample. The solid and dashed curves are explained in the text, and the dashed-dotted line represents the hadronic background.

LIFETIME MEASUREMENTS - GENERAL REMARKS

We now go to our lifetime measurements of τ, D^0 and B. The
first two of these are based on verticizing the decay tracks and
measuring the distribution of decay proper times (for τ this is
just proportional to the decay length since the τ is monoenergetic).
The procedure is discussed in adequate detail in the paper of
Jaros et al.[4] The B measurement is based on the observed distribution of impact parameters of high P_T leptons and has been discussed
in the paper of Lockyer et al.[5] Here I just want to make a few
general remarks:

(1) In all three cases the decays are identified through criteria
     which are <u>independent</u> of the decay times. We thus avoid the
     potential biases present in many of the techniques used for
     lifetime measurements, in which detection efficiency depends
     on time-of-flight.

(2) The measurement resolutions provided by the vertex detector
     are comparable to (τ, D^0) or broader than (B) the average
     time-of-flight signal. Our ability to get an accurate result
     is based on careful assurance that these resolution errors are
     random, and that their effect can be appropriately reduced with
     sufficient statistics.

(3) To obtain the best information from our data, we have used
     maximum likelihood fits of the experimental distributions to
     convolutions of theoretical distributions with expected reso-
     lutions. This procedure suffers from a potential bias - an
     underestimate of resolution will generally lead to an over-
     estimate of lifetime. We have taken account of this potential
     problem in our analysis procedure and have included in our
     systematic errors estimates of any residual contributions from
     this source.

THE τ LIFETIME

The τ lifetime is easily predicted from the standard formulas,

$$\tau_\tau = \frac{1}{\Gamma_e} B_e$$

where

$$\Gamma_e = \frac{G^2 M_\tau^5}{192\pi^3} = 0.627 \times 10^{12} \text{ sec}^{-1}$$

and B_e is the electron branching ratio. Taking for B_e our measured value of 17.6 ± 1.1%, we find for the expected τ_τ the value $(2.81 \pm 0.2) \times 10^{-13}$ sec if the coupling constant G is the same as in muon decay. The dominating uncertainty in this prediction is the error in the branching ratio. This expected lifetime corresponds to a mean decay length of about 650μ at our energy of 14.5 GeV per beam.

We base our measurement on the reconstruction of vertices of three-prong τ decays, and the determination of decay lengths between the relevant beam interaction points and these vertices. The detailed procedure has been discussed in the paper of Jaros et al.[4] Great care is exercised in the selection of τ decays where all three prongs are well measured and have at least two hits in the inner four layers of the vertex detector and at least one hit in the outer three layers. In addition, cuts on (1) total event energy, (2) total energy, momentum and invariant mass of the 3π system, and (3) mass of the 3π plus neutrals in the same hemisphere are used to ensure freedom from hadronic contamination and removal of events with converted electron pairs. Because the vertex detector measures only x, y information (in the plane perpendicular to the beam) the vertex reconstruction is done only in that plane, the known total momentum vector of the three pions then being used to convert decay length to three dimensions. The vertex fit χ^2 distribution has the expected shape for one degree of freedom and a cut at $\chi^2 < 4$ is made. Only events for which the calculated decay length error is less than 1.4 mm (about 75% of the events) are used in the final analysis. As discussed in reference 4, the position and shape of the beam distribution in x - y have been extensively studied.

To check for possible systematic biases, we have studied control samples of track triplets chosen from hadronic events to duplicate as closely as possible the kinematics of the τ decays. Unfortunately it is impossible to avoid completely the choice of tracks which may arise from charm decays, and therefore the average decay length of such a sample will always tend to have a small but finite positive value. Through insertion of known charm decay lifetimes in our Monte Carlo, we attempt to mimic as closely as possible the real situation. With appropriate choices of control triplets, we obtain mean effective decays lengths of about 100 μm, in both data and Monte Carlo. This agreement indicates that systematic offsets are under $20\sqrt{2}$ or about 30 μm.

Our decay length distribution, based on 423 three-prong decays, is shown in Fig. 3. To avoid the bias discussed in the previous section, we fit the data of Fig. 3 with a convolution in which the calculated decay length resolution is permitted to vary either through a scale factor which is a parameter of the fit or through the quadratic combination with the calculated resolution of an additional fixed error which is also a parameter of the fit. Both procedures yield the same mean decay lengths within 10μ and we average them.

After correction for an estimated 4% hadron contamination and 3.3% τ production through the two-photon process, we obtain a final mean decay length of 653 ± 56 μm. The mean τ energy after radiative corrections is 13.9 GeV, and our corresponding τ lifetime is

$$\tau_\tau = (2.80 \pm 0.24 \pm 0.30) \times 10^{-13} \text{ sec.}$$

Our systematic uncertainty has been given a fairly large value, because the above result is a preliminary one. We are still making checks which could shift it slightly, and we hope that we shall

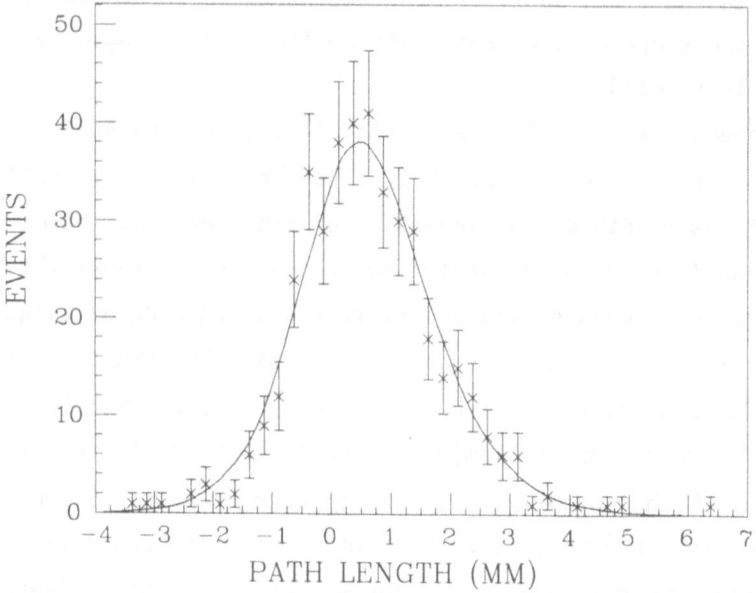

Fig. 3. τ decay length distribution. The solid curve is the
fit to the data.

soon be able to reduce the systematic error. Our preliminary result
is in excellent agreement with theory.

THE D^o LIFETIME

We identify D^o mesons through observation of the sequence

$$D^{*+} \rightarrow \pi^+ + D^o$$
$$\phantom{D^{*+} \rightarrow \pi^+ + D^o} \hookrightarrow K^- + \pi^+ \, ,$$

and choose events for which the D energy is at least 60% of the
beam energy to minimize the background. No attempt is made to
identify particles, and all tracks are tried as pions and kaons.
Oppositely charged Kπ pairs with invariant mass between 1.72 and
2.00 GeV/c^2 are considered as candidates and combined with additional pions of appropriate sign to study the distribution of
$(M_{D\pi} - M_D)$. Furthermore events are retained only if each of the

tracks which make up the D* candidate are well measured, have at least three vertex chamber hits, and can be fit to a vertex with reasonable χ^2. The π originating from the D* decay does not come from the real D^o decay vertex, but because of the low Q value of the decay it passes very close to that vertex and the success of the three-particle fit provides a check that none of the tracks have been scattered or mismeasured. The distribution of $(M_{D\pi} - M_D)$ after these cuts is shown in Fig. 4. The sample is taken to be those 27 events which fall between 143 and 149 MeV/c^2. We estimate that this sample includes two events from hadron contamination and one event from D* originating in B decay.

Decay lengths are calculated in the same manner as for the τ sample, and converted to proper times from the measured momenta. The mean lifetime is obtained from a maximum likelihood fit of the 27 events with appropriate corrections due to hadron background, B decay contributions, and possible underestimates of decay length errors. A non-D^o control sample is used to check possible systematic errors. Proper time distributions for the real data sample and for the control sample are shown in Fig. 5a and 5b respectively. Our final result for the D^o lifetime is $(4.2 \begin{smallmatrix} + 1.3 \\ - 1.0 \end{smallmatrix} \pm 1.0) \times 10^{-13}$ sec.

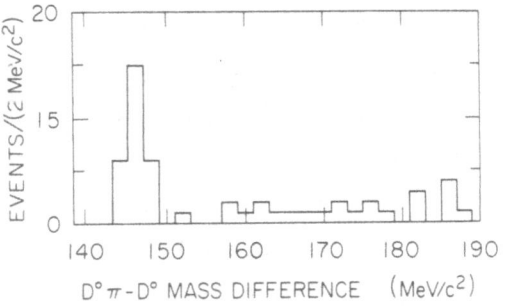

Fig. 4. The mass difference $M_{K\pi\pi} - M_{K\pi}$ for the selected sample.

71

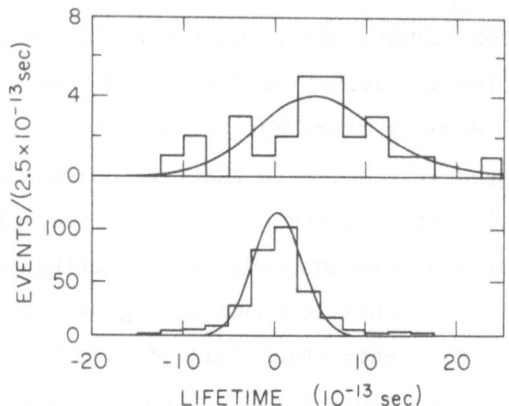

Fig. 5. The observed proper time distribution for (a) the 27
identified D^0 events; (b) the hadron control sample.
The curves are (a) the result of the fit described in
the text and (b) a Gaussian fit.

It is in good agreement with values from other experiments.

THE B LIFETIME

    Our analysis of B lifetime has been published,[5] and we have no
new results to report at this time. However, for completeness, I
briefly recapitulate the main ingredients of that analysis. We
identify heavy meson decays (bottom and charm) through detection
of electrons and muons in hadronic interactions, and set up $c - \bar{c}$
and $b - \bar{b}$ enriched samples through appropriate cuts on lepton total
momentum and transverse momentum relative to the event thrust axis.
We define for each lepton in the x-y plane an impact parameter of
its trajectory relative to the average beam position. We assume
that each heavy meson originates at the average beam position and
moves along the thrust axis in that direction which makes an acute

72

angle with the lepton momentum vector. If, with that convention, its apparent decay length is positive (negative) we assign a positive (negative) sign to the corresponding impact parameter. It is then an excess of positive impact parameters which provides a signal for a finite mean decay length for the heavy meson. The interpretation of the impact parameter distribution in terms of actual lifetimes involves consideration of cuts used in the sample enrichments, position resolution, backgrounds and the momentum spectra of the bottom and charmed mesons. Impact parameter distributions for the b and c enriched regions are shown in Fig. 6a and 6b. Fig. 6c shows the corresponding distributions for hadrons with a non-leptonic track which satisfies the cuts (other than lepton identification) required for the lepton in the b-enriched sample. We particularly note the large preponderance of positive impact parameters in Fig. 6a. From maximum likelihood fits (shown in Fig. 6) coupled with external information on charm lifetimes, we obtain our best estimate of the mean lifetime of B hadrons, averaged over the species populations weighted by semileptonic decay probabilities. Our result is,

$$\tau_B = (12.0 \; ^{+4.5}_{-3.6} \pm 3.0) \times 10^{-13} \; \text{sec.}$$

Since the impact of this result on the quark weak mixing matrix will be described elsewhere in these Proceedings, I do not discuss it here.

I do however want to mention that we are trying to improve this important lifetime measurement through two separate analyses presently in progress: (i) an improved lepton impact parameter analysis with increased statistics and more optimal procedures and (ii) a study of reconstructed vertices of hadrons produced in the b and c enriched samples. This latter procedure uses information which is largely independent of the lepton measurements, but whose interpretation requires substantial and careful Monte Carlo modeling. We hope that results from these studies will be available in the not too distant future.

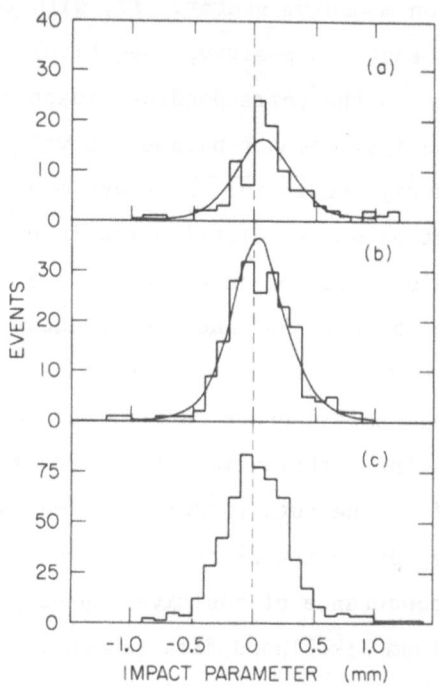

Fig. 6. Impact parameter distributions for (a) leptons in the
b - b̄ enriched sample, (b) leptons in the c - c̄ enriched
sample, (c) hadrons which satisfy all but the lepton cut
of the b - b̄ sample. The solid curves are results of
maximum likelihood fits.

REFERENCES

1. The members of the MARK II Collaboration are: G. S. Abrams,
   D.Amidei, A. R. Baden, A. M. Boyarski, J. Boyer, M. Breidenbach,
   P. Burchat, D. L. Burke, F. Butler, J. M. Dorfan, G. J. Feldman
   G. Gidal, L. Gladney, M. S. Gold, G. Goldhaber, L. Golding,
   G. Hanson, D. Herrup, R. J. Hollebeek, W. R. Innes, J. A. Jaros,
   I. Juricic, J. A. Kadyk, A. J. Lankford, R. R. Larsen,
   B. W. LeClaire, M. Levi, N. S. Lockyer, V. Luth, C. Matteuzzi,
   M. W. Nelson, R. A. Ong, M. L. Perl, B. Richter, M. C. Ross,
   P. B. Rowson, T. Schaad, H. Schellman, D. Schlatter,
   P. D. Sheldon, W. B. Schmidke, G. H. Trilling, C. de la Vaissiere,
   D. Wood, J. M. Yelton, and C. Zaiser.

2. W. Bacino et al., Phys. Rev. Lett. 42, 749(1975).

3. C. A. Blocker et al., Phys. Lett. 109B, 119(1982).

4. J. A. Jaros et al., Phys. Rev. Lett. 51, 955(1983).

5. N. S. Lockyer et al., Phys. Rev. Lett. 51, 1316(1983).

IMPROVED b LIFETIME MEASUREMENT FROM MAC*

W. T. Ford

Department of Physics

University of Colorado

Boulder, Colorado 80309

INTRODUCTION

Two recent publications, from the MAC[1] and Mark II[2] collaborations, have reported the somewhat surprising result that the lifetime
of particles made up of b quarks is in the 1 to 2 picosecond range, or
somewhat longer than the lifetimes of charm particles. Although the
charm decays are favored transitions while those of b particles depend
upon off-diagonal elements of the weak flavor mixing matrix, the
smallness of the b decay rates in face of the large available phase
space indicates that the off-diagonal elements are indeed very small.
The possibility for complete determination of the mixing matrix was
brought significantly nearer by the availability of the lifetime
information; what is needed now is to reduce the uncertainty of the
measurements, which was about 33% for both experiments. We describe
here an extension of the b lifetime study with the MAC detector,
incorporating some new data and improvements in the analysis.

THE MAC EXPERIMENT

The decay length for short-lived particles is inferred from the
distribution in the impact parameter, or distance of closest approach
to the beam, of tracks that have been identified as products of the
decay. Muons and electrons, which can be identified in the detector,
are used in the analysis; the method for isolating leptons born in b
decay is described below.

The MAC detector[3] was designed with emphasis on large acceptance
and the use of calorimeters to measure energy flow, identify electrons, and filter hadrons to facilitate muon identification. It

77

includes a cylindrical central drift chamber for tracking charged
particles which consists of ten layers of drift wires in the 5.7 kG
magnetic field of the surrounding solenoid coil. The layers, extending from 12 cm to 45 cm, each provide point measurement accuracy of
about 200 μm. The smallness of the inner radius helps the precision
of impact parameter measurements; the relatively short tracking
length, on the other hand, limits the precision because it leads to a
fairly large curvature error ($\delta p/p \simeq 0.065p$), which in turn affects
the extrapolation to the production point.

The drift chamber is surrounded by electromagnetic and hadron calorimeters, which provide energy-flow information (used to pick out multihadron events and to define their thrust axis) and identification of
electrons. Layers of lead interspersed with proportional wire chambers constitute the electromagnetic shower chamber, amounting to 16
radiation lengths of material. In the hadron calorimeter, layers of
steel alternate with proportional wire chambers, such that normally
incident particles traverse 91 cm of steel. The calorimeter steel in
both the central and endcap regions is magnetized to about 17 kgauss
by toroid coils. The entire calorimetric detector is surrounded by
drift chambers for muon identification and tracking. These chambers
determine the radial and axial components of the location and direction of particles penetrating the hadron calorimeters.

EVENT SELECTION

The parent sample for this analysis consists of approximately 75000
multihadron events having five or more charged prongs and calorimetric
energy flow consistent with production by single photon annihilation.
Cuts on the total energy, its component perpendicular to the beam, and
the net energy imbalance eliminate two photon annihilation
events[4],[5]. The muon (electron) sample corresponds to an integrated luminosity of 160 (127) pb^{-1} at a center-of-mass energy of
29 GeV. This sample is about 1.5 times as large as the one used in
ref. 1.

Considering first the muon selection, candidates were reconstructed
and momentum-analyzed by interpolation between isolated track segments
in the outer drift chambers and the primary event vertex, through the
toroidal magnetic field of the calorimeter, taking into account the
ionization energy loss of the particle in the calorimeter. It was
required that the resulting track be matched within typically 1° in
polar angle and 30% in momentum to a track reconstructed in the central drift chamber, and within 2 to 10 degrees to a segment reconstructed from the energy deposited in the central or endcap calorimeter. (The actual cuts were made on the appropriate χ^2, computed with
inclusion of all measurement errors and the effect of multiple scattering, and are momentum dependent.) It was required that the calorim-

eter pulse heights correspond to those of a single minimum-ionizing track.

An electron candidate is defined as a track in the central drift chamber associated with a shower in the electromagnetic shower chamber and with no significant energy deposition in the hadron calorimeter. Only tracks with momentum greater than 1.8 Gev/c and in the fiducial region $|\cos\theta| < 0.7$ are considered. The energy deposited in the calorimeters is measured in a 2° (azimuthal) × 5° (polar) cone around the direction of the track at the exit of the drift chamber and is required to follow a pattern typical of the development of an electromagnetic shower in our detector, as determined from non-radiative and radiative Bhabha events.

The next step is to find criteria for selecting those leptons that come from b decay, and to establish the purity of the resulting samples. For this purpose we use a Monte Carlo calculation which incorporates a version of the Lund string model to simulate the production and decay of hadrons[6] combined with code which traces in detail their interactions and the response of the detector[7]. We find that soft particles (those having less than about 2 GeV) are most likely to originate from one of the several background processes. In the case of the muons, the principal backgrounds are pion and kaon decay, and penetration of charged hadrons through the calorimeters with little or no interaction (punch-through). For the electrons, the contamination comes from e^+e^- pairs (photons converting in the vacuum pipe and Dalitz pairs), τ and ψ decays and hadron misidentification. About 50% of the electron pairs can be easily identified. The residual contamination after cuts comes mostly from hadrons having early interactions that produce a predominantly electromagnetic cascade.

Finally, the discrimination between the heavy b quarks and lighter c quarks that decay to leptons is achieved via the component of the lepton's momentum perpendicular to the thrust axis of the event ($p_\perp$). This is because a heavy hadron follows closely the direction of the primary quark which in turn is well approximated by the thrust axis. On decay, the heavy parent imparts a large transverse momentum to the daughter lepton. Fig. 1 shows the $p_\perp$ distribution for muons originating, according to the Monte Carlo, from b semileptonic decay compared with the distributions for c decay and for background (π, K decay and punchthrough). Similar distributions are found for the electron sample. The validity of the background calculations has been checked by comparing with observation in the case of three-prong τ lepton decays, for which all apparent muons and electrons must come from background processes. The specific kinematic cuts applied in the present analysis were total lepton momentum greater than 2 GeV/c and $p_\perp$ greater than 1.5 GeV/c. We find about 30% of both samples is background and about 18% comes from charm decays (see Table 1 below). These estimates are somewhat higher than those quoted in ref. 1, as a consequence of improved Monte Carlo statistics and improved knowledge of the b and c semileptonic branching ratios (see below).

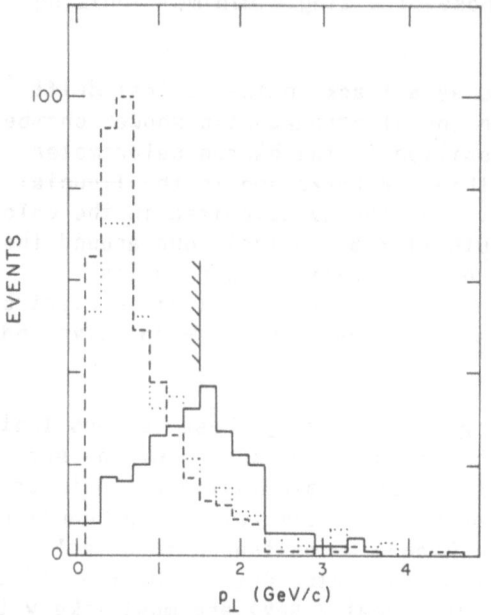

Fig. 1. Calculated $p_\perp$ distributions of contributions to the muon sample, for events having momentum greater than 2 GeV, from b (solid curve) and c (dashed curve) decays and from π/K and punchthrough background (dotted curve).

Fig. 2. Direction vectors and production and decay points relevant to heavy hadron leptonic decay.

IMPACT PARAMETER MEASUREMENT

The lifetime of the particles which decay to produce the observed leptons is inferred from the distribution in impact parameter of the lepton tracks with respect to the interaction point. In Fig. 2 are defined the flight path, ℓ, of the parent hadron (e.g., B meson), the directions of travel of the parent and lepton (e.g., muon), the decay angle ψ and impact parameter δ, all as projected onto a plane perpendicular to the beam axis. The quantity δ is defined as a signed quantity. That is, we assume that the parent travelled along the thrust axis in the direction that is consistent with the further assumption that the lepton went forward in the laboratory. The intersection of the lepton trajectory with the thrust axis then measures the decay path, which may come out negative because of measurement error or (rarely) because the lepton actually went backward in the decay. The signed impact parameter is this decay path times sinψ (with an additional factor for projection onto the x,y-plane), which is nice because ψ tends to shrink with the B meson's energy as 1/γ, compensating the relativistic time dilation γ factor in the decay path and so largely removing the dependence of δ on the B fragmentation function. Fig. 3 shows the distributions of δ for b's and c's as computed with the Monte Carlo without detector resolution. Taking the average over the sample distribution, the expression for ⟨δ⟩ is

$$\langle \delta \rangle = \langle \beta \gamma \sin\psi \sin\theta \rangle c\tau \equiv \alpha c\tau, \tag{1}$$

where θ is the polar angle of the impact parameter vector. The constant α, computed with the Monte Carlo program, is about 0.45 (0.15) for bottom (charm). It is confirmed to be very insensitive to the fragmentation function (e.g., α decreases by 11% as the average B momentum is changed from 0.8 to 0.5 times the beam momentum).

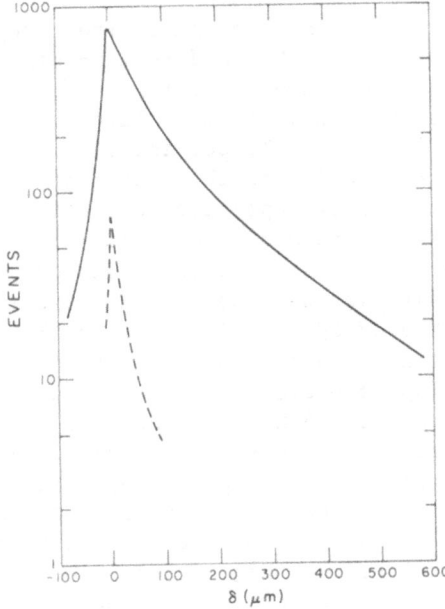

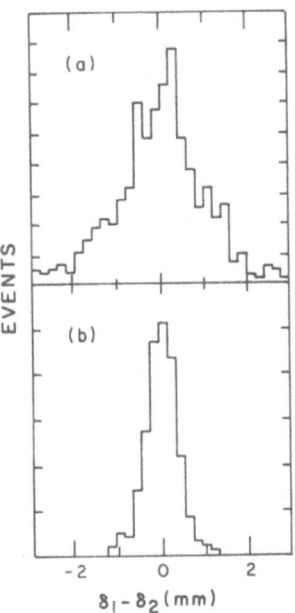

Fig. 3. Calculated impact parameter distributions for b (solid) and c (dashed) for an ideal detector.

Fig. 4. Separation between the two tracks in a mu pair event; (a) without, and (b) with beam momentum constraint.

The effect of approximating the B direction by the thrust axis (which has been accounted for in the calculation of α) is to cause a true positive δ to appear negative when the angle of the muon trajectory falls between those of the B flight path and the thrust axis, diluting the observed effect. Detailed calculations show that the resulting loss of sensitivity from this effect is negligible for b particles and about a factor of three for charm. On the other hand, for π and K decay and converted gamma ray backgrounds there is nearly complete cancellation. Thus quite fortuitously we get a much bigger effect in the impact parameter for the signal than for the backgrounds.

The precision of the measurement of δ is determined by the precision of the extrapolation of the lepton track reconstructed in the central drift chamber, and by the effective size of the beam interaction volume, including the effect of any uncorrected shifts with time of the beam position. The beam position for each run was determined by a simultaneous fit to all of the Bhabha events in that run. A plot of the results as a function of time was then used to establish beam position values for each block of runs between significant changes. From these data we find the effective rms beam size to be about 0.5 mm (horizontal) by 0.1 mm (vertical).

The uncertainty of the track extrapolation is computed from propagation of the point measurement error, as established from χ^2 distributions for track fits. It can also be measured directly from the separation near the beam between the two tracks in muon pair events, as illustrated by the histogram in Fig. 4(a). To see the effect of the curvature uncertainty on the extrapolation to the beam region, we have plotted in Fig. 4(b) the same two-track separation after the tracks have been refit with the known momentum of the muons imposed as a constraint on the track fits. In this case a factor of two or so improvement results. For the inclusive leptons used in the lifetime measurement, some improvement is achieved by refitting the tracks with the addition of such information as is known from devices external to the drift chamber, i.e., the toroid spectrometer momentum, which defines the expected curvature within about 30% for muons, and the shower chamber, which gives the electron energy within about .35 × $\sqrt{E}$ after we allow for the loss of part of the shower caused by cuts that carve out the shower energy belonging to a particular track. Overall the gain from these track refits is around 20% for the present event sample. This is new since the work of ref 1. Fig. 5 shows the distribution in the uncertainty of δ; the few events having uncertainty greater than 1 mm are now excluded from the sample.

Since the uncertainty differs among the events, the impact parameter distributions are constructed by weighting each event by its reciprocal squared error, so that the information contributed by each event is properly counted. These distributions are shown in Fig. 6 for the muon, electron and combined samples. Although the resolution is insufficient to let us see the exponential-like shape expected for an ideal detector (see Fig. 3), there is a clear indication, particularly in Fig. 6(c), of a shift of the distribution toward positive values of δ. The mean value for the combined sample is about 120 μm.

Since the beam is quite elongated in the horizontal relative to the vertical dimension, and in fact is known to move about more in that dimension also, it is of some interest to check that the above result is not an artifact of beam shifting by comparing events that see the

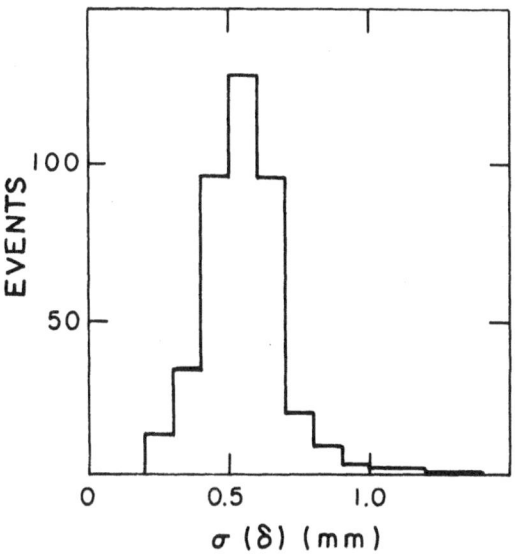

Fig. 5. Distribution of measurement uncertainties of the decay impact parameter, δ.

beam edge-on with those that do not. (The azimuthal symmetry of the event sample should insure against biases from this cause anyway.) We show in Fig. 7 the δ distribution of the combined sample divided into horizontal (±45°) and vertical subsamples. We see that although the numbers of events are about the same in these two samples, the sum of weights is nearly a factor of two larger, and the width of the distribution narrower for the horizontal events, as expected. The plot statistics show that ⟨δ⟩ is nearly the same for both muon samples, but for the electrons the vertical (poor beam precision) events give a negative mean value, which is offset in the full sample by a relatively large positive value for the horizontal events. We keep all events, letting the weights take care of the differences in quality. To the extent that there is a difference, it supports the conclusion that the positive ⟨δ⟩ is not caused by beam drift.

To understand the resolution function, including the issue of possible bias, a control sample was made from the parent multihadron sample by taking all charged particle tracks (including leptons) assigned to the primary vertex which pass the same momentum and $p_\perp$ cuts required for the data sample. Momentum and $p_\perp$ distributions of the surviving events are quite similar to those of the data sample. The treatment of these events differs from that of the lepton data only in that there is no external curvature information used in the track fits. The resulting weighted δ distribution for these events is shown in Fig. 8. There are a total of 18575 tracks; the mean value is 29±8 μm, and the median (see below) 23±7 μm. The Monte Carlo predicts

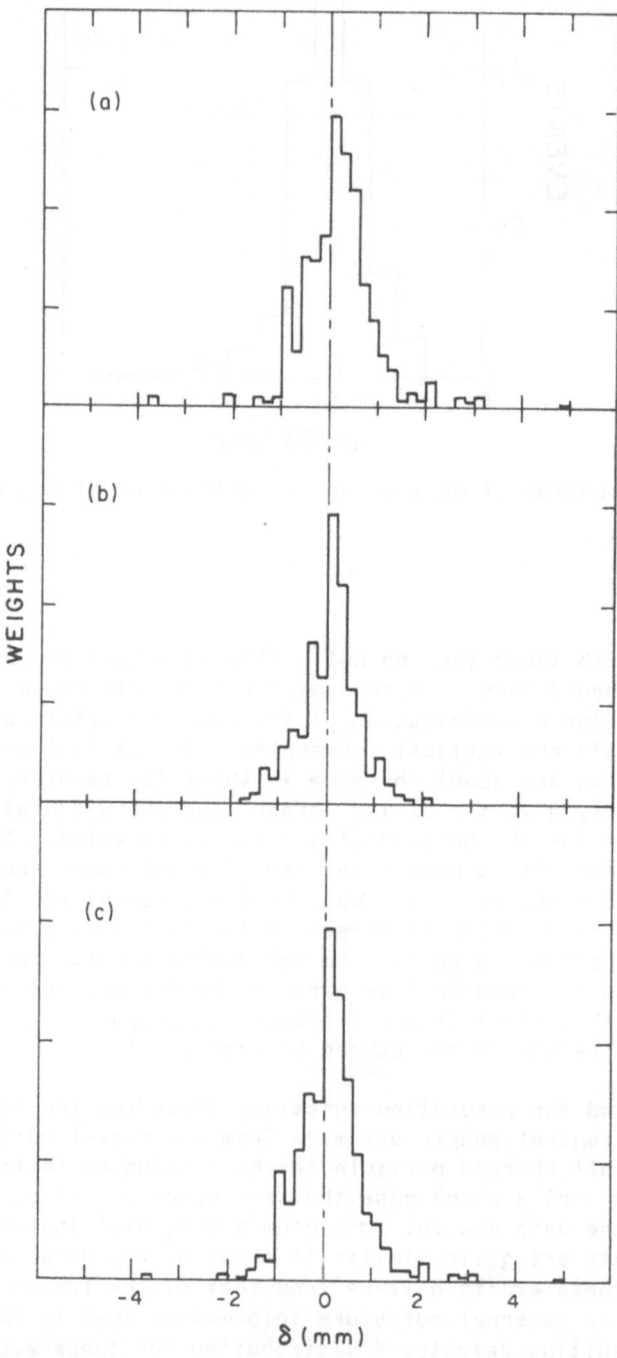

Fig. 6. Distributions of δ for (a) muons, (b) electrons, and (c) both.

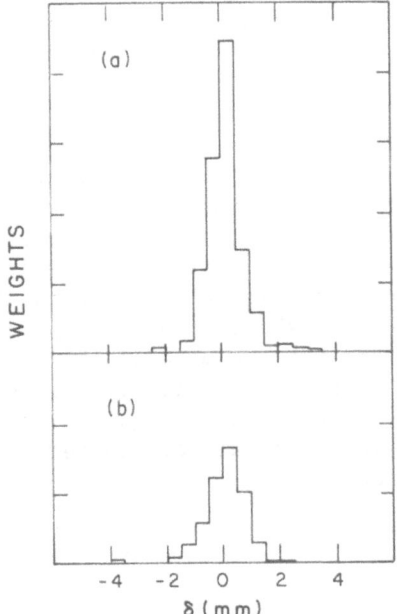

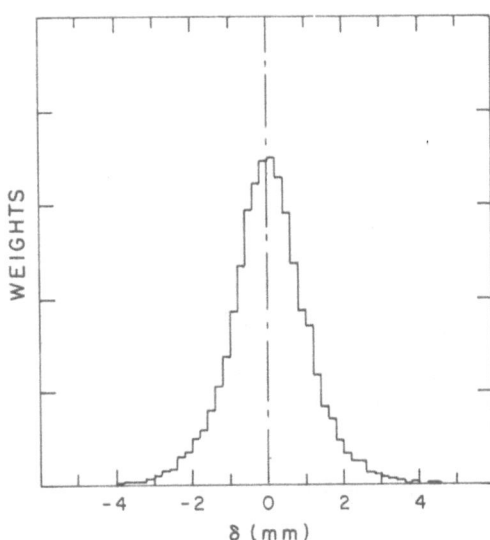

Fig. 7. Impact parameter distribution for the combined data sample: (a) horizontal, and (b) vertical events.

Fig. 8. Impact parameter distribution for the control sample.

a contribution from short-lived particles in the control sample of $(8 \ \mu m + 0.03c\tau_b) = (22 \pm 6) \ \mu m$ for $\tau_b = 1.6$ psec. We conclude that the resolution function is highly symmetric, contributing a mean-value shift of less than 10 μm or so; the final result below includes a component of 10 μm in the systematic error to account for this residual possible bias.

The effects of pions and kaons that decay before or just after entering the drift chamber must also be considered. We find that the δ distribution from early pion decays is very narrow and well centered; for the kaons the distribution is much broader, and there is a hint in Fig. 6(a) of a contribution at large |δ| from such decays in the data. These events add to the standard deviation of the distribution, but little to the mean, as we have confirmed by relaxing cuts to admit more decay events.

To avoid the problem that a few events at large |δ| may overly influence the mean, we have chosen in the present analysis to use the median value of δ instead to determine the lifetime. For smooth distributions the difference between mean and median is negligible when the resolution function is symmetric and wider than the unsmeared distribution, as it is here. Henceforth the symbol "⟨δ⟩" refers to the median value of the data distribution. The median value for the combined sample is 120 ± 28 μm, somewhat more than four standard deviations from zero.

It's perhaps worth a comment at this point that in considering how best to extract the answer for τ_b from these distributions we have elected not to use a maximum-likelihood fit. The advantage would be that in principle maximal use is made of the information in the shape of the distribution. There are two practical limitations, however. First is that for our detector the resolution strongly dominates the distribution so little information about the shape from the physics survives. The second is that any underestimate we make of the width of the resolution function can force the exponential (i.e., physics) component to be overestimated in the fit. Our method requires only that the resolution function be symmetric, which we have tested via the control sample study as described above.

RESULTS AND DISCUSSION

Eq. 1, which gives the b lifetime in terms of the average (or median) impact parameter, needs to be modified to account for the non-b components of the data sample:

$$\langle \delta \rangle = f_b \alpha c \tau_b + f_c \delta_c + f_{bg} \delta_{bg}, \tag{2}$$

where the subscripts b, c and bg refer to b- and c-flavored particles and light-quark background (excluding early decays to muons), respectively. Each factor f_i is the fraction of the sample corresponding to component i, and α is defined in equation (1). Here f_c includes cascades b→c→lepton. The calculated impact parameter for the control sample discussed previously is used to estimate $\delta_{bg} = (8 \mu m + 0.03 c \tau_b)$. For the muons we add to this a term $f_d \delta_d$ to account for π/K-decay occuring before the drift chamber. The values of all these parameters are given with their uncertainties in Table 1. The QCD and fragmentation parameters in the Monte Carlo used to calculate them were adjusted to fit measured hadron-production spectra[4], including energy-energy correlations[4] and the fragmentation of heavy flavor as measured with the momentum and $p_\perp$ distributions for inclusive leptons[5],[8],[9]. The b leptonic branching fraction used was[10] 12%. The charm leptonic branching fraction was taken to be[9],[11] 8%, and the lifetimes of charm particles used in the calculation, in units of 10^{-13} seconds, are[12] (D^0) 4.0; ($D^\pm$) 9.3; ($F^\pm$) 2.9 (population-weighted average = 5.5).

From Table 1 it can be seen that the expected values of $\langle \delta \rangle$ for charmed particle decays and for background, when weighted by the corresponding fractions, is of order 5-10 μm, compared with the observed values of 80 to 160 μm for the data samples. We conclude that the observed non-zero $\langle \delta \rangle$ in the data samples is to be attributed mainly

Table 1. Summary of parameters used to compute τ_b.

	μ sample	e sample
No. of events	238	160
f_b	0.53±0.07	0.52±0.07
f_c	0.18±0.04	0.18±0.04
f_{bg}	0.27±0.06	0.30±0.07
f_d	0.02±0.01	(0)
δ_c (μm)	24 ± 11	19 ± 11
δ_{bg} (μm)	8 ± 4	8 ± 4
δ_d (μm)	80 ± 46	-----
α	0.44±0.03	0.46±0.03
$\langle\delta\rangle$ (μm)	159 ± 39	83 ± 42
τ_b $(10^{-12}$ sec)	2.00±0.52	1.04±0.56

to the term proportional to τ_b in equation (2). This conclusion is
insensitive to the uncertainties in f_i's and δ_i's. Assigning a systematic uncertainty of 10 μm for $\langle\delta\rangle$ and combining with the other systematic errors quoted in Table 1 we find the lifetime values for the
two samples given in the last line of Table 1. Both are non-zero and
in reasonable agreement with each other. Combining the two we have
finally

$$\tau_b = (1.6 \pm 0.4 \pm 0.3) \times 10^{-12} \text{ sec,}$$

where the first error quoted is statistical, the second systematic.
Because the second and third term on the right-hand side of eq. 2 are
small, the systematic error in this measurement is primarily in the
scale factors in the first term of eq. 2. This result is consistent
with our previously published value[1] of
$(1.8 \pm 0.6 \pm 0.4) \times 10^{-12}$ sec, and with the Mark II value[2] of
$(1.2 +0.45 -0.36 \pm 0.3) \times 10^{-12}$ sec. The present result is not independent of our earlier one[1], but rather is an updating of it.

   With the assumption that systematic and statistical uncertainties
of the MAC and Mark II experiments may be combined as independent
quantities, and ignoring their non-Gaussian aspects, we obtain the
current world average

$$\tau_b = (1.4 \pm 0.4) \times 10^{-12} \text{ sec.}$$

The implications for flavor mixing will be discussed in several other
talks at this conference.

ACKNOWLEDGEMENTS

This work was supported in part by the Department of Energy, under contract numbers DE-AC02-81ER40025 (CU), DE-AC03-76SF00515 (SLAC), and DE-AC02-76ER00881 (UW), by the National Science Foundation under contract numbers NSF-PHY80-06504 (UU), NSF-PHY82-15133 (UH), NSF-PHY82-15413 and NSF-PHY82-15414 (NU), and by I. N. F. N.

REFERENCES

*MAC collaborators are: E. Fernandez, W. T. Ford, A. L. Read, Jr., J. G. Smith, Department of Physics, University of Colorado, Boulder, Colorado 80309; R. De Sangro, A. Marini, I. Peruzzi, M. Piccolo, F. Ronga, Laboratori Nazionali Frascati dell' I.N.F.N., Italy; H. T. Blume, H. B. Wald, Roy Weinstein, Department of Physics, University of Houston, Houston, Texas 77004; H. R. Band, M. W. Gettner, G. P. Goderre, B. Gottschalk,[a] R. B. Hurst, O. A. Meyer, J. H. Moromisato, W. D. Shambroom, E. von Goeler, Department of Physics, Northeastern University, Boston, Massachussets 02115; W. W. Ash, G. B. Chadwick, S. H. Clearwater, R. W. Coombes, H. S. Kaye, K. H. Lau, R. E. Leedy, H. L. Lynch, R. L. Messner, S. J. Michalowski,[b] K. Rich, D. M. Ritson, L. J. Rosenberg, D. E. Wiser, R. W. Zdarko, Department of Physics and Stanford Linear Accelerator Center, Stanford University, Stanford, California 94305; D. E. Groom, Hoyun Lee, E. C. Loh, Department of Physics, University of Utah, Salt Lake City, Utah 84112; M. C. Delfino, B. K. Heltsley, J. R. Johnson, T. L. Lavine, T. Maruyama, R. Prepost, Department of Physics, University of Wisconsin, Madison, Wisconsin 53706

1. E. Fernandez, et al., Phys. Rev. Letters 51,1022(1983).
2. N. S. Lockyer, et al., Phys. Rev. Letters 51,1316(1983).
3. MAC Collaboration, in Proceedings of the International Conference on Instrumentation for Colliding Beams, SLAC-250, Edited by W. Ash, 1982, p. 174; W. T. Ford, et al., Phys. Rev. Letters 49,106(1982); Roy Weinstein, in American Institute of Physics proceeding No. 98, Particles and Fields subseries, No. 29 (Nov., 1982); G. Gidal, B. Armstrong, and A. Rittenberg, Lawrence Berkeley Laboratory Report No. LBL-91 Supp., 1983.
4. B. K. Heltsley, Univ. of Wisconsin Report WISC-EX-83/233, 1983 (PhD thesis, unpublished).
5. E. Fernandez, et al., Phys. Rev. Letters 50,2054(1983).
6. T. Sjostrand, Comput. Phys. Commun. 27,243(1982); Lund Univ. Report LU TP-82-7 (1982, unpublished).
7. Electromagnetic showers are simulated by EGS, described in R. L. Ford and W. R. Nelson, SLAC Report No. SLAC-210, 1978. Hadronic cascades are simulated by HETC, described in T. W. Armstrong, in "Computer Techniques in Radiation Transport and Dosimetry", edited by W. R. Nelson and T. M. Jenkins. (Plenum Press, New York, 1980).

8.   D. Schlatter, SLAC-PUB-2982, 1982 (unpublished).

9.   M. E. Nelson, et al., Phys. Rev. Letters $\underline{50}$,1542(1983).

10. S. Stone, in Proceedings of the 1983 International Symposium on
     Lepton and Photon Interactions at High Energies, edited by
     D. G. Cassel and D. L. Kreinick, Cornell University, 1983
     (pp. 212-216, and ref. 24).

11. R. Brandelik, et al., Phys. Letters $\underline{70B}$,387(1977); R. Schindler,
     et al., Phys. Rev. D $\underline{24}$,78(1981); W. Bacino, et al.,
     Phys. Rev. Letters $\underline{43}$,1073(1979); $\underline{45}$,329(1980); J. M. Feller, et
     al., Phys. Rev. Letters $\underline{40}$,274,1677(1978).

12. G. Kalmus, in Proceedings of the 21st International Conference on
     High Energy Physics, Edited by P. Petiau and M. Porneuf (Paris,
     1982), p. C3-431, and references therein.

TRILLING:

Did you attempt to see what happens to your signal if you cut on the angle between thrust and lepton direction? The large angle events should contain all the signal, and the small angle events should show only very small signal.

FORD:

No, that's a good suggestion; we did not.

FRANZINI:

Could Trilling or Ford comment on the possible disagreement between the lifetime limit given by JADE and your present results?

TRILLING:

There is no real disagreement, but in fact I do not believe that the JADE result should be taken too literally since it involves what in my opinion is an incorrect statistical analysis. A correct analysis would give a higher limit.

FORD:

The size of the effect that I think Trilling refers to is roughly a factor $\sqrt{2}$ increase.

KLEINKNECHT:

What precision on the b lifetime do you expect with your new vertex detector around end of 1985?

FORD:

Depending how much luminosity we get, let's say a factor of 4 for measurement precision, which becomes maybe $4/\sqrt{2}$ given limited luminosity over one year. What I think is more important is we'll get measurement precision for each event that is comparable to the size of the effect, so we begin to see real effects in the spectrum shape. That would be more satisfying.

THE B AND ITS DECAYS

Paul Avery

Wilson Laboratory
Cornell University
Ithaca, NY   14853

INTRODUCTION

Considerable theoretical attention has been focussed on B meson decay in the last few years. Part of the reason for this interest is the recognition that B decay measurements provide strong tests of the standard model of electroweak interactions. For example, the B lifetime and the ratio $\Gamma(b \to u\ell\nu)/\Gamma(b \to c\ell\nu)$ measure quark mixing angles[1] (in the Kobayashi-Maskawa mixing scheme, for instance) that cannot be determined from the charm and strange sectors alone. Measurements of dileptons in B decay have also eliminated several classes of topless models from contention, while other "exotic" models, such as those that predict $b \to q\ell\ell$ or $b \to \overline{qq}\ell$, have been ruled out[2] by charged energy measurements and by the reconstruction of exclusive B decays.

In order to paint a unified picture of the experimental status of B decay, I have organized this talk along the following lines. In section 2 I give a brief description of the Υ system and present the first evidence for the fifth and sixth Υ resonances. The discussion on B decay commences in section 3 with a description of the spectator model and color separation of hadronic final states. These ideas are carried forward to section 4 in the discussion of inclusive properties of B decays. Topics include the semileptonic branching fraction, charged and neutral particle multiplicities, and kaon and baryon yields. Because of the recent progress in measuring exclusive B decay modes, I devote a substantial amount of space to this issue in section 5. I return to the spectator model in section 6 in order to discuss an important prediction of Quantum Chromodynamics (QCD) - namely color mixing suppression - and its experimental verification in $B \rightarrow \psi + X$ decays. Finally in section 7 I show how the branching fractions for two body decay modes of the B can be calculated. Some of these decay modes turn out to have surprisingly large rates.

In order not to overlap with other speakers of the conference, I will not discuss lepton results or mixing angles, except briefly, because those topics are adequately covered in the talk by J. Lee-Franzini in these proceedings. Similarly, I will leave out some interesting material concerning the B lifetime measurements of the MAC and MARK II collaborations, since they are described in the talks by W. Ford and G. Trilling elsewhere at this conference. At the time of writing this review, no machine other than CESR had collected any significant amount of information concerning the B decay; consequently the data I present here came exclusively from the CLEO and CUSB experiments at CESR.

2. THE T SYSTEM

The four resonances (see Fig. 1) comprising the T system have been extensively studied since 1979. The first three of these lie below the $B\overline{B}$ production threshold and are correspondingly narrow, with full widths measured in keV. The T(4S), in contrast, has a mass of 10.578 GeV and a width of 32 MeV. It lies above the threshold for open bottom (or beauty, if you like that sort of thing) production and, indeed, is found to decay almost 100% of the time

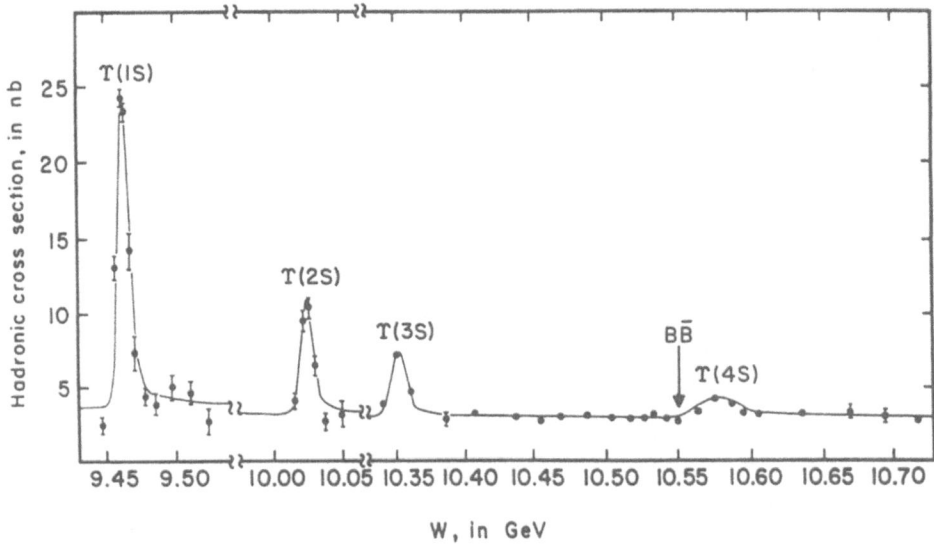

Fig. 1. The T system circa 1983.

into B mesons. The T(4S) is believed to be too light to decay into
BB̄* because the almost monochromatic photon from B* → Bγ decay has
not been seen at its expected energy of 50 MeV. In fact, the CUSB
group reports B[T(4S) → BB̄*] < .06 - .08 (90% C.L.) for photon
energies E_γ = 20-135 MeV.[3]

The last 6 months of data taking at CESR have indicated the
presence of at least one, and perhaps two or three, new T resonances
(Fig. 2) above the T(4S). Their parameters and those of the old T
system[4] are listed below:

	Mass (MeV)	FWHM (MeV)	Γ_{ee} (keV)
T(1S)	9459.9 ± 0.1	0.042 ± 0.03	1.26 ± 0.06
T(2S)	10023.8 ± 0.5	∿0.032	0.54 ± 0.03
T(3S)	10355.5 ± 0.5	∿0.027	0.40 ± 0.03
T(4S)	10578.0 ± 3.0	32 ± 8	0.32 ± 0.06
	10877 ± 6	90 ± 15 ± 20	0.20 ± 0.06
	11014 ± 3	50 ± 15 ± 20	0.08 ± 0.03

The values for the new resonances are preliminary and come from CLEO
data. The variations in R in this energy regime are rather

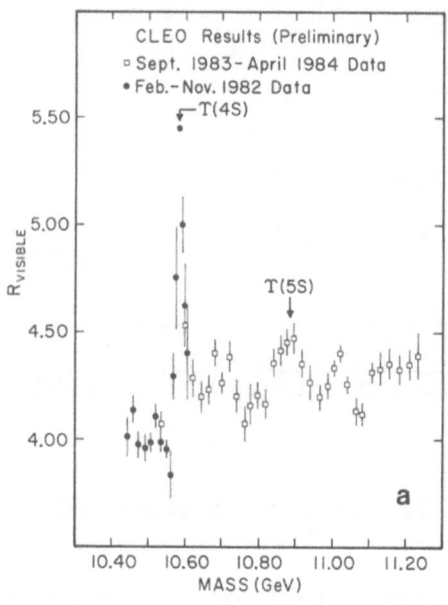

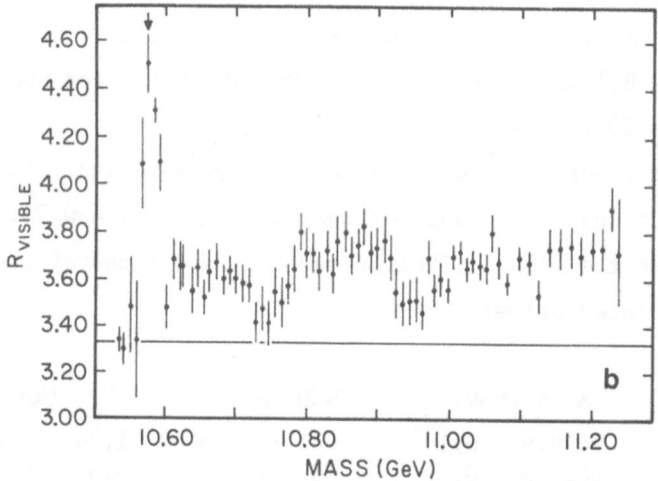

Fig. 2.   R. measurements in the region above the T(4S) for CLEO (a) and CUSB (b).

complicated and appear to be reminiscent of the undulations exhibited by the charmonium system near $D\overline{D}$ threshold.

3. THE SPECTATOR MODEL

Spectator processes (Fig. 3), in which the accompanying light quark or antiquark plays no role in the decay, are expected to comprise a larger fraction of B decays than of D decays. The reason for this is that the b quark is about 3 times heavier than the c quark and the spectator decay rate goes as the fifth power of the quark mass, faster than that of competing processes. Because of this relatively large mass, QCD corrections to the semileptonic and nonleptonic decay rates are held to the 15–25% level and the calculation of these corrections converges better for B decays than for D decays. For these reasons the B meson is regarded as being a relatively clean laboratory for testing the standard model.

Some b spectator decays are shown in Figure 3. The process can be thought of as a decay of the b quark into a c quark plus a virtual $\overline{W}$, with subsequent decay of the W into leptons or quarks (This visualization will become important later in this section when I discuss the D momentum spectrum). Note that theoretically one could substitute a u quark for the c quark in the decays pictured in Fig. 3, e.g. $b \rightarrow ue\nu$, $b \rightarrow u\overline{u}d$, $b \rightarrow u\overline{c}s$, etc. However, studies of the endpoint

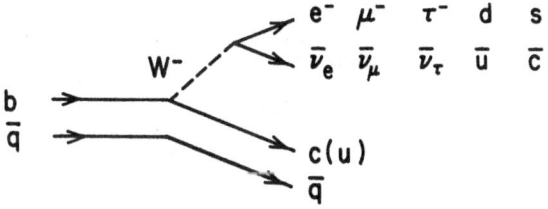

Fig. 3. The spectator model of B decay.

of the lepton momentum spectrum in B decay yield the following upper limits:[5]

CLEO $\Gamma(b \rightarrow u\ell\nu)/\Gamma(b \rightarrow c\ell\nu) < 0.04$ (90% C.L.)
CUSB $\Gamma(b \rightarrow u\ell\nu)/\Gamma(b \rightarrow c\ell\nu) < 0.055$ (90% C.L.)

which clearly rule out any significant contribution from $b \rightarrow u$ transitions. I will henceforth assume that $b \rightarrow u$ processes are negligible.

Using some rather simple kinematic arguments plus elementary QCD we can derive some important consequences of B spectator decay. First, since the virtual W is colorless the hadronic decay products of the W should evolve independently of those originating with the $c\bar{q}$ (c quark plus spectator antiquark) system. Some "color mixing" is expected to take place at the 10% level (as discussed in section 6), but it should be a minor effect. Second, the energy of the $c\bar{q}$ system is dominated by the energy of the c quark (2-3 GeV) because the spectator antiquark is light and has little Fermi motion inside the B. Third, this last argument implies that the mass of the $c\bar{q}$ system will be dominated by the rest mass of the c quark, so that the $c\bar{q}$ most frequently should end up as a D or a D* meson with perhaps 1 additional pion. Fourth, the first and second predictions imply that the momentum spectrum of the $c\bar{q}$ system should resemble the c quark momentum distribution in V-A decay and, furthermore, that it should be the same for both semileptonic and nonleptonic decays, since the W decays independently of the $b \rightarrow c$ vertex. The momentum spectrum will <u>not</u> resemble the phase space distributions expected for nonleptonic decays.

These predictions are supported by data taken at CLEO. In B semileptonic decay, $B \rightarrow X\ell\nu$, the shape of the lepton momentum spectrum indicates that $M_X \sim 2$ GeV while charged multiplicity measurements (discussed in section 4) show that the charged multiplicity for X is 2.8 ± 0.3. Both of these results strongly suggest that the $c\bar{q}$

system does indeed fragment into D or D* mesons with little accompanying pion production. Direct support for the charm momentum prediction comes from their study of the inclusive D° momentum spectrum in B decay.[6] In this analysis the D° signal for each momentum interval was obtained by fitting the $K^-\pi^+$ (and charge conjugate) mass distribution appropriate for the interval (Fig. 4). Instead of requiring kaon identification, they demanded that the kaon helicity angle satisfy $|\cos\Theta_k| < 0.8$ in order to reduce the combinatoric background. The resulting D° spectrum shown in Figure 5 agrees with the predicted momentum distribution for the c quark in $b \to ce\nu$ decay and strongly disagrees with a phase space distribution (dashed line in Figure 5). As a byproduct, CLEO also finds,

$$\# D^\circ/B \text{ decay} = 0.8 \pm 0.2 \pm 0.2.$$

An analysis of charged D* mesons produced in B decays yields a quite similar momentum spectrum (Fig. 5) and a total production rate of $0.35 \pm 0.08 \pm 0.12$ D^{*+} mesons per B decay. Assuming $N(D^{*+}) = N(D^{*\circ})$, we see that within large errors D* mesons dominate charm production in B decay.

4. INCLUSIVE PROPERTIES OF B DECAY

4.1 Leptons

Historically the observation of the large production of leptons at the $\Upsilon(4S)$ resonance clinched the argument that the $\Upsilon(4S)$ decayed into mesons containing a new quantum number which subsequently underwent weak decay. The decay into leptons is a pure spectator process as shown in Figure 1. Recently, several PEP and PETRA experiments[1] have joined the CESR groups in reporting B semileptonic branching fractions, albeit with larger systematic errors. For completeness I list below the most recent measurements[5] of electron and muon production (averaged over $\bar{B}^\circ$ and B^-) by CLEO and CUSB and leave a

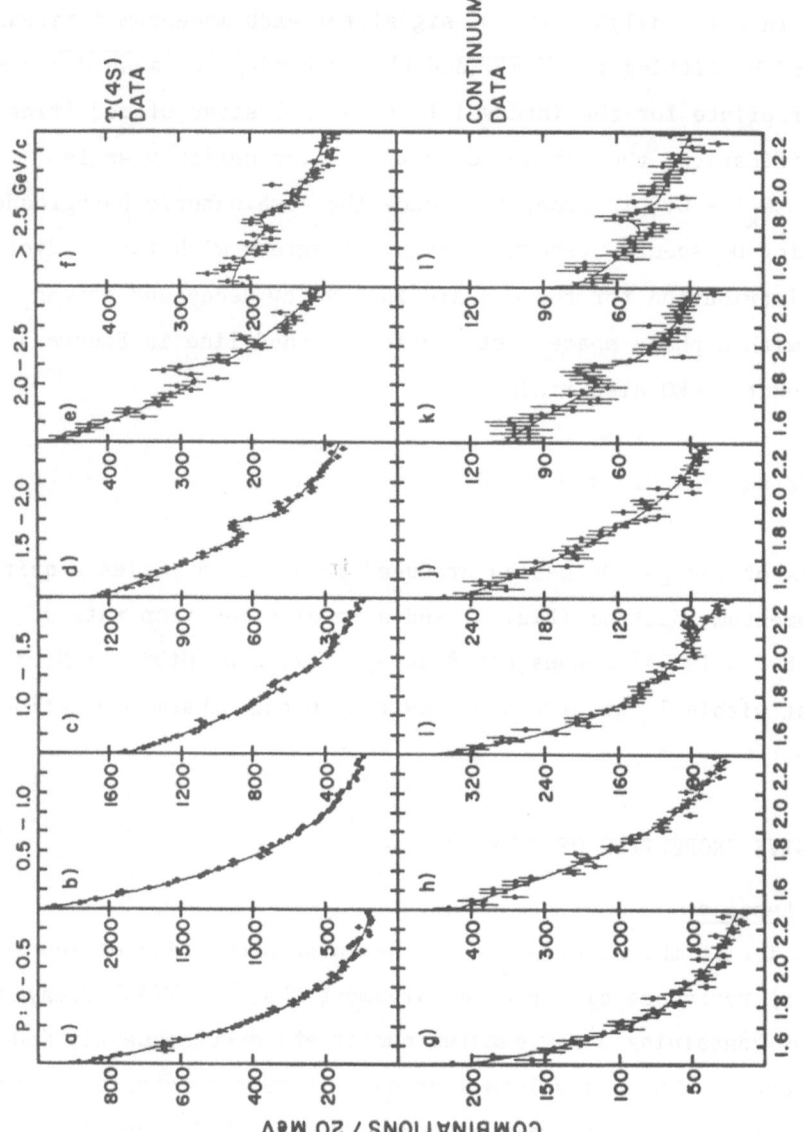

Fig. 4. $K^{\mp}\pi^{\pm}$ mass distributions on and off the $\Upsilon(4S)$ resonance for various $K\pi$ momentum intervals.

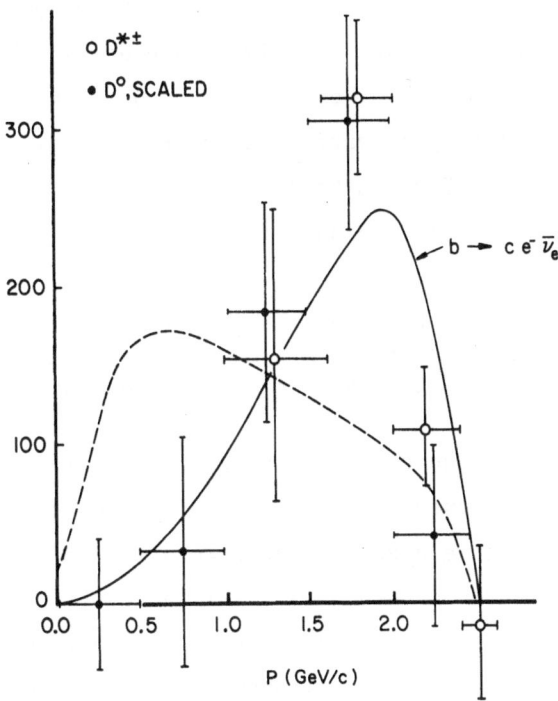

Fig. 5. The D° (solid points) and D*+ (open points) momentum spectra. The solid curve is the c quark momentum predicted by
b → ceν while the dashed line is the D momentum expected
from a phase space decay of the B.

discussion of the significance of these results to the paper of J. Franzini in these proceedings. The branching fractions are consistent with spectator model expectations:

$$B(B \to e + X) = .120 \pm 0.007 \pm 0.005 \qquad \text{CLEO}$$
$$B(B \to \mu + X) = .108 \pm 0.006 \pm 0.010 \qquad \text{CLEO}$$
$$B(B \to e + X) = .132 \pm 0.006 \pm 0.014 \qquad \text{CUSB}$$

4.2 Kaons

Charged kaons are identified by CLEO by both specific ionization (dE/dx) devices and time of flight counters in the momentum range $0.5 < P < 1.0$ GeV/c. They identify neutral kaons in their central drift chamber by measuring the characteristic decay $K_s \to \pi^+\pi^-$ in the momentum range $P > 0.3$ GeV/c. The results for total K production in B decay are listed below:

$$N_k = 1.45 \pm 0.10 \ / \ \text{B decay} \qquad \text{All}$$
$$N_k = 1.59 \pm 0.12 \ / \ \text{B decay} \qquad \text{Nonleptonic}$$
$$N_k = 1.00 \pm 0.20 \ / \ \text{B decay} \qquad \text{Semileptonic}$$

The semileptonic (and hence nonleptonic) kaon yield was obtained by studying a sample of events containing kaons and leptons.

These results have a simple interpretation in the spectator model. The semileptonic number is easily understood in light of the $b \to c \to s$ decay sequence which predicts 1 kaon per semileptonic decay. A prediction for the nonleptonic value can be obtained by adding the contributions of the decays $b \to c\tau\nu_\tau$, $b \to c\bar{u}d$, $b \to c\bar{c}s$, $b \to c\bar{u}s$ and $b \to c\bar{c}d$, appropriately weighted by phase space and Cabibbo angle factors. The τ semileptonic decay mode is included because it is experimentally classified as a hadronic decay. Assuming as for the semileptonic case 1 kaon per c quark or s quark, we predict 1.37 kaons per nonleptonic B decay. This means that all but 0.22 ± 0.12 kaons come from "valence" quarks; the rest are presumably produced by "ocean" $q\bar{q}$ pairs, especially $s\bar{s}$, which are thought to account for the relatively large charged multiplicity observed in nonleptonic B decay.

4.3 Multiplicity and Energy Fraction

The charged particle multiplicity has been measured by CLEO using a data set an order of magnitude larger than that employed for the 1982 paper.[7] The new numbers listed below do not differ significantly from the old values:

$$N_{ch} = 5.50 \pm 0.03 \pm 0.15 \qquad \text{All}$$
$$N_{ch} = 3.8 \pm 0.3 \qquad \text{Semileptonic}$$
$$N_{ch} = 6.1 \pm 0.3 \qquad \text{Nonleptonic}$$

An interesting interpretation of the above results can be made by adopting the $b \to c$ hypothesis and assuming a 3 to 1 predominance of D*'s to D's (average charged multiplicity of 2.6 ± 0.1).[8] Letting "D" represent the average D state, we find:

Average Semileptonic Decay: $B \to \text{"D"} + 0.2\pi^{\pm} + \ell^- + \nu$

Average Nonleptonic Decay: $B \to \text{"D"} + 3.4\pi^{\pm}$

For convenience, I have labeled as pions all charged particles produced in addition to the D or D*. This interpretation of the semileptonic charged multiplicity demonstrates rather directly the notion that the c plus spectator system produced either a D or a D* and little else, as discussed in section 3.

CLEO has also measured the photon multiplicity on and off the $\Upsilon(4S)$. The analysis uses the electromagnetic shower detectors located outside the magnet coil to identify photons of energy larger than 300 MeV. They find, after suitable corrections for acceptance, charged track overlap and the missing portion of the energy spectrum:

$$N_\gamma = 5.0 \pm 0.3 \pm 0.3 \; / \; B \text{ decay}$$
$$N_\gamma = 7.6 \pm 0.1 \pm 0.5 \; / \; \text{Continuum event}$$

These numbers are preliminary. Most of this multiplicity comes from
π° decay, although some contribution is expected from decays of D*,
F* and η mesons.

They have also measured the fraction of energy appearing either
as charged particles or as photons on and off the $\Upsilon(4S)$ resonance.
The results (the photon values are still preliminary) are shown
below:

Photons:	23.8 ± 1.7 ± 1.6%	B decay
Charged:	60.3 ± 0.7 ± 2.0%	B decay
Photons:	30.3 ± 0.6 ± 2.2%	Continuum
Charged:	61.2 ± 0.3 ± 2.0%	Continuum

The amount of missing energy, which presumably appears as neutrons,
K_L mesons or neutrinos, is larger for B decay (16%) than for
continuum production (8%), as expected from the large B semileptonic
branching fraction.

4.4 Baryons

One of the most interesting experimental observations in the
last year is that the B decays a significant fraction of the time
into baryons, with signals for both protons and lambdas seen by
CLEO.[9] Protons were identified in the momentum range $0.3 < P < 0.5$
GeV/c using dE/dx measurements in the inner proportional chamber and
in the interval $0.65 < P < 1.45$ GeV/c using time of flight and dE/dx
measurements made outside the coil. They detected lambdas in the
central drift chamber through the characterisitc decay $\Lambda \rightarrow p\pi^-$ for
$P > 0.45$ GeV/c. The analysis yeilds the following values for baryon
production in B decay (note that averages of particle and antiparticle rates are presented):

#p/B decay = 0.026 ± 0.006 ± 0.009	P > 0.30 GeV/c
#Λ/B decay = 0.022 ± 0.007 ± 0.004	P > 0.45 GeV/c

The proton value does not include those produced in Λ decay.

The data are presented in a momentum dependent form because
there is as yet no reasonable model of baryon production that permits an extrapolation to zero momentum. Nevertheless, an approximate
extrapolation can be attempted using the measured momentum dependence
to yield the approximate values:

#p/B decay ∿ 0.03
#Λ/B decay ∿ 0.03

To put these numbers in some sort of context, I have compared
them to (preliminary) results obtained in the nearby continuum
region and on the $\Upsilon(1S)$ resonance as shown below:

#p/Cont. event = 0.120 ± 0.005
#Λ/Cont. event = 0.040 ± 0.004

#p/$\Upsilon(1S)$ event = 0.22 ± 0.02
#Λ/$\Upsilon(1S)$ event = 0.105 ± 0.005

Interpretation of these results is difficult because there is
no model which quantitatively predicts baryon production in B decay.
Two possibilities are shown in Fig. 6. In the first[10] (Fig. 6a),
baryond production is described as a process in which a quark from
virtual W decay pairs up with the c quark to form a diquark while
the antiquark matches up with the spectator antiquark to make an
anti-diquark. Baryons are produced when an ocean $q\bar{q}$ pair is
"popped" from the vacuum. In the seond model baryons are produced
at the virtual W decay vertex by the creation of a diquark -
anti-diquark pair. Although the observation of charmed baryons
would distinguish between these two mechanisms, current data are not
quite up to the task.

Nevertheless, some tentative conclusions can be drawn from the
data. The Λ and proton momentum distributions (Fig. 7) fall essentially to zero around P = 1.0 GeV/c, indicating the presence of
particles other than the baryons in the decay. Also the Λ/p ratio

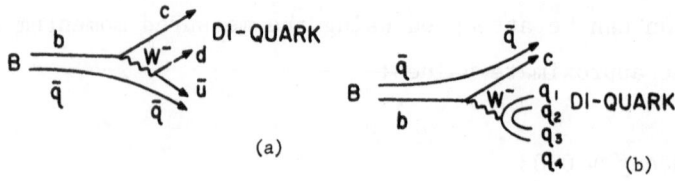

Fig. 6.   Two models of baryon production in B decay.

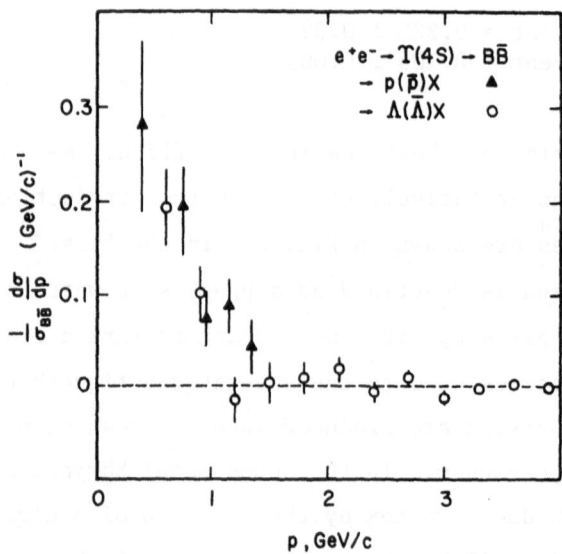

p , GeV/c

Fig. 7.   Momentum distributions measured by CLEO for protons and
lambdas produced in B decay.

appears to be approximately 1, although the supporting statistical evidence is not overwhelming. This observation is interesting because in both continuum and $T(1S)$ data the Λ/p ratio seems to be much smaller, and even in charmed baryon decay the Λ/p ratio is only about 0.4. We will have to wait for further data to resolve this point, but it is rather provocative.

5. EXCLUSIVE B DECAYS

5.1 First Observation

The discovery by CLEO of the exclusive decay channels of the B has been described elsewhere.[11] To summarize: D° mesons were detected through the $K^-\pi^+$ decay mode, where the kaon had to be positively identified by either time of flight or dE/dx measurements. D^{*+} mesons were identified by employing the familiar $D^{*+}-D^\circ$ mass difference technique, again using only the $K^-\pi^+$ decay mode of the D° (but without particle identification). Kinematic fitting of $D^\circ\pi^-$, $D^\circ\pi^+\pi^-$, $D^{*+}\pi^-$ and $D^{*+}\pi^-\pi^-$ mass combinations and their charge conjugates then yielded the mass distribution shown in Figure 8, which shows 18 events clustering around a mass of 5272 MeV. The masses of the charged and neutral B mesons obtained by this procedure (only D^* combinations were used in the mass determination because of the contamination of D° mesons by D^*'s) are:

$$M_{B^\circ} = 5274.2 \pm 1.9 \pm 2.0 \text{ MeV}$$
$$M_{B^-} = 5270.8 \pm 1.9 \pm 2.0 \text{ MeV}$$

The mass difference $\Delta M = 3.4 \pm 3.0 \pm 2.0$ MeV is consistent with Eichten's prediction[12] of 4.4 MeV, but it is also consistent with 0. The fact that the $T(4S)$ lies so close to $B\bar{B}$ threshold makes the relative production of charged to neutral B mesons sensitive to the exact value of ΔM. I will use Eichten's value of ΔM from now on in order to compute branching fractions. Phase space considerations

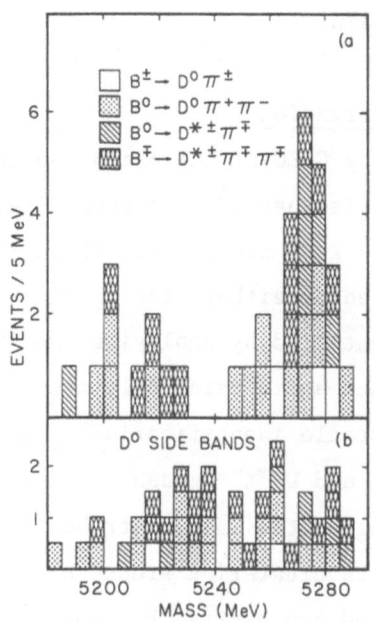

Fig. 8. Decay of the B into exclusive final states.

then imply that the charged to neutral B production ratio is 60/40
at the $\Upsilon(4S)$ resonance.

The branching fractions for the 4 decay channels listed above
are [all branching fractions in this paper assume the values
$B(D*^+ \rightarrow D^\circ \pi^+) = 0.60 \pm 0.15$ and $B(D^\circ \rightarrow K^- \pi^+) = 0.030 \pm 0.006$].

$$
\begin{array}{rcl}
B(\overline{B}{}^- \rightarrow D^\circ \pi^-) & = & 4.2 \pm 4.2\% \\
B(\overline{B}{}^\circ \rightarrow D^\circ \pi^+ \pi^-) & = & 13.0 \pm 9.0\% \\
B(\overline{B}{}^\circ \rightarrow D*^+ \pi^-) & = & 2.6 \pm 1.9\% \\
B(B^- \rightarrow D*^+ \pi^- \pi^-) & = & 4.8 \pm 3.0\%
\end{array}
$$

The large uncertainties in these values reflect the very poor
statistics available in each decay channel.

5.2 Measurements of $B^- \rightarrow D^\circ \pi^-$ Without Particle I.D.

Recently the CLEO group has managed to observe the decay
$B^- \rightarrow D^\circ \pi^-$ ($D^\circ \rightarrow K^- \pi^+$) without requiring that the kaon be positively
identified by the time of flight or dE/dx systems. They instead
demand that the kaon satisfy a helicity angle cut, $|\cos\theta_k| < 0.8$,
in order to reduce the combinatoric background. By foregoing particle
identification they obtain a larger D° signal (Fig. 9) at the expense
of a larger background under their $D^\circ \pi^-$ peak (Fig. 10). A combination was entered in the mass plot only if the energy E of the decay
products satisfied $-0.5 < E - E_{beam} < 0.25$ GeV. The asymmetric
energy cut ensures that $D*\pi^-$ decays, which feed down to $D^\circ \pi^-$, are
also included. No kinematic fitting was performed since the beam
constraint technique yields a similar mass resolution.

The signal of 12.3 ± 5.2 events contains contributions from the
decays $B^- \rightarrow D^\circ \pi^-$, $B^- \rightarrow D*^\circ \pi^-$ and $\overline{B}{}^\circ \rightarrow D*^+ \pi^-$, in which the $D*^\circ$ or $D*^+$
decays into a D° plus a slow pion or photon that is not detected.
CLEO interprets this signal under two different assumptions: (1) the
branching fractions for the above three decays are all equal and (2)
the signal is all due to the $D*$ decay modes. The branching fraction

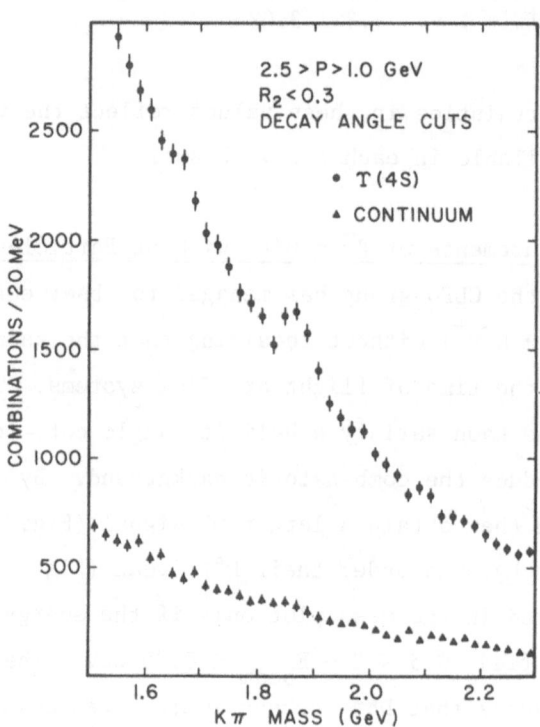

Fig. 9. $K^{\mp}\pi^{\pm}$ mass distributions for data taken on and off the
Υ(4S).

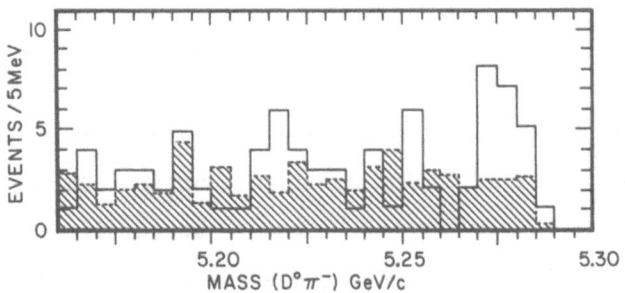

Fig. 10. $D^\circ\pi^- + \bar{D}^\circ\pi^-$ mass distributions without K identification. The shaded region is a background estimate obtained from D° sidebands.

for each of these assumptions is shown below:

$$D^\circ = D^* \quad B(\bar{B} \rightarrow D^\circ\pi^-) = 2.0 \pm 0.8 \pm 0.6\%$$
$$\text{All } D^* \quad B(\bar{B} \rightarrow D^{*\circ}\pi^-) = 4.0 \pm 1.7 \pm 1.0\%$$

5.3 $\underline{B \rightarrow D\pi \text{ and } D^*\pi \text{ from the Charged Particle Momentum Spectrum}}$

Because the B meson is nearly at rest when produced by the $\Upsilon(4S)$, nearly monochromatic decay products result when the B decays into two bodies. CLEO has exploited this fact by searching for the mono-energetic charged pion in the decays $B \rightarrow D\pi^\pm$ and $B \rightarrow D^*\pi^\pm$. The technique is particularly attractive because the charmed particle is never observed, one merely looks for a bump in the charged particle inclusive $x = 2p/\sqrt{s}$ distribution, where p is the momentum and $\sqrt{s}$ is the center of mass energy (twice the beam energy). In this technique charged and neutral B mesons are averaged together and the D and D* mesons cannot be individually resolved since the pion momentum is approximately the same for each.

The analysis proceeds as follows. First, the known charged lepton spectrum is subtracted from the x distribution. The portion of the spectrum above x = 0.25 is then fitted to a pion momentum distribution obtained by adding contributions from $D\pi$, $D2\pi$, $D3\pi$ and $D4\pi$ distributions (including the "Doppler" broadening caused by the small B motion in the lab). These momentum spectra are traced individually in Figure 11. The fit, shown in Figure 12, yields

$$1/2 \; [B(\overline{B}^\circ \to "D"\pi^-) + B(B^- \to "D"\pi^-)] = 1.3 \pm 0.3 \pm 0.3\%$$

where "D" refers to a 2 GeV charmed object (as discussed in sections 3 and 4), usually either a D or a D*. It should be pointed out that the assumption of a charged weak decay ($b \to c$, or $b \to u$) forces the charge of the pion in the above decay to be negative for b quarks and positive for $\overline{b}$ antiquarks.

5.4 $\overline{B}^\circ \to D*^+\pi^-$ from the D* Constraint

CLEO has exploited still another kinematic trick to measure the $D*^+\pi^-$ branching fraction of the $\overline{B}^\circ$ (of course, the charge conjugate mode is also seen). Only the "soft" π^+ resulting from $D*^+ \to D^\circ\pi^+$ decay and the "hard" π^- recoiling from the $D*^+$ need be observed in the process pictured schematically in Figure 13. If we call Θ the angle between the soft π and the (unobserved) D° and note that the energy of the B is equal to the beam energy, we obtain two constraints:

1. $E_{D^\circ} = E_{beam} - E_{\pi^-} - E_{\pi^+}$ $\qquad \to$ fix P_{D°
2. $(P_{D^\circ} + P_{\pi^+})^2 = M^2_{D*}$ $\qquad \to$ fix Θ

The ϕ angle of the reconstructed D* meson around the soft pion is still undetermined. Monte Carlo studies show that when the B is moving slowly, a best estimate for ϕ is obtained at the position

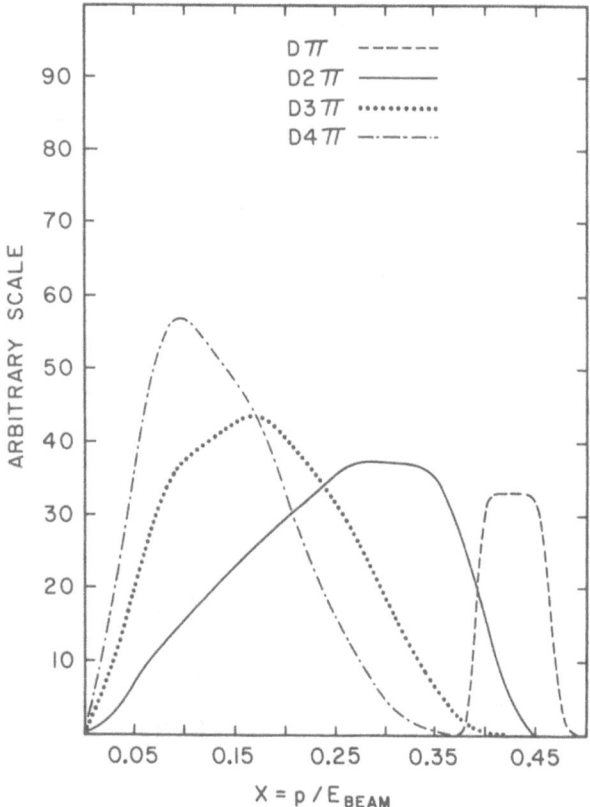

Fig. 11.   Charged particle momentum distributions expected from
phase space decay of the B meson.

that minimizes the $D*^{+}\pi^{-}$ invariant mass.   The mass resolution so
calculated rivals that obtained by more direct methods employing the
beam constraint, i.e. several MeV.   The mass distribution for data
taken on and off the $\Upsilon(4S)$ resonance is shown in Figure 14.   To be
included in this plot a combination had to satisfy the conditions
$E_{\pi^{+}} < 0.25$ GeV and $E_{\pi^{-}} > 2.3$ GeV.   The second requirement proved
necessary to remove background from high energy leptons resulting
from semileptonic B decays.   Fitting the enhancement yields a
signal of 41 ± 12 events which, when corrected for acceptance, yields
the preliminary value:

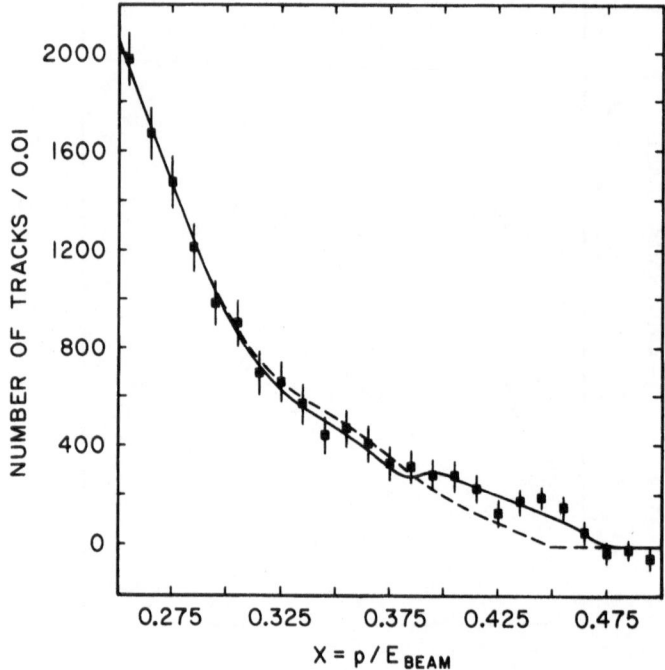

Fig. 12. Fit to the measured CLEO x distribution. The dashed curve
includes only the D2π, D3π, D4π contributions while the
solid line also includes the Dπ component.

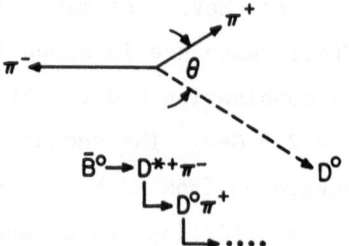

Fig. 13. The decay $\bar{B}^\circ \to D*^+\pi^-$, where the D° is unobserved.

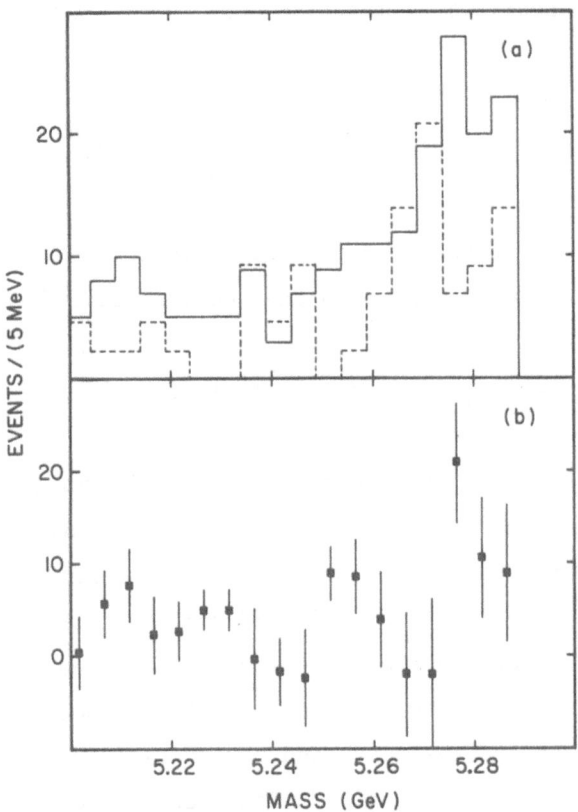

Fig. 14.   (a)   The D*+π⁻ + D*⁻π⁺ mass distribution using the D*
constraint for data taken on (solid line) and off
(dashed line) the Υ(4S).
(b)   Direct contribution of the Υ(4S).

$$B(\overline{B}^\circ \to D*^+\pi^-) = 2.1 \pm 0.6 \pm 0.6\%$$

6. COLOR MIXING AND B → ψ + X

Consider the spectator decay b → cūd. Using only naive weak interaction theory (i.e. ignoring QCD), we can write the effective hamiltonian as

$$H = G/\sqrt{2}(\overline{c}b)_L(\overline{d}u)_L \qquad\qquad (\overline{c}b)_L = \overline{c}\gamma^u(1 - \gamma^5)b$$

where I have suppressed the dependence on the K–M angles. QCD modifies this hamiltonian into the new effective form:

$$H^{QCD} = G/\sqrt{2}[(f_+{+}f_-)/2(\overline{c}b)_L(\overline{d}u)_L + (f_+{-}f_-)/2(\overline{c}u)_L(\overline{d}b)_L]$$

The f's are known under several names; I will refer to them as renormalization constants. Several authors[13] have calculated their values using the Leading Log approximation, sometimes to subleading order. Typical values are f_+ = 0.80 and f_- = 1.56.

The first term in the expression for H^{QCD} corresponds to "normal" color flow in which the c quark and spectator antiquark form a color singlet and the quark and antiquark from virtual W decay produce a second color singlet. The second term describes a "funny" color flow in which the antiquarks are switched (I shall henceforth refer to this latter situation as "color mixing"). Both of these processes are pictured in Figure 15. The formation of color singlets at the quark level is important because it is believed that hadron formation in each singlet takes place more or less independently of the other. The measured D° momentum spectrum discussed in section 3 supports this statement.

The decay B → ψ + X is particularly well suited for studying predictions of color mixing because there are no competing diagrams of the same order that can interfere with the process pictured in

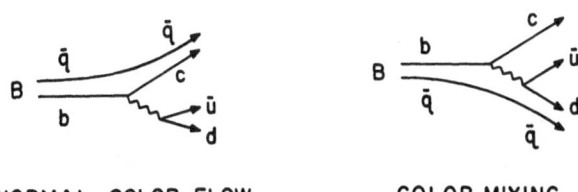

NORMAL COLOR FLOW   COLOR MIXING

Fig. 15.   Diagrams illustrating two different color flows possible
in b spectator decay.

Figure 16.   The CLEO group looked for ψ mesons in B decays by
examining the $\mu^+\mu^-$ and e^+e^- mass distributions in the vicinity of 3
GeV (Fig. 17).   Each of the two muons was required to be positively
identified while only one electron needed to be identified as such.
There is a small enhancement at the ψ mass in both distributions,
although CLEO prefers to quote the result as an upper limit:

$B(B \rightarrow \psi + X) < 1.6\%$ 90% C.L.

Taken as a signal, the enhancements correspond to a branching fraction
of $1.0^{+0.5}_{-0.4}\%$.

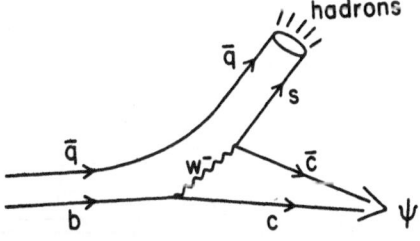

Fig. 16. $B \rightarrow \psi + X$ decay resulting from a color mixing process.

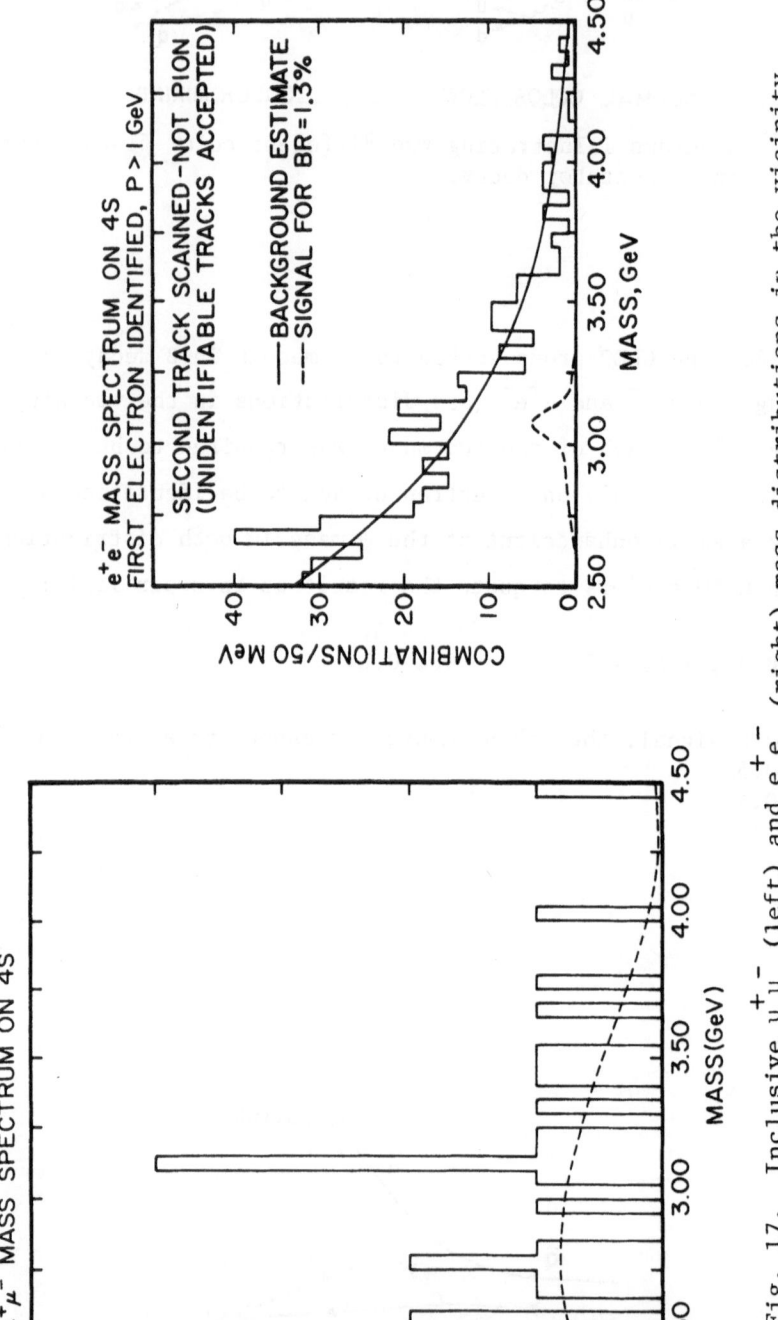

Fig. 17. Inclusive $\mu^+\mu^-$ (left) and e^+e^- (right) mass distributions in the vicinity of the ψ.

116

The theoretical rate calculated from H^{QCD} is $(2f_+-f_-)^2/9$ times the rate obtained by neglecting QCD. Unfortunately, a fortuitous cancellation in $2f_+-f_-$ makes the value extremely sensitive to the exact value of the renormalization constants. For the values mentioned above the branching fraction is 0.01%. Setting both f_+ and f_- to 1 yields a branching fraction of 1.5%.[14] Physically this latter case corresponds to allowing the c and $\bar{c}$ quarks to randomly match in color. Since ψ's are only produced when the colors match, a factor of 1/3 in the amplitude and 1/9 in the rate is introduced. Any deviation of f_+ and f_- due to QCD would make the rate even smaller. The conclusion we reach from this line of reasoning is that regardless of whether one believes the CLEO signal is real or not, there is substantial suppression of the color mixing decay $B \to \psi + X$.

7. TWO BODY DECAYS OF THE B

The spectator model can be used to predict quantitative rates for two body B decays of the form $B \to "D"X^-$, where "D" is the low mass charm system (usually either a D or a D*) discussed in section 3 and X^- is either a vector meson or a weakly decaying pseudoscalar particle. The charge of X is positive for $\bar{b}$ decay.

Let me proceed by using for an example the decay $B \to "D"\pi^-$. The amplitude for this process, pictured in Figure 18a, can be written:

$$M = G/\sqrt{2} \cdot \bar{c}\gamma^u(1-\gamma^5)b \cdot U_{bc} \cdot <\pi^-|A_\mu|0>$$

where c and b represent the spinors for the c and b quarks, U_{bc} is the K-M matrix coefficient and A_μ is the axial vector weak current. I have implicitly assumed both the factorizability of the amplitude into this form and vacuum saturation of the intermediate states. Both of these assumptions are universally employed in the literature.

Now the factor $<\pi^-|A_\mu|0>$ also appears in the amplitude for

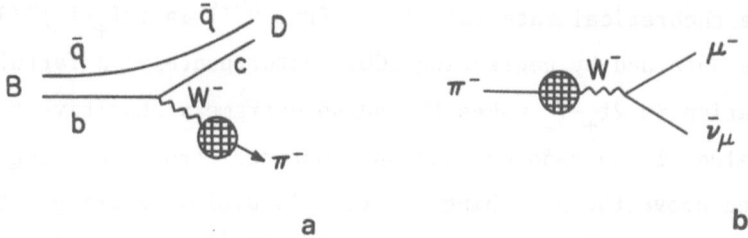

Fig. 18. Comparison of B → Dπ⁻ decay and π → μν decay.

π → μν decay (Fig. 18b) and can be expressed as:

$$\langle \pi^- | A_\mu | 0 \rangle = f_\pi \cos\Theta_c q_\mu$$

where Θ_c is the Cabibbo angle, q_μ is the π 4-vector and f_π is the pion decay constant whose value has been measured from the pion decay rate to be 0.132 GeV. The decay rate for B → "D"π⁻ can now be written:

$$\Gamma(B \to \text{"D"}\pi^-) = \Gamma_0 \cos^2\Theta_c \cdot 12\pi^2 \cdot (f_\pi^2/m_b^2) \cdot A(X_c^2, X_\pi^2) \cdot \lambda(1, x_D^2, x_\pi^2)$$

where $\Gamma_0 = (G^2 m_b^5/192\pi^3) U_{bc}^2$, $A(x,y) = (1-x)^2 - y(1+x)$, $\lambda^2(x,y,z,) = (x-y-z)^2 - 4yz$, and $X_\pi = m_\pi/m_b$, etc. The same formula applies to any decay of the form B → "D"X⁻, with appropriate substitutions for masses and Cabibbo factors, where X⁻ is a pseudoscalar meson.

A similar formula can be derived for the case in which X⁻ is a vector meson. In this case the amplitude can be written:

$$M = G/\sqrt{2} \cdot \bar{c}\gamma^u (1 - \gamma^5) b \cdot U_{bc} \cdot \langle X^- | V_\mu | 0 \rangle$$

where V_μ is the vector current. The factor $\langle X^- | V_\mu | 0 \rangle$ can be written as $f_M m_M \cos\Theta_c \varepsilon_\mu$, where f_M is the meson decay constant (obtained from its leptonic width), m_M is its mass and ε_μ is the polarization

4-vector. The rate for B → "D"V⁻ then becomes:

$$\Gamma(B \rightarrow \text{"D"}V^-) = \Gamma_o \cos^2\Theta_c \cdot 12\pi^2 \cdot (f_M^2/m_b^2) \cdot B(X_c^2, X_M^2) \cdot \lambda(1, X_D^2, X_M^2)$$

where $B(x,y) = (1+x-y)(1+x-2y)-4x$ and $X_M = m_M/m_b$ as before.

Formulas similar to the above equations were derived in 1971 by Y. S. Tsai[15] who successfully predicted the $\tau^- \rightarrow \nu_\tau \pi^-$ and $\tau^- \rightarrow \nu_\tau \rho^-$ branching fractions. B decay, however, is more complicated because of interfering diagrams and QCD corrections. One effect of QCD is to introduce a factor $(2f_+ + f_-)^2/9 \sim 1.11$ which increases the rates derived above by an amount that is fairly insensitive to the precise values of f_+ and f_- (f_+ decreases as f_- increases). A more serious problem arises from quantum interference with so-called "non-spectator" diagrams, such as those pictured in Figure 19. Calculations show[16] that these diagrams should not contribute much to the total B decay rate, although the interference with the "D"π^- and "D"ρ^- spectator diagrams might be substantial because of the small size of the latter's rates (typically a few per cent). More confidence can be placed on the predictions for charged B's because the non-spectator rates (Fig. 19) in this case are proportional to the square of U_{bu}, the K-M coefficient that describes the b → u coupling. This coupling constant has already been shown from CLEO and CUSB data to be small:[5]

$$|U_{bu}|^2/|U_{bc}|^2 < 0.02 \qquad\qquad 90\% \text{ C.L.}$$

and should therefore produce negligible interference with the spectator diagram.

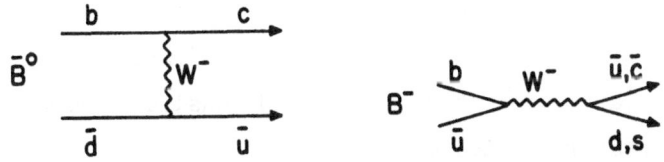

Fig. 19. Non-spectator diagrams of $\bar{B}^o$ and B^- decay.

The spectator rate predictions[17] for various 2 body B decays are listed below. In cases for which the decay constant f_M is not known, I have tried to use the best theoretical values together with extrapolations from the values of the neutral vector mesons (done in several ways) to estimate the branching fractions. The two approaches are in reasonable agreement. In any event, the dependence of the rate on f_M is explicity given.

Decay Mode	BR
"D"π^-	$1.89 \pm 0.25 \pm 0.13\%$
"D"F$^-$	$(3.78 \pm 0.39 \pm 0.25)\% \cdot (f_F/0.240)^2$
"D"ρ^-	$4.91 \pm 0.65 \pm 0.33\%$
"D"F$*^-$	$(5.34 \pm 0.76 \pm 0.36)\% \cdot (f_D/0.240)^2$

The first error comes from an assumed uncertainty of .15 GeV in the c and b quark masses and the second error reflects the B semileptonic branching fraction uncertainty (which is used to compute the total B decay rate).

Two points should be noted from the above table. First, the $B \rightarrow$ "D"F$^-$ and $B \rightarrow$ "D"F$*^-$ rates are relatively large and appear to constitute a large fraction of the total $b \rightarrow c\bar{c}s$ rate (estimated to be 13%, using reasonable quark masses). However, the predicted branching fractions are still somewhat too small to be observed in the current CLEO data sample. Measurement of these particular branching fractions would be especially important because there are no other competing diagrams of the same order to confuse the prediction. Second, the CLEO collaboration does see a branching fraction of about 2% for the "D"π channel, albeit with large errors. Better measurements of the "D"π and "D"ρ branching fractions (which has a moderately large predicted rate) would consitute important tests of these spectator predictions.

8. CONCLUSIONS

The experimental work of the last year or so in B meson decay strongly supports the conclusion that B decay is dominated by the dynamics of the spectator model. Data as dissimilar as the lepton momentum spectrum, the D° momentum spectrum and kaon yields are convincingly explained by simple applications of kinematics and QCD to the spectator model. Other phenomena, notably baryon production, have no satisfactory theoretical model to draw any significant conclusions from the data. Both experimental and theoretical work need to be done in this area.

The limits of applicability of the spectator model may be reached in trying to predict the absolute values of the branching fractions in two body B decays (after all, the corresponding predictions for the D system do not work so well), but further measurements in the near and intermediate future at CESR and DORIS will test the accuracy of these predictions with, one hopes, important consequences for QCD and the standard model.

REFERENCES

1. S. Stone, in "Proceedings of the 1983 Symposium on Lepton and Photon Interactions at High Energies", p.203, (1983);

2. A. Chen et al., Phys. Lett. 122B, 317 (1983);

3. R. D. Schamberger et al., Phys. Rev. D26, 720 (1982); The value quoted in the text comes from a manuscript submitted by the CUSB group to Phys. Rev. D;

4. M. G. D. Gilchriese, Cornell Preprint CLNS-83/593, 1983.
   R. K. Plunkett, Ph.D Thesis, Cornell University, 1983;
   R. Giles et al., Phys. Rev. D29, 1285 (1984);

5. A. Chen et al., Phys. Rev. Lett. 52, 1084 (1984);
   C. Klopfenstein et al., Phys. Lett. 130B, 444 (1983);

6. J. Green et al., Phys. Rev. Lett. $\underline{51}$, 347 (1983);

7. M. S. Alam et al., Phys. Rev. Lett. $\underline{49}$, 357 (1983);

8. R. Schindler et al., Phys. Rev. $\underline{D24}$, 78 (1981);

9. M. S. Alam et al., Phys. Rev. Lett. $\underline{51}$, 1143 (1984);

10. I. I. Bigi, Phys. Lett. $\underline{106B}$, 510 (1981);

11. S. Behrends et al., Phys. Rev. Lett. $\underline{50}$, 881 (1983);

12. E. Eichten, Phys. Rev. $\underline{D22}$, 1819 (1980);

13. G. Altarelli and L. Maiani, Phys. Lett. $\underline{52B}$, 351 (1974);
    G. Altarelli, N. Cabibbo and L. Maiani, Phys. Lett. $\underline{57B}$,
    277 (1975) and Nuc. Phys. $\underline{B88}$, 285 (1975);
    B. W. Lee and M. K. Gaillard, Phys. Rev. Lett. $\underline{33}$, 108 (1974);
    G. Altarelli et al., Phys. Lett. $\underline{99B}$, 141 (1981);
    R. Ruckl, submitted to Physics Reports;

14. Y. S. Tsai, Phys. Rev. $\underline{D9}$, 2821 (1971);

15. H. Fritzsch, Phys. Lett. $\underline{86B}$, 164 (1979);
    H. Fritzsch, Phys. Lett. $\underline{86B}$, 343 (1979);
    J. H. Kuhn, S. Nussinov and R. Ruckl, Z. Phys. $\underline{C5}$, 117 (1980);
    T. DeGrand and D. Toussaint, Phys. Lett. $\underline{89B}$, 256 (1980);
    M. B. Wise, Phys. Lett. $\underline{89B}$, 229 (1980);
    J. H. Kuhn and R. Ruckl, MPI-PAE/Pth 52/83, 1983;

16. J. P. Leveille, "B Decay Workshop", CLNS 51/505 (1981);

17. P. Avery, "Calculation of Two Body Decays of the B", Cornell
    Preprint CBX 84-13, 1984.

EXPERIMENTAL STATUS OF NEUTRINO MASS

Felix Boehm

California Institute of Technology
Pasadena, CA 91125

Introduction

Neutrino physics faces a number of interesting challenges, among them are the following fundamental questions: Are neutrinos massive? Are physical neutrinos pure or mixed states? Are neutrinos Dirac or Majorana particles? Evidence for neutrino mass and mixing would lend support to grand unified theories or other schemes beyond the standard electroweak model.

Neutrino mixing is described by a linear superposition

$$\nu_\ell = \Sigma U_{\ell i} \, \nu_i \qquad \begin{array}{l} \ell = e, \mu, \tau \\ i = 1, 2, 3 \end{array}$$

where ν_ℓ are weak interaction eigenstates, ν_i are mass eigenstates, and $U_{\ell i}$ are the mixing amplitudes. From astrophysics and cosmology, arguments can be made that the mass of neutrinos, summed over all flavors cannot exceed a value of about 100 eV.

What are the experimental possibilities allowing tests for neutrino mass and neutrino mixing? The present upper limits for the masses ($m\,(\nu_e) < 50$ eV, $m(\nu_\mu) < 0.5$ MeV, $m(\nu_\tau) < 160$ MeV) have been derived from kinematical studies of decay processes. Other phenomena testing mass and mixing are neutrino decay, searches for heavy neutrinos, and studies of neutrino oscillations. Neutrinoless

123

double beta decay emerges as an important tool for the quest of small Majorana masses.

Kinematic Tests – $\bar{\nu}_e$ Mass from ^{3}H Beta Decay

If neutrinos emitted in nuclear beta decay, such as in the low energy decay of ^{3}H $\rightarrow$ ^{3}He $+$ e$^-$ $+$ $\bar{\nu}_e$ have finite mass, m_ν, the study of the linearized beta spectrum is expected to show a change in slope near the endpoint. In that case there will be no electrons with energy greater than $E_0 - m_\nu$, where E_0 is the decay energy. Lubimov et al.[1] in 1980 have reported evidence for finite mass, $14 \leq m_\nu \leq 46$ eV. As subsequently pointed out by Simpson[2] their analysis ignores the finite line width of a calibration line. If corrected for this width the evidence disappears reducing their result to an upper limit, $m_\nu < 41$ eV. More recently, the ITEP group[3] has reported new results obtained with improved spectrometer resolution and improved signal-to-noise. Taking the spectrum of the final states in He as calculated by Kaplan[4] the new result is $m_\nu = 33 \pm 1.1$ eV. Concerns have been raised, however, over an apparent non-linearity of the energy spectrum as reported at the Brighton Conference.[3] Also, the reported endpoint of $E_0 = 18583 \pm 0.3$ eV is at variance with the value of the decay energy $E_0 = 18549 \pm 7$ eV determined[5] from the ^{3}H $-$ ^{3}He mass difference. The true endpoint of the electron spectrum (assuming a delta function resolution), according to Ref. 3, is expected to be at $18583-33 = 18550$ eV, in good agreement with Ref. 5, but at the same time indicating zero neutrino mass. Clearly independent verification of this important result is of paramount importance.

Neutrino Decay

If neutrinos have mass, the heavier ones could decay into the lighter ones. Neutrino decay has never been seen; however, if it were observed it would give information on the masses, and, because at least two neutrino flavors are involved, on neutrino mixing.

The radiative decays $\nu_2 \to \nu_1 + \gamma$ and $\nu_2 \to \nu_1 + e^+ + e^-$ are the most likely candidates. Their rates have been calculated and are reviewed in Ref. 6. The $\nu-\nu'\gamma$ rate is proportional to $|U_{\ell 2} U_{\ell 2}|^2$ characterizing the mixing between ν_2 and an intermediate virtual lepton ℓ, as well as between ν_1 and ℓ, in addition of being proportional to the fifth power of the mass of ν_2 and the fourth power of the mass m_ℓ. The calculated rate vs. ν_2 mass for $m_\ell = m_\tau$ and assuming full mixing ($|U_{\ell i}|^2 = 1$) is shown in Fig. 1. The $\nu_2-\nu_1 e^+ e^-$ rate is also given in that figure. Experimental limits are summarized by broken lines, whereby the transformation τ_γ (CM) $= (m_\nu/E_\nu) \tau_\gamma$ (LAB) has been taken into account. For example, the decay of reactor neutrinos from the Gösgen experiment [7] furnishes τ_γ (CM) ≥ 7 m_ν (eV) seconds. For $m_\nu = 30$ eV, we have $\tau > 210$ s, an upper limit about 10^{32} times lower than the calculated rate for full mixing. More pertinent, perhaps, is a comparison of the experimental and calculated $\nu-\nu'e^+ e^-$ limits for reactor neutrinos and for ν_τ's. The data lies above the calculated curve and thus can be interpreted as constraint for the mixing parameters $U_{\ell i}$. The most stringent constraint testing ν_τ decay comes from a recent CHARM experiment [9] to be described later at this Conference.

Admixture of Heavy Neutrinos

The preceding discussion is closely related to the question whether there exist heavy neutrinos admixed to the known light neutrino states. Experimental evidence points to the fact that a state ν_ℓ (such as ν_e) is predominantly composed of one light neutrino ν_i. If heavy neutrinos exist, their admixture must therefore be small, the branching ratio being given by $|U_{\ell i}|^2$.

Two body decays ($K \to \pi\nu$, $\pi \to e\nu$) offer sensitive tests to study these branching ratios. Each mass eigenstate is expected to manifest itself as a monochromatic peak in the lepton spectrum at some energy below the regular peak associated with the light neutrino, with an intensity given by $|U_{\ell i}|^2$. Spectroscopic

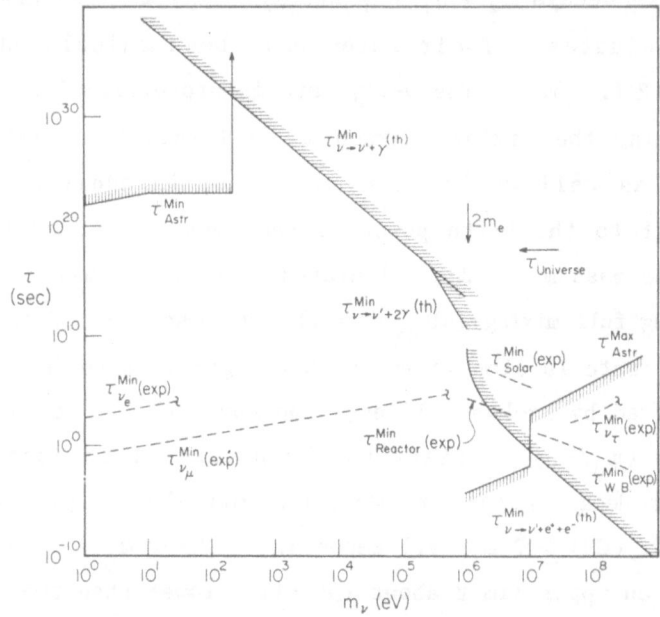

Fig. 1   Theoretical and experimental limits of neutrino lifetime:
τ (th)   is the calculated lifetime for maximum mixing,
where one assumes that τ is the heaviest charged lepton;
the horizontally shaded regions are allowed. The theoretical lifetime for 2γ decay is also shown. The τ_{Astr} are
boundaries based on astrophysical arguments; vertically
shaded regions are allowed. Experimental lower limits for
the dominantly coupled neutrinos extend to the upper
limits of the corresponding mass: τ_{ν_e} Ref. 7, τ_{ν_μ} Ref. 8,
τ_{ν_τ} Ref. 9, τ_{solar} Ref. 10, $\tau_{reactor}$ Ref. 11, and $\tau_{W.B.}$
Ref. 9, are limits for subdominantly coupled heavy
neutrinos.

measurements aimed at finding this peak have not revealed any
evidence and the limits for $|U_{\ell i}|^2$ for $\ell = \mu$ and $\ell = e$ from various
experiments are reviewed in Fig. 2.

The study of three body decays also has provided limits for these branching ratios (see Fig. 2). In a nuclear beta decay, for example, a heavy neutrino would show up as a discontinuity in the electron spectrum. Further limits for $|U_{\ell i}|^2$ come from oscillation experiments to be discussed below.

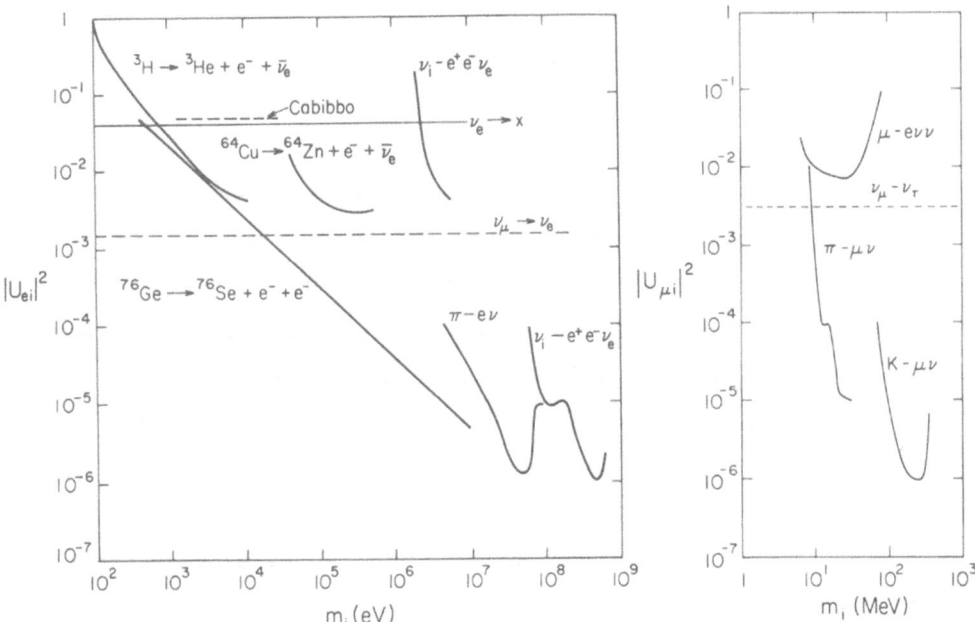

Fig. 2. Limits for mixing coefficients $|U_{\ell i}|^2$ and $|U_{\mu i}|^2$ describing the admixture of heavy neutrinos to the electron neutrino and the muon neutrino, respectively. The regions above the curves are excluded. The curves are based on the following references: $^3H \rightarrow {}^3He + e^- + \bar{\nu}$ Ref. 12, $^{64}Cu \rightarrow {}^{64}Zn + e^- + \bar{\nu}$ Ref. 13, $^{76}Ge \rightarrow {}^{76}Se + e^- + e^-$ Ref. 14, $\pi \rightarrow e\nu$ Ref. 15, $\nu - ee\nu$ Ref. 9, $\mu \rightarrow e\nu\nu$ Ref. 16, $\pi \rightarrow \mu\nu$ Ref. 17, $K \rightarrow \mu\nu$ Ref. 18.

Neutrino Oscillations

A large number of experiments to study neutrino oscillations have been carried out with low and high energy neutrinos. So far no evidence for oscillations has been found. Using a description in terms of two parameters, a mixing angle θ ($\sin\theta = U_{12} = -U_{21}$) and a mass parameter $\Delta m^2 = |m_2^2 - m_1^2|$, the probability for a transition from one neutrino state ℓ into another state ℓ' is given by

$$P(\ell-\ell') = \frac{\sin^2 2\theta}{2} \left[1 - \cos \frac{2.53 \; \Delta m^2 \; (eV)^2 \; L(m)}{E_\nu \; (MeV)} \right] .$$

The probability for ν_ℓ having disappeared is given by $1-P(\ell-\ell')$. Among the disappearance experiments, which test the effects of all oscillation channels, is the recent neutrino experiment at Gösgen[19,20]. This experiment which explores the rate of $\bar{\nu}_e$ induced reactions in a proton-rich liquid scintillator as a function of energy E_ν at various distances L has provided stringent limits for Δm^2 and $\sin^2 2\theta$ as shown in Figs. 3-5.

To detect the low energy antineutrinos the reaction $\bar{\nu}_e p - e^+ n$ has been utilized. The reaction is identified by a time-correlated e^+ and n signature. About 11,000 neutrino events were recorded each with the detector at positions of 38m and 46m from the reactor core. Backgrounds for each position were recorded during a one-month reactor-off period. Figure 3 shows the difference spectrum, reactor-on minus reactor-off for both positions together with a curve representing the expected spectrum for no oscillations[21].

In Fig. 4 the observed yields for both positions at Gösgen together with a previous result[22] from a 8.7 m position at ILL are displayed, in units of no oscillation yield, as a function of L/E_ν. As can be seen in Fig. 3 and 4 there is good agreement between experiments at various positions and also between experiments and expectation.

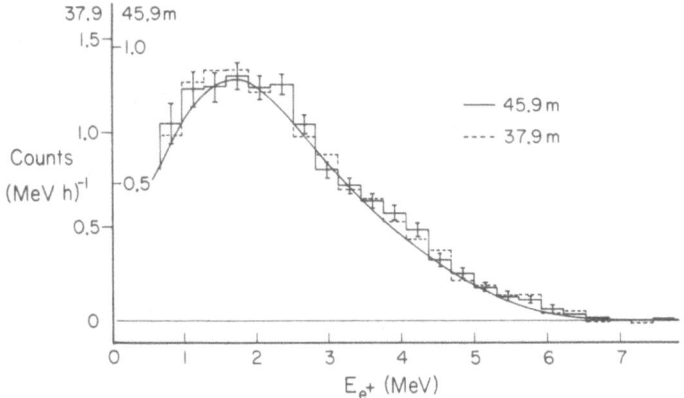

Fig. 3. Results of the Gösgen experiments. Positron spectra obtained by subtracting reactor-off from reactor-on spectra for the 38 m and 46 m experiments. The solid curve represents the predicted positron spectrum assuming no neutrino oscillations.

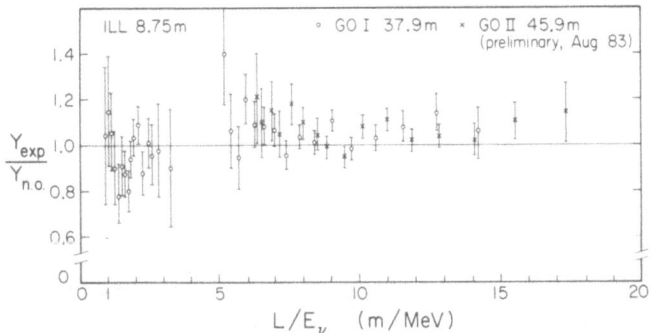

Fig. 4 Ratio of experimental to predicted (for no oscillations) positron spectra at 8.7 m, 38 m, and 46 m from the reactor core. The errors of the data points shown are statistical.

The data at 2 or 3 position can be analyzed without resorting to the no oscillation spectrum. The exclusion plots of Fig. 5 were obtained by considering the ratios of the data at 8.7 m, 38 m, and 46 m for each energy bin and fitting it to calculated ratios for various oscillation parameters. It is concluded that there are no oscillations with parameters greater than those to the right of the curves in Fig. 5.

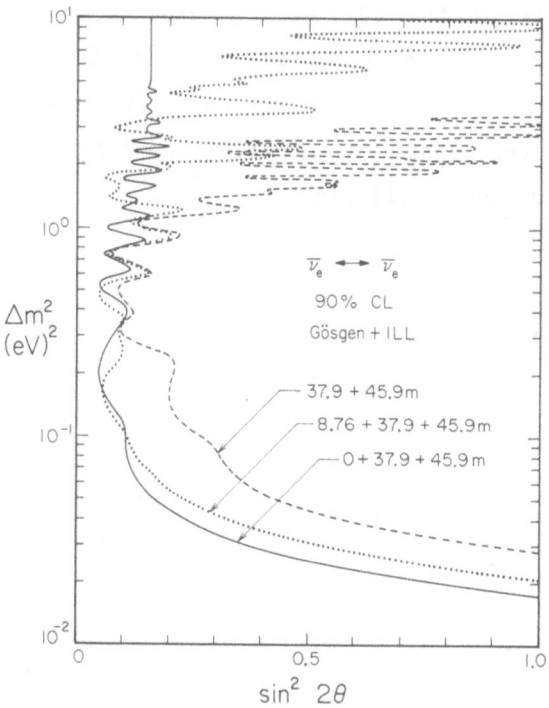

Fig. 5   Exclusion plot obtained from the Gösgen[19,20] and ILL[22] experiment at 90% confidence limit. The curve labelled 37.9 + 45.9 m refers to the limit obtained by using the 37.9 m and 45.9 m data only, for example. The 0 m label represents the on-line beta spectroscopic measurement.[21]

Finally, the important question should be addressed, where in the Δm^2 vs. $\sin^2 2\theta$ plane should one continue to search for oscillations? Unfortunately there is no guidance from theory whatsoever. As to the mixing angle, we can state that present limits are smaller than the Cabibbo angle. Figure 7 shows these limits, together with other possible dimensional guesses (lepton mass ratios). If the solar neutrino experiments are indeed telling us that neutrinos oscillate with large mixing angle, the Δm^2 values must lie between 10^{-2} and 10^{-10} eV^2, a region increasingly more difficult to explore.

Figure 6 summarizes the present limits for oscillations in several appearance and disappearance channels. (For references see caption).

Neutrinoless Double Beta Decay

Neutrinoless double beta decay of a nucleus $Z, (Z \rightarrow Z + 2 + e_1 + e_2)$ appears to be the most sensitive experimental method to explore neutrino mass. This process is forbidden by standard theory and, if observed, would signal violation of lepton number conservation which can be associated with non-zero Majorana neutrino mass or right-handed current.

Two-neutrino double beta decay $(Z \rightarrow Z + 2 + e_1 + \bar{\nu}_1 + e_2 + \bar{\nu}_2)$ which is expected to occur from standard theory may compete with the neutrinoless (0ν) process. Its study is of interest since it may help in estimating the value of the nuclear matrix element needed to analyze the 0ν decay.

Figure 8 illustrates the two processes. The characteristic signature of the 0ν decay is a monochromatic peak in the total energy spectrum of the electrons at the decay energy ε_0. 0ν decay could proceed by virtual neutrino exchange as shown in the figure. This is a two-nucleon mechanism: a neutron n_1 emits an electron e_1 and a neutrino, and the latter is absorbed by a neutron n_2 which then emits an electron e_2. The process can only proceed if the

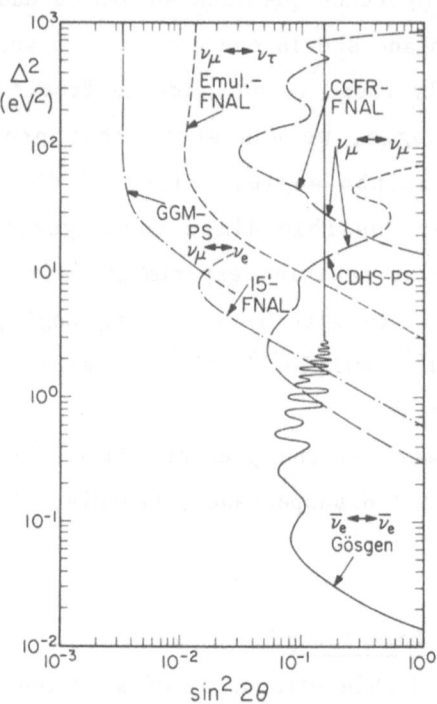

Fig. 6   Exclusion plot summarizing present status of experimental results for various oscillation channels. Some recent references are $\nu_\mu - \nu_e$: GGM Ref. 23, FNAL Ref. 24; $\nu_\mu - \nu_\tau$: FNAL Ref. 25; $\nu_\mu - \nu_\mu$:CCFR Ref. 26, CDHS Ref. 27, $\nu_e - \nu_e$: Gösgen Ref. 19,20.

neutrino is a Majorana particle ($\nu^M = \bar{\nu}^M$). In addition, the perfect helicity of the neutrino must be violated. This can be accomplished by a mass term m_ν (giving rise to a helicity violation m_ν/E) or by an explicit right helicity term, with amplitude η, in the charged lepton current. In the following discussion we assume that the 0ν process can be described by m_ν or η.

The transition probability in the 2ν process can be calculated[28] and is proportional to the phase space factor of the four fermions, $F_{2\nu}(\varepsilon_0)$, and to the square of the second order Gamow-Teller like matrix element, $|M_{GT}|^2$,

$$\Gamma^{2\nu} \sim F_{2\nu}(\varepsilon_0) \, |M_{GT}|^2 \, / < \Delta E_N + \varepsilon_0/2 + m_e >^2 ,$$

where ΔE_N is the average nuclear energy difference between the Gamow-Teller states in $Z + 1$ and the initial state in Z (Fig. 8). A rough estimate for the 2ν half-life yields $T_{\frac{1}{2}}^{2\nu} \simeq 10^{22\pm 2}(y)$, the uncertainty stemming from the matrix element M_{GT}.

Similarly, the 0ν rate is given by

$$\Gamma^{0\nu} \sim F_{0\nu}(\varepsilon_0) |M_{GT} / <r_{ij}> m_p|^2$$

where $F_{0\nu}(\varepsilon_0)$ is the two-fermion phase space factor containing the parameters m_ν and η, and the quantity $||^2$ is the Gamow-Teller matrix element divided by an average nucleon separation distance (the range of the virtual neutrino) measured in proton Compton wave length. A rough estimate gives $T_{\frac{1}{2}}^{0\nu} \simeq 10^{15\pm 2} \, \eta^{-2}$ (or m^{-2}) (y), where m is measured in units of electron mass m_e.

Selection rules could help distinguish between mass and right-hand current mechanisms. For the mass term, only 0^+-0^+ transitions are allowed, while for the current term one can have 0^+- 0^+, 1^+, 2^+. Thus, if a 0^+-2^+ branch is observed, i.e. a double beta branch into an excited state, it would give evidence for right handed currents.

Double beta decay has been studied in several nuclei both, by geochemical techniques (extraction of the daughter $Z + 2$ from the parent Z in an old ore) and with counters. Geochemical experiments can, of course, not distinguish between 0ν and 2ν decays. The results of some selected cases particularly suited for sensitive tests are now briefly discussed.

tured and is proportional to the phase space factor for three body
emission, F, i.e. ... see to the ...
nonrelativistic matrix element, |...|²

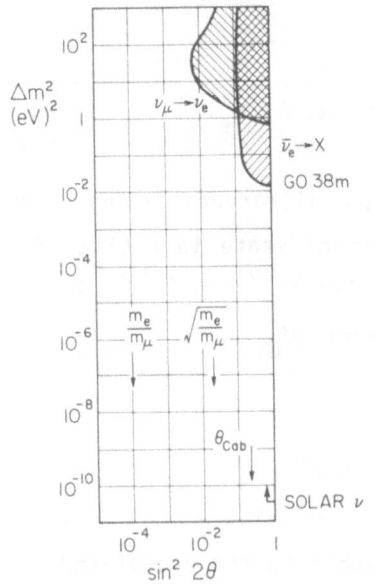

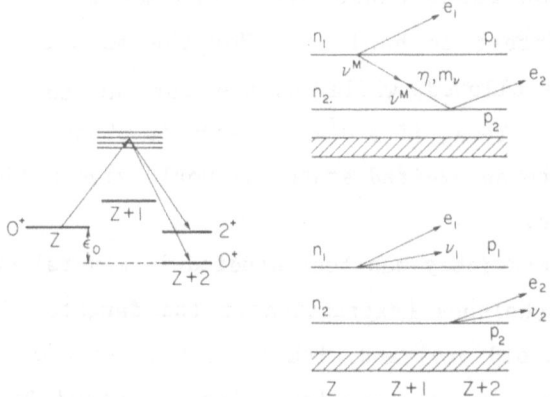

Fig. 7 Expanded Δm^2 vs $\sin^2 2\theta$ plane showing current experimental limits and some dimensional guesses.

Fig. 8 Illustration of double beta decay: (top-right) neutrinoless, (bottom-right) two-neutrino.

134

Te ratio: The uncertainty stemming from the nuclear matrix element can be eliminated by comparing two isotopes of the same element, in this case tellurium, presumed to have similar nuclear structure. The ratio of the half lives is then given by the ratio of the phase space factor, $F_{2\nu}$ and $F_{0\nu}$ for the two isotopes. Specifically, one finds for ^{130}Te and ^{128}Te the ratios $T_{\frac{1}{2}}^{2\nu}$ (130)/ $T_{\frac{1}{2}}^{2\nu}$ (128) = 1.8 x 10^{-4} and $T_{\frac{1}{2}}^{0\nu}$ (130)/$T_{\frac{1}{2}}^{0\nu}$(128) = 3 x 10^{-2} to 3 x 10^{-3} (depending on the values of m_ν and η). The geochemical work of Kirsten[29] gives $T_{\frac{1}{2}}$ (130)/$T_{\frac{1}{2}}$ (128) = (1.0 $\pm$ 1.1) x 10^{-4}, in agreement with 2ν decay, and thus consistent with the absence of 0ν decay. Upper limits for m_ν and η are: m_ν < 5 eV, η < 2 x 10^{-4}. The ratio of Kirsten, however, disagrees with another geochemical ratio obtained by Hennecke et al.[30], $T_{\frac{1}{2}}$ (130)/$T_{\frac{1}{2}}$ (128) = (6.3 $\pm$ 0.2) 10^{-4}. This result could be interpreted as requiring 0ν decay, with m_ν = 10 eV or η = 5 x 10^{-5}. Clearly no strong case can be made for or against lepton conservation until this discrepancy is resolved.

^{130} Te: The half life of the decay ^{130}Te $-$ ^{130}Xe averaged over the existing geochemical experiments[31] is 2.6 x 10^{21} y. Interpreted as 2ν decay this yields a nuclear matrix element $|M_{GT}$ (2ν)$|$ = 0.24. Haxton et al.[32] have calculated this decay and find $T_{\frac{1}{2}}$ $\simeq$ 1.7 x 10^{19}y, about 150 times shorter than observed. The situation is illustrated in Fig. 9. An estimate of the neutrino mass limit based on either value of the matrix element is also shown in the figure.

^{82} Se: Again, there exists a geochemical lifetime value[31] of about 1.5 x 10^{20} y. As illustrated in Fig. 10 the theoretical prediction[32] disagrees with the geochemical lifetime for 2ν decay. In addition, there is a cloud chamber experiment[33] which gives a 15 times shorter lifetime than the geochemical one. The experimental search for 2ν decay is continuing with an improved apparatus[34] and should help clarify the discrepancy. Also, for this transition there exists an experimental limit[35] of $T_{\frac{1}{2}}^{0\nu}$ < 3.1 x 10^{21} y. That

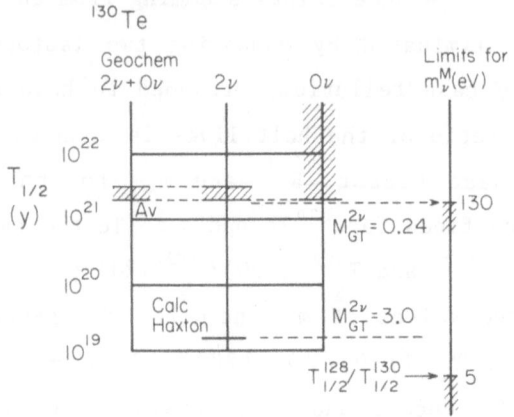

Fig. 9   The geochemical and calculated half-lives and inferred
neutrino mass limits for ^{130}Te (see text for explanations).

limit would imply that less than 5% of the geochemical rate is due
to 0ν decay, giving a neutrino mass limit $m_\nu < 32$ eV.   If the
argument were based on the cloud chamber data then $m_\nu < 12$ eV.

   ^{76}Ge: In the case of ^{76}Ge there now exist several sensitive
laboratory results giving tight bounds on $T_{\frac{1}{2}}^{0\nu}$.   Since there exists
no geochemical data from which to extract the matrix elements, one
has to rely on the calculations by Haxton et al.[32] possibly as modified by Doi et al.[31]   Using only the theoretical matrix elements of
Ref. 32 and the best current laboratory limit[36] for the $0^+ - 0^+$
transition of $T_{\frac{1}{2}}^{0\nu} > 3.7 \times 10^{22}$y, one obtains $m_\nu < 7$ eV. Recalling
the discrepancy, in the case ^{130}Te and ^{82}Se between the geochemical
and theoretical 2ν rates, one may, following Ref. 31, "scale down"
the theoretical matrix element by the factor corresponding to the
discrepancy in ^{82}Se (the nuclide closer to ^{76}Ge),   and   obtain
$m_\nu \leq 16$ eV.   The half-life limit for the $0^+ - 2^+$ branch   is
4×10^{21} y.[36]

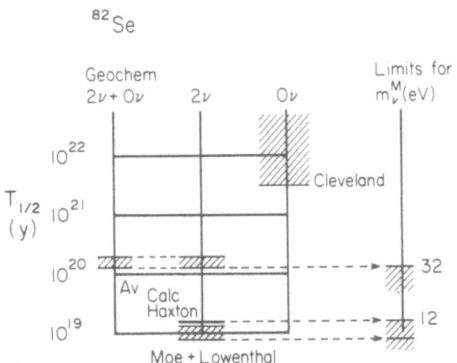

Fig. 10 The geochemical and calculated half-lives and inferred neutrino mass limits for ^{82}Se (see text for explanations).

A high resolution Ge detector is an ideal instrument for obtaining sensitive limits for 0ν decay. ^{76}Ge occurs in germanium with a natural abundance of 7.8%. Fiorini and his group pioneered the Ge experiments and the quoted best current upper limit has been reported by Bellotti et al.[36] from an experiment in the Mont Blanc tunnel. Other laboratories[37,38,39] have also reported results.

Figure 11 depicts the experimental arrangement of the Caltech experiment.[39] The Ge detector is shielded with Cu and Pb and surrounded by a radon tight can. A veto counter serves to reduce cosmic ray background.

The principal limitations for these experiments are detector size and, even more important, detector background. One of the principal sources of the background in the region of the decay energy ε_0 is the Compton contribution of the 2.6 MeV gamma ray accompanying ^{208}Tl decay, a ubiquitous natural contamination. In the Caltech experiment[39] this contamination has been virtually eliminated. Other background components come from cosmic rays.

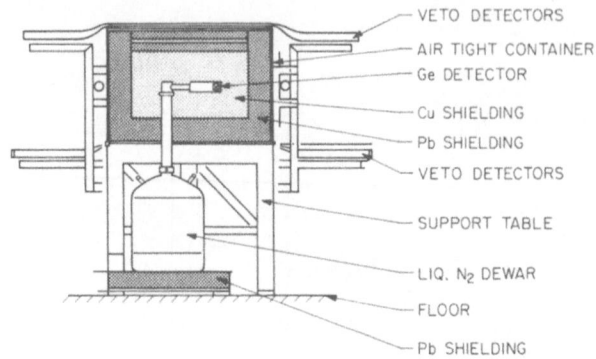

VETO DETECTORS
AIR TIGHT CONTAINER
Ge DETECTOR
Cu SHIELDING
Pb SHIELDING
VETO DETECTORS
SUPPORT TABLE
LIQ. N₂ DEWAR
FLOOR
Pb SHIELDING

Fig. 11 Ge detector setup for the Caltech double beta decay
         experiment.

They can be reduced by a veto system, as illustrated in Fig. 11.
However, high-energy bremsstrahlung and neutrons are not vetoed and
to reduce these components one must install the experiment in an
underground site, as Bellotti has shown.

Figure 12 illustrates a portion of the spectrum from the
Caltech experiment. After 3820 h of running time there is no

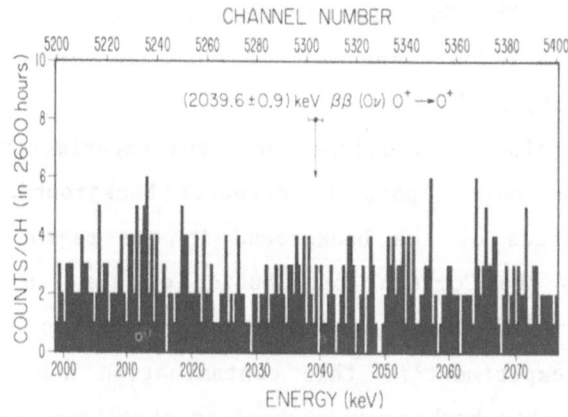

Fig. 12 High-resolution Ge spectrum near total decay energy
         ε_0 = 2040 keV.

evidence for a peak at 2.04 MeV. From the number of countes, $\underline{N}$, in a 3-keV interval (the detector resolution) and its fluctuation, $\sqrt{N}$, one obtains a 1σ limit for the 0ν lifetime of $T_{\frac{1}{2}}^{0\nu} > 1.9 \times 10^{22}$y.

As to the future of the Ge studies, it is safe to predict that ongoing efforts will stretch the sensitivity for $T_{\frac{1}{2}}$ to about $T_{\frac{1}{2}}^{0\nu} > 10^{23}$ y, which corresponds to a mass limit of $m_\nu < 10$ eV. To progress substantially below 10 eV, much larger sample sizes will be needed. The largest currently planned Ge experiments envision detectors of about 1000 cm^3, or 3×10^{24} atoms. In comparison, a large Xe time projection chamber (TPC) may have in excess of 10^{26} atoms of ^{136}Xe ($\varepsilon_0 = 2.5$ MeV). Both liquid Xe[40] and pressured gas Xe[41] TPC have been proposed. Depending on how well correlated electron tracks with energies up to 2.5 MeV can be identified, these detectors may allow the exploration of neutrino mass down to 1 eV or below.

Conclusion

At the present time the only indication for a non-zero neutrino mass comes from the ^{3}H experiment. A mass as large as 33 eV as suggested in this experiment would be difficult to reconcile with cosmological constraints, assuming some scaling between lepton and neutrino masses. This is illustrated in Fig. 13 which shows the cosmologically allowed and forbidden mass regions as well as the present experimental upper limits for neutrino mass. Speculating that the neutrino mass is proportional to the lepton mass, $m_{\nu\ell} \sim m_\ell$, and accepting $m_{\nu e} = 33$ eV, one finds ν_μ and ν_τ masses outside the cosmological bounds. If, on the other hand, $m_{\nu\tau} < 100$ eV, then it follows that $m_{\nu e} < 0.03$ eV. (This simple argument ignores mixing and other complications).

Among the most urgent experiments, therefore, is an independent study of the ^{3}H decay. Also, more sensitive double beta decay experiments have the capability of exploring Majorana neutrino mass down to 1 eV.

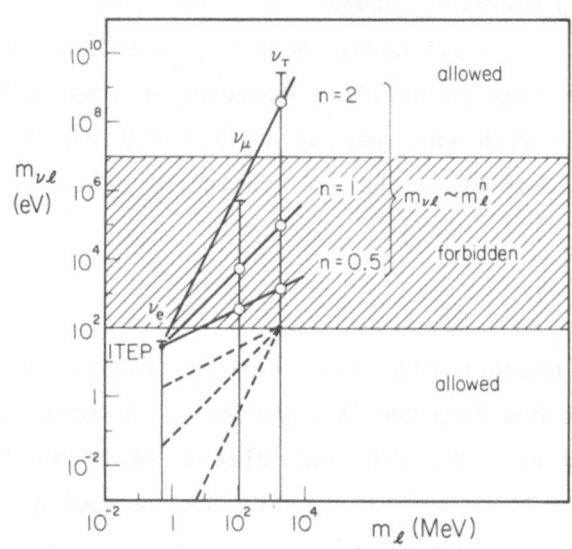

Fig. 13 Neutrino mass vs. lepton mass assuming a power law relationship. If the ν_e mass is taken from the recent ITEP experiment, the projected ν_μ and ν_τ masses are mostly in the region forbidden from cosmological considerations.

REFERENCES

[1] V.A. Lubimov et al., Phys. Lett. 94B, 266 (1980).

[2] J.J. Simpson, Proc. Int. Conf. Matter Non-Conserv., Ed. E. Bellotti, Serv. Doc. Lab. Frascati, Frascati, p. 279 (1983).

[3] S. Boris et al., paper presented at HEP 83, Brighton, July 1983.

[4] I.G. Kaplan et al., Phys. Lett. 112B, 417 (1982).

[5] L.G. Smith et al., Phys. Lett. 102B, 114 (1981).

[6] F. Boehm and P. Vogel, Ann. Rev. Nucl. Part. Sci., 34, to be published.

[7] F. Reines et al., Phys. Rev. Lett. 32, 180 (1974); see also Ref. 6.

[8] J.S. Frank et al, Phys. Rev. D 24, 2001 (1981).

[9] F. Bergsma et al., Phys. Lett. 128B, 361 (1983).

[10] D. Toussaint and F. Wilczeg, Nature 289, 777 (1981).

[11] K. Gabathuler et al., Phys. Lett. B, to be published; G. Zacek et al., Phys. Rev. D, to be published; P. Vogel, Phys. Rev. D, to be published.

[12] J.J. Simpson, Phys. Rev. D 24, 2971 (1981).

[13] K. Schreckenbach et al., Phys. Lett. 129B, 265 (1983).

[14] J.J. Simpson, Phys. Lett. 102B, 35 (1981).

[15] D.A. Bryman et al., Phys. Rev. Lett. 50, 1546 (1983).

[16] M.S. Dixit et al., Phys. Rev. D 27, 2216 (1983).

[17] R. Abela et al., Phys. Lett. 105B, 263 (1981).

[18] R.S. Hayano et al., Phys. Rev. Lett. 49, 1305 (1982).

[19] J.-L. Vuilleumier et al., Phys. Lett. 114B, 298 (1982).

[20] K. Gabathuler et al., Phys. Lett., to be published.

[21] K. Schreckenbach et al., Phys. Lett. 99B, 251 (1981).

[22] H. Kwon et al., Phys. Rev. D 24, 1097 (1981).

[23] J. Blietschau et al., Nucl. Phys. B 133, 205 (1978).

[24] N.J. Baker et al., Phys. Rev. Lett. 47, 1576 (1981).

[25] N. Ushida et al., Phys. Rev. Lett. 47, 1694 (1981).

[26] W. H. Smith et al., Proc. HEP 83, Brighton (1983).

[27] H.J. Meyer et al., Proc. HEP 83, Brighton (1983).

[28] See for example S.P. Rosen, Neutrino 81, Vol. II, p. 76, Univ. Hawaii, 1982.

[29] T. Kirsten et al., Phys. Rev. Lett. 50, 474 (1983).

[30] E. Hennecke et al., Phys. Rev. C 11, 1378 (1975).

[31] M. Doi et al., Progr. Theor. Phys. 69, 602 (1983) and references quoted therein; for ^{130}Te and ^{82}Se lifetimes see also F. Boehm, AIP Conf. Proc. 93, 321 (1982).

[32] W.C. Haxton, et al., Phys. Rev. D 25, 2360 (1982).

[33] M. Moe and D. Lowenthal, Phys. Rev. C 22, 2186 (1980).

[34] M. Moe, Neutrino 82, Vol. I, 231 (1982).

[35] B. Cleveland et al., Phys. Rev. Lett. 35, 737 (1975).

[36] E. Bellotti et al., Proc. HEP 83, Brighton (1983).

[37] F. Leccia et al., University of Bordeaux preprint (1983).

[38] F. T. Avignone et al., Phys. Rev. Lett. 50, 721 (1983).

[39] A. Forster et al., Phys. Lett. 138B, 301 (1984).

[40] H. Chen, private communication.

[41] A. Forster et al., TPC Workshop, Vancouver, June 1983; California Institute of Technology Report CALT-63-409.

Work supported by United States Department of Energy under Contract No. DEAT-03-81-ER40002.

RESULTS ON NEUTRINO OSCILLATIONS AND HEAVY NEUTRINO DECAYS

FROM THE CHARM COLLABORATION

CHARM Collaboration [1])
(CERN-Hamburg-Amsterdam-Rome-Moscow)

presented by L. Lanceri [2])
CERN
1211 Geneva 23, Switzerland

1. INTRODUCTION

The concept of flavour mixing, so fruitful in the quark sector of weak interactions, can be extended to the lepton sector [1]. In analogy with the Kobayashi-Maskawa quark mixing matrix, one can in-

---

[1]) Members of the collaboration are: F.Bersgma, J.Dorenbosch, M.Jonker, C.Nieuwenhuis (Amsterdam); J.V.Allaby, U.Amaldi, G.Barbiellini, L.Barone, C.Berger, A.Capone, Y.Eisenberg, W.Flegel, L.Lanceri, M.Metcalf, J.Panman, D.Perrel-Gallix, R.Plunkett, C.Santoni, F.Vannucci, K.Winter (CERN); I.Abt, J.Aspiazu, F.W.Büsser, H.Daumann, P.D.Gall, T.Hebbeker, F.Niebergall, K.H.Ranitzsch, P.Schütt, P.Stähelin (Hamburg); V.Gemanov, P.Gorbunov, E.Grigoriev, V.Kaftanov, V.Khovansky, A.Rosanov (Moscow); A.Baroncelli, B.Borgia, C.Bosio, M.Diemoz, U.Dore, F.Ferroni, E.Longo, L.Luminari, P.Monacelli, F.deNotaristefani, P.Pistilli, L.Tortora, E.Valente, V.Valente (Rome).
[2]) On leave of absence from INFN, Trieste, Italy.

troduce a unitary neutrino mixing matrix $U_{\alpha i}$ and express the weak flavour neutrino eigenstates ν_α (α = e,μ,τ) as superpositions of mass eigenstates ν_i (i = 1,2,3) :

$$|\nu_\alpha\rangle = \Sigma_i \, U_{\alpha i} \, |\nu_i\rangle \tag{1}$$

Necessary conditions for neutrino mixing are: (a) neutrinos must be massive, with at least two non-degenerate eigenstates having masses $m_i \neq m_j$; (b) the lepton flavour conservation law must be violated.

The idea of massive neutrino mixing is attractive from several points of view. Unlike photons, neutrinos are not required to be massless by any fundamental theoretical principle, and their mass-lessness appears unnatural in most grand unified models [2]. Massive neutrinos can help solving some cosmological and astrophysical problems (i.e. missing masses in galaxies, clusters and in the Universe [3]). There are experimental evidences, though not conclusive, in favor of massive neutrinos: the ^{3}He β-decay measurement performed at ITEP [4], and the solar neutrino flux problem [5].

However, no really firm prediction is offered by theory and the conceivable values for neutrino masses span many orders of magnitude. This fact opens a very wide field of experimental investigation. Many different techniques [6] are used to probe different mass regions, i.e. measurements of the kinematics of beta-decays and meson decays, the search for neutrinoless double-beta decays and the search for neutrino oscillations from a large variety of sources. Most of these methods are quite difficult and require control over large systematic uncertainties. In this respect, the detection of neutrino oscillations in high-energy accelerators neutrino beams, where cross-sections are larger and fluxes better understood, would be one of the cleanest signatures of massive neutrinos. We report here (section 2) on one such search performed at CERN by the CHARM Collaboration.

144

Upper limits on neutrino masses established in some experiments do not exclude large masses, in the MeV region, if the mixing angles are sufficiently small [7]. Such "heavy" neutrinos would be present in high-energy neutrino beams and would be detectable via their decay products. This kind of search was also performed at CERN by the CHARM Collaboration and is described in section 3.

2. NEUTRINO OSCILLATIONS : THE CERN PROGRAM AND THE CHARM RESULTS

The neutrino mass eigenstates or "stationary states" $|\nu_i\rangle$ evolve each in a different way in space and time. As a consequence, in a beam of neutrinos of energy E (in GeV) and initially pure flavour α (α = e,μ,τ,..), after a distance L (in km) a component of flavour β (β = e,μ,τ,.. $\neq \alpha$) develops, so that ν_β can be found with a probability :

$$P(\nu_\alpha \rightarrow \nu_\beta) = \sin^2 2\theta \, \sin^2(1.27\Delta m^2 L/E) \quad . \tag{2}$$

Δm^2 is the difference between the squares of the mass eigenvalues (in eV^2) and θ is the mixing angle in the usual two-flavour parametrization [1], assumed here for simplicity.

The detection of a variation in the flavour composition of a neutrino beam, or "neutrino oscillation", is therefore a clear signature for massive neutrinos. The squared mass eigenvalues difference Δm^2 and the mixing angle θ can be measured in such an experiment. The ratio L/E determines the range of masses to which the experiment is sensitive. The maximum sensitivity is achieved when $E/L \approx \Delta m^2$. When $E/L \gg \Delta m^2$ the effect becomes too small to be detected. For $E/L \ll \Delta m^2$ the finite sizes of the source and of the detector average out the oscillations: only the mixing angle can be measured. Table 1 gives a summary of the available neutrino sources and of the typical accessible ranges for Δm^2.
The sensitivity to the mixing angle is limited by the systematic uncertainties in the knowledge of the initial flux of the neutrino beam, for the experiments detecting neutrinos of flavour α and looking for the "disappearance" of a fraction of them. Experiments

TABLE 1

Sensitivity for various neutrino sources

Source	ν	Mean energy	L	Δm^2 (eV^2)
Accelerators	ν_μ	1-30 GeV	1000 m	1-30
Meson factories	ν_μ, ν_e	30 MeV	100 m	0.3
Reactors	$\bar{\nu}_e$	3 MeV	10 m	0.3
Atmospheric	ν_μ, ν_e	0.5 GeV	12000 Km	4×10^{-5}
Solar	ν_e	1-10 MeV	10^{11} m	10^{-10}

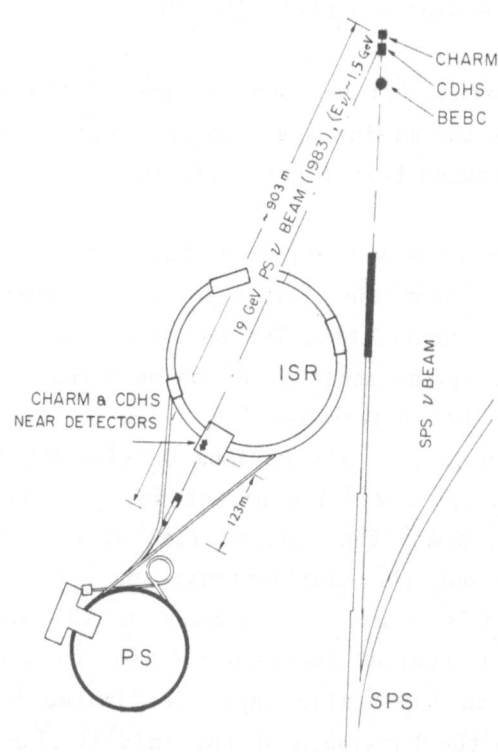

Fig. 1: Sketch of the neutrino beam derived from the CERN Proton Synchrotron and the BEBC, CDHS and CHARM detectors.

looking for the "appearance" of neutrinos of flavour β on the other hand may be limited by the backgrounds.

In 1982 a new neutrino beam line dedicated to oscillations experiments was constructed at CERN. As shown in Fig. 1, it pointed at the three existing neutrino detectors, BEBC, CDHS and CHARM, placed at a distance of about L ≈ 900 m from the ν source. The low energy neutrino beam ($\langle E_\nu \rangle$ ≃ 1.5 GeV) was produced by protons of 19.2 GeV/c, extracted from the Proton Synchrotron (PS), striking a beryllium target with no magnetic elements behind it. The divergence of the beam was much larger than the solid angles covered by the detectors, therefore the neutrino flux scaled approximately as $1/L^2$. Both CDHS and CHARM placed a second detector at a smaller distance from the target (L ≅ 123 for CHARM) to be independent of the absolute neutrino flux.

The CHARM fine-grained calorimeter [8] consists of subunits each formed by a layer of scintillators, a layer of streamer tubes, a 3x3 m² marble slab (8 cm thick), surrounded by an iron frame, and a layer of proportional drift tubes (PDT). Sixty such subunits, corresponding to a fiducial mass of approximately 120 tons, constituted the "far detector", at 903 m from the target. Eighteen subunits, corresponding to a fiducial mass of 36 tons, constituted the "close detector". The close detector was complemented by twenty 8 cm thick iron plates, interleaved with PDT layers, to increase the efficiency in the containment of long muon tracks ("muon catcher").

The trigger selected in each of the two detectors events in which at least three scintillator layers were hit. The data reported here correspond to 8.7 x 10¹⁸ protons on target, and were recorded in the Spring of 1983 from the two detectors simultaneously. The characteristics of the CHARM detector allowed us to perform both a ν_μ disappearance experiment, by measuring the rates of ν_μ induced charged current (CC) events in the two detectors, and to perform a ν_e appearance experiment by searching for candidates of ν_e induced CC events. (The estimated initial contamination of ν_e in the ν_μ beam was less than 0.5%).

To this end we selected contained events characterized by a clear isolated track originating in the detector volume, accompanied at most by some small hadronic activity at the origin (signature for a ν_μ CC event), or by a narrow and compact shower, compatible with the known behaviour of low energy electromagnetic showers in our detector (signature for a ν_e CC event). For each event the vertex, the track or shower direction and the projected length ℓ (in number of traversed subunits) were determined. In order to have equal efficiencies in the two detectors we considered in this analysis only events with $\ell \leq 15$ subunits; longer events, stopping in the muon catcher in the close detector, were analysed separately. Fiducial volume conditions (vertex in the marble of 15 subunits of the close detector and of 57 subunits of the far) and track angle limits eliminated background events due to traversing cosmic rays and beam-related entering muons. The remaining cosmics (3% in the far detector only, where they dominated the triggering rate) were statistically subtracted using the measured out-of-spill rate.

With these criteria 2043 events in the close detector and 270 events in the far detector were selected. Complicated events, having several prongs and/or diffused showers, were rejected by our criteria, but some neutral current (NC) events, characterised by a compact shower, were retained. According to a Monte Carlo simulation based on experimental bubble chamber results, a fraction of 20% NC events contaminated the selected sample.

2.1 The ν_e appearance search

The selected contained events were visually scanned searching for candidates of ν_e-induced quasi-elastic CC events, characterised by a single electromagnetic shower. The longitudinal and lateral profiles of the energy deposition in the scintillator layers were required to satisfy conditions [9] derived from the behaviour of low energy e.m. showers in our detector observed in a test beam, and from Monte Carlo simulations. 19 (66) candidates were found out of 270 (850) events scanned for the far (close) detector. The raw shower energy spectra of these events are shown in Fig.2.

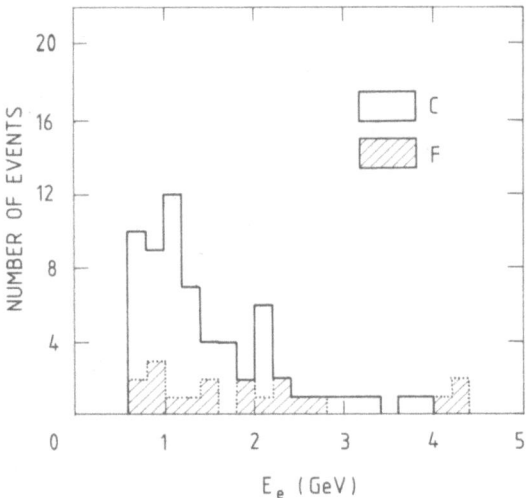

Fig. 2: Observed distributions of the electron energies of ν_e candidates in the close and in the far detector.

The ratio of efficiencies $\varepsilon_e/\varepsilon_\mu$ for finding ν_e CC events and ν_μ CC events was computed by Monte Carlo: $\varepsilon_e/\varepsilon_\mu = 1.00 \pm 0.12$. After correcting for these efficiencies and for the different lengths of the two detectors (a 7% effect), the ratios between events interpreted as ν_e and ν_μ CC events were:

close detector : $R_C = (\nu_e/\nu_\mu)_C = (9.2 \pm 1.1 \pm 1.1)\%$
far detector : $R_F = (\nu_e/\nu_\mu)_F = (8.9 \pm 2.0 \pm 1.0)\%$

The second error is systematic and accounts for the uncertainties in the efficiency computations; it is strongly correlated in the close and the far detector. The observed ratios can only be due to the contamination of NC events with an electromagnetic-like component, or to very fast oscillations at high Δm^2, resulting in an average effect in both detectors. By subtraction of the close and far detector results, one obtains an upper limit of 2.7% at 90% confidence level for the amount of $\nu_\mu \rightarrow \nu_e$ oscillation between the close and the far detector. The comparison of the two electron candidates energy spectra (Fig.2) by a maximum likelihood method

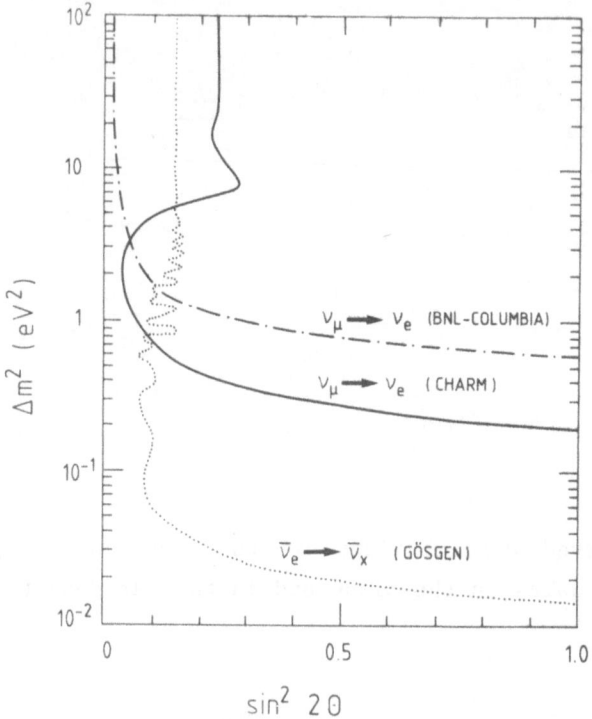

Fig. 3: The 90% confidence limit obtained in the appearance experiment $\nu_\mu \to \nu_e$ is compared with the best previously obtained limits for oscillations of the types: $\nu_\mu \to \nu_e$
(Baker et al., ref. 19) and $\bar{\nu}_e \to \bar{\nu}_x$ (J.L. Vuilleumier et
al., ref. 20).

taking into account the computed beam energy spectrum and detector
acceptances provides the 90% confidence level limit shown in Fig.3.

For full mixing ($\sin^2 2\theta = 1$) our best limit is $\Delta m^2 > 0.20$ eV^2
(at 90% c.l.) while our best sensitivity in the mixing angle is
$\sin^2 2\theta = 0.04$, at $\Delta m^2 = 2.0$ eV^2.

2.2. The ν_μ disappearance search

What about other neutrino flavours? Neutrinos associated with
the τ or possible heavier leptons are, in our beam, below the

threshold for the production of their lepton partners and only undergo neutral current interactions. A signature for the oscillations of muon neutrinos into these neutrino flavours is therefore
given by the measurement of a ratio of far to close detector CC
events rates different from the one expected in the absence of oscillations, and given in first approximation by the ratio of distances squared $(123/903)^2$. In contrast, the neutral current event
rates are independent of the presence of oscillations.

We searched for such effects in our samples of contained
events. We applied corrections for the different dead times (5.7%
in the close detector and 1.9% in the far detector) and for event
losses in the filter programs. Events with superimposed tracks or
spurious hits were in fact rejected; their frequencies (close detector: (11.3 ± 3.1)%; far detector: (7.0 ± 2.0)%) were determined
by scanning random samples of events of the two detectors.

We subdivided the events in bins of the projected length ℓ and
computed in each bin the observed ratio r_{obs} of the far to close
detector rates per unit fiducial mass. We then compared r_{obs} with
the value r_{exp} expected in the case of no oscillations. This value
was computed by a Monte Carlo program, simulating in detail [9] the
neutrino beam and taking into account our detector efficiencies.
Typically the finite size of the neutrino source induced a correction of about 4% in the ratio of rates to the naive $1/L^2$ scaling.

The results are shown in Fig. 4, where the ratio r_{obs}/r_{exp} is
plotted versus the projected length ℓ of the events. The average
accepted neutrino energy, computed in each bin by Monte Carlo, is
also shown. The fourth point, which appears in the figure, was obtained by a separate analysis of long events, with a relaxed containment condition in the close detector, where tracks were allowed
to leave the calorimeter and stop in the muon catcher. The errors
in Fig. 4 are statistical only. Adding all our systematic uncertainties in quadrature, we obtain a total of 5%. The major contribution (3.7%) came from the scanning corrections mentioned above.
About 1.5% was the uncertainty in the Monte Carlo calculation of
r_{exp}.

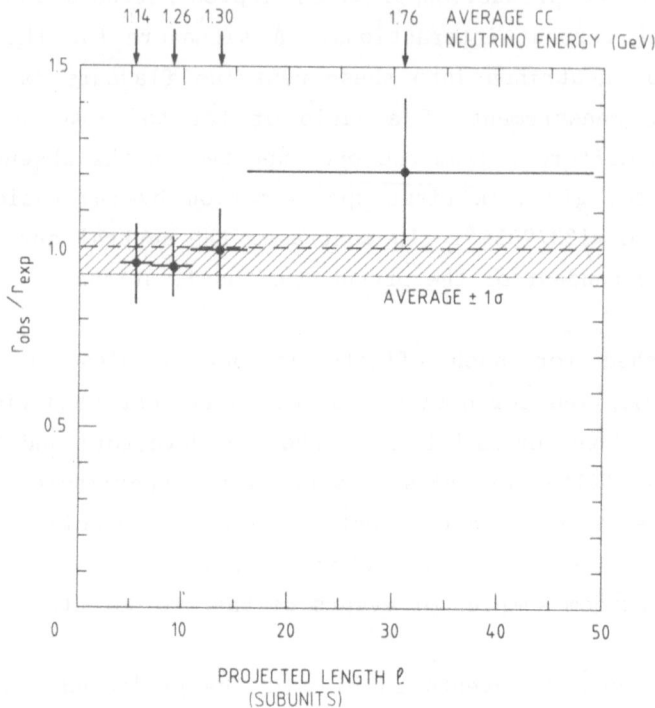

Fig. 4: The ratio r_{obs}/r_{exp} measured in the disappearance experiment is plotted versus the length ℓ (in number of subunits) of the events.

The results of this search are compatible with no disappearance of muon neutrinos. We compared the ratios r_{obs} to r_{exp} computed, in the hypothesis of oscillations, as functions of Δm^2 and $\sin^2 2\theta$. Taking into account the statistical and systematic errors, we derived the 90% confidence level limit shown in Fig.5. The best limits we obtain are $\Delta m^2 < 0.29$ eV^2 for full mixing and $\sin^2 2\theta < 0.18$ for $\Delta m^2 \approx 1.5$ eV^2. We checked that for an increase by 50% of the neutral currents contamination in the Monte Carlo calculation the best limit in Δm^2 increases by less than 0.02 eV^2.

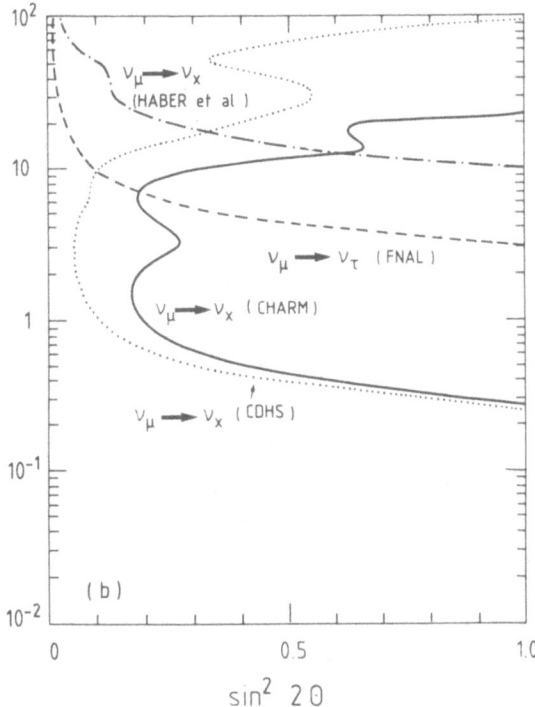

Fig. 5: The 90% confidence limit obtained in the disappearance experiment is compared with previous limits on ν_μ oscillations to ν_τ (Ushida et al., ref. 21) and $\nu_\mu \to \nu_x$ (CDHS Collab., ref. 12 and Haber et al., ref. 13).

2.3 Summary of neutrino oscillations from accelerators

The most recent results from experiments at accelerators are reviewed in Table 2, together with the sensitivities of some experiments which are still in progress or in preparation.

Former limits from high-energy accelerators have been reviewed in several conferences [10,11]: they were mainly obtained searching for the appearance of ν_e or ν_τ with only one detector exposed to ν_μ beams.

TABLE 2

Recent accelerator oscillation experiments

Lab.	Exp.t	Test	Δm^2 (90%CL, $\theta=\pi/4$)	$\sin^2 2\theta$ (90%CL)	Status
CERN	CHARM	$\nu_\mu \to \nu_x$	0.29-25 (eV)^2	0.18	completed
		$\nu_\mu \to \nu_e$	>0.20 (eV)^2	0.04	completed
	CDHS[12]	$\nu_\mu \to \nu_x$	0.26-90 (eV)^2	0.053	completed
FNAL	CCFR[13]	$\nu_\mu \to \nu_x$	10-900 (eV)^2	0.03	completed
CERN	BEBC	$\nu_\mu \to \nu_e$	>0.1 (eV)^2	0.02	in progress
BNL	E734,E775,	$\nu_\mu \to \nu_e$	>0.035 (eV)^2	0.002	in progress
	E776 [11]	$\nu_\mu \to \nu_x$	>0.30 (eV)^2		

The present generation of experiments has significantly improved the limits on the $\nu_\mu \to \nu_x$ disappearance, obtaining larger statistics and a better control over the beam fluxes by the simultaneous use of two detectors, and pushing E/L to the lowest available values.

At CERN, the CHARM collaboration has also searched for the $\nu_\mu \to \nu_e$ appearance, complementing the Gösgen reactor results [20] and improving the existing limits in the window $0.8 < \Delta m^2 < 3.5$ eV^2. Further improvements in this channel are expected from the BEBC experiment [14] at CERN, and from two BNL experiments, E775 and E776 [11].

Disappearance experiments at accelerators are approaching a sensitivity of order 1% to variations in the ν_μ flux. It will be hard to realize further improvements in the sensitivity to the mixing angle. On the other hand, extending the explored Δm^2 range implies ambitious projects of increasing the detector distance L far beyond the laboratory boundaries [15]. Alternative means to achieve this last goal will be provided by experiments at meson factories and with atmospheric and solar neutrinos.

3. A SEARCH FOR DECAYS OF HEAVY NEUTRINOS

It has been pointed out [7] that in leptonic meson decays $M \rightarrow \nu\ell$ ($M = \pi$, K, ...; $\ell = e$, μ, ...), heavy neutrinos should manifest themselves as secondary peaks in the centre-of-mass lepton momentum distribution, with rates $\sim |U_{\ell i}|^2$. No evidence has been found for such heavy neutrinos in π, K decay experiments [16,17].

Neutrinos with a mass larger than a few MeV would decay into a light neutrino and two electrons: $\nu_i \rightarrow \nu_e e^+ e^-$ with a lifetime (scaled from the μ lifetime τ_μ):

$$\tau_i = |U_{ei}|^{-2} (m_\mu/m_i)^5 \tau_\mu . \tag{3}$$

For neutrinos with mass larger than 110 MeV, also other decay modes are open ($\nu_i \rightarrow e\mu\nu$, $\nu_i \rightarrow e\pi$, ...). One can expect to detect heavy neutrinos in neutrino beams at accelerators. They may be produced by meson decays ($M \rightarrow \nu_i\ell X$) and then decay in the detector. The observed rate of $\nu_i \rightarrow \nu_e e^+ e^-$ events would be proportional to the product of mixing angles $|U_{\ell i}|^2 |U_{ei}|^2$.

The CHARM Collaboration has searched for two-electron final states due to heavy neutrinos using data collected in a proton beam-dump and in a wide-band neutrino beam exposure [18].

Decays of heavy neutrinos were searched for in an empty decay region ("decay detector", Fig. 6) of 35 m length and 3 x 3 m^2 cross-section, parallel to the main CHARM neutrino detector. No events were observed in the 400 GeV proton beam-dump neutrino beam exposure. A model-dependent upper limit on the mixing angle $|U_{ci}|^2$ was derived in the mass range 10-250 MeV using the following assumptions: i) the heavy neutrinos are ν_τ's produced in F decays (F $\rightarrow \nu_\tau \tau$), coupled to a single mass eigenstate ($|U_{\tau i}|^2 = 1$); ii) BR(F $\rightarrow \nu_\tau \tau$) = 0.03; iii) $\sigma(pp \rightarrow F\bar{F}X)/\sigma(pp \rightarrow D\bar{D}X) = 0.2$, the D production

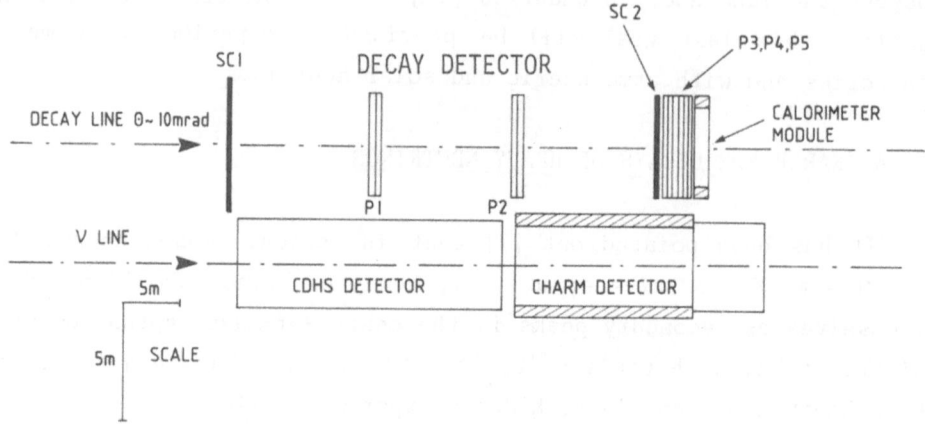

Fig. 6: Layout of the decay detector and of the main CHARM detector. SC1 and SC2 are scintillator planes. SC1 is used as
a veto counter. P1 to P5 are sets of 4 planes of proportional drift tubes. The calorimeter module is 6 radiation lengths deep.

being monitored by the prompt ν_μ rate measured at the same time in
the main detector.

The same results have also been analysed assuming that the
heavy neutrinos are produced by D decays. The $D\bar{D}$ production is
measured by the prompt ν_μ signal in the main detector and can be
extrapolated to the decay volume assuming, for the inclusive yield,
a dependence $(1-x)^4 \exp(-2p_T)$. We derived limits on $|U_{ei}|^2$ (from
$D \rightarrow \nu_i eX$; $\nu_i \rightarrow \nu_e e^+ e^-$) and on $|U_{ei} U_{\mu i}|$ (from $D \rightarrow \nu_i \mu X$; $\nu_i \rightarrow$
$\nu_e e^+ e^-$), extending the explored neutrino mass range up to ≈ 1750
MeV. Beam attenuation effects due to the opening of other decay
channels ($\nu_i \rightarrow \mu e \nu$, $\pi\mu$, πe, ...) become important at neutrino
masses > 1000 MeV for large mixing angles, and have been taken into
account.

Two-electron final states due to heavy neutrinos from π and K
decays were searched for in a sample of 2.7×10^6 (anti)neutrino
interactions collected in the main detector in a wide-band neutrino
beam exposure. After subtraction of the background due to known

156

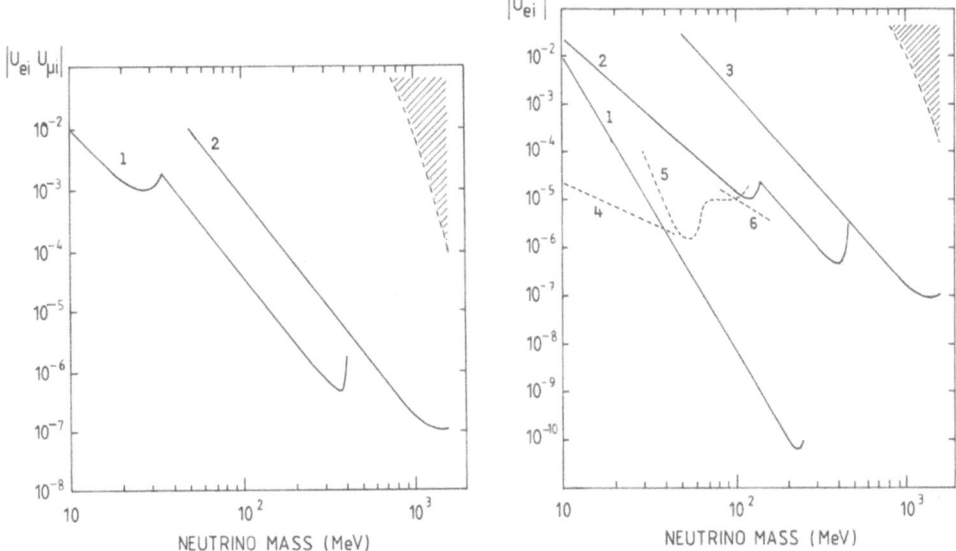

Fig. 7:   (a) 90% CL limits on $|U_{ei}U_{\mu i}|$ as a function of the neutrino mass from: (1) CHARM, Wide Band, π,K decay; (2) CHARM, Beam Dump, D decay.

(b) 90% CL limits on $|U_{ei}|^2$ from: (1) CHARM, Beam Dump, F decay; (2) CHARM, Wide Band, π,K decay; (3) CHARM, Beam Dump, D decay; (4) $\pi \to \nu e$ branching ratio [16]; (5) $\pi \to \nu e$ search for monoenergetic peaks [16]; (6) K $\to \nu e$ branching ratio [17].

sources, the number of candidates is compatible with zero.   Upper limits were set on $|U_{ei}|^2$ in the neutrino mass range 10-490 MeV and on $|U_{\mu i}U_{ei}|$ in the range 10-380 MeV.

The limits (90% confidence level) obtained by the CHARM Collaboration are summarised in Fig. 7.   The excluded regions are those above the solid curves, except the shaded areas corresponding to the sensitivity losses due to beam attenuation effects.   The dashed lines indicate limits from other experiments.

Recently both the main CHARM detector and the decay one have been exposed to the wide-band beam.   The combination of the higher beam flux and of the essentially background-free decay detector should allow us to improve our present limits.

157

REFERENCES

1.   B. Pontecorvo, Zh. Eksp. Teor. Fiz. 34 (1958) 247
[Sov. Phys. JETP 7 (1958) 172].
B. Pontecorvo, Zh. Eksp. Teor. Fiz. 53 (1967) 1725
[Sov. Phys. JETP 26 (1968) 984].

2.   L. Wolfenstein, this Conference.

3.   A. De Rújula, Nucl. Phys. A374 (1982) 619c.

4.   S. Boris et al., Proc. Int. Europhysics Conf. on High Energy
Physics, Brighton (UK), p.386 (1983).

5.   G. Zatsepin, Proc. Int. Conf. Neutrino '82, Balatonfüred,
Suppl., p. 53 (1982).

6.   F. Boehm, this Conference.

7.   R. E. Shrock, Phys. Rev. D24 (1981) 1232.

8.   A.N. Diddens et al., CHARM Collaboration, Nucl. Instr. and
Methods 176 (1980) 189.
M. Jonker et al., CHARM Collaboration, Phys. Scr. 23 (1981)
677-679.

9.   L. Lanceri, presented at the 4th Moriond Workshop on Massive
Neutrinos in Particle and Astrophysics, La Plagne, January
15-21, 1984.

10.   D.H. Perkins, Proc. Banff Summer Institute on Particles and
Fields, Banff(Canada), 1981 (Plenum, New York, 1983), p.379.

11.   H.H. Chen, Proc. 3rd Workshop on Grand Unification, Chapel
Hill, 1982 (Birkhäuser, Boston, 1982), p.206.

12.   F. Dydak et al., CDHS Collaboration, Phys. Lett. B134 (1984)
281.

13.   C. Haber et al., Fermilab. preprint, FERMILAB-CONF-
83/57-Exp. (1983).

14.   M. Baldo Ceolin et al. (Padova-Pisa-Athens-Wisconsin Collaboration), CERN/SPSC/P146 - PSCC/P33, 1980.

15.   F. Vannucci, presented at the same Conference as ref. 9.

16.   D.A. Bryman et al., TRIUMF preprint, TRI-PP-82-43, November
1982, Submitted to Phys. Rev. Lett.

17.   D. Berghofer et al., Proc. Int. Conf. on Neutrino Physics
and Astrophysics, Maui, Hawaii, 1981,vol.2, p.67.
K. Heard et al., Phys. Lett. 55B (1975) 327.
J. Heintze et al., Phys. Lett. 60B (1976) 302.

18. F. Bersgma et al., (CHARM Collaboration), Phys. Lett. 128B
     (1983) 361.
     C. Santoni, presented at the same Conference as ref. 9.
19. N.J. Baker et al., Phys.Rev.Lett. 47 (1981) 1576.
20. Caltech-SIN-TUM Collaboration, J.L. Vuilleumier et al.,
     Phys. Lett. 114B (1982) 298.
21. N. Ushida et al., Phys. Rev. Lett. 47 (1981) 1694.

DISCUSSION

LITTENBERG:
Doesn't the signal in $\nu_h \to e^+e^-\nu$ look just like $\gamma \to e^+e^-$ from π^0 created in neutral current interactions? How do you tell them apart?

LANCERI:
The Beam Dump decay experiment was dedicated to the search for decays of neutral objects and was essentially background-free. The decay region was delimited in air, whose low density implied a negligible background from neutrino interactions in the decay region.

On the contrary, the Wide Band Beam decay experiment was based on a re-analysis of existing data from a normal exposure of the massive CHARM calorimeter detector. The decay region in this case had to be devined inside the calorimeter itself. Therefore indeed there was a background of γ's from neutral current interactions of neutrinos in the decay region, which we had to subtract as follows.

The three main processes contributing events with an electromagnetic shower in our detector are
1) $\overset{(-)}{\nu}_\mu e \to \overset{(-)}{\nu}_\mu e$ (whose detection is the main aim of the WBB experiment;
2) Charged current events from the ν_e component of the beam;
3) Coherent π^0, γ production.

The $\nu_\mu e$ analysis extracts the $\nu_\mu e$ signal using two variables: E_{first} (energy deposited by the electromagnetic shower in the first scintillator layer after the event vertex) and $E^2\theta^2$ (where E is the shower energy and θ the angle with respect to the beam).

The $\nu_\mu e$ signal (1) appears at <u>small</u> $E^2\theta^2$ and <u>small</u> E_{first} (compatible with <u>one</u> minimum ionizing particle) and the background (3) is estimated using as a control region the complementary region in the variables E_{first}, $E^2\theta^2$. The e^+e^- pairs from ν_H decays kinematics are expected to appear overlapping with <u>small</u> opening angles in the forward region, <u>but</u> with E_{first} compatible with <u>two</u> minimum ionizing particles. Also in this case, coherent γ, π^0 background is extrapolated from the control region and subtracted: the result is compatible with zero with a rather large error (about ± 40 events). The limit on ν_H mass and mixing angle is coming out remarkably good even with this error since a large power of the ν_H mass ($m_{\nu_H}^5$) is entering in the expression giving the expected $\nu_H \to \nu_e e^+ e^-$ decay rate.

BOEHM:
When you showed the constraint labelled "solar" do you refer to the analysis by Wilcek? This analysis is based on ^8B neutrinos and thus is subject to some uncertainty. So is the assumed positron rate in space.

LANCERI:
Yes, indeed. I agree with you.

COMMENTS ON LIMITS ON ν MASS

Tso-Hsiu Ho

Institute of Theoretical Physics
Academia Sinica
P.O. Box 2735
Beijing, China

ABSTRACT

A summary of the present situation on the experimental determination of various neutrino masses, the direct experimental measurements, and the indirect estimations from cosmology and possible different scales of neutrino clusters is given. Certain theoretical questions concerning the interpretation of these experiments, especially the ITEP's tritium β-decay experiment and ^{158}Tb K-electron capture experiment, are discussed in some detail. Contradictions in the estimates on the ν mass from ITEP's experiments from cosmology, the neutrino clusters, and from other experiments are emphasized.

I) A SUMMARY OF PRESENT SITUATIONS ON THE EXPERIMENTAL DETERMINATIONS OF νe MASS

Professor Boehm [1] has just given a nice talk about this same subject, so I shall emphasize the related theoretical problems. For the purpose of convenience, the present experimental value of νe mass can be summarized in the following two tables:

TABLE I The neutrino mass from β-decay

Authors	Radioactive Sources	Mass of $\bar{\nu}_e$ (in eV)	Atomic $M(^3T)$–Atomic $M(^3He)$ in eV
Bergkvist (1972)[2]	T in $A\ell$	≤ 60	18651 ± 16
Lubimov et al. (1980)[3]	T in valine	$14 \leq m_{\nu e} \leq 46$	18594 ± 13
Boris et al. (1983)[4]	T in valine	33 ± 1.1 or > 20	18596 ± 2 or > 18592
Simpson (1981)[5]	T in Si (Li)	≤ 65	19573 ± 7

TABLE II The mass from electron capture

Authors	Nucleus	Q-value (in keV)	Estimated neutrino mass
Raun et al. (1983)[6]	^{193}Pt	56.6 ± 0.3	< 500 eV
Anderson et al. (1982)[7]	^{163}Ho	2.3 ± 1.0	< 1.3 keV
Yasumi et al. (1983)[8]	^{163}Ho	2.45 ± 0.08	$\ll 1.3$ keV
Bayer et al. (1983)[9]	^{157}Tb	62.9 ± 0.7 $\Rightarrow E_{\nu}^{L_1} = 0.174 \pm 0.75$	
Raghavan (1983)[10]	^{158}Tb	$E_{\nu}^{K}(1187) = 0.156 \pm 0.017$	≤ 70 eV

From Tables I and II, one can see that the ITEP's experiments[3,4] and Raghavan's experiments[10] are the most important. Hence, in the following I shall discuss in detail a few theoretical problems related to these two experiments.

II) SOME THEORETICAL PROBLEMS ON THE INTERPRETATION OF THE ITEP'S EXPERIMENTS

a) The importance of atomic or molecular structure corrections to the β-spectrum in the ITEP's experiments.

Suppose the radioactive source is free atomic tritium, then in the β-decay process

$$(^3\text{Te}^-) \longrightarrow (^3\text{He } e^-)^+ + e^- + \bar{\nu}_e , \tag{2.1}$$

the final state is a helium ion He^+ plus the e^- and antineutrino. The electron binding energy is 13.6 eV for the tritium atom and 54.4 eV for the ground state of He^+. If the mass of the neutrino is about 30 eV, while the difference between the binding energy is $3 \times 13.6 = 40.8$ eV, then the atomic or molecular corrections to the β-spectrum must be taken into account. Furthermore, owing to the disturbance of the change of the nuclear electric charge, the final atomic electron can be excited to different excited states, e.g., to 1S, 2S, 3S, ... and each accompanied with different transition probabilities. Thus, the final β-spectrum is a complicated superposition of β spectra of different end-point energy and different weight. The general form of the β spectra can be expressed as[11]

$$\frac{dN(E_\beta)}{d E_\beta} \propto \left\{ \sum_{n=1}^{m} <f_n|i>^2 (W_o + V_n - E_\beta)[(W_o + V_n - E_\beta)^2 - m_\nu^2]^{1/2} \right.$$

$$\theta(W_o + V_n - E_\beta - m_\nu) + (1 - \sum_{n=1}^{m} <f_n|i>^2)[(W_o + \bar{V}_{m+1} - E_\beta)^2 + \bar{V^2}_{m+1} \bar{V^2}_{m+1} - \frac{1}{2} m_\nu^2]$$

$$\left. \theta(W_o + \bar{V}_{m+1} - E_\beta - m_\nu) \right\} , \tag{2.2}$$

165

in which

$$W_o = M_i c^2 - M_f c^2 - m_e c^2 - E_R \tag{2.3}$$

is the end-point energy of the nuclei, M_i and M_f are the mass of the initial and final nuclei, E_R is the recoil energy,

$$V_n = E_i - E_{fn} , \tag{2.4}$$

is the corresponding atomic or molecular binding energy differ-ence, $\langle f_n | i \rangle^2$ is the transition probability to the different final atomic or molecular exciting states. In the case of the pure tritium atom, it can be accurately calculated:

$$\langle f_1 | i \rangle^2 = 0.702,$$

$$\langle f_2 | i \rangle^2 = 0.250,$$

$$\langle f_3 | i \rangle^2 = 0.013. \tag{2.5}$$

The term $(1 - \sum_{n=1}^{m} \langle f_n | i \rangle^2)$ is an estimation of the remainder, which can be approximated by the sum rule, i.e., under the assumption

$$\theta(W_o + V_n - E_\beta - m_\nu) \approx \theta(W_o + \bar{V}_{m+1} - E_\beta - m_\nu), \quad (n > m) , \tag{2.6}$$

one can carry out some averaging procedure. By using the formula

$$\bar{V}_n = \sum_n \langle f_n | i \rangle^2 V_n = \langle i | H_i - H_f | i \rangle , \tag{2.7}$$

$$\overline{v_n^2} = \sum_n \langle f_n | i \rangle^2 V_n^2 = \langle i | (H_i - H_f)^2 | i \rangle , \tag{2.8}$$

and the approximate formula

$$(W_o + V_n - E_\beta)[(W_o + V_n - E_\beta)^2 - m_\nu^2]^{1/2} \approx (W_o + V_n - E_\beta)^2 - \frac{1}{2} m_\nu^2, \tag{2.9}$$

one can sum up all the remainder and expressed in Eq. (2.2),

166

$$\overline{V_{m+1}} = \frac{\overline{V}_n - \sum_{n=1}^{m} <f_n|i>^2 V_n}{1 - \sum_{n=1}^{m} <f_n|i>^2} \quad , \tag{2.10}$$

$$\overline{V_{m+1}^2} = \frac{\overline{V_n^2} \sum_{n=1}^{m} <f_n|i>^2 V_n^2}{1 - \sum_{n=1}^{m} <f_n|i>^2} \quad . \tag{2.11}$$

In the case of atomic tritium, we have

$$V_n = \left(\frac{4}{n^2} - 1\right) \times 13.605 \text{ eV}, \tag{2.12}$$

$$\overline{V}_n = 27.21 \text{ eV}, \quad \overline{V_n^2} = 2 \times \overline{V}_n^2 = 1480.8 \text{ eV}^2. \tag{2.13}$$

If some people wish to use the approximate formulae, i.e., for m = 1,2,3,..., then the corresponding binding energy differences are

$$\overline{V}_2 = -4.839 \text{ eV}, \qquad \overline{V_2^2} = 1044.7 \text{ eV}^2,$$

$$\overline{V}_3 = -3.004 \text{ eV}, \qquad \overline{V_3^2} = 6486.1 \text{ eV}^2,$$

$$\overline{V}_4 = -44.01 \text{ eV}, \qquad \overline{V_4^2} = 8873.97 \text{ ev}^2. \tag{2.14}$$

In ITEP's first experiment [3], they used both the bare-nuclear-assumption and a two-level model of atomic tritium, i.e.,

$$\frac{dN(E_\beta)}{dE_\beta} \propto \{<f_1|i>^2 (W_o+V_1-E_\beta) [(W_o+V_1-E_\beta)^2 - m_\nu^2)]^{1/2}$$

$$+ (1-<f_1|i>^2)(W_o+V_1-E_1^*-E_\beta)[(W_o+V_1-E_1^*-E_\beta)^2 - m_\nu^2]^{1/2}\} , \tag{2.15}$$

with $\langle f_1|i\rangle^2 = 0.70$, $E_1^* = 43$ eV $\neq V_1 - \bar{V}_2 = 45.65$ to analyze their experiment they used $W_1 \equiv W_0 + V_1$ and m_ν as free parameters. Then they obtained

$$26 \text{ eV} \leq m_{\nu e} \leq 46 \text{ eV}, \quad \text{(two-level atomic model)} \qquad (2.16)$$

$$14 \text{ eV} \leq m_{\nu e} \leq 26 \text{ eV}, \quad \text{(bare nuclear assumption)} \qquad (2.17)$$

and finally they reached the conclusions on ν_e mass as

$$14 \text{ eV} \leq m_{\nu e} \leq 46 \text{ eV}, \quad \text{(99\% c.\ell.)} \qquad (2.18)$$

$$m_{\nu e} = 34 \pm 4 \text{ eV}. \qquad \text{(95\% c.\ell.)} \qquad (2.19)$$

However, this method of interpretation leads to some questions both from the authors themselves[3] and others[12,13]. The problems are:

   i) Is it possible to substitute the accurate formula by the two-level model?

   ii) Is it possible to substitute atomic tritium for the molecular valine ($C_5H_{11}NO_2$), i.e., involving 19 nuclei and 64 electrons?

In the following we shall discuss these two questions in some detail.

   b) The two-level model

   The essential step in the two-level model is putting the index "m" in the formula (2.2) as m = 1, i.e., assuming

$$\theta(W_o + V_n - E_\beta - m_\nu) \simeq \theta(W_o + \bar{V}_2 - E_\beta - m_\nu), \quad n > 1. \qquad (2.6')$$

This means that one must sum a lot of spectrums with the shape

$$(W_o + V_n - E_\beta)[(W_o + V_n - E_\beta)^2 - m_\nu^2]^{1/2}, \quad n > 1, \qquad (2.20)$$

in which the energy of the β-particles of different spectra n will

pass through a value zero, i.e., $E_\beta = W_0 + V_n - m_\nu$, and then
increase again, finally stopping at the value $\bar{E}_\beta = W_0 + V_2 - m_\nu$.
Thus, in the region 30 ~ 80 eV apart from the end-energy, the
average spectrum will deviate from the accurate one significantly.
If the interpretation of the experiment is quite insensitive to
this region, then the conclusion drawn from the two-level model
gives convincing results. However, in order to measure the mass
of neutrino up to a reliable accuracy, this region is just the
sensitive region where one should take care. This is the point
that one must keep in mind.

c) The molecular excitation correction

Ching, Chao and I have made a set of calculations for the
atomic or molecular structure correction to the β-spectra of the
tritium nuclei by using the variational wave functions of simple
atomic, ionic or molecular systems with great accuracy. Their
final results are given in Table III:

**Table III.a   The β-spectrum of tritium nuclei in simple
atomic or molecular systems**

Molecular	$\langle f_1|i\rangle^2$	V_1 (eV)	$\bar{V}_2$ (eV)	$\langle f_2|i\rangle^2$
T [11]	0.702	40.8	−4.84	0.250
T⁻ [16]	0.230	64.7	29.0	0.469
$T_2(\gamma_0{=}1.40$ a.u.$)$ [15]	0.582	49.1	4.09	0.170
$T_2(\gamma_0{=}1.44$ a.u.$)$ [12]	0.572	49.1	4.77	0.178

Table III.b The β-spectrum of tritium nuclei in simple atomic or molecular systems

Molecular	V_2 (eV)	$\langle f_3\|i\rangle^2$	V_3 (eV)	$\bar{V}_4$ (eV)	$\bar{V}_4^2$ (eV^2)
T [11]	0	0.013	−7.56	−44.0	8874.0
T$^-$ [16]	44.0	1.4×10^{-4}	41.7	5.60	1164.6
$T_2(\gamma_0=1.40$ a.u.$)$ [15]	21.7	0.072	14.7	−17.3	2664
$T_2(\gamma_0=1.44$ a.u.$)$ [12]	22.4	0.072	15.2	−16.6	2519

Kaplan et al.[13,14], using the Hartree-Fock-Roothan approximation, made an ab initio calculation for several complicated molecules, including valine. A few of their results, including those
for valine, are presented in Table IV.

Table IV The β-spectrum of tritium nuclei from valine and a few other molecules [13,14]

Molecule	$\langle f_1\|i\rangle^2$	V_1(eV)	$\bar{V}_2$(eV)
valine I [13]	0.635	47.1	−5.22
[14]	0.651	46.8	−6.22
valine II	0.629	48.3	−2.65
CH_3T	0.616	48.6	0.026
Na T	0.401	58.0	16.16

From Tables III and IV, one can see that the results of molecular corrections deviate from atomic tritium significantly with the value of $\langle f_1|i\rangle^2$ varying from 0.23 to 0.70, and the average value $\bar{V}_2$ varying from −4.84 to 29.0, therefore the atomic approximation to the complex molecular system is not very convincing. Furthermore, it has been pointed out in Refs. 12 and 14 that the branching ratio of the ground state transition of T_2 decay $\langle f_1|i\rangle^2$ is very sensitive to the accuracy of the wave functions selected and may vary from 0.73 to 0.572 for different wave functions. As to the complex molecules, what is the accuracy in such an approach? How many configurations can be practically taken into account? How are the excited levels calculated? It is unclear. This situation has even been emphasized by the authors themselves: "If the tritium β-spectrum in the source corresponds to the atomic tritium, then the estimated neutrino mass is described by the inequality (10) [i.e. Eq. (2.18) here]. However, since we can not prove such a conclusion, therefore it can not be regarded as a well-founded one".[3]

Thus, it is suggested that it is better to use the atomic tritium or it is more practical and easier to use molecular T_2 gas jet or a molecular crystal of T_2 as the radioactive source, since these sources can be calculated with sufficient theoretical accuracy[12].

d) The interpretation of the new ITEP experiment

Recently, the ITEP group published a new experimental result[4]. In this new experiment, the resolving power of the magnetic spectrometer increased from 45 eV to 20 eV, the background was reduced by a factor of 15! In their theoretical interpretations, they also include the corrections due to the width of the calibration lines, while in the old experiment these were neglected. This has been emphasized by Simpson[17] and Boehm[18], the inclusion of such corrections will change the old mass limit (2.18) down to

Table V   The interpretation of ITEP's new experiment with different theoretical formula

Theor. formula	Valine[14]	T_2[12]	T	bore nuclei
$m_{\nu e}$ (eV)	33±1.1	35.8±1	27.3±1.2	$6^{+4}_{-12} \approx 0$
E_0 (eV)	18583.2±0.3	18585±0.3	18580.3±0.3	18567.3±0.3

$$0 \text{ eV} \leq m_{\nu e} \leq 41 \text{ eV.} \qquad (99\%c.\ell.) \tag{2.21}$$

Since the mass limit of $\bar{\nu}_e$ is quite sensitive to the adopted model, ITEP uses different theoretical formulas to interpret their experimental results. These results are expressed in the Table V. According to the philosophy adopted by the old experiment the conclusion one can draw is that

$$6^{+4}_{-12} \text{ eV} \leq m_{\nu e} \leq 35.8 \text{ eV.} \tag{2.22}$$

However, owing to the reduction of the background, they can measure the spectrum 40 ~ 50 eV apart from the very end of the β-spectrum in certain accuracy. Since the average excitation energy $E_1^* \equiv V_1 - \bar{V}_2$ in the two-level model in general is around 50 eV, thus, "if one determines the E_0 parameter near the very end of the beta-spectrum in the energy interval 40 ~ 50 eV, then the resulting values should not be dependent on the assumption about the final state transitions"[4]. Table VI shows the values of E_0 resulting from the fit in the energy intervals of 40 and 50 eV under various assumptions.

Table VI   Different end energies determined by various theoretical formulae

Energy interval of the fit (eV)	Valine (eV)	Atom (eV)	Nucleus (eV)
18530 - 18570	18580±3.2	18579.2±3.1	18578.8±0.8
18520 - 18570	18579.9±2.1	18579.0±2.1	18578.2±1.9

Since the end-point energy E_0 derived in this way does not depend, within 2 eV, on the final state spectra, they thus get a model independent lower limit for E_0:

$$E_0 > 18575 \text{ eV} \quad \text{at 95\% c. } \ell. \tag{2.23}$$

The end-point energies determined in the bare nucleus case using two different ways differs by 11 ~ 12 eV, while the others differ only by 3 eV. Thus, the fitted value of the $m_{\nu e}$ in the bare nucleus case should be ignored and we then have a model independent limit of

$$m_{\nu e} > 20 \text{ eV} \quad \text{with 95\% c. } \ell. \tag{2.24}$$

However, one can see the end-region of the spectrum, i.e., 40 ~ 50 eV apart from the end-point energy (actually, it is 50 ~ 60 eV apart from the end energy E_0 = 18580 eV) is just the sensitive region where the two-level approximation is not very good. For example, one can see from Table III, the first excited level of T_2 is 49.1-22.1 = 27.0 eV, and for T^- is 20 eV, for atomic T is 40.8 eV. What is the value of first excited level of the final state of valine molecule? It is unknown at the present time. So it is difficult to believe that the fitted end-point energies given in Table VI are really model independent. However, as a series of calculations show, the ground to ground transition probabilities $<f_1|i>^2$ are always less than 100%, so it is unlikely that the results of the molecule structure correction will be identical with the bare nuclei case. Giving up the mass value $m_{\nu e} = 6^{+4}_{-12}$ may be reasonable, and plausibly the neutrino may have a finite mass. However, what the lower bound on neutrino mass really is at the present time still depends on the model, e.g., the two-level model assumed here.

III SOME THEORETICAL PROBLEMS RELATED TO THE ELECTRON CAPTURE EXPERIMENT

Raghavan's experiment [10] shows that in the following K-capture process

$$^{158}\text{TB} \longrightarrow {}^{158}\text{Gd}^h(1187 \text{ KeV}) + \nu_e,$$

the emitted neutrinos have an energy

$$E^K_\nu = 156 \pm 17 \text{ eV}. \tag{3.1}$$

It is well known that the electron capture rate is proportional to the phase factor

$$\lambda_K \propto E^{K2}_\nu \left(1 - \frac{m^2_\nu}{E^{K2}_\nu}\right)^{1/2}, \tag{3.2}$$

with

$$E^K_\nu = Q - 1187 \text{ KeV} - \epsilon_K, \tag{3.3}$$

and ϵ_K is the K-electron binding energy of the daughter atom. If $E^K_\nu \simeq 150$ eV and $m_{\nu e} \simeq 30$ eV, the presence of neutrino mass will give a 2% correction to the capture rate. Consequently, if the capture rate measurement can be made with precision $\leq 1\%$ and the value of E^K_ν can be determined up to 1%. Then it is not difficult to determine $m_{\nu e}$ up to the accuracy < 15 eV. However a single experiment to obtain the K/L ratio can not determine the two unknown values E^K_ν, $m_{\nu e}$ simultaneously. Thus, a set of

174

experiments to obtain the L_1/L_2, M_1/M_2 ratios should be done. An analysis of this problem, given by Ching et al.[19], shows that owing to the big energy difference $\epsilon_K - \epsilon_L \simeq 40$ KeV, the new experiments must be accurate to as high as $10^{-4} \sim 10^{-5}$, hence the calculated zero point wave function $\phi(0)$ must attain the same accuracy. It is shown by Erickson[20], also by Mohr[21], that the level shift from the radiative correction for high Z atoms is

$$\Delta E_{nj\ell} = \frac{4\alpha(Z\alpha)^4}{3\pi n^3} F_{nj\ell}(Z\alpha)\, m_e \, . \tag{3.4}$$

For $Z = 64$,

$$\Delta E_{1s} \simeq 100 \text{ eV},$$

$$\Delta E_{2s} \simeq 20 \text{ eV}, \tag{3.5}$$

while the lowest contribution is

$$\Delta E_{1s}^{(1)} \simeq -70 \text{ eV},$$

$$\Delta E_{2s}^{(1)} \simeq -7 \text{ eV}, \tag{3.6}$$

i.e. the high order correction is important. As pointed out by Ching et al.[19], the first order vacuum polarization will contribute to the level shift as

$$\Delta E_{1s}^{v.c.} \simeq -10 \text{ eV},$$

$$\Delta E_{2s}^{v.c.} \simeq -1 \text{ eV}, \tag{3.7}$$

while the corresponding change of zero point wave function $\Delta\phi(0)$ is $\Delta\phi(0)/\phi(0) \simeq 10^{-3}$. Thus it is expected that the radiative corrections can shift the value of $\phi(0)$ up to a few percent.

Unfortunately, at the present time it seems there doesn't exist a program which can carry out the atomic structure

calculation including the radiative corrections with high precision, so it would be most welcome to have such a program appear.

Alternatively, there is a well known result[22] that a
three-body decay will be more sensitive than a two-body decay in
determining the neutrino mass since for two-body decay $M \rightarrow \ell \, \nu_e$
one has

$$E_\ell = \frac{m_M^2 + m_\ell^2 - m_{\nu e}^2}{2 \, m_M} \, , \tag{3.8}$$

while for three-body decay $M \rightarrow N + \ell + \nu_e$,

$$E_\ell^{max} = \frac{m_M^2 + m_\ell^2 - (m_N + m_{\nu e})^2}{2 \, m_M} \, . \tag{3.9}$$

Consequently, the three-body decay has an advantage factor of
$\sqrt{(2m_N/m_\nu)} \approx 10^3 \sim 10^4$. Thus, the process of electron
capture is not really a most advantageous one. However, the
process of electron capture is sensitive to determining the Q
value or the end-pointenergy K^0_{max} of the γ-ray of a three-body
radiative electron capture, e.g., the process

$$^{158}Tb \quad \rightarrow \quad ^{158}Gd^h(1187 \text{ keV}) + \nu_e + \gamma \, , \tag{3.10}$$

with a statistical spectrum shape as

$$\frac{dW}{dk} \propto k(K^0_{max} - k) \sqrt{K^0_{max} - k)^2 - m_\nu^2} \, . \tag{3.11}$$

Thus, a combined study of the process (3.1) and (3.10) will be
easier than a single study[23] of process (3.10), i.e., a lower
statistical accuracy is sufficient to obtain the mass limit, since
in this situation, the K^0_{max} is not determined from the spectrum
(3.11) itself, but is known from (3.2).

IV CONSTRAINTS ON NEUTRINO MASS FROM COSMOLOGY AND NEUTRINO CLUSTERS

The constraints from cosmology have been known for a long time to be[24-26]

$$\sum_i m_{\nu i} \leq 200 \text{ eV} . \tag{4.1}$$

However, a set of newly determined and more accurate values of the helium abundance, $Y \leq 0.25$, is obtained[27] which leads to two important consequences:

a) $N_\nu = 3$,[23] while the old constraint is $N_\nu \leq 3$ or 4.

b) The age of the universe $t_0 \geq 13 \times 10^9$ yr. This is the consequence obtained from the analysis of the globular cluster, in which the helium content Y is a sensitive parameter in determining the age of the universe[29]. This newly determined lower bound of the age of the universe t_0, combined with the newly determined Hubble constant $H_0 = 52$ km/sec/Mpc,[30] will give a new upper bound for the deceleration parameter[31]

$$q_0 \equiv \frac{1}{2} \frac{\rho}{\rho_0} \leq 0.55 , \tag{4.2}$$

since

$$H_0 t_0 = f(q_0) \equiv \int_0^1 (1 - 2q_0 + \frac{2q_0}{x})^{-1/2} dx , \tag{4.3}$$

a well known theoretical result obtained from the standard model of cosmology[32]. Recently, an excellent review article about the deceleration parameter q_0 was given by Fang et al.[33], from which one can set up the limits for q_0 as

$$0.1 \leq q_0 \leq 4.0 , \tag{4.4}$$

and $q_B \equiv \frac{1}{2} \frac{\rho_B}{\rho_C}$ as

$$0.01 \leq q_B h_0^2 \leq 0.02 , \qquad (4.5)$$

which is obtained from the ^{2}H, ^{3}He and ^{7}Li abundance data. Then if $h_0 \equiv H_0/100$ km/sec/Mpc $= 0.52$ is adopted, one has

$$3.5 \times 10^{-31} \text{gm/cm}^3 \leq \rho_B \leq 7.1 \times 10^{-31} \text{ gm/cm}^3 . \qquad (4.6)$$

Thus, combining Eqs. (4.2), (4.4)-(4.6), one obtains a narrower limit[31]

$$1.7 \text{ eV} \leq \sum_i m_{\nu i} \leq 26 \text{ eV} \qquad (4.7)$$

if the neutrino-dominant universe is assumed. One can see this new limit is sharply in contradiction with the new ITEP's value $m_{\nu e} = 33$ eV, but still consistent with $m_{\nu e} > 20$ eV. However, Eq. (4.6) plus the ITEP's lower bound $m_{\nu e} > 20$ eV will lead to

$$m_{\nu \mu} + m_{\nu e} < 6 \text{ eV} \ll m_{\nu e} ! \qquad (4.8)$$

This result seems to contradict the theoretical speculation where people generally expect $m_\nu \propto m_\ell$ or $m_\nu \propto m_q$ in the same generation, also inconsistent with the neutrino oscillation experiments which show[22]

$$m_{\nu \mu} \approx m_{\nu \tau} \approx m_{\nu e} . \qquad (4.9)$$

Furthermore, if one believes that the inflationary universe concept is correct[34-36], then one gets

$$\sum_i m_{\nu i} = 24 \text{ eV} . \qquad (4.10)$$

From astrophysics, one also can obtain the limit of neutrino mass from the properties of the clusters of dark matter, provided the neutrino is the suitable candidate of the dark matter. If neutrinos have mass, they can form neutrino clusters by self-gravitating force. Tremaine and Gunn[37] show, owing to the extremely weak interactions between neutrinos, neutrinos will present themselves as an ideal Vlasov fluid, and as a consequence, the neutral density in the phase space is conserved. This permits the authors to set an upper bound for the neutrinos as

$$m_\nu \geq (101 \text{ eV}) (\frac{100 \text{ km}^{-1}}{\sigma})^{1/4} (\frac{1 \text{ kpc}}{Y_c})^{1/2} g_\nu^{-1/4} \quad , \quad (4.11)$$

where Y_c is the linear size of the cluster, σ the velocity of dispersion. For typical galaxies, one has

$$Y_c \leq 100 (\frac{0.50}{h_0}) \text{ kpc} \quad , \quad (4.12)$$

$$\sigma \simeq 100 \text{ km/sec} \quad . \quad (4.13)$$

Also if $g_\nu = 2$ is adopted, one obtains

$$m_\nu \geq 8 \text{ eV} \quad . \quad (4.14)$$

On the other hand, Ching et al.[38], starting from the Oppenheimer-Volkoff solution, obtained that if the degenerate neutrino halo is used to explain the behavior of rotation curve different from the Keplarian law up to 100 kpc, a neutrino mass limit can be deduced as

$$m_\nu \leq 9 \text{ eV} \quad . \quad (4.15)$$

If the scale extends to 1 Mpc,

$$m_\nu \leq 3 \text{ eV} \quad , \quad (4.16)$$

would be obtained instead. Neutrino mass limits can also be obtained from the estimation of the total mass in the galaxies. As analyzed by Ching et al.[39], the Oppenheimer-Volkoff solution predicted the existence of a heaviest stable neutrino star with the total mass as

$$M_{max} = 0.623 \left(\frac{2\pi hc}{G}\right)^{3/2} \frac{1}{8\pi m_\nu^2} \left(\frac{2}{g_\nu}\right)^{1/2} , \tag{4.17}$$

and if one assumes that the observed cluster of galaxies with the total mass of about 1×10^{16} $M_\odot$ mainly contributed from neutrinos, then one gets

$$m_\nu \leq 7.6 \text{ eV} . \tag{4.18}$$

If the total mass is of 10^{17} $M_\odot$, then one gets

$$m_\nu \leq 2.4 \text{ eV} , \tag{4.19}$$

instead. Another approach in the estimation of neutrino mass constraints is started from studying neutrino's Jeans mass. Many authors[40] show that before the age of the recombination, there exists a preferential cluster mass M_{pre}

$$M_{pre} = 7.15 \times 10^{15} \times \left(\frac{m_\nu}{10 \text{ eV}}\right)^{-3} \left(\frac{g_\nu}{2}\right)^{-2/3} M_\odot . \tag{4.20}$$

Since the neutrino Jeans mass must be less than the mass within the particle horizon M_n, which is about 1×10^{17} $M_\odot$, then one would get

$$m_\nu \geq 3.8 \text{ eV} . \tag{4.21}$$

On the other hand, if the mass of the so far observed largest structure of super cluster is about 10^{16} $M_\odot$, then by requiring that $M_{pre} \leq 10^{16}$ $M_\odot$, one gets

$$m_\nu \geq 9 \text{ eV} . \tag{4.22}$$

All these order of magnitude estimations lead to such concept that in order to explain the dark matter with the use of massive neutrinos in different scales such as galaxy and galaxy clusters, different neutrinos with different masses are needed.

V A Summary

In short, the above discussions can be summarized in the following:

    i)   The ITEP's value $m_{\nu e} = 33 \pm 1.1$ eV is in contradiction with cosmological constraints 1.7 eV $\leq m_{\nu e} \leq$ 26 eV.

    ii)   The results obtained from neutrino oscillation experiments with

$$m_{\nu\mu} \simeq m_{\nu\tau} \simeq m_{\nu e} \qquad (4.9)$$

are in contradiction with the estimated value from studying neutrino clusters, in which different neutrinos have different masses 3 ~ 9 eV are suggested.

    iii)   In addition, we have heard from Prof. Boehm in this conference, the Majorana neutrino mass $m_{\nu e} M$ is consistent with[1]

$$m_{\nu_e} M \simeq 0, \text{ or } m_{\nu_e} M \leq \text{ few eV.} \qquad (4.10)$$

Thus, this mass limit is also in contradiction with the definite lower mass limit

$$m_{\nu e} > 20 \text{ eV} \quad . \qquad (2.24)$$

Thus, the present situation seems puzzling to us and these three puzzles constitute the present situation regarding the neutrino mass problem!

REFERENCES

1.   F. Boehm, this Proceeding.

2.   K.E. Bergkvist, Nucl. Phys. B39 (1972) 317, 371.

3.   V.A. Lubimov, E.G. Novikov, V.E. Nozik, E.F. Tretyakov and V.S. Kosik, Phys. Lett. B94 (1980) 266; V.A. Lubimov, E.G. Novikov, V.E. Nozik, E.F. Tretyakov, V.S. Kosik and N.F. Myasoyedov, JETP (Russian ed.) 81 (1981) 1158.

4. S. Boris, A. Golutuin, L. Laptin, V. Lubimov, V. Nagovizin, E. Novikov, V. Nozik, V. Soloshenko, J. Tichomirov and E. Tretjakov, Proc. HEP 83, Brighton, (UK), 20-27 July (1983) 386.

5. J.J. Simpson, Phys. Rev. D23 (1981) 649.

6. H.L. Raun et al., Neutrino Mass and Gauge Structure of Weak Interactions, AIP Conf., Proc. No. 99, NY, AIP (1983).

7. J.V. Anderson et al., Phys. Lett. B113 (1982) 72.

8. E. Yasumi et al., papers submitted to the 1982 Neutrino Conference, Balaton, Hungary.

9. C.J. Beyer et al., Nucl. Phys. A408 (1983) 87.

10. R.S. Raghavan, Phys. Rev. Lett. 51 (1983) 975.

11. Ching Cheng-rui and Ho Tso-hsiu, Comm. Theor. Phys. 1 (1982) 11.

12. Ching Cheng-rui, Ho Tso-hsiu and Chao Hsiao-lin, Comm. Theor. Phys. 1 (1982) 267.

13. I.G. Kaplan, V.N. Smutny and G.V. Smelov, Phys. Lett. B112 (1982) 417.

14. I.G. Kaplan, V.N. Smutny and G.V. Smelov, JETP (Russian ed.) 84 (1983) 833.

15. Ching Cheng-rui and Chao Hsiao-lin, a new calculation on the β-spectrum of T_2, to be published in Comm. Theor. Phys.

16. Chao Hsiao-lin, On the β-spectrum of T^- ion, to be published in Comm. Theor. Phys.

17. J.J. Simpson, paper presented at ICOMAN 83, Frascati, Italy, January 1983.

18. F. Boehm, APS, Annual Meeting of the DPF, Virginia Polytechnic Inst. and State Univ., Blacksburg, VA, Sept. 15-17, 1983, Preprints CALT-63-413.

19. Ching Cheng-rui, Ho Tso-hsiu and Liu Jia-lin, to be published in Comm. Theor. Phys.

20. G.W. Erickson, Phys. Rev. Lett. 27 (1971) 780.

21. P.T. Mohr, Phys. Rev. Lett. 34 (1975) 1050.

22. P.H. Frampton and P. Vogel, Phys. Rept. 82 (1982) 339.

23. A. De Rujula, Nucl. Phys. B188 (1981) 414.

24. Ya. B. Zeldovich and R.A. Sunyaev, Astron. Zh. Pisma 6 (1980) 451.

25. C.R. Ching, Y.S. Wu, T.H. Ho, C.H. Chang and Z.L. Zou, Acta Astrophys. Sinica 1 (1981) 9.

26. M.K. Turner, "Neutrino 81", R.J. Cence, E. Ma and A. Roberts, editors, Univ. of Hawaii, (1981).

27. H.B. French, Lick Obs. Bull. No. 863 (1979; J. Lequeax et al., Astron. Astrophys. 80 (1979) 155; R. Dufour et al., Report at the Santa Cruz Workshop on Astronomy (1980); M. Peimbert and S. Torres-Peimbert, Astrophys. J. 203 (1976) 581.

28. B.E.J. Pagel, Phil. Trans. R. Soc. Lond. A30 (1982) 19; D. Kunth, Proc. of the 1st Moriond Astrophysics Meeting: Cosmology and Particles, ed. J. Audouze, P. Crane, T. Gaisser, D. Hegyi and J. Tran Thanh Van, (1981) 241.

29. D.N. Schramm, Phil. Trans. R. Soc. Lond. A307 (1982) 43.

30. A. Sandage, Astrophys. J. 256 (1982) 553.

31. Ching Cheng-rui and Ho Tso-hsiu, to be published in Physics Report.

32. S. Weinberg, Gravitation and Cosmology, Principles and Applications of the General Theory of Relativity, John Wiley, (1972).

33. Fang Li-zhi et al., Progress in Physics, Chinese edition, 3 (1983) 52; Proc. Acad. Sinica - Max-Planck Soc. Workshop on High Energy Astrophysics, Science Press, Beijing, China, Gordon and Breach Science Publisher S.A. (1983) 425.

34. A. Guth, Phys. Rev. D23 (1981) 347.

35. A.D. Linde, Phys. Lett. B108 (1982) 389.

36. A. Albrect and P.J. Steinhardt, Phys. Rev. Lett. 48 (1982) 1220.

37. S. Tremaine and J.E. Gunn, Phys. Rev. Lett. 42 (1979) 407.

38. C.R. Ching, T.H. Ho and Y.Z. Zhang, Comm. Theor. Phys. 2
     (1983) 1145; W.L. Huang, C.X. Xu, C.R. Ching and T.H. Ho,
     Comm. Theor. Phys. 2 (1983) 1039; Proc. Acad. Sinica -
     Max-Planck Soc. Workshop on High Energy Astrophysics, Science
     Press, Beijing, China, Gordon and Breach Science Publisher
     S.A. (1983) 467.

39. C.R. Ching, Y.S. Wu, T.H. Ho, C.H. Chang and Z.L. Zou,
     Preprint AS-ITP-002, (1981); Kexue Tongbao 27 (1981) 289.

40. See Ref. 31, and the references therein.

DISCUSSION

GOLDHABER:
What other candidates are there for the dark mass of the universe?

HO:
Some people consider the axion as a new candidate of the dark matter. However there is no direct evidence on the existence of this matter.

CHAU:
Could you please elaborate why different cosmological states require different neutrino masses?

HO:
According to the Oppenheimer-Volkoff theory of neutron stars, there exist a limiting mass. If the total mass of the condensed system is bigger than the limiting mass then gravitational collapse will occur. This kind of limiting mass is inversely proportional to the mass of the constituent. In the galaxies and cluster of galaxies, different scale of masses occur. Thus, it gives different limits to the mass of neutrino.

HAGELIN:
I am not entirely sure I have grasped the purport of your remarks: Are the astrophysical bounds on m_{ν_i} consistent among themselves? In particular, is the constraint $\sum_i m_{\nu_i} \leq 26$ eV

consistent with the requirement that neutrinos cluster at different scale?

HO:
Yes, the fact that astrophysics seems to require several different neutrino masses (for different flavors presumably) is still consistent with $\sum_i m_{\nu_i} \leq 26$ eV.

NEUTRINO MASSES AND MIXINGS

Lincoln Wolfenstein

CERN, Geneva
and
Carnegie-Mellon University, Pittsburgh, PA 15213

Abstract

   Gauge theories that contain right-handed partners for the usual neutrinos, such as SO(10), lead naturally to massive Majorana neutrinos. The possibility of obtaining massless neutrinos or massive Dirac neutrinos in such theories is discussed. The phenomenology of three flavors of massive neutrinos, including neutrino oscillations, decay kinematics, and double beta-decay, is reviewed. The theories of interest usually contain heavy neutral leptons in addition to the light neutrinos and some of the phenomenology related to these is summarized.

1. Theoretical Ideas about Neutrino Mass

   It is quite possible that all neutrinos are massless. Nevertheless the idea that neutrinos have a mass has fascinated experimentalists and theoreticians alike. The only experimental evidence for neutrino mass remains the spectrum of the H^3 decay of the Lubimov group[1] that gives $m_{\nu_e} \sim 30$ ev. No other experiments contradict this result nor do they provide direct or indirect confirmation. The standard electroweak theory has massless neutrinos but there is no fundamental reason why interactions beyond the standard model should not induce neutrino masses and indeed they are required in the minimal SO(10) grand unified theory. If neutrinos have masses one then expects Cabibbo-like mixing in the lepton sector. In this talk I will review some of the theoretical ideas about neutrino masses and mixings and the various types of experiments that are relevant.

The standard SU(2)xU(1) model has only left-handed neutrinos ν_{Li} so that neutrinos can not acquire a mass in the way the other fermions do. It is possible, however, for neutrinos, unlike other fermions, to acquire a Majorana mass by a mass term that connnects ν_{Li} to the right-handed anti-particle fields ν^c_{Rj}. Such a mass term does not in fact occur in the standard model because the interaction Hamiltonian contains as global symmetries the conservation of the lepton numbers L_e, L_μ, L_τ, whereas a Majorana mass term involves a lepton number change $\Delta L = 2$. An interesting possibility is that interactions much weaker than the standard electroweak interactions may cause $\Delta L = 2$ and thus the neutrino Majorana mass may be considered one of the few low-energy windows on such superweak interactions. One may be semi-quantitative by noting that the Majorana mass term also requires a change in the weak isospin $\Delta I_w = 1$ (ν_L with $I_3 = \frac{1}{2}$ is connected with ν^c_R with $I_3 = -\frac{1}{2}$). Thus if weak isospin violation occurs only via the vacuum expectation value (vev) v of Higgs doublets the[2]

$$m_\nu \sim (fv)^2/M$$

(1a)

where f is a Higgs coupling, the square is required to get $\Delta I = 1$ from a doublet, and the M in the denominator required by dimensional arguments corresponds to the mass scale of the $\Delta L = 2$ interaction. Since fv is the mass m_D of a normal Dirac fermion we can write (1a) as

$$m_\nu = m_D^2/M$$

(1b)

A detailed model is needed to know whether (and when) to use values between $m_e (\sim 10^{-3}$ Gev) and $M_t (>20$ Gev) for m_D. The striking feature of Eq. (1) is that the smaller m_ν the larger the mass scale of the new interaction we probe.

Models which have an intrinsic left-right symmetry and also a lepton-hadron symmetry yield in general massive neutrinos. The prototype is the grand-unified theory (GUT) SO(10). Such models have a set of right-handed neutral leptons N_{Ri} to go with ν_{Li} and the lepton-hadron symmetry has the consequence that there will be a Dirac mass term of order m_D linking the ν_{Li} with the N_{Rj}. This, of course, raised the problem of why the neutrino mass was not equal to m_D. The answer of Gell-Mann, Ramond, and Slansky[3] (GRS) was that N_R can acquire a large Majorana mass M (since they are SU(3)xSU(2)xU(1) singlets) so that the ν mix only very little with the N yielding as a light mass eigenstate.

$$\nu_{Ll} \cong \nu_{La} + (m_D/M)N^c_{La}$$

(2)

with

$$m_1 \sim (m_D/M)^2 M = m_D^2/M$$

It was in this context that Eq. (1b) was first derived. For the case of N generations such a model yields N light Majorana neutrinos $\nu_{Li},...$ with masses given by a generalization of Eq. (1b) and N heavy Majorana neutral leptons $N_{Ri}...$ with masses of the order of M. The mixings and mass hierarchy expected in SO(10) have been reviewed in many papers.[4]

The same general procedure can be carried out[5] in the simplest left-right symmetric theory SU(2)$_L$xSU(2)$_R$xU(1) (which, of course, is a subgroup of SO(10)). In this case the mass M is associated with $\langle \Delta_R \rangle$, where Δ_R is a Higgs triplet under SU(2)$_R$ and is the cause of the breaking of left-right symmetry. In such a model there is then a correlation between the scale $M \propto m_\nu^{-1}$ and the scale M_R of the right-handed intermediate bosons W_R. For interesting values of $W_R (\sim$ a few TeV) one tends to get a mass for ν_μ or ν_τ or both that is hard to reconcile with cosmological constraints I will discuss later.

One may ask whether within the general framework of SO(10) or similar models it is possible to have the neutrinos remain massless. Within SO(10) there are for one generation sixteen fundamental fermion fields forming the 16 representation. If one adds a seventeenth neutral fermion in the singlet representation (which could be needed in supersymmetry) and imposes a global lepton number symmetry one finds[6] that the resulting spectrum contains one massless neutrino and a heavy Dirac neutral lepton. For N generations the single lepton number symmetry guarantees N massless neutrinos.

Another question raised is whether it is possible to obtain Dirac neutrinos with a light mass in such a model. This is of particular interest with respect to the absence of neutrinoless double beta–decay. Recently it has been shown[7] that if one extends the fermions from 16 to 18 per generation by adding two singlets of SO(10) one can obtain N light Dirac neutrinos and N heavy Dirac neutral leptons.

I will explain briefly the basic ideas of these extended SO(10) models. Within standard SO(10) there are two two–component left–handed neutrino fields ν_L and n_L ($\equiv N_R^c$, the anti–particle of N_R) so that the mass matrix has the form (with left–handed particles defining the columns and the corresponding right–handed particles the rows)

$$
\begin{array}{cc}
 & \begin{array}{cc} \nu_L & n_L \end{array} \\
\begin{array}{c} \nu_R^c \\ n_R^c \end{array} &
\begin{pmatrix} m & m_D \\ m_D & M \end{pmatrix}
\end{array}
$$

where m_D is the Dirac mass and m, M are Majorana masses. If $m = M = 0$ the neutrino is a normal Dirac particle with mass m_D, which in SO(10), is of the order of the quark mass m_u. The GRS assumption was M is very large (GUT scale), $m \underset{\sim}{\sim} 0$ (since it requires $\Delta I_{weak} = 2$); the result is two Majorana particles with masses M and m_D^2/M. If we add a third neutrino field s_L (an SO(10) scalar) it is possible to maintain lepton number conservation with the mass matrix

	ν_L	n_L	s_L
ν_R^c	0	m_D	0
n_R^c	m_D	0	M
s_R^c	0	M	0

Inspection indicates one zero eigenvalue and one Dirac particle with mass $(m_D^2+M^2)^{1/2}$, which is approximately equal to M if M $\gg$ m. (Notice that in this notation a Dirac particle corresponds to equal and opposite mass eigenvalues as required by the zero trace of the matrix). The large value of M is assumed because it connects two SO(10) singlets.

If we add a fourth neutrino field σ_R we can assume two fields (v_L, s_L) with positive lepton number and two (n_L, σ_L) with negative lepton number yielding the mass matrix

$$
\begin{array}{c|cccc}
 & v_L & s_L & n_L & \sigma_L \\
\hline
\begin{array}{c} v^c_R \\ s^c_R \end{array} & & 0 & & M^T \\
\hline
\begin{array}{c} n^c_R \\ \sigma^c_R \end{array} & & M & & 0 \\
\hline
\end{array}
$$

$$
M = \begin{pmatrix} m_D & M_1 \\ m_D' & M_2 \end{pmatrix}
$$

where m_D, m_D' involve SU(2) breaking whereas M_1 or M_2 is an intermediate mass. The choice of Roy and Shanker[7] is $m_D' = 0$, $M_2 \gg M_1 \gg m_D$. The resulting mass spectrum has two Dirac particles with masses M_2 and $M_1 m_D / M_2$. All these considerations can be extended to three generations by replacing the elements of the mass matrix by 3x3 matrices.

Although the existence of right-handed partners for the v_{Li} required by left-right symmetric models is the most interesting theoretical argument for neutrino mass, it is possible to obtain a Majorana neutrino mass in SU(2)xU(1) models without them. The simplest possibility is the introduction of triplet Higgs particles with a vacuum expectation value v_3 much less than that of the usual Higgs doublet v_2. The triplet Higgs couples v_L to v^c_R with a coupling f so that it behaves as if it has lepton number L = 2. The neutrino mass is then $f v_3$. Two versions are possible (1) the violation of lepton number occurs in the Higgs Lagrangian; in this case typically these Higgs particles are heavy with mass m_H and the value of v_3 is of the order v_2^2 / m_H; this conforms to the general result of Eq. (1a). (2) The violation of the global lepton number occurs spontaneously when the triplet acquires its expectation value. This model due to Gelmini and Roncadelli[8] has many amusing consequences[9]. The most important is that in addition to the light neutrinos there is a massless Goldstone boson X called the majoron and a very light boson $\bar{H}_2^{\,\circ}$. Observation of these two new particles is very difficult; perhaps the most hopeful place is the decay $Z^\circ \rightarrow X + H_2^{\,\circ}$, which would be "seen" as $Z^\circ \rightarrow$ nothing and might be alternatively interpreted as two new flavors of neutrinos. This majoron model has the unique feature that there are no cosmological relic neutrinos; the only relics are the massless majorons.

2. Phenomenology of Three Massive Neutrinos

I now turn to the phenomenology associated with three massive neutrinos. I will assume CP invariance since evidence relative to CP violation in the mass matrix is very difficult to acquire. There are then three mass eigenvalues m_1, m_2, m_3 and three mixing angles which define the orthogonal matrix U that relates the mass eigenfunctions v_i to the flavor states v_a

$$v_a = \sum_{i=1}^{3} U_{ai} \, v_i \qquad a = e, \, \mu, \, \tau$$

(3)

In addition for the case of Majorana neutrinos there are two relative CP eigenvalues

$$\eta_{12}, \, \eta_{13} \, (\eta_{23} \equiv \eta_{12}\eta_{13}, \, \eta_{12} = \pm 1, \, \eta_{13} = \pm 1)$$

which have significant phenomenological implication. When the Majorana mass matrix is diagonalized by the matrix U, some of the eigenvalues come out relatively negative; these are those that have negative values of η_{ij}. Two degenerate Majorana neutrinos with opposite CP eigenvalues are equivalent to a Dirac neutrino. In this section I will concentrate on the case of three Majorana neutrinos, the case that is favored theoretically. The phenomenology associated with additional heavier neutral leptons will be discussed in the next section.

Five types of experiments are discussed:

1. Direct mass determination from decay kinematics.

2. Neutrino oscillation experiments provide correlated limits on mass differences and mixing angles.

3. Double beta–decay experiments provide information concerning the electron part of the neutrino mass matrix.

4. The search for rare decay branches (such as $\pi \rightarrow e +$ heavy neutrino) provides a correlated limit on neutrino masses and mixings.

5. The search for the decay of massive neutrinos also provides correlated limits on masses and mixings.

I will not attempt to review the experimental results, which except for the results reported here by Boehm, have been adequately reviewed in recent neutrino conference proceedings. I will, however, review some logical interrelations among the different types of experiments.

Because of the possibility of mixing it is not always clear what is meant by a determination of "the mass" of a type of neutrino. For the case of a study of decay kinematics (as the ^{3}H decay) there are two possibilities.

1) There are two or more significantly different mass eigenvalues. In this case neutrino oscillation experiments tell you there is little mixing for the relevant mass range. Thus the experiment simply determines the mass of one mass eigenstate. Experiments of type 4 looking for rare branches to heavy neutrinos can, of course, provide limits on the small parameter U_{ej}^2 for the probability that v_e includes a heavier v_j. This has been applied to the overall spectrum of ^{3}H by Simpson[10].

2) There are two or more almost identical eigenvalues. In this case one determines the essentially common mass.

For the case of neutrinoless double beta-decay the amplitude is proportional to the weighted sum

$$m_{\beta\beta}(v_e) = \sum_i (U_{ie})^2 \, \eta_{ij} m_i \equiv M_{ee}$$

(4)

where M_{ee} is the diagonal element of the original neutrino mass matrix that connects v_{eL} and v_{eR}^c. (This holds provided all m_i are relatively light; for masses well above 10 MeV the contribution to the sum is decreased because the range of the force associated with the neutrino exchange becomes comparable or smaller than relevant nuclear length parameters.) Thus neutrinoless double beta-decay does not provide a limit on the mass of v_e as measured in ^{3}H decay. A small admixture of a heavy neutrino with the opposite CP eigenvalue can cancel in Eq. (4) the contribution of the predominant low mass in v_e. Such a cancellation is not necessarily unnatural since there may be some symmetry that renders the diagonal value M_{ee} zero. After all, the mass matrix is in a theoretical sense more fundamental than the eigenvalues. It has recently been noted[11] "that if one of the neutrinos contributing significantly to the sum that gives $m_{\beta\beta}(\nu_e)$ has a mass much larger than 10 MeV, any cancellation that occurs will vary with the A of the nucleus so that a complete cancellation would not occur for both ^{78}Ge and ^{128}Te, for example. Of course, in this case, the result is <u>not</u> simply M_{ee} so that the cancellation in any case would appear somewhat accidental.

In the case of a heavy neutrino, say ν_3, which mixes with ν_e (or ν_μ) with a small probability $(U_{e3})^2$ (or $(U_{\mu3})^2$), experiments of type 4 or 5 are relevant. Rare decay branches of a system X (a nucleus, π, or K)

$$X \rightarrow e\nu_3$$

or

$$X \rightarrow \mu\nu_3$$

will be proportional to $(U_{e3})^2$ or $(U_{\mu3})^2$ and depend kinematically on the mass[12]. Results on $\pi \rightarrow e\nu_3$, $\pi \rightarrow \mu\nu_3$, and $K \rightarrow \mu\nu_3$ have been presented in the last couple of years[13].

A neutrino ν_3 which is sufficently massive may have a reasonable probability of decaying in a neutrino detector[14]. For masses much larger than m_e but less than m_μ the expected decay mode is

$$\nu_3 \rightarrow e^- + e^+ + \nu_e$$

with a probability proportional to $(U_{3e})^2$. For a beam resulting from the decay branch $\pi \rightarrow \mu + \nu_3$ one expects

$$\frac{\text{Decays}}{\text{Interactions}} \approx |U_{\mu3}|^2 \; |U_{e3}|^2 \; \frac{m_3}{E} \; \frac{(m_3/m_\mu)^5}{\sigma N c \tau_\mu}$$

(5a)

where τ_μ is the muon lifetime, σ is the ν-hadron cross-section and N is the number of target nuclei/cc. This gives roughly

$$\frac{\text{Decays}}{\text{Interactions}} \approx 10^8 \ (U_{\mu 3} U_{e 3})^2 \ (\frac{m_3}{m_\mu})^6 \ \frac{1}{(E(\text{Gev}))^2}$$

(5b)

In a recent experiment the CHARM group at CERN[15] found a limit of the order of 10^{-4} on this ratio corresponding to a limit of 10^{-3} to 10^{-6} on $(U_{\mu 3} U_{e 3})$ for values of m_3 varying from 20 to 300 MeV. A similar limit was obtained for $(U_{e 3})^2$ considering the beam to result from $\pi \to e \nu_3$ and $K \to e \nu_3$. If we identify ν_3 as ν_τ with $U_{\tau 3} = 1$ one can obtain much stronger limits on $|U_{e \tau}|^2$ in a beam dump experiment assuming ν_τ are produced via decays such as $F \to \tau \nu_\tau$. Thus the CHARM group[15] using a theoretical calculation of the ν_τ flux find from their beam dump experiment limits of 10^{-6} to 10^{-10} on $|U_{e \tau}|^2$ as $m(\nu_\tau)$ varies from 40 MeV to 250 MeV. It has been pointed out[16] that if $|U_{e 3}|^2$ is extremely small then for $m(\mu) > m(\nu_\tau) > 40$ MeV the predominant decay mode will be $\nu_3 \to \nu_\mu + \gamma + \gamma$ proportional to $|U_{\mu 3}|^2$. Amusingly in this range the two-photon decay mode is more probable than the one photon[16,17]. However once the charged lepton channel really opens (in our example, it would be $\nu_3 \to \mu^- + e^+ + \nu_e$) the two photon branching ratio becomes very small.

An example I have given previously[18] illustrates how the different types of experiments supplement each other. Consider only two generations with

$$\nu_e = \cos\theta \, \nu_1 + \sin\theta \, \nu_2$$

Let us assume the following data

H^3 decay:	$m(\nu_e) = 30$ ev
Double beta-decay:	$m_{\beta\beta}(\nu_e) < 15$ ev
Neutrino Oscillations, either:	a) $\sin^2 2\theta < .17$, Δm^2 arbitrary
	or
	b) $\sin^2 2\theta > .17$, $\Delta m^2 < 0.1$ ev^2.

What do we conclude? Either

a) $m_1 = 30$ ev

$\eta_{12} = -1$

$|m_1 \cos^2\theta - m_2 \sin^2\theta| < 15$ ev

which yields

45 ev $> (m_1 + m_2)\sin^2\theta > 15$ ev

Since $\sin^2\theta$ is small this corresponds to a good approximation to

45 ev $> m_2 \sin^2\theta > 15$ ev

(6)

With $\sin^2 2\theta < .17$

$m_2 > 300$ ev

As discussed above this conclusion has to be quantitatively modified (because of nuclear effects) for m_2 much above 10 MeV. This solution with a large m_2 and a small θ can in principle be tested by experiments of type 4 and 5. For example, the Bryman experiments[13] on $\pi \rightarrow e + \nu_2$ rule out Eq. (6) for ν_2 masses between 40 and 70 MeV.

b) $m_1 \underset{\sim}{\sim} m_2 = 30$ ev

with a mass difference

$|m_2 - m_1| < 0.1$ ev$^2/60$ ev $= 1.6 \cdot 10^{-3}$ ev

$\eta_{12} = -1$

and

$\cos 2\theta < \tfrac{1}{2}$ or $\theta > 30°$

The near degeneracy of masses with $\eta_{12} = -1$ strongly suggests, of course, that in this case ν_e is a Dirac particle; that is, $\theta = 45°$ and $m_2 = m_1$. One can distinguish small departures from the Dirac character since if $\theta \neq 45°$, double beta-decay is possible, and if $|m_1-m_2| \neq 0$ neutrino oscillations occur. For example, if $\Delta m^2 > 10^{-11} ev^2$ and $\theta = 45°$ there is a factor of 2 reduction in the solar neutrino flux. A possible alternative is what I have called (perhaps an unfortunate choice) a pseudo–Dirac particle[19]. This is the case in which the mass matrix appears to have conserved a lepton number but this number is not one of the lepton numbers L_e, L_μ, L_τ conserved by the weak Hamiltonian. Thus the mass eigenstates are found to be Dirac particles but they do not correspond to $\theta = 45°$ and so double beta-decay is allowed; furthermore when weak radiative corrections are included the Dirac particle is seen to be two Majorana particles with a small mass difference.

So far we have ignored astrophysical and cosmological constraints. Neutrinos ν_i with $m_i > 100$ ev (and masses less than 2 Gev) must be unstable to avoid giving too large an energy density (and therefore too short a lifetime) to the universe. A variety of cosmological and astrophysical arguments[20] require this lifetime to be of the order 10^2 to 10^4 sec or less for m_ν less than 100 MeV and well above 1 MeV assuming the decay is $\nu_i \to \nu_e + e^+ + e^-$ due to the mixing U_{ei}. A new somewhat more severe cosmological constraint ($\tau_\nu < 10^2$ sec for the whole range from 5 to 200 MeV) based on the survival of primordial helium and deuterium has been given by Sarkar and Cooper[21]. The correlated constraints of $|U_{ei}|^2$ and m_i from double beta decay and experiments of type 4 and 5 discussed above yield lifetimes that rule out values of m_ν less than 20 MeV if we want to satisfy the cosmologists. It is pointed out by Sarkar and Cooper that if you accept the published analysis of the CHARM beam dump experiment then their cosmological argument just rules out values of $m(\nu_\tau)$ all the way up to the experimental limit of 200 MeV. They conclude therefore $m(\nu_\tau)$ is less than 100 ev. While this conclusion must be considered as very tentative it does show that future beam dump experiments coupled with more information on F production and the decay $F \to \tau + \nu_\tau$ could eliminate the possibility of a large value for $m(\nu_\tau)$. For $m_i < 1$ MeV the decay mode $\nu_i \to \nu + \nu + \bar\nu$ has been considered generated by some exotic interaction. Of particular interest is the possiblity in a scheme with a neutrino mass hierarchy that ν_μ might have a mass in the range 100–500 kev. In fact it is difficult[22], if not impossible, to build a viable model of this sort consistent with cosmological constraints.

197

3. Heavy Neutral Leptons

Most of the theoretical models that yield massive neutrinos (as well as the example of a left-right model[6] that gives massless neutrinos) also contain heavy neutral leptons N_a. In the minimal models the mass of the neutrinos ν_a and their mixing $U_{\nu N}$ with N_a are described by Eqs. (1) and (2) so that

$$M_a \sim m_{Da}^2/m_{\nu a}$$

(7a)

$$|U_{\nu N}|_a^2 \sim m_{Da}^2/M_a^2 = m_{\nu a}/M_a$$

(7b)

In the models of Refs. 6 and 7 these constraints do not obtain. In the case of three generations flavor mixing must also be taken into account; we will ignore this for simplicity.

If we assume $m_{\nu a}$ is close to its experimental upper limit for each generation a we find values of M of the order 10 to 100 Gev. On the other hand if we accept the cosmological limit $m(\nu_\tau) < 100$ ev and assume a single mass scale for M we have M of the order 10^8 Gev or larger. The mass range $M_a < 100$ Gev with mixings $|U_{\nu N}|_a^2 > 10^{-8}$ may conceivably be experimentally observable[23]. In general this corresponds to the ideas that the right-handed scale defining $m(W_R)$ is not more than 10 to 100 times $m(W_L)$. A limit $m(W_R) < 20\ m(W_L)$ was established[24] based on the contribution of W_R in the box diagram contributing to $m(K_s) - m(K_L)$. In a model designed to explain CP violation[25] using the $SU(2)_L \times SU(2)_R \times U(1)$ model it is necessary that W_R be close to this limit. The mass range $M \gg 100$ Gev is probably inaccessible for any prospective experiments.

Heavy neutral leptons N_a can decay

$$N_a \rightarrow \ell_a + \bar{\ell} + \nu$$

(8a)

$$N_a \rightarrow \ell_a + \text{hadrons}$$

(8b)

where ℓ_a is a charged lepton with a probability proportional to $|U_{\nu N}|_a^2$. In the models under discussion they may also decay via W_R into the mode (8b) but not (8a) with a probability proportional to $\{m(w_L)/m(w_R)\}^4$. Lifetimes for N_e and N_μ are of the order[23]

$$\tau(N) = (10^{-11} \text{ to } 10^{-12} \text{ sec}) \times (m_N/\text{Gev})^{-5.18} \times \{(U_{\nu N})^2$$
$$+ (m(W_L)/m(w_R))^4\}^{-1}$$

Considering the range from 1 to 100 Gev with $(U_{\nu N})^2$ from 10^{-4} to 10^{-8} we span a lifetime range from a millisecond to 10^{-18} sec. Thus detection of N decays may involve decay paths too short to observe, decays that can be identified by short gaps in the production target, or decays that can be observed in a downstream detector as in the CHARM experiment discussed above 15. A variety of possible experiments have been recently analyzed by Gronau, Leung, and Rosner[23].

For masses above 10 Gev the most likely production mechanisms are

(1) ν + (nucleon) $\rightarrow$ N + X

proportional to $(U_{\nu N})^2$ via the neutral current

(2) $W \rightarrow \ell + N$
     $Z \rightarrow \bar{\nu} + N$

with branching ratios of the order $0.1|U_{\mu N}|^2$ for $m_N < \frac{1}{2}Mw$. At LEP one might hope to find an N of mass up to 50 Gev if $|U_{\nu N}|^2 \sim 10^{-4}$ to 10^{-5}.

(3) $e^- + p \rightarrow N + x$

due to the W_R coupling. The possibility of finding N in this way at HERA was discussed at this meeting by Soergel.

If there is a sizeable amount of flavor mixing among the N_a then even for large masses there may be a significant rate for $\mu \rightarrow e + \gamma$. For large mixing and $m(N_a) \gg m_w$, one finds[6]

$$\frac{\Gamma(\mu \rightarrow e \gamma)}{\Gamma(\mu \rightarrow e \nu \nu)} \approx 10^{-3} (U_{\nu_\mu N})^2 (U_{\nu_e N})^2$$

4. Conclusion

In spite of extensive efforts our knowledge of neutrino masses and mixings is very limited. Constraints on these come from a wide variety of laboratory experiments and from cosmology. Theoretical guidance is very limited, the major motivation for neutrino mass comes from left-right symmetric theories which naturally, but not necessarily, lead to massive Majorana neutrinos. Grand unified theories lead to a mass hierarchy, which coupled with cosmological constraints, suggest that only ν_τ has a mass above 1 ev and this lies below 100 ev. In this case the most likely experimental frontier is $\nu_\mu - \nu_\tau$ neutrino oscillations.

ACKNOWLEDGMENTS

I wish to thank the Theory Division of CERN for its hospitality. This work was supported in part by the U.S. Department of Energy and by a John Simon Guggenheim Memorial Foundation fellowship.

REFERENCES

1. V. Lubimov et al, Phys. Lett 94B, 266 (1980); Proc. Int. Europhysics Conf. on High Energy Phys., Brighton, 1983 (Rutherford Appleton Laboratory, Chilton, U.K.), p. 386.
2. S. Weinberg, Phys. Rev. Lett. 43, 1566 (1979); Phys. Rev. D22, 1694 (1980).
3. M. Gell-Mann, P. Ramond, and R. Slansky, private communication (1977) and in Supergravity, eds. P. van Nieuwehuizen and O. Freedman (North Holland, 1979), p. 317; T. Yanagida in Proc. Workshop on Unified Theory and Baryon Number in the Universe, eds. A. Sawada and A. Sugamoto, KEK, 1979.
4. L. Wolfenstein in Proc. of Neutrino Mass Mini-Conference, eds. V. Barger and D. Cline (Univ. of Wisconsin, 1980); D. Chang and P. B. Pal, Phys. Rev. D3113 (1982); and references therein.
5. R. N. Mohapatra and G. Senjanovic, Phys. Rev. Lett. 44, 912 (1980); Phys. Rev. D23, 165 (1981).
6. D. Wyler and L. Wolfenstein, Nuc. Phys. B218, 205 (1983).
7. M. Roncadelli and D. Wyler, Phys. Lett. 133B, 325 (1983); P. Roy and O. Shanker, Tata Institute preprint TIFR TH/83-35 (1983).
8. G. B. Gelmini and M. Roncadelli, Phys. Lett. 99B, 411 (1981).
9. H. Georgi, S. L. Glashow, and S. Nussinov, Nucl. Phys. 193B, 297 (1981); G. B. Gelmini, S. Nussinov, and M. Roncadelli, Nucl. Phys. 209B, 157 (1982).
10. J. J. Simpson, Phys. Rev. D24, 2971 (1981).
11. S. Petcov, P. Rosen and A. Halprin, Phys. Lett. 125B, 335 (1983).
12. R. E. Shrock, Phys. Rev. D24, 1232 (1981).
13. T. Yamazaki in Proc. 12th Int. Conf. on Neutrino Physics and Astrophysics, Balatonfured, Hungary 1982; D. A. Bryman, et al, Phys. Rev. Lett. 50 1546 (1983).

14. M. Gronau, Phys. Rev. D28, 2762 (1983) and references therein.
15. F. Bergsma et al, Phys. Lett. 128B, 361 (1983).
16. J. F. Nieves, Phys. Rev. D28, 1664 (1983).
17. R. Ghosh, Phys. Rev. D29, 493 (1984).
18. L. Wolfenstein, Proc. Neutrino Mass and Gauge Structure of Weak Interactions Mini-Conference, Wisconsin, 1982.
19. L. Wolfenstein, Nuc. Phys. B186, 147 (1981).
20. See, for example, M. S. Turner in Proceedings of Neutrino 81 (R. J. Cence, E. Ma and A. Roberts, eds.) University of Hawaii, 1981, Vol. 1, p. 95.
21. S. Sarkar and A. M. Cooper, to appear in Proceedings ESO/CERN Symp. "Large Scale Structure of the Universe" Geneva 1983.
22. P. B. Pal, Nucl. Phys. B227 (1983) 237-251.
23. M. Gronau, C. N. Leung, and J. Rosner, Fermilab preprint EFI 83/63; C. N. Leung and J. Rosner, Phys. Rev. D28, 2205 (1983); and references therein.
24. G. Beall, M. Bander, and A. Soni, Phys. Rev. Lett. 48, 848 (1982).
25. Darwin Chang, Nucl. Phys. B214, 435 (1983).

DISCUSSION

CHAU:

In your conclusion, you emphasize that there is no compelling reason for $m_\nu \neq 0$. But, maybe more importantly is that there is no compelling reason for $m_\nu = 0$. Put another way, if there is a spin 1/2, $m_\nu = 0$ particle, what is the good theoretical reason for it, (e.g. gauge invariance $\leftrightarrow$ zero photon mass)?

WOLFENSTEIN:

I did mention at the beginning that the global conservation of lepton number which yields $m_\nu = 0$ does not appear fundamental, so that is a motivation for considering non-zero mass of the Majorana type. However, I do not find as a compelling argument that everything which is not fundamentally forbidden must occur.

HIGGS MASS IN SALAM-WEINBERG THEORY*

M.A.B. Bég

The Rockefeller University
1230 York Avenue
New York, New York 10021

ABSTRACT

The Salam-Weinberg theory, of electroweak interactions, is discussed under the premise that the pure $\lambda\phi^4$ theory is trivial in four dimensions. Arguments underlying this premise are reviewed; it is shown that the requirement of consistency leads to an upper bound of about 125GeV for the Higgs-mass, if the mass of the top-quark is less than 80GeV and the theory is realized in the canonical way.

1. INTRODUCTION

I shall discuss the consistency of the canonically realized Salam-Weinberg theory[1] under the premise that the pure $\lambda\phi^4$ theory is a trivial field theory in four dimensions[2]. By "canonically realized" I shall mean (a) that the gauge group underlying

---

* Invited talk at Europhysics Conference held at Erice, Sicily, March 5-11.

electroweak and strong interactions is $U(1) \otimes SU(2)_L \otimes SU(3)_C$ and (b) that the Higgs mechanism which reduces the symmetry of the electroweak interactions to $U(1)_{EM}$, the electromagnetic gauge group, is triggered with the help of the usual elementary spin-0 fields, to wit: four real fields that can be combined into a complex doublet, so that three are devoured to produce massive $W^{\mp}$ and Z and one manifests itself as a physical particle even[3] under CP. It is obvious that this construct faces serious problems of consistency if the pure $\lambda\phi^4$ theory is trivial; the usual interpretation, wherein one presumes that the ϕ-sector by itself produces the Goldstone bosons needed for the Higgs mechanism, must be abandoned. We may, however, demand that the coupled Higgs--gauge-field system be non-trivial; for a wide range of quark masses, this leads to an upper bound of about 125 GeV for the mass of the physical Higgs boson[4].

2. SOME RIGOROUS RESULTS

Before I explain how the Higgs mass can be bounded, let me review the status of the input-premise: the triviality of the $\lambda\phi^4$ theory in the real world. I should emphasize that, strictly speaking, we are dealing here with an article of faith; a rigorous demonstration of the triviality of the continuum theory is not yet in hand. Nonetheless, one may predicate one's faith on some results that have been derived with a measure of rigor; these are listed below. An essential ingredient in many of them is the postulate that one may move from Euclidean space formulation of the theory to Minkowski space (real time) without let or hinderance; in technical jargon, the solutions of the field equations are supposed to be in accord with the Osterwalder-Schrader axioms.[5]

(i) Triviality, if there, can not be proven in any finite order of perturbation theory.

This almost obvious result can be put on a firm basis using the work of Glimm and Jaffe.[6]

(ii) The renormalized coupling constant lies in a bounded interval:[7]

$$0 \leq \lambda_{Ren.} < \lambda_{Max.} \text{ for } d \leq 4 \tag{1}$$

where d is the dimensionality of the space in which the theory is defined.

A necessary condition for triviality is thus satisfied.

(iii) For the theory in the symmetric phase, triviality for $d > 4$ has been established by Aizenman[8] and Frohlich[9].

(iv) For $d = 4$, Frohlich has noted that triviality can be established if Z_3 -- the wave-function renormalization constant-- vanishes.[9]

(v) The continuum limit of the latticized theory is trivial, or consistent with triviality, in all existing calculations.[10] When other non-perturbative calculational techniques are available, notably the 1/N expansion for the O(N)-symmetric theory, triviality again follows in the limit of infinite cut-off.[11]

This is the most compelling reason for believing that the theory is indeed trivial. In the following we[4] accept this result; our apologia is that if it turns out that there is no disease, our cure -- while deprived of a sound raison d'etre -- will do no harm.

3. **HEURISTIC CONSIDERATIONS**

To get some insight into the nature of the problem, and the manner in which it can lead to an upper bound for the Higgs mass, let me sketch a heuristic non-proof of triviality.

The renormalization group invariant or running coupling constant, λ, may be written at the one-loop level[12] as

$$\bar{\lambda}(t)^{-1} = \lambda_{Ren.}^{-1} - (3/2\pi^2) \cdot \ln(p/m_w) \tag{2}$$

where $2t = \ln (p^2/m_w^2)$, p is the momentum variable customarily used in renormalization group calculations and m_w, the mass of the W-boson, has been used to introduce a mass scale.

Now stability requires that $\bar{\lambda}(t)$ be positive semi-definite; hence $p_{max.}$, the largest admissible momentum in the theory, must be such that

$$\lambda_{Ren.} \leq 2\pi^2/3 \ln (p_{max.}/m_w). \tag{3}$$

If $p_{max} \to \infty$, $\lambda_{Ren.} \to 0$ and we get a trivial theory; if non-triviality is achieved with some definite mechanism corresponding to a finite cut-off, eq.(3) can be converted into an upper bound for the Higgs mass via[1]

$$m_H^2 = m_w^2 \cdot 8 \lambda_{Ren.} \sin^2\theta_w/e^2 \tag{4}$$

where θ_w is the electroweak angle and e is the electric charge.

Needless to say, the introduction of cut-offs that can not be pushed to infinity destroys the mathematical structure of a field theory, reducing it to a kind of phenomenology. While introduction of ad hoc cut-offs is not a defensible proposition, effective cut-offs can arise naturally in grand unified theories or theories which bring gravity into the picture via, for example, embedding of electroweak theory in a local supersymmetric theory. In such theories, however, the triviality problem acquires a new dimension; I have nothing to say about it at this time. Let us return, therefore, to our original mission: investigation of the consistency of the canonically realized theory.

4.   A FRAMEWORK FOR RESOLUTION

The framework which we[4] (Constantinos Panagiotakopoulos, Alberto Sirlin and I) offer for resolution of the conundrum may be outlined as follows:

(i) We propose that the coupled Higgs--gauge-field sector is non-trivial.

(ii) To uphold (i), we require that $y\left[\equiv\bar{\lambda}(t)/\bar{g}_1(t)^2\right]$ does not diverge or become unduly large for "large" t.

Here g_1 is the coupling constant for the U(1) factor in the strong-electroweak group (with g_2 and g_3 defined in an analogous way); the precise meaning of "large" will become clear in the following.

Observe that (ii) may be deemed to be a necessary condition for consistency; if it is not satisfied, the ϕ-sector would decouple from the rest of the Lagrangian; arguments for the triviality of this sector would then go through.

(iii) To implement (ii) we demand that y be driven to an ultra-violet stable fixed point of the renormalization group.

That y must indeed be of reasonable magnitude if it is to make it to a fixed point will be demonstrated below. As noted in ref. 12, $\bar{\lambda}(t)/\bar{g}_2(t)^2$, the other ratio that one may be tempted to examine, can not go to a fixed point.

5. RENORMALIZATION GROUP ANALYSIS

The renormalization group equations for the various couplings are:

$$\frac{d\bar{g}_i}{dt} = \beta_i(\bar{g}_j, \bar{\lambda}, \bar{G}) \tag{5a}$$

$$\frac{d\bar{\lambda}}{dt} = \beta_\lambda(\bar{g}_j, \bar{\lambda}, \bar{G}) \tag{5b}$$

$$\frac{d\bar{G}}{dt} = \beta_G(\bar{g}_j, \bar{\lambda}, \bar{G}) \tag{5c}$$

where i and j (=1,2 or 3) label the gauge group and G is a generic fermion-Higgs Yukawa coupling. At the one loop level[12]

$$\beta_i = \varepsilon_i \, b_i \, g_i^3 / 16\pi^2 \tag{6}$$

where $b_i > 0$, $\varepsilon_1 = +1$, $\varepsilon_2 = \varepsilon_3 = -1$. (The notation presumes that there are not too many fermions!) Before we write down the one-loop β-functions for the other couplings[12], we introduce variables:

$$x \equiv \bar{g}_2^2/\bar{g}_1^2 \; ; \; \zeta \equiv \ln\left[\bar{g}_1^2/\bar{g}_1^2(t=0)\right] ; \; z \equiv \bar{G}^2/\bar{g}_1^2 \; ; \; u \equiv \bar{g}_3^2/\bar{g}_1^2$$

with G now identified as the Yukawa coupling associated with the top quark. (The effect of the other fermions in the standard model, on our analysis, is negligible.)

Note that x, ζ, and u are simply related by virtue of eqs.(6). We have:

$$u(x) = (b_1/b_3) \, Cb_2 \, x/\left[b_1 + x \, b_2(1-C)\right] \tag{7a}$$

$$x = (b_1/b_2) \, C' \, \exp(-\zeta)/\left[1-C' \, \exp(-\zeta)\right] \tag{7b}$$

where C and C' can be determined from our knowledge of the coupling constants at t=0; the actual values, C ≈ 0.664 and C' ≈ 0.622, are such that the mappings in eqs. (7a,b) are non-singular. Thus the domain of ζ, $0 \leq \zeta < \infty$, corresponds to the interval $\tan^{-2}\theta_w \geq x > 0$. Note further that, at the tree level,

$$y(t=0) = m_H^2/(8m_w^2 \, \tan^2\theta_w) \tag{8a}$$

$$z(t=0) = m_t^2/(2m_w^2 \, \tan^2\theta_w) \tag{8b}$$

The renormalization group equations for the quartic and Yukawa couplings[12] can be put in the form:

$$\frac{dy}{dx} = -\frac{192y^2 - 8y(3+2b_1+9x-12z)+3(1+2x+3x^2)-48z^2}{16x(b_1+b_2x)} \tag{9a}$$

$$\frac{dz}{dx} = -\frac{z(9z-2b_1-16u(x))}{2x(b_1+b_2x)} \tag{9b}$$

(Some electroweak contributions, small in comparison with $2b_1 +$ $16u(x)$, have been neglected in eq.(9b)).

Both equations are of the Riccati type[13]. The second can be solved exactly; using ζ as the independent variable, we have

$$z(\zeta) = \frac{z(0)\exp\left[\chi(\zeta)\right]}{1+z(0)\frac{-9}{2b_1}\int_0^\zeta \exp\left[\chi(\zeta)\right]d\zeta} \qquad (10)$$

where $\chi(\zeta) = -b_1^{-1}\int_0^\zeta (b_1 + 8u)d\zeta$.

Observe that if $z(0)$ is arbitrary, there will be a ζ such that $z(\zeta)$ blows up. To have a solution that tends smoothly to the fixed point $z=0$, as $\zeta \to \infty$, it is necessary to remove this singularity by requiring that $z(0)$ be bounded from above; this translates into an upper bound M ($\simeq$ 168 GeV) on the mass of the top quark.

Eq. (9a) can be solved analytically only for small x and z; this is adequate, however, to show that there exists a non-singular solution which is driven to the fixed point $y* = \left[1 + (2/3)b_1 - \{(1+(2/3)b_1)^2 - 4\}^{1/2} \right]/16$ as $x \to 0$, if the initial value of y is bounded from above. To proceed further, and actually calculate this bound for physical values of x, it is necessary to integrate eqs. (9a,b) numerically; this was done using a computer program based on the Runge-Kutta method[13], with the following values of the input parameters: $b_1 =(20/9)n_g + 1/6 = 6.83333$; $b_2 = 22/3 - (4/3)n_g - 1/6 = 3.16666$, $b_3 = 11-(4/3)n_g = 7$, x-initial $\equiv x_0 = 3.54545$, u-initial $\equiv u_0 = 10.0987$. These numerical values correspond to n_g (number of generations) = 3, $\sin^2\theta_w=0.22$, $\Lambda_{\overline{MS}}= 0.1$GeV and initialization at momentum $m_w \approx 81$GeV. ($\Lambda_{\overline{MS}}$ is the usual QCD parameter[1]).

6. RESULTS

The computer generated solutions of eqs. (9a,b) are consistent with the basic fact that if the theory is to be in the domain of attraction of an ultra-violet stable fixed point, its parameters can not be arbitrary; in particular, the Higgs mass must lie in a bounded interval. We distinguish three cases:

(A) If m_t, the mass of the top quark, is less than a determinable value, $m*$, then

$$0 < y \ (t=0) \leq Y_{max.}(m_t); \ (m_t < m*) \tag{11}$$

With the aforestated values of the input parameters, we find that $m* \simeq 80\text{GeV}$ and that $Y_{max.}$ varies slowly with m_t and approximately equals 1.02; this number leads to

$$m_H \lesssim 125\text{GeV};(m_t < 80\text{GeV}) \tag{12}$$

(B) If m_t exceeds $m*$, there is also a non-trivial lower bound for y-initial which arises from the requirement that y remain positive definite for all t:

$$Y_{min.}(m_t) \leq y(t=0) \leq Y_{max.}(m_t); \ (m_t > m*) \tag{13}$$

In this case the upper and lower bounds vary quite sharply with m_t.

(C) If m_t equals M ($\simeq 168\text{GeV}$), the maximal value allowed in our formulation, the upper and lower bounds on $y(t=0)$ coalesce into one; the Higgs mass is then determined rather than bounded:

$$m_H \simeq 175\text{GeV}; \quad (m_t \simeq 168\text{GeV}) \tag{14}$$

These features of the solution can be seen clearly in Fig. 1, where the bounds on Higgs mass are plotted as a function of the top-quark mass. Observe the manner in which the critical point at $m_t = m*$, in the family of solutions that make it to the fixed point while maintaining their positivity, manifests itself.

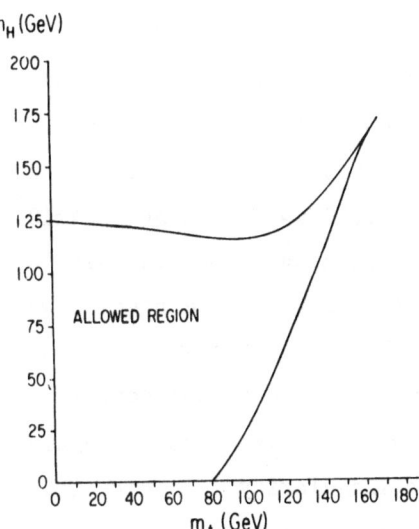

Fig. 1. Upper and lower bounds on the Higgs-boson mass plotted as a function of the top-quark mass. [From ref. 4; the lower bound of Weinberg and Linde ($m_H > 7$ GeV; see, for example, ref. 1) is not depicted].

7. QUESTIONS OF PRINCIPLE

Before we proceed to a critique of our analysis let us note that what was promised at the outset has been delivered. First, the requirement of a reasonable magnitude for y is indeed met if it goes to a fixed point. Thus for m_t = 31GeV, $y(x_0)$ = 1.02, we find that $0.027 < y(x) < 1.02$ for $5 \times 10^{-5} < x < x_0(=3.54545)$. This range of y may be compared to the value at the fixed point: $y = y^* \simeq 0.023$. Second, if $y(x_0)$ is such that the fixed point is not attained, a catastrophic blow-up can in fact occur; for example, with $y(x_0)=1.10$ we find that $y \sim 4 \times 10^8$ for $x \sim 0.16$.

There is a shortcoming in our formulation which, paradoxically enough, permits us to give a precise meaning to the words "large t". The problem, if it may be so described, stems entirely from our retention of only one-loop contributions to the various β-functions. For g1, this leads to a singularity similar to the Landau-Pomeranchuk singularity[14] in pure QED. Crossing this singularity would take us into an Alice-in-Wonderland world where probabilities can acquire negative values. Our judicious choice of variables, however, comes to the rescue; $x \to 0$, or $\zeta \to \infty$, corresponds to approaching this singularity but takes us no further; that is to say, all momenta are cut off at a value determined by the position of the ghost pole. The numerical value of this cut-off ($\simeq 4 \times 10^{41}$ GeV) is quite large; that it is actually finite is, therefore, perhaps irrelevant. A more important difficulty is the breakdown of the loop expansion for β_1 in the neighborhood of the ghost pole. We may reasonably expect that perturbation theory will be good if t is such that $\alpha_1(t)$ ($\equiv \bar{g}_1^2/4\pi$) $\lesssim 0.1$; this corresponds to $p_{max.} \lesssim 4 \times 10^{37}$GeV-- large enough to permit us to assume that its finiteness has no effect on the physics we study -- and $t_{max.} \lesssim 82$. Thus t must be deemed to be "large" if it is 0(100).

We have treated the electroweak system as an isolated system, the canonical theory as a fundamental theory. An alternate viewpoint would be to regard the Salam-Weinberg theory as an effective low energy theory, the fundamental theory being a grand unified theory or, better still, a super unified theory which embraces gravity as well. Such theories, however, would also suffer from problems of consistency; the nature of these problems -- more complex than anything we have discussed -- has not yet been clarified. Pending their elucidation and resolution, one may impose upon the Salam-Weinberg theory a provisional constraint, a consistency requirement of sorts, that the theory be not required to handle momenta in excess of say the Planck mass. This leads to an upper bound on the Higgs mass which, for m_t <80GeV, is approximately $\sqrt{2}$ times larger than the one derived above. Indeed all alternate theoretical approaches to the problem of Higgs mass, that are known to us, lead to upper bounds larger than ours.[15].

8.   CONCLUDING REMARK

The message to real physicists, our experimental colleagues, is that it may be worthwhile to look for the elusive Higgs particle at masses $\sim$ 10GeV - 150GeV. This is a very interesting mass range in that it is also the abode of pseudo-Goldstone bosons in the hypercolor scenario[16]. Should any semi-weakly coupled spin-0 particle be found in this range, it would be an important discovery. Investigation of the detailed properties of such a particle could lead to a clarification of the nature of the Higgs mechanism and illuminate the fundamental structure of electroweak theory.

ACKNOWLEDGEMENTS

This work was supported in part by the U.S. Department of Energy under Contract Grant No. DE-AC02-81ER40033B.000.

REFERENCES

1. For a recent review, see: M.A.B. Bég and A. Sirlin, Phys. Rep. **88**, 1 (1982).

2. K.G. Wilson, Phys. Rev. **B4**, 3184 (1971).
K.G. Wilson and J. Kogut, Phys. Reports **12C**, 78 (1974).
See also: G. Parisi, nucl. Phys. **B100**,368 (1975).

3. That the CP property may provide a means of identifying a true Higgs has been noted by A. Ali and M.A.B. Bég, Phys. Lett. **103B**, 376 (1981).

4. M.A.B. Beg, C. Panagiotakopoulos and A. Sirlin, Phys. Rev. Lett. **52**, 883 (1984).

5. K. Osterwalder and R. Schrader, Commun. Math. Phys. **42**, 281 (1975). See Also: V. Glaser, ibid. **37**, 257 (1974).

6. J. Glimm and A. Jaffe, Phys. Rev. Lett. **33**, 440 (1974).

7. J. Glimm and A. Jaffe, Ann. Inst. Henri Poincare, **22**, 97 (1975).

8. M. Aizenman, Phys. Rev. Lett. **47**, 1 (1981).

9. J. Frohlich, Nucl. Phys. B200 [FS4], 281 (1982).

10. K.G. Wilson and J. Kogut, ref. 2; G.A. Baker Jr. and J. Kincaid, Phys. Lett. **42**, 1431 (1979); J. Stat. Phys. **24**,469(1981).
G.A. Baker Jr., L.P. Benofy, F. Cooper and D. Preston, Nucl. Phys. **B210**, 273 (1982).
C.M. Bender, F. Cooper, G.S. Guralnik, R. Roskies and D.H. Sharp, Phys. Rev. **D23**, 2976 (1981); ibid **D23**, 2999 (1981); ibid **D24**, 2772 (E) (1982). B. Freedman, P. Smolensky and D. Weingarten, Phys. Lett. **113B**, 481 (1982).

11. W.A. Bardeen and M. Moshe, Phys. Rev. **D28**, 1372 (1983). See also: G. Parisi, Nucl. Phys. **B100**, 368 (1975); S. Coleman, R.Jackiw and H.D. Politzer, Phys. Rev. **D10**, 2491 (1974).

12. D.J. Gross and F. Wilczek, Phys. Rev. **D8**, 3633 (1973). See also: T.P. Cheng et.al., Phys. Rev. **D9**, 2259 (1974).

13. See, for example, E.L. Ince, Ordinary Diffential Equations (Dover, New York, 1956).

14. L.D. Landau, in Niels Bohr and the Development of Physics (McGraw Hill, New York, 1955). However, the interpretation generally accepted is due to Bogoliubov. See: N.N. Bogoliubov and D.V. Shirkov, Introduction to the Theory of Quantized Fields (Interscience, New York, 1959), Sect. 43.2.

15. B.W. Lee, C. Quigg and H.B. Thacker, Phys. Rev. $\underline{D16}$, 1519 (1977); N. Cabibbo, L. Maiani, G. Parisi and R. Petronzio, Nucl. Phys. $\underline{B158}$, 295 (1979); R. Dashen and H. Neuberger, Phys. Rev. Lett. $\underline{50}$, 1897 (1983); D.J.E. Callaway, CERN preprint TH.3660 (1983). [The viewpoint in the last paper seems, at least superficially, similar to ours; the analysis, however, is very different and the results differ.]
C. Whitmer, Princeton preprint (1984).

16. For a review of the subject, and extensive references to the literature, see: M.A.B. Bég and A. Sirlin, ref. 1.

DISCUSSION

CHAU:
Could you comment on other upper bounds on the Higgs boson mass?

BÉG:
Almost all bounds are based on some formulation of the principle
that weak interactions lend themselves to a perturbative
treatment. To my knowledge, the first such bound is due to
B.W. Lee, C. Quigg and H. Thacker [Phys. Rev. D16, 1519 (1977];
they found that m_H < 1 TeV. As I stated in my talk, Dashen
and Neuberger -- who offer a cure for the $\lambda\phi^4$-triviality problem
of the Salam-Weinberg theory, different from ours -- also obtain
an upper bound of the order of a TeV.

GOLDHABER:
Can you say something about other quark masses?

BÉG:
Quarks with masses of the order of a few GeV have a negligible
effect on our analysis; only the top quark can play a role. The
answer to your question is therefore: No.

QUARK MIXING FROM HEAVY QUARK SEMILEPTONIC DECAYS

Juliet Lee-Franzini

SUNY at Stony Brook

Stony Brook, N.Y.

ABSTRACT

Heavy quark semileptonic decay data: lifetime, branching ratio and the electron spectrum are analyzed in a self consistent way. Values for the $|V_{ub}|$ and $|V_{cb}|$ elements of the mixing matrix are determined to be <0.0051 at 90% c.l. and $= 0.044 \pm 0.005$ respectively. Combining these results with nucleon and hyperon β-decay data the K-M angle s_3 is found to be < 0.022 at 90% c.l. and s_2 to be bounded between 0.044 and 0.066, and the mixing matrix to be almost diagonal.

INTRODUCTION

In the standard model[1] of the electroweak interaction, with $SU(2)_L \times U(1)$ as the gauge group, the quark current coupled to the charged gauge bosons (W's) has the form

$$J_\alpha \propto (\bar{u}, \bar{c}, \bar{t})\gamma_\alpha(1-\gamma_5)\begin{pmatrix} d' \\ s' \\ b' \end{pmatrix} \quad (1)$$

where d',s' and b' are the charge $-1/3$ weak isospin partners of u, c and t respectively. The primed states are mixtures of the mass eigenstates d, s and b defined by a unitary matrix V as:

$$\begin{pmatrix} d' \\ s' \\ b' \end{pmatrix} = \begin{pmatrix} V_{ud} & V_{us} & V_{ub} \\ V_{cd} & V_{cs} & V_{cb} \\ V_{td} & V_{ts} & V_{tb} \end{pmatrix}\begin{pmatrix} d \\ s \\ b \end{pmatrix} \quad (2)$$

217

This matrix can in general be expressed in terms of three angles and a phase. The original parametrization of Kobayashi and Maskawa of the mixing matrix was[2]:

$$V_{KM} = \begin{pmatrix} c_1 & s_1 c_3 & s_1 s_3 \\ -s_1 c_2 & c_1 c_2 c_3 - s_2 s_3 e^{i\delta} & c_1 c_2 s_3 + s_2 c_3 e^{i\delta} \\ -s_1 s_2 & c_1 s_2 c_3 + c_2 s_3 e^{i\delta} & c_1 s_2 s_3 - c_2 c_3 e^{i\delta} \end{pmatrix} \qquad (3)$$

where $c_1 = \cos \theta_1$, $s_1 = \sin \theta_1$ etc. The three angles can be chosen in the first quadrant and $0 < \delta < 2\pi$.

Maiani[3] was the first to note the decreasing cross generation mixing and proposed an alternate parametrization where the Cabibbo angle retains its original definition and in general mixing between generations is more transparent. Wolfenstein[4] and, most recently, Chau and Keung[5] have proposed representations more convenient for studying CP violation. In the following, the K-M angles are predominantly used, because of historic familiarity.

The main aim of this paper is to derive the constraints on V_{ub} and V_{cb} from data on weak decays of B mesons, use them in conjunction with the known values of $V_{ud} = 0.9737 \pm 0.0025$ [6] and $V_{us} = 0.231 \pm 0.003$ [7] to obtain the range of allowed K-M angles, and finally make estimates of the magnitude of possible $B^0 \overline{B}^0$ mixing. The analysis is preceded by a recapitulation of the experimental evidence for the standard six quark model and followed by an attempt to pinpoint the arenas for future experimentation.

CHECKING THE SIX QUARK MODEL

Flavor Changing Neutral Currents

Since the weak isospin partner of the b' has thus far not shown its presence in experimental searches, it is reasonable to question if the b' is a left handed weak isospin singlet. Peskin and Kane[8] have shown that such models would imply flavor changing neutral currents (FCNC) such that $b \to s$ or d via the emission of a Z^0 which produces a lepton pair, and that this rate would be greater or equal to one eighth of the normal semileptonic rate. Experimentally, a pair of unlike sign leptons can result from the simultaneous semileptonic decay of a B$\bar{\text{B}}$ pair (parallel decays), or from sequential semileptonic decays of the B and its daughter D meson (cascade decays). CLEO[9] recently reported 59.7±11.1 $\mu^+\mu^-$ and e^+e^- pairs from B$\bar{\text{B}}$ decays while they expect 57.9 events from parallel and cascade contributions. The resulting limit for the branching ratio (BR) for B $\rightarrow$ X$\ell^+\ell^-$ is < 0.3% at 90% c.l., which is at least a factor of five smaller than the Peskin and Kane limit of (1/8)BR(B $\rightarrow$ X$\ell\nu$) = 1.5%. Other groups have also imposed limits: MAC reports a limit of < 0.6% at 90% c.l., and Mark J and JADE each report a limit of < 0.7% at 95% c.l.[10]. Thus, one can safely conclude that FCNC's are absent in b decays and that the top exists.

Exotic Couplings

The standard model predicts b decays into $q\bar{q}q$ and $\ell\ell\bar{q}$ in definite ratios such that the energy carried away by neutrinos (missing energy) is on the average $\approx$ 6.5% of the total c. of m. energy[11]. Exotic couplings were proposed such that b's decay predominantly into $\ell\ell q$ or $\ell q\bar{q}$, resulting in missing energies of the order of 33% and 17% respectively[12]. Furthermore, such models also imply different correlations of the charged multiplicity versus the fraction of energy carried by the charged particles with respect to the standard model. CLEO[13] and CUSB[11], by a combination of measurements excluded the above mentioned exotic couplings by at least 90% c.l. Finally, there had been proposals where the t and b quarks were placed in a right handed weak isospin doublet[14], in which case the b decays would not be depressed by the $|V_{ij}|^2$ coefficients, and the B meson would have a very short lifetime, of the order of 10^{-15} seconds. The recent measurements of the B lifetime by MAC[15] and Mark II[16] certainly exclude that possibility.

SPECTATOR MODEL FOR SEMILEPTONIC DECAYS

Introduction

At high energy e^+e^- colliders, heavy quarks are most often

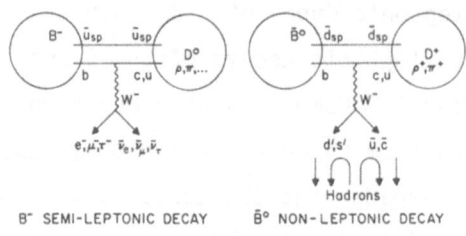

B⁻ SEMI-LEPTONIC DECAY   B̄° NON-LEPTONIC DECAY

Figure 1 The Spectator Model

produced bound with light quarks in the form of heavy flavored
mesons (B's, D's...) which decay subsequently into hadrons
(nonleptonic decays) or into leptons and hadrons (semileptonic
decays). At the parton level, the flavor changing weak decay of
the initial heavy quark $Q_i \to q_f$ proceeds via its emission of a W
which then produces a quark or lepton pair. The color neutrality
of the W argues that no, or little, interaction exists between the
quarks that it produced and the light quark companion (in the
meson) which combines with the flavor changed q_f in the final
hadronization process. The passive role of the accompanying quark
is implied in its appellation: the spectator quark, $\bar{q}_{sp}$. Figure 1
depicts, for typical B semileptonic and nonleptonic decays, this
conceptual two step model, which is referred to as the "spectator
model".

The model's validity for semileptonic decay is immediately
apparent semiquantitatively because it provides a rationale for
predicting the leptonic rate of any Cabibbo allowed process (for
ex. $c \to e\nu s$) by scaling up from the muon decay rate using the
factor $(m_Q/m_\mu)^5$, where m_Q is a reasonable effective quark mass.
Thus, if $m_c \approx 1.5$ GeV, $\Gamma(D^+ \to e\nu X)$ is $\approx 2.5 \times 10^{11}$ sec^{-1}, a value
not far from experimental measurements[17]. For nonleptonic decays
the model predicts that the final state meson composed of the
flavor changed quark and the spectator antiquark, should have a
momentum spectrum characteristic of three body decays. Recently
this has been confirmed by CLEO[18] in their measurement of the D
meson momentum spectrum from B inclusive decays.

Our main interest in this paper is to extract $|V_{ub}|$ and $|V_{cb}|$ values from experimentally measured B decay parameters: the lifetime τ_B, the semileptonic branching ratio BR(B → evX) and the β decay spectrum end point. The spectator model in its present formulation by Altarelli et al[19], which includes soft gluon radiation correction and fermi motion of the bound quarks, was used to extract $|V_{cs}|$ from D semileptonic data (as we shall see) with some success. It proves ideal for application in the B system because of the massiveness of the b quark and the sharpness of the electron spectrum, and most of our discussion will be in that context.

Lepton Spectrum from Free Quark Decay

Consider a free b quark of mass M_b, β decaying at rest into an electron ($m_e \approx 0$), ν and a c or u quark (with $m_q = m_c$ or m_u). The electron energy distribution in terms of $x = 2E_e/m_b$, which is bounded by $1-\epsilon^2 = x_M$ where $\epsilon = m_q/M_b$, is given by:

$$d\Gamma_b^0/dx = (G_F^2 M_b^5/96\pi^3)[x^2(1-\epsilon^2-x)^2/(1-x)^3][(1-x)(3-2x)+(3-x)\epsilon^2] \quad (4)$$

This spectrum computed for $m_u=0$, $m_c=1.72$ GeV and $M_b=5.04$ GeV is shown in figure 2. The b→u spectrum is in fact identical to the muon β decay spectrum prior to any radiative corrections. The b→c curve shows the effect of a massive quark in the final state. The total rate for b→u is therefore ~ 2 or 3 times of that for b→c (for equal couplings). The two spectra are well separated near their end points. Figure 3 shows the experimental electron spectrum, measured by CUSB[20] at the T''' resonance peak energy, where ~30% of the cross section is due to $e^+e^- \to T''' \to B\bar{B}$ and 70% of the cross section is due $e^+e^- \to$ hadrons. The B's are produced with β=v/c of 0.08. Note that a sharp drop off of the electron spectrum (at ~ 2.2 GeV) is evident even in this non background subtracted sample. The experimentally measured electron contribution from continuum events is shown by the solid curve in the figure.

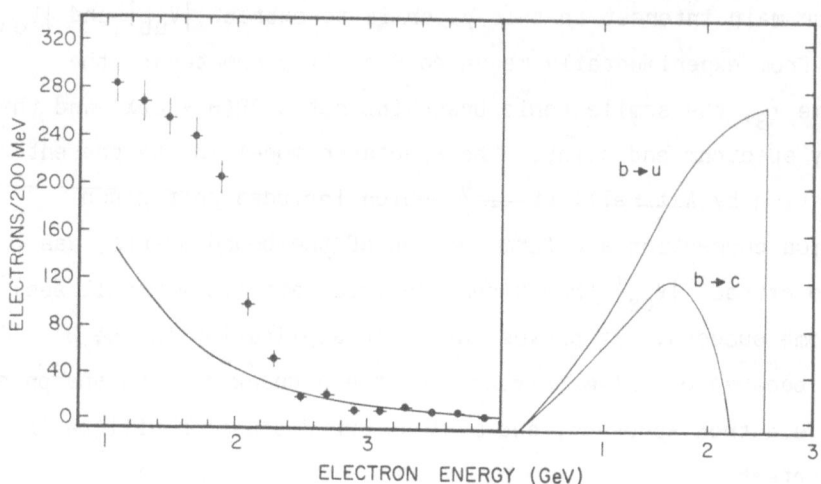

ELECTRONS/200 MeV

ELECTRON ENERGY (GeV)

Figure 2. Free b quark β decay lepton Figure 3. CUSB electron
          spectum.   spectrum on the
                                                  Υ'''.

The corresponding free c quark leptonic decay spectrum is

$$d\Gamma^0_c/dx = (G_F^2 M_c^5/16\pi^3)[x^2(1-\epsilon^2-x)^2/(1-x)] \qquad (5)$$

This distribution is peaked around the middle, see figure 4 where

the data are from D decays produced at the ψ'' with β=0.14, as

measured by DELCO[21] at SPEAR. Because of the shape of the e

spectrum, the determination of the spectrum end point is

considerably harder for c→s,d than b→c,u. The total decay rate

from integration of either equation (4) or (5) from 0 to x_M is

$$\int(d\Gamma^0/dx)dx = (G_F^2 M_Q^5/192\pi^3)[1-8\epsilon^2+8\epsilon^6-\epsilon^8-24\epsilon^4\ln\epsilon] \qquad (6)$$

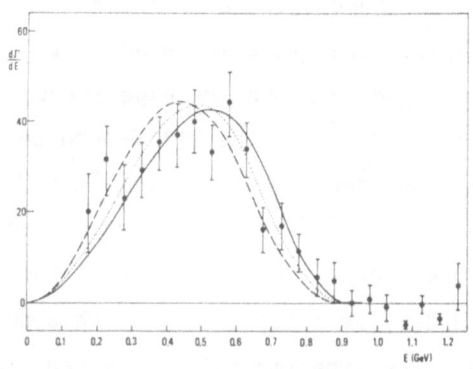

Figure 4 DELCO electron spectrum on the ψ''.

where $M_Q = M_b$ (or M_c), $\varepsilon = m_q/M_Q$. The quantity in the square
bracket is often (incorrectly) called "phase space factor" and
referred to as $I(\varepsilon_1,0,0)$ in the literature[22], where the zeroes
refer to $m_e/M_Q \approx 0$ and $m_\nu/M_Q = 0$. For nonleptonic decays, the
$I(\varepsilon_1,\varepsilon_2,\varepsilon_3)$'s for $Q \rightarrow q_1 q_2 q_3$ is best obtained by numerically
integrating its generic form, which is quoted, among other places,
in the paper by Cortes et al[23]. This paper, incidentally, gives
several definite integrals for $I(\varepsilon_1,\varepsilon_2,\varepsilon_3)$, some of which seem to
contain misprints[24]. However, it is not obvious at this juncture
which quark masses should be used in equation (6), a point which
will be resolved in a consistent way later.

Soft Gluon Radiation Corrections

The initial and final state quarks (Q, q_f) can radiate soft
gluons, in much the same way initial and final state leptons
radiate photons during β decay transition. Figure 5 list five of
such diagrams: the first three are virtual corrections in the
decay $Q \rightarrow q_f \ell \nu$ and the last two are real corrections due to gluon
bremsstrahlung[25]. The QCD corrected spectrum is given by
$$d\Gamma^0/dx = d\Gamma^0/dx \times [1-(2\alpha_s/3\pi)G(x,\varepsilon)] \tag{7}$$
where $d\Gamma_0/dx$ are given by equations (4) and (5), $\alpha_s = \alpha_s(M_Q/2)$,
computed from the two loop formula for $\Lambda_{MS} \approx 200$ MeV[26], and with
$G(x,\varepsilon) = G^{bremss}(x,\varepsilon) + G^{virtual}(x,\varepsilon)$.

Both the virtual correction and the gluon bremsstrahlung
corrections have a regular and a divergent part. Fortunately, the
divergent part of these two processes cancel exactly. Also, while
each $G(x,0)$ separately diverges in the limit $\varepsilon=0$, their sum has a
finite limit whose explicit form is given by Corbo[25]. The effect
of this singularity is avoided by giving the final quark some mass
such that $\varepsilon \neq 0$.

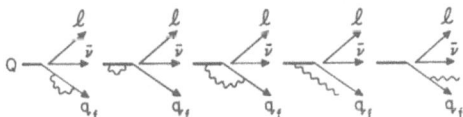

Figure 5 Diagrams responsible for soft gluon radiations.

Finally, as $x \rightarrow x_M$, recall $x_M = 1-\varepsilon^2$, $G(x,\varepsilon)$ has a logarithmic divergence of the form $\ln^2(x_M-x)$ which threatens to plunge $d\Gamma/dx$ of equation (7) to negative values. To quote Cabibbo et al[25], "...the origin of this divergence is well known: at $x \approx x_M$ the infrared divergences in the vertex corrections cannot be compensated by the bremsstrahlung terms which vanish because of phase space...". The cure is equally well known, one needs to smear away the critical region of $x \approx x_M$ by averaging $d\Gamma/dx$ over some finite Δx around $x \approx x_M$, either by giving the quark some fermi motion, or by folding in an experimental resolution function. Since in the analysis of lepton spectra, we do either or both of the above, the singularities can be ignored. However, since $G(x,\varepsilon)$ is essentially a constant function and has some finite value over the remainder of the x range, we can factor out $G(x,\varepsilon)$ to obtain its correction to the total rate. In particular, $\int G(x,\varepsilon)dx = G(\varepsilon)$ is tabulated in the literature[27] and is plotted in figure 6. Henceforth for the rate calculations we will use $f(\varepsilon) = [1 - (2\alpha_s(M_Q/2)/3\pi)G(\varepsilon)]$ for the QCD correction factor to the decay rate. Its maximum effect, for $\alpha_s \approx 0.25$, is to depress the semileptonic rate by 20% when $\varepsilon \approx 0$.

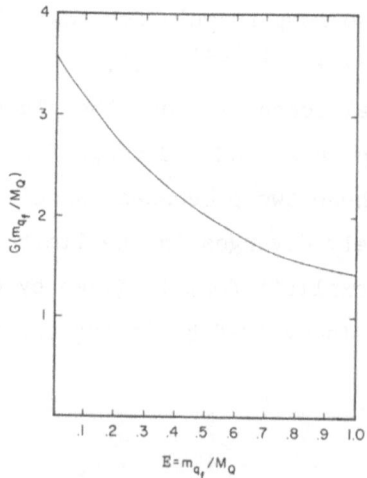

Figure 6 $G(\varepsilon)$ vs ε.

Bound State Effects, Doppler Smearing, Experimental Resolutions

In the spectator model as depicted in figure 1, the decaying quark Q is assumed to be moving in the bound $(Q\bar{q}_{sp})$ system and the spectator antiquark $\bar{q}_{sp}$ to recoil with opposite momentum and definite mass m_{sp}, ultimately combining with q_f to form the recoiling system X with mass M_X. The Fermi motion distribution $\phi(|p|)$ is taken as a gaussian $\phi(|p|)=(4/\sqrt{\pi}p_F^3)\exp(-|p|^2/p_F^2)$. By energy and momentum conservation, then Q has a mass:

$$W_Q = \sqrt{[M_{B,D}^2 + m_{sp}^2 - 2M_{B,D}\sqrt{(|p|^2+m_{sp}^2)}]}$$

where $M_{B,D}$ are the actual B and D masses, i.e. masses of the original bound states. The lepton spectrum from the decay of the bound state is $d\Gamma_{Bound}/dE=\int\phi(|p|)p^2dp[d\Gamma_Q(W,P,E)/dE]$ where the integral is to be taken from 0 to the maximum allowed value of $|p|$, and $d\Gamma_Q/dE$ is the energy spectrum of the lepton from the decay in flight of a quark of mass W and momentum p, which is obtained by boosting the distributions given in equations (4) and (5).

In addition, since the B's and D's are produced with a β of 0.08 or 0.14 respectively, Doppler smearing also has to be folded in to obtain the lepton spectrum from a bound state decaying in flight. Finally, the experimental resolution should be folded into the theoretical spectra before fitting to data. In fact, DELCO[21], CUSB[20] and CLEO[28] all have experimental resolutions whose effects are negligible compared to the Doppler smearing.

Summary of Formulae for Extraction of K-M Matrix Elements

From the above we can write the semileptonic width for the decay of a pseudoscalar meson P_Q containing a heavy Q into a final hadronic state containing a quark q as

$$\Gamma(P_Q \rightarrow e\nu X_q) = (G_F^2/192\pi^3)\langle M_Q^5\rangle I(\langle\varepsilon\rangle)f(\langle\varepsilon\rangle) \tag{8}$$

using (the appropriate averages of) M_Q and m_q obtained from fitting the measured electron spectra as outlined above.

The partial width for semileptonic decay is given by

$$\Gamma(P_Q \to e\nu X_q) = BR(P_Q \to e\nu X_q)/\tau_{P_Q} = \sum |V_{q_1 Q}| (G_F^2/192\pi^3)<M_Q^5>I(\epsilon_i)f(\epsilon_i) \quad (9)$$

where we include the mixing angles and sum i over the appropriate flavors. We further define

$$R_B = \Gamma(B \to X_u \ell\nu)/\Gamma(B \to X_c \ell\nu) = |V_{ub}|^2 I(\epsilon_u)f(\epsilon_u)/|V_{cb}|^2 I(\epsilon_c)f(\epsilon_c) \quad (10),$$

the corresponding R_D is obtained by substituting u→d, c→s. Then,

$$\Gamma(B \to X_{all} e\nu) = (1+R_B)\Gamma(B \to X_c e\nu), \quad \Gamma(D \to X_{all} e\nu) = (1+R_D)\Gamma(D \to X_s e\nu) \quad (11).$$

To summarize, using $G_F = 1.166 \times 10^{-5}$ GeV^{-2} [29] and M_b in GeV,

$$|V_{cb}|^2 = [BR(B \to Xe\nu)/\tau_B] \times [K_{cb}/(1+R_B)], \quad (12)$$

$$K_{cb} = 2.88 \times 10^{-11} \sec/[<M_b>^5 I(\epsilon_c)f(\epsilon_c)] \quad (13)$$

$$|V_{ub}|^2 = R_B |V_{cb}|^2 [I(\epsilon_c)f(\epsilon_c)/I(\epsilon_u)f(\epsilon_u)] \quad (14).$$

The corresponding expressions for $|V_{cs}|^2$ and $|V_{cd}|^2$ are obtained by substituting B→D, c→s and u→d.

EXTRACTING $|V_{cs}|$ and $|V_{cd}|$ from D SEMILEPTONIC DECAYS

An attempt to determine K_{cs} was done by Alterelli et al[19]. Figure 4 shows their computed spectra and the DELCO D→eXν data[21]. The dotted curve ($p_F=150$ MeV/c, $m_{sp}=150$ MeV, $m_s=300$ MeV) has the lowest χ^2 of the six combinations of various parameters that the authors have tried. In fitting the spectrum for the best value of m_s (which really correspond to an average of mostly m_s and some admixture of m_d) we found two other combinations of p_F and m_{sp} which are equally acceptable to the data. These three possible fits are are presented in Table I. All three have as input the spectrum given by equation (11), using $M_D=1.867$ GeV (average of M_D^+ and M_D^0), $P_{beam}(D)=0.26$ GeV. We note incidentally that all three fits yield $M_X \approx 600$ MeV, consistent with the picture that kaons are almost always present in the the hadronic system recoiling against the electron. Delco[21] had reported that R_D is $\approx 6\%$.

Table I K_{cs} and $|V_{cs}|$ from DELCO Lepton Spectrum

m_{sp} (MeV)	300	150	≈ 0
p_F (MeV/c)	≈ 0	150	300
χ^2 (10 d.o.f.)	7.2	7.2	6.1
$\langle m_c \rangle$ (GeV)	1.57	1.63	1.52
$\langle m_s \rangle$ (GeV)	0.5+0.15−0.05	0.32+0.09−0.06	0.23+0.08−0.7
End point (GeV)	0.7+0.02−0.05	0.78+0.01−0.02	0.74+0.01−0.013
m_X (MeV)	630	600	580
rms spread (MeV)	82	105	234
K_{cs} 10^{-12}sec	4.18+0.5−0.03	4.26+0.8−0.33	5.56+0.69−0.48
$|V_{cs}|$	0.81 ± 0.14	0.82 ± 0.15	0.933 ± 0.15
$|V_{cd}|$	0.04 ± 0.04	0.035 ± 0.035	0.05 ± 0.05

The three values for K_{cs} listed in Table I differ by up to 30%. A better measurement of the lepton spectrum is needed in order to understand whether this instability is due to the model or to the dynamics of the system. At present, the D semileptonic rates are equally uncertain. To evaluate $|V_{cs}|^2$ I used an average BR from the two SPEAR experiments[21,30]: BR($D^{\pm,0} \to eX\nu$)=(10±3)%, and an average of $\tau(D^+)$ and $\tau(D^0)$ as given by Reay's compilation[17]: $\tau(D)$=(6.05±0.54)×10^{-13} sec. One notes that while the values of the matrix elements are reassuringly consistent with those obtained from unitarity considerations (see later) and those obtained from ν induced heavy quark production analyses[31], we are far from having an independent precision determination from the semileptonic data.

EXTRACTING $|V_{ub}|$ and $|V_{cb}|$ from B SEMILEPTONIC DECAYS

B-Meson Mass

B meson studies barely began three years ago. The first evidence of open b-flavor was through the observation of a sudden increase in cross section of electrons with energy greater than 1 GeV on the Υ''' resonance peak, indicating β decay of a new heavy

quark[32]. Sometime elapsed before b-flavored mesons (B's) were
reconstructed through exclusive hadronic channels by CLEO[33],
yielding $M(B^o)$=5274 ±1.9±2 MeV and $M(B^{\pm})$=5270.8±2.3±2.0 MeV. the
possibility that these B mesons were B*'s (excited states of B's
with $m(B^*)-m(B)<m_{\pi}$) was excluded by CUSB[34] with their null result
on the Υ search on the T''' where the B*'s would be expected to
dominantly decay via B*→B+Υ. The relative fraction of charged to
neutral B's produced at this resonance was inferred using a
theoretical mass difference of $M(B^o)-M(B^{\pm})$=4.4 MeV and threshold
effects to be 40%:60%. For the computations we use an average B
mass value of M_B = 5272 ± 2 MeV.

B-Meson Lifetime

    The B meson lifetime has been recently measured by two groups
at PEP. For the present calculation I combined the latest
MAC[15] measurement as reported by Ford of τ_B=1.6±0.4±0.3 psec with
the Mark II[16] value of 1.2+0.45-0.36±0.3 psec to obtain
τ_B=1.4±0.3±0.3 psec. I believe that the systematic uncertainties
(the second error) should be averaged and not combined with
statistical errors in quadrature or otherwise. Incidentally, as
we shall see, at present it is the error on this measurement of ≈
22% which dominates the uncertainty in the determination of the
mixing matrix elements.

BR(B→Xℓν)

    Leptonic branching ratios have been measured at PEP and
PETRA, as well as at CESR where the copious production of B's at
the T''' allow higher statistical accuracies. The average BR(B→e
orμ,νX) from the high energy colliders is 11.8±1.2 %[10]. The CESR
results are included the following table:

	CLEO[28]	CUSB[20,36]	<CESR>
BR(B→eνX)	12.0±0.7±0.5	13.2±0.8±1.0	12.6±0.5±0.7
BR(B→μνX)	10.8±0.6±1.0	11.2±0.9±1.0	11.0±0.5±1.0
BR(B→e or μ,νX)	<e or μ>		11.8±0.36±0.75

World Average BR(B→e or μ, νX)   (11.8±0.35±0.75)%

This error of ≈5% contributes negligibly at present to the
uncertainty in the determination of the $|V_{c\ or\ u,b}|$'s.

Limits on R_B

    Limits on the ratio R_B were obtained at CESR by examining the
decay lepton spectra as shown in figures 7 (CUSB) and 8 (CLEO).
We had noted that for a free b quark decay, the end point of the
lepton spectrum from the b→u transition is well separated from
that of the b→c transition (figure 2). This difference remains
virtually intact in the bound-b (B) semileptonic decays: curve B
versus curve A in figure 7, dotted curves versus dashed ones in
figure 8. The curves were computed using the same parameters:
M_B=5.272 GeV, $P_{beam}(B)$=400 MeV/c, p_F=150 MeV/c, m_{sp}=150 MeV,
except for curve B (dotted curve) m_u was set to 150 MeV (to avoid
end point singularities) and for curve A (dashed curve) m_c was set
to 1.7 GeV, a mass preferred by the data. The areas of the b→u
curves were reduced by ≈1/3 to fit on the same graph as the b→c
curves which were normalized to the data. These calculations give
$M(X_u)$=850 MeV with rms spread of 250 MeV which is consistent with
X being physical π's, ρ's, A_2's etc. and $M(X_c)$≈2.0 GeV with the
rms spread of only 90 MeV, consistent with X being 50%D and
50%D*'s. Curve C (dot-dash curve) is due to β decay of the
daughter D meson.

    Both data sets are consistent with there being no b→u
transitions. By comparing these predicted distributions with the
data, both groups place limits on R_B, CUSB[20] had a limit of 5.5%
at 90% c.l. from their electron spectrum, CLEO[28] a 4% limit at
90% c.l. from their combined e and μ data. To obtain the limit
given by all the CESR results, I combined the likelihood functions
given to me from both groups, it is shown in figure 9 and yields
R_B < 0.03 at 90% c.l., <0.038 at 95% c.l.

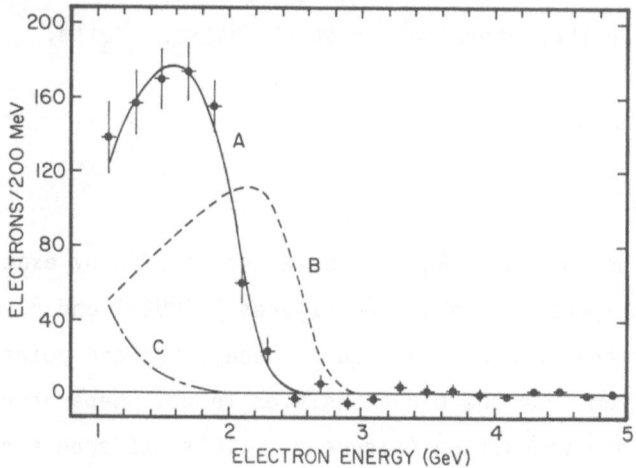

Figure 7 CUSB B → eνX spectrum, see text for curves.

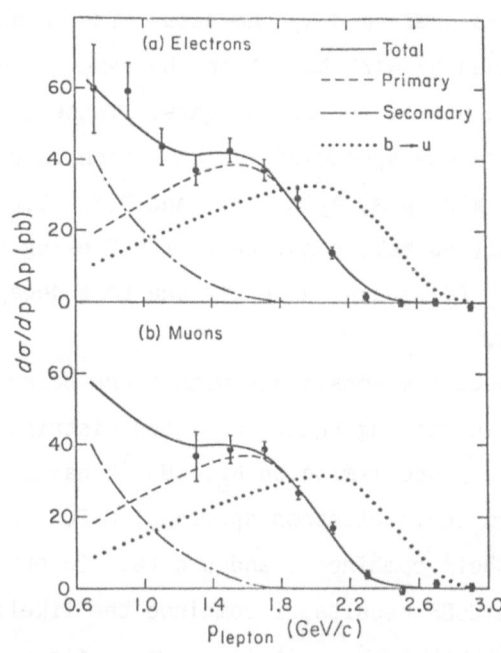

Figure 8 (top) CLEO B → eνX spectrum, (bottom) B → μνX spectrum.

Table II. K_{cb} from CUSB Lepton Spectrum

m_{sp} (MeV)	300	150	≈ 0
p_F (MeV)	≈ 0	150	300
χ^2 (for 18 d.o.f.)	11.1	11.6	9.8
$\langle m_b \rangle$ (GeV)	4.97	5.04	4.93
$\langle m_c \rangle$ (GeV)	1.64 $\pm$ 0.06	1.74 $\pm$ 0.06	1.61 $\pm$ 0.05
End Point (GeV)	2.214 $\pm$ 0.020	2.219 $\pm$ 0.020	2.202 $\pm$ 0.016
m_X, variance (GeV)	2.01, 0.033	1.98, 0.089	1.90, 0.227
K_{cb} 10^{-14} sec	2.351 $\pm$ 0.107	2.353 $\pm$ 0.134	2.412 $\pm$ 0.113

K_{cb} from the Lepton Spectrum

The $\langle M_b \rangle$ and effective m_c which enter into the evaluation of $|V_{ub}|^2$ and $|V_{cb}|^2$ were obtained by fitting the CUSB spectrum to various combinations of m_{sp} and p_F and selecting the ones with the minimum χ^2. Three representative cases are presented in Table II.

We note that in contrast with range of K_{cs} values shown in Table I, the K_{cb} values are much better defined. This is partly due to a better determination of the electron spectrum endpoint, and partly to a compensation between the $\langle M_B \rangle^5$ and $I(\varepsilon)$ factors. The error on K_{cb} is at present dominated by the uncertainty in the fitted value of the c quark mass $\approx 3.5\%$ (in contrast to $\approx 35\%$ for the D case). The effective end point of the electron spectrum

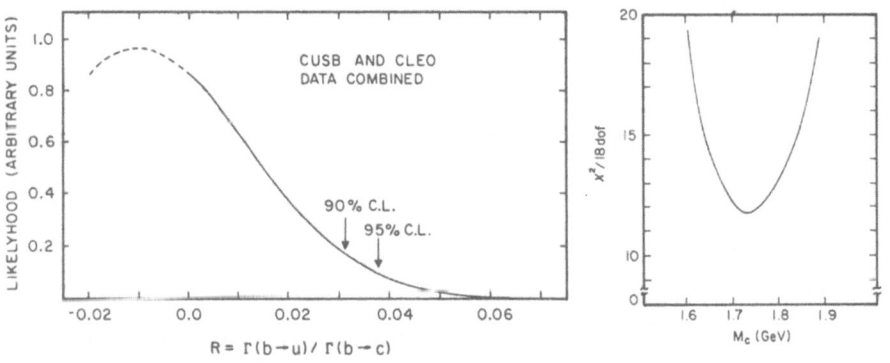

Figure 9 Combined likelihood for CUSB and CLEO data.

Figure 10 CUSB χ^2 vs m_c

$[(M_Q{}^2 - m_q{}^2)/2M_Q]$ is given for both cases in the tables. Figure 10 shows a typical χ^2 versus m_c plot. For all computations we use $K_{cb} = (2.353 \pm 0.134) \times 10^{-14}$ sec.

$|V_{cb}|$ and $|V_{ub}|$

Substituting into equations (12) to (14) all the relevant numbers we obtain:

$$|V_{cb}|^2 = 0.001925 \pm 0.000431, \qquad |V_{cb}| = 0.0439 \pm 0.0049 \qquad (15)$$
$$|V_{ub}|^2 < 0.0000260 \text{ at } 90\% \text{ c.l.} \qquad |V_{ub}| < 0.0051 \text{ at } 90\% \text{ c.l.} \qquad (16).$$

The statistical error on the $|V_{ub}|$ limit is 10% of its value, and comes mainly from the uncertainty in τ_B. In addition,

$$|V_{ub}|/|V_{cb}| < 0.116 \text{ at } 90\% \text{ c.l.} \qquad (17),$$

where the statistical error on this limit of $\approx 3.5\%$ comes from the uncertainty of m_c from the lepton spectrum.

In anticipation of more precise lifetime and R_B measurements,

$$\text{we write} \quad |V_{cb}|^2 = (2.777 \pm 0.179) \times 10^{-15} \text{sec}/[(1+R_B)\tau_B] \qquad (18)$$

$$\text{and} \quad |V_{ub}|^2 = R_B(1.228 \pm 0.086) \times 10^{-15} \text{sec}/[(1+R_B)\tau_B] \qquad (19)$$

LIMITS ON K-M ANGLES

s_3

From the K-M matrix (equation 3) we see $s_3 = |V_{ub}|/s_1$. Using $s_1 = 0.231 \pm 0.003$ [7] and $|V_{ub}| < 0.0051$, we obtain $s_3 < 0.022$ at 90% c.l. and $c_3 \approx 1$ to better than $1/10^3$. The statistical uncertainty on s_3 is 10% of its value.

s_2

Similarly we have $c_1{}^2 c_2{}^2 s_3{}^2 + s_2{}^2 c_3{}^2 + 2c_1 c_2 s_3 s_2 c_3 c_\delta = |V_{cb}|^2$. Setting $c_3 = 1$, and $c_2 = 1$ (which we shall see is a valid assumption), we can solve $s_2 = \pm\sqrt{(|V_{cb}|^2 - s_\delta{}^2 c_1{}^2 s_3{}^2)} - s_3 c_1 c_\delta$. Buras[37], among others argue that only the + solution is a physical one. We further obtain $s_2 = |V_{cb}|[\sqrt{(1 - 8.06 R_B s_\delta{}^2)} - 2.84\sqrt{R_B} c_\delta]$, by using $c_1 = 0.9737 \pm 0.0025$ [6]. In figure 11 we show this relation evaluated for $\delta = 0$ to π, the curve is for $R_B = 0.03$ and the line for $R_B = 0$. We

232

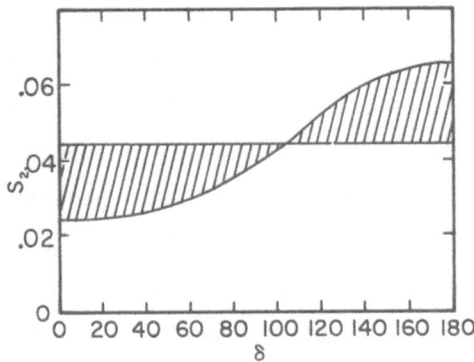

Figure 11 s_2 vs δ, the allowed region is shaded.

note that s_2 is bounded between 0.022 and 0.044 for $\delta = 0$ to about 105° and between 0.044 and 0.066 for δ's greater than 105°. This implies that $c_2 \approx 0.995$ to 0.9995.

Gilman and Hagelin[38] have shown that the sign of the CP impurity parameter ϵ limits δ to lie between 0 and 180 degrees. We can use their equation to further constrain the region allowed for s_2. Using $|\epsilon| = 2.27 \times 10^{-3}$ [29] and the fact the s_2 and s_3 are small, they write 4.1×10^{-1} GeV^2 $= (B_K/0.33)s_1{}^2 s_2 s_3 s_\delta \times \{\}$ where $\{\}$ equals $m_c{}^2[-0.7+0.81\ln(m_t/m_c)+0.6(m_t/m_c)^2 s_2(s_2+s_3 c_\delta)]$, and B_K is the correction for using vacuum intermediate states, which these days can vary between 0.33 and 1 depending on the method of computation. Superimposing this condition we find that for m_t's $< m_W$ and B_K in the abovenamed range, δ lies in the second quadrant.

MIXING MATRIX

Using c_1, s_1 and the bounds on s_2 and s_3 determined above, we can determine the absolute values of all the elements of the mixing matrix, this is equivalent to using the unitary conditions and we obtain:

$$
|V| =
\begin{matrix}
0.9737 \pm 0.0025 & 0.231 \pm 0.003 & <0.0051 \\
0.231 \pm 0.003 & 0.972 \pm 0.002 & 0.044 \pm 0.005 \\
<0.015 & 0.043 \pm 0.001 & >0.999
\end{matrix}
\quad (20)
$$

One notes that this matrix is almost diagonal, and there is very little mixing between the second and third generation of quarks. It is also interesting that the values deduced from unitarity for $|V_{cd}|$ and for $|V_{cs}|$ are very close to the ones resulting from analyses of neutrino induced charm quark production[31]. This may well imply that either there are only three generations, or that as expected from present experience, that the fourth generation mixes little or none with the other three.

Wolfenstein[4], and Chau and Keung[5] have recently proposed new parametrizations of the K-M matrix which capitalize on the above feature. In the latter's parametrization all first order CP violation effects are proportional to a product $X_{CP} = s_x s_y s_z s_\phi$ (which is equal to $s_1^2 s_2 s_3 s_\delta$ in the K-M notation), which when we substitute in the experimental values is $<5.2 \times 10^{-5} s_\phi$.

B-$\bar{\text{B}}$ MIXING

Given the above matrix we can estimate in the usual manner (using box diagrams) the amount of B-$\bar{\text{B}}$ mixing expected for B_d's (bound $b\bar{d}$) and B_s's (bound $b\bar{s}$). Following Gilman and Hagelin[38]'s notation, we write:

$$\Delta M = 2|M_{12}| = n_{QCD} G_F^2 f_B^2 B_B m_B m_t^2 / 6\pi^2 \times |V_{tb}^2 V_{td}^{*2}| \quad \text{or} \quad |V_{tb}^2 V_{ts}^{*2}|$$

where $n_{QCD} \approx 0.85$, f_B is the form factor of the B (which we take to be $\approx f_K$) and B_B is the bag factor for the B which is expected to be ≈ 1. The mixing parameter r_2 which is measured experimentally by counting by the number of same sign dileptons divided by all dileptons can be approximately given as $(\Delta M/\Gamma)^2/2$. It is $\approx 3.2 \times 10^5 \times (s_1 s_2 c_\delta)^4 \times (m_t/35 \text{GeV})^4$ for the $B_d \bar{B}_d$ system. This small number gets multiplied by a factor of about 56 for the $B_s \bar{B}_s$ system where $|s_1^2 s_2^2|$ is replaced by $|s_2 + s_3 e^{i\delta}|^2$ and r_2 becomes thereby hopefully measurable.

The present experimental status on $B_d \bar{B}_d$ mixing is such that CLEO can only rule out complete mixing[9]. They see 10.4 ± 6.1 like sign dileptons while expecting 17 from parallel and cascade decays and 93.2 ± 11.7 unlike sign dileptons while expecting 18 from parallel and cascade decays. The interpretation of these results

234

depends on the relative lifetimes of the neutral and charged B's,
as well as on their relative proportion produced on the T'''.
Assuming equal lifetimes and 60%:40% fraction, CLEO expects 20%
like sign dileptons and 80% unlike sign dileptons which leaves
them with -0.09±0.09 excess same sign dileptons, hence they rule
out complete mixing.

We are at present hunting for possible sources of resonant
B_s production. In figure 12 we see the CUSB scan at CESR above
the T''' region. We see a bump around 10.86 GeV which could be
the T(5S), and its spacing from the lower resonance suggests that
it could be above the $B_s\bar{B}_s$ threshold. Both groups, CLEO and CUSB,
are looking for enhanced s quark production on this structure.
This is a first step towards possible future studies on
$B_s\bar{B}_s$ mixing.

CONCLUSION AND FUTURE OUTLOOK

We have shown that by using the B lepton spectra and B
lifetime measurements, $|V_{cb}|$ and (the limit on) $|V_{ub}|$ are
determined to 10% of their value, see eqs 15 and 16. Their ratio
(eq. 17) and the coefficients dependent on $\langle M_b \rangle$ and effective
m_c are determined to ~ 3.5% accuracy (eqs 18,19). The K-M matrix
is now almost determined, except for the phase δ which most likely
lies in the second quadrant, see eq. 20. The most immediate
improvements will come from better B lifetime measurements
forthcoming from PEP and PETRA. Prospects for long range B

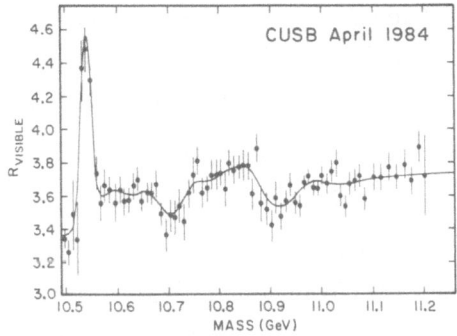

Figure 12 $R_{visible}$ vs E_{cm}.

studies at CESR have been discussed by P.Franzini at this conference.

ACKNOWLEDGEMENTS

In the past three months the author had the pleasure of many discussions with theoretical colleagues, from A. Buras, L.L. Chau, J. Hagelin, W.Y. Keung, L. Maiani, R. Rückl to L. Wolfenstein on their work. She thanks P. Franzini and B. Gittelman for detailed discussions on the CUSB and CLEO analyses, C. Klopfenstein for help in calculations, and R. Poling for the CLEO likelihood data. Finally, she thanks R.D. Schamberger and P.M. Tuts for invaluable help in preparing the present paper. This work was supported in part by the National Science Foundation.

REFERENCES

1. S.L. Glashow, Nucl. Phys. 22 (1961) 579; S. Weinberg, Phys. Rev. Lett. 19 (1967) 1264; S.L. Glashow, J. Iliapoulos and L. Maiani, Phys. Rev. D2 (1970) 1285.

2. M. Kobayashi and T. Maskawa, Prog. Theor. Phys. 49 (1973)652.

3. L. Maiani, Phys. Lett. 62B (1976) 183.

4. L. Wolfenstein, Phys. Rev. Lett. 51 (1983) 1945.

5. L.L. Chau and W.Y. Keung, BNL preprint, 1984.

6. R.E. Shrock and L.L. Wang, Phys. Rev. Lett. 41 (1978) 1692 and 42 (1979) 1589.

7. M. Bourquin et al., Rutherford Lab preprint. Latest results were reported by Siebert at this conference.

8. G.L. Kane and M.E. Peskin, Nucl. Phys. B195 (1982)29.

9. R. Giles et al., in Proc. of 1983 Int. Symp. on Lepton and Photon Interactions at High Energies, ed. D.G. Cassel and D.L. Kreinick, Cornell Univ. (1983) 902.

10. S. Stone, see ref.9, 203.

11. P. Franzini, Jour. de Phys. Coll. c-3 (1982) 114.

12. E. Derman, Phys. Rev. D19 (1979) 317; H. Georgi and S.L. Glashow, Nucl. Phys. B167 (1980) 173.

13. A. Chen et al., Phys. Lett. 122B (1983) 317.

14. V. Barger et al. Phys. Rev. D24 (1981) 1328.

15. E. Fernandez et al., Phys. Rev. Lett. 51 (1983) 1022.

16. N.S. Lockeyer et al. Phys. Rev. Lett. 51 (1983) 1316.

17. N. Reay, see ref. 9, 244.

18. J. Green et al., Phys. Rev. Lett. 51 (1983) 347.

19. G. Altarelli et al., Nucl. Phys. B208 (1982) 365.

20. C. Klopfenstein et al., Phys. Lett. 130B (1983) 444.

21. W. Bacino et al., Phys. Rev. Lett. 43 (1979) 1073.

22. H.B. Thacker and J.J. Sakurai, Phys. Lett. 36B (1971) 103.

23. J.L. Cortes et al. Phys. Rev. D25 (1982) 188.

24. P. Franzini, private communication.

25. N. Cabibbo et al., Nucl. Phys. B155 (1979) 93;
    G. Corbo, Nuc. Phys. B212 (1983) 99.

26. G. Altarelli, Phys. Rep. 81 (1982) 1.

27. N. Cabibbo and L. Maiani, Phys. Lett. 79B (1978) 109.

28. A. Chen et al., Phys. Rev. Lett. 111B (1984) 1084.

29. Particle Data Group, Phys. Lett. 111B (1982) 1.

30. R. Schindler et al., Phys. Rev. D24 (1981) 78.

31. K. Kleinecht and B. Renk, Z. Phys. C15 (1982) 19.

32. See P. Franzini and J. Lee-Franzini, Phys. Rep. 81 (1982)
    239 and references therein.

33. S. Behrends et al., Phys. Rev. Lett. 50 (1983) 881.

34. R.D. Schamberger et al., Phys. Rev. D26 (1982) 720;
    see ref. 9, 894.

35. E. Eichten, Phys. Rev. D22 (1980) 1819.

36. G. Levman et al., Phys. Lett. (1984) to be published.

37. A.J. Buras et al. Preprint MPI-PAE/PTh 7/84.

38. F.J. Gilman and J.S. Hagelin, Phys. Lett. 133B (1983) 443;
    SLAC-PUB-3226 (1983).

39. P. Colic et al. Phys. Rev. D26 (1982) 2286;
    J. Donaghue et al., Phys. Lett. 119B (1982) 412;
    N. Cabibbo, at this conference.

DISCUSSION

HAGELIN:
You mentioned that the radiative corrections to the momentum
spectrum from soft gluons is small by analogy with
electromagnetic radiative corrections in μ decay. But in μ decay
(and presumably also here) the radiative corrections to the
momentum spectrum are huge (although they largely cancel in the
total rate). If the radiative corrections are 20%, as you say,
for the total rate, they could be ten times larger at the
end-point of the electron spectrum. How might this impact your
analysis of $\Gamma(b \to u e \nu)/\Gamma(b \to c e \nu)$ based on the momentum spectrum?

LEE-FRANZINI:
No, they do not change the spectrum very much as Cabibbo et al.
had shown with a complete algebraic analysis, especially if one
gives just a little mass to the emitted quark.

PHAM:
I can comment on this point. Let's write the electron-spectrum
partial decay width in the form

$$\frac{d\Gamma}{dE} = \frac{1}{n-4} g^D(E) + g^F(E)$$

where the infra-red divergent part $g^D(E)$ is splitted out as the
pole term $\frac{1}{n-4}$, n is the space-time dimension. The finite part is
$g^F(E)$. Although the integrated width $\int dE\, g^D(E)$ is exactly
equal to zero (beautiful cancellation between loops contributions
and soft-gluon ones), the partial width $g^D(E)$ is not zero.
However this divergent-part is substantially large in a small
range of E near the lower point, therefore the shape $\frac{d\Gamma}{dE}$ in the
whole range of E might not be strongly affected by this infrared
divergence.

NEW RESULTS ON HEAVY FLAVOR PRODUCTION FROM PETRA

M. Pohl

Inst. für Hochenergiephysik

Eidgen. Techn. Hochschule

Zürich, Switzerland

ABSTRACT

I report on recent results obtained by experiments at the e^+e^- storage ring PETRA. Data collected at c.m.s. energies between 30 and 46.2 GeV are analyzed to single out $e^+e^- \rightarrow c\bar{c}$ ($b\bar{b}$) tagged by the semileptonic decay of the quarks into muons. The transverse momenta of these muons with respect to the quark direction as well as the event shape allow for an efficient separation of the sample into its b and c contents and thus for a measurement of the c and b semileptonic branching ratios. The longitudinal momentum spectrum gives information about their fragmentation properties. Charm quark identification is also obtained by observing high momentum D^* mesons among its fragmentation products. Both methods allow for a measurement of the forward-backward charge asymmetry in b and c production, found to be in agreement with the usual assignment of weak isospin for these quarks and thus indicating that the b quark is member of a doublet. The absense of flavor changing neutral current b decays also supports the hypothesis that a top quark should exist. No indication of $t\bar{t}$ resonance production is however observed at PETRA up to $\sqrt{s} = 45.2$ GeV.

1. SEMILEPTONIC CHARM AND BOTTOM DECAYS

1.1 Event Selection

To identify the production of heavy flavors, events of the type

$$e^+e^- \to \mu^{\pm} + hadrons \tag{1}$$

are used, preferentially coming from the semileptonic decay of c and b flavored particles in the final state. The event selection is based on the sample of identified hadronic events[1,2]. Chambers outside an absorber serve to select candidate muons. It is then required that no "kinks" bigger that the spatial resolution of the apparatus including multiple Coulomb scattering in the absorber are seen on the trajectory. The statistics of the event selection are shown in tab.1 for the MARK J and TASSO analysis as an example.

The determination of background in the sample is based on a full cascade simulation of hadronic showers[3] built up from experimentally measured elementary cross sections for hadronic interactions and extensively tested with experimental data. As an example, the simulation correctly predicts, within errors, the amount of muons cut by the "no kink" requirement.

A rough indication of the composition of the sample at this stage of the analysis is given in tab.2. Note that the fraction of non-prompt background is momentum dependent. The reason for the low

Tab. 1: Event selection statistics for the semileptonic charm and bottom decay candidates

Experiment	$\langle\sqrt{s}\rangle$ GeV	$\int L\, dt$ pb^{-1}	hadronic events	muon candidates	no kinks
MARK J	34.6	76	25000	2170	962
TASSO	34.5	75	21553	2025	1136

Tab. 2: Approximate composition of the sample of inclusive muon
candiates with "no kinks".

Experiment	prompt c,b decay	background
MARK J	68%	32%
TASSO	35%	65%

background fraction in the MARK J experiment is essentially the short
decay path to the first absorber material.

1.2 Flavor Separation

Most efficient for the separation of the sample into portions
enriched in their b and c contents is the transverse momentum $p_\perp$
of the muons with respect to the quark direction. This direction is
approximated by the thrust axis of the hadronic event component[4]
(see fig.1). Fig.2 shows the distribution $dN/dp_\perp$ observed with
MARK J together with the result of the Monte Carlo calculation
corresponding to the final analysis (see section 1.3). The histo-

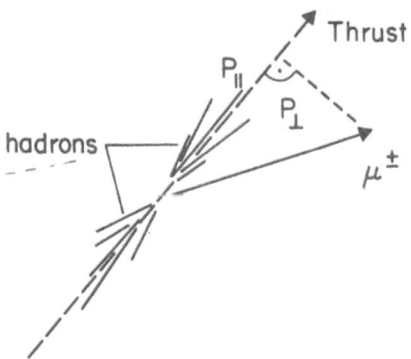

Fig. 1: Definition of event axis and muon parallel $(p_{//})$ and transverse momenta $(p_\perp)$ in $e^+e^- \rightarrow \mu^\pm$ + hadrons.

grams show the transverse momentum spectra of c→μX and b→μX separately, and the sum of both plus the abovementioned background contributions. Below a transverse momentum of 1.2 GeV/c, the sample consists of about 60% c→μX, 10% b→μX and 30% background. For larger momenta it is enriched in b decays (50%) with an admixture of c (25%) and less background (25%). A small admixture of cascade decays b→c→μX is contained in the above numbers.

In the MARK J analysis, additional separation power is obtained from an analysis of the event shape by using the thrust variable[1] T, a measure of the width of the hadronic energy flow around the event axis. Fig.3 shows the thrust distribution of the two subsamples mentioned above ($p_\perp < 1.2$ GeV/c and $p_\perp \geq 1.2$ GeV/c) compared to the distribution for all hadronic events (solid line). While the distri-

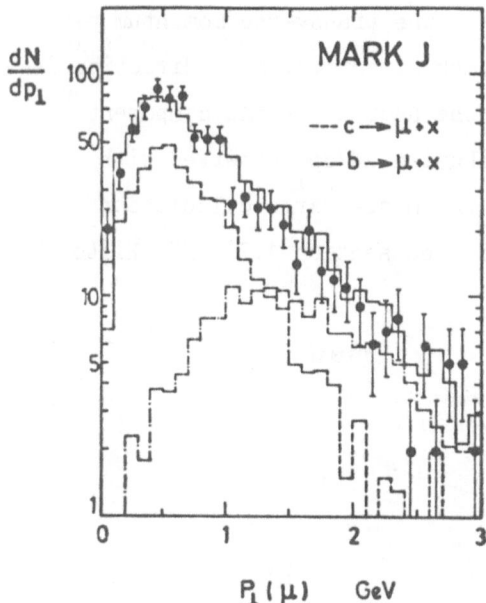

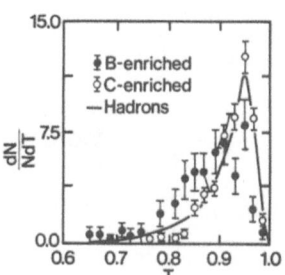

Fig. 2: Muon $p_\perp$ distribution from MARK J compared to MC prediction for prompt c, prompt b decays and all muons (solid histogram).

Fig. 3: Thrust distribution from MARK J for all hadron events, events with μ of $p_\perp< 1.2$ GeV/c and $p_\perp \geq$ 1.2 GeV/c.

bution of the c-enriched sample closely follows the one for all (u,
d, s, c and b) events, thrust for the b-enriched sample is substantially broadened by the higher mass of the primordial final state.

Fig. 4 shows the distribution of the longitudinal muon momentum $p_{//}$ (along the thrust axis) from TASSO in three different $p_\perp$
bins ($p_\perp$ < 0.6 GeV/c, 0.6 < $p_\perp$ < 1.2 GeV/c, 1.2 < $p_\perp$), after background substraction. Data are compared to the Monte Carlo calculation
(see sec. 1.3) and the b contribution is given separately (hatched
histogram). It is seen that at high transverse momentum, the b contents of the sample increases dramatically. It is obvious also that
at high $p_\perp$ the longitudinal momen-
tum distribution substantially
hardens, indicating a hard fragmentation of the b quark.

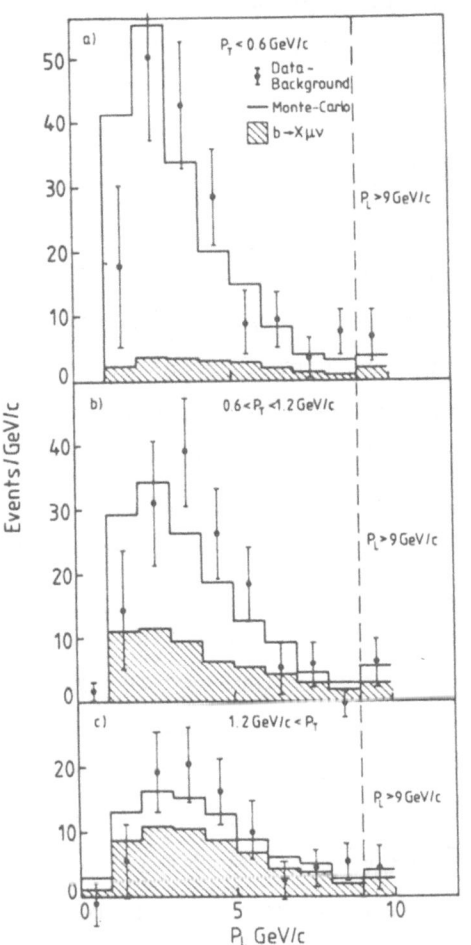

Fig. 4: Distribution of longitudinal momenta from TASSO for three
different $p_\perp$ bins, corrected for
non-prompt background. Data are
compared to MC predictions for
prompt b decay (hatched histogram)
and all prompt muons (solid histogram).

243

1.3 Semileptonic Branching Ratios and Fragmentation Functions

While it would thus be possible to obtain an efficient flavor separation by using simple cuts in $p_\perp$ and/or thrust as done by other groups[5], it turns out that an increase in the statistical significance of the results can be obtained by a multidimensional analysis. For that purpose, the MARK J group subdivides the sample into 8x8x8 bins in the variables $p_\perp$, T and $p_{//}$, thus measuring $d^3\sigma/dp_\perp dp_{//}dT$. The TASSO group uses a 3x3 bin distribution in $p_\perp$ and $p_{//}$, thus $d^2\sigma/dp_\perp dp_{//}$. As shown above, $p_\perp$ and thrust are efficient for flavor separation and thus for a measurement of the decay rate, whereas the longitudinal momentum is a measure of the fractional energy carried by the decaying heavy meson and thus sensitive to fragmentation parameters.

A spectator model for heavy quark decay[6,7] is used to predict the shape $1/N\, d^3N/dp_\perp dp_{//}dT$ $(1/N\, d^2N/dp_\perp dp_{//})$ for prompt muons from b and c decay, including QED radiative corrections[8] to order $\alpha3$, QCD corrections[9] to order α_s^2 and full detector response simulation. In the same way, the shape of the background distributions from π/K decay and punch-through is calculated. The Monte Carlo predictions are then fitted to the experimental distributions using the semileptonic branching ratios $B(c\to\mu X)$ and $B(b\to\mu X)$ and the fragmentation functions as parameters. In the MARK J analysis, the background rate is left free in the fit, while TASSO fixes it using the events rejected by the "no kink" requirement as an additional constraint. Note that while the cascade decay $b\to c\to\mu X$ is of course included in the calculation for b decay, its contribution is <u>not</u> included in the b branching ratio quoted. Note also that the conversion of muon rates into branching fractions is evidently based on the assumption that the production rate for primordial $c\bar{c}$ and $b\bar{b}$ final states is correctly predicted by QED and QCD including all known radiative corrections.

The semileptonic branching ratios extracted from the data in this way are given in tab.3. The first error quoted is statistical, the second systematic. The systematic errors were estimated [1,2] by variations of the fitting method and sample definition, as well as variations of the parameters of the spectator model. They also cover uncertainties in the simulation procedure for the background determination. In particular, the MARK J errors cover a 50% deviation of the punch-through rate from the value determined in the fit and predicted by the Monte Carlo, and a 20% change in the rate of non-prompt muons from π/K decay. Fig.5 compares these measurements to data obtained by other experiments [18] and to branching ratios measured for $c\to eX$, $b\to eX$. All results are in agreement within statistical and systematic errors and compatible with e/μ universality of heavy quark decays. Note however that comparison to data obtained on resonances is not straight forward, since the breakdown of the average final state into different mesons need not be the same as in continuum production.

Tab. 3: Semileptonic branching ratios obtained by MARK J and TASSO.

Experiment	$B(c\to\mu X)$	$B(b\to\mu X)$
MARK J	$0.115\pm0.010\pm0.017$	$0.105\pm0.015\pm0.013$
TASSO	$0.082\pm0.012^{+0.02}_{-0.01}$	$0.117\pm0.028\pm0.01$
CELLO	$0.123\pm0.029\pm0.039$	$0.088\pm0.024\pm0.035$

Simultaneously, fragmentation parameters for b and c are extracted. The parameters quoted in tab.4 correspond to a parametrization of the fragmentation function $f_Q(z)$ for flavor Q in the form proposed by Peterson et al. [10]

$$f_Q(z) = 1/z \, (1 - 1/z - h_Q^2/(1-z) \,)^{-2} \tag{2}$$

$$z = (E_M + p_M) / (E_Q + p_Q) \qquad\qquad (3)$$

where E_M and E_Q are the energies of the heavy meson and its parent quark and p_M, p_Q their momenta along the quark direction. h_Q is a fragmentation parameter related to the quark and meson masses.

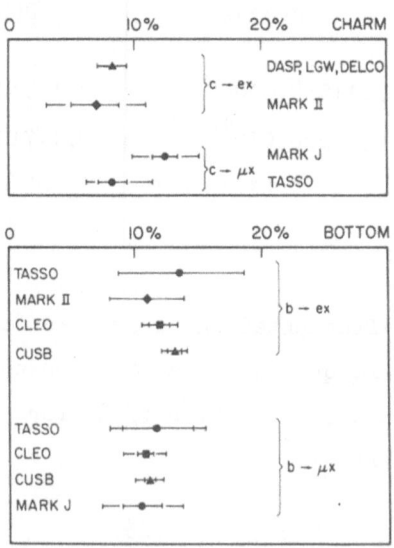

Fig. 5: A survey of results for the semileptonic branching ratios of charm and bottom quarks[18].

Tab. 4: Fragmentation parameters as measured by MARK J and TASSO

Exp.	$<z_c>$	$<z_b>$	$|h_c|$	$|h_b|$
MARK J	$0.46\pm0.02\pm0.05$	$0.75\pm0.03\pm0.06$	$0.8\pm0.1\pm0.2$	$0.15\pm0.03\pm0.05$
TASSO	$0.77^{+0.05+0.03}_{-0.07-0.11}$	$0.85^{+0.10+0.02}_{-0.12-0.07}$	$0.08^{+0.07+0.16}_{-0.05-0.03}$	$0.05^{+0.13+0.07}_{-0.05-0.02}$

The groups agree on a very hard fragmentation of the b quark, as evidenced by the high average fractional energy z retained by the heavy meson after fragmentation. This has also been observed by other groups (see fig.6)[18], using b→μX as well as b→eX. Agreement between the charm fragmentation parameters is still fair, when taking into account systematic errors. More data on charm fragmentation are reported in section 2.2.

Independent of the above functional form (2), the fragmentation functions f_c and f_b can be determined as a function of z by intro-

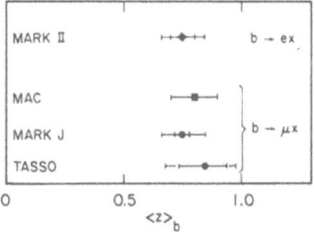

Fig. 6: A survey of the average fractional energy $<z_b>$ retained by B mesons in the fragmentation process.

247

ducing them directly as parameters in the fit. Fig.7 shows the re-
sultant measurement by MARK J with 10 bins for each function, compared to parametrization (2). The average values obtained from this
procedure ($<z_c>$ = 0.46±0.05, $<z_b>$ = 0.74±0.10) as well as the shape
of the fragmentation functions agree well with the parametrization
results.

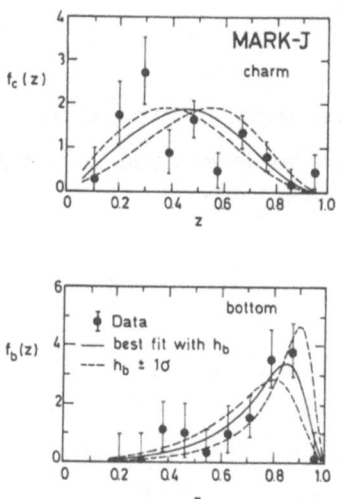

Fig. 7: Charm and bottom fragmentation functions from MARK J, compared to parametrization (2) with h_Q as given in tab.4 (solid
curves) and ±1σ variation in h (dashed curves).

2. LEADING CHARMED PARTICLES

2.1 Event Selection

A second method for the identification of charm production in e^+e^- reactions is based on the selection of hadronic events with leading (high momentum) charmed mesons (D^* or F). The decay channels used for mass reconstruction are

$$D^* \rightarrow D^o$$
$$\hookrightarrow K\pi, K\pi\pi, K\pi\pi\pi \tag{4}$$

$$F \rightarrow \phi\pi$$
$$\hookrightarrow K^+K^- \tag{5}$$

To reduce combinatorial background and to assure that the charm quark is not picked up from the sea during fragmentation, a typical requirement is that the momentum fraction

$$x = p_M/p_Q = 2p_M/\sqrt{s} \tag{6}$$

be bigger than $\simeq 0.4$. A particularly clean signal is obtained by using the mass difference

$$\Delta M = M_D^* - M_D \simeq m_\pi \tag{7}$$

This difference is plotted in fig.8 for the decay channels (4) from TASSO data[11]. A clear signal of in total 119 D^* is seen above a background of 26 events. Fig. 9 shows the mass distribution of K^+K^- candidates for all events and for combinations with $x_{KK}>0.4$.[12] A clear ϕ signal at the correct mass ($M_\phi=1023\pm2$ MeV/c^2) is observed in high momentum pairs. When the selection is restricted to K^+K^- combinations inside the ϕ region ($M_\phi -0.015 < M_{KK} < M_\phi+0.015$GeV/c^2), the high momentum $KK\pi$ combinations ($x_{KK\pi} > 0.4$, fig. 10a) show a mass peak at the F position ($M_F = 1.975\pm0.009\pm0.010$ GeV/c^2). The

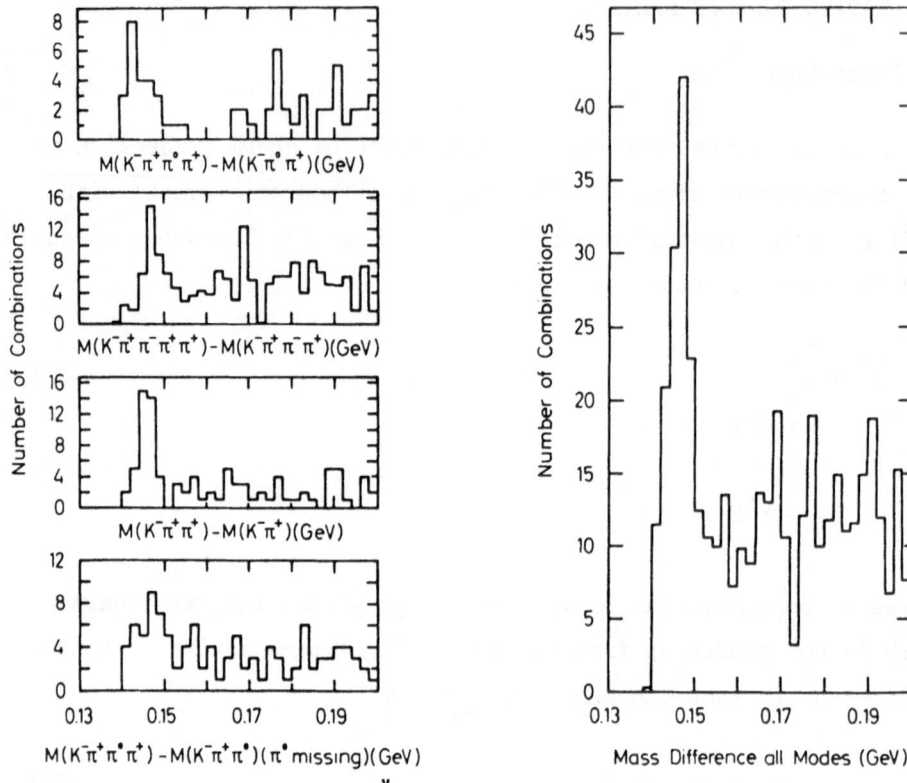

Fig. 8: Mass difference M(D*)-M(D) as measured by TASSO for all D* decay channels used.

mass plot of combinations from a close side band (1.05 < M_{KK} < 1.10 GeV/c^2) shows no peak (fig. 10b). A fit to the data in fig.10a with a gaussian resonance and a second order polynomial background yields a signal of 49±14 events.

2.2 Charm Fragmentation Properties

The observed charmed mesons can be used to directly measure the fragmentation function $f_c(z)$. Fig.11 shows a compilation of data from various experiments[18] using D* mesons. General agreement is observed, but there is certainly room for an improvement of the statistical significance. Fig.12 compares results for $<z_c>$ obtained by diffe-rent methods from νN scattering and e^+e^- annihilations. The data

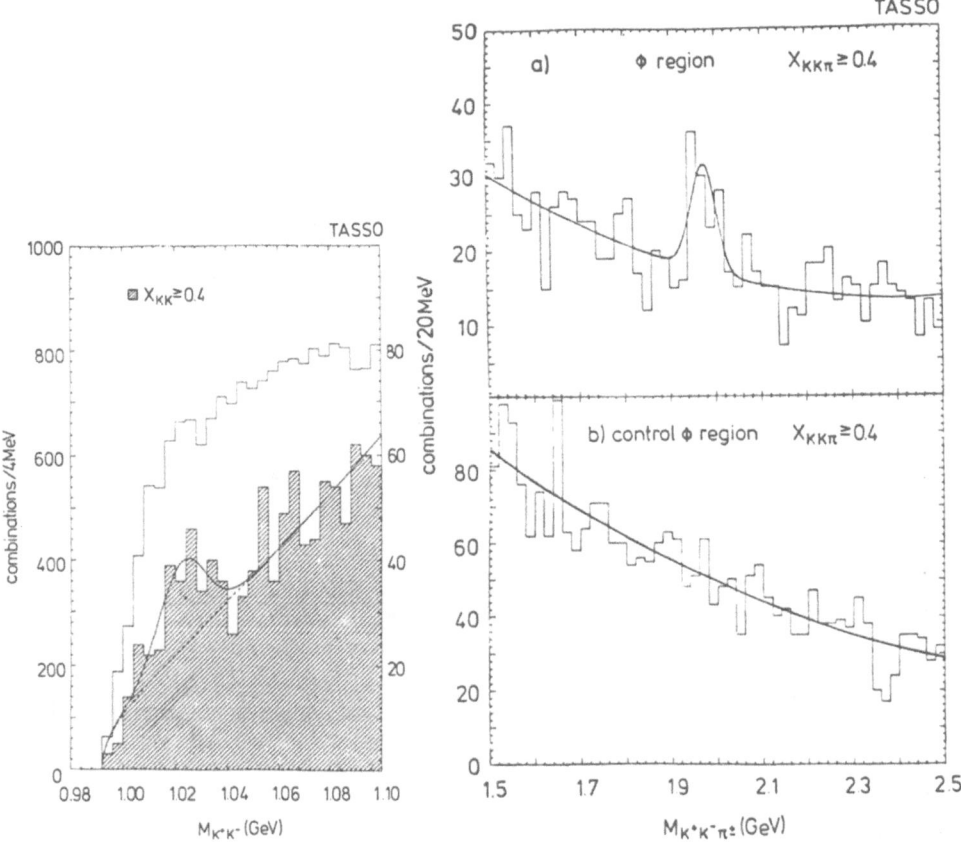

Fig. 9: Mass distribution of K^+K^- candidates for all pairs and pairs with x>0.4 (hatched). The curves indicate the fitted signal and background.

Fig. 10: Mass distributions of $KK\pi$ pairs with x>0.4 and KK mass regions as given in the text. The curves indicate the fitted signal and background.

cluster around $<z_c> \simeq 0.6$, although the scatter is still large due to statistical and systematic uncertainties. The observed momentum fractions of F mesons[12] are compatible with this conclusion.

In addition to the fragmentation function, the selection of inclusive D^* production provides a sample of charm quark jets promising some more insight into the fragmentation process of heavy quarks.

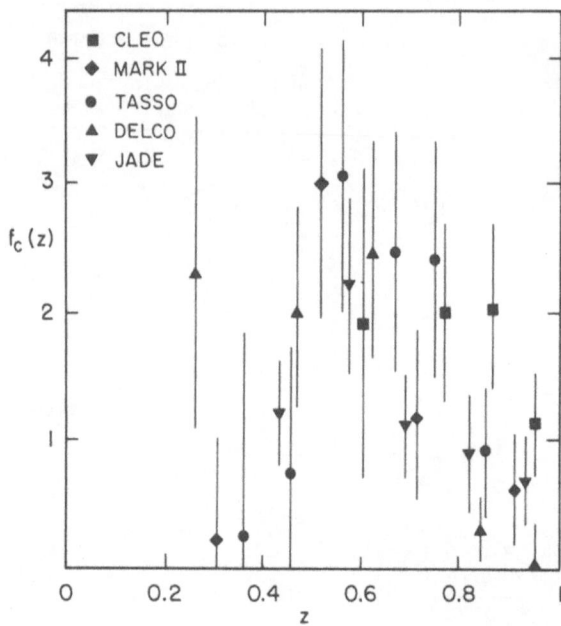

Fig. 11: Compilation of data on the charm fragmentation function
obtained using D* production.

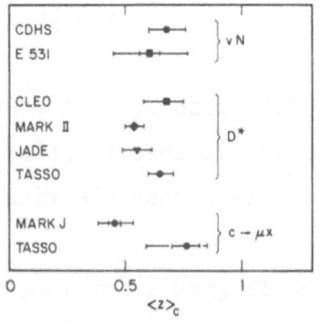

Fig. 12: A survey of data on the fractional energy $<z_c>$ retained by
C mesons after fragmentation.

As sketched in fig.13, the remaining hadrons in the jet where the
D^* is observed ("D^* jet") give information about the remainder of
the fragmentation after the charmed meson was branched off. The jet
opposite to the D^*, which is unbiased by the selection, represents
an average "charm jet" at the beam energy. It is interesting to study
these two kinds of jets and to compare them to jets produced by light
quarks (u, d, s). This has been done by the TASSO collaboration[13].

Fig.14 shows inclusive particle spectra ($x = 2p/\sqrt{s}$, rapidity y,
$p_\perp^2$) and event shape parameters (thrust, sphericity) of charm jets compared to the average (u, d, s, c and b) at the same c.m.s. energy.
No significant differences are observed, nor are any to be expected
since all jets (except for a small admixture of b) are produced far
away from threshold ($\sqrt{s} \gg 2M_c$) and differences of the fragmentation
functions are washed out by the weak charm decay. In contrast, the
inclusive properties of the remaining cascade on the D^* side do differ

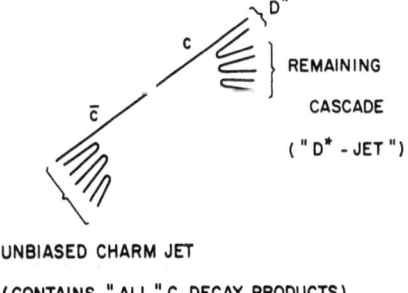

Fig. 13: Sketch of the division of $c\bar{c}$ events tagged by a D^* into an
unbiased c jet opposite to and the cascade along the D^*.

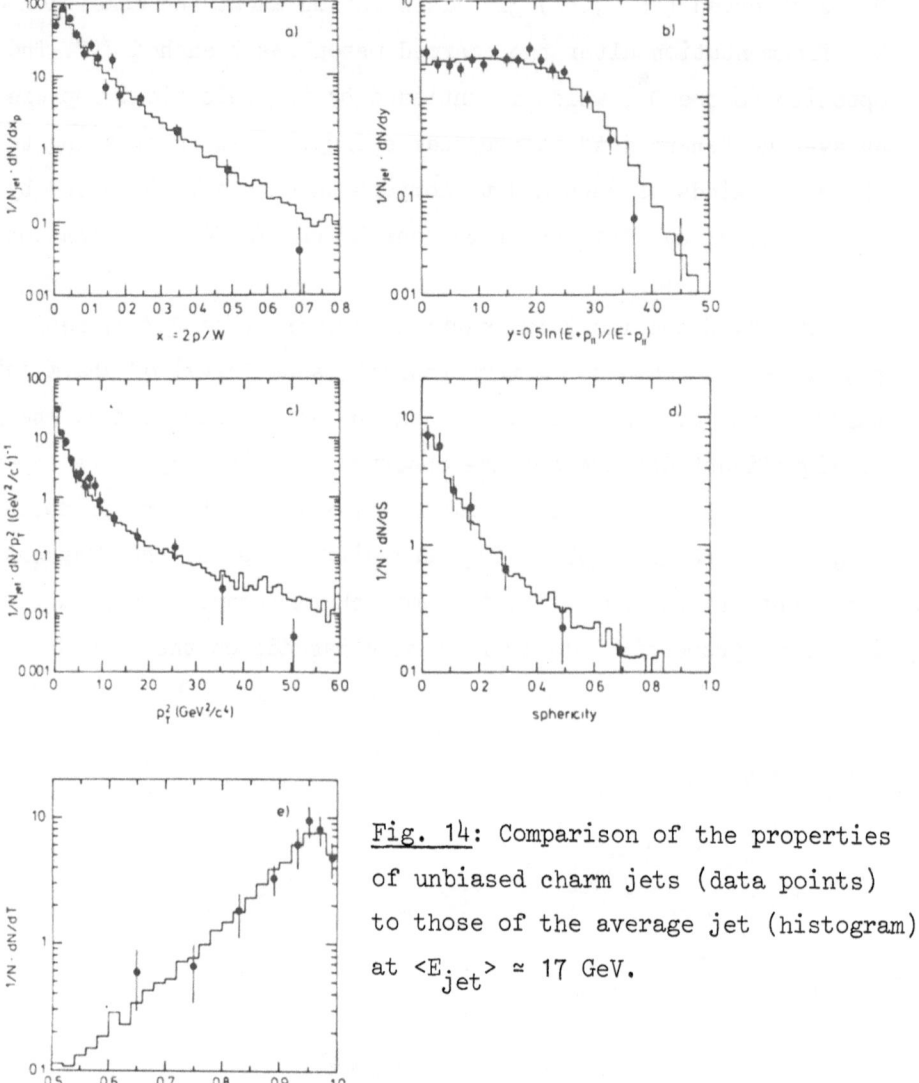

Fig. 14: Comparison of the properties of unbiased charm jets (data points) to those of the average jet (histogram) at $\langle E_{jet} \rangle \simeq 17$ GeV.

from those of the average jet at $\sqrt{s} \simeq 34$ GeV as evident from fig.15. This difference, however, is entirely due to the different jet energies, since the D^* jet has an average energy of only about 10 GeV. It is thus appropriate to compare it to the average jet at $\sqrt{s} = 14$ GeV, with which its properties agree well.

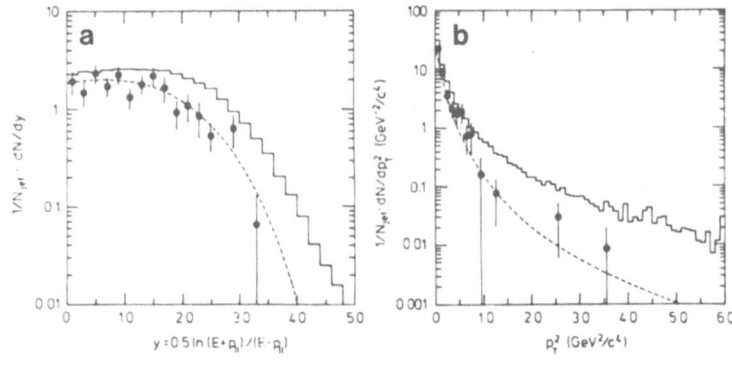

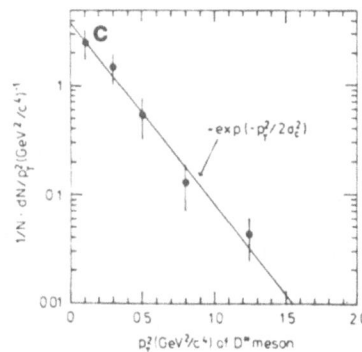

Fig. 15a,b: Properties of the D* jet compared to the average jet at $<E_{jet}>$ $\simeq$ 17 GeV (solid histogram) and $<E_{jet}>$ = 7 GeV (dashed curve).

Fig. 15c: Transverse momenta of D*'s with respect to the event axis.

Moreover, the transverse momentum spread σ of particles around the jet axis, as determined from the $p_\perp^2$ distribution of fig.15c,

$$\sigma_c = 0.36 \pm 0.02 \quad GeV/c \tag{8}$$

is the same as the one obtained from all produced quarks

$$<\sigma_q> = 0.36 \pm 0.01 \quad GeV/c \tag{9}$$

Also the rate of gluons radiated from charm quarks, as measured by determining the strong coupling constant $\alpha_s(c)$ from the transverse momentum distribution in the event plane[13], shows no deviation from its value for all other flavors

$$\alpha_s(c) / \alpha_s(all\ flavors) = 1.00 \pm 0.20 \pm 0.20 \tag{10}$$

255

In summary, a jet originating from a charm quark at high energies looks the same as any other jet when the decay products of the charmed meson are included. The only difference observed is the energy fraction transferred to the charmed meson which is substantially higher than for light quarks and mesons. After the charmed meson has been branched off, there is again no difference seen between the remainder of the cascade and a light quark jet at the same energy. These observations are in obvious agreement with cascade type fragmentation mechanisms as assumed in the aforementioned Monte Carlo models[6,7].

3. ELECTROWEAK PROPERTIES OF HEAVY QUARKS

After this detour into the strong interactions of heavy quarks we return to their weak interaction properties, this time with respect to the weak neutral current. As extensively discussed in the context of leptonic reactions $e^+e^- \to \mu^+\mu^-$, $\tau^+\tau^-$ [14] electroweak interference leads to a forward-backward charge asymmetry for fermions f

$$A_{f\bar{f}} = \frac{\#f(\cos\theta>0) - \#f(\cos\theta<0)}{\#f(\cos\theta>0) + \#f(\cos\theta<0)} \qquad (11)$$

with the scattering angle θ defined with respect to the e^- direction. For quark pair production it is given by

$$A_{q\bar{q}} \simeq -3/2 \times g_A(e)\, g_A(q)\, /\, Q_q \qquad (12)$$

with g_A the axialvector coupling constant of electrons or quarks to the weak neutral current and the quark charge Q. The pole term χ is

$$\chi = \frac{G_F}{\sqrt{8}\ \pi\alpha} \ \frac{s\cdot M_Z^2}{s-M_Z^2} \quad \simeq \quad -0.3 \qquad (13)$$

at PETRA energies and with a Z^o mass $M_Z \simeq 95$ GeV/c^2 as recently

256

measured[15]. In the standard electroweak model[16], where the left handed quarks are arranged in doublets of weak isospin, all axial vector couplings are fixed. We thus expect the rather high asymmetries quoted in tab.5. To measure these, in addition to a determination of the quark flavor, we need to tell a quark jet from an antiquark jet. In fact, while the thrust axis of a $q\bar{q}$ event remembers the quark direction, the charge of prompt muons remembers the quark charge:

$$e^+e^- \to c\bar{c} \qquad\qquad\qquad e^+e^- \to b\bar{b} \qquad\qquad (14)$$
$$\;\vert_{\to\mu^+X} \qquad\qquad\qquad\qquad \vert_{\to\mu^-X}$$

Thus the quantity $-Q_\mu\cos\theta_T$ corresponds to the antiquark direction for $c\bar{c}$, to the quark direction for $b\bar{b}$ production. In the multidimensional analysis described in section 1.3, the longitudinal momentum $p_{//}$ is therefore replaced by $-Q_\mu\cos\theta_T$, keeping the flavor separating quantities $p_\perp$ (and T). The fragmentation functions are fixed to the values obtained previously and a fit performed to obtain $B(c\to\mu X)$, $B(b\to\mu X)$, $A_{c\bar{c}}$ and $A_{b\bar{b}}$. It is no surprise that the branching fractions come out the same as before, since the flavor separation remains essentially unchanged. The asymmetry results are given in tab.6.

Of course, observing leading charmed particles $(D^{*\pm})$ also gives information about the quark charge, since the D^* contains the pri-

Tab. 5: Electroweak asymmetries for $q\bar{q}$ production expected in the standard model

Flavor	Q_q	$g_A(q)$	$A_{q\bar{q}}$
u,c	2/3	1/2	- 14%
d,s,b	-1/3	-1/2	- 25%

mordial $\overset{(-)}{c}$ quark. The advantage of this method is the very small background ($\approx$ 4% from cascade b decays), its only draw-back beeing the small statistics available. Fig.16 shows the angular distribution of D* events from TASSO and JADE. A deviation from a symmetric $(1+\cos^2\theta)$ distribution is observed at the 1.5σ level. The resulting asymmetries $A_{c\bar{c}}$, determined by a fit to the functional form

$$dN/d\cos\theta \sim 1 + 8/3\ A\ \cos\theta\ + \cos^2\theta \tag{15}$$

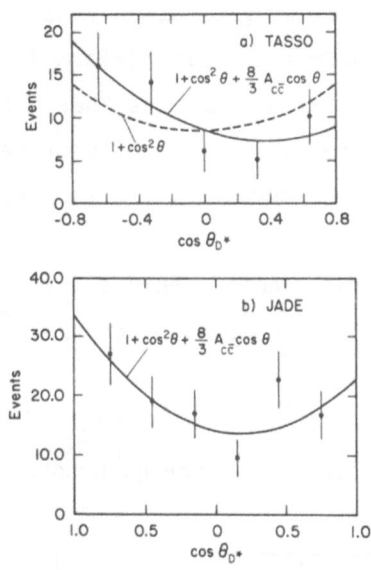

Fig. 16: Angular distribution of D* mesons from a) TASSO, b) JADE. solid curves are best fits to the form (15). Dashed curves are the QED prediction.

are given in tab.5. Averaging the results and assuming $g_A(e) = -\frac{1}{2}$ as determined in νe scattering[14], we can convert these measurements into values for the quark axial couplings

$$g_A(c) = + 0.59 \pm 0.16$$

$$g_A(b) = - 0.48 \pm 0.13$$

(16)

where we expect $+\frac{1}{2}$ and $-\frac{1}{2}$, respectively, from the standard settings of weak isospin. The data thus clearly indicate that the b quark should be the "down" member of a weak isodoublet, and that its "up" member, the t quark, is missing.

Tab. 6: Experimental data from PEP[18] and PETRA on the electroweak asymmetry for a) $c\bar{c}$ and b) $b\bar{b}$ production

Experiment		$A_{c\bar{c}}$ (%)	expected (%)
JADE	D*	− 14± 9	− 14
MARK J	c→μX	− 17± 9	− 14
TASSO	D*	− 13±10	− 14
HRS	D*	− 8± 9	− 9
	D⁰	− 11±14	− 9
TPC	D*	− 31±12	− 9
	c→μX	− 22±13±10	− 9

Experiment		$A_{b\bar{b}}$ (%)	expected (%)
JADE	b→μX	− 26± 9	− 26
MARK J	b→μX	− 15±22	− 25
TASSO	b→μX	− 38±28	− 27
MAC	b→μX	− 7± 9	− 12
TPC	b→eX	− 38±37	− 19

Tab. 7: Experimental upper limits on flavor changing neutral current decays of the b quark.

Experiment	95% CL upper limit on $B(b \to \mu^+\mu^- X)$
CLEO[18]	0.35 %
JADE	0.7 %
MARK J	0.7 %

In fact, if there were no t quark and the b were in an isosinglet, neutral current decays changing b flavor, e.g.

$$b \to s \, (d) \, \mu^+\mu^- \qquad\qquad\qquad (17)$$

would occur at the 1% level[17]. Searches for a significant rate of opposite charge muon pairs inside the same jet were unsuccessful, yielding the stringent limits on the branching ratio $B(b \to \mu^+\mu^- X)$ quoted in tab.7.

We thus have two independant pieces of evidence, the weak iso-spin of the b quark and the absence of b flavor changing neutral currents, that the t quark should indeed exist.

4. SEARCH FOR HIGH MASS RESONANCES

4.1 $t\bar{t}$ Resonances in $e^+e^- \to$ hadrons

It is thus obviously one of the prime tasks of the worlds highest energy e^+e^- storage ring to search for the production of the top quark. This has been done at PETRA by an energy scan, recently exten-

ded to $\sqrt{s}$ = 45.2 GeV and still continuing to even higher energies.
When a $t\bar{t}$ bound state is produced, we expect a narrow resonance to be
observed in $e^+e^- \rightarrow$ hadrons, thus a pronounced bump to appear in the
hadronic crossection, normalized to the pointlike QED cross section
σ_o for $e^+e^- \rightarrow \mu^+\mu^-$

$$R = \frac{\sigma (e^+e^- \rightarrow \text{hadrons})}{\sigma_o} \qquad (18)$$

The resonance strength, the integrated cross section under the bump,
depends on the product of the resonance's width into e^+e^-, Γ_{ee}, and
its branching ratio into hadrons B_h. Extrapolating Γ_{ee} from the
scaling law

$$\Gamma_{ee} / Q^2 = \text{const.} \qquad (19)$$

valid for low energy resonances (ρ, ω, J/ψ, Υ) and assuming a high
hadronic branching ratio ($B_h \simeq 70\%$), we roughly expect for a $t\bar{t}$ resonance ($Q_t = 2/3$)

$$\Gamma_{ee} B_h \simeq 3.5 \text{ keV} \qquad (20)$$

Fig.17 shows the combined data[19] on the hadronic cross section R from
all four PETRA experiments in the range of the recent scan ($40 \leq$ s
≤ 45 GeV), compared to the expectation for a $t\bar{t}$ resonance. No experiment observes a significant bump in R, 95% CL upper limits on $\Gamma_{ee}B_h$
are given in tab.8. Note however, that with these data, the production
of a resonance of 1/3 charged quarks cannot be excluded. In the combined data, the best candidate for a narrow resonance appeared at
$\sqrt{s} \simeq 44$ GeV. A later remeasurement at this energy, however, gave no
confirmation of a high cross section, as shown in fig.17.

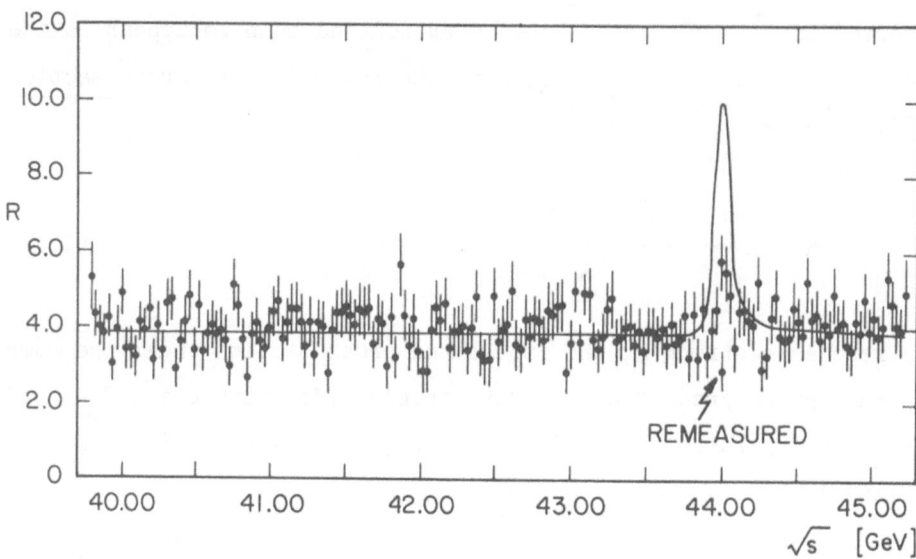

Fig. 17: Combined data[19] on the cross section for $e^+e^- \rightarrow$ hadrons from CELLO, JADE, MARK J and TASSO. The curve shows the expectation for a $t\bar{t}$ resonance with $\Gamma_{ee}B_h = 3.5$ keV.

Tab. 8: 95% CL upper limits for the strength $\Gamma_{ee}B_h$ of a narrow resonance in $e^+e^- \rightarrow$ hadrons.

Experiment	$\sqrt{s}$ for best resonance candidate (GeV)	95% CL upper limit for $\Gamma_{ee}B_h$ (keV)
CELLO	43.15	2.7
JADE	42.49	2.3
MARK J	44.00	2.9
TASSO	42.64	2.4

4.2 Generalized Resonance Search

An obvious generalization of the above method is to give up the restriction to hadrons and to simulaneously analyze all channels:

$$e^+e^- \rightarrow \text{hadrons, } e^+e^-, \mu^+\mu^-, \tau^+\tau^-, \gamma\gamma \tag{21}$$

The motivation for this is not only to prevent missing a $t\bar{t}$ resonance because of peculiar decay properties. It can also shed some light on the existence of scalar particles X, proposed[20] to explain the unexpectedly high rate of radiative Z^o decays, observed at the CERN $p\bar{p}$ collider[21], by a decay

$$Z^o \rightarrow X \tag{22}$$
$$\quad |_{\rightarrow 1^+1^-}$$

Inorder to explain the high branching ratio of this decay, as well as the observed kinematics, these models need a high width of X into e^+e^- ($\Gamma_{ee} \simeq 2$ MeV) and a mass overlapping with the PETRA energy range ($40 \leq M_X \leq 50$ GeV/c^2). It would thus be produced at PETRA, if its mass is below $\simeq 45$ GeV. It would contribute a resonance type structure to the cross section for final state i with a strength

$$S_i = \int \sigma_i(W) \, dW = (2s+1) \, 2\pi^2/M_X^2 \, \Gamma_{ee} B_i \tag{23}$$

with c.m.s. energy $W=\sqrt{s}$ and spin s of the X particle. Summing over all final states, we obtain (for n_l kinds of leptons)

$$\Sigma \, S_i = (n_l S_l + S_\gamma + S_h) \simeq (2s+1) \, 2\pi^2/M_X^2 \, \Gamma_{ee} \tag{24}$$

a direct limit on Γ_{ee}, independent of the detailed decay properties. Fig.18 shows cross section data from MARK J for $40 \leq \sqrt{s} \leq 45$ GeV, together with the best fits for a narrow resonance in each channel. 95% CL upper limits on resonance strengths and $\Gamma_{ee} B_i$ are given in

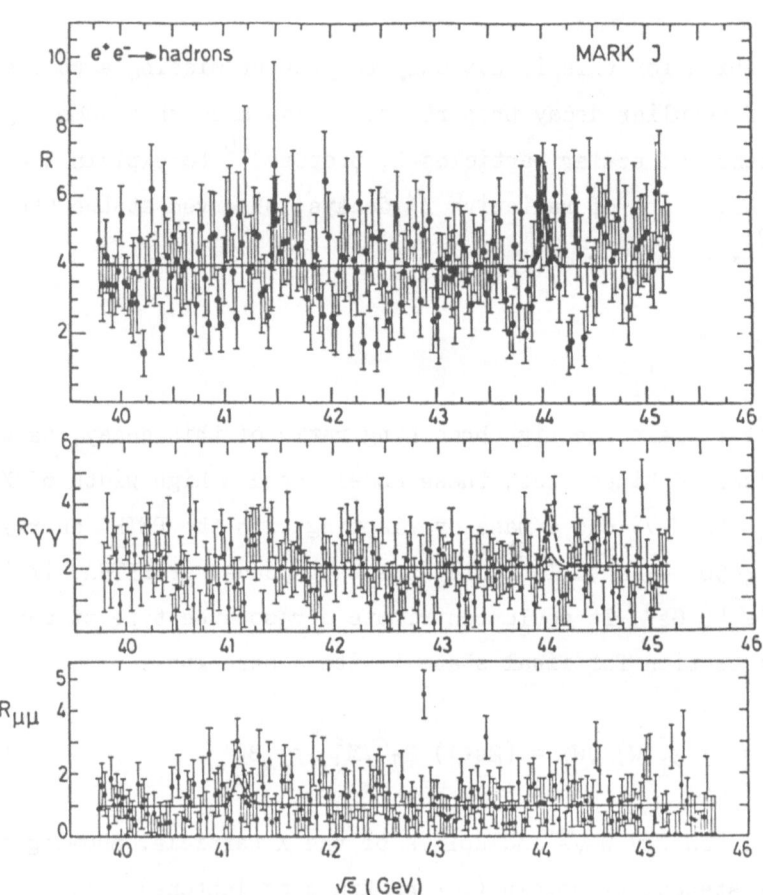

Fig. 18: Cross section measurements from MARK J for $e^+e^- \to$ hadrons, $\gamma\gamma$ ($|\cos\theta|<0.8$) and $\mu^+\mu^-$ ($|\cos\theta|<0.8$). Curves indicate best fit (solid) and 95% CL upper limit (dashed) for a narrow resonance.

tab.9. The sum of all channels yields

$$(2s+1)\ \Gamma_{ee} \simeq (2s+1)\ \Gamma_{ee}\ (B_h + B_\gamma + 6B_\mu) < 39 \text{ keV} \qquad (25)$$

for $39.79 < M_X < 45.22$ GeV/c^2. The upper limit is thus two orders of magnitude smaller than what is needed to explain the radiative Z^o decays.

Even if M_X were slightly higher than the maximum $\sqrt{s}$, a change of the angular distribution of $e^+e^- \rightarrow e^+e^-$ would be seen due to γ-X-interference. Fig.19 therefore shows the relative cross section for large angle bhabha scattering as a function of $\sqrt{s}$ from MARK J, together with the expectation[22] for a scalar of mass $M_X = 46$ and 48 GeV/c^2. No deviation from QED is seen, ruling out $40 \leq M_X \leq 48$ GeV/c^2 at the 95% CL. Experiments will soon reach a sensitivity up to masses of 50 GeV/c^2.

Another explanation for the radiative Z^o decays would be the existence of a high mass e* with electron quantum numbers, such that Z$^o \rightarrow e\ e^* \rightarrow e\ e\ \gamma$, in which case M_{e^*} would have to be around 75 GeV. Again, one would see a deviation in the $e^+e^- \rightarrow \gamma\gamma$ angular distribution, through exchange of a virtual e* in the t channel (fig.20), provided

Tab. 9: 95% CL upper limits on the integrated cross sections and resonance strength for production of X with spin s.

final state	$\int \sigma_i dW$ (MeV nb)	$\Gamma_{ee} B_i (2s+1)$ (keV)	
		MARK J	CELLO
hadrons	34	8.7	
$\gamma\gamma$	14.5	3.7	2.6
$\mu\mu$	20.7	4.5	5.6
sum	$(2s+1)\ \Gamma_{ee} < 39$ keV		

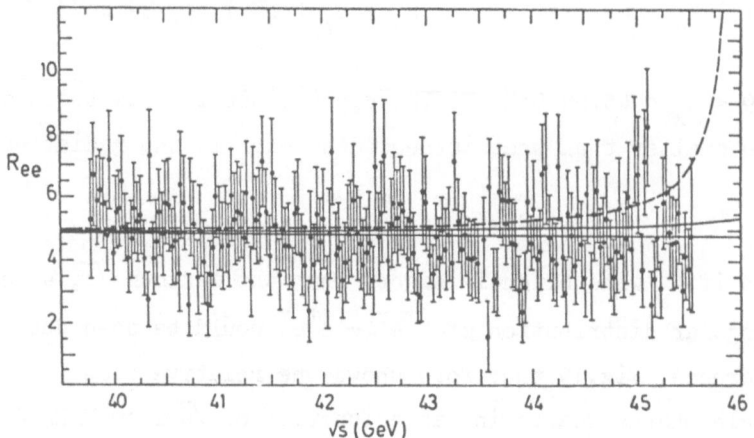

Fig. 19: Cross section measured by MARK J for bhabha scattering ($|\cos\theta|<0.5$) compared to the expectation from QED (straight line), a scalar particle with mass 46 GEV/c^2 (dashed curve) and 48 GeV/c^2 (solid curve).

that the e* → γe coupling λe is large enough. Fig. 21 shows the angular distribution for e$^+$e$^-$ annihilation into two photons from CELLO and MARK J at high energies. No deviation from the QED expectation is observed. Fig. 22 gives the 95% lower limiton M_{e*} as a function of λ. Again it is seen that sensitivity to masses for the order of 75 GeV/c^2 will soon be reached.

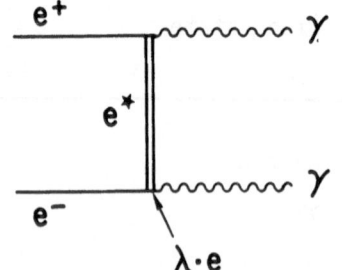

Fig. 20: Feynman graph for t channel e* exchange in e$^+$e$^-$ → γγ.

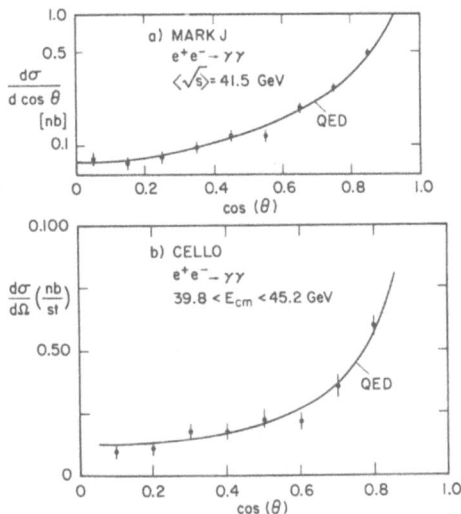

Fig. 21: Angular distribution for $e^+e^- \to \gamma\gamma$ from MARK J and CELLO, compared to the QED prediction.

5. CONCLUSIONS

Impressive data on the properties of heavy flavors c and b have been collected using e^+e^- colliding beams. From the recent PETRA data we conclude:

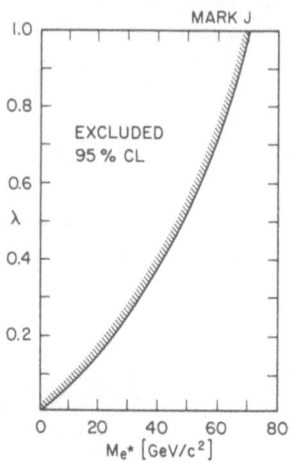

Fig. 22: 95% CL upper limit on the e* mass from MARK J as a function of λ~

The semileptonic branching ratios of the average C and B meson final stat produced in the continuum are B(c→μX) 10% and B(b→μX) 12%.

The longitudinal fragmentation of c and b quarks is hard, with an average energy fraction retained by the C (B) meson of $<z_b>≈0.6(<z_b>≈0.75)$.

The transverse fragmentation of c is normal and a high energy c jet, including the c decay products, looks like a light quark jet. The gluon radiation rate from c is also not abnormal.

The standard model weak isospin assignments for c and b are in agreement with data. In particular, the asymmetry of $b\bar{b}$ production and the absence of b-flavor changing neutral currents indicate that b should be in a doublet with the t quark, yet to be discovered.

The mass of a $t\bar{t}$ resonance is larger than 45.2 GeV/c^2. The existence of a lower lying resonance of $1/3$ charge quarks is not ruled out by existing scan data; it is, however, improbable, since no new threshold for continuum heavy quark production is observed up to this energy.

No "composite scalar X" with a high width into e^+e^- exists with mass between 40 and 45 GeV/c^2. PETRA experiments will soon reach sensitivity to masses as high as $M_X ≈ 50$ GeV/c^2 and $M_{e*}≈ 75$ GeV/c^2.

268

ACKNOWLEDGEMENTS

I gratefully acknowledge the hospitality of the DESY Directorate, especially Professors V. Soergel and P. Söding, during my stay at Hamburg. I wish to thank the members of the CELLO, JADE and TASSO collaborations, in particular Prof. G. Flügge, Drs. L. Becker and D. Lüke, as well as my colleagues from the MARK J group for providing me with data prior to publication. I also thank Prof. A. Böhm, F.P. Poschmann and Dr. G. Viertel for useful discussions. Finally, thanks are due to Prof. L.L. Chau for having managed to organize such a hot meeting in such chilly surroundings.

REFERENCES

1) MARK J Collaboration, B. Adeva et al., MIT-LNS Report 131 (1983), to be published in Phys. Rep.
   MARK J Collaboration, B. Adeva et al., Phys. Rev. Lett. 51 (1983) 443

2) TASSO Collaboration, M. Althoff et al., DESY 83-121, to be published in Zeitschr. für Physik C

3) H. Fesefeldt, Simulation of Hadronic Showers and Measurement of Muons, RWTH Aachen preprint PITHA(1984), to be published

4) To reduce the distortion of the measured event axis by hard gluon bremsstrahlung, the MARK J group rejects flat events with a cut on the oblateness of the broad jet[1] $O_b > 0.3$

5) For a review of earlier data see: H. Schneider, Semileptonic Decays of c and b quarks, Invited talk tiven at the International Conference on High Energy Physics, Brighton, 1983, Karlsruhe preprint KfK 3643 (1983)

6) A. Ali et al., Phys. Lett. 93B (1980) 155 and Nucl. Phys. B168 (1980) 409
   For an excellent review on the theory of quark decays see also: R. Rückl, Weak Decays of Heavy Flavors, Habilitationsschrift, Munich Univ. (1983)

7) B. Anderson et al., Phys. Lett. $\underline{94B}$ (1980) 211

   T. Sjöstrand, Computer Phys. Comm. $\underline{27}$ (1982) 243 and $\underline{28}$ (1983) 229

8) F.A. Berends and R. Kleiss, Nucl. Phys. $\underline{B177}$ (1981) 237, Nucl. Phys. $\underline{B186}$ (1981) 22; F.A. Berends, R. Kleiss and S. Jadach, Inst. Lorentz Leiden Preprint (1981); R. Kleiss, Univ. of Leiden Thesis (1981), unpublished

9) R.K. Ellis et al., Nucl. Phys. $\underline{B178}$ (1981) 30

10) C. Peterson et al., Phys. Rev. $\underline{D27}$ (1983) 105

11) TASSO collaboration, M. Althoff et al., DESY 83-010 (1983), to be published

12) TASSO collaboration, M. Althoff et al., DESY 83-119 (1983), to be published

13) TASSO collaboration, M. Althoff et al., DESY 84-005 (1984), to be published

14) For a review, see e.g. A. Böhm, Electroweak Interference, Plenary talk at the Int. Conf. on High Energy Physics, Brighton 1983, DESY 83-103 (1983)

   B. Naroska, Electroweak Interference, Neutrino Electron Scattering and Related Topics, Rapporteur's talk at the 1983 International Symposium On Lepton and Photon Interactions at High Energies, Cornell, DESY 83-111 (1983)

15) G. Arnison et al., Phys. Lett. $\underline{126B}$ (1983) 398

   P. Bagnaia et al., Phys. Lett. $\underline{129B}$ (1983) 130

16) S.L. Glashow, Nucl. Phys. $\underline{22}$ (1961) 579

   S. Weinberg, Phys. Rev. Lett. $\underline{19}$ (1967) 1264, Phys. Rev. $\underline{D5}$ (1972) 1412

   A. Salam, in Elementary Particle Theory, N. Svartholm Edt., Stockholm 1968, p. 361

   S.L. Glashow, J. Iliopoulos and L. Maiani, Phys. Rev. $\underline{D2}$ (1970) 1285

17) V. Burger, W.Y. Keung and R.J.N. Phillips, Phys. Rev. <u>D24</u> (1981) 1328

   H. Georgi and S.L. Glashow, Nucl. Phys. <u>B167</u> (1980) 173

   G.L. Kane and M.E. Peskin, Nucl. Phys. <u>B195</u> (1982) 29

18) For a review of PEP data see L. Barbaro-Galtieri, these proceedings; for a review of CESR data see J. Lee-Franzini, these proceedings

19) The combined measurements are compiled by D. Lüke with the consent of all experiments to include preliminary data

20) M. Veltman, Univ. of Mich. Preprint UMHE 83-22 (1983); M.J. Duncan and M. Veltman, UMHE 84-1 (1984); G. Gounaris, R. Kögeler and D. Schildknecht, Univ. of Bielefeld, Preprint Bi-TP 83/17 (1983); W. Hollik, F. Schrempp and B. Schrempp, DESY 84-3 (1984) F.W. Bopp et al., Si84-3 (1984)

21) G. Arnison et al., CERN-EP/83-162 (1983)

22) See W. Hollik, B. Schrempp and F. Schrempp, ref. 20

23) Similar results have been obtained by the CELLO collaboration, H.-J. Behrend et al., to be published

DISCUSSION

TRILLING:
In analysis of high p_T leptons showing broadened thrust distributions, is the thrust not kinematically related to the observation of a particle with large p_T? Furthermore the choice of a high p_T particle may bias the hadronic background in favor of the three jet events.

POHL:
There is some kinematic correlation, but when you look at the jet opposite to the μ, you still see broadening of thrust, indicating that a non-negligible fraction of the effect comes from the b mass. I forgot to mention that in order to prevent mis-measurement of the quark direction due to hard gluons, events with a flat energy flow ($O_B > 0.3$) have been cut out.

BARBARO-GALTIERI:
Is any other detector, besides TASSO, looking for an F signal?

POHL:
In order to reduce combinatorial background, you have to use particle identification to veto against pions and protons in the $\phi \to K^+K^-$. Since you are looking for high momentum kaons, you can however not require positive kaon identification for time-of-flight measurement or Cerenkov counters. Only the TASSO particle identification system seems to be set up to do this job.

NEW RESULTS ON FLAVOR PRODUCTION AT PEP

A. Barbaro-Galtieri

Lawrence Berkeley Laboratory
University of California
Berkeley, CA 94720

1. INTRODUCTION

This report includes results from five PEP detectors: DELCO, HRS, MAC, MARK II and TPC. All, except the TPC, are presently taking data at PEP. The TPC is being upgraded: a new superconducting coil is being installed and other improvements are being implemented.

The results discussed here are either new or improved since the Cornell Conference. Other results from PEP have been reported at this Conference by George Trilling (MARK II) on τ, D^0, and b lifetimes and a new limit on the tau neutrino mass, and by Bill Ford on the MAC measurement of the b lifetime.

Section 2 will discuss new results on Particle Searches and a limit on neutrino generations. Section 3 includes new data on weak couplings of c and b quarks. In Section 4 various new results on hadron production are reported. Section 5 summarizes the new results. All data were obtained in e^+e^- collisions with total energy $\sqrt{s}$=29 GeV.

2. PARTICLE SEARCHES

2.1 Fractionally Charged Particles (TPC)

The TPC collaboration has already reported new upper limits on a search for Q=±4/3 particles [1]. They now have new results on Q=±1/3, ±2/3 particles [2]. Simultaneous measurements of curvature and ionization loss (dE/dx) for each particle produced in the reaction e^+e^-→hadrons allows the identification of particles with charges different from unity.

Figure 1 shows the scatter plot of dE/dx versus momentum for tracks with more than 80 ionization samples produced in 12000 hadronic events. The average charged multiplicity is 12.6 tracks per event. The dE/dx resolution is 3.7%. The bands populated by known Q=1 particles are clearly visible in Fig. 1. The search for fractionally charged particles is made in two regions outside these bands, (a) between 4 and 8 KeV/cm energy loss and (b) between 24 and 40 KeV/cm. A few tracks fall in region (b), but they are easily rejected as candidates by simple requirements: i) track must point to the primary vertex (reject deuterons and tritons from secondary interactions), ii) ionization must be due to a single track (reject overlapping tracks). For the total sample of 29000 hadronic events, corresponding to $\int Ldt=77$ pb^{-1}, no candidates are found.

The upper limits for |Q|=1/3, 2/3 production are shown in Fig. 2 as a function of quark mass. The limits are mass and model dependent since the acceptance of the apparatus is a sharp function of the momentum of the particle. Results for two such models are shown in the figure, as well as limits obtained by other experiments [3] for the reaction e^+e^-→q$\bar{q}$ X.

2.2 Selectron Search and Neutrino Counting (MAC)

The MAC collaboration has new results on their supersymmetric electron search. As previously reported [4] they can extend the selectron mass search to values above the total e^+e^- energy by using the process represented in the diagram of Fig. 3a. The radiative annihilation of e^+e^- into two photinos [5] provides sensitivity, depending on the luminosity, to selectron masses as high as, e.g., 60 GeV at $\sqrt{s}=29$ GeV.

The signature for this reaction is provided by a single gamma in the apparatus and nothing else. The search can only be made above a value of γ energy, E_{min}, where the QED background cannot contribute. The background is due to processes like $e^+e^-\gamma$, $\mu^+\mu^-\gamma$, and $\gamma\gamma\gamma$ where the two leptons or two of the photons escape along the beam pipe. A good acceptance for this process is therefore dependent upon the ability to detect tracks as close as possible to the beam pipe. The variable used is E_T, the transverse energy of the detected photon. The QED background extends to a value of E_T which is related to the smallest veto angle for the photons and leptons of the QED reactions.

The measured E_T spectrum is shown in Fig. 3d. This was obtained with a veto angle $\theta_v \simeq 5^o$, improved from the previous analysis [4] for which $\theta_v = 9^o-12^o$ (Fig. 3c). The integrated luminosity for the new analysis is 34 pb^{-1}. No candidate events are found for $E_T > 3$ GeV. For the old analysis, the minimum γ energy was 3 GeV and the search region was for $E_T > 4.3$ GeV. The lower limits, for the combined

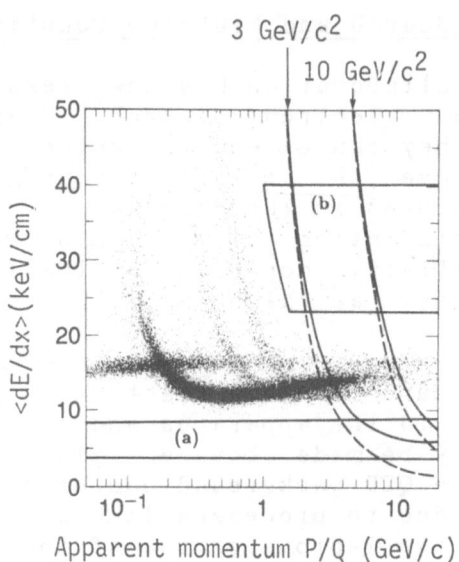

Fig. 1.   Ionization loss vs momentum for tracks in jets
           as measured by the TPC detector [1].   Regions
           (a) and (b) are search regions for   Q=1/3,   2/3
           particles.   The lines indicate expected <dE/dx>
           curves for   Q=1/3   (dashed)   or   Q=2/3   (solid)
           particles with masses of 3 and 10 GeV.

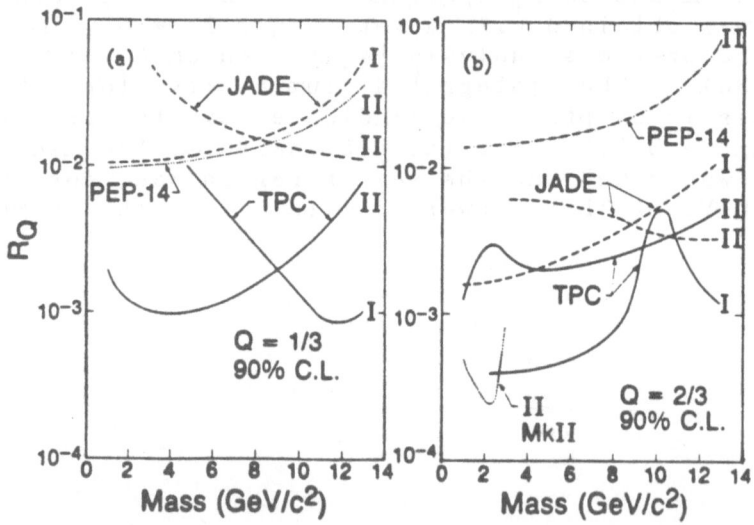

Fig. 2.   Upper   limits   reported   by   the   TPC   [2]   on
           production   of   particles   with   Q=1/3 (a), and
           Q=2/3 (b).   Results from MARK II, PEP14 and
           JADE   are also shown [3]. Curves labelled I use
           a momentum distribution $dN/dp \simeq p^2/E$,   curves   II
           use $dN/dp \simeq (p^2/E) \exp(-3.5E)$, E in GeV.

276

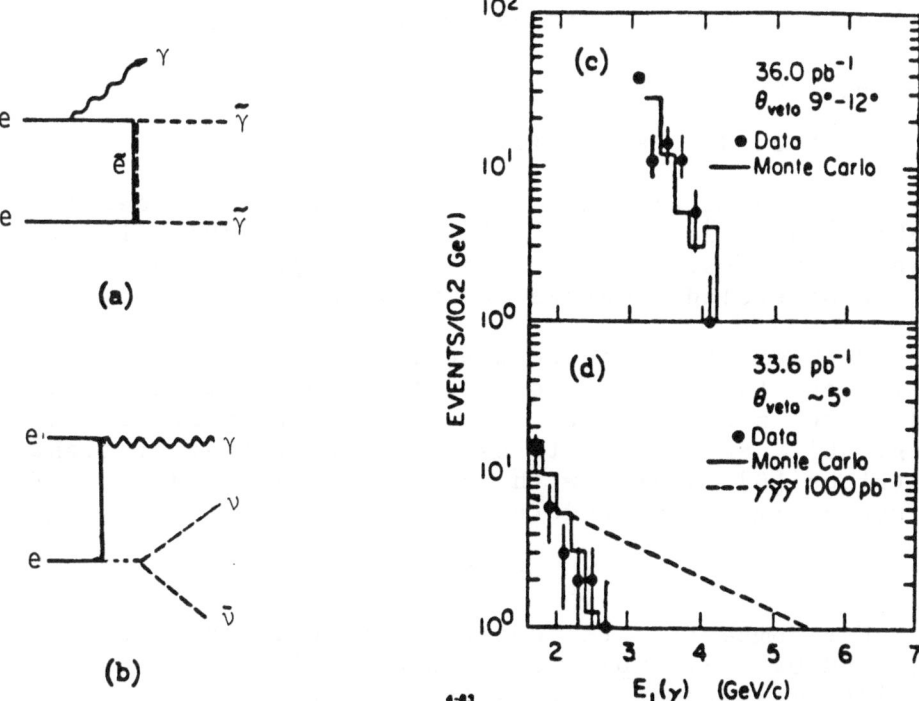

Fig. 3. (a) Radiative (e^+e^-) annihilation into $\tilde{\gamma}\tilde{\gamma}$, used for selectron search, (b) diagram used for neutrino counting. (c) MAC's measured spectrum of transverse energy, for events with one γ only. Solid curve: spectrum of QED background processes. The veto angle was $9^\circ-12^\circ$, (d) same as (c) with veto angle $\simeq 5^\circ$. Dashed curve: expected spectrum for $e^+e^- \to \gamma\tilde{\gamma}\tilde{\gamma}$.

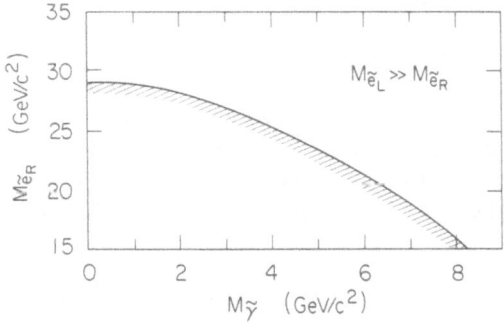

Fig. 4. The region of excluded selectron masses for massive photinos (MAC [4]).

luminosity (73 pb^{-1}) are as follows:

$$m_{\tilde{e}} > 29 \text{ GeV} \quad 90\%\text{CL for } m_L \gg m_R, \; m_{\tilde{\gamma}} = 0$$

$$m_{\tilde{e}} > 35 \text{ GeV} \quad 90\%\text{CL for } m_R = m_L, \; m_{\tilde{\gamma}} = 0$$

where m_L and m_R are the masses for the left-handed and right-handed selectron. For heavy photinos, limits lower than these are expected [5]. Figure 4 shows the region of excluded masses for $m_L \gg m_R$.

The same data can be used to calculate an upper limit on the number of neutrino generations. The radiative process $e^+e^- \rightarrow \gamma\nu\bar{\nu}$ via a virtual Z^o (Fig. 3b) can be used to estimate the number of light neutrinos [6]. Again, with no candidates found a 90% CL upper limit can be determined:

$$N_\nu < 43$$

For comparison the upper limits set by the UA1 [7] and UA2 [8] experiments in $\bar{p}p$ collisions from the Z^o width measurement are as follows:

$$\text{UA1} \quad \Gamma(Z^o) < 8.5 \text{ GeV} \quad N_\nu < 34$$
$$\text{UA2} \quad \Gamma(Z^o) < 11 \text{ GeV} \quad N_\nu < 47$$

The UA1 group [7] also estimates the number of neutrino generations from the lower limit of the $Z^o \rightarrow e^+e^-$ branching ratio to be $N_\nu < 18$.

3. WEAK COUPLINGS OF b AND c QUARKS

In e^+e^- annihilation heavy quarks are produced via one photon or Z^o intermediate states. At PEP and PETRA energies the interference between these two production mechanisms is expected to produce a measurable asymmetry in the distribution of the quark production angle with respect to the beam direction.

In lowest order QED, and assuming only one Z^o, the forward-backward asymmetry can be written as follows:

(1) $\qquad A = (F-B)/(F+B) = -3/2\chi(a_e \cdot a_f)/Q_f$

where Q_f is the charge of the final quark considered, a_e and a_f are the axial coupling constants for the

278

electrons and the final quark, and, away from the Z^o pole, χ can be approximated with the expression

(2) $\chi = \dfrac{G_F}{8\pi\alpha\sqrt{2}} \dfrac{sM^2}{s-M^2_{Z^o}} = -.042$ for $s = \sqrt{29}$ GeV

The standard model predicts $a_e = a_b = -1$, $a_c = +1$.

Two methods have been used to measure this asymmetry: a) use D^* and D^o mesons, products of the c quark, and b) use leptons that are decay products of hadrons produced in the hadronization of c and b quarks.

3.1 $\underline{D^* \text{ and } D \text{ as Tag for c Quarks (HRS and TPC)}}$

The charm quark is easily tagged by identifying D^* events using the decay

(3) $D^{*+} \rightarrow D^o \pi^+$ (and c. c.)

with subsequent D^o decay into $K^-\pi^+$, $K^-\pi^+\pi^+\pi^-$, $K^-\pi^+\pi^o$ etc. The decay (3) has a small Q value (5.8 MeV) and the D^*–D mass difference exhibits a narrow peak at the pion mass. The background from u, d, and s quarks will be very small, whereas the contribution from b quark can be calculated to be, depending on the spectrum of D's from bottom quarks, at most

$$b/c = Q_b^{\,2}/Q_c^{\,2} = 1/4$$

This method, originally used by the MARK I collaboration [9], has been used extensively in e^+e^- collisions [10-12].

New results from PEP since the Cornell Conference are from TPC [14] and HRS [11]. The TPC results are shown in Fig. 5. Here particle identification is used to identify the kaon, in high momentum $(D^o\pi^+)$ combinations, i.e. $z = 2E_{D^*}/\sqrt{s} > 0.5$. The $D^{*\pm}$ peak includes 120 events, obtained with 77 pb^{-1} of data. The decay modes used are $D^o \rightarrow K^-\pi^+$, $K^-\pi^+\pi^o$ and $K^-\pi^+\pi^+\pi^-$ and the charge conjugate states. For the asymmetry measurement (Fig. 5b) only the first decay mode was used in order to minimize the background. The value of the asymmetry is $A = -.20\pm.13$, where -0.09 is expected.

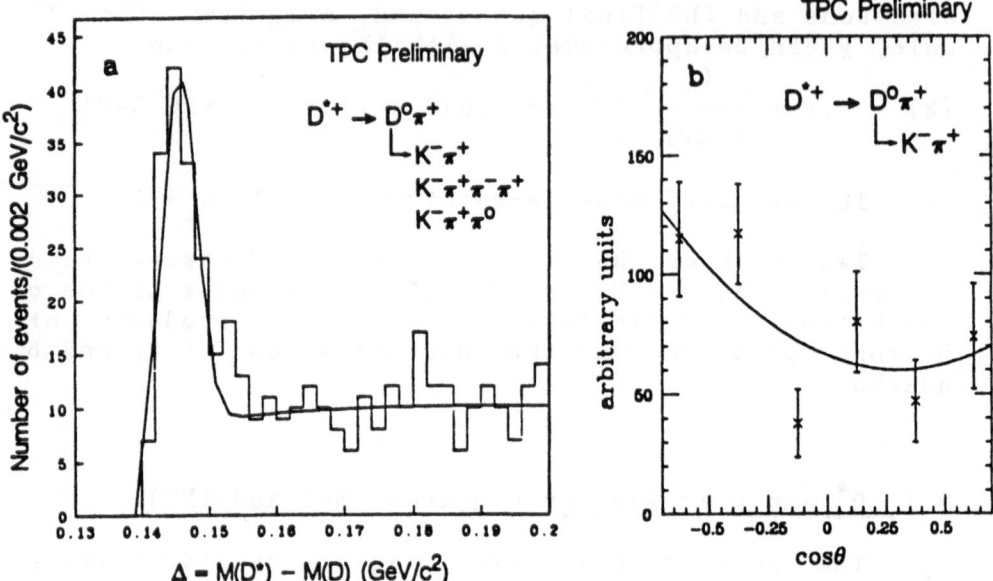

Fig. 5. TPC results on $D^{*\pm}$ production and asymmetry. (a) $D^{*+} - D^{0}$ mass difference (and c.c. states). The curve is a fit to a Gaussian shape ($\sigma=3$ MeV) and a smooth background. (b) Asymmetry for events with $D^{0} \rightarrow K^{-}\pi^{+}$ (and c.c. states).

New results from HRS [11] on D^{*} production and asymmetry are shown in Fig. 6. Here no particle identification is used, the good momentum resolution ($\delta p/p=0.1\%p$ at high momentum) being sufficient to reduce backgrounds. The sample analyzed corresponds to 120 pb^{-1} of integrated luminosity. Figure 6a shows the mass difference for the $D^{0} \rightarrow K^{-}\pi^{+}$ decay mode, with the requirement $z > 0.4$. Figure 6b shows the mass difference for another D^{0} decay mode and Fig. 6d shows the asymmetry for the combination of three decay modes: $D^{0} \rightarrow K^{-}\pi^{+}$, $K^{-}\pi^{+}\pi^{0}$, $K^{-}\pi^{+}\pi^{+}\pi^{-}$. The asymmetry from the whole sample is $A = -0.08\pm0.09$.

HRS [11] has also measured the asymmetry using inclusive $D^{0} \rightarrow K^{-}\pi^{+}$ production. Because of the good momentum resolution, a D^{0} peak is present in the $K\pi$ invariant mass (no particle identification is used), when the following two criteria are applied: i) require $|\cos\theta_{K}| < 0.7$, where θ_{K} is the angle between the kaon and the D direction in the $K\pi$ rest frame, ii) require $z > 0.5$. Table I shows the asymmetry obtained in this case; Fig. 6c shows the $K^{-}\pi^{+}$ (and c.c.) invariant mass distribution for the 120 pb^{-1} sample.

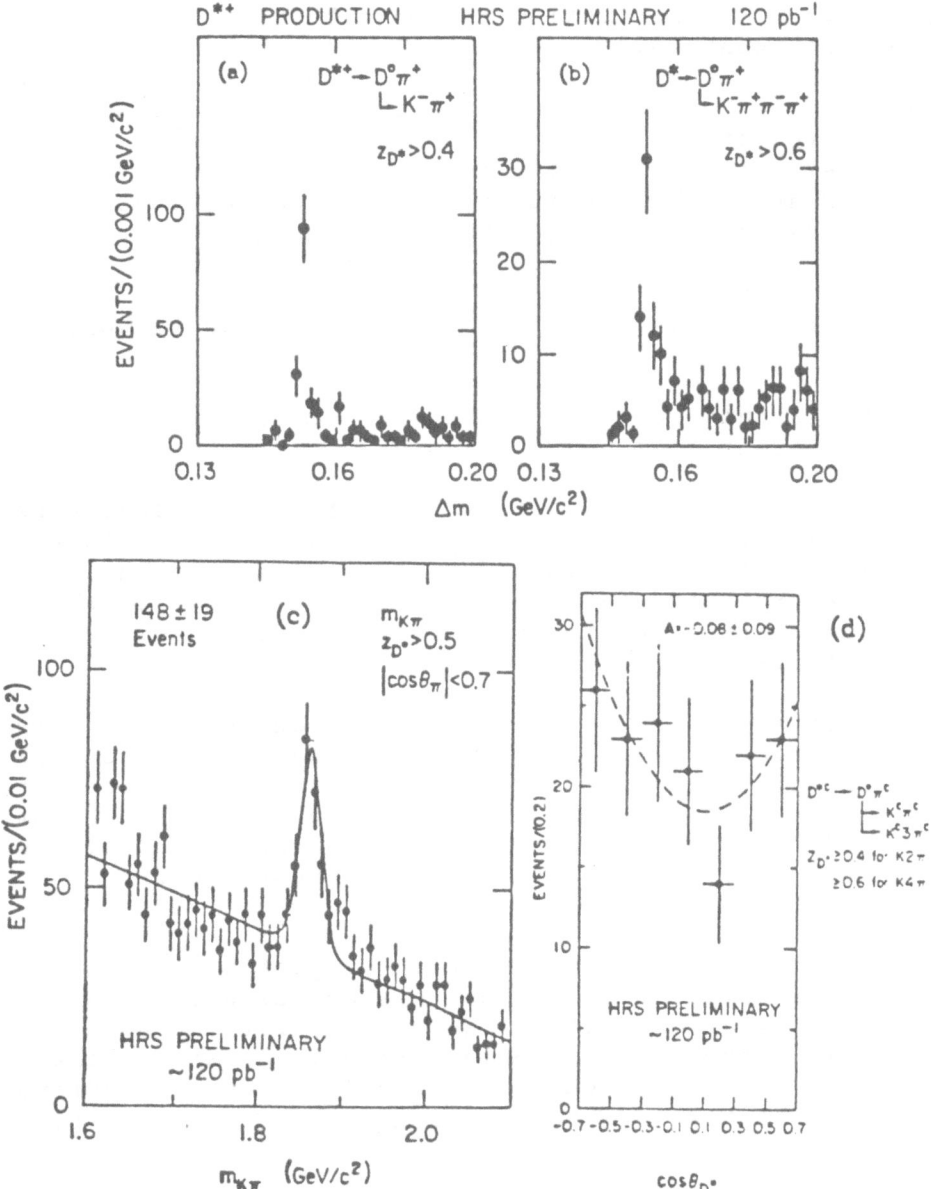

Fig. 6. HRS results on $D^{*\pm}$ production and asymmetry [11]. (a) D^{*+}-D^0 mass difference for $D^0 \rightarrow K^- \pi^+$, (b) mass difference for the $D^0 \rightarrow K^- \pi^+ \pi^+ \pi^-$ decay, (c) the $K^- \pi^+$ invariant mass distribution, and (d) asymmetry for the complete $D^{*\pm}$ sample.

3.2 Leptons as a Tag for b and c Quarks (TPC)

This method uses semileptonic decays of hadrons carrying charm or bottom quantum numbers to tag c and b quarks. The assumption is that the thrust axis of the jet containing the lepton gives the direction of the produced quark. In addition, the charge of the lepton is used to distinguish quark jets from antiquark jets. The expected asymmetry at $\sqrt{s} = 29$ GeV (34 GeV) is -9% (-14%) for c quarks and -19% (-26%) for b quarks. The asymmetry is defined to be positive if there are more quarks than antiquarks produced along the direction of the incident electron.

There are various types of backgrounds to contend with:

a. Background due to hadrons misidentified as electrons or muons.

b. For the electron case, source of backgrounds are real electrons from γ conversion pairs (in the material between the e^+e^- interaction point and the tracking detector) and Dalitz pairs for which one of the electrons has too low a momentum to be detected.

c. Background from the other heavy quark. This is important because $c \rightarrow \ell^+$ whereas $b \rightarrow \ell^-$, so the background from the other quark will tend to reduce the asymmetry. It is therefore important to select c or b enriched samples and to estimate the contribution from the other heavy quark.

Two variables p and p_T with respect to the thrust or sphericity axis of the jets are used to separate the samples. In addition, an overall fit including as variables the semileptonic branching ratios, the z ($z=2E/\sqrt{s}$) distributions of the produced parent hadrons and the backgrounds has to be made before asymmetries are measured.

Many groups have published results using this technique [15-16]. New results come from TPC [17] which analyzed heavy quark decays into electrons. Electrons are identified by combining the information from the electromagnetic calorimeter (HEX) and the energy loss in the TPC. This allows a good hadronic

rejection. The hadron misidentification probability
is 0.003%–0.3%, depending on p and p_T. The major
source of background is due to electrons from photon
conversions in the material (0.2 r.1) in front of the
TPC. Figure 7 shows the electron spectra for the
samples with $p_T < 1$ GeV/c and $p_T > 1.0$ GeV/c. Results
of the overall fit to the data will be discussed in
Sec. 4.1. For the asymmetry measurement, restricted
samples have been chosen in order to reduce
backgrounds. The c–enriched sample comprises 270
events with $p > 1.5$ GeV/c and $p_T < 1.0$ GeV/c; the b
enriched sample includes 80 events with $p > 1.5$ GeV/c
and $p_T > 1.0$ GeV/c. The contributions from $c \rightarrow e$
decays, $b \rightarrow c \rightarrow e$ decays, $b \rightarrow e$ decays and backgrounds are
$(57 \pm 8)\%$, $(7 \pm 1)\%$, $(12 \pm 2)\%$ and $(24 \pm 6)\%$ in the
c–enriched sample and $(20 \pm 3)\%$, $(4 \pm 1)\%$, $(65 \pm 6)\%$
and $(11 \pm 3)\%$ in the b–enriched sample, respectively.
Figures 7c–d show the background subtracted and
acceptance–corrected distributions of the angle
between the incident e^- and the thrust axis. The
results of the fits are shown in Table I.

3.3   Summary

Table I summarizes all the results obtained at
PEP and PETRA. Here the systematic and statistical
errors quoted by each experiment have been added in
quadrature.

We notice the following;

a.   values of $a_e a_c$ obtained from D^0, D^*, e and μ agree
     with each other.

b.   If we assume $a_e = -1$ as expected we obtain for the
     average over all experiment the values:

$$a_c = 1.25 \pm 0.34 \qquad \text{Standard Model: } a_c = 1$$
$$a_b = -1.03 \pm 0.24 \qquad \text{Standard Model: } a_b = -1$$

In conclusion, the measured axial coupling
constants for charm and bottom agree with the Standard
Model predictions within $\simeq 30\%$.

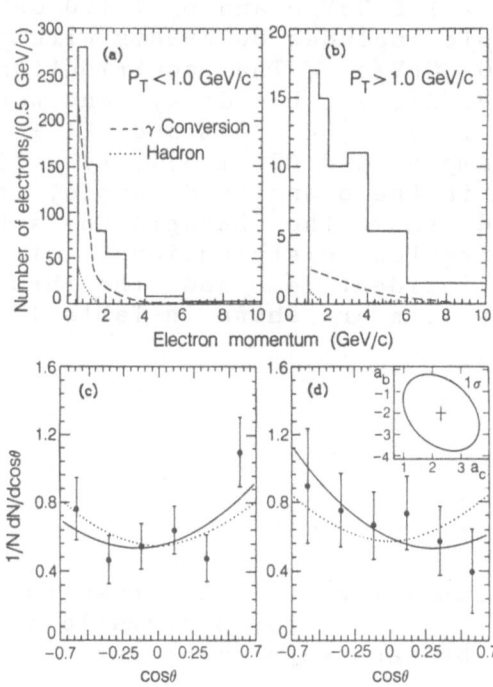

Fig. 7.   TPC results on prompt electron spectrum and
          asymmetry [17]. (a), (c) charm enriched sample;
          (b), (d) bottom enriched sample. Dashed and
          dotted curves in (a), (b) show the background.
          Dotted curves in (c), (d) show the $1+\cos^2\theta$
          distributions. The insert in (d) is the equal
          likelihood contour in the a_c-a_b plane.

Table I. Results on heavy quark axial coupling constants, for electrons, a_e and for b and c quarks, a_q. Here A is the measured asymmetry (see text).

Exp.	Quark	Ref.	$\sqrt{s}$ (GeV)	A (%) Measured	A (%) Expected	$a_e \cdot a_q$
	charm					
HRS	D*	11	29	-8±9	-9	-0.8±0.9
HRS	D$^{\circ}$	11	29	-11±14	-9	-1.2±1.3
TPC	D*	14	29	-20±13	-9	-2,1±1.4
TASSO	D*	11	34	-13±10	-14	-0.9±0.7
JADE	D*	12	34	-27±14	-14	-1.9±1.0
AVERAGE	D*					-1.21±0.43
TPC	e	17	29	(-22±16)[b]	-9	-2.3±1.7
MAC	μ	15	29	(-5±11)[a]	(-3)[a]	-1.6±3.6
MARK J	μ	16	35	-17±9		-1.2±0.6
AVERAGE	e,μ					-1.33±0.56
AVERAGE	charm					-1.25±0.34
	bottom					
TPC	e	17	29	(-38±37)[b]	-19	2.0±1.9
MAC	μ	15	29	(-7±9)[a]	(-12)[a]	0.6±0.7
MARK II	μ,e	15	29	(-28^{+13}_{-11})[b]	-19	$1.5^{+0.7}_{-0.6}$
MARK J	μ	16	35	-15±22	-25	0.6±0.9
TASSO	μ	16	34	-38±28	-27	1.4±1.0
JADE	μ	16	34	-26±9	-26	1.0±0.3
AVERAGE	bottom					1.03±0.24

a. Background not subtracted.
b. Derived quantity, not quoted by the authors.

4. HADRON PRODUCTION

Hadron production in e^+e^- collisions provides a simple source of jets that allow the study of the process of hadronization. The quark-parton model provides a simple explanation for the observed jet structure: e^+e^- annihilation first produces $q\bar{q}$ pairs which later fragment into hadrons producing the observed jets. The hadronization process, however, is not that simple; there are still a number of unanswered questions that need investigation. Detailed experimental information is essential.

Over the past few years a number of models have been developed to study the fragmentation of quarks and gluons into hadrons. In these models, quarks and gluons fragment into hadrons either "independently" from other partons [18], or in "color singlet strings" [19] or in "parton showers" based on perturbative QCD [20-22]. Some Monte Carlo generators using these different models have been made available to experimentalists and comparisons with data can be made. In this review the comparison will be mostly made with the LUND Monte Carlo [19].

In this section results obtained from PEP since the Cornell Conference [23] on heavy quark fragmentation, particle content of jets, baryon production and flavor correlations will be reported.

4.1 Heavy Quark Fragmentation

Direct information on heavy quark fragmentation is more accessible from the data than for light quarks. Tagging of heavy quarks can be done through D^*, D^o and prompt leptons, as discussed in Sec. 3. Since the c and b quarks are heavy, $c\bar{c}$ and $b\bar{b}$ pairs are unlikely to be created in the hadronization process. The suppression factor is model dependent, but in first approximation it can be assumed that the heavy mesons will usually contain a first rank (primary) quark.

Experimental results have shown that the fragmentation function of heavy quarks is peaked at high energy. The heavy hadron is very likely to carry a large fraction $(z=2E/\sqrt{s})$ of the available energy. This was anticipated by Suzuki and Bjorken [24]. A simple form for the fragmentation function of heavy quarks has been suggested by Peterson et al. [25]:

$$(4) \qquad D(z) = \frac{A}{z} \left(1 - \frac{1}{z} - \frac{\epsilon_Q}{(1-z)} \right)^{-2}$$

where A is a constant and $\epsilon_Q = m_q^2/m_Q^2$

Here Q and q indicate the heavy and light quark that make the heavy hadron. The experimental results are usually compared to this form.

286

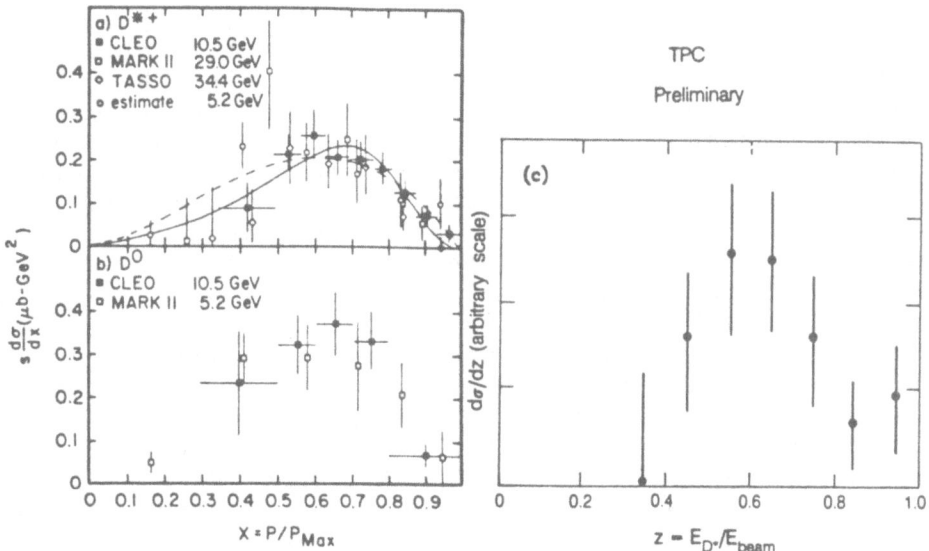

Fig.8. Results on D* fragmentation function. (a) Inclusive cross section as a function of x for D*+ as compiled by CLEO [10], including results from the MARK II [10] and TASSO [11]. (b) D⁰ cross sections from CLEO [10] and MARK II [26]. (c) The TPC preliminary data on D* fragmentation [14].

A.   Charm fragmentation from D* (TPC)   New results on D* fragmentation are reported by TPC [14]. The D*+-D⁰ mass difference is shown in Fig. 5 and details are given in Sec. 3.1. The fragmentation function is shown in Fig. 8c. A partial compilation of results from other experiments, on D* and D⁰ is shown in Fig. 8a,b taken from Avery et al. [10]. Here data from CLEO [10], TASSO [11], MARK II at 29 GeV [10], and MARK II at 5.2 GeV [26] are shown in a plot of the scaled cross section $sd\sigma/dx$. Values of <z> and of ϵ are shown in Table II. All data agree within errors.

B.   Results from prompt leptons (TPC, DELCO).   As discussed in Sec. 3.2, the prompt leptons in hadronic events can be used as tag of c and b quark fragmentation into hadrons. From the lepton spectrum it is possible to infer the spectrum of the hadrons produced from heavy quarks and measure their z distributions.

New results from PEP are from DELCO(e) [27] and TPC(e,μ) [17,14]. The TPC electron identification and data have been discussed in Sec. 3.2. In the DELCO experiment, electron identification is made primarily with a threshold Cerenkov counter, although extra rejection of background is done with a shower counter. Two separate experiments were carried out by filling the Cerenkov counter with isobutane (range 0.5 to 2.5 GeV/c) and then with nitrogen (range 0.5-5.5 GeV/c). Results from the two experiments agree within statistics. Backgrounds to the electron signal are due mostly to γ conversions and misidentified hadrons (total of 28% and 18% for the two different experimental conditions). Figure 9a shows the electron cross section for the DELCO measurements along with those for TPC [17] and MARK II [15]. Figure 9b shows the DELCO background substracted p_T distribution. This illustrates how effective the cut at P_T = 1 GeV is in separating the charm enriched from the bottom enriched samples.

The TPC muon results [14] are shown in Fig. 10. The TPC muon system covers 98% of the solid angle with four layers of proportional chambers and a total iron thickness varying between 0.67 and 1.25 meters of steel. For this analysis only part of the solid angle is used and the muon momentum cutoff is at 2 GeV/c. The background from kaon decays is reduced by using the dE/dx particle identification in the TPC. The sample of 982 muon candidates includes a background from punch-through and decays of 38%. Figure 10 shows the background subtracted c and b enriched samples.

The results obtained by the different groups are summarized in Table II. The procedure used is to make an overall fit to the data in bins of p and p_T (including background) with four free parameters: semileptonic branching ratios for b and c quarks, ϵ_c and ϵ_b (see Eq. 4). From the values of ϵ one can then calculate the <z> for each quark. Both ϵ and <z> are reported in Table II.

C.   <u>Summary</u>.   Many experiments have measured heavy quark fragmentation parameters. They are summarized in Table II. The results are in good agreement within errors in the measured values of <z>. The values of ϵ show a larger variation, but, since D(z) varies slowly with ϵ, this does not imply a variation in shape between the many experiments.

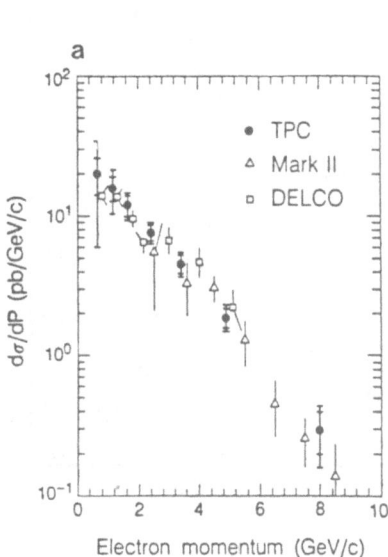

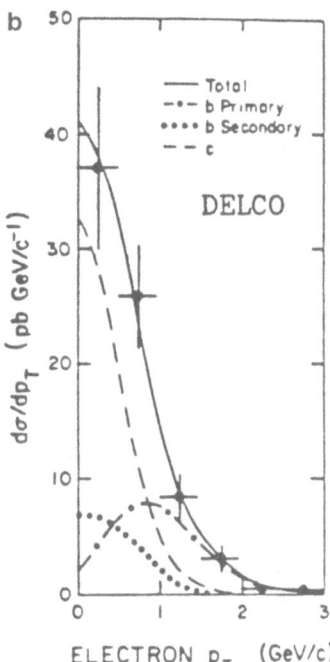

Fig.9.   (a) Prompt electron spectrum from the TPC [17],
DELCO [27] and MARK II [15].   (b) DELCO's
background subtracted p_T distribution of direct
electrons.   Curves   represent   different
contributions to the spectrum.

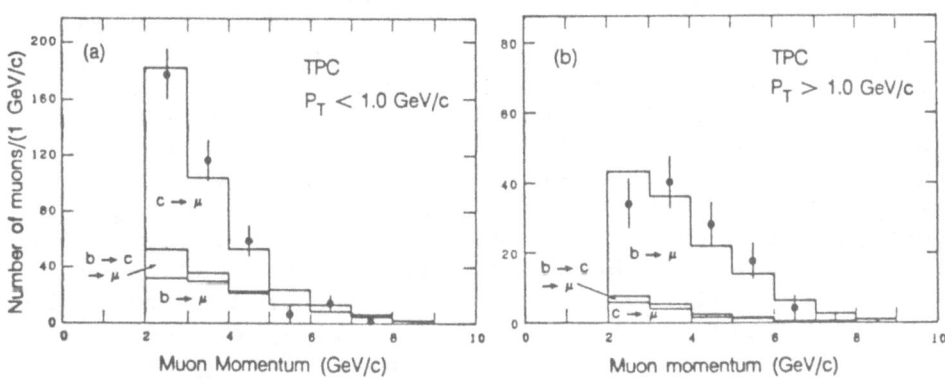

Fig.10.   TPC background subtracted spectra of direct
muons (preliminary) [14].   (a) Charm-enriched
sample, (b)   bottom-enriched   sample.   The
different   contributions   to   the spectrum shown
are the result of a fit.

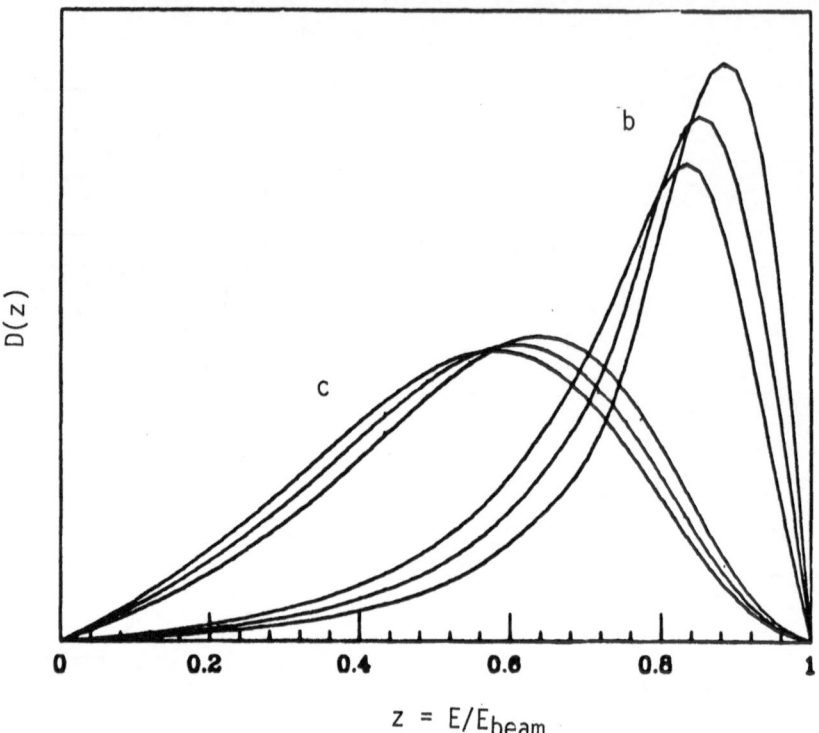

$$z = E/E_{beam}$$

Fig. 11. The Peterson et al. [25] form for the charm and bottom fragmentation function, calculated with $\epsilon_c = 0.25 \pm 0.05$ and $\epsilon_b = 0.017 \pm {}^{0.010}_{0.007}$ (see text).

Can the results from the different experiments be averaged? This seems to be somewhat inappropriate because some experiments, i.e. MARK J [29] and MAC [15], use a different variable $z = (E+p_L)/(E_Q+p_Q)$, rather than $2E_h/\sqrt{s}$. TASSO [29] used both and gave results for c and b quarks that differ by (5–8)% for the two parameterizations. Another difficulty is due to radiative corrections: DELCO quotes two values (with and without radiative corrections) that differ by (2–4)%. With caution then, we can average the results. For charm we average $<z>_c$ as obtained from the D* data, without including the CLEO results, because the value of z_{min} is very different from the others. This average (see Table II) agrees very well with the average from the e, μ data. For ϵ_c we also average the D* data. For the bottom quark, the errors on the measured ϵ_b are very asymmetric, so we choose to average $<z>_b$ and derive ϵ. The results are as follows:

$$<z>_c = 0.57 \pm 0.01 \qquad \epsilon_c = 0.25 \pm 0.05$$

$$<z>_b = 0.78 \pm 0.03 \qquad \epsilon_b = 0.017 \pm 0.010$$

the errors are the ones obtained in the weighted

290

Table II. Heavy quark fragmentation results, from e^+e^- experiments. In all cases systematic and statistical errors have been added in quadrature for this summary. Average values to be taken with caution (see text).

		$\sqrt{s}$	$<z>$[a]	ε	Ref.
Charm					
MARK II	D*	29	0.58±0.06		10
CLEO	D*	10.5	[0.63±0.02][b]	[0.10±0.02][b]	10
TASSO	D*	34.4	0.58±0.04	0.18±0.07	11
HRS	D*	29	0.56±0.02	0.29±0.10	28
HRS	D^o,D^+	29		0.36±0.12	28
TPC	D*	29	0.58±0.02	0.28±0.11	14
Average	D*		0.57±0.01	0.25±0.05	
MARK J	μ	34.6	0.46±0.05	(0.80 ± 0.22)[c]	28
TASSO	μ	34.4	$0.71^{+0.06}_{-0.13}$	$0.006^{+0.055}_{-0.006}$	29
TASSO	e	34.4	$0.55^{+0.11}_{-0.09}$	$0.23^{+0.39}_{-0.18}$	30
TPC	μ	29	0.55±0.08	$0.26^{+0.30}_{-0.14}$	14
DELCO	e	29	0.66±0.06	$0.071^{+0.076}_{-0.038}$	27
Average μ,e			0.56±0.03		
Bottom					
MARK II	e	29	0.79±0.09	$0.015^{+0.032}_{-0.016}$	15
MARK II	μ	29	0.73±0.18	$0.043^{+0.235}_{-0.054}$	15
MARK J	μ	34.6	0.75±0.07	(0.15 ± 0.06)[c]	28
MAC	μ	29	0.80±0.10	$0.008^{+0.037}_{-0.008}$	15
TASSO	μ	34.5	$0.81^{+0.05}_{-0.10}$	$0.0025^{+0.0310}_{-0.0028}$	29
TASSO	e	34.5	0.75±0.08	$0.022^{+0.042}_{-0.020}$	30
DELCO	e	29	0.76±0.06	$0.025^{+0.034}_{-0.015}$	27
TPC	c	29	0.83±0.06	$0.0065^{+0.0204}_{-0.0042}$	17
TPC	μ	29	0.74±0.06	$0.033^{+0.042}_{-0.023}$	14
Average Bottom			0.78±0.03	$(0.017^{+0.010}_{-0.007})$[d]	

[a] All experiments, except MARK J and MAC, use $z=2E_h/\sqrt{s}$ (see text).
[b] Values not included in the average.
[c] This is the quoted value for $\sqrt{\varepsilon}$.
[d] Value obtained from $<z_b>$.

average and do not reflect the problems discussed above. So these values are used only to show qualitatively the difference between the two fragmentation functions (Fig. 11). The best way to obtain an average would be to fit all the data together, rather than average the individual results.

4.2 Inclusive Particle Production

Two types of data can be obtained by studying production of different particles in jets: particle yields and particle spectra.

Particle yields provide information on multiplicities of $q\bar{q}$ pairs pulled from the vacuum, on the type of pairs (i.e. u:d:s ratio), on the ratio of diquark to quark pairs (relevant to one model of baryon production) and also on how many vector-vs-pseudoscalar particles (V/P ratio) are produced in the hadronization phase.

Particle spectra provide information on the fragmentation function, $D(x)$. The spectra are usually presented in distributions like

$$D(x) = (1/\sigma\beta) \ d\sigma/dx$$

with $x = 2E_h/\sqrt{s}$ (different variables are used by different experiments, $x_p=2p/\sqrt{s}$ is often used) and β is the velocity of the particle. $D(x)$ represents then the probability that a particle is produced at x in a unit interval of x.

In this section new results on single particle production and on resonance production will be presented.

A. Single Particle Production (HRS,TPC). A large fraction of the particles produced in a jet are pions that are mostly decay products of resonances. The charged pion multiplicity has been measured by TASSO [31] and now by TPC [32] to be about 10.5 out of 13 charged particles. New results discussed here are on charged particles (HRS [33] and TPC [32]), on γ and π^o production (TPC [14]), and on Λ and Ξ (TPC [14]).

Figure 12a shows the HRS inclusive charged particle spectrum (obtained in a sample of $\int Ldt=20pb^{-1}$) compared with earlier results from TASSO

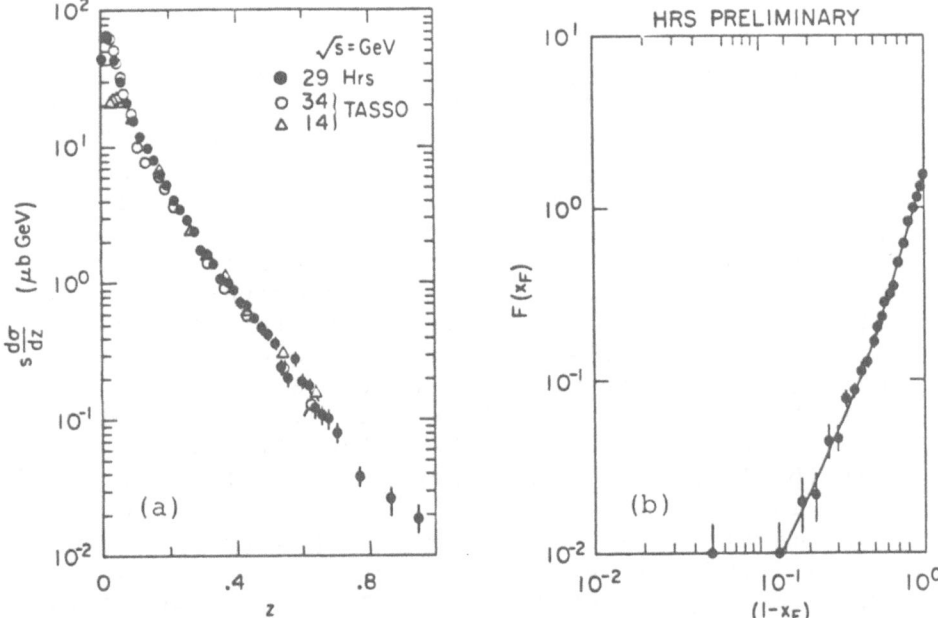

Fig. 12.   Inclusive cross sections for charged hadron
production measured by HRS [33]. (a) Scaled
cross section versus $z=2p/\sqrt{s}$, compared with
TASSO [34], (b) same data plotted in terms of
$x_F=2p_L/\sqrt{s}$, p_L being the momentum component
along the thrust axis. The vertical axis is
$F(x_F)=(1/\sigma\pi)(2E/\sqrt{s})(d\sigma/dx_F)$.

[34] at 14 and 34 GeV. The agreement is very good in
the overlap region. The HRS data extend to higher z
values. The same data are shown in Fig. 12b, plotted
in a variable more appropriate to study longitudinal
fragmentation properties. The variable is x_F =
$2p_L/\sqrt{s}$, where p_L is the component of the momentum of
the particle along the thrust axis. The data suggest
a flattening off of the spectrum at large values of x_F
that the authors attribute to production of hadrons
from light quarks. Monte Carlo studies indicate that
hadrons that are decay products of particles
containing heavy quarks contribute mostly in the
region with $x_F < 0.6$. The curves represent separate
fits to the data in the ranges x_F=0.1 to 0.6 and
x_F=0.6 to 0.9 to the function $(1-x_F)^n$. Values of n =
3.36 ± 0.01 and $n = 2.16 \pm 0.35$ were obtained for the
two regions. These values of n support the hypothesis
that light quarks are the major contributors at large
x. Another question that HRS may be able to answer
with their full sample of data is the behavior for

293

x→1, a region that so far has been inaccessible to other experiments.

TPC has recently published the normalized cross sections versus x for $\pi^{\pm}$, $K^{\pm}$, p and $\bar{p}$ [32]. The dE/dx measurements of the TPC (Fig. 13) are used to identify particles in jets. At low momentum simple counting in the π,K,p, bands will provide the number of particles of the different species. In the region of ambiguity with the electron band, electron subtraction is done by estimating the electron contamination by smooth fits to the spectrum outside the region of ambiguity. In the relativistic rise region the electron subtracted spectrum is fit to a sum of three distributions (for π,K,p) by maximum likelihood method (see inset in Fig. 13). The final corrected distributions for the three types of particles are shown in Fig. 14. The curves are calculated using the LUND model [19]. The baryon production mechanism included in this model assumes that a fraction of the times a diquark anti-diquark pair is produced in the color field rather than a quark antiquark pair. The diquark pair then combines with a quark to form a baryon. The parameters used are as follows: s/u probability ratio, $P(s)/P(u)=0.3$, and diquark to quark probability, $P(qq)/P(q)=0.09$. The integrated distributions give the following multiplicities:

$$
\begin{aligned}
N_{\pi^{\pm}} &= 10.7 \pm 0.6 \\
N_{K^{\pm}} &= 1.35 \pm 0.13 \\
N_{p,\bar{p}} &= 0.60 \pm 0.08
\end{aligned}
$$

The preliminary inclusive π^{o} differential cross section from the TPC [14] is shown in Fig. 15a. The electromagnetic calorimeter (HEX) [35] of the TPC facility consists of six modules, outside the 1.4 r.l. conventional magnet coil. Each module consists of 40 layers of lead-fiberglass-aluminum laminates with gold-plated tungsten wires in the 6 mm thick gas gap. It operates in Geiger discharge mode, in a 1 atm argon ethyl-bromide (4%) gas mix. The photon energy resolution is 16% at 1 GeV. Photons can also be reconstructed from γ conversion pairs in the 0.2 r.l. of material ahead of the TPC. The two spectra agree with each other within errors. Photons from the HEX are used to reconstruct π^{o}s at momenta $P_{\pi o} > 1$ GeV/c.

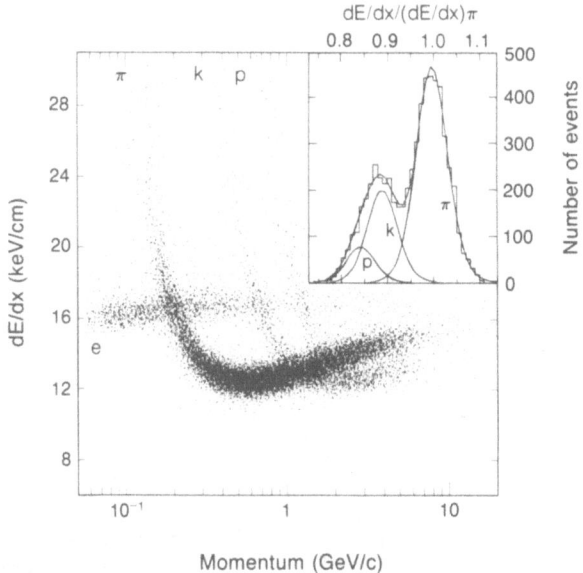

Fig. 13.   The TPC dE/dx measurement versus momentum [32].
Clean bands for π, K, p are seen. The inset
shows the ratio to the dE/dx expected for
pions, for tracks with 3.5<p<6.0 GeV/c.

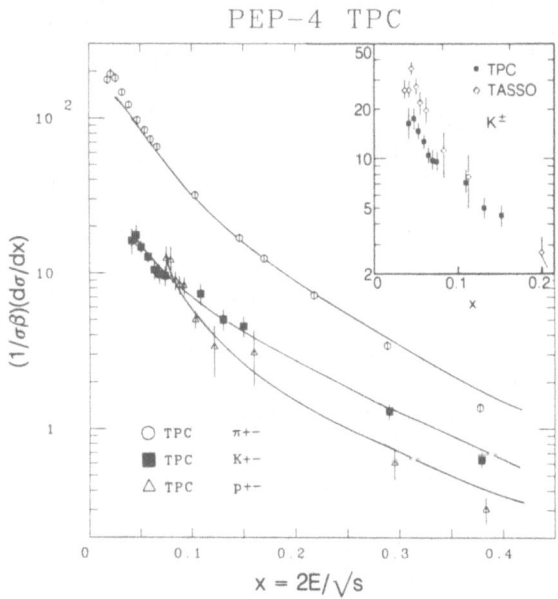

Fig. 14.   TPC normalized cross section versus x for
pions, kaons and protons at $\sqrt{s}$=29 GeV [32].
The inset shows the comparison with TASSO [31].
The curves are calculated using the LUND Model.

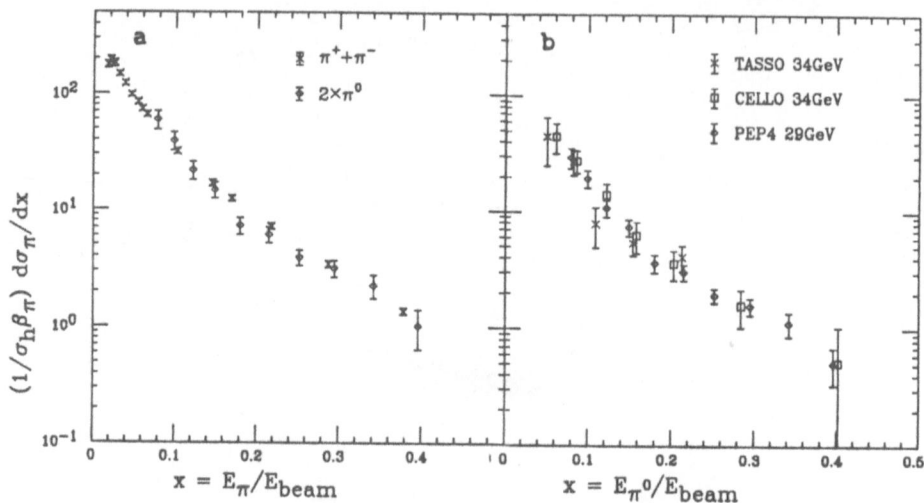

Fig. 15. Neutral pion normalized cross section measured by the TPC [14]. (a) Comparison with TPC charged pion data [32]. (b) Comparison with TASSO [36] and CELLO [37].

The measured π^0 spectrum, compared with TASSO [36] and CELLO [37] is shown in Fig. 15b. The comparison with TPC's measured charged pion spectrum [32] is shown in Fig. 15a. The ratio $2\sigma(\pi^0)/(\sigma(\pi^+) + \sigma(\pi^-)) = 0.92 \pm 0.14$ indicates a very good agreement between the two spectra. The total energy going into photons is measured to be $E_\gamma = 7.4 \pm 0.7$ GeV. The neutral pion energy is $E_{\pi^0} = 6.2 \pm 0.8$ GeV, to be compared with the charged pion energy of $E_{\pi\pm} = 12.1 \pm 0.7$ GeV. The measured π^0 multiplicty (integrated over the whole spectrum) is found to be $N_{\pi^0} = 5.3 \pm 0.7$ to be compared with $N_{\pi\pm} = 10.7 \pm 0.6$ discussed above.

The TPC Collaboration also reports results on Λ and Ξ production [14]. TASSO first observed Ξ production with a sample of 14 events [38]. The TPC results are shown in Fig. 16. The Λ sample (Fig. 16a) selection criteria are based on kinematics of the Λ decay into $\pi^- p$, on flight distance and on requirements that the dE/dx measurements (Fig. 13) for the pions and protons are consistent with the expected values for the assigned mass hypothesis. The detection efficiency for Λ reaches a maximum of 15% at 2 GeV/c and decreases to several per cent at 10 GeV/c. A total of 570 Λs and $\bar{\Lambda}$s are obtained. The Ξ sample was selected with criteria based on kinematics of the $\Xi \rightarrow \Lambda\pi$ decay, for a momentum $P_\Xi > 1$ GeV/c, and on identification of the pion by dE/dx measurement. The efficiency was about 2%. Figure 16b shows the Ξ^- signal of 16±4 events, the curves show the result of a fit with a smooth background and a Gaussian ($\sigma=18.7\pm4.3$ MeV). The shape of the background was obtained by fitting the sample of "wrong sign combinations," i.e., $\Lambda\pi^+$ and $\pi^-\bar{\Lambda}$ events (not shown).

Figure 16c shows the differential cross sections for protons, lambda and Ξ^- as measured by the TPC. The curves are again obtained with the LUND model [19]. The values of the relevant parameters are:

$$R = P(qq)/P(q) = 0.085$$
$$r = (P(su)/P(ud))*(P(d)/P(s))=0.4$$

where $P(qq)$ and $P(q)$ are the probabilities to produce a diquark or a quark pair, respectively. For r, the extra strange-diquark suppression factor, the flavors are explicitly indicated. The values of these parameters are different from the default values of the LUND program ($R=0.09$, $r=0.2$), but they are not the result of an overall fit. Particle yields obtained by integrating the distributions of Fig. 16c over the entire energy region, are

$$N_{p,\bar{p}} = 0.60 \pm 0.08$$

$$N_{\Lambda,\bar{\Lambda}} = 0.216 \pm 0.013 \pm 0.018$$

$$N_{\Xi,\bar{\Xi}} = 0.025 \pm 0.009 \pm 0.006$$

where both statistical and systematic errors are included (for the first measurement the two errors have been added in quadrature). The Λ and Ξ results are in agreement with the TASSO results [38], shown in Table III.

B. <u>Resonance Production (MARK II, TPC, DELCO)</u>. The investigation of resonance productin in e^+e^- annihilation provides important information on parton fragmentation. Most of the single particles we have discussed in the previous paragraphs are decay products of resonances, and therefore they are one more step removed from the original partons. The fraction of vector mesons produced, V/P ratio, is an

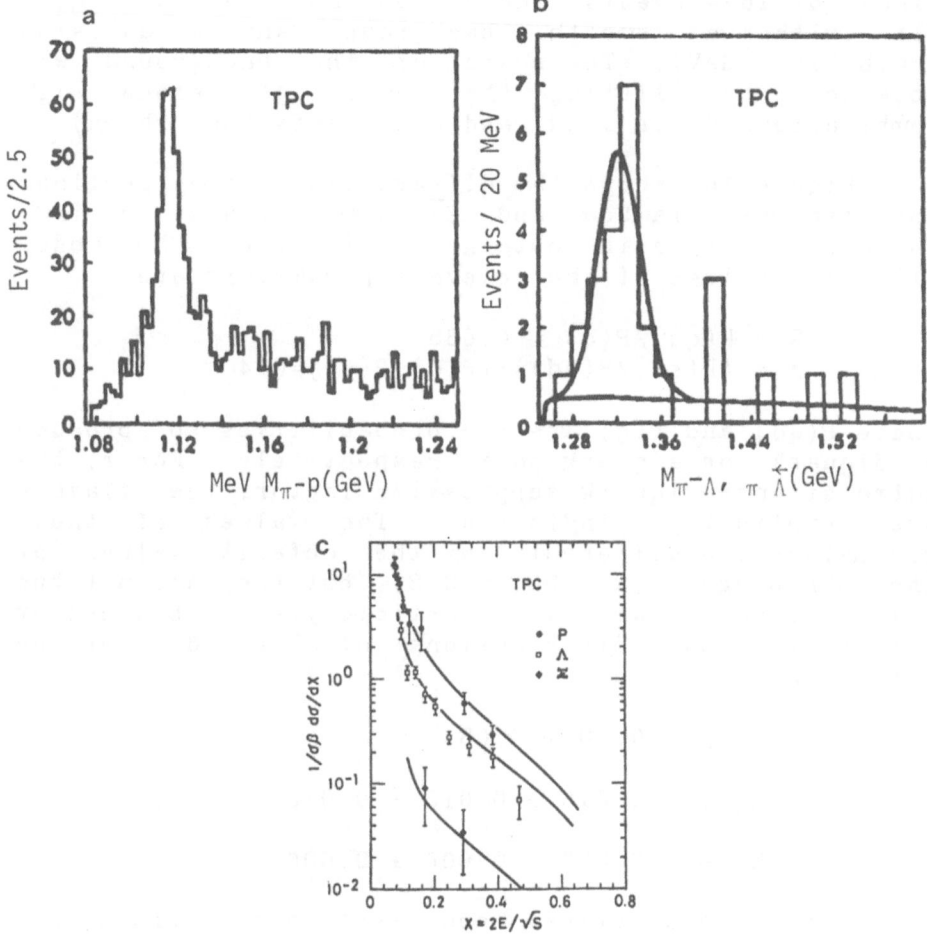

Fig. 16. Inclusive Λ and Ξ production (TPC, prelim.). (a) The π⁻p invariant mass showing the Λ signal, (b) the Λπ⁻ (and c.c.) spectra showing the Ξ signal, and (c) inclusive p,Λ,Ξ cross sections. The curves are from the LUND Model.

important parameter in many models [18–22], therefore, it is important to determine it from the data. In addition, the mass spectrum of hadrons produced in the hadronization process can be approximately predicted in some models and therefore it is important to try to measure it.

New results on ρ and K^* production are reported by MARK II [39], on K^* and ϕ production by TPC [14,40] and on ϕ production by DELCO [41].

The MARK II data [39] used to study ρ^o and K^{*o} production are shown in Fig. 17. No particle identification is used, so each track is considered alternatively to be a pion or a kaon. All possible $\pi^+\pi^-$ pairs are plotted in Fig. 17a; all possible $K^\pm\pi^\mp$ pair are shown in Fig. 17b. The $\pi\pi$ plot includes contributions from K^{*o} where the $K^\pm$ has been misidentified as a pion. The $K\pi$ plot includes contributions from K_s, ρ, ω where one of the pions has been misidentified as a kaon, as well as contribution from K^{*o} where the K and π are given the wrong mass assignment. Using only pairs with P > 1 GeV/c, a simultaneous fit to both distributions was made, adding contributions from background (a 4th order polynomial), K_s, ω, ρ and K^*. The shapes for these states are those obtained from Monte Carlo; the normalizations are left free. After background subtraction, the results are shown in Fig. 17c,d, where each contribution is drawn separately. The ρ^o and K^{*o} yields were obtained from this fit to be $N_\rho = 6255\pm780$, $N_K*=3635\pm565$. After acceptance correction for each state they obtain

$$\sigma(K^*)/\sigma(\rho) = 0.87 \pm 0.18 \pm 0.22$$

The number of ρ and K^* per event are:

$N_\rho = 0.410\pm0.051\pm0.067$ for $0.07<x<0.7$

$N_K* = 0.356\pm0.056\pm0.095$ for $0.07<x<0.7$

where the first error is from statistics, the second from systematics. The ρ^o differential cross section, shown in Fig. 17e, agrees with earlier measurements by TASSO [42]. The TASSO yield/event shown in Table III, is integrated over the whole x range.

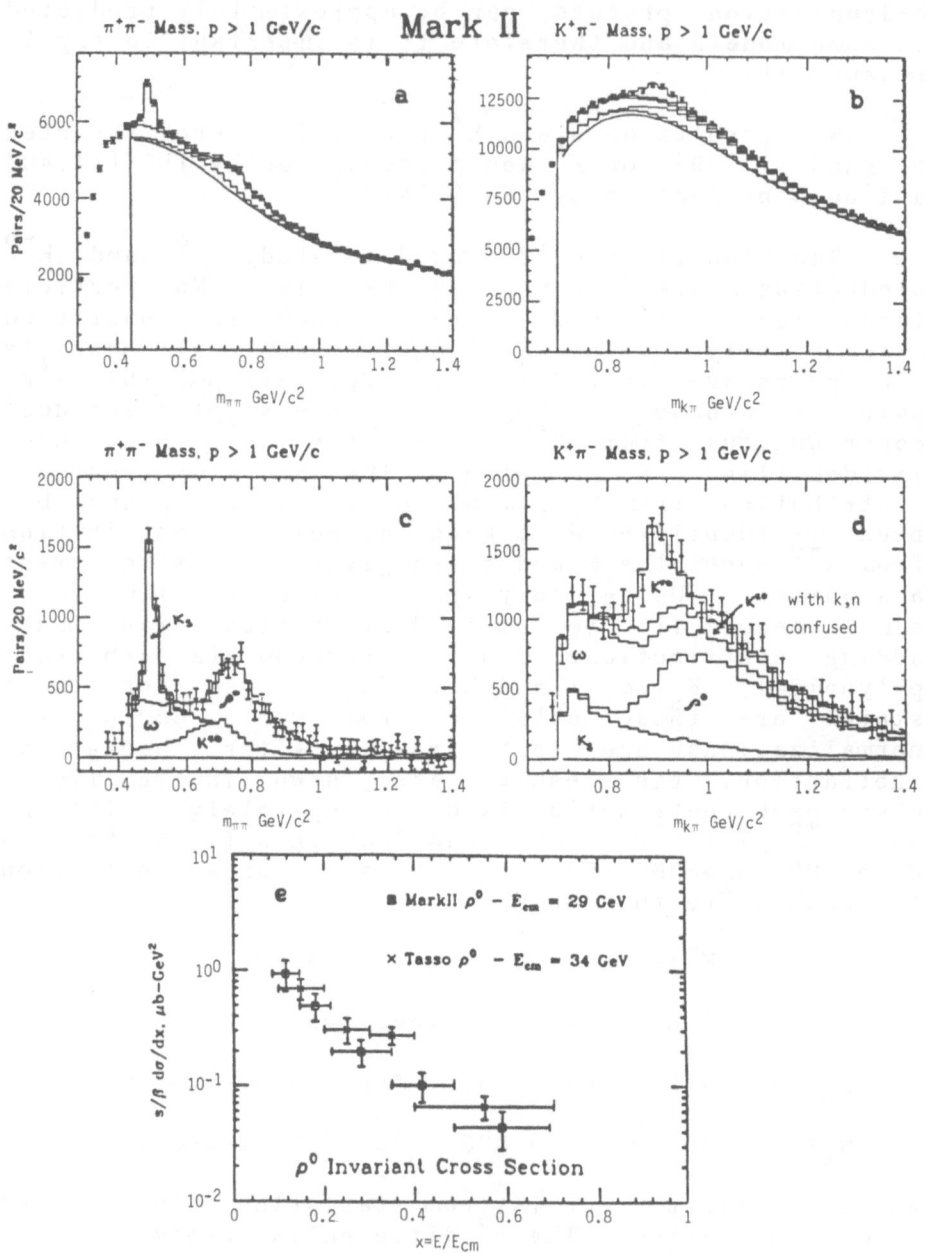

Fig. 17. Inclusive ρ and K^{*0} production reported by MARK II [39]. (a), (b) $\pi^+\pi^-$ and $K^+\pi^-$ spectra respectively; (c) and d), as (a) and (b), with background subtracted. (e) Inclusive ρ^0 cross section versus x, compared with TASSO [42].

The TPC results on K^* and ϕ production [14,40] are shown in Fig. 18. Here particle identification is used and the resonances appear as peaks in the $K^-\pi^+$ and K^+K^- invariant mass, respectively. The kaons are identified unambiguously at low momenta (Fig. 13). At other momenta and in the overlap regions a method based on the probability that a track be a kaon is used [40]. From the measured dE/dx and momentum for each track and the empirically determined dE/dx curves for each particle type (i = e,π,K,p) a χ^2 is calculated for each track hypothesis. A probability for the track to be of type i is calculated from the χ^2, and then weighted by the measured charged particle fractions [32] as function of momentum. A probability w > 70 % is then required for a kaon to be identified, assuring a purity of 70% on the average for the kaon sample. For the K^* analysis, a similar procedure is also used for the pion identification.

In a 69 pb^{-1} sample of data analyzed this way the $K\pi$ and KK invariant mass plots of Fig. 18a-c are obtained. Fits with the sum of a Breit–Wigner shape and a smooth background to the "unlike signs" distributions yield 2250±120 K^* events and 62±11 ϕ events. The r.m.s. width for the ϕ peak in Fig. 18b is 6.2±0.4 MeV, consistent with the detector resolution. The normalized inclusive cross sections versus x ($2E/\sqrt{s}$) for K^* and ϕ production are shown in Fig. 18e. The TPC results for K^* (preliminary) [14] and for ϕ [40] produced per event are as follows:

$$N_{K^*} = 0.39 \pm 0.04 \pm 0.05 \qquad 0.1<x<0.8$$

$$N\phi = 0.084 \pm 0.013 \pm 0.018 \qquad 0.<x<1.$$

The curves shown have been calculated with the LUND model [19] using s/u = 0.3 and V/P = 1. They agree with the data reasonably well.

The DELCO results on ϕ production [41] are shown in Figs. 18d and 18e. A sample of 92 pb^{-1} of data has been used. The kaons are identified above 2.5 GeV/c by a Cerenkov counter filled with isobutane with pion threshold at 2.5 GeV/c. Figure 18d shows the invariant K^+K^- mass distribution obtained when both tracks are identified by the Cerenkov counter. An enhancement is observed at the ϕ mass, containing 26 ϕ with an estimate background of 14 events. The width of the peak is σ=10 MeV, consistent with the

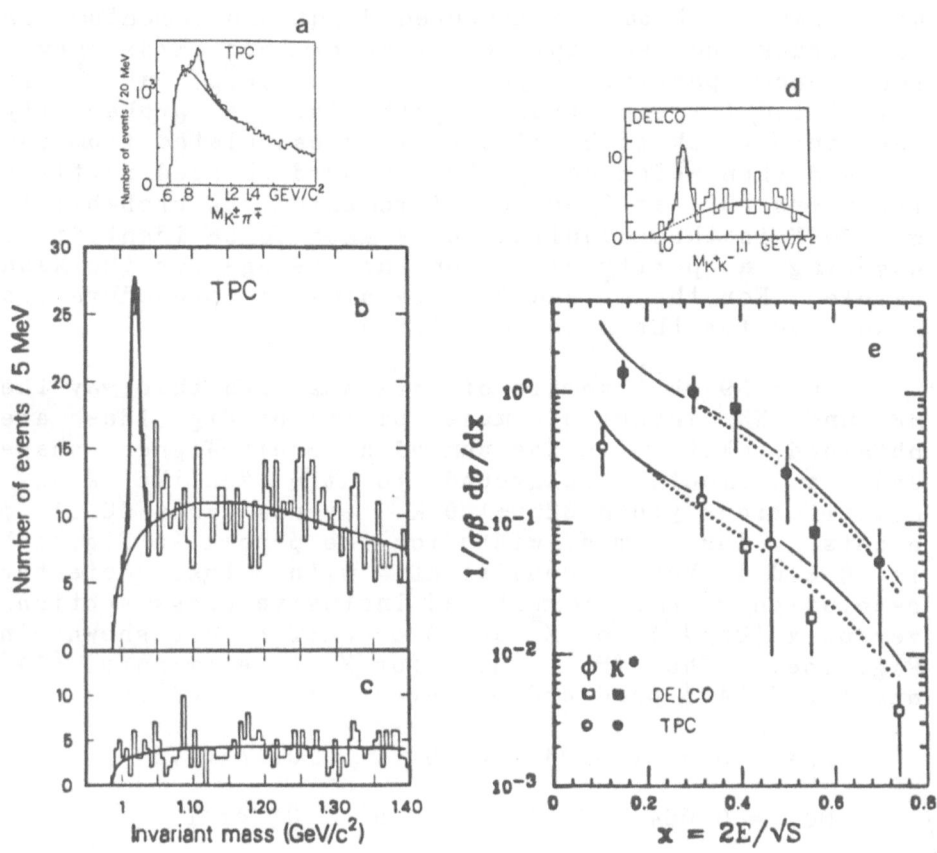

Fig. 18. Vector meson production data. (a) $K^{\pm}\pi$
invariant mass from the TPC [14]. (b) and (d)
K^+K^- invariant mass from the TPC [40] and DELCO
[41] respectively. (c) Same charge KK mass
distribution from the TPC. (e) Normalized
inclusive cross sections for K^* and ϕ
production. The curves are calculated with the
"Symmetric" LUND Model (solid lines) and with
the "Standard" LUND Model (dashed lines).

Table III. Particle yield measured at PEP and PETRA in e^+e^- Collisions in the $\sqrt{s}$ = 29-34.5 GeV range.

Particle	Particles/Event	Experiment	Ref.	Remarks
$\pi^\pm$	10.7 ±0.6	TPC	32	
	10.3 ±0.4	TASSO	31	
π^0	6.1 ±2.0	TASSO	36	
	5.2 ±1.8	CELLO	37	
	5.3 ±0.7	TPC	14	
η	0.72±0.10±0.18	JADE	46	
ρ^0	0.73±0.06	TASSO	42	
	0.410±0.051±0.067	MARK II	39	x=0.1-0.7[a]
$K^\pm$	1.35±0.13	TPC	32	
	2.0 ±0.2	TASSO	31	
$K^0+\bar{K}^0$	1.6 ±0.1	TASSO	43	
	1.45±0.08±0.15	JADE	44	
$K^{*0}+\bar{K}^{*0}$	0.39±0.04±0.05	TPC	14	x=0.1-0.8
	0.356±0.056±0.095	MARK II	39	x=0.07-0.7
ϕ	0.084±0.013±0.018	TPC	40	
D^0	(0.48±0.17)[b]	HRS	11	x > 0.5
D^+	(0.21±0.09)[b]	HRS	11	x > 0.5
$D^{*\pm}$	0.31±0.08±0.11	TASSO	11	x > 0.3
$D^{*\pm}$	(0.25±0.09)[b]	HRS	11	
$F^\pm$	(0.015±0.003±0.005)/B	TASSO	45	x>0.3 B=BR(F→$\phi\pi$)
$p+\bar{p}$	0.60±0.08	TPC	32	
	0.8 ±0.1	TASSO	31	
$\Lambda+\bar{\Lambda}$	0.216±0.013 0.018	TPC	14	
	0.28±0.04	TASSO	43	
Ξ^-	0.025±0.009±0.008	TPC	14	
	0.026±0.008±0.009	TASSO	38	

[a]In this x interval the TASSO result agrees very well with the MARK II.

[b]Calculated by us, using R_h = 4.0.

resolution. The region of x covered is 0.35<x<0.8. The values obtained for the normalized cross section are plotted in Fig. 18e and are in agreement with the TPC data.

C. Summary. New results on single particle and resonance production from PEP add information to the wealth of data collected so far in e+e- annihilation on jet production.

Table III is a compilation of the published measurements on particle yields and the new results presented here on $K^{\pm}$, $\pi^{\pm}$, π^{o}, p, Λ and Ξ. Information on resonant states is particularly important because resonances are more directly related to the quantum numbers and the production mechanisms of the original partons. The new results reported here on ρ^{o}, K^{*o} and ϕ production are also included in Table III.

The inclusive x distributions for charged particles reported by HRS (Fig. 12) shows a flattening off at large x, providing more information on light quark fragmentation. The x distributions for charged particles reported by the TPC (π,K,p,Λ,Ξ in Figs. 14 and 16) are in agreement with expectations from the LUND Model with parameters

$$u/s = 0.3 \qquad\qquad qq/q = 0.085$$
$$V/P = 1 \qquad\qquad (sd/ud)*(d/s) = 0.4$$

The LUND Model [19] also agrees well with the K^{*o} and ϕ distributions from the TPC, shown in Fig. 18.

4.3 Characteristics of Proton Production in Jets (TPC)

Baryon production in quark and gluon fragmentation provides important information for the study of jet formation. Baryons, because of their large mass, experience little momentum degradation in resonance decays, thus they carry information that closely reflects the distribution of primary fragmentation products.

The TPC group has done a detailed study of proton production [47]. The data are shown in Fig. 19. Figure 20a shows the TPC measured fraction [32] of different charged particle species in jets at $\sqrt{s} = 29$ GeV. The general trend is that at high momentum the kaon and proton fractions (f_K and f_p respectively) increase, whereas the pion fraction decreases. This is well described by the LUND Model [19]. The increase of f_p with momentum is shown in more details in Fig. 19a, where now the "Standard" LUND is used with the value $qq/q = 0.075$. The dependence of f_p on p_T, transverse momentum to the sphericity axis, is shown in Fig. 19b-e for different ranges of total momentum p. Here again a sharp increase with p_T is observed. The same effect has been observed by the EMC collaboration in deep inelastic lepton-nucleon scattering [48]. For momenta above 3.7 GeV/c the p_T dependence flattens out.

The high value of f_p at large p_T and the large f_p seen in T decays [49] has stimulated speculations that proton production increase is related to gluon emission [50]. The TPC results can be explained with an alternative mechanism, higher proton production in gluon jets not being necessary to explain the data [47]. The alternative mechanism is simply that f_p is constant at the level of primary produced hadrons and the apparent increase in f_p with p and p_T is due to the effects of resonance decays. Resonances produce a large number of low energy pions that dilute the proton fraction in jets. The dashed lines in Fig. 19 show the fraction of protons among the primary hadrons as calculated by the "Standard" LUND Model. Here f_p appears to be constant with p_T, whereas among all the particles (solid line) f_p has a strong p_T dependence. Only at large values of p (Fig. 19e), where resonance decay effects are suppressed, the flat p_T dependence survives.

Another aspect of proton production studied by the TPC group is the p_T distribution of protons at a fixed rapidity. Figure 20b shows the proton cross section $(1/\sigma)d\sigma/(dydp_T^2)$ for the $|y|<1$ region. The corresponding pion cross section is also shown for comparison. The proton p_T distribution can be represented by a Gaussian shape with $\sigma = 0.55 \pm 0.04$ GeV/c. The full lines represent the prediction of the LUND Model that has been tuned to describe the mean p_T of the observed mesons. It reproduces the large p_T of the protons with no additional assumptions. The protons are likely to reflect the p_T distribution of primary hadrons, whereas resonance decays soften the observed pion distributions.

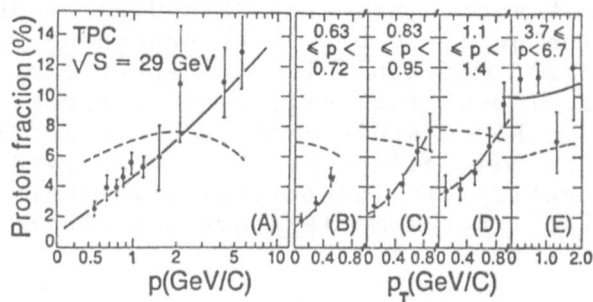

Fig. 19. Fraction of protons among charged particles, measured by the TPC [47], (a) as a function of momentum, (b)-(e) as function of p_T for fixed total momentum. Curves are calculated with the LUND Model for primary hadrons (dashed) and all hadrons (solid).

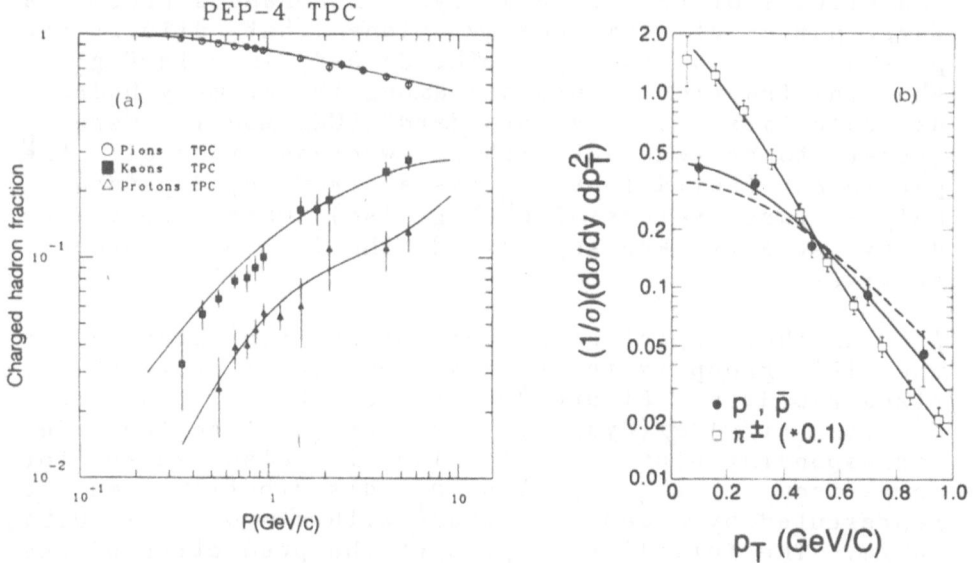

Fig. 20. TPC proton production studies [32,47]. (a) Charged particle fractions as function of momentum, (b) p_T spectra for fixed rapidity ($|y|<1$) for protons and pions.

The proton p_T distribution can also be used to test baryon production mechanisms. If a proton is made out of a diquark and a quark (both of which have the same p_T distribution) the proton mean p_T is expected to be $\simeq \sqrt{2}$ $(p_T)_q$. If the proton is made out of three independent quarks (all three having the same mean p_T), the mean proton p_T is expected to be $\simeq \sqrt{3}$ $(p_T)_q$. Included in Fig. 20b are curves obtained with the LUND Monte Carlo for the diquark-quark case (full lines) and with a modified version of the program emulating the three independent quark hypothesis (dashed lines). The diquark-quark model for the proton fits the data very well. The three independent quarks model fits the data also, although it appears to be slightly disfavored.

In conclusion, new data on proton production reported by the TPC [47] points out the effects of resonance decays in the apparent increase of proton production at large p_T. To establish whether this increase is also due to increased proton production in gluon jets, more data are needed and careful studies must be done to disentangle kinematic effects. The study of p_T distributions of protons shows a broader distribution for protons than for pions, this effect can also be explained by resonance decays. Finally with the present statistics, the data cannot distinguish between a baryon production mechanism with three independent quarks and a diquark-quark mechanism, although the latter seems to be favored by the data.

4.4 Two Particle Correlations (TPC)

The study of two particle correlations in e^+e^- annihilation provides information on the hadronization process. The initial state is a massive virtual photon that decays into a $q\bar{q}$ pair (plus possibly a hard gluon) and then in the color field stretched between the two original quarks, other quark pairs are created. At a later time these quarks are transformed into observable hadrons. Correlations between particles that are far apart in rapidity (LRC) provide information on the initial partons and can test flavor compensation. Correlations between particles that are close in rapidity space (SRC) provide information on local conservation of quantum numbers.

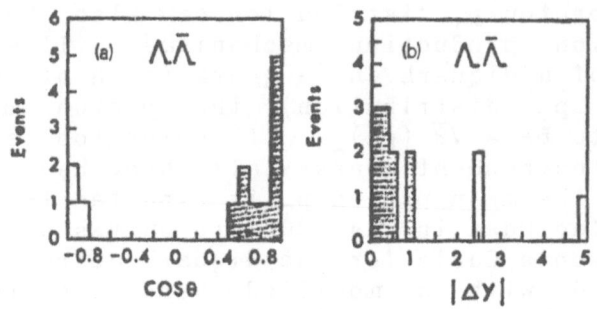

Fig. 21.　TPC preliminary results on Λ-Λ̄ correlations
[14].　　(a)　Opening angle between the Λ and Λ̄.
(b) Distribution of rapidity difference between
the Λ and Λ̄.

The TPC collaboration is reporting Λ-Λ̄
correlation studies and, using particle
identification, presents results on flavor
correlations: $\pi\pi$, KK and Kπ [14].

A. <u>Baryon Correlations</u>.　Baryons are made of
three quarks in a color singlet state. In order to
understand the mechanism of baryon production, it is
important to know if baryon number is conserved
locally, i.e., baryon pairs are produced close in
rapidity space or if baryon number is conserved
globally, i.e., there is no definite correlation
between the baryons produced in a jet. TASSO [51] and
JADE [52] have reported data on pp and p̄p correlations
that show evidence for local baryon compensation.

The TPC collaboration reports new data on Λ-Λ̄
correlations.　In the sample of Λ and Λ discussed in
Sec. 4.2-A there are fourteen events with two entries.
They are divided as follows:

Λ-Λ̄　　　13 events
Λ-Λ　　　 1 event
Λ̄-Λ̄　　　 0 events

Background contribution to the Λ-Λ̄ sample is estimated

to be 2.2±0.7 events as calculated by the TPC detector simulation that uses the LUND Monte Carlo generator. The distribution of the opening angle between the pair is shown in Fig. 21a. Ten of the thirteen pairs have a small opening angle, which is interpreted to mean that they are produced in the same jet. For each Λ a rapidity is calculated using the sphericity axis as the jet axis. The distribution of $|\Delta y|$, the rapidity difference between the Λ and $\overline{\Lambda}$, is shown in Fig. 21b. On the average, for the ten pairs produced in the same jet, $|\Delta y|$ = 0.4± 0.3. This again shows evidence for local baryon compensation.

B. <u>Flavor Correlations</u>. Because of particle identification over a large solid angle, the TPC can study correlations between particles of known flavor. Kaon and pion identification is done using the method discussed in Sec. 4.2-B. For this analysis the requirements on purity are more stringent, and tighter cuts are imposed on the data. The final kaon sample has purity >75% for all rapidity values, whereas the pion sample has purity >90%.

The two particle charge correlations reported by the TASSO collaboration [53] showed evidence for the charged nature of primary partons. A similar technique is used by the TPC collaboration to study $\pi\pi$, KK and Kπ correlations. For each particle in a jet, a rapidity is calculated with respect to the sphericity axis. Then a test particle "a" at a given rapidity y' is chosen and for all other particles of type "b" at rapidity y, a charge density is calculated

$$D(ab)=(N_b(\text{opp. Q as "a"}) - N_b(\text{same Q as "a"}))/N_a$$

The values of D(ab) versus y, for a test particle in the rapidity interval y'=1.5 to 4, are plotted in Fig. 22 for the three different cases: $\pi\pi$, KK, πK. The distributions are corrected for acceptance, particle misidentification and radiative effects, using our detector simulation in conjunction with the LUND generator [19]. The corrections are a function of rapidity as well as of the angle of the sphericity axis with respect to the beam line. Error bars in the distributions include the systematic uncertainties associated with these corrections.

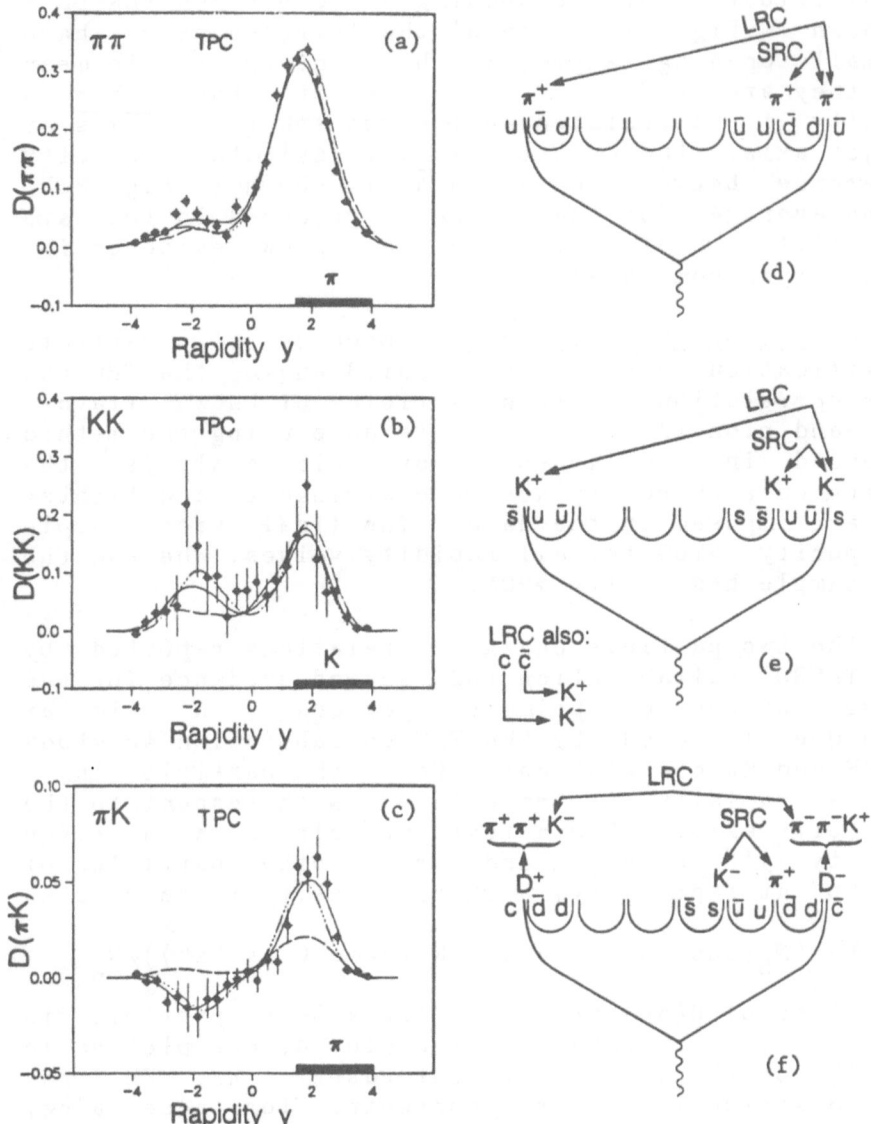

Fig. 22. Flavor tagged charge density, $D(ab)$, versus rapidity y for (a) $\pi^+\pi^-$, (b) K^+K^- and (c) $\pi^\pm K^\mp$. The test particle has $y'=1.5-4.0$. The solid (dashed) curves show the expected correlation from the LUND Model with (without) c and b quarks. The dot–dashed curves were calculated with the Webber Model [20]. (d)–(f) Show some mechanisms responsible for LRC and SRC.

The $\pi\pi$ correlations in Fig. 22a are similar to the charge correlations reported by TASSO [53] and by Drijard et al. [54] in an ISR experiment. There is a large peak (SRC) at about y=2 due to resonance decays and local charge conservation. A LRC is evident as well; this is evidence for charged primary partons. Integration of the distribution shows that the $\pi^{\pm}$ charge is compensated by another pion (95±2)% of the time.

The KK distribution in Fig. 22b is qualitatively similar to that of Fig. 22a for $\pi\pi$. However, for the KK distribution the relative strengths of the SRC and the LRC are almost equal, whereas for the $\pi\pi$ case the SRC is stronger than the LRC. For the SRC the only known resonance is the ϕ, which, as seen in Table III, is produced at a significant rate and contributes to the effect along with local strangeness compensation. The LRC are due to primary $c\bar{c}$ and $s\bar{s}$ quarks, as indicated in Fig. 22e.

Finally, the Kπ correlation of Fig. 22c illustrates the effect of heavy quark decays. The LRC shows a significant same sign charge correlation. This can be explained with charm production as illustrated in Fig. 22f. The curves drawn in Fig. 22c are the predictions of the "symmetric" LUND model (solid line), of the same model without heavy c and b, quarks (dashed line) and of the Webber [20] model (dot-dashed line). It is clear that most of the observed SRC and LRC are due to heavy quarks. The Webber model and the LUND Model agree with the data for all cases: $\pi\pi$, KK and Kπ. It is not possible to distinguish between these two models with the information contained in these graphs.

In conclusion, short and long range flavor correlations have been observed by the TPC collaboration. Short range correlations indicate that quantum numbers are locally conserved. Long range correlations show that the primary quarks carry flavor and that heavy quarks produce sizable effects.

5. SUMMARY

New results from TPC, DELCO, HRS, MAC and MARK II have been presented.

Limits on production of $Q = 1/3$, $2/3$ particles in jets have been improved by about an order of magnitude by the TPC.

A new limit on the mass of the supersymmetric electron has been set by MAC at 35 GeV for zero photino mass.

From the electroweak interference term, new values of the axial weak coupling constant to charm and bottom have been reported by HRS and TPC. The world average of these type of measurements yield (see Table I):

$$a_c = 1.25 \pm 0.34 \qquad a_b = -1.03 \pm 0.24$$

where the Standard Model predicts $a_c = +1$ and $a_b = -1$.

New results on heavy quark fragmentation have been reported by DELCO and TPC. The Peterson et al. [25] form of the fragmentation function fits the data. The ratio of the ϵ parameters obtained from an average of the data is:

$$\epsilon_b / \epsilon_c = 0.07 \pm 0.04$$

in agreement with what is expected from the ratio of the squared masses for charm and bottom ($\cong 0.10$).

New data on jet composition and particle spectra have been reported by HRS, TPC, MARK II, and DELCO. Production of Ξ has been observed which, in conjuction with Λ and proton information, can test some of the assumptions in baryon production models. New data on yields and spectra of K^*, ρ, and ϕ add information relevant to the understanding of the s/u, V/P, and Φ/K^* production ratios.

TPC has made some detailed studies of proton production and emphasized the effect of resonance decays in the apparent increase of the proton fraction with momentum and transverse momentum. The proton p_T distribution is measured to have a Gaussian shape with $\sigma = 0.55 \pm 0.04$ GeV/c.

New results have been presented on two particle correlations, that are expected to provide deeper understanding of the fragmentation process. Local baryon compensation has been observed by the TPC in

$\Lambda-\bar{\Lambda}$ production. Finally, flavor correlations have been studied by the TPC in e^+e^- annihilation. Both short range and long range flavor compensation are observed in the $\pi\pi$, KK, and Kπ charge-weighted rapidity correlations.

This work was supported by the Department of Energy under contract No. DE-AC03-76SF00098.

REFERENCES

1. H. Aihara et al. (TPC), Phys. Rev. Lett. 52,168 (1984).
2. H. Aihara et al. (TPC), Search for Q=2/3, 1/3 Particles Produced in e^+e^- Annihilation, Lawrence Berkeley Laboratory Report LBL-17551 (March 1984), to be published.
3. J. Burger, Proc. Int Symposium on Lepton and Photon Interactions at High Energies, edited by W. Pfeil (Univ. of Bonn, 1981), p. 115, includes JADE results. J.M. Weiss et al. (MARK II), Phys. Lett. 101B, 439 (1981). M.C. Ross et al. (PEP-14), Phys. Lett. 118B, 199 (1982).
4. M. Piccolo (MAC), Proc. Summer Institute on Particle Physics, Ed. by P. McDonough, SLAC (1983) p. 673. See also, S. Yamada, Proc. Int. Symposium on Lepton and Photon Interactions. at High Energies, ed. by D. Cassell and D. Kreinick, Cornell University (1983), p. 525.
5. P. Fayet, Phys. Lett. 117B,460 (1982). J. Ellis and J.S. Hagelin, Phys. Lett. 122B, 303 (1983). T. Kobayashi and M. Kuroda, Report TPR-83-22 (1983), to be published. J.D. Ware and M.E. Machacek, Northeastern Univ. Report NUB-2626 (1984), to be published.
6. E. Ma and J. Okada, Phys. Rev. Lett. 41,287(1978); K.J.F. Gaemers, R. Gastmans et al., Phys. Rev. D19, 1605 (1979).
7. G. Arnison et al. (UA1), Phys. Lett. 126, 398 (1983); B. Sadoulet et al. (UA1), Proc. Int. Symp. On Lepton and Photon Interactions at High Energies, ed. by D. Cassel and D. Kreinick, Cornell University (1983), p. 27.
8. P. Bagnaia et al.(UA2), Phys. Lett. 129B,130 (1983).

9. G. Feldman et al., Phys. Rev. Lett. 38,1313(1977).
10. J.M. Yelton et al. (MARK II), Phys. Rev. Lett. 49
    430 (1982). P. Avery et al. (CLEO), Phys. Rev.
    Lett. 51, 1139 (1983).
11. M. Althoff et al. (TASSO), Phys. Lett. 126B, 493
    (1983); also Ref. 13 for asymmetry measurements.
    S. Ahlen et al. (HRS), Phys. Rev. Lett. 51, 1147
    (1983) and private communication.
12. W. Bartel et al. (JADE), presented to the Cornell
    Conference, see Ref. 13.
13. B. Naroska, Proc. Int. Symp. on Lepton and Photon
    Interactions at High Energies, ed. by D. Cassel
    and D. Kreinick, Cornell Univ. (1983), p. 97.
14. H. Aihara et al. (TPC), preliminary results.
15. M.E. Nelson et al.(MARK II), Phys. Rev. Lett. 50,
    1542(1983); M.E. Nelson, LBL-16724 (1983), see
    also Ref. 13. E. Fernandez et al. (MAC), Phys.
    Rev. Lett. 50, 2054 (1983), and H.S. Kaye,
    Stanford Linear Accelerator Center Report SLAC-262
    (1983), see also Ref. 13 for weak couplings.
16. B. Adeva et al. (MARK J), Phys. Rev. Lett. 51, 443
    (1983), see also Ref. 13. M. Althoff et al.
    (TASSO), DESY 83-121 (1983), to be published.
    W. Bartel et al. (JADE), unpublished, see Ref. 13
    for weak couplings.
17. H. Aihara et al. (TPC), "Prompt Electron
    Production in e⁺e⁻ Annihilation at 29 GeV,"
    LBL-17545 (March 1984), to be published.
18. P. Hoyer et al. , Nucl. Phys. B161, 349 (1979);
    A. Ali et al., Phys. Lett. 93B, 155 (1980).
19. B. Andersson, G. Gustafson, G. Ingelsman and
    T. Sjostrand, Physics Reports 97, 31 (1983); also
    T. Sjostrand, Comp. Phys. Com. 27, 243 (1982), and
    28, 229 (1983). Unless explicitly stated, we use
    the "Symmetric" LUND Model (JETSET 5.2). When
    referring to "Standard" LUND we mean JETSET 4.3.
20. B.R. Webber, CERN preprint TH-3713 (1983);
    G. Marchesini and B.R. Webber, CERN preprint
    TH-3525 (1983).
21. R.D. Field, S. Wolfram, Nucl. Phys. B213,65(1983).
22. T.D. Gottschalk, Nucl. Phys. B214, 201 (1983);
    also Caltech preprints CALT-68-1052, 1059 (1983).
23. J. Dorfan, Proc. of the 1983 Int. Symp. on Lepton
    and Photon Interactions, ed. by D. Cassel and
    D. Kreinick, Cornell Univ. (1983), p. 686.
24. M. Suzuki, Phys. Lett. 71B, 139 (1977).
    J.D. Bjorken, Phys. Rev. D17, (1978).

25. C. Peterson et al., Phys. Rev. <u>D27</u>, 105 (1983).
26. G. Goldhaber et al. (MARK II), Phys. Lett. <u>69B</u>, 503 (1977).
27. D.E. Koop et al. (DELCO), Phys. Rev. Lett. <u>52</u>, 970 (1984).
28. J.M. Weiss (HRS), Proc. Summer Inst. on Particle Physics, Ed. by P. McDonough, SLAC (1983), p. 643.
29. B. Adewa et al. (MARK J), Phys. Rev. Lett. <u>51</u>, 443 (1983). M. Althoff et al. (TASSO), DESY 83-121 (1983).
30. M. Althoff et al. (TASSO), preliminary results, quoted by J. Dorfan, Ref. 23.
31. M. Althoff et al. (TASSO), Z. Physik <u>C17</u>, 5 (1983).
32. H. Aihara et al. (TPC), Phys. Rev. Lett. <u>52</u>, 577 (1983).
33. D. Bender et al. (HRS), "Study of Quark Fragmentation at 29 GeV: Global Jet Parameters and Single Particle Distributions," ANL-HEP-PR-84-09 (1984), to be published.
34. R. Brandelik et al. (TASSO), Phys. Lett. <u>114B</u>, 65 (1982),
35. H. Aihara et al. (TPC), Nucl. Inst. and Meth. <u>217</u>, 259 (1983).
36. R. Brandelik et al. (TASSO), Phys. Lett. <u>108B</u>,71 (1982).
37. H.J. Behrend et al. (CELLO), Z. Physik <u>C20</u>, 207 (1983).
38. M. Althoff et al. (TASSO), Phys. Lett. <u>130</u>, 340 (1983).
39. H.M. Schellman et al.(MARK II), preliminary result.
40. H. Aihara et al. (TPC), "Φ Meson Production in e^+e^- annihilation at 29 GeV," LBL-17616 (March 1984), to be published.
41. DELCO Collaboration, preliminary result.
42. R. Brandelik et al. (TASSO), Phys. Lett. <u>117B</u>, 135 (1982).
43. As reported in S.L. Wu, DESY 83-007 (1983) and Proc. Summer Inst. on Particle Physics, Ed. by A. Mosher, SLAC (1982), p. 555.
44. W. Bartel et al.(JADE), Z. Physik <u>C20</u>,187 (1983).
45. M. Althoff et al. (TASSO), Phys. Lett. <u>136B</u>, 130 (1984).
46. W. Bartel et al. (JADE), Phys. Lett. <u>130B</u>, 454 (1983).
47. H. Aihara et al. (TPC), "Characteristics of Proton Production in Jets From e^+e^- Annihilation at 29 GeV," LBL-17705 (April 1984), to be published.

48. J.J. Aubert et al. (EMC), Phys. Lett. <u>135B</u>,225 (1984).

49. H. Albrecht et al., Phys. Lett. <u>102B</u>,291(1981); also CLEO Collaboration, as reported by J. Dorfan, Ref. 23.

50. G. Schierholz, M. Teper, Z. Physik <u>C13</u>, 53 (1982).

51. M.Althoff et al. (TASSO), Z. Physik <u>C17</u>, 5 (1983); M.Althoff et al., Report DESY 84-4 (to be published).

52. W. Bartel et al.(JADE), Phys. Lett. <u>104B</u>,325 (1981).

53. R. Brandelik et al. (TASSO), Phys. Lett. <u>100B</u>, 357 (1981).

54. D. Drijard et al. (ACCDHW), Nucl. Phys. <u>166B</u>, 233 (1980).

DISCUSSION

MACHACEK:

One must be careful in quoting bounds on selectron masses when photinos are produced in the final state. Such bounds as quoted apply only for photinos with masses < 1 GeV/c^2 while models predict photino masses as large as 8-10 GeV/c^2, in which case the bounds must be relaxed.

BARBARO-GALTIERI:

Yes, you are correct. The limits I showed are for photino masses ~ 0. For larger photino masses the selectron masses lower limit gets smaller. Thank you for bringing that point up.

GOLDHABER:

It would be interesting to investigate the relative production of di-quarks (e.g. uu(4/3e) and dd(2/3e)) by e^+e^- by detecting both protons and neutrons. Is this feasible?

BARBARO-GALTIERI:

It is very difficult to detect neutrons with the present detectors. The di-quark model, (the model that substitutes a quark from a vacuum with a di-quark pair that then combines with another quark to make a baryon), has a probability for the di-quark to be produced proportional to the di-quark mass and not to the charge so dd and uu would be equally probable if $m_u \sim m_d$.

TRILLING:

The ε_c values for PEP experiments are all higher than the average you quoted. What was the CLEO value which brought the average down?

BARBARO-GALTIERI:

The CLEO value was $\varepsilon_c = 0.10 \pm 0.02$. The TASSO value is $\varepsilon_c = 0.18 \pm 0.07$. The PEP results are larger than these two. The overall average is 0.12 ± 0.02 due to the fact that the CLEO average has small errors. If I leave out the CLEO value, the average is 0.25 ± 0.05.

KLEINKNECHT:

What is the total multiplicity of hadrons at your energy (to be compared with the 0.6 ± 0.1 protons/event)?

BARBARO-GALTIERI:

The total charged multiplicity is 13, to which you have to add ~ 6 for π^0.

HADRO AND PHOTOPRODUCTION OF HEAVY FLAVOURS

G. Bellini

CERN - Geneva - Switzerland

Dipartimento di Fisica dell'Università - Milano - Italy

Istituto Nazionale di Fisica Nucleare - Sezione di
Milano - Italy

ABSTRACT

The present situation of hadro and photoproduction of heavy
flavours is summarized and the more recent experimental results are
discussed. The main characteristics of the experiments now on the
floor or in preparation are also briefly reviewed.

1. INTRODUCTION

The physics of the heavy flavours is very young and still
unexplored for a large part. But, even for the aspects already
studied, the experimental data are few and mostly consisting of
inclusive measurements. Actually it is very hard to disentangle
from the inclusive data experimental results which be model independent. Then the open problems are several.

In my talk I will concentrate my attention on the aspects not
yet settled, which received further improvements by recent experimental results. In the first part I will discuss parameters,
Branching Ratios, total cross sections and production mechanisms
of some charmed particles; in addition I will spend few words on
what we can expect from the experiments now on the floor. The
second part includes the cross sections of Beauty hadro and photoproduction and a discussion on the four experiments now in preparation at the fixed target machines[1].

2. NEW RESULTS ON CHARMED PARTICLES

The more recent results on charmed particles concern the F meson, the charmed baryons and the D decay Branching Ratios.

2.1. THE MASS OF THE F MESON

In Table I the parameters of the F meson, measured in the more recent experiments, are summarized.

TABLE I

F MESON

Experiment	Decay modes	Mass (MeV)	Lifetime $(10^{-13}$ sec)
WA4[2] (γ; 20-70 GeV) (Ω)	$\eta 3\pi$ $\eta' 3\pi$ $\eta \pi^{\pm}$	2020±12±20	–
WA57[3] (upgraded Ω)	$\eta \pi^{\pm}$	2017±13	–
NA1[4] (γ; 80-150 GeV) (live target)	$K^- K^+ \pi^{\pm}\pi^0$ (7 eV) $\eta\pi^+\pi^-\pi^{\pm}\pi^0$ (9 eV)	2050±30	$5.0^{+5.0}_{-2.5}$
NA11[5] (hadrons– 175 and 200 GeV)	$K^- K^+ \pi$ (6 events; 3 events are ambiguous with Λ_c)	1975±4	3.2±1.8
E531[1d] (ν)(emulsion)	$K^+ \bar{K}^0 \pi^- \pi^+$ (1 eV) $K^+ K^- \pi^0 \pi^0$ (1 eV) $\pi^+ \pi^- \pi \pi^0$ (2 eV) $K^+ \bar{K}^0$ (1 eV) $\phi\pi$ (1 eV) $\phi\pi^+\pi^0$ (2 eV)	1938±14	$2.6^{+1.2}_{-0.8}$±0.2
CLEO[1d] (CESR)	$\phi\pi^{\pm}$	1970±5±5	–

Three experiments show mass values larger than 2 GeV, but two of them make use of the same apparatus, the CERN Omega Spectrometer, even if upgraded in WA57. These results are obtained essentially by studying the (η+nπ) channels, with the exception of the NA1 experiment where the $K^-K^+\pi^\pm\pi^0$ channels are also analyzed, but without any constraint on the K^+K^- invariant mass.

The NA11 experiment obtained 1957±4 MeV for the F mass value, in agreement with the E531 and CLEO results; these three experiments study mostly the (φ+nπ) decay channels.

If we take into account also the old DASP results which found M-2.03±0.06 GeV for the F decaying into ηπ±, I should be tempted to conclude that experiments studying (η+nπ) decay modes obtain values larger that 2 GeV, while the results of experiments analyzing (φ+nπ) channels are in favour of a mass lower than 2 GeV.

This disagreement could be a consequence of biases introduced by the different cuts adopted for the analysis of the different channels, but also the event misinterpretation could play an important role. Just as an example, a wrong π-K identification of the D-decay products gives origin to a large mass around 2020 MeV!

Among the different experiments at fixed target machines I would like to mention NA11, which uses a technique based on a vertex detector, designed with six planes of Silicon microstrip chambers. The charge-division read out of these devices allows a precision of ∿5 μm in measuring the track coordinates in the plane, making possible a good reconstruction of the secondary verteces (see Fig.1).

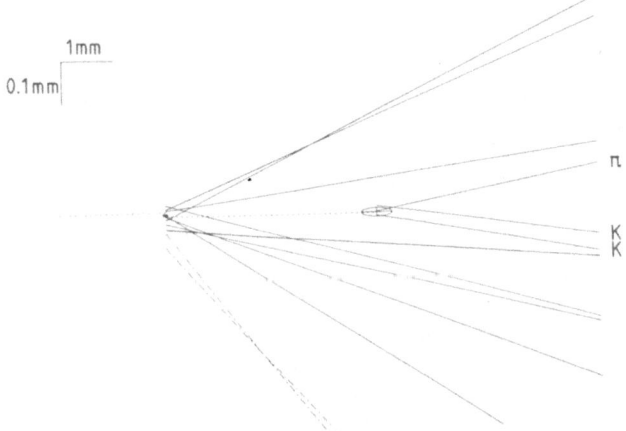

Fig.1. Display of the vertex region for an event including a decay
F→KKπ(exp. NA11). The dashed-dotted lines represent tracks
seen in the Silicon microstrip detector, but not in the
forward spectrometer.

The ($\eta+n\pi$) and ($\phi+n\pi$) decay channels are the Cabibbo-favoured final states, suggested by the current spectator quark model. They are studied in several experiments, as shown in Table I.

Some indications in favour of the existence of multipion decay modes, as expected from the annihilation diagram, have been obtained from the E531 and NA1 experiments. In E531 two events show a secondary vertex, which can be interpreted as a short lived particle decaying in the $\pi^+\pi^-\pi^-\pi^0$ channel; their masses are consistent with a value of 1980 MeV. NA1 data show a bump in a 5 charged π's channel (see Fig.2), with an invariant mass peaked around 2020 MeV; its width is larger than the experimental resolution and therefore, even in this case, the presence of the F in this channel cannot be considered as firmly established.

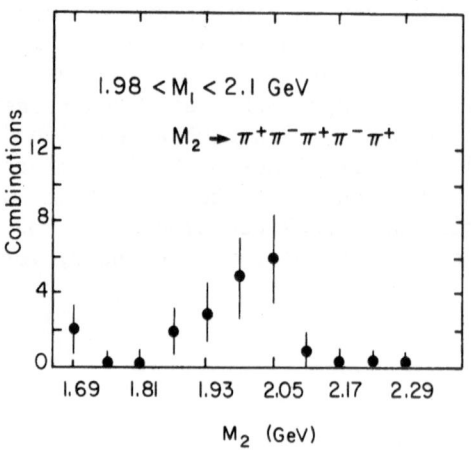

Fig.2. Invariant mass of the $\pi^+\pi^-\pi^+\pi^-\pi^+$ system when the associated mass of the F^- candidate falls in the interval: 1.98 – 2.1 GeV (exp.NA1).

Before closing this paragraph concerning the F meson, I would like to mention a result obtained by a French-Soviet Collaboration using the hydrogen bubble chamber Mirabelle[6]. This result, not so recent, but forgotten often, concerns the evidence of a clear $\phi\pi^+(\to K^+K^-\pi^+)$ signal, 7 standard deviations over the background, at 2145±10 MeV of invariant mass (Γ=25 MeV). This evidence (see Fig.3) is obtained after applying the following cuts: M(K^+K^-) constrained in the 1017-1022 MeV interval; x_F (K^+K^-)$\geq$ 0.2, which corresponds to the kaon fragmentation region; $P_T(K^+K^-\pi^+)$ >500 MeV/c. The x_F cut is introduced because ϕ is shown to be mostly produced in the kaon fragmentation region.

The interpretation of this signal as a state consisting of ($c\bar{s}$) quarks, like the F+ meson, is reinforced by the absence of any evidence in the $\phi\pi^-$ channel. This can be explained taking into

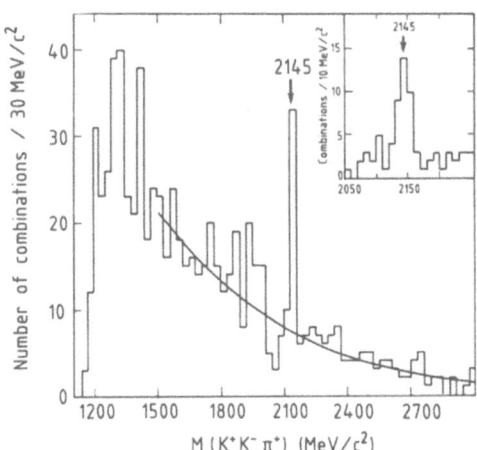

Fig. 3. $\phi\pi^+$ mass distribution for $x_F(K^+K^-) \geqslant 0.2$ and $P_T(K^+K^-\pi^+) > 500$
MeV/c (see text;ref.6). At the top of the figure the peak
region is plotted with smaller bins.

account the number of quarks in common between the incident and the
outgoing states. The incident K^+ is a $(u\bar{s})$, F^+ a $(c\bar{s})$ and F^- a $(\bar{c}s)$
state.

The interpretation of this signal, which has been confirmed
also by the BEBC data obtained with a 70 GeV K^+ beam, is still
enigmatical.

2.2. CHARMED BARYONS

The experimental knowledge of the charmed baryons was recently
improved by the evidence of A^+ (cus) and T^0 (css) production,
obtained at CERN with a 135 GeV/c Σ^- beam (exp. WA62[7]).

The signal concerning the A^+ baryon has been found studying
the $p\pi^-K^-\pi^+\pi^+$ channel, with the $p\pi^-$ constrained into the Λ^0 mass
region. π, K and p are identified, in principle; particles not
identified either as K^- nor as Λ decay products are considered as
π. The mass value is placed at 2460 MeV with 23 MeV of FWHM, which
corresponds to the experimental resolution of the apparatus. The
peak (fig.4) contains 82±16 events above a background of 147±5,
corresponding to ~6 standard deviations and to a $\sigma \cdot B =$ 5.3±2.0
μb/Be nucleus. The sample of 82±16 events belongs to the ($\Lambda K^{*0}(892)$
π^+) channel for a 40 % and to the ($\Sigma^{+*}(1385) K^-\pi^+$) for a 30 %.

Possible reflections due to π^--K^-, π^+-K^+ and $K^--\bar{p}$ misidentification have been studied and then have been excluded. No evidence
has been found in the $pK^-K^-\pi^+\pi^+$ and $pK^-\bar{K}^0\pi^+$ channels.

A preliminary lifetime measurement gives $5.0^{+3.0}_{-2.0} \cdot 10^{-13}$ s[8].

The evidence for the T^0 production[8] is restricted to four

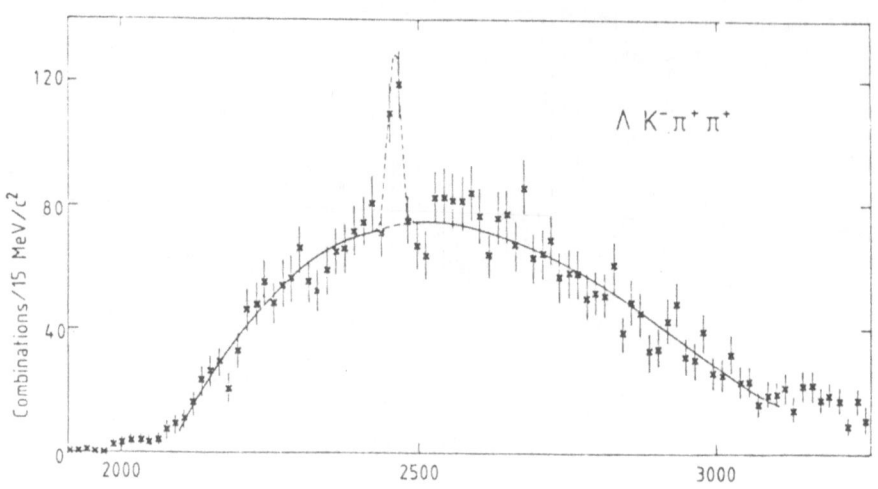

Fig.4. ($\Lambda K^- \pi^+ \pi^+$) effective mass containing 4002 combinations made
from 3352 events. The full line is a polynomial of order 3;
the dotted line is a gaussian with a width equal to the
experimental resolution.

events of the channel $\Xi^- K^- \pi^+ \pi^+$ with the $K^- \pi^+$ mass in the $K^*(892)$
region, which are clustered together. If one takes into account
that the $u\bar{u}$, $d\bar{d}$, $s\bar{s}$ production scales as $u\bar{u}$: $d\bar{d}$: $s\bar{s}$ = 1:1: $\frac{1}{10}$ and
that the acceptance of the apparatus for Ξ^- production is smaller ($\sim\frac{1}{2}$)
than for Λ^0, the sample reduction from 82 A^+ events to 4 T^0 events
seems reasonable.

Still in the domain of the charmed baryon physics, peaks
corresponding to $\overline{\Lambda}_c$ and $\overline{\Sigma}_c^{--}$ have been isolated by the CDHW Collaboration working at the ISR SFM . The small mass difference between
the ground states (D, Λ_c) and the excited states (D^*, Σ_c) have
been exploited to reduce the combinatorial background, which
otherwise seems quite overwhelming. The signals shown in Fig. 5
were obtained using only events with K and p fully identified and
by requiring $\Delta m = m (K^+ \bar{p} \pi^- \pi^-) - m (K^+ \bar{p} \pi^-) < 200$ MeV.

2.3. BRANCHING RATIOS

The main results on the Branching Ratios concern new measurements of the two, three and four prong decay of the D mesons.
In Table II the B.R. recently found by the LEBC-EHS Collaboration[10] for some D^+ decay modes are compared with previous measurements of Mark II. Actually the B.R. obtained by the EHS-LEBC
Collaboration are normalized to the total three and four prong
decay rates; the absolute Branching Ratios are obtained by means
of the B.R. for the three and four prong channels measured at
Spear.

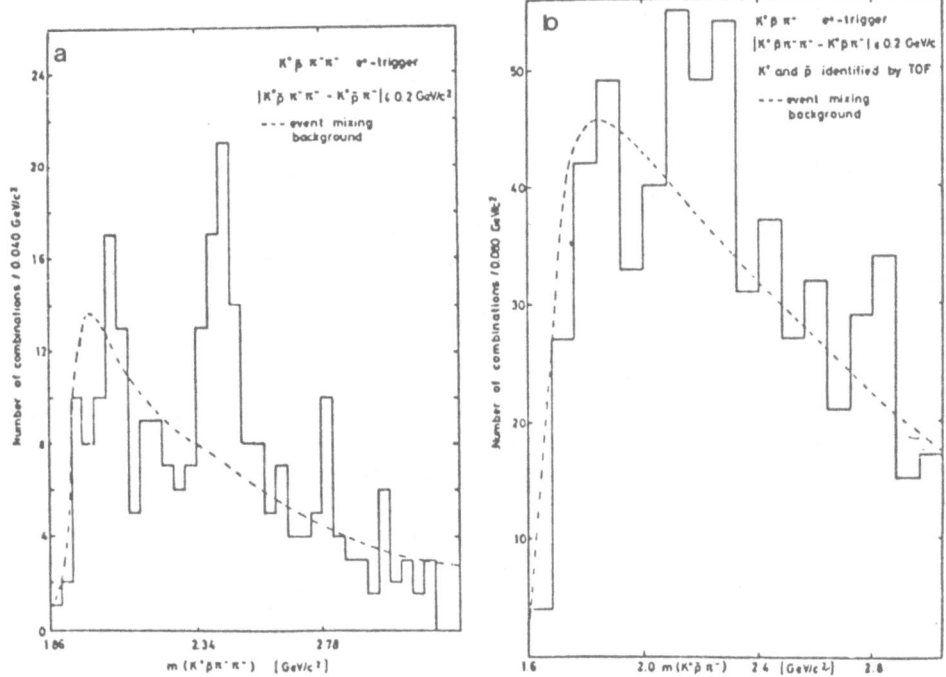

Fig. 5. $K^+\bar{p}\pi^-\pi^-$ and $K^+\bar{p}\pi^-$ mass distribution (CDHW Coll.[9]). The signals indicate production of $\bar{\Sigma}_c^{--}$ (a) with the subsequent decay into $\bar{\Lambda}_c^- \pi^-$ (b) see text.

TABLE II

Channel	Branching Ratios	
	LEBC-EHS	Mark II
$D^+ \to K^- \pi^+ \pi^+$	14^{+6}_{-5}	6.3 ± 1.5
$D^0 \to K^- \pi^+$	13^{+11}_{-7}	3.0 ± 0.6
$D^0 \to K^- \pi^- \pi^+ \pi^+$	10 ± 4	8.5 ± 2.1

The EHS apparatus is not able to reconstruct the π^0's and to identify π, K and protons. Nevertheless, following Montecarlo simulations, possible reflections of F and Λ_c decays and of D mesons decaying to final states including neutral particles have been found to give a contribution of at most a fraction of event in the D peak.

The acceptance of the EHS apparatus covers only the positive X_F region.

The new values obtained by the LEBC-EHS Collaboration are definitively larger than the Mark II measurements; they push down the total cross sections which were measured from these decay modes.

A second important lot of Branching Ratios were recently remeasured by the Tagged Photon beam Spectrometer Collaboration at FNAL[11]; they studied the ratio between $D^0 \to K^- \pi^+ \pi^0$ and $D^0 \to K^- \pi^+$, and the contribution of the $K^{*-} \pi^+$, $K^{*-} \pi^0$, $K^- \rho^+$ final states to the $K^- \pi^+ \pi^0$ decay mode. Combining these measurements with the current average value of the $D^0 \to K^- \pi^+$ B.R., the absolute values quoted in Table III are obtained. These new results are fully compatible with the old data of Mark II, with the exception of the $(K^- \rho^+)$ decay B. R., which is lower in the T.P.S. measurement.

I would like to recall that the ratios:

$$\frac{B(D^0 \to K^{*-} \pi^+)}{B(D^0 \to \overline{K}^{*0} \pi^0)} \text{ and } \frac{B(D^0 \to K^- \rho^+)}{B(D^0 \to \overline{K}^0 \rho^0)}$$

have to be consistent with 2 for a $I = 1/2$ final state. While the first ratio is $\simeq 2$ for both the T.P.S. and Mark II results, the second one is slightly nearer to 2 in the new measurement.

The sample of D^0's used by the Fermilab TPS has been obtained looking for the events $\gamma + p \to D^{*+} +$ anything, where $D^{*+} \to \pi^+ + D^0$ and D^0

Table III

Channel	Branching Ratios	
	T.P.S.	Mark II
$D^0 \to K^- \pi^+ \pi^0$	10.3 ± 3.7	–
$\to K^- \rho^+$	$3.2^{+2.3}_{-1.8}$	7.2 ± 3.0
$\to \overline{K}^{*0} \pi^0$	$0.9^{+1.4}_{-0.9}$	$1.4^{+2.3}_{-1.4}$
$\to K^{*-} \pi^+$	$3.4^{+3.9}_{-2.8}$	$3.6^{+0.6}_{-0.1}$
$\to \overline{K}^0 \rho^0$	–	$0.1^{+0.6}_{-0.1}$

decays either to $K^-\pi^+$ or to $K^-\pi^+\pi^0$. The selection is carried out: first applying the cut $|M(K^-\pi^+(\pi^0))-M_D| < 60$ MeV, then plotting $\Delta M = M(K^-\pi^+\pi^+(\pi^0)) - M(K^-\pi^+(\pi^0))$ and accepting only the events contained in the clear peak which is shown at $\Delta M \simeq 0.1454$ GeV. This peak is due to $D^{*+} \rightarrow \pi^+ D^0$.

3. CROSS SECTIONS AND PRODUCTION MECHANISMS

Until now the experimental estimations of the total cross sections for charmed particle production are few and not very reliable because they suffer for many theoretical assumptions. The procedures for measuring the total cross sections include:

i)   Large corrections for the small acceptance and efficiency of the experimental apparatus. These corrections strongly depend on the production and decay mechanisms, which are assumed.

ii) Extrapolations from single state production cross sections and few decay channels. Then relative production rates and Branching Ratios are needed.

iii) An extrapolation from nuclear target (when used) to Hydrogen (an A^1 dependence of the cross sections are currently used).

3.1. PHOTOPRODUCTION CROSS SECTIONS

In Fig.6 some of the more important evaluations of the total production cross sections of charmed particles are reported and compared with the photon-gluon fusion model (full line[1d].).

In the low energy region the data refer to real incident photons. The more recent results concern the WA58 data [12] (black points),which make use of emulsions associated to the CERN Omega Spectrometer. The selection is based upon the identification of a secondary vertex in the emulsion, without requiring a full reconstruction of the event. The data are corrected for the scanning only and not for the spectrometer inefficiencies. Therefore these cross sections are probably underestimated.

The cross sections for $D^{*\pm}$ production (black squares) are measured by the FNAL TPS[11], following the procedure described in 2.3.. The ratio between these values and the total cross sections is < 0.25 for the full energy range which has been explored.

In the high energy region the cross sections are calculated for incident virtual photons. The data reported in Fig.6 concern the European Muon Collaboration[13] and the FNAL Muon Collaboration[14].

As an example of the procedure used to calculate these cross sections I report here the more significant steps of the selection carried out by the E.M.C.. Dimuon and trimuon events are measured to obtain $d^2\sigma(\mu N \rightarrow c\bar{c}X)/dQ^2 d\nu$. The dimuon and trimuon data are mutually consistent and the ratio of their yields in the overlap region is consistent with the known semileptonic B.R. of the D mesons. Using the current Hand convention, the $\sigma(\gamma^* N \rightarrow c\bar{c}X)$ are extracted and the real photoproduction of open charm is obtained

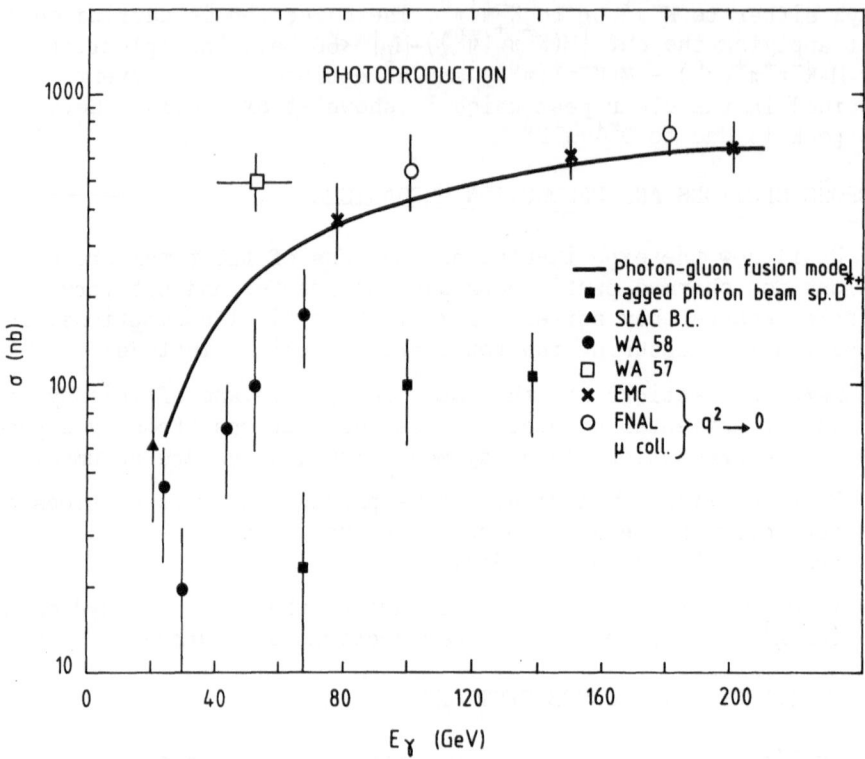

Fig. 6. Photoproduction cross sections per nucleon for charmed
particles (see text).

by extrapolating the $\sigma(\gamma^*N)$ to $Q^2=0$ [Q^2= 4-momentum transfer to the
virtual photon]. At each value of $\nu = p \cdot q/M$ [q = 4-vector of the
virtual photon] a fit is performed on the Q^2 dependence using the
form

$$\sigma_\gamma(Q^2) = \frac{\sigma_\gamma(Q^2=0)}{(\frac{1+Q^2}{M^2})^2}$$

where $\sigma_\gamma(q^2=0)$ and M are free parameters in the fit. Corrections
for acceptances and cuts are calculated using the photon-gluon
fusion model, which gives a good representation of the data
[Q^2, ν, P_T^2 distributions in the measured ranges]. The form exp
(AZ_D) is used as fragmentation function; a fit performed to the
shape of the dimuon Z_μ distribution in the [0.-0.4 range] gives
the best value A=1.6±0.6 [$Z_D = P_D^{lab}/P_{ch._quark}^{lab}$; $Z_\mu = E_\mu^{lab}/\nu$;
E_μ=decay muon energy in the Lab. system]. Finally in the range
$4M_C^2 < s < 4M_D^2$, 1/6 of the cross section was assumed to be lost to
bound charm (J/ψ) production and the rest to be open charm. The
average semileptonic B.R. of the D is taken = 8.2%.

328

The general trend of the cross section vs E_{inc} seems to be consistent with the previsions of the photon gluon fusion model.

3.2. D* PHOTOPRODUCTION

Measurements on the D* photoproduction are carried on by the T.P.S. Collaboration[11]. The energy of the incident photons falls in the range 40-160 GeV. The events are divided in two samples: the single recoil proton events, which show only one charged track in the vertex detector surrounding the target (diffractive like events), and the target fragmentation triggers, with many recoiling tracks.
The main results obtained from the single recoil proton events can be summarized as follows:

i) The production angle of the D* in the photon fragmentation system shows a distribution strongly anisotropic (see Fig.7). In other words the P_T of the D* is much smaller than expected for an isotropic decay of the forward state.

ii) The forward state is not simply $D^*\overline{D}$ or $D^*\overline{D}^*$, but includes extra particles. $D^*\overline{D}^*$ or $D^*\overline{D}$ are produced alone only in the 11±7 % of the events. Therefore the state produced is not usually a low mass $c\bar{c}$ state.

From the target fragmentation events the upper limit $\sigma(\gamma p \to \overline{D^{*-}}\Lambda_c)$ <60 nb is measured. This limit is lower than the values obtained at the CERN Omega Spectrometer, with an average photon beam energy of ∿50 GeV; the lower energy probably favours the production at the slower vertex.

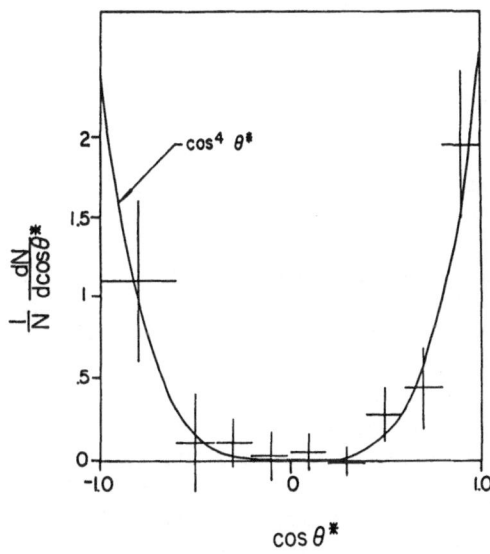

Fig. 7. Distribution of the D* production angle in the photon fragmentation system CMS (exp.T.P.S.; ref.11).

3.3. PROPERTIES OF THE CHARMED PARTICLES HADROPRODUCED

a) A dependence

An A^α dependence of the total production cross section on the nuclear target atomic weight is normally assumed. I would like to recall that in the incoherent production of non-charm particles, α is a strong function of x_F. Therefore the acceptance limitations of the apparatus and the actual x_F behaviour strongly influence the measurements of the A dependence.

The J/ψ absorption cross section was measured using the vector dominance relation

$$\frac{d\sigma}{dt} \, (\gamma p \rightarrow Vp) = \frac{\alpha}{16 \cdot \pi} \cdot \frac{4\pi}{f_V^2} \cdot \sigma_T^2 \, (Vp \rightarrow Vp)$$

$\sigma(\psi N)$ was calculated from the measured values $|d\sigma/dt|_{t=0}$, neglecting the real part of the forward scattering amplitude and using the lepton-pair partial width $\Gamma_{\ell\ell}$=4.8 GeV[15]. A very small $\sigma(\psi N)$ was found ($\approx$1 mb), which leads to a linear dependence on A(α=1), confirmed also by the ratios of yields in Be and Ta[16].

For the open charm states the comparison of the $(D\bar{D})$ yields in LEBC (H)[10] and in BIBC (A=17.1)[17] gives α=1. This conclusion is valid only at $x_F > 0.25$, which corresponds to the acceptance range of the apparatus.

The A dependence will be probably clarified by the FNAL E400 experiment[18], which uses a structured target, consisting of a 400 µm thick W layer, followed by 10 Silicon wafers, 200 µm thick, followed again by a 4000 µm Be sheet. The Si telescope decides in which part of the target the interaction has taken place. It is so possible to compare the different rates in exactly the same experimental conditions.

b) x_F behaviour

The data obtained with π^- beams show large discrepancies. The LEBC-EHS experiment (E_{inc} = 360 GeV)[19] and the CFRS Collaboration (E_{inc} = 270 GeV-Fe target)[20] find two different behaviours for D^+, $\bar{D}^0$ and D^-,D^0 production. If the dependence on x_F is parametrized using the formula $(1-x_F)^n$, n=6±3 for D^+, $\bar{D}^0$ (consistent with a central production) and =1±1 for D^-, D^0 (showing a leading effect).

These results are not confirmed by BIBC (E_{inc} = 340 GeV)[1d] and NA11 (E_{inc} = 175-200 GeV)[21], which do not observe slope change. They obtain n$\approx$1 for all data.

For incident protons we expect a strong x-dependence because the D-mesons do not share any valence quark with the proton. In fact the CFRS beam dump experiment (350 GeV p beam; Fe target) obtains n=5.0±0.8, assuming only D production[22], and the BCF Collaboration at the ISR ($\sqrt{s}$=63 GeV) has indications in favour of n$\approx$3 for D^+ production[23].

330

On the other hand the LEBC EHS (350 GeV p beam) obtains n=1.8±0.8 for D production[24]. This result is amazing because the x_F behaviour for D production is in this way even flatter than what it was found for the K^0_S production at high energy (n≈5), though the K^0 has a d-quark in common with the proton. Nevertheless the LEBC data are confirmed indirectly by WA62 (135 GeV Σ^- beam; Be target) which gives n=1.7±0.7 for the A^+ production[7].

3.4. TOTAL CROSS SECTIONS FOR HADROPRODUCTION OF CHARMED STATES

a) $\underline{\pi \text{ beams}}$

The total hadroproduction cross sections suffer from large uncertainties because most of the experiments measure only a tiny fraction of the total cross section. Just as an example, the LEBC-EHS experiment detects a fraction of the total cross section corresponding to ∿7 %; in addition the particle identification is very weak in this apparatus.

The acceptance of NA11 is few orders of magnitude lower than the EHS acceptance, mostly due to the trigger requirements, which demand a semileptonic decay for one charmed particle, and to the off-line analysis, which selects a particular decay mode for the associated charmed state, in addition to the geometrical ineffi-ciency of the apparatus.

Therefore the total cross sections obtained from these data are strongly dependent on the assumptions adopted for the x_F behaviour, for the correlations between $c\bar{c}$ particles and for the Branching Ratios.

Using the new LEBC-EHS Branching Ratios, discussed in 2.3., the total cross sections for $D\bar{D}$ production are:

$$\sigma(D\bar{D}) = 12.2^{+7.5}_{-3.8} \text{ μb from LEBC-EHS (360 GeV } \pi^-)$$

$$\sigma(D\bar{D}) = 13.4\pm6 \text{ μb at 175-200 GeV}$$

and $\qquad = 8.3\pm6$ μb at 120 GeV from NA11 (π^- beam).

These results agree with the prompt single muon experiment at Fermilab (278 GeV π^-)[20], which gives:

$$\sigma(D) = 8.2\pm0.9 \text{ μb}$$

$$\sigma(\bar{D}) = 9.5\pm0.7 \text{ μb}$$

assuming 8 % for the average semileptonic B.R. and 60 % $\bar{D}\rightarrow K\mu\nu$, 40 % $\bar{D}\rightarrow K^*\mu\nu$.

All these three experiments detect only the positive x_F region.

Total cross sections for $D^{*\pm}$ production are measured by NA11 at two different energies: 120 and 175-200 GeV. The D^* yield is very near to the D production at both energies ($\sigma(D^*)/\sigma(D)=0.9^{+3.1}_{-0.6}$).

b)   p beams

I would like to divide the data obtained with p beams in two parts: the first concerns the experiments carried out at the CERN SPS/FNAL energies, the second at the ISR energy.

The LEBC EHS experiment (360 GeV incident p), using the new B.R., find, for the $D\bar{D}$ production cross section, $15.5^{+8.2}_{-4.6}$ μb[10].

This value can be compared with the results of the beam dump experiments, recently analyzed in a very coherent way by Aziz and Gurtu[25]. Three typical experiments are compared: the prompt neutrino data of the CDHS and FMOWW Collaborations, collected at a 400 GeV p beam with Cu and W targets, and the prompt muon experiment CFRS, carried out at a 350 GeV p beam with a Fe target. The acceptances are very different for the different experiments: 1.8 mrad for the CDHS apparatus and 38 mrad for FMOWW and CFRS.

The total cross sections of the reactions

$$p+A \rightarrow (\Lambda_c\bar{D}) + X$$

and

$$p+A \rightarrow (D\bar{D}) + X$$

are investigated; the critical part of the procedure concerns the choice of the semileptonic B.R. and of the X and A dependences.

The D and Λ_c semileptonic B.R. measured at SPEAR were used (10.0±3.2 % for average D^+/D^0 and 4.5±1.7 % for Λ_c) and equal $\Lambda\pi\ell\nu(K\pi\ell\nu)$ and $pK\ell\nu$ ($K^*\pi\ell\nu$) decay rates for Λ_c(D) were assumed.

The A and x dependences were chosen in analogy with the strange particle production, ΛK and $K\bar{K}$. Therefore the baryon-meson production was considered a quasi-diffractive process, whereas the meson-meson production was treated as a central process (see Table IV).

The cross sections so obtained range from 11±4 μb (CFRS) to 29±9 (CDHS) for $D\bar{D}$ and from 17±6 (CFRS) to 44±13 (CDHS) for $\Lambda_c\bar{D}$. The values obtained for $\sigma(D\bar{D})$ are not in contrast with the LEBC data, but they could also agree with the cross section obtained by means of the old Mark II B.R.($\sigma(D\bar{D}) = 40^{+16}_{-7}$ μb).

Before to leave the SPS energies I would like to spend few words about two "anomalous" cross sections[16] connected with the signal seen at Mirabelle (see 2.1.) and with the A^+ production

TABLE IV

$D\bar{D}$		A^1	$(1 - |x|)^7$
	Λ_c		$(1-|x|)\cdot\{a_1+q_2\cdot x^2\cdot\exp(-a_3\cdot x^2)\}$
$\Lambda_c\bar{D}$		$A^{2/3}$	
	$\bar{D}$		$(1-|x|)^3$

detected by WA62 (see 2.3). Assuming 0.1 for the $\phi\pi$ decay mode B.R. of the 2145 MeV signal, the $\sigma \cdot B = 11\pm3$ µb measured by Mirabelle corresponds to a total production cross section of 110 µb.

The $\sigma \cdot B = 5.3\pm2.0$ µb/Be nucleus measured for the A^+ production gives a total cross section >50 µb/nucleon if we adopte the very conservative assumption that the $A^+ \to \Lambda K\pi\pi$ B.R. is 0.1 and the A and x_F dependences are well reproducted by the expression A^1 and $(1-x_F)^{1.7}$.

These cross sections are evidently "out of range" in comparison with the cross sections measured for the $D\bar{D}$ and $\Lambda_c\bar{D}$ production.

At the ISR energies three different experimental approaches for the measurement of the charm cross section can be compared: high p_T electrons, dilepton events, direct detection of charmed states.

For the high p_T electrons, it is assumed that most of these leptons come from decays of charmed particles. The measurement of the corresponding cross section depends on a lot of assumptions concerning the fragmentation functions, the decay distributions, the type of particles and the Branching Ratios. The cross section, so estimated, falls in the interval 200-300 µb.

The dilepton data are recently analyzed by Fischer and Geist[26]. Taking into account that the semileptonic decay of charms and beauty can give contribution to the dilepton production also at large effective masses, they adjust the charm and beauty cross sections to saturate the lepton pair $d\sigma/dM$ below 4 GeV. Using reasonable assumptions on the $D\bar{D}$ correlation length, the main inclusive distributions (as the x_F behaviour, the $<p_T>$ dependence on mass and s, the lepton angular distributions) can be reproduced and cross sections $\sim$100 µb at the ISR and $\sim$10 µb at the SPS energies are found for charm production.

Concerning the experiments which search for a direct evidence of charmed particles, I would mention the result of the BCF Collaboration[23]. Using for $D\bar{D}$ a $(1-x_F)^3$ dependence (the central production seems favoured) and a flat distribution for Λ_c (together with a $(1-x_F)^3$ for the associated $\bar{D}$), the following cross sections are obtained: $\sigma(D\bar{D}) \simeq 850$ µb and $\sigma(\Lambda_c^+\bar{D}) \simeq 200$ µb. Any other assumption on the x dependences tend to increase these values. The CDHW Collaboration[9], working at the same energies, obtain a $\sigma(cc)$ ranging from 300 to 900 µb.

The present situation is summarized in Fig.8. The cross sections found at the SPS/FNAL energies are clustered around few tens of µb, if we forget the two "anomalous" measurements concerning the F^* and A^+ particles. At the ISR energies the data are strongly spread out. In general more model independent measurements are needed.

In the low energy region, where the comparison is possible, charms seem produced with the same rate at π and p beams. This disagrees with what we expect from processes dominated by fusion mechanisms, where the charm production pion-induced would be much larger than the production proton-induced.

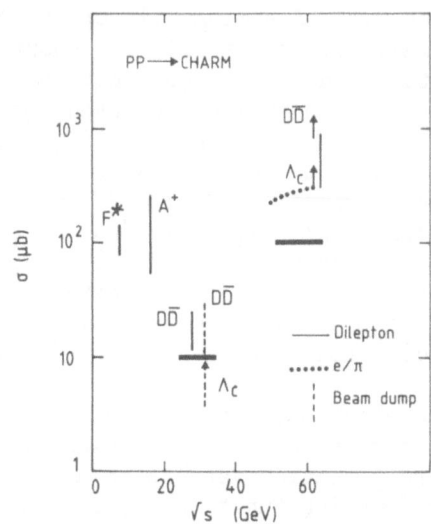

Fig. 8. Cross sections per nucleon of charm production with proton beams. F* refers to the 2145 MeV peak mentioned in 2.1..

4. EXPERIMENTS SEARCHING FOR CHARMS

At the CERN SPS three experiments are now working on charm physics: NA11, NA1 and NA14.

NA11 uses the Silicon microstrip technique described in 2.1.. This experiment is expected to provide good measurements of $D^{\pm}$, D^0, F (and perhaps Λ_c) parameters and lifetimes.

NA1 is now taking data at a photon beam of 200 GeV top energy using a new live target, which consists of a Germanium bulk detector, followed by a telescope of Silicon layers, 200 μm thick[27] (Fig.9).

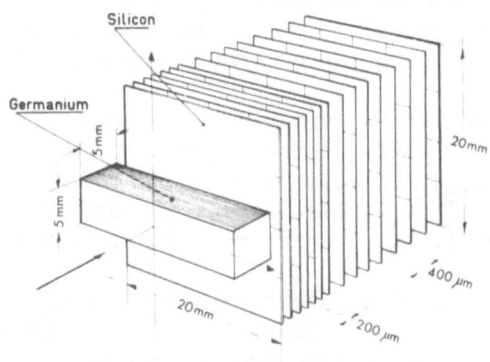

Fig. 9. The live target used by the NA1 experiment.

40 strips (50 μm wide, 100 μm pitch) are deposited on the top face of the Ge block, while an electrode is diffused on the bottom[28]. If the beam enters the region between the opposite electrodes at a zero angle with respect to their planes, the spatial development of the events can be followed by means of the strip signals. The use of the Germanium allows to improve the granularity of the detector, keeping a good signal-to-noise ratio; compared with the Silicon, the energy loss in Germanium is larger by a factor 2 and the corresponding pulse height is 2.5 time higher. NA1 is expected to provide good measurements of D^0, F and perhaps Λ_c parameters and lifetimes and to give some information on the Branching Ratios.

In the NA14 experiment the vertex detector uses the two techniques of live targets and Silicon microstrip chambers[29]. A telescope of 30 Silicon layers, 300 μm thick, is followed by 8 planes of microstrip chambers, 50 μm of pitch. The Silicon detectors are divided in 24 sections each, 300 μm width, 200 μm apart: it is so possible to work with a very intense beam if its spot is large enough to share the flux among the different sections. A photon beam produced by $2 \cdot 10^8$ electrons/sec could be used without dramatic pile-up problems. The microstrip chamber read out does not make use of the charge division method, as in NA11 experiment; therefore a precision of ∿50 μm in the plane has to be expected. In this experiment the Silicon microstrip chambers are used in order to select the events for which it is worthwhile to look for secondary vertices in the live target.

At Fermilab machine, in addition to the TPS Collaboration[11], which apply conventional techniques, the E400 experiment[18] is on the floor; this experiment collected already a very high statistics with a 350 GeV neutron beam, using a structured live target which consists of four Silicon detector telescopes. Changes of multiplicity between adjacent telescopes select possible charm decays, near to the production point. This experiment, now still running with the target described in 3.3., is expected to give good information on B.R., production mechanisms and cross sections.

5. HADRO AND PHOTOPRODUCTION OF BEAUTY STATES

No positive signals for Beauty production have been obtained until now at fixed target machines. The upper limits measured with hadron and photon beams are strongly model dependent because of the large corrections due to the very small acceptances of the apparatus. A central or nearly central production is generally assumed as generation mechanism and a linear dependence on the nuclear target atomic weight is adopted.

The only experiment of the "peak hunting" type is WA11[30], carried out with a 190 GeV π^- beam and a Be target; B mesons are searched for studying the $\psi K(n\pi)$, ψKK, $\psi\phi$ channels with a muon pair in the final products. 7 nb is the measured as upper limit of the cross section$^{(\circ)}$

NA19 used emulsions, exposed to a 350 GeV π^- beam[31]; the Beauty candidates were tagged by the presence of 3 muons in the final states. The sensitivity of the experiment was very low and no Beauty events were found. The upper limit of the cross section is 90 μb for a Beauty lifetime of $\sim 10^{-3}$s, which grows up to ~ 180 μb if the Beauty lifetime is $\sim 10^{-12}$s.

Finally the European Muon Collaboration looks for same sign dimuon and trimuon events with a 250 GeV μ^+ beam and an Fe target. Three wrong sign trimuon events were found, two including 3 μ^+ and one consisting of 1 μ^+ plus 2 μ^-. The corresponding cross sections would be 5.0 ± 5.0 nb[32].

At the ISR the only positive result remains the Λ_B^0 signal found by the BCF Collaboration[23], which decays into $pD^0 \pi^+ (\to pK^+\pi^+\pi^-)$. In the mass interval assumed for the Λ_B (5.35–5.5 GeV) the x_F distribution is nearly flat. Corrections for the efficiency and trigger cuts were calculated assuming that:

i) the $\bar{B}$ state produced in association with the Λ_B^0 is an anti-beauty-flavoured meson;

ii) the longitudinal and transverse momentum for Λ_B^0 is the same as for Λ_c^+;

iii) the Λ_B^0 decay mechanism follows the leading baryon conditions:

$$x_L(pD^0\pi^-) \geqslant 0.32$$

$$|y(pD^0\pi^-)| \geqslant 1.4$$

In addition the B.R. for the reaction $\bar{B}$ state $\to e^+ +$ anything was assumed to be $\simeq (13 \pm 6)\%$.

The measured cross section is $\sigma(\Lambda_B^0) = 120^{+82}_{-50}$ μb, which is at least two orders of magnitude higher than the cross section obtained analyzing the dilepton events with the same technique described in 3.4.[26].

5.1. THE PRESENTLY APPROVED PROGRAMME FOR BEAUTY SEARCH AT THE CERN SPS AND FNAL TEVATRON

The two experiments on Beauty search now on the floor at the CERN SPS, WA71[33] and WA75[34], use emulsions, exposed to a 350 GeV π^- beam. The triggers are different: WA71 triggers on multiplicity jumps (which would correspond to charm decays) between two Silicon telescopes placed just after the emulsion; WA75 asks an high p_T muon or multimuon. In both the experiments the emulsion region,

---

(°) The cross sections of this paragraph are recalculated taking into account up to date Branching Ratios (B(B→ψX)=1 %; B(B→μX)=12 %; B(B→D)=100 %; B(D→μX)=8 %)[16].

where one has to look for, is restricted by means of a vertex detector consisting of Silicon microstrip chambers, 50 µm of pitch. The total sensitivity is <1 Beauty event/nb for WA71 and ≈1.5 Beauty event/nb for WA75.

The two experiments in preparation at the FNAL Tevatron seem more promising.

E653[35], planned for the 600 GeV p beam, designed a set-up consisting of: emulsions, 15 planes of microstrip Silicon chambers, a forward spectrometer including a particle identifier, γ detectors, an hadron calorimeter, muon hodoscopes. The trigger requirement is a single muon with large momentum and the selection procedure is based upon the identification of a charm sample. The microstrip chambers of the vertex detectors are read-out with the charge division method and achieve a precision on the plane of ∿15 µm;

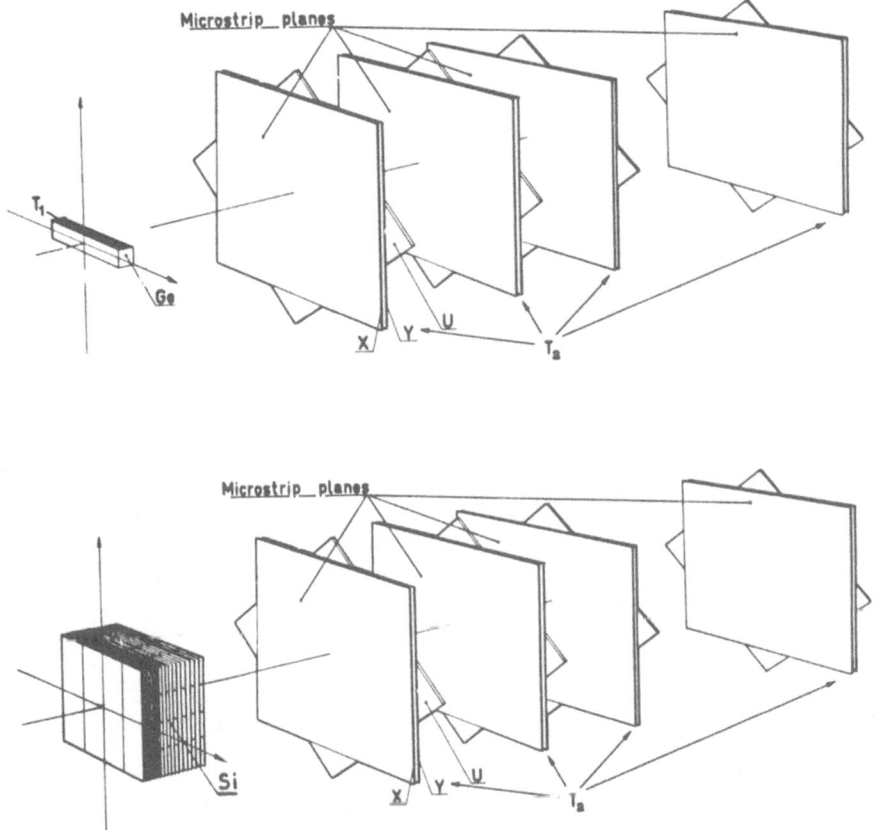

Fig. 10. The vertex detectors designed for the Fermilab E687 experiment.

they are used to search for tracks intersecting with the muon track at a point removed from the production vertex. The total sensitivity is ∿1 Beauty event/nb, but the Beauty production cross section is expected to grow up rapidly with the incident energy.

The E687 experiment[36] will run, at least at the beginning, with a γ beam with 500 GeV of top energy. The set-up includes: a live target, a high resolution tracking system, placed near to the vertex and consisting of 12 Silicon microstrip chambers, 25-50 μm of pitch, a magnetic spectrometer which assembles drift chambers, three C counters, electromagnetic and hadron calorimeters. The live target will be either a Ge bulk detector, 50 μm of pitch, useful to search for particle with very short living time (10^{-14}-$5 \cdot 10^{-13}$s), or a Si detector telescope, which covers lifetimes >$5 \cdot 10^{-13}$s. The vertex detector of the experiment is sketched in Fig.10.

In this experiment the identification of Beauty flavoured states is based upon the detection of two short decays following each other, in addition to the selection of the topology and mass. The total sensitivity is expected to be ∿100 Beauty events/nb with $\tau_B \simeq 5 \cdot 10^{-14}$s and ∿700 events /nb for $\tau_B \simeq 10^{-12}$s.

REFERENCES

1. The following review papers:
   a) W. Geist and S. Reucroft, Proc. of the Workshop on SPS Fixed Target Physics, CERN 83-02 Vol. II: 190 (1983);
   b) J. Sacton, Proc. of the Workshop on SPS Fixed Target Physics, CERN 83-02 Vol.II: 222 (1983),
   c) S. Reucroft, XXIV Int. Symposium on Multiparticle Dynamics, Lake Tahoe, California, preprint CERN-EP/83-155 (1983);
   d) N.W. Reay, Proc. of the Int. Conference "Physics in Collision III", Como - Italy, Edit. Frontières p.223 (1983).

   represented a good source of information in preparing this invited talk.
2. D. Aston et al., Phys. Lett. 100B: 91 (1981) and Nucl. Phys. B189: 205 (1981).
3. M. Atkinson et al., preprint CERN-EP/82-128 (1982).
4. S.R. Amendolia et al., contributed paper nb.0742 to the 21st Int. Conf. on High Energy Physics, Paris (1982); G. Bellini, Proc. of the Fifth Int. Conf. "Novel Results in Particle Physics, Vanderbilt, ed. A.I.P. p.105 (1982).
5. R. Bailey et al., preprint CERN-EP/84-18 (1984).
6. I.V. Ajinenko et al., Phys. Lett. 95B: 451 (1980).
7. S.F. Biagi et al., Phys. Lett. 122B: 455 (1983).
8. K.P. Streit, private communication.
9. D. Drijard et al., contributer paper to the High Energy Physics Conference, Brighton - U.K. (1983).
10. M. Aguilar-Benitez et al., Phys. Lett. 135B: 237 (1983).

11. J. Bronstein et al., presented by M.S. Witherell at the Como Workshop "Search for Heavy Flavours" published on the Proc. of the Int. Conf. "Physics in Collision III", Como - Italy, Edit. Frontières p.317 (1983).
    D.J. Summers et al., preprint Fermilab - Pub.83/84 - Exp. (1983)
    K. Sliwa et al., preprint Fermilab - Pub.83/96 - Exp.732.516 (1983).
12. M.I. Adamovich et al., preprint: CERN-EP/83-183 (1983) and CERN-EP/83-184 (1983).
13. J.J. Aubert et al., Nucl. Phys. B213: 31 (1983).
14. A.R. Clark et al., Phys. Rev. Lett. 45: 1455 (1980) and Phys. Rev. D24: 55 (1981).
15. A.M. Boyarski et al., Phys. Rev. Lett. 34: 1357 (1975).
16. U. Camerini et al., Phys. Rev. Lett. 35: 483 (1975).
17. A. Badertsher et al., Phys. Lett. 123B: 471 (1983).
18. P. Coteus, presented at the Como Workshop "Search for Heavy Flavours" published on the Proc. of the Int. Conf. "Physics in Collision III", Como - Italy, Edit. Frontières p.451 (1983).
19. M. Aguilar-Benitez et al., Phys. Lett. 123B: 98 (1983).
20. A. Bodek et al., U. Rochester preprints: COO 3065-324, 338 (1983).
21. R. Bailey et al., Phys. Lett. 132B: 237 (1983).
22. A. Bodek et al., U. Rochester, preprint COO 3065-337 (1983).
23. M. Basile et al., presented by G. D'Alì, Proc. of the Europhysics Study Conference "The Search for Charm, Beauty and Truth at High Energies", Erice (1981), Ed. Plenum p.467 (1983).
24. M. Aguilar-Benitez et al., Phys. Lett. 123B: 103 (1983).
25. T. Aziz and A. Gurtu, preprint TIFR-BC 83-02 (1983) and references herein.
26. H.G. Fischer and W.M. Geist, preprint CERN-EP/83-30 (1983).
27. Frascati, Milano, Pisa, Torino, Trieste Coll., CERN/SPSC/82-33 SPSC/P170/Add 1 (1982).
28. G. Bellini et al., Phys. Rep. 83: 1 (1982).
29. D. Websdale, Proc. of the Europhysics Study Conf. "The Search for Charm, Beauty and Truth at High Energies", Erice (1981), Ed. Plenum p.287 (1983).
30. R. Barate et al., Phys. Lett. 121B: 429 (1983).
31. J.P. Albanese et al., Phys. Lett. 122B: 197 (1983).
32. J.J. Aubert et al., Phys. Lett. 106B: 419 (1981).
33. M. Adamovich et al., CERN/SPSC/81-18 (1981).
34. J.P. Albanese et al., CERN/SPSC/81-69 (1981).
35. N.W. Reay, Proc. of the Europhysics Study Conference "The Search for Charm, Beauty and Truth at High Energies", Erice (1981), Ed. Plenum p.237 (1983).
36. E687 Proposal and addenda, Fermilab (1981, 1982).

DISCUSSION

TRILLING:
Can you clarify the normalization of the LEBC and TPS branching ratio measurements?

BELLINI:
The LEBC experiment measures the ratios between the $D^+ \rightarrow K^-\pi^+\pi^+$, $D^0 \rightarrow K^-\pi^+$, $D^0 \rightarrow K^-\pi^-\pi^+\pi^+$ and the three prongs and four prongs decays of the $D^{\pm}$ and D^0, respectively. The absolute Branching ratios are obtained using the 3-prongs ($D^{\pm}$) and the 4-prongs (D^0) Branching ratios obtained at SPEAR.

The Tagged Photon Beam Spectrometer measures the ratio $(D^0 \rightarrow K^-\pi^+\pi^0)/(D^0 \rightarrow K^-\pi^+)$. For the absolute Branching ratio of $D^0 \rightarrow K^-\pi^+$, we still have to use the SPEAR data.

TRILLING:
Is there not evidence for multipion decays of F from emulsion?

BELLINI:
There are two events in the $\pi^+\pi^-\pi^-\pi^0$ decay mode obtained by the FNAL E531 experiments.

GENTILE:
We have three Λ_c events from NA27 which are confirmed from particle identification. For NA16 branching ratio kinematics fits have been made to all events. This procedure includes an additional constraint (p_T balance in decay) which is violated, in general, if a neutral particle is missing.

BELLINI:
Yes, I agree. As I have already said, you have calculated with Monte Carlo simulations that reflections of F and Λ_c decays and D-meson decays to final states containing neutral particles could give contributions of at most a fraction of event in your D peak.

STECH:
The measurement of multipion decays of the F^+ is of great importance for the understanding of nonleptonic decays. The result presented appears to give the first direct evidence that helicity suppression is overcome and annihilation diagrams are indeed important. Can you extract a branching ratio for this decay made from your data?

BELLINI:
In principle, yes, but the corrections for acceptances and efficiencies have to be done yet.

REVIEW ON MEASUREMENTS OF CHARM LIFETIME

Kiyoshi Niu

Department of Physics
Nagoya University
Nagoya 464, Japan

ABSTRACT AND INTRODUCTION

The study of charm particle lifetime has progressed enormously in these years, from chaos to the point where measurements are consistent within rather large errors. Before summarizing the present data, the early results from cosmic ray studies are briefly reviewed. To go further, higher statistics study will be urgently needed in which errors of the order of a few percent should be aimed at.

EARLY COSMIC RAY RESULTS

Experimental study of charmed particle has now a history over a dozen of years. Fig. 1 shows the progress in the accumulation power of the directly observed charm decays per experiment. There are three lines with different slopes. One is from cosmic ray experiments, the next is from accelerator experiments with only emulsions, and the last one is from the experiments with hybridized apparatus.

A pioneering work on observation of charm particles[1] had been carried out in the cosmic ray field as was the case of the strange particle study many years ago. It was several years in advance of the commencement of the "charmed age" in the accelerator physics community.

The discovery of one event of 10 TeV range, containing a pair creation and decays of particles with lifetime around 10^{-13} sec and mass of $2 \simeq 3$ GeV was reported at the 12th International Cosmic Ray Conference at Hobert, Australia, in 1971. Fig. 2 shows the sketch of the essential part of the event. A pair of kinks was observed 1.4 cm and 4.9 cm down from the primary interaction vertex, and a

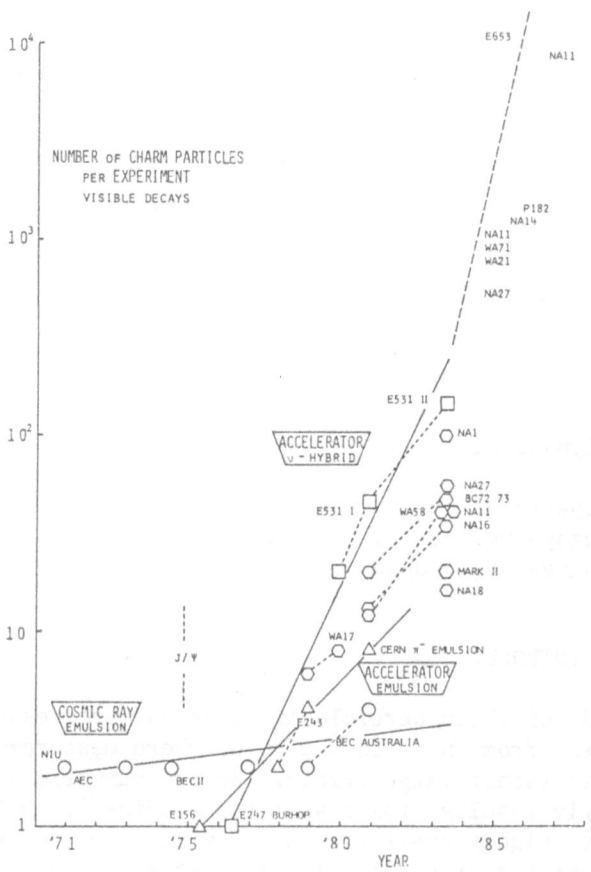

Fig. 1 The progress in the accumulation power of
the directly observed charm decays per experiment.

coplanarity relation among a parent and a daughter of a kink with a
π^{o} meson suggested a two body decay of a new particle.

Just after the discovery, this particle was pointed out to be
a charmed particle by a Japanese theoretical group[2]. A certain
excitement was created in the cosmic ray physics community, but the
field of charm physics lay domant in the high energy physics community until the later discovery of the J/Ψ[3] in 1974. The reason may
be, the work was carried out in far far east where no remarkable
high energy experimental physics activity was yet recognized at that
time, and also they used cosmic rays and the emulsions.

From technical point of view, however, the detector used by them
was not a pure emulsion stack but so called the emulsion chamber.
It was quite suitable to detect and analyse a new particle with lifetime in the order of 10^{-13} sec which was produced by primaries of 10
TeV range. It was a complex detector consisting of an emulsion tar-

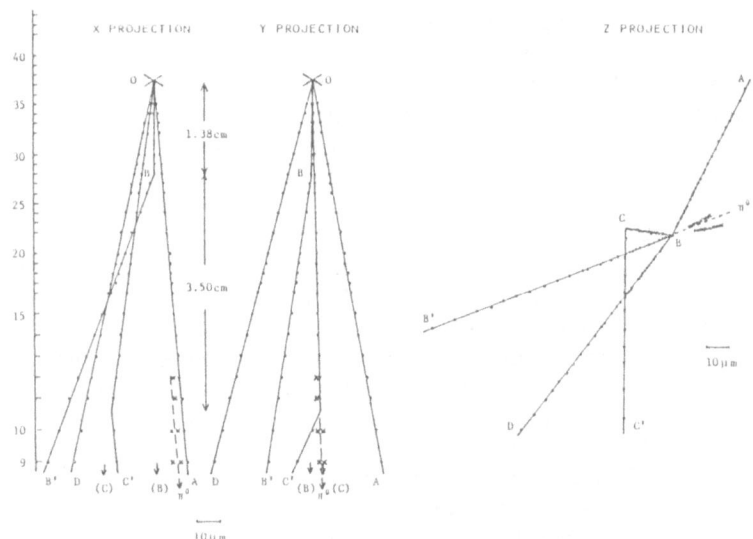

X PROJECTION Y PROJECTION Z PROJECTION

Fig. 2 The first cosmic ray event which shows a pair
production and decay of charm particles observed in 1971.

get and a lead-emulsion sandwich as an energy momentum analyser as
is shown in Fig. 3. Charged particle momentum could be analysed by
the relative scattering method with M.D.M. of TeV/c range, and the
energy of γ rays or prompt electrons could be measured by the analysis of cascade showers induced by them in the analyser part of the
detector. Moreover, these cascade showers and charged particles in

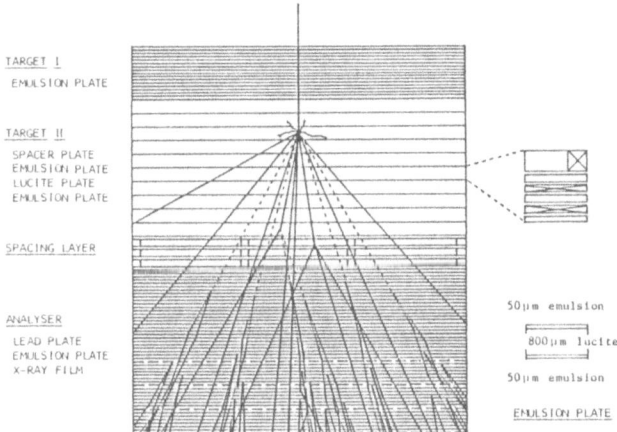

TARGET I
EMULSION PLATE

TARGET II
SPACER PLATE
EMULSION PLATE
LUCITE PLATE
EMULSION PLATE

SPACING LAYER

ANALYSER
LEAD PLATE
EMULSION PLATE
X-RAY FILM

50 μm emulsion
800 μm lucite
50 μm emulsion
EMULSION PLATE

Fig. 3 Configuration of the emulsion chamber which was
used to study charmed particles in cosmic ray experiments.

343

the forward cone could be traced back not only to the primary vertex
but also to the secondary vertex a few cm away from the origin.

Adoption of this type of the emulsion chamber enabled us to discover charmed particle decays, and the extention of this technique
has been essential for the success in the later study of charm lifetimes by means of the emulsion counter hybrid apparatus.

Using the emulsion chambers, additional example of charmed particles had been gradually accumulated including a beautiful example
observed by other Japanese group[4]. Number of events per experiment,
however, grew only slightly in the cosmic ray experiment as is shown
in Fig. 1. But, just after the discovery of J/Ψ particle, about 20
charm particles were accumulated including those from reanalyses of
older events published by others.

At the 14th International Cosmic Ray Conference at Munnich in
1975, a report was made pointing out the lifetime difference of a
factor of $\simeq 3$ between neutral and charged component of charm particle[5]. Fig. 4 is the one which was presented at that conference.
Including this, following 3 conclusions listed in Table I were deduced from the charmed particle observation in the cosmic rays up to
that time. Again, it was too early to give an excitement in the
physics community, and a letter reporting these results was refused
by the referee of some Journal. The third conclusion is, however,
confirmed by the up-to-date data. The first one is also found to be
consistent with the data accumulated by the hadro-production experiments in the accelerator energy region.

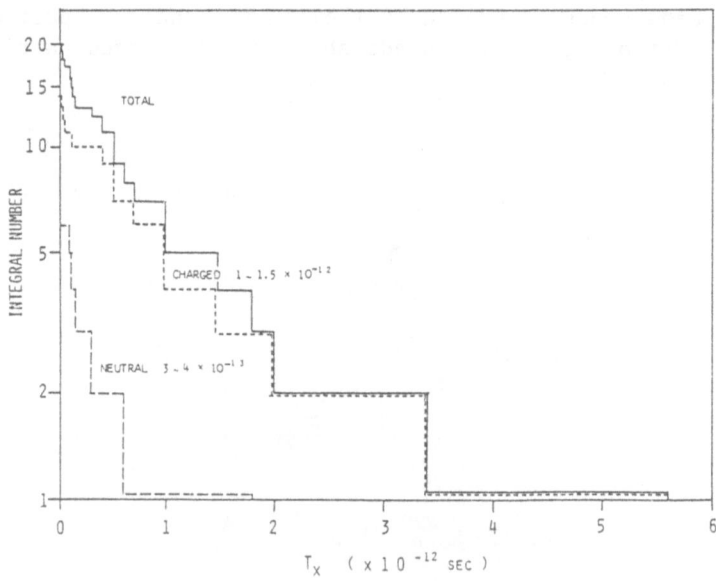

Fig. 4   Distributions of lifetime of charm
particles observed in cosmic ray experiments.

344

Table I   Three conclusions from short-lived particle observation
           in Cosmic Ray experiments.   (in 1975)

A) Production Rate; 1 pair / 20 - 40 inelastic events

B) Cascade Decay;   copious

C) Lifetime; $\tau_\pm > \tau_0$, $(1 - 2)\times10^{-12}$: $(3 - 4)\times10^{-13}$
                    (charged)   (neutral)

STATUS OF CHARM STUDIES AT ACCELERATOR ENERGIES

   After J/Ψ, the main hunting ground of charmed particles has been
moved to the accelerator laboratories. The situation seems quite
similar to that case of the strange particle study. There is, however, a big difference between old transition and new transition.
The emulsions was replaced by the bubble chambers along with the old
transition, because of long flight path of strange particles, and
because of a big difference of the analysing power between the two
methods. In the case of charm particle study, emulsion still remained as the most powerful detector even after the transition. This is
because emulsion was the only detector with resolving power of one
micron which is indispensable to observe directly both production and
decay vertices of charmed particles as is shown in Fig. 5.

   Japanese cosmic ray emulsion group, therefore, took part in the
accelerator experiment with improved design of the emulsion chambers.
Accumulation speed of charmed particle decays has been increased by
the use of accelerated beams, as is shown in Fig. 1.

   The speed has also much accelerated by the adoption of high
energy neutrinos as a producing agent, and by the introduction of a
hybrid apparatus initiated by Prof. Burhop's group[6]. This is because
neutrino is the most effective producing agent of charmed particle
in the sense of production rate per interaction, and because the
spatial resolution of emulsion is complemented by a powerful down-

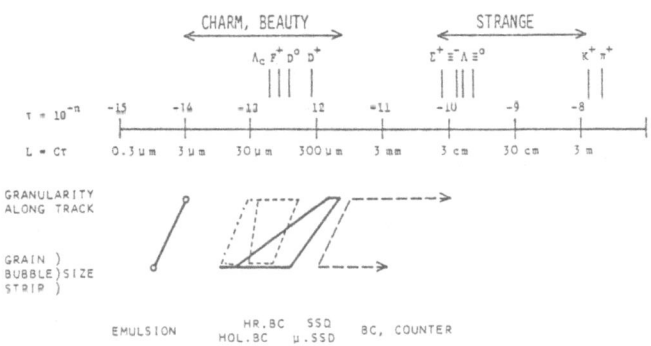

Fig. 5   Comparison of the resolving power of track detectors.

Table II   New emulsion techniques for hybrid experiment.

1) <u>Vertical emulsion plate</u>, as a 'super multi-layer counter telescope' with submicron resolution.

2) <u>Tracing back method of secondary tracks</u> predicted by counters as a highly efficient locating technique both for primary and secondary vertices.

3) <u>Fresh emulsion sheet as a low background interface</u> between emulsion target and counter vertex detector.

4) <u>Computer aided semi-automatic scanning machine</u> to speed up drestically location time of the events in emulsion target.

stream detector to minimize the scanning labour in the emulsion.

The most successful experiment along this line has been carried out at Fermilab as the Experiment E-531[7], in which Japanese emulsion group again made an essential contribution with drastically renewed emulsion techniques. There are 4 points in the new method as listed in Table II . Among them, the fourth one was the most essential to overcome a miserablly low analysing power which had been the weakest point of the old emulsion technique. The computer aided semi-automatic scanning machine has changed completely the way of emulsion work. Style of new emulsion work became more or less comparable to the bubble chamber film analysis.

Due to introduction of such new emulsion techniques on one hand, and due to introduction of high resolution bubble chamber techniques on the other hand, the study of charm lifetimes has taken rapid strides in these years. Most of the experiments dedicated to measure charm lifetimes in the past few years became using of large acceptance spectrometers, for example the one shown in Fig. 6, differing mainly in their choice of particle beams and types of vertex detectors used to see decays. The most recent experiments are summarized briefly in Table III, where thsy have been listed according to their using of detectors. In the following, brief explanations are given for each experiment.

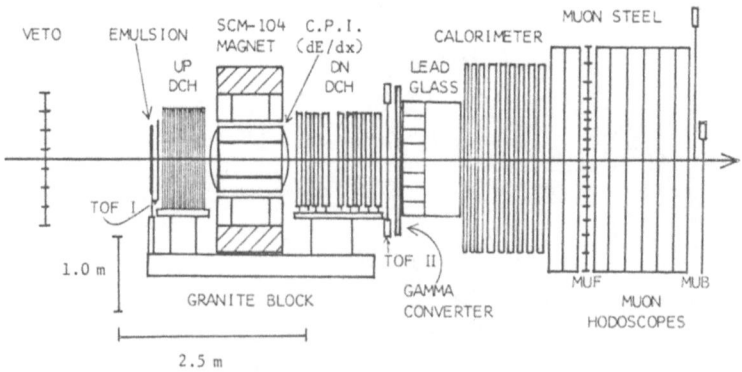

Fig. 6   A hybrid emulsion counter spectrometer used in E-531.

Table III   List of recent experiments studying charm lifetimes.

Type name (beam)	t_{min} $(10^{-13} s)$	Efficiency	Background	## Decays
Emulsions				
E 531 (ν)	0.05	90%	< 1%	140
WA 58 (ν)	0.05	60%, rapidly varying for neutrals.	< 2%	40
Bubble Chambers				
NA 16 (π,p) (LEBC)	2	10% – 60% with t.	< 2%	35
NA 18 (p)	2	≃ 11%, no neutrals detected.	23%	16
NA 27 (π,p) (HOLEBC)	0.8	same as LEBC.	< 2%	55
BC72/73 (γ)	3	99%, for > 500 μm.	≃ 10%	43
$e^+ e^-$ collider				
MARK II (e^+e^-)	1	Low, only D^* → selected modes.	≃ 10%	27
Silicon, electronic				
NA 1 (γ)	2	Low, only certain decay modes.	≃ 20%	110
NA 11 (γ) (μ strip)	1	Low, only certain $c\bar{c}$ modes.	≃ 14%	40

E-531[7]; They used a hybrid emulsion spectrometer in a neutrino
beam. The downstream spectrometer shown in Fig. 6 catches particles
over a wide solid angle, momentum analysing and tagging both charged
and neutral particles. The new emulsion techniques mentioned just
before have been successfully utilized firstly in this experiment.
They exposed 23 liters of Fuji emulsion in the first run and 35
liters in the second run. Analysis of the latter is almost finished.
The scanning technique of a single minimum track predicted by the
spectrometer back into the emulsion shown in Fig. 7 was devised which
gave rise to 90% track finding efficiency. This technique is used
for finding charm decays mainly for events in the vertical target,
60% of the total volume in the first run, but all of the events from
the second run. Calibration of scanning efficiency was done by finding electron pairs, and was proved to be flat down to 5 cm from the
production vertex as shown in Fig. 8. Event yields and the present
status of lifetimes from E-531 are shown in Table IV. Fig. 9 shows
a reconstructed Λ_c decay found in a vertical target of E-531.

WA-58[8]; The WA-58 collaboration exposed single emulsion pelli-
cles 0.6 mm thick at an angle of 5° with respect to the CERN tagged
photon beam. Event finding was done by volume scanning 45 mm^3 centered on the vertex prediction from the OMEGA spectrometer. The suc-

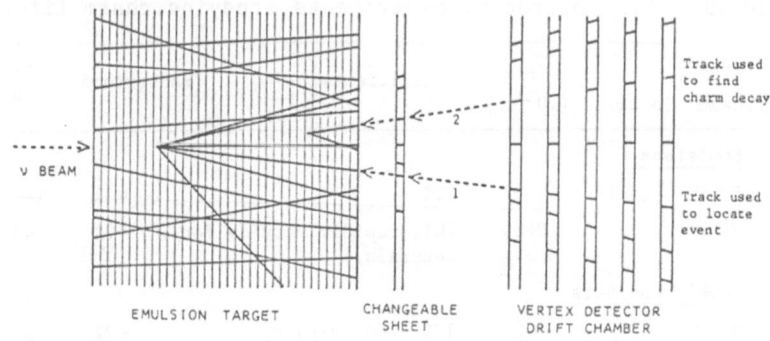

Fig. 7   Follow (Scan) back method of predicted
minimum ionized tracks to find events and decays.

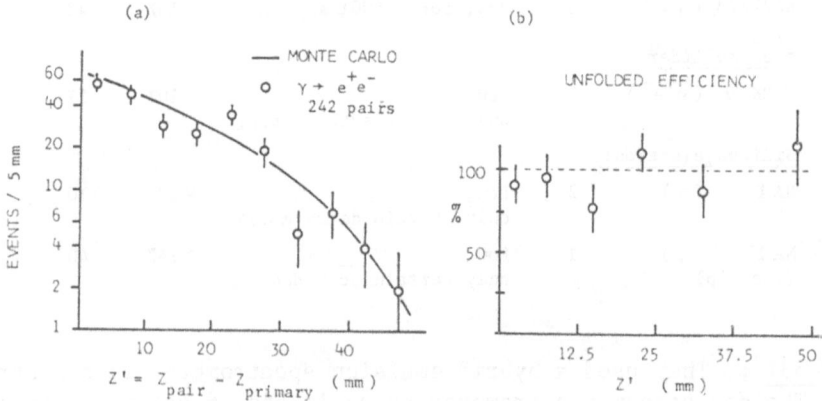

Fig. 8   a) Comparison of predicted and measured distributions of
elctron pairs.   b) Unfolded efficiency for finding electron pairs
by the scan back method.

Table Ⅳ   Event yield and present status of lifetimes in E-531.

	Predicted	Searched	Found	Charm	K^O, Λ^O
Phase I	1821	1821	1248	50	4
Phase Ⅱ	4163	3844	2990	102	12

Particle	# Decays	Lifetime (x 10^{-13} sec)
D^O	56	$3.3 \pm {}^{0.5}_{0.4} \pm 0.25$
$D^{\pm}$	11	$11.5 \pm {}^{7.5}_{3.5}$
$F^{\pm}$	8	$2.6 \pm {}^{1.2}_{0.8} \pm 0.2$
Λ_c^+		$2.3 \pm {}^{1.0}_{0.8} \pm 0.2$

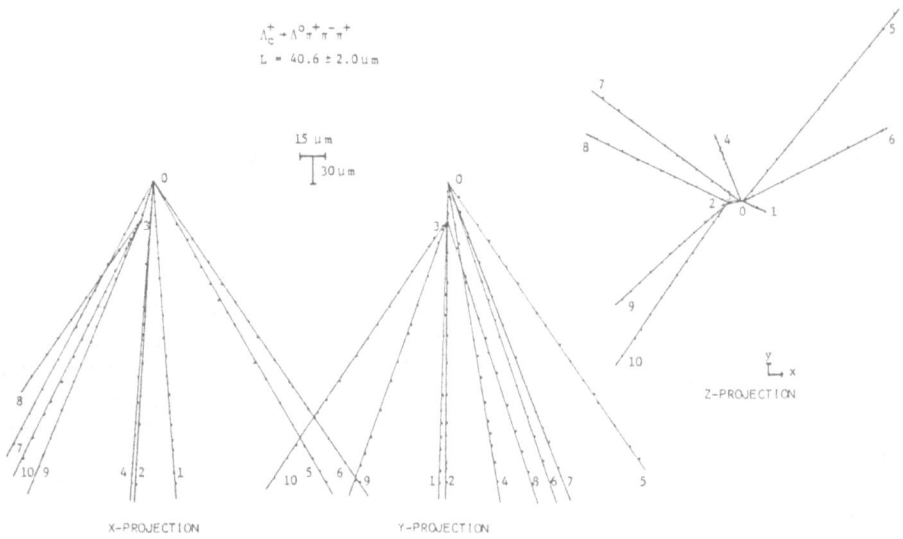

$\Lambda_c^+ \rightarrow \Lambda^0 \pi^+ \pi^- \pi^+$

$L = 40.6 \pm 2.0 \mu m$

15 μm

30 μm

X-PROJECTION

Y-PROJECTION

Z-PROJECTION

Fig. 9   A reconstructed Λ_c^+ decay in a vertical target in E-531.

cess rate is 50%. Charged decay candidates are found first by following charged tracks to the edge of the emulsion. Following length
is typically 3 mm. If a decay candidate is not found, a volume scan
for neutrals is done for 550 μm, otherwise the volume scan continues
to maximum 1,000 μm. In total 60 decays have been found, of which
42 are used for lifetime determinations.

NA-16[9]; NA-16 (LEBC-EHS collaboration) employs the small hydrogen bubble chamber LEBC built specifically for the study of charm
decays. Bubble size of 45 μm and bubble density of 80 blobs/cm are
achieved by running at higher pressure and temperature than conventional condition, and by using a short delay of 300 μs between the
interaction and the firing of the flash lump. The European Hybrid
Spectrometer EHS is coupled with LEBC for charged particle momentum
analysis and for π^0 reconstruction. Information is available from
the dE/dx chamber ISIS for dientifying some tracks.
     After subtructing background and unreconstructed events from
the observed sample, a cut in decay length which is smaller than the
computed minimum detectable length was applied to get unbiased sample. A total of 52 charm decays, 23 neutral and 28 charged and 1
ambiguous having kinematic fits with 3 or 2 degrees of freedom have
been used in the lifetime analysis.

NA-27[10]; The EHS collaboration has been pursuing a new experiment NA-27 at CERN. They are using the high resolution hydrogen
bubble chamber HOLEBC as a vertex detector in which resolved bubble
diameter of $15 \simeq 20$ μm is obtained. The final version of the EHS
consisting of two bending magnets, and associated wire and drift

chambers is used for downstream track reconstruction and identification. It provides momentum measurement to an accuracy of 1% or better over the momentum range of interest. From the preliminary study of the data they are reporting the lifetime results on D^o, $D^{\pm}$ and Λ_c^+ deduced from 16 four prong decays and 36 three prong decays.

NA-18[11] ; NA-18 group used the BIBC (Berne Infinitesimal Bubble Chamber) vertex detector in a 340 GeV/c π^- beam at CERN. It is a 6.5 cm long freon chamber. Because of the heavy liquid fill, the bubble density achieved is higher, 300/cm for minimum ionizing particles, and the optical system resolved individual bubbles down to 30 μm diameter with high contrust in the dark field. The downstream detector employs a streamer chamber for charged track reconstruction. No particle identification is available nor is there a π^o detector, so that events are considered for analysis only if the decay products are all charged and the event is completely constrained. To get bias free data for lifetimes, they corrected each decay length using minimum detectable decay length. Twenty one charmed meson candidates have been identified and used in lifetime analysis.

BC-72[12] ; In this experiment, the SLAC 40 inch hydrogen bubble chamber was modified for studying charmed decays by running at 29°K and by adding a single high resolution camera whose flash was triggered ealier than 3 conventional cameras. The resolution obtained is 70 bubbles/cm averaging 55 μm in diameter. A 20 GeV backward scattered nearly monochromatic photon beam was chosed for the experiment. The down stream spectrometer is the SLAC Hybrid Facility, which contains 2 Cerenkov counters for particle identification, and an electromagnetic shower detector comprised of hodoscopes and lead glass blocks. The hadronic interactions in the fiducial volume passing the 3 cuts are considered for lifetime analysis. Cuts are due to maximum impact distance larger than 110 μm, the second maximum impact distance larger than 40 μm, and decay length longer than 500 μm. Recent result include 2.4 million pictures with 72 visible multiprong charm decays in 62 events. Events passing lifetime cuts include 51 decays, 22 neutral, 21 charged and 8 ambiguous. Lifetimes are quoted for 43 clear events.

MARK II[13]; The MARK II detector with particle spectrometer using a large solenoidal coil provides charged particle tracking and electron identification over 65%, and muon identification over about 45% of the solid angle. As a vertex detector a high precision drift chamber is located just outside the beam pipe. In conjunction with the main tracking chamber, it measures trajectories near the interaction point with 100 μm accuracy in the plane perpendicular to the beam. The D^o lifetime is measured by finding distance of the D^o decay vertex from the average beam position, for a sample of D^o coming from D* decay. The average beam position is known to ± 20 μm vertically and horizontally. The beam size is 480 μm × 65 μm, and the beam position is stable.

D^0 and $\overline{D}^0$ are selected by changing pairs of these tracks whose invariant mass is between 1.72 and 2.00 GeV/c^2 when one track is assingned a pion mass and the other a kaon mass. These D^0 candidates are combined with a third track of appropriate charge, and the mass difference $\Delta M = M_{K\pi\pi} - M_{K\pi}$ is studied. Twenty seven events which are clustering in the vicinity of $\Delta M = 145.4$ MeV/c^2 comprises the $D^{*+} \rightarrow D^0\pi^+$ sample in which 1.5 background events are estimated. The D^0 lifetime is deduced by fitting the proper lifetime distribution for these 27 events with a maximum likelihood technique using the convolution of Gausian resolution function and an exponential distribution as a fitting function.

NA-1 [14] ; The NA-1 group used an active silicon target to study charm lifetimes. The target consists of fourty 300 μm thick silicon wafers, 14 mm in diameter, separated by gaps of about 100 μm. Charged particles leave the equivalent of 90 Kev in each Si wafer compared to a r.m.s. noise level of 30 KeV. The event with pulse hight signature consisting of an interaction of photon beam followed by two charmed decays can be seen in this target. The spectrometer deployed contains 4 magnets for charged track momentum analysis and 5 sets of photon detectors using either lead scintillator hodoscopes or lead glass absorvers. Two Cerenkov counters separate pions and kaons from 5 to 21 GeV/c.

Unlike that of the preceeding experiments, the charm event has to be selected firstly by reconstruction in the spectrometer. They picked up those events with pulse hight steps in the target showing the leveling at least two minimum ionizing particles equivalent which hold more than 4 wafers after the interaction, and those with a step in pulse hight of $\Delta n = 2$, 4 or 6 again holding at least 4 silicon wafers. Since the association of decay length in the target and charmed particles reconstructed in the spectrometer is ambiguous, the average Lorentz boost is attributed to both charmed particles. Because the analyses required coherent charmed pair productions, this approximation is believed to contribute little to the error in lifetime. Ninty eight $D^{\pm}$ decays and 8 $F^{\pm}$ decays are used in lifetime estimation.

NA-11 [15] ; This group has become the first to use successfully microstrip silicon detectors in the study of lifetimes of charm particles. By making use of the 5 micron r.m.s. resolution of these precision devices, they were able to reconstruct secondary vertices in the high energy hadron interactions. In the down-stream was placed the ACCMOR spectrometer which consisted of two stage magnetic spectrometers with two large multicell Cerenkov counters and a large area shower counter. Proportional chambers were used on-line to provide an effective mass trigger.

Of millions of 200 GeV/c π^-·Be reactions, at first those events with clean single electron signature and with an identified $K^{\pm}$ were analysed in the vertex telescope. After rejecting background due to electron pairs and so on, they searched for events where kaon and one

to three pions from a good vertex which is significantly displaced
from the primary vertex. A clean sample of 23 neutral and 13 charged
D mesons has been obtained. Recently, they also have measured the
mass and lifetimes of charm particles decaying into Kππ using the
same apparatus and similar analysing method. They obtained 3 uniquely identified F mesons, 5 D mesons with Cabbibo forbidden decay, and
4 possible Λ_c baryons.

LIFETIME SUMMARY

Present summary of lifetime measurements for each charm state
are shown in Fig. 10, and in Table V. In the following, some brief
discussions for each lifetime are given.

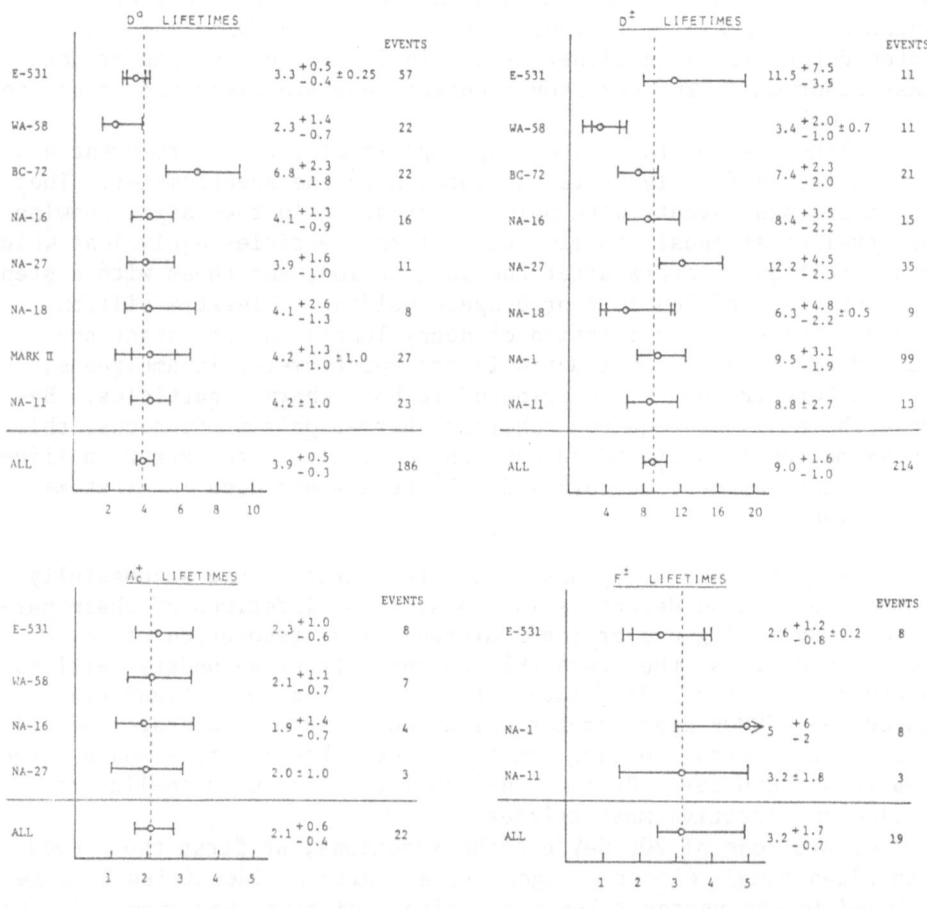

Fig. 11   Summary of lifetimes of charm particles. (× 10^{-13} sec).

Table V World average of charm lifetimes.

Particle	Lifetime ($\times 10^{-13}$ sec)
D^O	$3.9 \pm ^{0.5}_{0.3}$
$D^\pm$	$9.0 \pm ^{1.6}_{0.8}$
$F^\pm$	$3.2 \pm ^{1.7}_{0.7}$
Λ^+_c	$2.1 \pm ^{0.6}_{0.4}$

$\underline{D^O}$; The world average is $\tau(D^O)=(3.9 \pm ^{0.5}_{0.3}) \times 10^{-13}$ sec. Six of 8 experiments presented overlap with the world average, but the spread in values seems to be slightly more than might be expected from statistics alone. Fig. 11 shows a scatter plot of average charged daughter impact parameter versus proper decay time for 30 fully constrained D^O decays from E-531. Twenty nine of the events cluster closely about the line $\delta = ct$, as does the thirtieth when a backward π^O is included. This means that the average impact parameter is a useful quantity which may be used to differentiate between the O-C solutions occuring, for example, in all semi-leptonic decays. Using this correlation for unconstrained decays, a semi-logarithmic differential plot of the number of D^O decays is displayed in Fig. 12, as a function of proper decay time for 56 events taken from E-531. From this, single lifetime of $(3.3 \pm ^{0.5}_{0.4}) \times 10^{-13}$ sec is deduced.

$\underline{D^\pm}$; The world average is $\tau(D^\pm)=(9.0 \pm ^{1.6}_{0.8}) \times 10^{-13}$ sec. Six of 8 measurements have error which overlap with the world average. But, the contamination from short-lived F and Λ_c particles may decrease the apparent lifetime in most of the experiments. An integral decay spectrum for charged 3 prong decays taken from E-531 data is shown

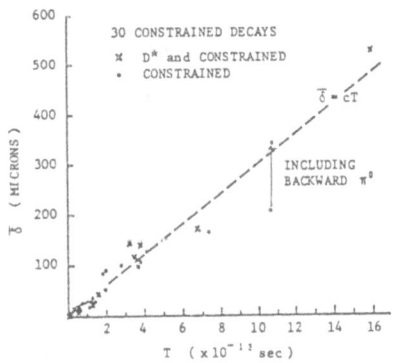

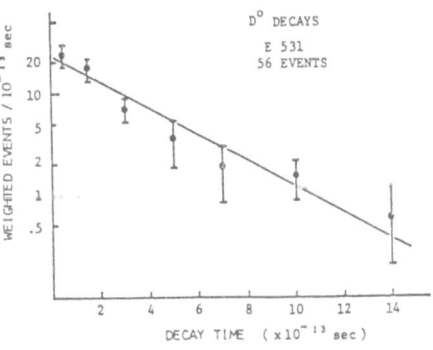

Fig. 11 Average charged track impact parameter $\bar{\delta}$ versus proper decay time for 30 constrained D^O decays from E-531.

Fig. 12 Differential time distribution for D^O events from E-531.

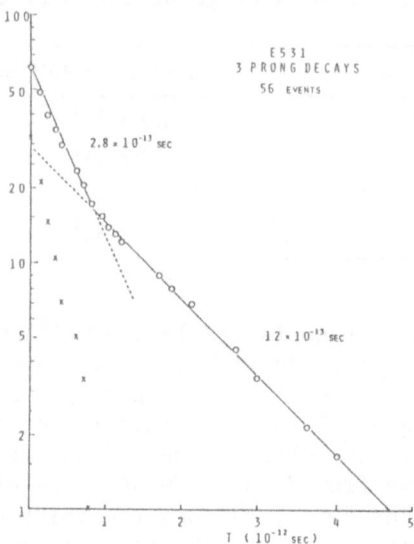

Fig. 13   Integral time distribution for E-531 charged 3 prong decays showing clear evidence for populations with two different lifetimes.

in Fig. 13.   The break between the short lived F and Λ_c decays and longer lived $D^{\pm}$ is quite evident at approximately 7×10^{-13} sec. Obviously, decays with ambiguities in the assignment of the parent particle are a potential source of bias in the lifetime.

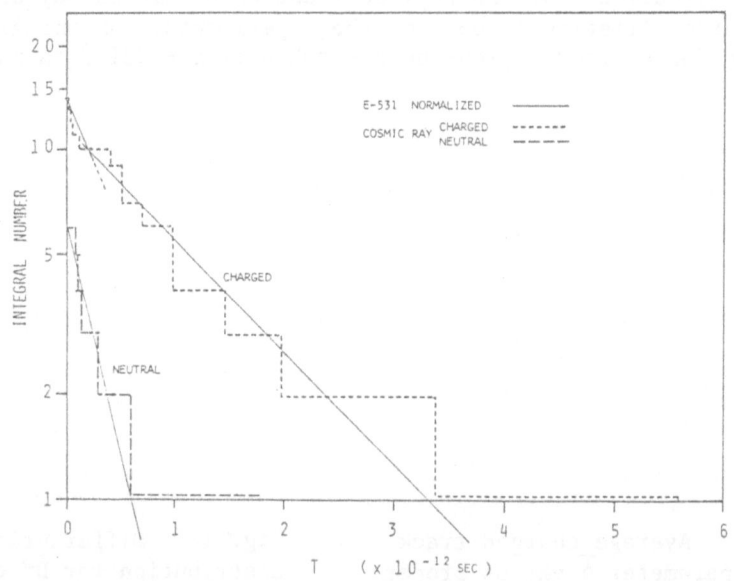

Fig. 14   Comparison of the decay time spectra of charged and neutral charm particles from E-531 and those from cosmic rays.

$\tau(D^{\pm})/\tau(D^{O})$ ratio; Using the two world averages, the ratio of charged to neutral D meson lifetimes is computed to be 2.31 ± 0.4. The ratio is about 3 standard deviation larger than one. Here, one should remind the early result of lifetime ratio from cosmic ray observation mentioned in the first part of this paper. Fig. 14 shows the comparison of the decay time spectra of charged and neutral charm particles from E-531 and those from cosmic rays. Agreement is quite satisfactory, and this shows the confidence on earlier cosmic ray results on charm lifetimes.

Λ_c^+; The world average is $\tau(\Lambda_c^+) = (2.1 \pm ^{0.6}_{0.4}) \times 10^{-13}$ sec.

$F^{\pm}$; The F lifetime has not been settled yet. E-531 has 8 F decays including those decaying into all pions and those containing a ϕ meson. The mass value based on all 8 decays is $1994 \pm 15 \pm 25$ MeV. This value is in substantive agreement with the value of $1970 \pm 5 \pm 5$ MeV obtained by the CLEO group[16]. NA-1 group has 8 F candidates, 3 with an $\eta \pi^+ \pi^- \pi^{\pm} \pi^0$ decay mode, and 5 with $K^- K^+ \pi^{\pm} \pi^0$ decay mode. The measured mass value is 2050 ± 30 MeV. Recently, NA-11 group has got 6 F candidates of $K^+ K^- \pi^{\pm}$ decay mode. The measured mass value is 1975 ± 4 MeV, but 3 of them are ambiguous coming near to the Λ_c mass when replacing one of K with p. These are excluded in estimating the F lifetime.

The world average of F lifetime is $\tau(F^{\pm}) = (3.2 \pm ^{1.7}_{0.7}) \times 10^{-13}$ sec. However, there is some danger in combining lifetime results from experiments with differing F masses.

The second generation experiments to measure charm lifetimes have met with increasing success. Errors to the charm lifetimes are gradually decreasing with the growing statistics. At present, D^O and $D^{\pm}$ lifetimes are known to within 15%, Λ_c and F lifetimes are known within a factor of two. The precise lifetime determinations may differentiate between the various weak decay models given in Fig. 15, when combined with precise knowledge of the decay branching ratio. In case that of the $D^{\pm}$ lifetime is longer than that of the D^O, the effect could come from suppression of $D^{\pm}$ rate or enhancement of the D^O rate, perhaps through the exchange diagram, which is permitted only for neutral decays. The annihilation diagram is Cabbibo favoured only for F decays. If it contributes strongly, F mesons could have a shorter lifetimes than D mesons, and F branching ratio for decays into various final states could be affected.

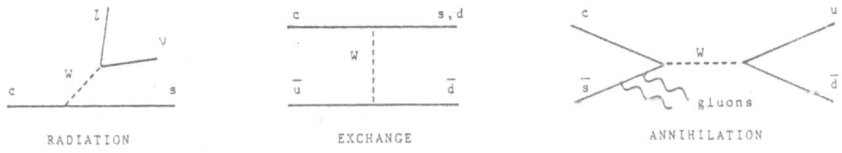

Fig. 15   Cabbibo favoured diagrams for charm decay.

Combining obtained lifetimes with semileptonic branching ratios, the following absolute semi-leptonic decay rates are obtained for each charm state[17]. These agree within rather large errors both with each other and with 2.6×10^{11} /sec predicted by the standard model.

$$\Gamma_{SL}(D^0 \rightarrow e\nu X) = (1.4 \pm 0.8) \times 10^{11} \text{ /sec}$$

$$\Gamma_{SL}(D^{\pm} \rightarrow e\nu X) = (2.4 \pm 0.6) \times 10^{11} \text{ /sec}$$

$$\Gamma_{SL}(\Lambda_c^+ \rightarrow e\nu X) = (2.0 \pm 1.1) \times 10^{11} \text{ /sec}$$

The ratio of charged to neutral D meson lifetimes is more than 3 standard deviation larger than one. Apparently, the D^0 decay rate is enhanced beyond that predicted by the simplest spectator radiative weak decay model. F and Λ_c lifetimes are not well known, whether the annihilation diagram plays a strong role must await future experiments. Error of the order of a few percent should be aimed at in the near future experiments.

FUTURE PROSPECTS

Several groups have been started or preparing the experiments, the aim of which is to collect samples of $10^3 \approx 10^4$ clean events, really adequate to a detailed study of the lifetimes and the production mechanisms. Table VI shows the list of those experiments.

Of course, in these high statistics study, a clear view of the vertex is important for both to measure the lifetime of different charm particles, and to get an unambiguous identification of charm events out of the large multibody production. The better the space resolution of the vertex detector, the larger the sample of identified events. For the vertex detectors other than the emulsions, therefore, a great effort has been devoted to the improvement of the space resolution. On the other hand, more than one order level up of the measuring and analysing speed in the emulsion target has been intended.

Defining the resolution σ as the capability of identifying separately two tracks close to each other, the new, small, high-rate bubble chambers are approaching a limit $\sigma \approx 15$ μm by means of smaller bubbles and of the holographic read-out, which decouples the depth of focus from the resolution and allows to afford a larger number of tracks per picture than with standard optics.

Solid state detectors are developed along two complementary lines, as is shown in Fig. 16. Active targets of smaller and smaller longitudinal granularity are being built to cover decay times down to a few$\cdot 10^{-14}$ sec, while micro-strip chambers of increasing electrode density become feasible with the miniuarization of the read-out electronics. Solid state detectors are the basis of the new experimental programmes of NA-1, NA-11 and NA-14.

Table VI   List of experiments to study charm and beauty particles.

Name	Apparatus (Beam)	Yield	Data Taking
CERN			
NA 1	Active Ge target + FRAMM. ($70 < E_\gamma < 175$ GeV)	10^3 D, F	1983
NA 14	Active target +μstrip + Spectr. ($100 < E_\gamma < 200$ GeV)	$*10^3$ D,F,Λ_c	1984/85
P 182	HOLEBC + EHS. ($100 < E_\gamma < 200$ GeV)	10^3 D, F	1984/85
NA 11	Active target +μstrip + ACCMOR. (200 GeV/c π^-)	10^3 D 10^4 D	1985 1987
NA 27	HOLEBC + EHS. (360 GeV/c P)	500 D 50 F, Λ_c	1983/84
WA 71	Emulsion +μstrip + Ω. (350 GeV/c π^-)	10^3 Charm * Beauty	1983/(84)
WA 21	BEBC. (wide band ν)	$*10^2$ Decays	1983/84
FNAL			
E 653	Emulsion +μstrip + Spectr. (800 – 1000 GeV/c P/π^-)	$2 \cdot 10^2$ Beauty $*10^4$ Charm	1984 – 86

To conduct a high statistics charm study in the experiment E-653 at Fermilab, a new emulsion technique has been developed by Nagoya emulsion group[18]. This is an extention of the technique successfully used in the E-531 collaboration. By means of that technique we have attained analysing power of 1,000 neutrino events per year, and we

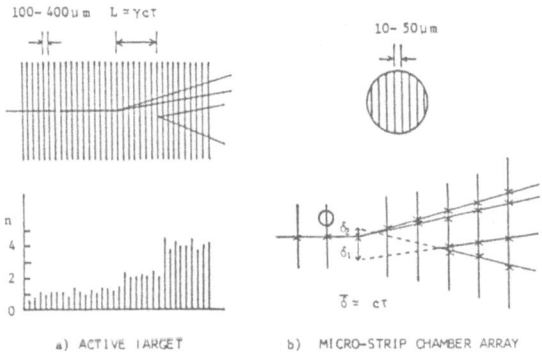

Fig. 16   Two ways of using solid state detectors for measuring the decay path.

have succeeded to get more than 150 charmed particles in a reasonable time. The analysing power of this level is not enough to conduct new experiment. When 100 liters of emulsion is exposed to the hadron beam with density of $10^5/cm^2$, 3×10^8 interactions are recorded in the emulsion, among which 6×10^5 charm pairs and 1,200 Beauty pairs may be recorded assuming cross section of 25 μb and 50 nb, respectively. Samples are enriched by on-line, selecting events with high P_T muon, and off-line, by picking up secondary vertex with triggered muon. About 2×10^4 charm pairs and 100 beauty pairs are concentrated in nearly 10^5 hadron events. This number is still too many to deal with the technique so far used.

New emulsion techniques developed to solve the problem are in two ways. One is a method to save time needed to locate events predicted by down stream micro-strip vertex detector. The idea is to use minimodule with area of 3 cm × 3 cm, cut from a mother module of 24 cm × 24 cm, and sticking all 25 plates from the same minimodule on the same supporting plate as is shown in Fig. 17. Then tedious plate by plate changing to follow the track is converted to a simple x - y displacement of a same supporting plate. By this, tracking speed from one plate to the other will be drastically reduced.

The other is using a new machine to pic up automatically a single track or a group of tracks with specified penetrating (dip and azimuthal) angle, out of the jungle of the noisy tracks in the heavily exposed emulsion plates. This machine is called as the automatic track selector which is a special type of the image processor. As is shown in Fig. 18, only images of the focussed grains in every microscopic field at 16 successive depth separated 20 μm each other are digitized and recorded on 16 individual frame memories with 512 × 512 pic-cells when the focal plane of the objective lens is shifted from the surface to the bottom of the emuslion plate. All of 16 frame informations are real time processed, and those tracks with 16 specified angles are selected and picked up separately in one action.

By means of the new emulsion techniques, we expect to be able to reach the analysing power 10 times higher than the present one,

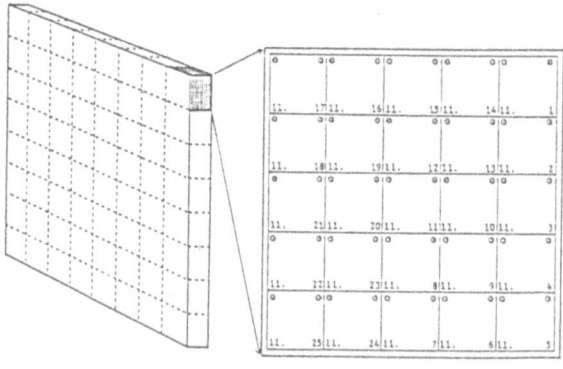

Fig. 17   Minimodule cut from a mother module and sticked on a plate.

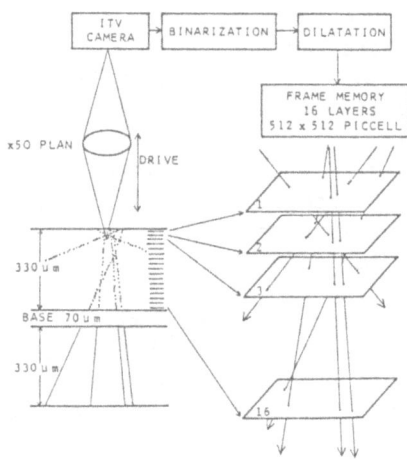

Fig. 18　Multi-layer image processor as an automatic track selector in the vertical emulsion target.

as is shown in Table VII.　The same technique is now being applied to the beauty decay search experiment WA-75 at CERN, and it seems to be working.

Emulsions may continue to remain as the detector with highest resolution in the high statistics study of charm particles by ideal hybridization of the new emulsion techniques and the micro-strip solid state detectors which can predict primary and sedondary vertices with accuracy of 10 μm.

Table VII　Analysing power of the emulsion events.

Task	Experiment (Used Machine)	E-531 (Semi-auto)	WA-75　E-653 (Full-auto)
Location of an event, Decay scan to 1 mm.		90 min.	5 min.
Follow down, Angle measurement, Charged decay search to 6 mm.		130 min.	12 min.
Scan back, Neutral decay search.		60 min.	10 min.
Analysing time / event		280 min.	27 min.
Measurable events / Machine·year		500	5,000 (25,000 location only)
No. of Machine used.		2 Semi-auto.	3 Full-auto + 10 Semi-auto.

In near future, thousands of charm decays will be accumulated by the experiments listed in the table Ⅵ. In addition to measuring lifetimes presisely, these experiments will measure production cross sections and decay modes for charms in a relatively bias-free and model independent way.

Exciting new physics will be explored. Certainly new ground state charmed particles remain to be discovered, especially in the baryon sector, and the spectroscopy of excited states remains to be elucidated. The all leptonic decay $F \to \tau \nu_\tau$ should be seen, yielding an improved limit for the tau neutrino mass as well as a clean measurement of the F form factor. $D^o - \bar{D}^o$ mixing will be searched for with a sensitivity at least 10 times better than can be done at present. Besides those, the direct study of weakly decaying beauty particles will commence, with as yet unknown results.

REFERENCES

1. K.Niu et al; Prog.Theor.Phys.,46,(1971),1644.
2. T.Hayashi et al; Prog.Theor.Phys.,47,(1972),280, and 1998.
3. J.J.Aubert et al; Phys.Rev.Lett.,33,(1974),1404.
   J.E.Augustin et al; Phys.Rev.Lett.,33,(1974)1406.
   C.Bacci et al; Phys.Rev.Lett.,33,(1974),1408.
   G.S.Abrams et al; Phys.Rev.Lett.,33,(1974),1453.
4. H.Sugimoto et al; Prog.Theor.Phys.,53,(1975),1541.
5. K.Hoshino et al; Conf.Papers,14th Int.Cosmic Ray Conf.(Munnich), 7,(1975),2442, and K.Niu; Proc.19th Int.Conf.High Energy Phys. (Tokyo),(1978),447.
6. E.H.S.Burhop et al; Phys.Lett.,65B,(1978),299.
7. N.Ushida et al; Phys.Rev.Lett.,45,(1980),1049 and 1053. ibid,47,(1981),1694, ibid,48,(1982),844, ibid,51,(1983),2362. Phys.Lett.,121B,(1983),287 and 292.
8. M.I.Adamovich et al; Phys.Lett.,140B,(1984),119.
9. M.Aguilor-Benitez et al; Phys.Lett.,122B,(1983),312.
10. E.H.S. Collaboration; Moriond Conf.,(1984).
11. A.Badertscher et al; Phys.Lett.,123B,(1983),471.
12. K.Abe et al; Phys.Rev.Lett.,48,(1982),1526, and SLAC-PUB-3271, (1983).
13. N.Lockyer; SLAC-PUB-3245, J.A.Jaros; Proc.Phys.in Collision-3 and Workshop,Search for Heavy Flavours,(Como),(1983),405.
14. E.Albini et al; Phys.Lett.,110B,(1982),339. S.R.Amendolia et al; CERN-EP/82-200.
15. R.Bailey et al; CERN-EP/84-18.
16. A.Chen et al; Phys.Rev.Lett.,51,(1983),634.
17. N.W.Reay; Proc.Phys.in Collision-3 and Workshop,Search for Heavy Flavours,(Como),(1983),223.
18. K.Niwa; Uchusen Kenkyu (in Japanese),27,(1984),140, and K.Niu; Proc.Phys.in Collision-3 and Workshop,Search for Heavy Flavours,(Como),(1983),475.

THE EXPERIMENTAL SITUATION IN D MESON DECAY

David Hitlin

California Institute of Technology
Pasadena, California 91125

INTRODUCTION

It has been clear for some time that the simple light quark spectator model of charm quark decay[1] does not provide a satisfactory explanation of the experimental situation. It is not clear, however, which specific changes to this picture are necessary. These changes, proposed by numerous authors, fall into two broad categories. The first approach assumes the fundamental correctness of the spectator model. That is, the dominant mechanism is thought to be decay of the charmed quark via emission of a W, with subsequent decay of the W into quark pairs. The light quark component of the meson is merely a spectator. The failures of this simple picture, that is, the non-equality of $D°$ and D^+ lifetimes and the erroneous prediction of suppression of the decay $D° \rightarrow \overline{K}°\pi°$ are dealt with by two basic modifications:

1) The change of the two QCD couplings f_+ and f_- from their calculated values.[2] These coefficients have now been calculated by renormalization group techniques not only in leading log approximations but, recently, in the next-to-leading log order. The next order changes are, in fact, small, reinforcing the correctness of the leading log values ($f_+ \cong .7$, $f_- \cong 1.9$, for six fermions and a mass scale of ~ 2 GeV) which have been in use for several years. Nonetheless, in order to account for the experimental facts within this context, it is necessary to postulate that f_- is, in fact, very much larger than f_+. Since the two spectator diagrams in D^+ decay lead to the same final quark state, they can

interfere. If $f_- >> f_+$, this interference can be destructive, reducing the D^+ decay rate and lengthening the D^+ lifetime. Thus, in this picture the D^+/D° lifetime difference is ascribed to an increase in the D^+ lifetime, with the D° and F^+ lifetimes occurring at values one would estimate by scaling from muon decay by $(m_\mu/m_c)^5$.

2) The second approach attributes the shorter D° lifetime to the importance of additional (W exchange) amplitudes, occurring only in D° decay.[3] These are, naively, suppressed by helicity conservation at the light quark vertex. Either through explicit radiation of soft gluons, or through the gluon component of the quark wavefunction, the W exchange diagram is then enhanced. In this picture, the D^+ lifetime would occur at the "normal" value while the D° lifetime (and perhaps the F^+ lifetime through similarly enhanced W annihilation graphs) would be shortened. Since the W exchange process leads to $I = 1/2$ final states in hadronic D° decay, whereas the spectator process produces both $I = 1/2$ and $I = 3/2$ final states, reinforcement of $I = 1/2$ configurations would be indicative of the importance of exchange diagrams.

Cabibbo-suppressed D decays provide an additional puzzle. Namely, they should occur at a rate $\sim \tan^2\theta_c$ or 5% of the corresponding Cabibbo-allowed decay. The Mark II has found, however, that the rate for $D^\circ \rightarrow K^+K^-$ is more than three times that for $D^\circ \rightarrow \pi^+\pi^-$. Many explanations[4] have been advanced for these different branching ratios, but a choice among the different classes of explanation requires the measurement of additional Cabibbo-suppressed modes.

THE MARK III DETECTOR

The puzzling situation created by the $D^\circ/D^\pm$ lifetime ratio and the other unexpected D branching ratios found by the Mark II detector at SPEAR, motivated the construction of a new spectrometer, the Mark III, which was specifically optimized for the reconstruction of exclusive final states, such as those produced in $\psi'' \rightarrow D\bar{D}$ decay. Properties of the Mark III[5] which make it a good match to the study of charmed meson weak decays are:

1) Good solid angle coverage: 93% of 4π for charged particles and 95% for photons.

2) Reduction of multiple Coulomb scattering of charged particles, producing a momentum resolution below 1 GeV/c of 1.5-2.5%.

3) Excellent reconstruction efficiency for low energy photons, achieved by placing the barrel shower counter inside of the solenoidal magnet coil. Usable efficiency extends to 50 MeV. Energy resolution for photons is $\sigma(E) = 0.18\sqrt{E}$ (GeV), while angular resolution is 10 mrad.

4) Good particle identification capability. Pions and kaons are separated over 80% of 4π by a time-of-flight system with a resolution of 180 psec. Electron/pion separation is done using the time-of-flight system and the shower counter. A dE/dx system for $\pi/K/p/e$ separation in the $1/\beta^2$ region provides additional discrimination, but is not used in the data discussed below. Muon/pion separation above 600 MeV/c is accomplished by using the flux return iron as a hadron absorber.

The Mark III was installed at SPEAR in the fall of 1981. Two main physics topics have dominated data-taking to this point. A sample of 2.7×10^6 produced J/ψ's has been studied for evidence of glueball production, and ~9000 nb^{-1} has been accumulated at the ψ'' resonance ($\sqrt{s} = 3.768$ GeV) for the study of D meson decays. This paper will discuss (largely qualitatively) initial preliminary results from the first 5000 nb^{-1} of the ψ'' data sample. It is not possible, at this time, to present definitive experimental results on D meson decays, but an attempt will be made to indicate the improvement in understanding of the mechanism of charm decay that can be expected from the Mark III data.

THE EXPERIMENTAL SITUATION

The most striking evidence that the simple light quark spectator model of charm decay is inadequate is, of course, the non-equality of D° and D^+ lifetimes. Lifetimes of the $D°$, D^+, F^+ and Λ_c^+ have been measured directly in emulsion, bubble chamber and active target electronic experiments, with the charmed particles produced by cosmic rays and hadron, photon and neutrino beams at accelerators. In e^+e^- annihilation, direct measurements of the $D°$ lifetime have been made, as well as measurements of the ratio of $D°$ and D^+ lifetimes by determination of the ratio of semileptonic branching ratios. As the accompanying article by Niu deals with the subject of charmed particle lifetimes in great detail, only the results of Niu's compilation[6] will be quoted here:

$$\tau(D°) = (3.9 \pm 0.4) \times 10^{-13} \text{ sec}$$
$$\tau(D^+) = (8.2 ^{+1.3}_{-0.9}) \times 10^{-13} \text{ sec}$$
$$\tau(F^+) = (3.2 ^{+1.4}_{-0.8}) \times 10^{-13} \text{ sec}$$
$$\tau(\Lambda_c^+) = (2.2 ^{+0.7}_{-0.4}) \times 10^{-13} \text{ sec}$$

The experimental numbers entering into those averages are subject to systematic errors, given that efficiencies for detached two- and three-prong vertices may be different, and scanning biases must be accounted for. The ratio method, employed by both Mark II[7] and DELCO[8] at SPEAR is subject to different systematics. The DELCO measurement, of the ratio of events containing one and two electrons, depends crucially on knowledge of the electron detection efficiency. The method employed by Mark II (and Mark III), that of measuring the number of electrons of correct sign recoiling against reconstructed $D°$ and D^+ hadronic decays ("tags") at the ψ'' resonance where D mesons are produced in pairs, depends on knowledge of the electron misidentification probability. These two experiments yield the results:

$$\tau(D^+)/\tau(D°) > 4.3 \text{ at } 95\% \text{ CL} \qquad \text{(DELCO)}$$

$$= 3.1^{+4.6}_{-1.4} \qquad \text{(Mark II)}$$

The Mark II result is consistent with the results of individual measurements, while the DELCO result is difficult to reconcile with others.

The Mark III measurement is not complete as of this writing, so we must limit ourselves here to a discussion of the expected precision of the result. First, the question of the statistics of the tags. The Mark II used a total of just less than 800 tagged D's. In the 5000 nb^{-1} of Mark III data presented herein, there are more than 2100 tags in the six decay modes shown in Table 1. The full sample will contain ~3800 events in these modes. Beam constrained mass plots of these six modes are shown in Figure 1. The

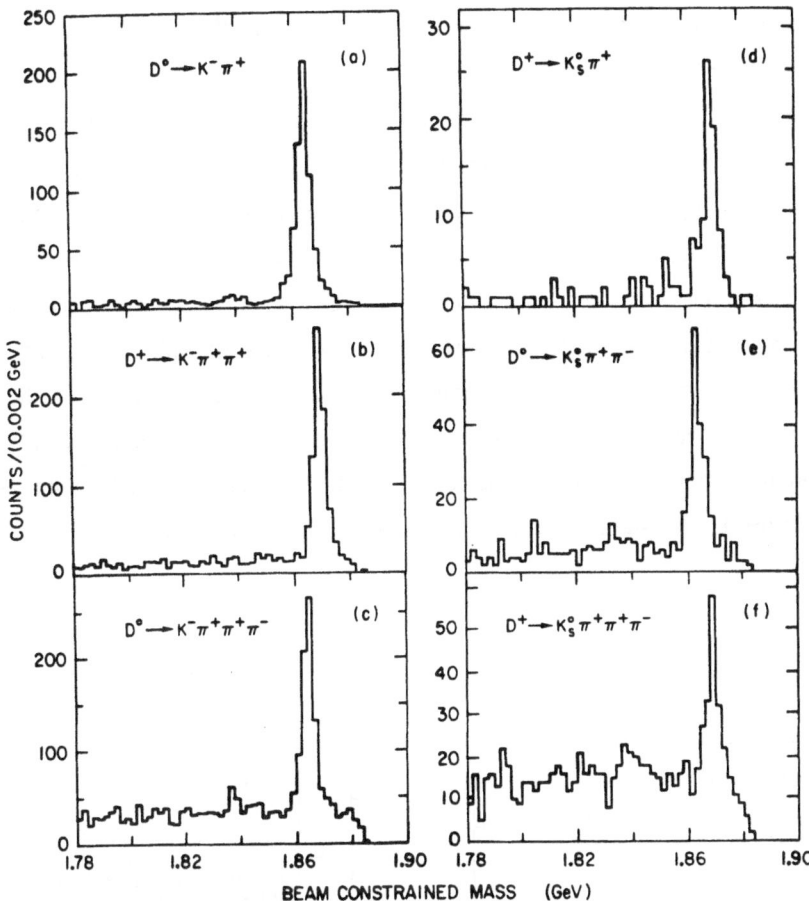

Fig. 1.   Beam constrained mass distributions for the 5000 nb^{-1} data sample of ψ'' data. (a) $K^-\pi^+$, (b) $K^-\pi^+\pi^+$, (c) $K^-\pi^+\pi^+\pi^-$, (d) $K_s^\circ\pi^+$, (e) $K_s^\circ\pi^+\pi^-$ and (f) $K_s^\circ\pi^+\pi^+\pi^-$.

tag sample is thus increased by a factor of ~5. At least as important as this increase in statistics is the improvement in $\pi \to e$ misidentification probability, which is done in the Mark III by a combination of time-of-flight and electromagnetic shower shape techniques. Using a sample of pions from K_s° decay and a sample of electrons from radiative Bhabhas, this misidentification probability has been measured to be better than 3% over the momentum spectrum of electrons from D decays. An additional improvement of a factor of 10 is expected by use of dE/dx measurements in the inner layer of the drift chamber. This combination of improvements is expected to lead to a measurement of the

D lifetime ratio to a precision of better than 20%, comparable to the precision of the current world average, in a single measurement.

Inspection of Table 1 also indicates that, with the exception of the $\overline{K}^\circ\pi^+\pi^+\pi^-$ mode, the Mark III results are in excellent agreement with the branching ratio measurements of the Mark II.[7]   Figures 2 and 3 show beam constrained mass distributions for four additional D decay modes from the partial Mark III sample. The $D^\circ \to \overline{K}^\circ\pi^\circ$ and $D^+ \to \overline{K}^\circ\pi^+\pi^\circ$ modes were seen by the Mark II with 8 ± 4 and 9 ± 5 events, respectively. Figure 2 shows that these modes are confirmed by the Mark III with 25 and 105 signal events. The $\overline{K}^\circ\pi^\circ$ mode is particularly important, as this mode is expected, in the light quark spectator model, to be suppressed by a factor of $\sim$20 with respect to $K^-\pi^+$, by a combination of coupling constants and color mismatch. A large $\overline{K}^\circ\pi^\circ$ branching ratio is thus confirmed by the Mark III data. The modes $D^+ \to K^-\pi^+\pi^+\pi^\circ$ and $D^\circ \to \overline{K}^\circ\pi^+\pi^+\pi^-\pi^-$

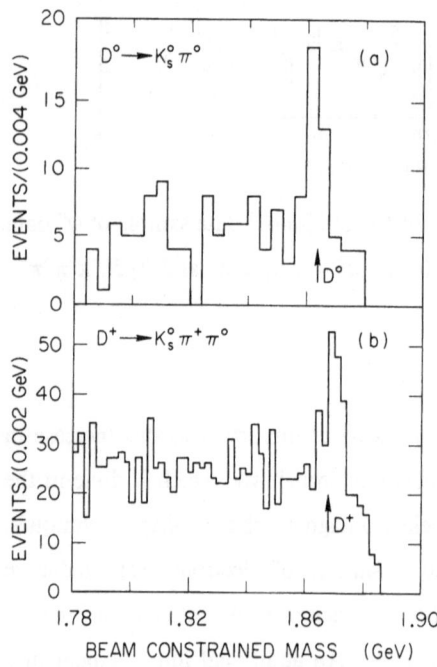

Fig. 2. Beam constrained mass distributions for (a) $K_s^\circ\pi^\circ$ and (b) $K_s^\circ\pi^+\pi^\circ$.

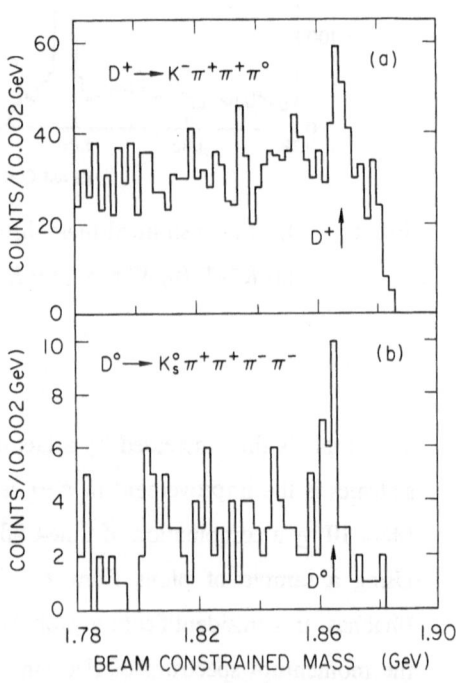

Fig. 3. Beam constrained mass distributions for (a) $K^-\pi^+\pi^+\pi^\circ$ and (b) $K_s^\circ\pi^+\pi^+\pi^-\pi^-$.

(Figure 3) are new; both are present with substantial branching ratios. In general, the Mark III observes decay modes containing π°'s with a much larger efficiency than the Mark II, due to the improvement in low energy photon detection efficiency.

To this point, D decay branching ratios at the ψ'' have been extracted by dividing the measured $\sigma_D \cdot BR(D \rightarrow \overline{K}(n\pi))$ by the measured $D\overline{D}$ production cross section at the ψ'' resonance, on the assumption that D° and $D^\pm$ are produced at the ψ'' equally except for a threshold factor due to the $D^\circ - D^\pm$ mass difference. The D semileptonic branching ratios, however, are measured absolutely, by counting correct sign electrons recoiling against reconstructed hadronic D° or $D^\pm$ hadronic decays. With the improvement in statistics and detection efficiency of the Mark III, it will be possible to use "double tags", that is events in which two hadronic decays are reconstructed, to also measure several of the hadronic decay branching ratios in an absolute manner, albeit with reduced statistical precision. In this manner, the uncertainties of the D production cross section can be removed, facilitating normalization of hadronic and photoproduction cross sections for charmed particles.

The are two other approaches to the study of hadronic charm decays which can shed light on the details of the decay process. One way to distinguish between models proposed to deal with the failures of the light quark spectator model is to search for signature decay modes. For example, the decay $D^\circ \rightarrow \varphi^\circ \overline{K}^\circ$ is forbidden in the spectator model (modulo OZI violation) but is allowed in the W exchange model at an expected level of $2 - 5 \times 10^{-1}$ of the $D^\circ \rightarrow \overline{K}^\circ \rho^\circ$ rate.[9] The Mark III is sensitive to this decay in the mode $D^\circ \rightarrow \overline{K}^\circ K^+ K^-$, which indeed shows a significant D° signal (Figure 4). Since the D° is produced with a unique momentum at the ψ'', a cut of ± 50 MeV about this momentum serves to remove background. Figure 5 displays the $K_s K^+ K^-$ invariant mass plotted against $K^+ K^-$ mass after such a cut. It will be seen that there is a significant cluster of events at low $K^+ K^-$ mass and at the D° mass, whereas the phase space distribution of $K^+ K^-$ is expected to peak much higher. Since the Mark III resolution on the φ is much narrower than the observed $K^+ K^-$ distribution around the φ mass, these events cannot come solely from $D^\circ \rightarrow \overline{K}^\circ \varphi^\circ$. Observation of even a small number of such events would indicate a significant $D^\circ \rightarrow \overline{K}^\circ \varphi^\circ$ branching ratio, since detection efficiency of this mode is a few percent due to the high probability of decay of slow $K^\pm$. We are unable to make a more precise statement at this time, but a detailed study of the full data sample should allow, at a minimum, the establishment of a useful limit on this decay mode.

367

Table 1. Mark III Data (5000 nb^{-1} sample) at the ψ''

| Decay Mode | Mark II | | Mark III | | |
	Events	Efficiency	Efficiency	Events Expected	Events Seen
$K^-\pi^+$	263 ± 17	0.39	0.48	511 ± 33	523
$K^-\pi^+\pi^+$	239 ± 17	0.22	0.41	704 ± 50	666
$K^-\pi^+\pi^+\pi^-$	185 ± 18	0.095	0.23	708 ± 69	602
$\overline{K}^\circ\pi^+$	36 ± 7	0.09	0.16	99 ± 19	69
$\overline{K}^\circ\pi^+\pi^-$	32 ± 8	0.04	0.10	124 ± 31	163
$\overline{K}^\circ\pi^+\pi^+\pi^-$	21 ± 9	0.04	0.09	203 ± 68	97

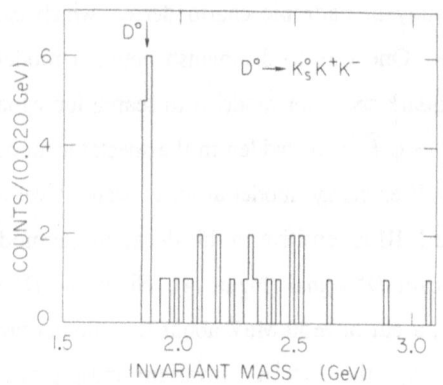

Fig. 4. Invariant mass distribution for $K_s^\circ K^+K^-$.

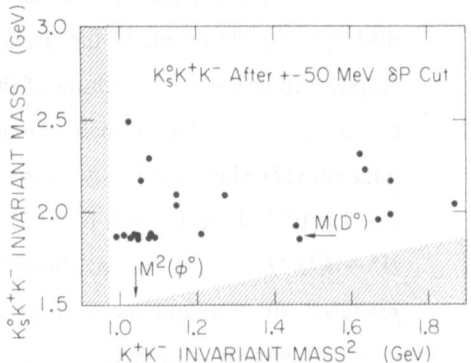

Fig. 5. $M^2(K^+K^-)$ vs. invariant mass of $K_s^\circ K^+K^-$, with a cut of ± 50 MeV about the momentum of the D°.

The second approach to the mechanism of the decay process centers on the study of the isospin content of the hadronic final state. While the light quark spectator process leads to $I = 1/2$ and $I = 3/2$ final states, the W exchange process in D° decay leads only to $I = 1/2$ states. Thus a measurement of the isospin content of the final state can be analyzed to determine the importance of exchange diagrams. This idea is most cleanly formulated in the context of pseudoscalar-vector final states in two triangle relations between D decay amplitudes:

$$A(D^\circ \to K^- \rho^+) + \sqrt{2}A(D^\circ \to \overline{K}^\circ \rho^\circ) - A(D^+ \to \overline{K}^\circ \rho^+) = 0$$

and

$$A(D^\circ \to K^{*-}\pi^+) + \sqrt{2}A(D^\circ \to \overline{K}^{*\circ}\pi^\circ) - A(D^+ \to \overline{K}^{*\circ}\pi^+) = 0$$

In terms of measured quantities, these relations can be summarized as follows:

$$\left[\sqrt{B(D^\circ \to K^{*-}\pi^+)} - \sqrt{B(D^+ \to \overline{K}^{*\circ}\pi^+)(\tau(D^\circ)/\tau(D^+))} \right]^2$$

$$\leqslant 2\, B(D^\circ \to \overline{K}^{*\circ}\pi^\circ) \leqslant$$

$$\left[\sqrt{B(D^\circ \to K^{*-}\pi^+)} + \sqrt{B(D^+ \to \overline{K}^{*\circ}\pi^+)(\tau(D^\circ)/\tau(D^+))} \right]^2$$

and

$$\left[\sqrt{B(D^\circ \to K^- \rho^+)} - \sqrt{B(D^+ \to \overline{K}^\circ \rho^+)(\tau(D^\circ)/\tau(D^+))} \right]^2$$

$$\leqslant 2\, B(D^\circ \to \overline{K}^\circ \rho^\circ) \leqslant$$

$$\left[\sqrt{B(D^\circ \to K^- \rho^+)} + \sqrt{B(D^+ \to \overline{K}^\circ \rho^+)(\tau(D^\circ)/\tau(D^+))} \right]^2 .$$

The interest in these triangle relations stems from the decomposition of the $D \to PV$ amplitudes, which we will generically call A_{+-}, A_{00} and A_{0+}, into $I = 1/2$ and $I = 3/2$ amplitudes. Including the possibility of final state interactions,[10] these can be written as:

$$A_{+-} = \left[\sqrt{\frac{2}{3}} A_{1/2} e^{i\delta_{1/2}} + \sqrt{\frac{1}{3}} A_{3/2} e^{i\delta_{3/2}} \right] V_{cs} V_{ud}^{*} A(D \rightarrow PV)$$

$$A_{00} = \left[-\sqrt{\frac{1}{3}} A_{1/2} e^{i\delta_{1/2}} + \sqrt{\frac{2}{3}} A_{3/2} e^{i\delta_{3/2}} \right] V_{cs} V_{ud}^{*} A(D \rightarrow PV)$$

$$A_{0+} = \sqrt{3} A_{3/2} e^{i\delta_{3/2}} V_{cs} V_{ud}^{*} A(D \rightarrow PV)$$

It has not been possible, thus far, to check the triangle relations or to perform the isospin decomposition, since only two $D \rightarrow K^{*}\pi$ and two $D \rightarrow K\rho$ branching ratios have been measured. The experimental situation[7,11,12] is summarized in Table 2. It will be noted that only upper limits exist for $D^{\circ} \rightarrow \overline{K^{*}}^{\circ}\pi^{\circ}$, and that no data on $D^{+} \rightarrow \overline{K}^{\circ}\rho^{+}$ has been published. While it is not yet possible to present fully analyzed data on these modes, the Mark III has been able to measure a sufficient number of $D \rightarrow K\pi\pi$ modes to permit the extraction of all six $D \rightarrow K^{*}\pi$ and $D \rightarrow K\rho$ amplitudes. These are indicated by the symbol $\sqrt{}$ in Table 2.

Recall that the W exchange diagram, available only to the D°, leads to an $I = 1/2$ final state. In the limit of a purely $I = 1/2$ final state, A_{0+}, which has not to this point been measured, vanishes. Thus, in this (unrealistic) limit, the simpler ratio

$$\frac{B(D^{\circ} \rightarrow K^{*-}\pi^{+})}{B(D^{\circ} \rightarrow \overline{K^{*}}^{\circ}\pi^{\circ})} = \frac{B(D^{\circ} \rightarrow K^{-}\rho^{+})}{B(D^{\circ} \rightarrow \overline{K}^{\circ}\rho^{\circ})} = 2$$

obtains.

While both Mark II and E516 find ratios for the $K^{*}\pi$ modes which are consistent with 2, the experiments differ on the $K\rho$ modes. This difference resides in the $D^{\circ} \rightarrow K^{-}\rho^{+}$ branching ratio, although, the results, having large errors, are not inconsistent.

With the Mark III data, it will be possible to check the full triangle relation for both PV channels. As an example, a few of these channels are exhibited below, for the Mark III partial sample of 5000 nb^{-1}. The $K^{-}\pi^{+}\pi^{\circ}$ channel, shown in Figure 6, is the source of the differing Mark II and E516 measurements. The Dalitz plot, obtained with a cut of ± 5 MeV about the D° mass, contains 480 events on a background of 400, is shown in Figure 7a. The $\pi^{+}\pi^{\circ}$ projection (Figure 7b) indeed shows a substantial $K^{-}\rho^{+}$ signal. There is also evidence for both $K_{s}^{\circ}\pi^{\circ}$ and $K^{*-}\pi^{+}$ in this channel.

Table 2. $D \to K^*\pi$ and $D \to K\rho$ Measurements

Channel	Final State	Branching Ratio (%)			Observed by
		Mark II	E516	NA11	Mark III
$D^\circ \to K^{*-}\pi^+$	$K^-\pi^\circ\pi^+$	3.6 ± 1.3	$3.4^{+3.9}_{-2.8}$	---	✓
	$\overline{K^\circ}\pi^-\pi^+$	---	---	---	✓
$D^\circ \to \overline{K^{*\circ}}\pi^\circ$	$K^-\pi^+\pi^\circ$	$1.4^{+2.3}_{-1.4}$	$0.9^{+1.4}_{-0.9}$	---	✓
	$\overline{K^\circ}\pi^\circ\pi^\circ$	---	---	---	---
$D^+ \to \overline{K^{*\circ}}\pi^+$	$K^-\pi^+\pi^+$	<3.7 at 90% C.L.	---	<1.0 at 90% C.L.	✓
	$\overline{K^\circ}\pi^\circ\pi^+$	---	---	---	✓
$D^\circ \to K^-\rho^+$	$K^-\pi^+\pi^\circ$	$7.2^{+3.0}_{-3.1}$	$3.2^{+2.3}_{-1.8}$	---	✓
$D^\circ \to \overline{K^\circ}\rho^\circ$	$\overline{K^\circ}\pi^+\pi^-$	$0.1^{+0.6}_{-0.1}$	---	---	✓
$D^+ \to \overline{K^\circ}\rho^+$	$\overline{K^\circ}\pi^\circ\pi^+$	---	---	---	✓

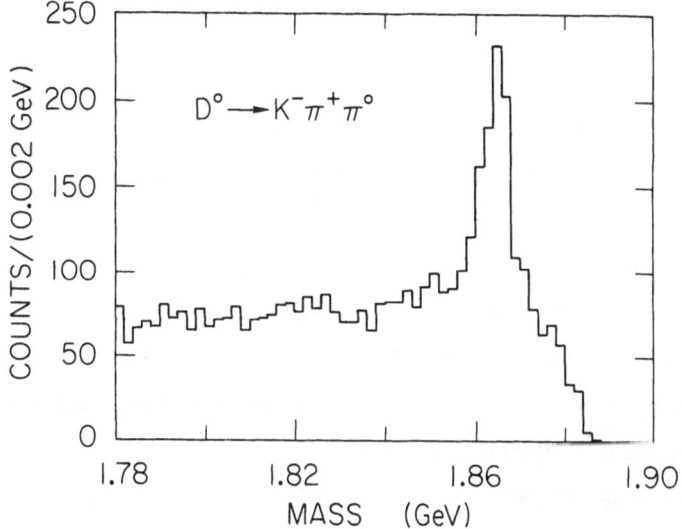

Fig. 6.   Beam constrained mass distribution for $K^-\pi^+\pi^\circ$.

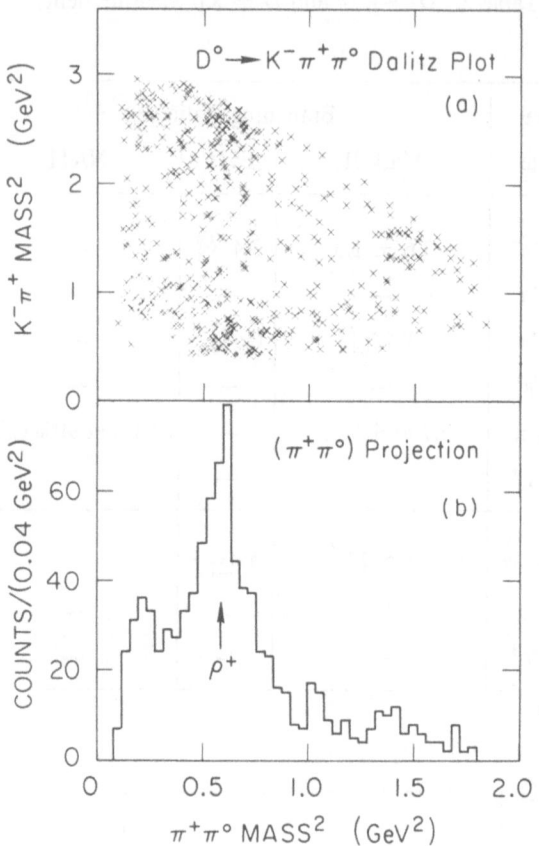

Fig. 7.   (a) Dalitz plot for $D^\circ \to K^-\pi^+\pi^\circ$.

(b) $\pi^+\pi^\circ$ projection of the Dalitz plot.

The $K^-\pi^+\pi^+$ mass distribution is shown in Figure 1.   The Dalitz plot with a ± 5 MeV cut around the D^+ mass is shown in Figure 8a.   It contains 720 events with a background of 75. Figure 8b shows $K^-\pi_1^+$ projection, which exhibits a clear $K^{*\circ}\pi^+$ signal.

The $K_s^\circ\pi_1$ vs. $\pi^+\pi^-$ shown in Figures 9 and 10, respectively. While a strong $K^{*\circ}\pi^+$ signal is seen, the $\overline{K}^\circ\rho^\circ$ signal is quite weak.

The last mode, $\overline{K}^\circ\pi^+\pi^\circ$, shown in Figure 2 is the other D decay mode whose Dalitz plot analysis has not been published. The Dalitz plot, containing 92 events on a background of 115 is shown in Figure 11a. The $\pi^+\pi^\circ$ projection (Figure 11b) appears to show a $\overline{K}^\circ\rho^+$ component.

These $D \to K\pi\pi$ distributions are currently being analyzed by the maximum likelihood method to extract all six PV rates. Upon completion of this work, it will be

372

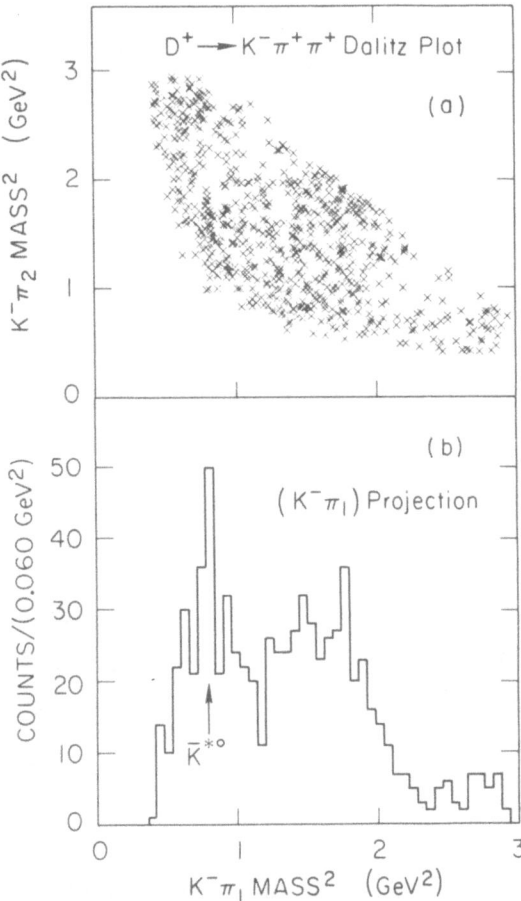

Fig. 8. (a) Dalitz plot for $D^+ \to K^-\pi^+\pi^+$.

(b) $K^-\pi_1$ projection of the Dalitz plot.

possible to check the complete triangle relations, enabling us to go beyond the crude test for an $I = 1/2$ final state.

One of the most surprising results to emerge from the Mark II study of D decays was the observation that the Cabibbo-suppressed decay modes $D^\circ \to K^-K^+$ and $D^\circ \to \pi^-\pi^+$ did not occur with equal rates.[13] The naive expectation was that both decays should happen at a rate $\tan^2\theta_c$ of $D^\circ \to K^-\pi^+$ or ~5%. The measured ratio $B(D^\circ \to K^-K^+)/B(D^\circ \to \pi^-\pi^+) = 3.4$ sparked a great deal of speculation. Explanations advanced included unexpectedly large SU(3) violation, an unexpectedly large role for Penguin diagrams, the existence of a charged Higgs' meson or final state interactions.

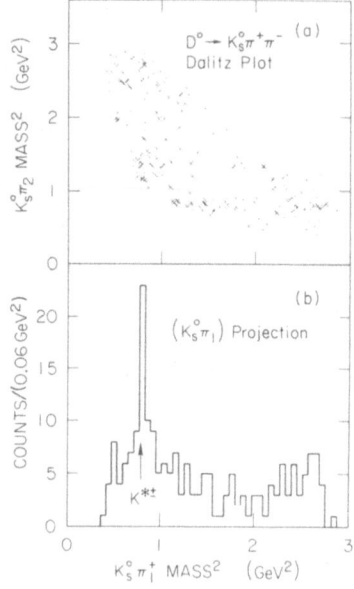

Fig. 9. (a) Dalitz plot for $D^\circ \to K_s^\circ \pi^+ \pi^-$. (b) $K_s^\circ \pi$ projection of the Dalitz plot.

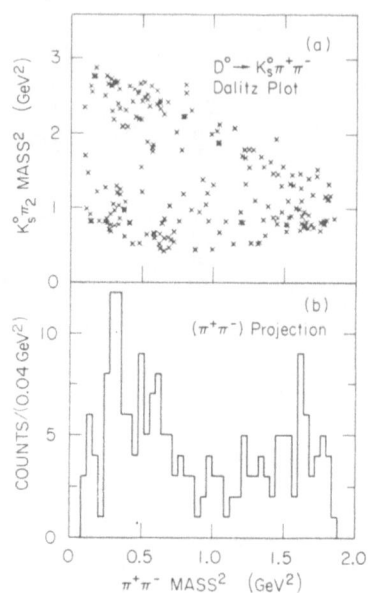

Fig. 10. (a) Dalitz plot for $D^\circ \to K_s^\circ \pi^+ \pi^-$. (b) $\pi^+ \pi^-$ projection of the Dalitz plot.

The first contribution of the Mark III to the puzzle of the Cabibbo-suppressed modes consists of a confirmation of the surprising experimental result. Figure 12 shows the invariant mass of the allowed $K^- \pi^+$ and suppressed $K^- K^+$ and $\pi^- \pi^+$ channels plotted against ΔP, the difference between the expected (unique) D° momentum from ψ'' production and the observed momentum. After a cut of ± 40 MeV in ΔP, the three invariant mass projections are shown in Figure 13. The $K^- \pi^+$ channel shows the expected strong signal. The $K^- K^+$ channel has a strong D° signal, accompanied by a clearly separated peak at 1985 MeV caused by pions misidentified as kaons by the time-of-flight system. The $\pi^- \pi^+$ channel shows a smaller D° signal, with a lower satellite caused by $K \to \pi$ misidentification. A gaussian fit to the $K^- \pi^+$ peak yields $\sigma = 19$ MeV and 453 ± 25 events above background. With gaussians of mass 1864 MeV and width 19 MeV fit to the $\pi^- \pi^+$ and $K^- K^+$ distributions, together with gaussians centered on the misidentification peaks with $\sigma = 21$ MeV due to shifted D° momentum, 14 ± 5 $\pi^- \pi^+$

374

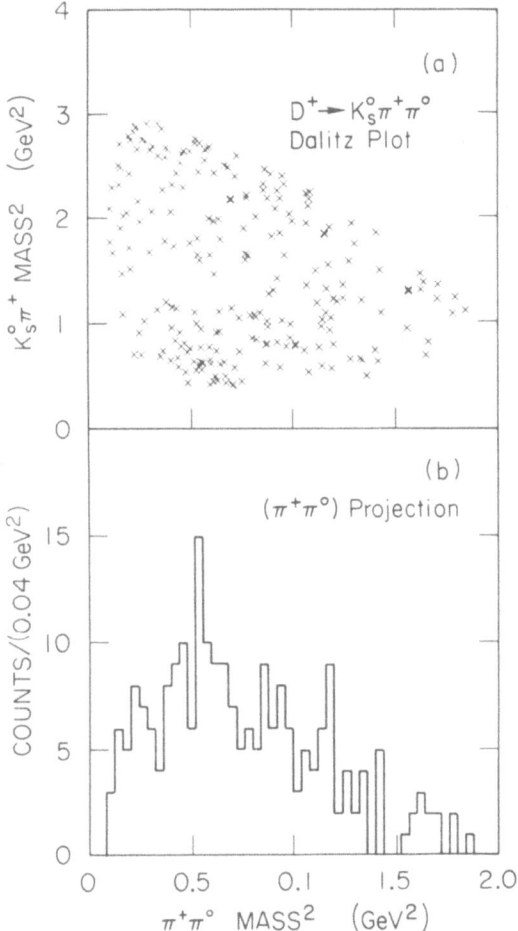

Fig. 11.　(a) Dalitz plot for $D^+ \rightarrow K_s^\circ \pi^+ \pi^\circ$.

(b) $\pi^+\pi^\circ$ projection of the Dalitz plot.

events and 52 ± 8 K^-K^+ are found above background. Corrections of the order of 20% must be made for the differing efficiency of the three modes, but it is clear that the Mark III data supports a conclusion that indeed $B(D^{\bullet} \rightarrow K^-K^+)$ is three times as large as $B(D^\circ \rightarrow \pi^-\pi^+)$.

With the experimental result confirmed, the solution to the puzzle of Cabibbo-suppressed decays lies in the measurement of other decay modes. The current experimental situation on Cabibbo-suppressed hadronic decay modes of the D is summarized in Tables 3 and 4, which include recent limits on all neutral decay modes set by the Crystal

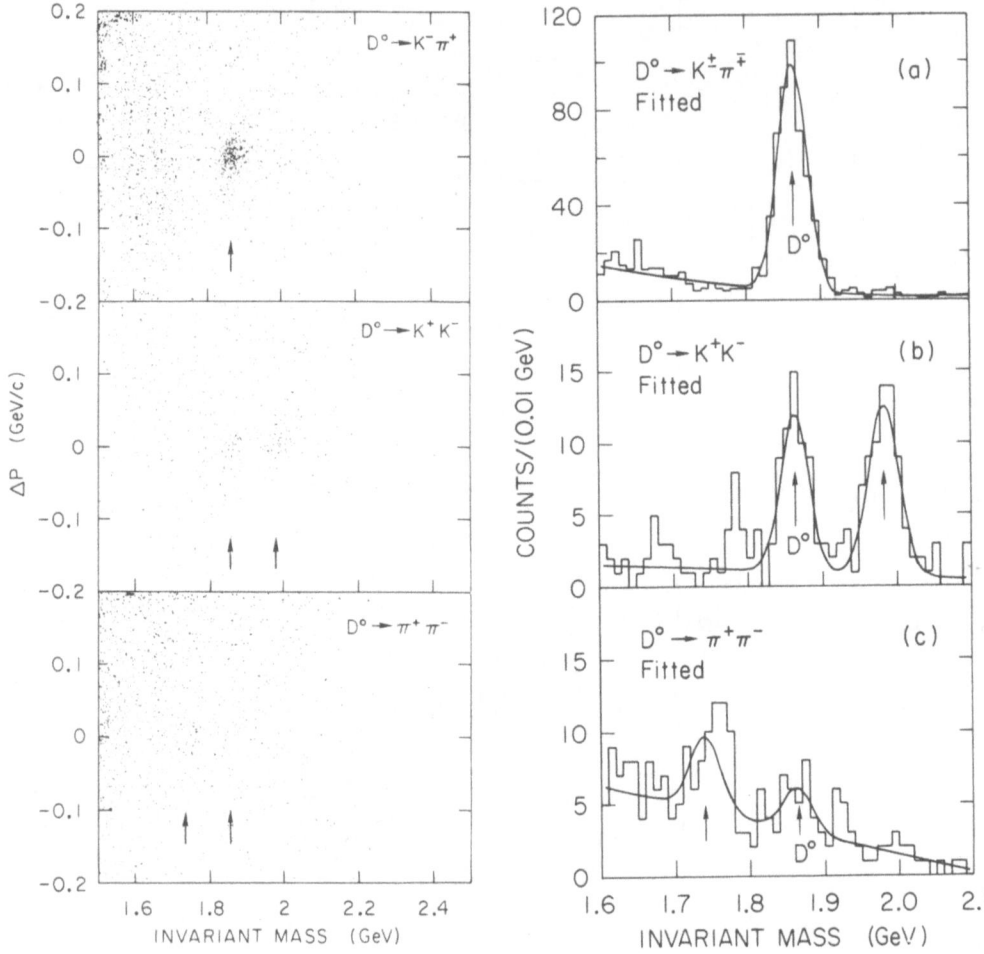

Fig. 12. Invariant mass distributions vs. deviation from expected D° momentum for (a) $K^-\pi^+$, (b) K^+K^- and (c) $\pi^+\pi^-$.

Fig. 13. Invariant mass projections, with fits to Gaussians, for (a) $K^-\pi^+$, (b) K^+K^- and (c) $\pi^+\pi^-$.

Ball.[14]   In Table 3, we again denote by $\sqrt{}$ marks the modes which are observed by the Mark III. A sufficient number of modes are observed to permit an analysis of these decays in terms of the possible contributing amplitudes and, in combination with related

Cabibbo-allowed decays, the measurement of all the two body decay amplitudes in D decay. A particularly important decay mode is $D^+ \to \pi^+\pi^\circ$, since the ratio

$$\frac{B(D^+ \to \pi^+\pi^\circ)}{B(D^+ \to \pi^+\bar{K^\circ})} = 1/2\left|\frac{V_{cd}}{V_{cs}}\right|^2$$

should allow determination of $|V_{cd}/V_{cs}|$ free of final state interaction effects, both final states being exotic. Since the $\pi^+\pi^-$ and K^+K^- modes could have substantial final state corrections, the $\pi^+\pi^\circ$ mode may be the most important link in our understanding of the puzzle of the Cabibbo-suppressed D decays.

A more direct understanding of the V_{cd}/V_{cs} ratio would, of course, come from the observation of the rate for Cabibbo-suppressed semi-leptonic decays, such as $D^\circ \to \pi^-e^+\nu$ or $D^+ \to \pi^\circ e^+\nu$. With the data sample of the Mark III, however, while a few examples of these decays are likely to be seen, the statistics are such that a useful measurement of the branching ratios is unlikely. It will be possible, however, to reconstruct of the order of 100 $D \to K e\nu$ and $D \to K^*e\nu$ events recoiling against hadronic decays. The Dalitz plots of these decays will be analyzed, providing, for the first time, information on the q^2 dependence of the form factors in these semi-leptonic decays.

Table 3. Cabibbo-Suppressed Hadronic Decays of D Mesons

MODE $\left(\dfrac{\text{Suppressed}}{\text{Allowed}}\right)$	Mark II Measurement	Observed by Mark III
$\pi^-\pi^+/K^-\pi^+$	0.033 ± 0.015	✓
$K^-K^+/K^-\pi^+$	0.113 ± 0.030	✓
$\pi^\circ\pi^+/\bar{K^\circ}\pi^+$	<0.30 at 90% C.L.	✓
$\bar{K^\circ}K^+/\bar{K^\circ}\pi^+$	0.25 ± 0.15	✓
$\pi^-\pi^+\pi^+/K^-\pi^+\pi^+$	<0.084 at 90% C.L.	✓
$K^-K^+\pi^+/K^-\pi^+\pi^+$	<0.14 at 90% C.L.	✓
$\pi^-\pi^+\pi^-\pi^+/K^-\pi^+\pi^-\pi^+$	<0.21 at 90% C.L.	---

Table 4. Limits on Neutral $D°$ Decays (Crystal Ball)

Mode	Branching Ratio at 90% C.L. (%)
$\pi°\pi°$	< 0.28
$\eta°\pi°$	< 0.74
$\eta°\eta°$	< 1.0
$\pi°\pi°\pi°$	< 1.6
$\eta°\pi°\pi°$	< 6.6
$\overline{K}°\pi°$	< 7.6

CONCLUSIONS

This paper has briefly reviewed the experimental situation in the study of charmed meson decay. There are quite a number of ongoing experiments which will improve our understanding of charmed particle lifetimes and branching ratios. An interim status report on the analysis of the Mark III ψ'' data sample has been presented. At the time of this writing, the analysis is incomplete, but some qualitative results are intended to show that the Mark III data will allow a substantial advance in our understanding of the mechanism of charmed particle decay by means of the following measurements:

1) Improved precision on many $D°$ and D^+ branching ratios

2) Measurement of several new allowed and Cabibbo-suppressed branching ratios

3) Study of the Dalitz plots of all the allowed pseudoscalar-vector decays, leading to a qualitative understanding of the $I = 1/2$ and $I = 3/2$ fractions in the final states

4) A substantially improved measurement of the $D°/D^+$ lifetime ratio by study of the semi-leptonic branching ratios

5) Measurement of absolute branching ratios for several hadronic decay modes by means of double-tag events.

6) Understanding of the puzzle of the Cabibbo-suppressed decays by measurement of several new branching ratios

7) Study of the Dalitz plots of D semi-leptonic decays.

REFERENCES

1. J. Ellis, M. K. Gaillard, and D. V. Nanopoulos, Nucl. Phys. B100:313 (1975).

2. B. Guberina, S. Nussinov, R. D. Peccei, and R. Rückl, Phys. Lett. 89B:111 (1979);

   Y. Koide, Phys. Rev. D20:1739 (1979);

   K. Jagannathan and V. S. Mathur, Phys. Rev. D21:3165 (1980);

   N. Deshpande, M. Gronau and D. Sutherland, Phys. Lett. 90B:431 (1980).

   G. Altarelli, G. Curci, G. Martinelli and R. Petrarca, Phys. Lett. 99B:141 (1981) and Nucl. Phys. B187:461 (1981).

3. M. Bander, D. Silverman and A. Soni, Phys. Rev. Lett. 44:7 (1980);

   W. Bernreuther, O. Nachtmann and B. Stech, Z. Phys. C4:257 (1980);

   S. P. Rosen, Phys. Rev. Lett. 44:4 (1980);

   H. Fritzsch and P. Minkowski, Phys. Lett. 90B:455 (1980).

4. V. Barger and S. Pakvasa, Phys. Rev. Lett. 43:812 (1979);

   H. Fritzsch and P. Minkowski, Nucl. Phys. B171:413 (1980);

   I. Bigi, Phys. Lett. 90B:177 (1980);

   D. Sutherland, Phys. Lett. 90B:173 (1980);

   L. F. Abbott, P. Sikivie and M. B. Wise, Phys. Rev. D21:768 (1980);

   M. Suzuki, Phys. Rev. Lett. 43:818 (1979).

5. D. Bernstein et al., SLAC-PUB-3222 (1983), to appear in Nucl. Instrum. and Methods;

   Members of the MARK III Collaboration are R. M. Baltrusaitis, D. Coffman, G. Dubois, J. Hauser, D. G. Hitlin, J. D. Richman, J. J. Russell, and R. H. Schindler, California Institute of Technology; K. O. Bunnell, R. E. Cassell, D. H. Coward, S. Dado, K. F. Einsweiler, L. Moss, R. F. Mozley, A. Odian, J. R. Roehrig, W. Toki, F. Villa, N. Wermes, and D. E. Wisinski, Stanford Linear Accelerator Center; D. E. Dorfan, R. Fabrizio, F. Grancagnolo, R. P. Hamilton, C. A. Heusch, L. Koepke, W. Lockman, R. Partridge, J. Perrier, H. F. Sadrozinski, T. L. Schalk, A. Seiden, and A. Weinstein, University of California at Santa Cruz; J. J. Becker, G. T. Blaylock, B.

Eisenstein, G. Gladding, S. A. Plaetzer, A. L. Spadafora, J. J. Thaler, B. Tripsas, A. Wattenberg, and W. J. Wisniewski, University of Illinois, Champaign-Urbana; J. S. Brown, T. H. Burnett, V. Cook, C. Del Papa, A. L. Duncan, P. M. Mockett, A. Nappi, J. C. Sleeman, and H. J. Willutzki, University of Washington, Seattle.

6. K. Niu, these Proceedings.

7. R. Schindler et al., Phys. Rev. D24:78 (1981).

8. W. Bacino et al., Phys. Rev. Lett. 45:329 (1980).

9. I. Bigi and M. Fukugita, Phys. Lett. 91B:121 (1980).

10. H. J. Lipkin Phys. Rev. Lett. 44:710 (1980).

11. D. J. Summers et al., Phys. Rev. Lett. 52:410 (1984).

12. R. Bailey et al., Phys. Lett. 132B:237 (1983).

13. G. Abrams et al., Phys. Rev. Lett. 43:481 (1979).

14. R. Partridge, Ph.D. Thesis, California Institute of Technology, 1984 (unpublished).

DISCUSSION

TRILLING:

In determining the lifetime ratio from semileptonic branching ratios, one must be sure that only Cabibbo favored decays are counted.

HITLIN:

I believe you are referring to the fact that the phase space corrections for Cabibbo-suppressed semileptonic decays of the D^0 and D^+ differ by more than the phase space corrections to the Cabibbo allowed decays of the two charmed mesons. The actual inclusive weak couplings do not differ. It would be lovely to contemplate a measurement of sufficiently high precision for this correction to provide a limitation.

LITTENBERG:

Has anyone proposed studying the W annihilation diagrams in F decay via semileptonic decay? Presumably one could distinguish the annihilation component via the absence of which contain s-quarks (e.g K's, η's) accompanying the leptons.

HITLIN:

Yes. It has been pointed out that in semileptonic decay of the F meson via the annihilation graph, the source of hadrons can only be some form of color singlet gluon emission from the s-quark.

KLEINKNECHT:

How many years of running would you need to get 100 Cabibbo-forbidden leptonic D decays ($D \rightarrow \pi e\nu$)?

HITLIN:

SPEAR runs fifty percent of the time for high energy physics and fifty percent for synchrotron radiation. We thus get about 15 weeks of beam per calendar year. At the present luminosity of ~ 2×10^{30} cm^{-2} sec^{-1}, it would take perhaps five years. There are plans to install a minibeta scheme, patterned after that of Doris II, in the summer of 1984. This is expected to raise the luminosity by a factor of 2 to 4, depending on the beam energy.

HAGELIN:

Wouldn't it be more conservative to compare inclusive strange final states with inclusive non-strange final states in order to check the expected pattern of Cabibbo suppression? Comparing exclusive modes like $K\pi$ versus $\pi\pi$ does not manifestly take into account SU(3) breaking arising from the relative ease of pulling light quarks out of the vacuum compared with strange quarks.

HITLIN:

That is true. There is however an experimental problem in accurately defining inclusive channels as you describe.

STECH:

The Cabibbo suppressed <u>exclusive</u> decays may show form factor effects (c.f. Fritzsch et al.). In the annihilation process, the slower strange quarks are more likely to form $K\bar{K}$ than the faster down quarks from $\pi\pi$.

TRILLING:

Has the F been seen in non-ϕ $K^+K^-\pi$ as well as in $\phi\pi$?

HITLIN:

I believe the NA11 events are in $K^+K^-\pi$ non-resonant, and $\phi\pi$.

SUPERGRAVITY AND PARTICLE PHYSICS: MODELS

WITH SPONTANEOUS SUPERSYMMETRY BREAKING

S. Ferrara

CERN
Geneva, Switzerland

Supersymmetric quantum field theories are those Lagrangian field theories which possess a set of conserved spinor currents:

$$J^i_{\mu\alpha}(x) \, , \quad \partial^\mu J^i_{\mu\alpha}(x) = 0 \qquad \begin{pmatrix} \alpha = \text{spinor index} \\ \mu = \text{vector index} \\ i = 1 \ \ldots \ N \end{pmatrix} \qquad (1)$$

such that the corresponding charges

$$Q^i_\alpha = \int d^3x \, J^i_{0\alpha}(x) \qquad\qquad (Q^i_\alpha \ \text{Hermitian}) \qquad (2)$$

are constant of motion

$$\frac{d \, Q^i_\alpha}{dt} = 0 \qquad\qquad (3)$$

The basic anticommutation relation

$$\{Q^i_\alpha, Q^j_\beta\} = (\gamma^\mu C)_{\alpha\beta} \, P_\mu \, \delta^{ij} + (\text{central charges}) \qquad (4)$$

(C charge conjugation matrix)

shows the deep interplay between the supersymmetry and the space-time Poincare symmetry. The operator P_μ is the displacement operator of the Poincare algebra. The interactions and particle content of supersymmetric theories strongly depend on the total number of spinor charges N.

From the representations of N-extended supersymmetry on massless one-particle states one gets the following multiplet structure

$$\lambda_{max}, \ \lambda_{max} - \frac{1}{2}, \ \ldots, \ \lambda_{max} - \frac{K}{2}, \ \ldots, \ \lambda_{max} - \frac{N}{2} \qquad (5)$$

$$+ \text{ (PCT conjugate states with } \lambda_{max} + \frac{N}{2} - \lambda_{max})$$

in which λ_{max} is the massless state of maximal (positive) helicity inside the multiplet. From Eq. (5) it follows

$$\lambda_{max} \geq \frac{N}{4} \ \left(\frac{N+1}{4} \ \text{ for odd } N\right) \qquad (6)$$

so that renormalizable theories ($\lambda_{max} \leq 1$) require $N \leq 4$. If one includes gravitational interaction with the graviton being identified with the state $\lambda_{max} = 2$, then one gets the bound $N \leq 8$.

Supersymmetric theories improve the ultraviolet behavior of conventional quantum field theories. There are examples of ultraviolet finite four-dimensional quantum field theories (N=4 Yang-Mills and N=2 Yang-Mills with soft-breaking terms[2]). There are also examples of supersymmetric theories of gravity (supergravity) which have a milder quantum behavior than Einstein theory, and N=8 supergravity is considered to be the most appealing candidate theory for a completely unified theory of all fundamental interactions[3].

Supersymmetry seems to be the only known symmetry which protects scalar masses from huge radiative corrections due to the so called non-renormalization theorem[4] (naturalness principle of small numbers). This fact led to the speculation that supersymmetry may solve the so-called hierarchy problem of conventional G.U.T.'s, i.e., the stability of $M_W/M_X \ll 1$ in perturbation theory[5-6]. An outstanding question is how to embed the observed fundamental interactions in supersymmetric theories.

The gauge group of observed interaction should be

$$G \supset G_0 = SU_C(3) \times SU_L(2) \times U(1) \tag{7}$$

There are two ways to unify G with the algebraic structure given by Eq. (1):

$$[G, Q_\alpha^i] = 0$$

$$[G, Q_\alpha^i] \neq 0 \tag{8}$$

The second possibility is only possible in extended super-gravities since Q_α^i can only be not inert under a gauge symmetry in the context of local supersymmetry[6].

On the other hand the elementary gauge fields of extended supergravities can at most gauge an orthogonal group SO(8) so there is no room for the gauge group G_0.[7] Only dynamically generated gauge symmetries like SU(N) (N $\leq$ 8) can accomodate, in this scenario, the minimal low energy gauge interactions SU(3) $\times$ SU(2) $\times$ U(1).[8]

The first possibility, with Q_α^i inert under G, requires N $\leq$ 4 and if one furthermore assumes that quarks and leptons belong to supersymmetric multiplets at low-energies then N $\leq$ 2. In fact only for N $\leq$ 2 the supersymmetry algebra admits representations with spin 1/2 and spin 0 particles in arbitrary representations R of the gauge group G.

The basic multiplets of N=1 supersymmetry are:

e_μ^a, $\psi_{\mu\alpha}$ (gravitino), spin 2, 3/2)   graviton multiplet

V_μ^A, λ_α^A (gauginos) (spin 1, 1/2) (A = 1...dim G) gauge
                                                                        multiplet

ψ_L^i, z^i (complex scalar partners of left handed fermions) (9)
            i = 1 ... dim R (spin 1/2, 0) matter multiplet

The basic multiplets of N=2 supersymmetry are:

e^a_μ, $\psi_{\mu\pm}$, B_μ (spin 2, 3/2, 1)   (graviton multiplet)

$(V^A_\mu, \lambda^A_\pm, z^A)$ (spin 1/2, 0)   (vector multiplet)

ψ^i_L, ψ^i_R, z^i_1, z_{2i} (spin 1, 1/2, 0)   (matter multiplet)
$$(10)$$

From the structure of the spin 1/2, 0 multiplet (hypermultiplet)
it follows that in N=2 supersymmetry it is unavoidable to have,
in the same multiplets of known fermions, mirror fermions in
complex conjugate representations of the gauge group G.[6]

The connections of supersymmetric theories and elementary
particle interactions is only possible if supersymmetry is broken
in some way. In order not to lose the improved ultraviolet
behavior peculiar to supersymmetric theories and not to further
lose the predictivity of these theories the most natural way is
to demand that supersymmetry be spontaneously broken. Globally
supersymmetric theories can also be broken by soft breaking terms
of at most dimension three.[9] However when supersymmetry is made
local, as it always happens when it is coupled to gravitation,
explicit soft breaking terms are not allowed any more. There is
an analogous counter-part in the breaking of internal symmetries
when they are promoted to gauge symmetries.

Let us confine ourselves to the spontaneous breakdown of N=1
supersymmetry. Spontaneous supersymmetry breaking is
characterized by the condition

$$Q_\alpha |0> \neq 0 \qquad\qquad (11)$$

In field theory Eq. (11) implies that there is some spin-1/2
spinor field $\psi_\alpha(x)$ such that

$$<0| \delta_\varepsilon \psi_\alpha(x) |0> \neq 0 \qquad\qquad (12)$$

under an infinitesimal supersymmetry transformation. Eq. (12) in
turn implies

$$< 0|\{Q_\alpha, \psi_\beta(x)\}|0 > = \varepsilon_{\alpha\beta} < 0|0(x)|0 >$$

$$= \varepsilon_{\alpha\beta} < 0|0(0)|0 > \neq 0 \qquad (13)$$

where $O(x)$ is same scalar operator of dimension two. Its vacuum expectation value determines the scale of supersymmetry breaking M_S

$$<0|0(0)|0> = M_S^2 \qquad (14)$$

The field $\psi_\alpha(x)$ is the Goldstone fermion of spontaneously broken supersymmetry (goldstino). If ψ_α is a different mixture of several spin 1/2 fields then M_S is given by

$$M_S = \left(\sum_i |h_i|^2 + \frac{1}{2} \sum_A D^{A\,2} \right)^{1/4} \qquad (15)$$

when h_i, D^A are dimension 2 operators, functions of the scalar fields of the theory, with non-vanishing v.e.v. The fields h_i, D^A play the role of the auxiliary components of the scalar and vector multiplets defined by Eq. (9), needed to have an off-shell realization of the supersymmetry algebra in field theory. In global supersymmetry the N=1 supersymmetric most general renormalizable gauge interaction is described by the following Lagrangian[5]

$$\mathscr{L}_{YM} = -\frac{1}{4}(F_{\mu\nu}^A)^2 - \frac{1}{2}\bar{\lambda}^A \slashed{D} \lambda^A - \frac{1}{2}\bar{\chi}_{Li} \overleftrightarrow{\slashed{D}} \chi_R^i$$

$$-\frac{1}{2}|D_\mu z_i|^2 - 2|f^i|^2 - \frac{1}{2}\left(-\frac{1}{2}\tilde{g}\, z^{*i}T_i^{Aj}z_j + \xi^A\right)^2$$

$$+ \left(f^{ij}\chi_{Li}\chi_{Lj} - i\,\tilde{g}\, T_j^{\alpha k}z_k \bar{\lambda}_R^A \chi_R^j + h.c. \right)$$

$$+ \frac{i}{4}\theta_A \left[F_{\mu\nu}^A \tilde{F}_{\mu\nu}^A - \frac{1}{2}\partial_\mu (\bar{\lambda}^A \gamma_5 \gamma_\mu \lambda^A) \right] \qquad (16)$$

in terms of the analytic function (superpotential)

$$f(z) = \eta^i z_i + m^{ij}z_i z_j + g^{ijk} z_i z_j z_k \qquad (17)$$

which must be G-invariant and at most of degree three for renormalizability. $\tilde{g}$ is the gauge coupling constant $(\tilde{g} \rightarrow (\tilde{g}_A)$

for a non-simple group G) and ξ^A are Fayet-Iliopoulos terms[10], only possible for Abelian U(1) factors of G. The possible connection of N=1 spontaneously broken supersymmetry and the gauge hierarchy problem implies that particles participating to the observed (non-gravitational) low-energy interactions should form approximate N=1 supermultiplets with some other (yet unobserved) partners with mass splitting of order 100GeV-1TeV[11]. This means that the intramultiplet mass splitting

$$\delta \ m_i^2 = g_i \ M_S^2 \tag{18}$$

expressed in terms of the Goldstino coupling g_i and the super-symmetry breaking scale M_S should be

$$g_i \ M_S^2 \simeq 0(M_W^2) \tag{19}$$

Eq. (19) is the basic connection of broken supersymmetry to the low-energy world.

In model building using global supersymmetry there are essentially two ways for satisfying Eq. (19):

$$g_i \sim 0(1) \quad M_S \sim 0(M_W) \quad \text{(Fayet-Iliopoulos breaking)} \tag{20}$$

$$g_i \sim 0\left(\frac{M_W}{M}\right) \quad M_S \simeq 0\left((M_W M)^{1/2}\right) \quad \text{(Intermediate scale} \tag{21}$$
$$\text{breaking - geometric hierarchy)}$$

The two methods are uniquely selected by the requirement that the scalar superpartners of ordinary quarks and leptons should be much heavier than their fermionic counterparts with masses in the 100 GeV-1 TeV range. This is only possible with a D-type breaking induced by a new U(1) gauge interaction or with a radiative breaking induced by a hidden sector which at the tree level spontaneously breaks supersymmetry through the O'Raifeartaigh mechanism. In global supersymmetry these two mechanisms avoid unwanted tree level mass formulae of the type

$$\sum_J (-)^{2J} (2j+1)m_J^2 = 0 \tag{22}$$

which would be in open conflict with experiment if other colored or charged states (mirror fermions) are not introduced in the theory.

In the radiative breaking scenario radiative mass differences are of the type[11]

$$\delta m \sim (g)^n \frac{M_S^2}{M} \quad \text{when M is a large scale } (M_X \text{ or } M_P) \quad (23)$$

connected to the heavy particle masses which circulate in loop-graphs and couple the hidden sector to the observable sector of the theory. However since $1/M \sim O(k)$, in which k is the gravitational coupling constant the simplest and most natural way to obtain a mass splitting as given by Eq. (21) is through supergravity couplings where it naturally occurs that[12,13,14]

$$\delta m_i \simeq O(k M_S^2). \quad (24)$$

The phenomenon yielding to Eq. (24) is the super Higgs effect[12] which now we discuss.

When supersymmetry is promoted to a gauge symmetry a globally supersymmetric matter system is necessarily coupled to the graviton multiplet given in Eq. (9).

The gauge field of local supersymmetry is a spin 3/2 mass-less field $\psi_{\mu\alpha}(x)$, which in absence of supersymmetry breaking is supposed to describe a massless spin 3/2 particle, the gravitino. If supersymmetry is spontaneously broken the super Higgs phenomenon occurs, namely the Goldstino is eaten up by the gravitino, which becomes massive with mass[15]

$$m_{3/2} = \sqrt{\frac{8\pi}{3}} \frac{M_S^2}{M_P} \quad (25)$$

the gravitino mass depends on M_S quadratically. If $M_S \simeq O\left((M_W M_P)^{1/2}\right)$ then $m_{3/2} \simeq O(M_W)$. Under this circumstance supergravity effects can become important in the region of the Fermi scale.

The possibility of having broken supersymmetry with vanish-ing cosmological constant in N=1 supergravity coupled to Yang-

Mills and matter multiplets is due to the fact that the gravitational effects modify the scalar potential. This is best understood by the fact that scalar potential is a function of the auxiliary fields (h_i, D^A, u) of matter as well as of the graviton multiplet[12]

$$V = V(h_i, D^A, u) \qquad (u = \text{complex scalar field}) \qquad (26)$$

The overall scalar potential is

$$V = - e^{-\mathscr{G}} (\mathscr{G}^i (\mathscr{G}^{-1})_i^{\;j} \mathscr{G}_j + 3)$$

$$+ \frac{1}{2} \tilde{g}^2 \, \text{Re} \, f_{AB}^{-1} (\mathscr{G}^i T_i^{Aj} z_j)(\mathscr{G}^k T_k^{B\ell} z_\ell) \qquad (27)$$

$(\mathscr{G}^i = \frac{\partial \mathscr{G}}{\partial z_i}, \;\; \mathscr{G}_i = \frac{\partial \mathscr{G}}{\partial z^{*i}})$ $\mathscr{G} = \mathscr{G}(z, z^*)$ is a real gauge invariant

function $(\mathscr{G}^i T_i^{Aj} z_j = z^{*i} T_i^{Aj} \mathscr{G}_j)$ which defines the Kahler scalar

potential related to the scalar kinetic terms

$$\mathscr{G}^i_{\;j} \, D_\mu z_i \, D^\mu z^{*j} \qquad (28)$$

and $f_{AB}(z)$ is an analytic function related to the Yang–Mills part of the Lagrangian ($f_{AB} = f_{BA}$)

$$- \frac{1}{4} \, \text{Re} \, f_{AB} F^A_{\mu\nu} F^B_{\mu\nu} + \frac{i}{4} \, \text{Im} \, f_{AB} F^A_{\mu\nu} \tilde{F}^B_{\mu\nu} \qquad (29)$$

If we split the function $\mathscr{G} = - J - \ell n |f(z)|^2$ in which $f(z)$ is the superpotential, then the overall scalar potential becomes

$$V = e^J \left((J^i f + f^i)(J^{-1})_i^{\;j} (J_j f^* + f_j^*) - 3|f(z)|^2 \right)$$

$$+ \frac{1}{2} \tilde{g}^2 \, \text{Re} \, f_{AB}^{-1} (J^i T_i^{Aj} z_j)(J^k T_k^{B\ell} z_\ell) \qquad (30)$$

Spontaneous supersymmetry breaking occurs when the Lagrangian contains a bilinear spin 3/2–spin 1/2 term

$$- \bar{\psi}_{\mu R} \gamma^\mu \psi_L + \text{h.c.} \qquad (31)$$

which in turn defines the goldstino mode

$$\psi_L = e^{-\mathscr{G}/2} \mathscr{G}^i \chi_{Li} - \frac{i}{2} D^A \lambda^A_L \qquad (32)$$

The latter can be rotated away via a finite local supersymmetry transformation.

If $\langle e^{-\mathscr{G}/2}g_i\rangle$, $\langle D^A\rangle$ are not vanishing at $V = 0$, $V_{z_i} = 0$ then we have the super Higgs effect in flat space.

The gravitino mass is given by:

$$m_{3/2} = \langle e^{-\mathscr{G}/2}\rangle \tag{33}$$

The simplest models for particle physics based on spontaneously broken supergravity used a "minimal coupling" of a N=1 Yang-Mills system to N=1 supergravity[5]. This is defined by a flat Kahler metric and canonical Yang-Mills terms:

$$\mathscr{G}^i{}_j = -\frac{1}{2}\delta^i{}_j \; ; \quad f_{AB} = \delta_{AB} \tag{34}$$

In this case the following mass relation holds[12]

$$\text{Supertrace } M^2 = \sum_J (-)^{2J}(2J+1)m^2{}_J = (N-1)\left(2m^2{}_{3/2} - k^2 D^{A2}\right)$$

$$- 2\,\tilde{g}_A D^A \text{Tr}T^A \quad (\text{N = number of chiral multiplets}) \tag{35}$$

The first term in the right-hand side of Eq. (35) shows that each scalar particle of any chiral multiplet, with the exception of the goldstino partners, receives on the average a gravitational mass shift $m^2{}_{3/2}$. If $m_{3/2} \simeq O(M_W)$ the scalar partners of quarks and leptons have masses $O(m_{3/2})$.

To obtain the most general low-energy effective Lagrangian in spontaneously broken supergravity one usually splits the chiral multiplets in two sets $z_i = (x_i,y)$ in which $\langle x\rangle \simeq O(M_p)$, $\langle y\rangle/\langle x\rangle \ll 1$. Then it is assumed that the Kahler metric J defined by Eq. (29) has the following structure in y

$$J = \frac{1}{2}y_i\Lambda^i{}_j(\zeta,\zeta^*)y^{*j} + B^i(\zeta,\zeta^*)y_i + B^*{}_i(\zeta,\zeta^*)y^{*i} \tag{36}$$

in which $\zeta = kx$. Moreover $f(\zeta,y)$, the overall superpotential, must be a finite polynomial of at most degree three in y with ζ-dependent coefficients, such that

$$f(\zeta, y=0) \approx O(M^2) \tag{37}$$

Under the previous assumptions in the limit $M_P \to \infty$, $m_{3/2}$ fixed, the z-fields (hidden sector) responsible for the super Higgs effect decouple from the low-energy fields y and the low-energy effective Lagrangian becomes[5]

$$\mathscr{L}_{Broken} = \mathscr{L}_{SUSY}(f) + m_{3/2}(c^A \bar{\lambda}_L^A \lambda_L^A + h.c.)$$

$$- m_{3/2} \; Re \; h(y) - \frac{1}{2} m^2_{3/2} \; y^{*j} \Lambda^i_j y_i \tag{38}$$

$h(y)$ is an analytic G-invariant polynomial of degree 3. The superpotential f contains ordinary Yukawa couplings for quarks, leptons and Higgs chiral multiplets. In minimal $SU(3) \times SU(2) \times U(1)$ models they are

$$Q = (3,2,\tfrac{1}{6}), \quad u^c = (\bar{3},1,-\tfrac{2}{3}), \quad d^c = (\bar{3},1,\tfrac{1}{3})$$

$$L = (1,2,-\tfrac{1}{2}), \quad e^c = (1,1,1)$$

$$H = (1,2,-\tfrac{1}{2}), \quad H^c = (1,2,\tfrac{1}{2}), \quad Y = (1,1,0) \tag{39}$$

in which Y is an overall singlet needed in order to trigger the $SU(2) \times U(1)$ breaking in models based on tree-level symmetry breaking[16].

The result given by the effective Lagrangian in Eq. (38) indicates that in the limit $M_P \to \infty$, $m_{3/2}$ fixed, spontaneously broken supergravity reduces to a globally supersymmetric theory with soft-breaking terms controlled by the supersymmetry order parameter $m_{3/2}$.

Let us now comment on the sector which determines the sponteneous breakdown of local supersymmetry. The simplest example is given by the Polonyi model with linear superpotential

$$f(z) = \mu/k \; (z+\beta) \tag{40}$$

in which the z-singlet field is the scalar partner of the would be goldstino. If one fine-tunes the parameter β to be given by $\beta = 1/(2-\sqrt{3})/k$, then the supergravity potential $V(z,z^*)$ has a

The unsatisfactory point in this choice of the superHiggs sector of the theory is the fact that one has to fine-tune the β-parameter in order to get vanishing vacuum energy and one has to choose $\mu \simeq O(M_W)$ in order to explain the mass hierarchy in the observable sector, namely the small (logarithimic) renormalization of M_W under (non-gravitational) radiative corrections.

Recently a class of theories have been proposed in which these problems can be circumvented. These theories, generally denoted as no-scale models, have the property of having a flat potential in the hidden z-sector[17]. A flat potential means a vanishing z-potential

$$V(z) \equiv 0 \quad \text{with non-vanishing superpotential } f(z). \quad (41)$$

In the case of a single field z, this is obtained by the unique solution

$$\mathscr{G} = \log\left(\phi(z) + \phi^*(z^*)\right)^3 \quad (42)$$

in which $\phi(z)$ is an arbitrary analytic function. The gravitino mass

$$m_{3/2} = e^{-\mathscr{G}/2} = \left|\phi(z) + \phi^*(z^*)\right|^{-3/2} \quad (43)$$

is undetermined in this case since $\langle z \rangle$ is not fixed.

The class of Kahler metrics induced by Eq. (43) correspond to a non-compact manifold with constant curvature SU(1,1)/U(1).

$$R_{zz^*} = \partial_z \partial_{z^*} \log \mathscr{G}_{zz^*} = \frac{2}{3} \mathscr{G}_{zz^*} \quad (44)$$

The interesting point of no-scale models is due to the fact that they may lead to a dynamical determination of the gravitino mass[18] of the form

$$m_{3/2} = e^{-0(1)/\alpha} M_P \quad (45)$$

due to logarithmic radiative corrections of the renormalizable sector of the theory which remove the tree level degeneracy of the potential. A relation like (45) may explain the hierarchy between $m_{3/2}$ and M_P!.

A natural condition in order for this phenomenon to happen is that

$$\text{SuperTrace } M^2 = \sum_{J=0}^{1} (-)^{2J}(2J+1)m^2_J = 0 \qquad (46)$$

Interestingly enough, recently[19] a class of models has been found in which $\mathcal{G}^i_k$ is not flat but still Eq. (46) is fulfilled. They correspond to Kahler manifolds of the type $U(N,1)/U(N) \times U(1)$ with the Ricci tensor satisfying

$$R^i_j = \frac{N+1}{3} \mathcal{G}^i_j \qquad (47)$$

For N=1 this reduces to Eq. (43).

SUPER HIGGS EFFECT IN N=2 SUPERGRAVITY

Spontaneous breakdown of local supersymmetry can be studied for higher N-extended theories. For N=2 theories this is particularly relevant in view of the fact that recently a larger class of ultraviolet finite theories have been constructed[2]. These theories are N=2 globally supersymmetric Yang-Mills theories with soft-breaking terms of special type. The natural question arises whether such terms can be relics of supergravity couplings at low-energies or just happened for the case of N=1 local supersymmetry.

Here we would like just to give the most general form of the N=2 scalar potential for an arbitrary number of vector multiplets defined by Eq. (10). Let us denote by z^A the scalar partners of the N-2 vector fields V_μ^A. The coupling of N=2 vector multiplets to N=2 supergravity is entirely specified[20] by an analytic function $f(z^A)$. If the vector fields V_μ^A gauge a group G, then f has to be gauge invariant i.e.

$$f_A \, c^{ABC} \, z_B = 0 \qquad (48)$$

c^{ABD} being the structure constants of the group. An additional SO(2) group can be gauged as well. This is the gauge symmetry which rotates the two gravitinos.

In general, if there are Abelian vector multiplets in the theory, the SO(2) gauge group can be gauged by a linear combination of the vector field B_μ of the supergravity multiplet and the other Abelian vector fields[21]:

$$g_o B_\mu + g_i V_\mu^i \qquad (49)$$

We will consider here the case $g_i = 0$. If $g_i \neq 0$ the scalar potential can always be transformed in a form such that $g_i = 0$.[22] The scalar fields z^A parametrized a Kahler manifold with Kahler metric

$$J = - \ln Y \qquad (50)$$

$$Y = \frac{1}{2} \left[f + f^* + \frac{1}{2} (f^A - f^{*A})(z^{*A} - z^A) \right] \qquad (51)$$

The chiral part of the scalar potential, in terms of $N=1$ variables, is given by the following $\mathscr{G}$ function[23]

$$\mathscr{G} = \ln Y - \ln 8 \, g_o^2 \qquad (52)$$

so that it is not vanishing only for $g_o \neq 0$. If the vector multiplicity gauge a group G which commutes with the supersymmetry generators, then the overall potential is given by Eq. (27) with

$$f_{AB}(z) = - \frac{1}{4} \frac{\partial^2 f}{\partial z_z \partial z_B} \qquad (53)$$

We define as a "minimal" coupling the special case in which

$$f(z) = 1 - z^{A^2} \qquad \frac{\partial^2 f}{\partial z_A \partial z_B} = - \frac{1}{2} f_{AB} \qquad (54)$$

This corresponds to the Kahler metric

$$J = \ln (1 - |z^A|^2) \qquad (55)$$

i.e. a σ-model over the Grassmannian manifold $U(1,n)/U(n) \times U(1)$.

All the extrema of the $N=2$ scalar potential can be classified in this case[24]. There is a unique extremum for which $\langle z^A \rangle = 0$ and supersymmetry is unbroken in anti-de Sitter space. There are other extrema for which $\langle z^A \rangle \neq 0$ which break supersymmetry in anti-de Sitter space.

For unitary groups $SU(N)$ the symmetry breaking pattern is found to be

$$SU(N) \rightarrow SU(m_1) \times SU(m_2) \cdots SU(m_N) \times U(1)^{p-1} \qquad (56)$$

in which $m_1, \ldots m_N$ are all sets of integers such that $\sum_{n=1}^{N} n \, m_n = N$ and p is the number of non-zero m_n. For $N \geq 7$ the breaking $SU(N) \rightarrow SU(3) \times SU(2) \times U(1)$ is possible. For any $N \geq 3$ the maximal unbroken residual symmetry is $SU(N-2) \times U(1)$. One may consider the possibility of a solution for which the vacuum energy is zero at the extremum of the potential. This is impossible for f(z) as given by Eq. (54).

If we consider the simplest situation of a single scalar field z but arbitrary f(z) one finds the general mass formula[21]

$$m_A^2 + m_B^2 = 0 \quad \text{at } V_z = 0, \quad V = 0, \qquad (57)$$

which in turn implies that $m_A = m_B = 0$. Moreover it has been shown that such a point is never a local minimum for any choice of f(z). Therefore the only solution with vanishing cosmological constant is a flat potential[22]

$$V \equiv 0. \qquad (58)$$

Interestingly enough there are just two solutions of Eq. (60) which break supersymmetry. They are particular cases of Eq. (43) and correspond to the following choices of f(z),[22]

$$f_1(z) = i(az+b)^3 + i \, P_2(z)$$

$$f_2(z) = (az+b)^{3/2} + i \, P_2(z) \qquad (59)$$

in which a,b are real coefficients and $P_2(z)$ is an arbitrary polynomial of second degree with real coefficients. The first solution has a generalization to many multiplets

$$f(z^A) = i \, d_{ABC} \, z^A z^B z^C \qquad (60)$$

and gives the same Kahler metric which is obtainable from coupling N=2 supergravity to n-1 multiplets in D=5 dimensions.[25]

REFERENCES

1. See for instance: P. Fayet and S. Ferrara, Phys. Rep. 32, 251 (1977); A. Salam and J. Strathdee, Fortschz. Phys. 68, 189 (1981); J. Wess and J. Bagger, "Supersymmetry and Supergravity" (Cambridge University Press).

2. P. Howe, K. Stelle and P.C. West, Phys. Lett. 124B, 55 (1983); A. Parkes and P.C. West, Phys. Lett. 127B, 353 (1983).

3. For a review see i.e.: P. van Nieuwenhuizen, Phys. Rep. 68, 191 (1981).

4. J. Wess and B. Zumino, Phys. Lett. 49B, 52 (1974); J. Iliopoulos and B. Zumino, Nucl. Phys. B76, 310 (1974); J. Ferrara, J. Iliopoulos and B. Zumino, Nucl. Phys. B77, 413 (1974); M.T. Grisaru, M. Rocek and W. Siegel, Nucl. Phys. B159, 429 (1979).

5. For a review see i.e.: R. Barbieri and S. Ferrara, Surveys in High Energy Physics 4, 33 (1983); H.P. Nilles, Univ. of Geneva preprint UGVA-DPT 1983/12-412 (to appear in Phys. Report).

6. E. Witten, Nucl. Phys. B188, 513 (1981).

7. M. Gell-Mann, talk given at the Washington APS Meeting (1977).

8. E. Cremmer and B. Julia, Phys. Lett. 80B, 48 (1978), Nucl. Phys. B159, 141 (1979); J. Ellis, M.K. Gaillard, L. Maiani and B. Zumino, in Unification of the Fundamental Particle Interactions (eds. S. Ferrara, J. Ellis and P. van Nieuwenhuizen) (Plenum Press, New York, 1980) p. 69.

9. L. Girardello and M.T. Grisaru, Nucl. Phys. B194, 65 (1982).

10. P. Fayet and J. Iliopoulos, Phys. Lett. 51B, 461 (1974).

11. For recent reviews see: S. Ferrara, Int. Europhysics Conf. on High Energy Physics, Brighton, U.K. (eds. J. Guy, C. Costain, Rutherford Appleton Lab.) p. 522.; same proceedings: P. Fayet, p. 33 and D.V. Nanopoulos, p. 38.

12. E. Cremmer, S. Ferrara, L. Girardello and A. Van Proeyen, Phys. Lett. 116B, 231 (1982); Nucl. Phys. B212, 413 (1982).

13. R. Barbieri, S. Ferrara and C.A. Savoy, Phys. Lett. 119B, 343 (1982).

14. A. Arnowitt, A. Chamseddine and P. Nath, Phys. Rev. Lett. $\underline{49}$, 970 (1982).

15. S. Deser and B. Zumino, Phys. Rev. Lett. $\underline{38}$, 1433 (1977).

16. P. Fayet, Proc. 21st. Int. Conf. on High Energy Physics, Paris, 1982 (eds. P. Petian and M. Porneuf); J. Phys. $\underline{43}$, Suppl. 12, C3-673, (1982).

17. E. Cremmer, S. Ferrara, C. Kounnas and D.V. Nanopoulos, Phys. Lett. $\underline{133B}$, 61 (1983).

18. J. Ellis, A.B. Lahanos, D.V. Nanopoulos and K. Tamvakis, Phys. Lett. $\underline{134B}$, 429 (1984); J. Ellis, C. Kounnas and D.V. Nanopoulos, CERN preprints TH-3768 (1983), Th-3874 (1984), TH-3848 (1984).

19. N. Dragon, M.G. Schmidt and U. Ellwanger, Heidelberg preprint HD-THEP-84-10 (1984).

20. B. de Wit, P.G. Lauwers, R. Philippe, Su, S.Q. and A. Van Proeyen, Phys. Lett. $\underline{134B}$. 37 (1984).

21. B. de Wit and A. Van Proeyen, NIKHEF-H/84-4.

22. E. Cremmer, J.P. Derendinger, B. de Wit, S. Ferrara, L. Girardello, C. Kounnas, A. Van Proeyen (in preparation).

23. J.P. Derendinger, S. Ferrara, A. Masiero and A. Van Proeyen, CERN preprint TH. 3813 (to appear in Phys. Lett. B).

24. J.P. Derendinger, S. Ferrara, A. Masiero and A. Van Proeyen, Phys. Lett. $\underline{136B}$, 354 (1984).

25. M. Gunaydin, G. Sierra and P.K. Townsend, Ecole Normale Superiore preprint LPTENS 83/32 (1982) (to appear in Nucl. Phys. B); same authors Cambridge DAMPT preprint (1984).

DISCUSSION

TERAZAWA:

I just want to point out that supersymmetry is not only the possibility to explain the dominance of kaon channels in proton decay. Our composite model is also a model in which kaon channels are enhanced in proton decay compared to the $e^+\pi^0$ channel.

LITTENBERG:

Are there gravitinos with $m \simeq m_W$ floating around left over from primordial times? Presumably they could decay into photon + photino for example. Maybe we could detect them.

FERRARA:

These are bound on the photonino mass coming from cosmology. However these bounds can be removed in the scenario of the inflationary universe.

CHAU:

You stopped your discussion at $N = 2$. Could you comment on higher N theories?

FERRARA:

In higher N theories the spin 1/2 states of the elementary multiplets cannot be identified with quarks and leptons just because they do not have the right quantum numbers with regard to $SU(3) \times SU(2) \times U(1)$. Quarks and leptons should possibly emerge as composite states of spin 0 - spin 1/2 constituents.

WALI:

I think it is a bit too strong a statement to say that the minimal SU(5) or some other grand unified theory is ruled out. There are great uncertainties in evaluating low energy matrix elements. Generally one has to make other assumptions not implied by GUTS. For instance, if we include vector mesons and follow the old-fashioned calculations of current algebra using effective Lagrangians, we can suppress $\pi^0 e^+$ mode relative to ωe^+ or $\rho^0 e^+$ mode in the minimal SU(5).

FERRARA:

I think that minimal SU(5) model is not completely out but it is probably out. The detection of proton decay in strange channels (νK) in my opinion could be an indication for low energy supersymmetry.

TRAMPETIC:

Would you comment anything about magnetic monopoles and their effects from the point of the SUSY?

FERRARA:

Monopoles are present also in SUSY GUTS. What could possibly change is that production rate and the way they catalyze proton decay.

THE EXPERIMENTAL SEARCH FOR SUPERSYMMETRY*

G.L. Kane

Randall Laboratory of Physics
University of Michigan
Ann Arbor, MI   48109

ABSTRACT

Various ways that experimental evidence for supersymmetry[1] might appear are considered. Possible signatures are discussed in detail. Recent events reported at the CERN collider are examined as possible evidence for supersymmetry.

INTRODUCTION

There has been great interest in supersymmetry for a decade, and increasingly in the past few years as a possible symmetry of nature that might be found experimentally. After considerable study by a number of people, it appears that one can say with confidence that either we can find evidence for supersymmetry or become convinced that it is absent at all energies accessible to experiment.

If supersymmetry were unbroken every particle would have a partner differing by half a unit of spin but identical otherwise-- in particular, with the same mass. Since this is obviously not so, the supersymmetry must be broken. Although there are a

401

number of models with a broken supersymmetry, none is yet compelling, so the mass spectrum we expect is essentially unknown. In interpreting experimental searches we can get around that problem by searching for all masses; the absence of a signal implies a range of masses is excluded. Below we will summarize present limits.

In the near future there are opportunities to discover evidence for supersymmetry at existing machines, and new possibilities at each new facility. In particular, some recently reported[2,3] events at the CERN pp collider are what is expected if supersymmetric partners are being produced.

SPECTRUM AND COUPLINGS

In a supersymmetric theory every ordinary particle has a partner with all quantum numbers identical except spin, which differs by 1/2 unit. More complicated theories can be written, but they will reduce to this situation by the time they confront experiment.

The particles, the partners, and a useful nomenclature are shown in Table 1.

One can get a minimal set of vertices by taking any standard model vertex, replacing particles in pairs by their supersymmetric partners, and making the minimal change in Lorentz structure. In particular, the coupling strengths remain the same, apart from an occasional $\sqrt{2}$. So qualitative interactions are easy to deduce. Some additional interactions occur, but all the ones obtained that way will be present.

For the charginos and neutralinos there is mixing among the weak eigenstates, so the mass eigenstates do not have definite SU(2) quantum numbers. We label them to some extent in terms of how they interact. However, for simplicity we will refer to the lightest neutralino as a $\tilde{\gamma}$ and the lightest chargino as a $\tilde{w}^{\pm}$ even though these may not interact like a photon or a W.

TABLE 1

Supersymmetric partners

Normal Particles	Weak Interaction Eigenstates		Mass Eigenstates		Mass Eigenstates with Specific Couplings	
	Symbol	Name	Symbol	Name	Symbol	Name
$q=u,d,s,$ c,b,t	$\tilde{q}_L, \tilde{q}_R$	scalar-quark	$\tilde{q}_1, \tilde{q}_2$	scalar-quark		
$\ell=e,\mu,\tau$	$\tilde{\ell}_L, \tilde{\ell}_R$	scalar-lepton	$\tilde{\ell}_1, \tilde{\ell}_2$	scalar-lepton		
$\nu=\nu_e, \nu_\mu, \nu_\tau$	$\tilde{\nu}$	scalar-neutrino	$\tilde{\nu}$	scalar-neutrino		
g	$\tilde{g}$	gluino	$\tilde{g}$	gluino		
$W^\pm$	$\tilde{W}^\pm$	wino			$\tilde{w}^\pm$	wino
H_1^+	$\tilde{H}_1^+$	higgsino	$\tilde{\chi}_{1,2}^\pm$	charginos	$\tilde{h}^\pm$	higgsino
H_2^-	$\tilde{H}_2^-$	higgsino			$\tilde{\omega}_1, \tilde{\omega}_2$	wiggsino
γ	$\tilde{\gamma}$	photino			$\tilde{\gamma}$	photino
Z^0	$\tilde{Z}^0$	zino			$\tilde{z}$	zino
H_1^0	$\tilde{H}_1^0$	higgsino	$\tilde{\chi}_i^0$	neutralinos	$\tilde{h}_1, \tilde{h}_2$	higgsino
H_2^0	$\tilde{H}_2^0$	higgsino			$\tilde{\zeta}_1, \tilde{\zeta}_2$	zigosino
$\binom{W^3}{B}$	$\binom{\tilde{W}^3}{\tilde{B}}$	$\binom{\text{wino}}{\text{bino}}$			$\binom{\tilde{w}_0}{\tilde{b}_0}$	$\binom{\text{wino}}{\text{bino}}$

This table gives the nomenclature for supersymmetric partners. For the partners of bosons, mixing generally occurs among the weak eigenstates, so one needs different symbols and names for weak and for mass eigenstates. Couplings of weak eigenstates are determined by the theory, but couplings of mass eigenstates depend on the amount of mixing; for mass eigenstates the names and symbols are either generic $(\tilde{\chi}_i)$ or reflect the couplings. As always, Z^0 and γ are linear combinations of W^3, B [$\gamma = \cos\theta_w B + \sin\theta_w W^3$, $Z = -\sin\theta_w B + \cos\theta_w W^3$], with the same combination of $\tilde{W}^3, \tilde{B}$ giving $\tilde{Z}^0, \tilde{\gamma}$.

INTERACTIONS AND DECAYS

Which decays dominate depend on the (unknown) masses. Two-body decays usually dominate if allowed. The main modes are as follows:

$$\begin{aligned}
\tilde{\nu} &\to \tilde{\gamma}\nu \\
\tilde{q} &\to q\tilde{g},\ q\tilde{\gamma} \\
\tilde{\ell}^{\pm} &\to \ell^{\pm}\tilde{\gamma} \\
\tilde{w}^{\pm} &\to \ell^{\pm}\tilde{\nu}_{\ell},\ \nu\tilde{\ell}^{\pm},\ q\bar{q}\tilde{g},\ q\bar{q}\tilde{\gamma} \\
\tilde{g} &\to q\bar{q}\tilde{\gamma},\ g\tilde{\gamma} \\
\tilde{z} &\to \ell^{+}\ell^{-}\tilde{\gamma},\ q\bar{q}\tilde{\gamma},\ q\bar{q}\tilde{g}
\end{aligned} \tag{1}$$

In a collider detector the $\ell^{\pm}$ would appear as hard charged leptons, the $q,\bar{q}$ as jets. Photinos must interact[4] by exciting a heavy $\tilde{q}$, so their interaction cross sections are small (numerically of order a ν cross section) and they escape collider detectors. $\tilde{\nu}$ escape collider detectors or decay into things that escape.

CURRENT LIMITS

Here I will briefly summarize the present limits on supersymmetric partners masses--that is, if they were lighter than these masses they would have been observed. Detailed caveats and references are given[5] in ref. 1.

In the talk of A. Galtieri at this meeting recent results were presented from the MAC Collaboration for scalar-electrons, $\tilde{e}$, based on the absence of the reaction $e^{+}e^{-} \to \gamma\tilde{\gamma}\tilde{\gamma}$ by $\tilde{e}$ exchange. The results exclude a region which is very crudely given by $\tilde{m}_{e} + 4\tilde{m}_{\gamma} \lesssim 30$ GeV, depending on assumptions about left- and right-handed $\tilde{e}$ masses. PETRA (and PEP) results[5,1] from all detectors require $\tilde{m}_{\mu}$, $\tilde{m}_{\tau} \gtrsim 17$ GeV from the absence of $e^{+}e^{-} \to \tilde{\mu}^{+}\tilde{\mu}^{-}$, $e^{+}e^{-} \to \tilde{\tau}^{+}\tilde{\tau}^{-}$.

Scalar-neutrinos are only limited[6] by the absence of the decays $\tau \to \ell\tilde{\nu}_{\ell}\tilde{\nu}_{\tau}$ and $\mu \to e\tilde{\nu}_{\mu}\tilde{\nu}_{\ell}$, so the $\tilde{\nu}_{i}$ masses must be large enough to exclude such decays kinematically; $\tilde{m}_{\nu\tau} > m\tau$ and $\tilde{m}_{\nu\mu} > m_\mu$ would suffice, for example.

Scalar-quarks of either charge probably must be heavier than about 17 GeV. This conclusion is based on a more complicated chain of reasoning.[1] If there were scalar-quarks lighter than about 3 GeV, scalar-quarkonium "ψ" states would have been formed,[7] and are not. From about 3-17 GeV for $Q = 2/3$ scalar-quarks one would have observed events $e^+e^- \to \tilde{q}\bar{\tilde{q}}$, $\tilde{q} \to jet+\tilde{\gamma}$, giving missing energy, and acolinear and acoplanar jets, when photinos escape; such events are not observed.[5,1] If scalar-quarks are relatively light compared to m_w, and if $\tilde{m}_u \neq \tilde{m}_d$, one would expect[1] noticeable isospin violation in strong interactions, so probably $(\tilde{m}_u-\tilde{m}_d)/(\tilde{m}_u+\tilde{m}_d) \ll 1$. However, if gluinos are lighter than scalar-quarks, probably one can only conclude that $\tilde{m}_q \lesssim 3$ GeV.

Essentially no limits exist on neutralinos. Whatever their masses, they would not have been discovered yet. The situation is interesting[8]; even if the production cross section for $\tilde{\chi}^0$'s were large, considerable energy is carried away in their decays, and the events would have been cut away in earlier searches. Recent efforts are just reaching the levels of interesting cross sections for these events.

Eventually the associated production of neutralinos, $e^+e^- \to \tilde{z}\tilde{\gamma}$ or $q\bar{q} \to \tilde{z}\tilde{\gamma}$, followed by $\tilde{z} \to f\bar{f}\tilde{\gamma}$ ($f\bar{f}$ are any quark or lepton pair), will provide[8] one of the most stringent restrictions on the possibility of detecting supersymmetry, or a good way to find and study it. That is because these processes have excellent experimental signatures and reasonable production cross sections, and the neutralino mass eigenstates generally include two that are rather light compared to the scale of supersymmetry breaking.

For charginos the limits depend[8,5] on the $\tilde{\gamma}$ mass and on what decay modes dominate. Although charginos can be easily produced in $e^+e^- \to \tilde{\chi}^+\tilde{\chi}^-$, if photinos are heavy they carry off too much energy for the event to be an acceptable signal, and if $\tilde{\chi}^\pm \to \ell^\pm \times \tilde{\gamma}$ does not have at least a typical branching ratio there may not be a good signature. Thus charginos would probably have been discovered if they were lighter than about 17 GeV, but under some conditions that limit could be evaded.

Finally, for gluinos the situation is complicated.[9] Beam dump experiments, where $pp \to \tilde{g}+X$, $\tilde{g} \to \tilde{\gamma}+X$, and $\tilde{\gamma}$ interacts in the dump detector, imply $\tilde{m}_g \gtrsim$ few GeV; however, there are caveats. The $\tilde{g}$ lifetime must be in a certain range for the experiments to be sensitive. One must include the effect that even massless $\tilde{g}$ which are bound in a hadron will decay.[10] The apparent absence of $\psi \to \gamma + \tilde{n}$, where $\tilde{n}$ is a $\tilde{g}\tilde{g}$ (gluinoball) state, gives a lower limit[11] which seems to exclude massless g under any circumstances.

Further experiments and analysis will sharpen the above limits, or help determine where some evidence for supersymmetric partners might be found.

COLLIDER SIGNATURES FOR SUPERSYMMETRY

Now let us see what would happen if supersymmetric partners were being produced at a hadron collider. The various states would decay as in eq. 1. Assume the masses are such that $\tilde{m}_g > \tilde{m}_q \gtrsim \tilde{m}_w > \tilde{m}_\nu$. Then the dominant modes would be $\tilde{g} \to jj\tilde{\gamma}$, $\tilde{q} \to j\tilde{\gamma}$, $\tilde{w} \to \ell\tilde{\nu}_\ell$, where j is a quark jet, ℓ a charged lepton, and $\tilde{\gamma}$ and $\tilde{\nu}$ escape the collider detector.

The dominant processes would be associated production,

$$q\bar{q} \to \tilde{g}\tilde{w}$$

$$qg \to \tilde{q}\tilde{w}$$

$$q\bar{q} \to \tilde{g}\tilde{\gamma}$$

$$qg \to \tilde{q}\tilde{\gamma}$$

$$qg \to \tilde{g}\tilde{q} \ .$$

Consider each of these to see what events might arise, and how they would look in a collider detector. For example,

$$\tilde{g} + \tilde{w} \rightarrow q\bar{q}\tilde{\gamma} + e\tilde{\nu} \rightarrow jje P_T$$

where P_T stands for missing transverse momentum, due to the combined effect of the two undetected particles. Similarly,

$$\tilde{q} + \tilde{w} \rightarrow je P_T$$

$$\tilde{g}\tilde{q} \rightarrow jjj P_T$$

$$\tilde{g}\tilde{\gamma} \rightarrow j P_T$$

$$\tilde{g}\tilde{\gamma} \rightarrow jj P_T \ .$$

Events of the kinds $jje^\pm P_T$, $je^\pm P_T$, and $j P_T$ are currently being reported[2,3] at the CERN collider. If some of these events are indeed the production of supersymmetric partners there are several immediate conditions that must be satisfied. First, since the missing P_T is due to two undetected particles, it must not be that of a single massless particle so $E^2 - P^2 \neq 0$. Second, sometimes the two particles go in directions where they give little missing P_T, so there must also be events of the type $jje^\pm$, $je^\pm$ with little missing P_T. Third, there must be many events of the type $jjj P_T$ from $\tilde{g}\tilde{q}$ production (see rates below). Fourth, the rates must be right, and given the masses the rates are calculable in a supersymmetry theory.

One obtains[12] the results in Table 2. The cross sections are evaluated with masses $\tilde{m}_g = 80$ GeV, $\tilde{m}_q = \tilde{m}_w = 40$ GeV, suggested by the kinematical structure of the events; careful Monte Carlo analysis of the combined topology and rates will allow us to see more precisely what masses are relevant. We put $\sqrt{s}$ = sum of final masses plus 40 GeV, roughly corresponding to the peak of the cross section. The integrated luminosity of the experiments is 116 nb^{-1}, so a little over one event is expected for each

Table 2

Produced Pair	$\hat{\sigma}(10^{-35}$ cm$^2)$
$\tilde{W}_{e\tilde{\nu}} + \tilde{g}$	$4(s_+^2 + s_-^2) + 5\, s_+ s_-$
$\tilde{W}_{e\tilde{\nu}} + \tilde{q}$	$1.4(s_+^2 + s_-^2)$
$\tilde{g} + \tilde{\gamma}$	6
$\tilde{q} + \tilde{\gamma}$	1.5
$\tilde{q} + \tilde{g}$	380

10^{-35} cm^2 of cross section. The quantities s_+ and s_- are $\sin\phi_+, \sin\phi_-$ and measure the mixing of the $W^{\pm}$ partner and the $H^{\pm}$ partner into the mass eigenstates. In typical models $1/2 \lesssim s_+ \simeq s_- \lesssim 1$ for the lightest chargino state. Structure function effects have not yet been included; they are expected to have little effect since the typical structure function is about unity at these x, Q^2 values. We see that the event rates are about right to interpret a few events as possible production of super-symmetric partners! A crucial prediction is that there must be many events of the type $jjjP_T$, which have not yet been studied.

If it turns out that these events are not consistent with a supersymmetric interpretation, we obtain[12,13] extremely strong limits on masses, since several of the masses must be made heavier than the 40-80 GeV range considered here.

FURTHER POSSIBILITIES

Two additional methods to search for production of supersymmetric partners should be emphasized.

(1) If the masses are suitable, which occurs in many models, one can have[14,8] $W^{\pm} \to \tilde{w}^{\pm} + \tilde{\gamma}$, followed by $\tilde{w}$ decay while $\tilde{\gamma}$ escapes. As before these are really mass eigenstates although we refer to them as $\tilde{w}, \tilde{\gamma}$ for simplicity. If $\tilde{w}^{\pm} \to \ell^{\pm} \nu \tilde{\gamma}$ by any channel, the signature is a single charged lepton and P_T. While this is in principle detectable, the background[15] from $W \to e\nu$, $W \to \tau\nu$ and $\tau \to \ell\nu\bar{\nu}$, and semileptonic decays of heavy quarks, makes separating a signal very difficult. Perhaps better is $\tilde{w}^{\pm} \to q'\bar{q}\tilde{\gamma}$, so the signature is a one-sided event with two jets (or a single fat jet if $q\bar{q}$ overlap) and P_T; there may be less background for such a mode. When hundreds of $W^{\pm}$ decays can be studied one will find this mode or be able to say that $\tilde{m}_w + \tilde{m}_\gamma \gtrsim m_w$, which puts very strong constraints on models.

(2) As remarked above in the discussion on neutralinos, the associated production of neutralinos, $f\bar{f} \to \tilde{z}\tilde{\gamma}$, followed by $\tilde{z} \to \bar{f}f\tilde{\gamma}$ (where $f\bar{f}$ is any $q\bar{q}$ or $\ell^+\ell^-$ pair in either production or decay) gives a one-sided event with a good signature and a reasonable cross section. Again, $\tilde{z}$ and $\tilde{\gamma}$ are really mass eigenstates that are mixtures of weak interaction eigenstates, though we use familiar names for simplicity. The properties of the mass matrix of the neutralinos are such that there are usually some eigenvalues that are light compared to the scale of supersymmetry breaking. [The determinant of the mass matrix has a factor $\tan^2\theta_w$ and some additional suppressions in some mass regions, so if one eigenvalue is of order the scale of supersymmetry breaking Λ, another is order $\Lambda\tan^2\theta_w \sim \Lambda/5$.] Thus eventually neutralino associated production at any high luminosity collider will probe TeV or higher scales of supersymmetry breaking, and provide a

definitive test of whether supersymmetry will be realized in nature on the weak scale.

CONCLUSIONS

Two points should be emphasized:

(a) Perhaps some supersymmetric partners have been observed at the CERN collider. Both the event topologies and the rates are consistent with this possibility. Some easily testable predictions must be satisfied if this interpretation could be correct.

(b) If supersymmetric partners exist on the scale of $\tilde{m} < 1$ TeV they can eventually be detected (if we construct the right machines), and one can definitely exclude their presence if they are not detected.

REFERENCES

*Research supported in part by U.S. D.o.E.

1. For a recent review along the lines of this article, and references to earlier literature, see H.E. Haber and G.L. Kane, UM HE TH 83-17, to be published in Physics Reports. See also C.A. Savoy, in Beyond the Standard Model, Proc. of the 18th Rencontre de Moriond, 1983, edited by J. Tran Thanh Van (Editions Frohtieres, France, 1984); P. Nath, R. Arnowitt and A.H. Chamseddine, Northeastern preprint NUB#2613, lectures given at the 1983 Summer Workshop on Particle Physics, Trieste; I. Hinchliffe and L. Littenberg, in Proc. of the 1982 DPF Summer Study of Elementary Particle Physics and Future Facilities, p. 242, edited by R. Donaldson, R. Gustafson and F. Paige (American Physical Society, New York, 1982); P. Fayet, in Proc. of the 21st International Conf. on High Energy Physics, Paris, p. C3-673, edited by P. Petiav and M. Porneuf (Les Editions de Physique, France, 1982);

410

D.V. Nanopoulos, A. Savoy-Navarro and Ch. Tao, Proc. of the Supersymmetry vs. Experiment Workshop, to be published in Physics Reports; J. Ellis, Proc. of the 11th SLAC Summer Institute on Particle Physics, p. 239, ed. by Patricia McDonnough, SLAC report no. 267.

2. P. P. Bagnaia et al., UA2 Collaboration, CERN EP/84-40.

3. G. Arnison et al., UA1 Collaboration, CERN EP/84-42.

4. P. Fayet, Phys. Lett. 86B (1979) 272.

5. See also S. Yamada, Proc. of the 1983 International Symposium on Lepton and Photon Interactions at High Energies, ed. D.G. Cassel and D.L. Kreinick, Cornell University Press, and "A Summary of Recent Results from MARK J," MIT LNS Report 131.

6. G.L. Kane and W. Rolnick, Nucl. Phys. B232 (1984) 21.

7. C. Nappi, Phys. Rev. D25 (1982) 84.

8. J.-M. Frere and G.L. Kane, Nucl. Phys. B223 (1983) 331.

9. See ref. 1 and also S. Dawson, E. Eichten, and C. Quigg, FERMILAB-Pub-83/82-THY.

10. E. Franco, Phys. Lett. 124B (1983) 271.

11. T. Goldman and H.E. Haber, Los Alamos preprint.

12. H.E. Haber and G.L. Kane, UCSC-TH-169-84.

13. See also G. Alterelli and B. Mele, "Search for Gauginos at the CERN p̄p Collider, preprint, April 1984.

14. S. Weinberg, Phys. Rev. Lett. 50 (1983) 387. R. Arnowitt, A.H. Chamseddine, and P. Nath, Phys. Rev. Lett. 50 (1983) 1232; Phys. Lett. 129B (1983) 445. P. Fayet, Phys. Lett. 133 (1983) 363. J. Ellis, J.S. Hagelin, D.V. Nanopoulos, and M. Srednicki, Phys. Lett. 127B (1983) 233. B. Grinstein, J. Polchinski, and M.B. Wise, Phys. Lett. 130B (1983) 285.

15. D.A. Dicus, S. Nandi, W.W. Repko, and X. Tata, Phys. Rev. D29 (1984) 67.

DISCUSSION

GOLDHABER:

You assume that the photino is absolutely stable. Could you foresee a super-unified theory where it would decay into ordinary particles?

KANE:

That is possible, by giving the scalar-neutrino a vacuum expectation value. Although this violates lepton number conservation, it is not very restricted by data because the predicted violations are hard to see. It has been studied by Aulakh and Mohapatra, Hall and Suzuki, Lee and Frere and myself.

FRANZINI:

How do you get both $\tilde{g} \tilde{w} \rightarrow e^{\pm} jj$ with missing p_T and $g w \rightarrow e^{\pm}$ with no missing p_T?

KANE:

Since there are two missing particles ($\tilde{\gamma}$ and $\tilde{\nu}$) these momenta add sometimes and cancel sometimes. The relative rates depend on experimental cuts.

LEE-FRANZINI:

Keung has written a paper on searching for gluinoballs on the Υ, which seems promising. CLEO has a missing energy search on the $\Upsilon' \rightarrow \chi_b$, I don't know the limit. Given your present interpretation what is the mass of the lightest particles: photinos and gluinos?

KANE:

I agree with your comments. There is also a paper on $\Upsilon \rightarrow \gamma + \eta_g$ by Goldman and Haber. Gluinos have to be rather heavy for the present interpretation to be correct, as described in the talk. If photinos and gluinos get mass radiatively, then one expects $m_{\tilde{\gamma}} \simeq (C\alpha/\alpha_s)m_{\tilde{g}}$ where C is a color factor of order 1-2. Then $m_{\tilde{\gamma}}$ is of order 10-15 GeV.

TERAZAWA:

I, of course, prefer my interpretation of the UA2 anomalous W jet events by excited gauge bosons. But, provided that your interpretation is right, how could you determine the masses of the gluino and the scalar quark with two particles missing?

KANE:

The masses of $\tilde{w}$, $\tilde{g}$, $\tilde{q}$ will be rather well fixed by the kinematical constraints on cross sections and on jet momenta and angles. Other masses are constrained but more weakly. Other experiments will see or constrain them further. Fixing photino and scalar-neutrino mass will, as you suggest, be difficult.

RÜCKL:
You assume in your analysis that one can reliably calculate the hadronic production of heavy SUSY particles such as scalar quarks or gluinos. However similar calculations of heavy quark production are known to fail. What is your opinion on this difficulty?

KANE:
I think such calculations always fail conservatively; the perturbative result is less than the reported one. So I expect the perturbative SUSY cross sections to be safe lower limits.

WOLFENSTEIN:
It seems somewhat paradoxical that a 70 GeV gluino can be detected by UA2 but one cannot limit the gluino mass in the range above 5 GeV by other experiments.

KANE:
That is partly a reflection on experimental detectors and on experimenter's priorities. At the CERN collider a limit of about 40 GeV can be set on the gluino mass, and will be soon, I think. The limit must be set on a statistical basis, while a signal can turn up in a few events, so one can look further if one is lucky with a signal.

MORSE:
SUSY requires several Higgs bosons, both charged and neutral. Is it possible that we will now get a contribution to CP violation from the Higgs sector?

KANE:
Not from the minimal Higgs set. Since we do not understand what constitutes simple Higgs physics, it could happen that the Higgs sector is more complicated in a supersymmetric theory and it could then contribute to CP violation.

SEGRE:
What are the events seen by UA2 and what is the SUSY interpretation?

KANE:
The events are described in Peter Hansen's talk, and I have interpreted them as production of W + g and W + q.

KLEINKNECHT:
If your interpretation of the UA2 events is right, why should they peak at an invariant transverse mass of 150 GeV? Should there be any sharp peak?

KANE:

I would expect a broader peak (of width ~ 30 GeV) just for kinematical reasons: there is a high threshold for production of massive particles on the low side, and there is a strong cutoff on the high side because of both phase space and the falling structure functions.

KLEINKNECHT:

How do you explain a common mass for the 3 UA2 events? Maybe Hansen should answer.

HANSEN:

If Gordie's interpretation is right, then it was wrong of me to assume the electron and neutrino in these events to come for a (normal) W. The transverse masses are consistent with the W hypothesis, though somewhat higher than you would expect from the M_T^W distribution.

HORIZONTAL SYMMETRIES AND FLAVOR MIXINGS

IN LEFT-RIGHT SYMMETRIC MODELS*

Rabindra N. Mohapatra[†]

Department of Physics and Astronomy
University of Maryland
College Park, MD. 20742

ABSTRACT

We review the prospects for calculating flavor mixings in weak interactions using continuous or discrete Horizontal symmetries. A model obtained by imposing Z_4 symmetry on an $SU(2)_L$ x $SU(2)_R$ x $U(1)_{B-L}$ electroweak model gives an interesting mass matrix for three generations. Its predictions for mixing angles and the top quark mass are discussed. The Z_4-model also provides an interesting resolution of the strong CP-problem.

§1.   INTRODUCTION

Despite the spectacular successes of the $SU(2)_L$ x $U(1)$ x $SU(3)_c$ model to describe electroweak and strong interaction processes at low energies, an important aspect of quark-lepton physics remains unexplained. This has to do with the proliferation of quarks and leptons beyond the ones (i.e. u,d,ν_e,e) important for our daily existence. There are two more families (or

---

*Invited talk presented at the Europhysics Workshop on "Flavor Mixing in Weak Interactions" held in Erice from March 5-10, 1984 (to be published in the proceedings).

[†]Work supported by a grant from the National Science Foundation.

generations) of fermions, i.e. (c, s, ν_μ, μ^-) and (t, b, ν_τ, τ^-) (to be called respectively as second and third generations), which behave very similarly to the first generation as far as <u>all</u> their inter-actions go; however, they have different masses and a definite pattern of mixings among them. This is the problem we will address in this review from different angles proposed in the last few years within the unified gauge theory framework. Before proceeding to this discussion, let us separate the three aspects of the problem:

(a) Why are there three generations or if there are more than three, how many are there? Phenomenologically, assuming asymptotic freedom up to a given energy implies that the number of generations, $N_g \leq 8$.

(b) There is a hierarchical pattern between the masses of the different generations (by mass of quark, we will mean the current quark masses); for instance, for the first generation, $m_u \approx 5$ Mev, $m_d \approx 8$ Mev ; $m_e \approx .5$ Mev and $m_{\nu_e} \leq 10$–20 ev with an average mass $\langle m \rangle_1 \approx 5 \times 10^{-3}$ Gev ; for the second generation, $m_c \approx 1.5$ Gev , $m_s \approx 160$ Mev , $m_\mu \approx 100$ Mev and $m_{\nu_\mu} \leq .5$ Mev implying $\langle m \rangle_2 \sim 1$ Gev ; and for the third generation, $m_t \leq 30$ Gev , $m_b \approx 4.5$–5 Gev , $m_\tau \approx 1.8$ Gev , $m_{\nu_\tau} \leq 250$ Mev implying $\langle m \rangle_3 \approx 10$–20 Gev .

(c) Finally, there also appears to be a definite nearest neighbor pattern in the mixings between generations if we assume three generations. In terms of the mixing angles that parametrized charged current weak interactions, a typical[1] set of values are (in magnitude)

$$U \approx \begin{pmatrix} .9733 \pm .0024 & .231 \pm .003 & .005 - .015 \\ .231 \pm .003 & .9715 \pm .0011 & .043 - .055 \\ .014 - .033 & .043 - .055 & .997 - .999 \end{pmatrix}$$

$$(1)$$

Looking at these various mixing parameters, it becomes clear that, they are correlated with the constituent quark masses between the generations involved in the mixing. For instance, the Cabibbo angle θ_c , that mixes the d and s-quarks can roughly be written as

$$\theta_c \approx \sqrt{m_d/m_s} \qquad (2)$$

This relation was derived in the left-right symmetric models by

various authors[2] several years ago. A similar pattern of relations can also be inferred for other mixing angles given in eqn.
(2). This would suggest that a complete understanding of the
family mixings would have to involve the dynamics of weak interactions and the origin of quark-lepton masses. Although some
clues to the mass hierarchy problem emerges from the study of
supersymmetric (composite) models of quarks and leptons, the true
picture is far from clear. In this paper, we will therefore
content ourselves with a more modest approach and seek possible
symmetries between different generations (or horizontal symmetries) that can constrain the parameters enough to relate the
mixing angles to the quark masses. Identification of such a
symmetry may take us a step closer to understanding the dynamics
of weak interactions much the same way as SU(3) flavor symmetry of
Gell-Mann and Ne-eman helped to clarify the underlying structure
of strong interaction physics, starting from nucleons via SU(3)
symmetry to quarks, then to color degree of freedom and finally to
Quantum-Chromo-Dynamics. The smoking gun in the case of strong
interactions was the approximate mass degeneracy among baryons and
among mesons. Our hope is that the analogue of this for electroweak physics may be the quark lepton mass hierarchies and flavor
mixings.

§2. MASS MATRICES IN GAUGE THEORIES

Before proceeding to discuss the constrained mass matrices
that can lead to predictions for flavor mixing angles, we will
count the number of arbitrary parameters in various gauge theories. For N_g generations, ignoring CP-violation, the number of
mixing angles that parametrize the weak current is $N_g(N_g-1)/2$ and
the number of quark masses is $2N_g$ leading to a total number of
observables $\frac{1}{2}N_g(N_g + 3)$. For three generations, there are nine
observables. So, to be able to derive relations between these
parameters, we need a mass matrix, which has fewer parameters.

In a gauge theory, the fermion masses vanish prior to spontaneous breaking of gauge symmetries due to the chiral properties
of weak interactions. Subsequent to spontaneous breakdown of
gauge symmetries, the Yukawa couplings of the theory lead to non-
zero fermion mass matrices. In general, we can write,

$$\mathcal{L}_Y = \sum_{a,b,i} h_{abi} \, \bar{\psi}_{aL} \, \phi_i \, \psi_{bR} + \text{h.c.} \tag{3}$$

where a,b are generation indices and i denotes the number of
Higgs multiplets. They lead to fermion mass matrices of the form:

$$M_{ab}^{(u)} = \sum_i h_{ab,i} \langle \phi_i^{(u)} \rangle$$

$$M^{(d)}_{ab} = \sum_i h_{ab,i} <\phi^{(d)}_i>$$

where $\phi^{(u)}_i$ and $\phi^{(d)}_i$ denotes the component of the ith Higgs multiplet that couple to the $I_{3W} = \pm \frac{1}{2}$ (i.e. up-like or down-like) components of weak SU(2)-multiplet. Now, we are ready to count the number of arbitrary parameters that characterize $M^{(u)}_{ab}$, $M^{(d)}_{ab}$, in various unified gauge theories.

(a) <u>SU(2)$_L$ x U(1) model</u>: For the minimal SU(2)$_L$ x U(1) model, which has only one Higgs doublet, $M^{(u)}$ and $M^{(d)}$ are arbitrary real N_g x N_g matrices (ignoring CP-violation), which have each N^2_g arbitrary elements. (For $N_g = 3$, we have a total of 18 free parameters, which is twice the number of observables for 3 generations). This result remains true for an arbitrary number of Higgs multiplets.

(b) <u>Minimal SU(5) model</u>: In the minimal SU(5) model,[3] the fermions are assigned to 10 and $\bar{5}$ representations of the group and the Higgs multiplet responsible for generating fermion masses transforms as a 5 -dimensional representation of SU(5). In this case, $M^{(d)}$ is completely arbitrary and therefore involves N^2_g-parameters as in the standard model. On the other hand, the $M^{(u)}$ is a symmetric[4] matrix due to group theoretic constraints, with $N_g(N_g + 1)/2$ arbitrary parameters. This reduces the total number of arbitrary parameters for three generations from 18 (in the standard model) to 15. This is, however, larger than the nine observables. In fact, to fix the leptonic mass spectrum, a 45 - dimensional Higgs multiplet is needed, which destroys the symmetry of $M^{(4)}$, thus increasing the number of arbitrary parameters to 18 as in the standard model.

(c) <u>Left-Right Symmetric Models</u>:[5] The quarks and leptons in this model are assigned to left-right symmetric doublets under the SU(2)$_L$ x SU(2)$_R$ x U(1)$_{B-L}$ gauge group as follows:

$$Q_L \equiv \begin{pmatrix} u_L \\ d_L \end{pmatrix} \quad (\frac{1}{2} , 0, + \frac{1}{3})$$

$$Q_R \equiv \begin{pmatrix} u_R \\ d_R \end{pmatrix} \quad (0, \frac{1}{2} , + \frac{1}{3}) \tag{5}$$

and similarly for the leptons. The Higgs multiplet, ϕ responsible for giving mass to the fermions belong to the ($\frac{1}{2}$, $\frac{1}{2}$, 0) representation of the gauge group. In general, ϕ is a reducible representation and has a "charge"-conjugate multiplet

$\tilde{\phi} \equiv \tau_2 \phi^* \tau_2$ associated with it. Both ϕ and $\tilde{\phi}$ transform as follows under the gauge transformation

$$\phi \rightarrow U_L \phi U_R^+$$

and $\quad \tilde{\phi} \rightarrow U_L \tilde{\phi} U_R^+$ \hfill (6)

The invariant Yukawa coupling is given by:

$$\mathcal{L}_Y = \sum_{a,b} (h_{ab} \bar{Q}_{La} \phi Q_{R,b} + \tilde{h}_{ab} Q_{La} \tilde{\phi} Q_{Rb} + \text{h.c.} . \hfill (7)$$

If under left-right transformation, $Q_L \longleftrightarrow Q_R$, it follows that,[6] if $h, \tilde{h}$ are real

$$h_{ab} = h_{ba} \quad \text{and} \quad \tilde{h}_{ab} = \tilde{h}_{ba} \hfill (8)$$

if they are complex

$$h = h^+ \quad \text{and} \quad \tilde{h} = \tilde{h}^+ \hfill (9)$$

Ignoring CP-Violation, we then find that, eqn. (8) implies equality of left- and right-handed mixing angles and the number of free parameters is $N_g(N_g + 1)$, which for $N_g = 3$ is 12 . Thus, it is also not enough to predict the flavor mixing angles but, with the imposition of additional horizontal symmetries provides a more advantageous starting point over the standard model. We will focus in the rest of the article on this.

§3. HORIZONTAL SYMMETRIES AND CONSTRAINED MASS MATRICES

Since the gauge interactions provide the basic electro-weak dynamics of unified gauge theories, the largest horizontal group must be such as to leave it invariant. In the case of the standard model, the invariance group of the gauge interactions is $SU(2)_L \times U(1) \times G_1$ where

$$G_1 = U(N_g)_L \times U(N_g)_R^{(+)} \times U(N_g)_R^{(-)} \hfill (9)$$

where the superscripts $+ , -$ stand for the charge $\frac{2}{3}$ and $-\frac{1}{3}$ quarks respectively. In the case of the $SU(5)$ group, the corresponding group G_5 (which is the analog of G_1) is

$$G_5 = U(N_g)_L^{(+)} \times U(N_g)_L^{(-)} . \hfill (10)$$

For the Left-Right symmetric models, on the other hand, we have,

$$G_{LR} = U(N_g)_L \times U(N_g)_R \qquad\qquad (11)$$

The horizontal groups we choose will be subgroups of G_{LR}. One is then left with the following possibilities:

(1) Local horizontal symmetry

(2) Global continuous horizontal symmetry

(3) Discrete horizontal symmetry.

The general strategy in all these models will be to have the fermions and the Higgs fields transform non-trivially under the horizontal symmetry groups. The parameters $h_{ab,i}$ in the Yukawa couplings, then, get related to each other, thereby reducing the number of arbitrary parameters in the theory.

In case (1), which has been discussed in many papers,[7] restricted forms of mass matrices may be obtained. We will not discuss these models here, since, we do not believe any of them leads to a sufficiently interesting mass matrix with realistic weak interaction properties. However, one common feature of these models worth emphasizing is that they lead to flavor changing neutral current processes such as $K_L^0 \to \mu e$, $\mu\mu$, etc. through the exchange of the horizontal gauge bosons. The stringent experimental limits on these processes imply that the mass of the horizontal gauge boson must be more than 10^5Gev or so.[7]

Coming to global horizontal symmetries, attempts to consider such models were not made for a long time for the reason that once the symmetry is broken spontaneously, as it has to be to construct realistic models, it would lead to real massless Goldstone bosons and it was feared that such bosons will be in conflict with strict observational limits on long range forces. But it was shown by Chikashige, Peccei and this author[8] that the flavor conserving Goldstone boson couplings are always of γ_5-type and therefore lead to spin dependent long range forces in the non-relativistic limit. The experimental constraints on the strength λ of such long range forces is, $\lambda \lesssim 10^{-6}$ where λ is an appropriately defined dimensionless coupling constant and is very weak. This allows for scales of spontaneous breaking for arbitrary continuous global symmetries to be as low as a Tev, thereby raising the possibilities that such symmetries can be probed by existing particle physics experimental facilities. The work of ref. 8 was concerned with the breaking of global lepton number symmetry but could easily be extended to other such symmetries such as horizontal symmetry as has recently been done.[9] Again, the constrained fermion mass matrices that emerge are not realistic enough. An interesting phenomenological consequence is the process $K \rightarrow \pi + \chi$ where χ is the Goldstone boson corresponding to breaking of the horizontal global symmetry. This process is predicted to occur with a branching ratio $\sim 10^{-12}$ so that improved K-decay experiments could reveal the existence of such continuous symmetries.

Finally, let us discuss the discrete horizontal symmetries. For the purpose of this discussion, we will assume three generations and first, consider the minimal Higgs system needed to implement the symmetry breakdown[10] i.e. the left-right symmetric triplet Higgs $\Delta_L(3,1,-2) \oplus \Delta_R(1,3,-2)$ and the ϕ and $\tilde{\phi}$ both of which transform as $(\frac{1}{2}, \frac{1}{2}, 0)$ under $SU(2)_L \times SU(2)_R \times U(1)_{B-L}$. The details of symmetry breaking have been presented elsewhere[10] and we will not repeat it here. However, we simply wish to point out that to have a realistic fermion masses and mixings both ϕ and $\tilde{\phi}$ multiplets are essential. At the first stage of symmetry breaking, we have

$$\langle \Delta_R^o \rangle = V_R \neq 0 \quad \text{and} \quad \langle \Delta_L^o \rangle = 0 \tag{12}$$

thus reducing the symmetry from $SU(2)_L \times SU(2)_R \times U(1)_{B-L}$ down to the standard model symmetry $SU(2)_L \times U(1)_Y$. At the second stage, we require

$$\langle \phi \rangle = \begin{pmatrix} \kappa & 0 \\ 0 & \kappa' \end{pmatrix} \tag{13}$$

so that the $SU(2)_L \times U(1)_Y \rightarrow U(1)_{em}$. As far as the fermion masses go, at the first stage only the right-handed neutrinoes acquire a heavy Majorana mass of order m_{W_R} and at the second stage, all the charged fermions acquire mass. So, in the study of quark masses and mixings, it suffices to consider only ϕ and $\tilde{\phi}$ and discuss their transformation properties under the discrete horizontal symmetry group.

We will separate the discussion into two main parts: (a) first, we consider the abelian discrete groups such as Z_n and (b) then, we study some non-abelian groups such as the permutation groups S_n $(n \geq 3)$. In each case, we will search for various possible ways of assigning fermions and Higgs bosons to irreducible representations of the group so as to get interesting mass matrices. We will note that, for the minimal Higgs system (ϕ and $\tilde{\phi}$), only the cyclic group Z_4 gives an interesting mass matrix and we will duscuss it in sec. 4.

(a) Cyclic (Z_n) Groups: We first remind the reader that the

elements of Z_n are $e^{\frac{2\pi i}{n} m}$, $m = 1,\ldots n$ and since it is an
abelian group, all its representations are one dimensional. Also,
it has n distinct representations. For the Z_2 group, the most
constrained mass matrix, we get is of the form:

$$
\begin{pmatrix}
0 & a & 0 \\
a & 0 & b \\
0 & b & 0
\end{pmatrix}
\qquad (14)
$$

which is unrealistic since it implies $\sum_f m_f = 0$ for up and down
type quarks separately.

Let us now turn to the Z_3-group, whose elements are $e^{\frac{2\pi i}{3} n}$,
$n = 1,2,3$. There are the following possibilities: either ϕ is
invariant under the group or it transforms as another representation of the group. If it transforms as a singlet, both ϕ and
$\tilde{\phi}$ can couple to the same set of fermions. In this case, either
one gets (i) either $\sum_f m_f = 0$ for up or down quarks separately as
before (ii) or one generation completely decoupled from the other
two or (iii) an arbitrary mass matrix without any interesting
constraint. Clearly both cases (i) and (ii) are unphysical.

For Z_n , $n \geq 5$, the results are similar to the Z_3
case. The case of Z_4 horizontal symmetry leads to an extremely
interesting mass matrix[11,12] for three generations and will be
discussed in sec. 4.

(b) <u>Permutation Groups</u>: Now, we turn to the implications of
permutation groups S_3 and S_4 for the fermion mass matrices.
For the $SU(2)_L \times U(1)$ models, their implications have been
studied by several authors.[13] Here, I will present only some
preliminary results for the $SU(2)_L \times SU(2)_R \times U(1)_{B-L}$ models.[14]

We first present some remarks on the properties of permutation groups; they are non-abelian groups with total number of
elements equal to n! for S_n . Moreover, we are interested only
in those S_n-groups, which have {2} and {3}-dimensional representations. This resticts us only to S_3 and S_4 groups. Furthermore, we will assign the fermions to non-singlet representation,
since otherwise, we will reproduce either the results of the Z_2
case mentioned earlier or have a totally unconstrained mass
matrix. We will only summarize the results and for readers
interested in details, we refer to ref. 14.

(a) The first case we consider assigns ϕ (and hence $\tilde{\phi}$) to singlet representations under the group. In this case, one can choose all fermionic fields to transform as singlets or one singlet and another doublet. In the first case, the mass matrix is completely arbitrary; in the second case, they are diagonal leading to vanishing mixing angles.

(b) A somewhat more interesting case is to assign $(\phi,\tilde{\phi})$ to doublets of S_3 and S_4 . Let us discuss S_3 case first. The only interesting case is to assign two generations (Q_2,Q_3) to doublets and the remaining Q_1 one to a singlet. The general form of Yukawa coupling can be written as:

$$\mathcal{L}_Y = h_1 (\bar{Q}_{2L} \phi Q_{3R} + \bar{Q}_{3L} \phi Q_{2R}$$

$$+ \bar{Q}_{2L} \tilde{\phi} Q_{2R} - \bar{Q}_{3L} \tilde{\phi} Q_{3R}) + h.c.$$

$$+ h_2 (\bar{Q}_{1L} \phi Q_{2R} + \bar{Q}_{1L} \tilde{\phi} Q_{3R}) + h.c. \tag{15}$$

This leads to mass matrices which are traceless and symmetric, thus implying that

$$m_u + m_c + m_t = 0 \quad \text{and} \quad m_d + m_s + m_b = 0 \tag{16}$$

This is unacceptable.

Coming to the S_4 case, we can assign all three generations to 3 –dim. representation of S_4 . This leads to the following unique Yukawa coupling:

$$\mathcal{L}_Y = h[\frac{1}{\sqrt{2}} (\bar{Q}_{1L} \phi Q_{2R} + \bar{Q}_{2L} \phi Q_{1R}) + \frac{1}{\sqrt{6}} (\bar{Q}_{1L} \tilde{\phi} Q_{1R}$$

$$+ \bar{Q}_{2L} \tilde{\phi} Q_{2R} - 2\bar{Q}_{3L} \tilde{\phi} Q_{3R})] + h.c. \tag{17}$$

This again leads to symmetric traceless matrices and is therefore unacceptable.

From examining these simple cases, we find that, except for Z_4-horizontal symmetry (to be discussed shortly), in other cases, the Higgs sector must be extended by including additional $\phi(\frac{1}{2} , \frac{1}{2} , 0)$ multiplets to arrive at acceptable fermion mass matrices. In that situation, one has a lot more freedom and it is not possible to give any compact description. However, we present one interesting scenario with extended Higgs sector, that gives an interesting mass matrix.

Consider the Cyclic group Z_{2N_g} as the horizontal symmetry group. Let us assume the quark generations to transform as follows:

$$Z_{2N_g} : \quad Q_a \rightarrow e^{\frac{i\pi a}{N_g}\gamma_5} Q_a$$

$$\phi_{a,b} \rightarrow e^{-\frac{i\pi}{N_g}(a+b)} \phi_{a,b}$$

where $a,b = 1,\ldots N_g$. $\qquad\qquad\qquad\qquad\qquad$ (18)

If we admit into the theory Higgs multiplets of type $\phi_{m,m+1}$ for $m = 1,\ldots N_g-1$ and ϕ_{N_g,N_g} , then, they lead to the mass matrix of the following form:

$$
\begin{pmatrix}
0 & m_{12} & & & \\
m_{12} & 0 & m_{23} & & \\
& m_{23} & 0 & m_{34} & \\
& & m_{34} & 0 & \\
& & & & \ddots \\
& & & & & m_{N_g,N_g}
\end{pmatrix}
$$

$\qquad\qquad\qquad\qquad\qquad\qquad\qquad\qquad\qquad\qquad\qquad$ (19)

This mass matrix has been well-studied in the literature.[2,15] It leads to the following mixing pattern between generations, i.e.

$$V_{a'a+1} \approx \sqrt{\frac{m_a^{(-)}}{m_{a+1}^{(-)}}} - \sqrt{\frac{m_a^{(+)}}{m_{a+1}^{(+)}}}$$

$\qquad\qquad\qquad\qquad\qquad\qquad\qquad\qquad\qquad\qquad\qquad$ (20)

where the a stands for the generation number and + , − stand for + 2/3 and −1/3 quark of the particular generation. . For instance, this will imply $V_{cd} \approx V_{us} \approx \sqrt{\frac{m_d}{m_s}} \approx$ .22 which is in

424

rough agreement with presently known values: $V_{us} \approx .231 \pm 0.003$ obtained from an analysis of semileptonic decays.[16] For the

b-c and t-s mixing, this implies: $V_{bc} \approx \sqrt{\dfrac{m_s}{m_b}} - \sqrt{\dfrac{m_c}{m_t}} \approx .01 -$

.049 for m_t = 45 to 65 Gev . The present experimental values[10] are: .043 − .055 . This model can, therefore, be tested once the t-quark is discovered.

§4. Z_4 HORIZONTAL SYMMETRY AND PREDICTIONS FOR QUARK MIXINGS

We now describe the constraints on the quark mixings in the case of the cyclic horizontal symmetry Z_4 . The elements of Z_4 can be represented by (1,i,−1,−i) and we will assign the left- and right-chirality quarks of each family to transform under the group as follows:

$$Z_4: \qquad Q_{1L}, Q_{3L} \longrightarrow Q_{1L}, Q_{3L}$$

$$Q_{2L} \longrightarrow -Q_{2L}$$

$$(Q_{1R}, Q_{3R}) \longrightarrow i(Q_{1R}, Q_{3R})$$

$$Q_{2R} \longrightarrow -iQ_{2R}$$

$$\phi \longrightarrow -i\phi$$

$$\tilde{\phi} \longrightarrow +i\tilde{\phi} \qquad\qquad (21)$$

We will work with the minimal Higgs system[10] for the model. The imposition of this discrete system constrains the Higgs potential and leads to a minimum[17] for

$$\langle \phi \rangle = \begin{pmatrix} \kappa & 0 \\ 0 & \kappa' e^{i\alpha} \end{pmatrix} \qquad\qquad (22)$$

The Yukawa coupling for this case, can be written as:

$$\mathcal{L}_Y = \sum_{a,b=1,3} \left[g_{ab} \, \bar{Q}_{aL} \phi Q_{bR} + g_{22} \bar{Q}_{2L} \phi Q_{2R} \right.$$

$$+ h_{12} (\bar{Q}_{1L} \tilde{\phi} Q_{2R} + \bar{Q}_{2L} \tilde{\phi} Q_{1R})$$

$$\left. + h_{23} (\bar{Q}_{2L} \tilde{\phi} Q_{3R} + \bar{Q}_{3L} \tilde{\phi} Q_{2R}) \right]$$

$$+ \text{h.c.} \qquad\qquad (23)$$

We choose the theory to conserve CP prior to spontaneous symmetry breaking, which implies g and h to be real. We can make a rotation in the (1,3) space to diagonalize $(g_{ab})_{diag.} \equiv (g_{11}, g_{33})$ and this simply redefines h_{12} and h_{23}. On substituting eqn. (22), this leads to the following structure for up and down quark mass matrices:

$$
M_u = \begin{pmatrix} g_{11}\kappa & h_{12}\kappa' e^{-i\alpha} & 0 \\ h_{12}\kappa' e^{-i\alpha} & g_{22}\kappa & h_{23}\kappa' e^{-i\alpha} \\ 0 & h_{23}\kappa' e^{-i\alpha} & g_{33}\kappa \end{pmatrix} \quad (24)
$$

$$
M_d = e^{i\alpha} \begin{pmatrix} g_{11}\kappa' & h_{12}\kappa e^{-i\alpha} & 0 \\ h_{12}\kappa \, e^{-i\alpha} & g_{22}\kappa' & h_{23}\kappa e^{-i\alpha} \\ 0 & h_{23}\kappa e^{-i\alpha} & g_{33}\kappa' \end{pmatrix} \quad (25)
$$

We see that, if we ignore CP-violation ($\alpha = 0$) there are seven parameters. We should therefore be able to obtain relations between mixing parameters and quark masses. To diagonalize thesse matrices, we first assume $g_{33} \gg g_{22}, h_{23} \gg g_{11}, h_{12}$. In this approximation, we can first diagonalize the lower 2×2 matrices:

$$
M_u^0 = \begin{pmatrix} 0 & & \\ & g_{22}\kappa & h_{23}\kappa' e^{-i\alpha} \\ & h_{23}\kappa' e^{-i\alpha} & g_{33} \end{pmatrix} \quad (26)
$$

and

$$
M_d^0 = \begin{pmatrix} 0 & & \\ & g_{22}\kappa' e^{+i\alpha} & h_{23}\kappa \\ & h_{23}\kappa & g_{33}\kappa' e^{i\alpha} \end{pmatrix} \quad (27)
$$

Since these are symmetric matrices, we can diagonalize them as follows:[6]

$$U \, M_u U^T K_u = D_u$$

$$\text{and} \quad V M_d V^T K_d = D_d \tag{28}$$

where $K_{u,d}$ are unitary diagonal matrices. Eqn. (26) and (27) are diagonalized for

$$U^{(o)} = \begin{pmatrix} 1 & & \\ & \cos\theta_u & \sin\theta_u e^{i\beta_u} \\ & -\sin\theta_u e^{-i\beta_u} & \cos\theta_u \end{pmatrix}$$

$$V^{(o)} = \begin{pmatrix} 1 & & \\ & \cos\theta_d & \sin\theta_d e^{i\beta_d} \\ & -\sin\theta_d e^{-i\beta_d} & \cos\theta_d \end{pmatrix}$$

and

$$K = \begin{pmatrix} 1 & & \\ & 1 & \\ & & e^{i\gamma_{u,d}} \end{pmatrix} \tag{29}$$

where $\tan\beta_u = \tan\beta_d = \dfrac{g_{33} - g_{22}}{g_{33} + g_{22}} \tan\alpha$

$$\text{and} \quad \tan 2\theta_u = \frac{2h_{23}\omega}{g_{22} - g_{33}}$$

$$\tan 2\theta_d = \frac{2h_{23}}{\omega(g_{22} - g_{33})} \tag{30}$$

with $\omega = \kappa'/\kappa$.

One can also obtain the eigenvalues, which give the quark masses. The relative mixing between second and third generation can, then, be written as:

$$\cos 2(\theta_u - \theta_d) = \frac{(m_c^2 - m_t^2)^2 \omega^2 + (m_s^2 - m_b^2)^2}{(1 + \omega^2)(m_c^2 - m_t^2)(m_s^2 - m_b^2)} \; . \tag{31}$$

One can also find the value of ω near $\alpha = 0$ and $\pi/2$:

$$\alpha = 0 \; , \quad \omega^2 = \left(\frac{m_s - m_b}{m_c + m_t} \right)^2$$

and $\quad \alpha = \dfrac{\pi}{2} \; , \quad \omega^2 = \left(\dfrac{m_s - m_b}{m_c - m_t} \right)^2 \tag{32}$

Also one finds that

$$\omega \simeq \left(\frac{m_b}{m_t} \right) \ll 1 \tag{33}$$

Eqn. (33) then implies that the mass matrix for up-quark sector is almost purely diagonal. This can be seen from eqn. (30), which implies:

$$\tan 2\theta_u = \omega^2 \tan 2\theta_d \ll \tan 2\theta_d \tag{34}$$

The mixing angles in charged currents will therefore be given almost entirely by the mixing angles in the down quark sector. We further find that,

$$(i) \quad \text{for } \alpha \approx 0 \; , \; m_t \le \frac{m_b m_c}{m_s} \left(1 - 2 \left(\frac{m_s}{m_b} \right) \right) \tag{35}$$

whereas

$$(ii) \quad \text{for } \alpha \approx \frac{\pi}{2} \; , \; m_t \ge \frac{m_b m_c}{m_s} \left(1 + 2 \left(\frac{m_s}{m_b} \right) \right) \tag{36}$$

Eqn. (35) implies $m_t(m_c) \le 45$ Gev , which when extrapolated to m_t , will lead to a lower value:

Equation (ii) implies $m_t(m_c) \geq 51$ Gev .

Using eqn. (32) in eqn. (31), we predict

$$V_{cb} \approx V_{ts} \approx \sqrt{\left| \frac{m_s}{m_b} - \frac{m_c}{m_t} \right|} \tag{37}$$

This is a very sensitive function of m_t and will predict $V_{cb} \approx .06$–$.08$ for $m_c \approx 1.3$ Gev and 0–$.054$ for $m_c = 1.5$ Gev . Thus, a precise determination V_{cb} from B-decays would restrict the value of m_t and thus, the discovery of the top quark would then test this model. It will also dictate a value of α , thus predicting CP-violation parameters, which we do not discuss here.

Next, we pass on to diagonalize the full mass matrix following ref. (12). We choose, the full unitary matrix to be

$$U = U^{(1)}U^{(0)} \tag{38}$$

where

$$U^{(1)} = \begin{pmatrix} 1 & X_{1u} & X_{2u} \\ -X_{1u} & 1 & 0 \\ -X_{eu} & 0 & 1 \end{pmatrix} \tag{39}$$

and similarly for V . We then determine,

$$h_{12} \approx \sqrt{m_d m_s}$$

and

$$X_{1d} \approx + \sqrt{\frac{m_d}{m_s}} \, e^{i\alpha}$$

$$X_{2d} \approx - \sqrt{\frac{m_d}{m_s}} \cdot \left(\frac{m_s}{m_b}\right)^2 e^{i(\alpha - \beta + \gamma_d)}$$

$$X_{1u} \approx + \sqrt{\frac{m_d}{m_s}} \left(\frac{m_s^2}{m_c m_b}\right) e^{i\alpha}$$

$$X_{2u} \approx 0 \tag{40}$$

We thus see that, the prediction for Cabibbo angle θ_c is in agreement with observation.

§5. A SOLUTION TO THE STRONG CP-PROBLEM IN THE Z_4-MODEL

In this section, we wish to point out that, there exists a very interesting solution to the strong CP-problem in the Z_4-model described in the previous section. The strategy[18] is to use exact left-right symmetry prior to spontaneous symmetry breaking to set $\theta = 0$ and then require hermitean mass matrices to arise <u>naturally</u> so that

$$\text{Arg Det } M_u = \text{Arg Det } M_d = 0 \tag{41}$$

This would then imply that $\theta_{tree} = 0$ naturally. Since, we would like the theory to have weak CP-violation, we will choose the Yukawa couplings in eqs. (23) to be complex. Left-Right symmetry would then imply that,

$$\mathcal{L}_Y = \sum_{a,b,c} g_{ab} \bar{Q}_{aL} \phi Q_{bR} + g_{22} \bar{Q}_{2L} \phi Q_{2R}$$

$$+ h_{12} \bar{Q}_{1L} \tilde{\phi} Q_{2R} + h_{12}^* \bar{Q}_{2L} \tilde{\phi} Q_{1R}$$

$$+ h_{23} \bar{Q}_{2L} \tilde{\phi} Q_{3R} + h_{23}^* \bar{Q}_{3L} \tilde{\phi} Q_{2R} + \text{h.c.} \tag{42}$$

where $g = g^+$ and g_{22} is real.

As before, we can diagonalize g_{ab} and the matrix h after this rotation also retains its hermiticity. Now, if we can show that the Higgs potential has a solution with $\alpha = 0$, for a range of parameters, we would obtain hermitean M_u and M_d i.e.

$$M_u = \begin{pmatrix} g_{11}\kappa & h_{12}\kappa' & 0 \\ h_{12}^*\kappa' & g_{22}\kappa & h_{23}\kappa' \\ 0 & h_{23}^*\kappa' & g_{33}\kappa \end{pmatrix} \tag{43}$$

and $M_d = M_u(\kappa \longleftrightarrow \kappa')$. $\tag{44}$

Thus, $\theta = 0$ naturally at the tree level. It is easy to see that requirement of left-right symmetry and hermiticity makes <u>all</u> couplings appearing in the potential real. As a result, we find that at the minimum of the potential, $\langle\phi\rangle$ is real[19] for a range of values of the potential. Thus, the model also provides a natural solution to the strong CP-problem. In higher orders, a finite θ will arise and we find it to be small.

Finally, I wish to point out that, the Z_4-symmetry is actually a subgroup of a $U(1)$-symmetry, which combines the Peccei-Quinn symmetry and a horizontal symmetry. One could therefore embed the Z_4-symmetry in this $U(1)$ symmetry to provide an alternative mechanism to solve the strong CP-problem. To make this discussion more explicit, we need to define the $U(1)$-charge for each generation:

$$U(1): \quad \phi \rightarrow e^{i\theta}\phi$$

$$Q_a \rightarrow e^{i\gamma_5\theta+4i(a-1)\theta} Q_a \tag{45}$$

where $a = 1,2,3$. This symmetry is respected by the Higgs potential if two terms $(\text{Tr}\ \phi\tilde{\phi}^\dagger)^2$ and $\text{Tr}\ \phi\tilde{\phi}^\dagger\phi\tilde{\phi}^\dagger$ are dropped from it. It is of course broken spontaneously by vacuum, which leads to the axion. To make this consistent with the negative results of the axion searches, a gauge singlet η , which is non-singlet under $U(1)$ needs to be introduced. This η-field will have a vacuum expectation value of order 10^{12} Gev so that the axion is hidden,[20] and is consistent with astrophysics. We will elaborate on this aspect of the model in a separate publication.[21]

§6. CONCLUSION

In summary, we have explored the possibility that, discrete horizontal symmetries may provide an insight into the nature of flavor mixing between different generations. We find one particular group, Z_4 , which leads to extremely interesting predictions for mixing angles: some of them are $V_{cd} \approx V_{us} \approx \sqrt{\dfrac{m_d}{m_s}}$ and $V_{cb} \approx V_{ts} \approx \sqrt{\left|\dfrac{m_s}{m_b} - \dfrac{m_c}{m_t}\right|}$. Thus, a more accurate measurement of V_{cb} along with the discovery of t-quark would provide a test of this model. We also find that the model provides two interesting solutions to the strong CP-problem -- one, by using the discrete symmetry alone; the other by embedding the Z_4 in a chiral $U(1)$ group that combines the Peccei-Quinn symmetry with the horizontal symmetry. It will be interesting to look for embedding of this model into the $SO(10)$ grand unified theory.

REFERENCES

1. See the talk by K. Klenknecht at this conference. Some other recent references include: A.J. Buras, W. Skominski and H. Stegger, MPI-PAE/PTh 77/83 (1983). K. Klenknecht and B. Renk, Phys. Lett. 130B, 459 (1983). L.L. Chau and W.Y. Keung, Phys. Rev. D (to appear). For an earlier review, see L.L. Chau, Phys. Rep. 95c, 1 (1983).

2. F. Wilczek and A. Zee, Phys. Lett. 70B, 41 (1977); S. Weinberg, Festschrift for I.I. Rabi, Trans. NY ser. II 38 (1977); H. Fritzsch, Phys. Lett. 70B, 436 (1977).

3. H. Georgi and S.L. Glashow, Phys. Rev. Lett. 32, 438 (1974).

4. R.N. Mohapatra, Phys. Rev. Lett. 43, 893 (1979).

5. J.C. Pati and A. Salam, Phys. Rev. D10, 275 (1974); R.N. Mohapatra and J.C. Pati, Phys. Rev. D11, 566, 2558 (1975); G. Senjanović and R.N. Mohapatra, Phys. Rev. D12, 1502 (1975).

6. R.N. Mohapatra, F.E. Paige and D.P. Sidhu, Phys. Rev. D17, 2642 (1978).

7. R.N. Mohapatra, Phys. Rev. D9, 3461 (1974); R.N. Mohapatra, J.C. Pati and L. Wolfenstein, Phys. Rev. D11, 3319 (1975); C.L. Ong, Phys. Rev. D19, 2738 (1978); F. Wilczek and A. Zee, Phys. Rev. Lett 42, 421 (1979); J. Chakrabarti, Phys. Rev. D20, 2411 (1979); Y. Chikashige, G. Gelmini, R. Peccei, and M. Roncadelli, Phys. Lett. 94B, 499 (1980); A. Davidson, M. Koca and K.C. Wali, Phys. Rev. Lett. 43, 92 (1979).

8. Y. Chikashige, R.N. Mohapatra and R. Peccei, Phys. Lett. 98B, 265 (1981).

9. G. Gelmini, S. Nussinov and T. Yanagida, Trieste Preprint (1982); D.B. Reiss, Phys. Lett.   ; F. Wilczek, Phys. Rev. Lett.49, 1549 (1982).

10. R.N. Mohapatra and G. Senjanovic, Phys. Rev. Lett. 44, 912 (1980) and Phys. Rev. D21, 165 (1981).

11. R.N. Mohapatra and G Senjanović , Phys. Lett. 73B, 176 (1977).

12. G. Ecker, W. Grimus and W. Konetschny, Nucl. Phys. B177, 489 (1981).

13. E. Derman and H. Tsao, Phys. Rev. D20, 1207 (1979), Y. Yamanaka, H. Sugawara and S. Pakvasa, Phys. Rev. D25, 1895 (1982).

14. R.N. Mohapatra and J. Sucher, in preparation.

15. V.P. Nair, K.C. Wali and L. Michel, Syracuse Preprint (1984).

16. M. Bourquin, et. al., Rutherford Lab. Preprint (1983).

17. For an analysis of the Higgs potential and study of the conditions under which eqn. (22) represents a minimum, see, G. Branco and R.N. Mohapatra, Univ. of Maryland Preprint (1983).

18. M.A.B. Bég and H.S. Tsao, Phys. Rev. Lett. <u>41</u>, 278 (1978); R.N. Mohapatra and G. Senjanović, Phys. Lett. <u>79B</u>, 283 (1978).

19. As has been shown in ref. 17, for this situation to occur, certain coupling parameters must be of order $(m_{W_L}/m_{W_R})^2$ and clearly, if m_{W_R} is very, very heavy, the model would become unnatural. For present lower bounds of m_{W_R} of order of 1-2 Tev , the relevant coupling parameters are of order $\approx 10^{-3}$, which is an acceptable fine tuning.

20. For a review and references, see J.E. Kim, Proceedings of the Asia Pacific Conference, World Scientific Publishing, Singapore (1983).

21. R.N. Mohapatra, in preparation.

DISCUSSION

BÉG:
What cure do you offer for the strong CP problem in your version
of the SO(10) scheme?

MOHAPATRA:
We have not looked into the strong CP-problem in this new version
of SO(10), and at the moment we have none.

KANE:
Please clarify how you cure the problem of generating enough
baryon asymmetry in your new SO(10) model.

MOHAPATRA:
In the old SO(10) version, the suppression of $(n_B - n_{\bar{B}})/n_\gamma$ was
due to a factor $(m_{W_R}/M_U)^2$ arising from the fact that parity and

parity and $SU(2)_R$ were broken at the same scale. In our
version, the scale of parity breaking is the GUT scale M_U
regardless of what M_W is and this means that the additional
suppression is absent.

CHAU:
Would you survey for us what is the current experimental limit on
m_{W_R}?

MOHAPATRA:
The most stringent limit on the W_R mass comes from
consideration of the left-right box graph contribution to the
$\Delta S = 2$ $K_L - K_L$ mass difference. This new contribution adds a
factor of -430 $(m_{W_L}/m_{W_R})^2$ to the old Gaillard-Lee left-right
contribution and, implies that $m_{W_R} \geq 1$ TeV or so. This is

assuming left and right gauge couplings and mixing angles to be
equal. Once this is given up, the limit becomes weak.

KLEINKNECHT:
If the question was on experimental limits for the W_R boson
mass, there is a limit from the LBL- TRIUMF μ decay experiment
$m_{W_R} > 370$ GeV.

MOHAPATRA:
This limit is dependent on the assumption that neutrinos are
Dirac particles.

WEAK MIXING ANGLES IN UNIFIED THEORIES

Kameshwar C. Wali

Physics Department
Syracuse University
Syracuse, N.Y. 13210

ABSTRACT

The replication of families or generations with identical electroweak interactions in the low energy region presents a deep mystery. It is the historic e-µ puzzle in a considerably extended form. Horizontal or intragenerational interactions and the consequent mixing angles govern a wide range of phenomena concerning both old and new particles. In this review, I discuss the recent theoretical ideas and two specific models pertaining to horizontal symmetries.

INTRODUCTION

We learned the other day from Professor Juliet Lee Franzini that there is indirect, but compelling evidence for the existence of top-quark. This follows from the study of b-quark couplings which are consistent with the standard six quark model. There is no evidence for any new or flavor changing neutral couplings (see Franzini's talk in these proceedings). Consequently we must assume that the top quark exists. We then have three generations or families of quarks and leptons. As far as the standard $SU(2)_L \times U(1)$ electroweak theory is concerned, the three families have identical

quantum numbers, and hence the description of their electromagnetic
and their low energy (<100 GeV) weak interactions proceeds along
identical lines. They behave, in this respect, as though they are
exact copies of each other except in their mass spectrum reminding
us of the historic e-μ puzzle.

The e-μ puzzle by itself, however, was more of a theoretical
and somewhat of an academic nature. There were few experimental
consequences outside the domain of Quantum Electrodynamics. The
generation puzzle or the phenomenon of "superfluous replication"
(according to Glashow) presents itself in a much more extended form
with consequences covering a wide range of experimental phenomena
associated with the physics of the new flavors. For instance, the
currently believed quark masses display a bewildering hierarchy.
The generalized Cabibbo-type angles and phases (Kobayashi-Maskawa
angles and phases), which come into play when one transforms the
gauge eigenstates into mass eigenstates, govern the charm, top,
bottom decays, weak CP-violation and give rise to quantitative
estimates of the rare decay modes. The number of flavors, especially the number of light neutrinos, is believed to have important
implications in astrophysics. Thus generally speaking, a whole
range of phenomena depend on our understanding of the generation
puzzle to which the standard theory has no answer. It is necessary
to go beyond the standard theory and consider new interactions, the
so-called "Horizontal" or "Intragenerational" interactions. A
complete theory that attempts grand unification should incorporate
the generation structure and the new interactions in a non-trivial
way and predict the form of the mass matrices so that some of the
above questions can be answered. Without such a theory, there are
too many arbitrary parameters.

In recent years a great deal of effort has been invested by
several authors in the study of such interactions and the possible
symmetries (horizontal symmetries). In what follows I shall first
describe in a qualitative way the basic ideas underlying such
attempts. Then after a brief review of the formalism that leads to

436

the definition of the weak mixing angles, I shall discuss two very
recent models which illustrate the general ideas and which allow
one to calculate the mixing angles in terms of a relatively fewer
number of parameters.

GENERAL IDEAS

Recent attempts to go beyond the standard theory fall into two
major categories: Electroweak Interactions and Grand Unified
Theories.

Electroweak Interactions

As a first step, extend the electro/weak interaction group
$SU(2)_L \times U(1)$ by adding a horizontal group factor G_H,

$$SU(2)_L \times U(1) \rightarrow G_{EW} \sqsupset SU(2)_L \times U(1) \times G_H.$$

The group G_H which characterizes the intragenerational interactions
is chosen to be discrete or continuous. The part of the Lagrangian,
that includes the kinetic energy and gauge interaction terms ex -
hibits certain discrete symmetries. Permutation symmetry among the
various generations is an obvious example. Extended to Higgs
sector, such symmetries reduce the number of free parameters, restrict the form of the mass matrices and lead to relations between
the mixing angles and fermion masses. Discrete symmetries have also
the advantage that they do not give rise to Goldstone bosons when
the symmetry is broken spontaneously. However, choice of discrete
symmetries is itself quite arbitrary. There is also no possibility
of incorporating such symmetries either in the framework of a grand
unified theory or as a part of a dynamically broken symmetry scheme.
In contrast, continuous symmetries are more in the spirit of gauge
theories. They can, at least in principle, originate in a grand
unified theory or dynamically broken symmetry schemes. Consequently,
several gauge groups have been proposed. They all have varying
degrees of qualitative successes.

Additional gauge symmetries, however, bring along with them

additional gauge and Higgs mesons. They produce, in general, flavor changing neutral currents leading to constraints that have to be imposed from outside. Therefore it is not surprising that quantitative details have not been worked out in such models.

A particularly interesting class of models is based on left-right symmetry. We shall hear from Professor Mohapatra about this class and its present status.

Grand Unified Theories

The main drawback of the approach based on the considerations of purely electroweak interactions is that the number of generations is not predictable. The same drawback is shared by the currently popular grand unified theories based on SU(5), SO(10), or the exceptional group E_6. They all are single generation schemes. They are exact analogues of the standard electro/weak interaction theory but for the fact that they include strong interactions. The number of generations is still a mystery within the framework of such theories. They have nothing to say about the intragenerational interactions. It is necessary to go beyond these theories as well.

Grand unification raises some basic questions. What are the basic constituents? Are they quarks and leptons as we "see" them in the low energy region? Or, is the generation puzzle, a manifestation of subquark, subleptonic structure? Are the basic constituents fermions or supersymmetric multiplets, consisting of both fermions and bosons? Lacking answers to these questions at present, we proceed on the assumption that quarks and leptons along with the currently observed generation structure persist all the way to a grand unification mass scale. Subsequently, we look for multi-generational grand unified theories which incorporate the generation structure in a non trivial way.

Multi-generational grand unification schemes can be classified into two main categories:

1) those in which the conventional Higgs mechanism is used to break the symmetry

2) those in which dynamical symmetry breaking is invoked, necessitating the introduction of hypercolor or technicolor.

In both classes of theories, one starts with certain assumptions which have become, so to say, the rules of the game. These are a) no exotic fermions such as color singlet fractionally charged ones or quarks other than triplets b) renormalizability c) no superfluous replication d) the representation be complex e) the number of generations be $\geqslant 3$ f) the theory be asymptotically free. In the case of class (1) theories there is no solution satisfying all the above criteria if the grand unification group G is simple. If we relax one or more assumptions, solutions are possible. For example SU(11) provides a solution with exactly three generations if we take the anomaly free combination of representations [4]+[8]+[9]+[10] of SU(11) to represent the fundamental fermions. ([n] denotes a representation with n SU(11) indices all antisymmetrised.) There are, however, altogether 561 fermions in the above representation, of which only 45 survive, giving the three families in the low energy region. The others can be made to acquire superheavy masses. Because of the large number of fermions, the model does not satisfy the criterion of asymptotic freedom. Likewise if we relax the requirement (c) and allow a replication of representations in the choice of anomaly free combination, one can obtain a model based on SU(9) with three generations, whose only virtue is that it contains a fewer number of fermions to start with than the SU(11) model. The number of generations in such an approach is not really predicted; solutions with any number of generations can be found.

Next, if we consider the technicolor alternative, we are required to assume that

$$G \supset [SU(3)_C \times SU(2) \times U(1)] \times G_{TC},$$

where G_{TC}, the group that describes technicolor must be bigger than $SU(3)_C$ in order to provide superstrong forces which form condensates in the TeV region. If $G_{TC} \equiv SU(n)$, then $n \geqslant 4$. With this proviso, if we examine simple groups, there is no solution for SU(N) groups for any N. In the case of orthogonal groups of the class O(4n+2), SO(18) provides a solution with three generations, but the technicolor group is Sp_4 which is only as big as SU(3). The smallest representation of the next possible candidate SO(22), contains 1024 fermions! With this huge number of fermions, asymptotic freedom is lost.

Thus simple groups do not appear to provide a satisfactory answer for a multigenerational, grand unified theory with or without technicolor. In fact there is a theorem which, with the previously stated assumptions, demonstrates that SU(5) and SO(10) are almost unique as unitary and orthogonal group candidates for grand unification. If simple groups are thus ruled out, the next simplest possibility is a semi-simple group structure of the form G×G, G×G×G and so on. Discrete symmetries can be imposed to assure that there is only one gauge coupling constant instead of more than one as would be the situation in the general case. In the class of models based on SU(N)×SU(N), SU(5)×SU(5) is the minimal multigenerational model which has five generations. With the assumptions listed before, one can restrict such models to 5⩽N⩽10. Such models do have some attractive features, but very few details have been explored as yet to come to any definitive conclusions. Also incorporation of the technicolor alternative is not possible in such models. For the latter purpose one has to go to the G×G×G structure.

A grand unified theory based on the semi-simple group structure $SO(10)_V \times SO(10)_H$ has some very attractive features. The "vertical" $SO(10)_V$ describes each family of ordinary fermions and hence incorporates automatically all the good features of the grand unified theory based on $SO(10)_V$. The "horizontal" $SO(10)_H$ contains hypercolor, extended hypercolor, and horizontal interactions. The irreducible representation $(\underline{16},\underline{10})+(\underline{10},\underline{16})$, an unusual combination of spinor and vector representations, is almost uniquely selected out. The ensuing particle spectrum and quantum numbers allow the existence of all the needed interactions, unlike other hypercolor models

440

where one has to introduce such interactions from outside. Further, the model satisfies the criteria for renormalizability and asymptotic freedom with a non-trivial generation structure of four conventional fermionic families. SU(4) emerges as the unique unitary group describing hypercolor. There appear to be no serious difficulties concerning flavor changing neutral currents. This is due to a fortuitous combination of group theoretic features of the model, which constrain the effective exchanges of single bosons in their couplings to K and D mesons. However, the model has the unsatisfactory feature of requiring more than one primary Higgs mechanism to cause the initial symmetry breaking. The discrete symmetry which fixes the group structure and the fermionic representation has to be broken right away! Nonetheless the model has many realistic features and merits further study.

MASS MATRICES AND MIXING ANGLES

The gauge-invariant part of the Lagrangian $\mathcal{L}_Y$, which gives rise to mass matrices is that which contains the Yukawa-type fermion-Higgs couplings,

$$\mathcal{L}_Y = \tilde{\psi}_i \, \Gamma_{ijk} \, \psi_j \phi_k + \cdots \, ,$$

where ψ_i, ϕ_k represent the fermion and Higgs fields respectively and Γ_{ijk} are coupling constants which may be real or complex. When the symmetry is broken spontaneously by giving vacuum expectation values to the appropriate ϕ's, we obtain fermion mass matrices that belong to different charge sectors. For N-generations, they are N×N complex matrices whose matrix elements are expressed in terms of the Yukawa couplings Γ_{ijk} and the vacuum expectation values which are, in general, complex. Thus as these parameters are unconstrained to a large extent, the mass matrices are arbitrary unless there are additional symmetries which restrict their form.

A complex matrix M can be diagonalized by a bi-unitary transformation,

$$U_L M U_R^\dagger = M \text{ (diagonal)}$$

where U_L and U_R are the unitary matrices that diagonalize the Hermitian matrices $MM^\dagger$ and $M^\dagger M$ respectively, that is,

$$U_L MM^\dagger U_L^\dagger = U_R M^\dagger M U_R^\dagger = (M \text{ (diagonal)})^2$$

Now if $U_L^{(d)}$ and $U_L^{(u)}$ are such arbitrary matrices that diagonalize the down (charge $-\frac{1}{3}$) and up (charge $\frac{2}{3}$) quark mass matrices, the generalized Cabibbo–Kobayashi–Maskawa matrix that characterizes the charge changing gauge interactions is U_c, where

$$U_c = U_L^{(u)} U_L^{(d)\dagger}$$

By suitably redefining the phases of the quark fields, we can make the first row and the first column of U_c real leaving specified number of physically meaningful phases in the remaining elements of U_c. In the case of three generations, U_c has the form

$$U_c = \begin{bmatrix} U_{ud} & U_{us} & U_{ub} \\ U_{cd} & U_{cs} & U_{cb} \\ U_{td} & U_{ts} & U_{tb} \end{bmatrix}$$

$$= \begin{bmatrix} c_1 & s_1 c_3 & s_1 s_3 \\ -s_1 c_2 & c_1 c_2 c_3 - s_2 s_3 e^{i\delta} & c_1 c_2 s_3 + s_2 c_3 e^{i\delta} \\ -s_1 s_2 & c_1 s_2 c_3 + c_2 s_3 e^{i\delta} & c_1 s_2 s_3 - c_3 c_3 e^{i\delta} \end{bmatrix},$$

where $c_i = \cos\theta_i$, $s_i = \sin\theta_i$, $\theta_1, \theta_2, \theta_3$ are the mixing angles. δ is the non-trivial phase. If it vanishes, then there is no CP-violation due to charged weak gauge boson interactions.

TWO SPECIFIC MODELS

As examples of the two broad categories mentioned earlier, we shall consider two recent models. One is based on a discrete permutation symmetry superimposed on the standard electro/weak theory. The other uses a global, axial U(1) symmetry in conjunction with the SO(10) grand unification scheme of a single generation.

442

Permutation Symmetry S_4

The electro/weak group structure is augmented by S_4, the symmetry group of permutations on four objects, so that

$$G_{EW} = SU(2)_L \times U(1) \times S_4.$$

Why S_4? The proponents of this class of models have three families in mind. It would appear that for N generations, the natural candidate would be S_N with N Higgs doublets coupled to the fermions. It turns out, however, that this first guess is not favored by experiments. In such a class of models, one quark doublet is always decoupled from the rest, and hence, in the case of three generations b cannot decay by W-exchange. As it can also be shown that b also cannot decay via non-leptonic decay modes, S_3 is clearly not suitable for three generations.

In the S_4 model for three generations, the <u>three</u> left-handed doublets of fermions and <u>three</u> Higgs doublets are assigned to <u>3</u> representation of S_4. The right-handed singlets belong to <u>1+2</u> representation. In addition to the three Higgs doublets, there is a Higgs singlet in the model, which belongs to S_4 singlet. It is this singlet which is the source of CP-violation in the model.

After the minimization of the Higgs potential and the usual transformation to mass eigenstate basis, the following results emerge,

a) Mass relations involving both quarks and leptons. For instance, $m_c/m_t \approx m_\mu/m_\tau$, also $m_c/m_t \approx m_s/m_b$. Hence m_t is expected to lie in the range 25 GeV to 35 GeV.

b) With some approximations, the Kobayashi-Maskawa matrix elements are given in terms of fermion masses. For instance, $|U_{bc}| = |U_{ts}| \approx \frac{2}{\sqrt{3}} (m_\mu/m_\tau)$, $|U_{ub}| = |U_{td}| = \frac{\sin\theta_c}{\sqrt{3}} (m_\mu/m_\tau)$. The predicted values are in good agreement with recent experimental findings on the lifetime τ_B of B-meson.

c) No CP-violation due to gauge boson exchanges. Only source of CP-violation is due to flavor-changing Higgs exchange. CP-violation is the superweak type.

443

d) There are flavor changing neutral currents leading to rare decays. The rates for such decays, electric dipole moments of different quarks and leptons and mixing parameters for K, D, B, and B_S are all calculated.

Thus, although the model has some arbitrariness in the choice of the discrete symmetry group and the choice of representations (appear to be quite special for three generations), it makes several predictions which will be tested soon. It is worth noting that the model predicted a long life-time for the B-meson long before it was experimentally known.

Model based on $SO(10) \times U(1)_{PQ}$

From the earlier discussion in section 2, it is fair to say that, at present, there is no multi-generational grand unified theory which predicts the number of generations and allows one to calculate the mixing angles in terms of relatively few input parameters. It is, therefore, worthwhile to explore an alternate approach to the generation problem. It may very well be that the number of generations is large. The low energy families we see are due to subquark structure. At the quark-lepton level, we then should seek a meaningful formalism which is capable of dealing with the new physics mentioned earlier. Such a formalism should be relatively insensitive to the number of generations, that is, the results in the case of N generations should undergo only small corrections when a heavier (N+1)th generation is added. A model that satisfies these criteria emerges by combining single generation grand unification group G with a global, axial U(1) symmetry. The latter symmetry was introduced by Peccei and Quinn (hence $U(1)_{PQ}$) to solve the strong CP-violation problem. Within the framework of grand unification schemes, the axion that arises when $U(1)_{PQ}$ is broken becomes a "phantom" or "invisible" axion.

444

Such axions have astrophysical consequences and appropriate constraints have to be imposed on the models. Assuming that this can be done, one is lead to examine the semi-simple group structure $G \times U(1)_{PQ}$, where G is a single generation grand unification group, $SU(5)$, E_6, or $SO(10)$. In such models, $U(1)_{PQ}$ plays a dual role. While it serves to eliminate the strong CP-violation problem, it also acts as horizontal flavor symmetry, yielding $U(1)_{PQ}$ assignments that distinguish the different generations and lead to restrictions on the form of mass matrices. For multigenerational models, $SO(10)$ is the favored single generation candidate. Hence, we shall confine our attention to $SO(10) \times U(1)_{PQ}$.

Let the left-handed, spin 1/2 chiral states of each fermion family ψ_L^i (i=1,2,3,...N) belong to a spinorial 16-dimensional representation of $SO(10)$. Under $U(1)_{PQ}$, let them transform as

$$\psi_L^i = e^{ix_i a} \psi_L^i \, ,$$

where a is a continuous parameter and x_i ($x_i \neq x_j$, $i \neq j$) are what one might call "Peccei-Quinn" family quantum numbers. In order to obtain mass matrices that have nondegenerate, non-vanishing eigenvalues (which is what is dictated by the presently known quark mass spectrum), at least two Higgs multiplets are necessary. If their Peccei-Quinn assignments are h and h' respectively, then one can show from simple arguments that

$$x_k = \frac{1}{4} (h+h') + \frac{1}{4} (-1)^{N-k} (2N-2k+1)(h-h') .$$

Furthermore, in the global $U(1)$ case, h+h'=0 so that there is a nonvanishing anomaly present in order that the theory is renormalizable. Thus we have for N generations a unique set of PQ assignments,

$$x_N = \frac{1}{2}h, \; x_{N-1} = -\frac{3}{2}h, \; x_{N-2} = \frac{5}{2}h, \ldots x_1 = (-1)^{N-1}(2N-1)h$$

In the special case of three generations,

$$x_3 = \frac{1}{2}, \ x_2 = -\frac{3}{2}, \ x_1 = \frac{5}{2},$$

where we have assumed, without any loss of generality, h=1.

The above quantum numbers combined with the hierarchy in quark masses lead to a novel type of composite picture in which the lighter generations are more composite than the heavier ones. Electron is more composite than muon. One can show qualitatively that the mass hierarchy is such that log m varies linearly with respect to the generation index. We shall not go into these details here as our main concern is the weak mixing angles. In the $SO(10) \times U(1)_{PQ}$ model, we assume that the two Higgs multiplets are a 10 (complexified) and a 126. Then the quark mass matrices are complex, symmetric matrices with the generic form

$$M = \begin{bmatrix} 0 & Ae^{i\alpha} & 0 \\ Ae^{i\alpha} & 0 & Bei^{\beta} \\ 0 & Be^{i\beta} & Ce^{i\gamma} \end{bmatrix}$$

which can be written as

$$M = PXP,$$

where

$$P = \mathrm{diag}(e^{i(\alpha-\beta+\gamma/2)}, \ e^{i(\beta-\gamma/2)}, \ e^{i\gamma/2}),$$

and

$$X = \begin{bmatrix} 0 & A & 0 \\ A & 0 & B \\ 0 & B & C \end{bmatrix}.$$

Thus X is a real symmetric matrix with every element being positive. A real orthogonal matrix 0 diagonalizes X,

$$0X0^T = \mathrm{diag}(m_1, -m_2, m_3),$$

where $0 < m_1 < m_2 < m_3$. They are the values of the current quark masses. The most important feature of the model is that both X and 0 can be expressed in terms of m_1, m_2, m_3,

$$A = \left[\frac{m_1 m_2 m_3}{m_1 - m_2 + m_3}\right]^{1/2}, \quad B = \left[\frac{(m_3 + m_1)(m_3 - m_2)(m_2 - m_1)}{m_1 - m_2 + m_3}\right]^{1/2},$$

$$C = m_3 - m_2 + m_1$$

$$0 = \begin{bmatrix} \left[\frac{m_2 m_3 (m_3 - m_2)}{(m_3 - m_1)(m_2 + m_1)C}\right]^{1/2} & \left[\frac{m_1 (m_3 - m_2)}{(m_3 - m_1)(m_2 + m_1)}\right]^{1/2} & -\left[\frac{m_1 (m_2 - m_1)(m_3 + m_1)}{(m_3 - m_1)(m_2 + m_1)C}\right]^{1/2} \\ \left[\frac{m_1 m_3 (m_3 + m_1)}{(m_3 + m_2)(m_2 + m_1)C}\right]^{1/2} & -\left[\frac{m_2 (m_3 + m_1)}{(m_3 + m_2)(m_2 + m_1)}\right]^{1/2} & \left[\frac{m_2 (m_3 - m_2)(m_2 - m_1)}{(m_3 + m_2)(m_2 + m_1)C}\right]^{1/2} \\ \left[\frac{m_1 m_2 (m_2 - m_1)}{(m_3 + m_2)(m_3 - m_1)C}\right]^{1/2} & \left[\frac{m_3 (m_2 - m_1)}{(m_3 + m_2)(m_3 - m_1)}\right]^{1/2} & \left[\frac{m_3 (m_3 - m_2)(m_3 + m_1)}{(m_3 + m_2)C(m_3 - m_1)}\right]^{1/2} \end{bmatrix}$$

Then, if $X^{(d)}$ and $X^{(u)}$ correspond to the mass matrices $M^{(d)}$ and $M^{(u)}$ in the down- and up-charge sectors,

$$M^{(d)} = P^{(d)} X^{(d)} P^{(d)}, \quad M^{(u)} = P^{(u)} X^{(u)} P^{(u)},$$

where $P^{(d)}$ and $P^{(u)}$ are the corresponding diagonal, pure phase matrices. If $O^{(d)}$ and $O^{(u)}$ are the desired orthogonal matrices that diagonalize $X^{(d)}$ and $X^{(u)}$,

$$O^{(d)} X^{(d)} O^{(d)T} = \text{diag}(m_d, -m_s, m_b)$$

$$O^{(u)} X^{(u)} O^{(u)T} = \text{diag}(m_u, -m_c, m_t),$$

where $m_d, \ldots, m_t$ denote the masses of the indicated quarks, then the Cabibbo-Kobayashi-Maskawa matrix in the charged current interaction, namely,

$$U_c = U_L^{(u)} U_L^{(d)\dagger}$$

is given by

$$U_c = Q[O^{(u)} P^{(u)*} P^{(d)} O^{(d)T}] R,$$

with the matrix in the rectangular parentheses determined completely by the parameters that enter in the mass matrix (vacuum expectation values, coupling constants) and Q and R being two arbitrary diagonal pure phase matrices. They reflect the arbitrary phases of the quark fields. We define

$$P = P^{(u)*}P^{(d)} = \text{diag}(e^{i\phi_1}, e^{i\phi_2}, e^{i\phi_3}),$$

$$c_{ij} = \sum_k O^{(u)}_{ik} O^{(d)}_{jk} \cos \phi_k,$$

$$s_{ij} = \sum_k O^{(u)}_{ik} O^{(d)}_{jk} \sin \phi_k,$$

and choose Q and R so that U_c reduces to the standard Kobayashi-Maskawa form. Then the required mixing angles $\theta_1, \theta_2, \theta_3$ and $\text{Im}(U_c)_{12}$ that is related to the weak CP-violation parameter are given by

$$\cos \theta_1 = (U_c)_{11} = \sqrt{c_{11}^2 + s_{11}^2},$$

$$-\sin \theta_1 \cos \theta_2 = (U_c)_{21} = -\sqrt{c_{21}^2 + s_{21}^2}$$

$$\sin \theta_1 \cos \theta_3 = (U_c)_{12} = \sqrt{c_{12}^2 + s_{12}^2}$$

$$-\sin \theta_2 \sin \theta_3 \sin \delta = \text{Im}(U_c)_{22}$$

$$= \frac{(s_{11}c_{22}+c_{11}s_{22})(c_{12}c_{21}-s_{12}s_{21})-(c_{11}c_{22}-s_{11}s_{22})(s_{12}c_{21}+c_{12}s_{21})}{\cos\theta_1 \cos\theta_2 \cos\theta_3 \sin^2\theta_1}$$

From the above equations, it follows that the mixing angles and CP-violation depend on the six quark masses and two phase angle differences. As only the topquark mass is unknown, the model contains only three unknown parameters - the top quark mass m_t and two phase differences, say

$$\alpha = (\phi_1-\phi_2), \quad \beta = (\phi_2-\phi_3) \quad (\text{then } \phi_1-\phi_3 = \alpha+\beta).$$

The mixing angles and $\text{Im}(U_c)_{22}$ are given by

$$\cos^2\theta_1 = K_0 + K_1\cos\alpha + K_2\cos(\alpha+\beta) + K_3\cos\beta,$$

$$\sin^2\theta_1\cos^2\theta_2 = L_0 + L_1\cos\alpha + L_2\cos(\alpha+\beta) + L_3\cos\beta,$$

$$\sin^2\theta_1\cos^2\theta_3 = N_0 + N_1\cos\alpha + N_2\cos(\alpha+\beta) + N_3\cos\beta,$$

$$\cos\theta_1 \cos\theta_2 \cos\theta_3 \sin^2\theta_1 \, \text{Im}(U_c)_{22}$$

$$= A_1 \sin\alpha + A_2 \sin(\alpha+\beta) + A_3 \sin\beta$$

$$+ \sin\alpha \, (A_4 \cos(\alpha+\beta) + A_5 \cos\beta)$$

$$+ \sin(\alpha+\beta) \, (A_6 \cos\alpha + A_7 \cos\beta)$$

$$+ \sin\beta \, (A_8 \cos\alpha + A_9 \cos(\alpha+\beta)),$$

where K_i, L_i, N_i, and A_i are all given functions of masses. Suppose we take the standard typical values for the quark masses, namely,

$$m_d = 7.5 \text{ MeV} \quad m_s = 150 \text{ MeV} \quad m_b = 5000 \text{ MeV}$$

$$m_u = 5 \text{ MeV} \quad m_c = 1250 \text{ MeV} \quad m_t > 30 \text{ GeV},$$

and compute these functions for varying values of m_t. One finds that $K_0, \ldots N_3$ are very slowly varying functions of m_t. With the help of these numbers one can study the choice of other parameters in the model (see for details, Nair, Michel and Wali, Ref. 5).

The most stringent requirement arises from the very well determined Cabibbo angle (see Kleinknecht in these proceedings),

$$\sin\theta_1 = 0.231 \pm 0.003.$$

First of all, it rules out the choice $\alpha=\beta=0$, which would have implied that $\text{Im}(U_c)_{22}=0$. In other words, there would have been no weak CP-violation in the conventional way due to gauge bosons in the charged interactions. One then had to appeal to the Higgs sector for CP-violation. Secondly, the study of the numbers shows that K_1 contributes most dominantly, K_2 and K_3 being relatively of no significance to the value of θ_1. The value $\alpha=90°$ leads to

$$\sin\theta_1 = 0.2264$$

for $m_t = 30 - 100$ GeV. This is remarkably close to the lower limit $\sin\theta_1=0.228$ set by experiments. The variation of $\sin\theta_1$ from 0.228-0.234 allows the variation of α from 91.5° to 97°. We take $\alpha=94°$ to yield

$$\sin\theta_1 = 0.231$$

independent of m_t when it is varied from 30 GeV to 100 GeV. Having fixed α this way, we vary β to set limits on its variation for various values of m_t. For this purpose, we assume the limits (K. Kleinknecht and B. Renk, Phys. Lett. 130B (1983) 459),

$$0.015 < \sin\theta_2 < .09,$$

$$\sin\theta_3 < .04,$$

which include the recent experimental results

$$|(U_c)_{bc}| = 0.053 \begin{array}{c} + .010 \\ - .009 \end{array}$$

on B-meson lifetime. The results show that for m_t=30–100 GeV, one can find β such that all the experimental constraints are satisfied quite well.

Finally, we come to the CP-violation effect predicted by the model. The K^0-$\overline{K}^0$ transition matrix M_{12} from the standard relevant box graph is given by (see for instance, Ling-Lie Chau, Physics Reports 95, 1 (1983))

$$M_{12} = - \frac{G_F^2 M_W^2}{16\pi^2} (\sum_{i,j=u,c,t} \lambda_i \lambda_j A_{ij}) M_{1,2,vac} B,$$

where

$$M_{12,vac} = \frac{-4}{3} f_K^2 m_K$$

with $f_K \cong 1.23 \, m_\pi$, is the vacuum insertion contribution and B is a constant characterizing the deviation of the vacuum-insertion calculation from unity. The other quantities appearing in the above equation are defined by Ling-Lie Chau. We calculate the quantity

$$M = \sum \lambda_i \lambda_j A_{ij}$$

for m_t = 30–100 GeV. Inserting the value of $M_{12,vac}$,

$$\text{Im } M_{12} = -(.114 \times 10^{-13} \text{ MeV}) \times cB \text{ ImM},$$

where $c = 1.0223 \times 10^7$. In order that we do not conflict with experiments, $cB \ \text{Im} \ M \lesssim 1$.

The main points of our results are summarized in Table 1. The most stringent limits on β are provided by the limits on $|(U_c)_{bc}|$. We note that the values $m_t > 90$ GeV are excluded, both from the constraint on $(U_c)_{bc}$ and $cB \ \text{Im} \ M \lesssim 1$, if we take $B \cong \frac{1}{3}$. For each value of m_t, an allowed range of β emerges. In this range, $\sin\theta_2$, $\sin\theta_3$, $\sin\delta$, and $c \ \text{Im}M$ are slowly varying, increasing functions of β. We have given in Table 1 the values of the above mentioned quantities for the end points of the allowed range of β. It is worth noting that $(U_c)_{bc}$ involves a specific combination of sines and cosines of all the mixing angles and the Kobayashi-Maskawa phase δ. By restricting the absolute value of this matrix element, the model predicts, for all the investigated values of m_t, limits for $\sin\theta_2$ and $\sin\theta_3$,

$$.038 \lesssim \sin\theta_2 \lesssim .057, \quad .011 \lesssim \sin\theta_3 \lesssim .022.$$

These limits are more stringent than the current phenomenological constraints. The model also predicts $\sin\delta$ to lie between .909 $\lesssim \sin\delta \lesssim$ .999. Thus, for each value of the top-quark mass, the complete Kobayashi-Maskawa matrix is known within certain limits. Consequently, the model provides a rich body of results that can be compared with experiments, once the value of m_t is known.

A variant of the model which includes 120 representation of SO(10) and therefore is somewhat more general has been proposed by Stech (Phys. Lett. 130B (1983) 189. See also his talk in these proceedings).

Summarizing, it appears that the generic form for the quark mass matrix discussed here seems to provide a good and satisfactory description of low energy parameters including the new piece of information concerning B-meson lifetime. Such a form for the mass matrix was suggested a long time ago by Fritzsch (Phys. Lett. 73B (1978) 317, Nucl. Phys. B155 (1979) 189) from heuristic

TABLE 1

m_t in GeV	β	$\underline{\sin\theta_2}$	$\underline{\sin\theta_3}$	c Im M	$\sin\delta$
30	9°–16°	.0397–.0567	.0177–.0225	1.013–2.101	.9994–.9994
40	13°–20°	.0379–.0572	.0160–.0214	1.019–2.552	.9862–.9927
50	13°–20°	.0387–.0560	.0148–.0200	1.055–2.797	.9123–.9852
60	10°–19°	.0382–.0563	.0132–.0192	1.064–3.233	.9360–.9790
70	3°–17°	.0385–.0566	.0109–.0185	0.9731–3.666	.9092–.9740
80	0°–14°	.0457–.0568	.0120–.0177	1.686–4.055	.9383–.9698
90	0°–9°	.0523–.0564	.0138–.0166	2.961–4.259	.9563–.9653
100	–	–	–	–	–

By restricting $|(U_c)_{bc}|$, we find allowed ranges for β, $\sin\theta_2$, $\sin\theta_3$, $\sin\delta$, and c Im M. Typical values for the absolute values of KM matrix elements,

m_t=30 GeV, β=13° m_t=40 GeV, β=16° m_t=60 GeV, β=15°

$$\begin{bmatrix} .973 & .231 & .005 \\ .231 & .972 & .053 \\ .011 & .052 & .996 \end{bmatrix} \begin{bmatrix} .973 & .231 & .004 \\ .231 & .971 & .052 \\ .011 & .051 & .999 \end{bmatrix} \begin{bmatrix} .973 & .231 & .004 \\ .231 & .971 & .054 \\ .011 & .053 & .998 \end{bmatrix}$$

considerations. But it can be derived within the framework of a grand unified theory combined with Peccei-Quinn symmetry as horizontal flavor symmetry. Within the framework of grand unified theories the model is the most economical one as it does not introduce new symmetries. $U(1)_{PQ}$ needed to avoid strong CP-violation is used as the flavor symmetry. The canonical structure of the resulting mass matrices allows one to eliminate several unknown parameters of the Higgs sector and express the mixing angles in terms of quark masses and two phases of the vacuum expectation values. When the top quark mass m_t is known, the model determines almost uniquely the KM matrix. m_t also plays a crucial role in the weak CP-violation predicted by the model. If $m_t < 40$ GeV, the predicted CP-violation parameter ε is small compared with the experimental value. But the model has a rich Higgs structure bound to produce CP-violating interaction.

CONCLUDING REMARKS

The generation puzzle is indeed a mystery which has to be reckoned with in order to have more complete unified theories. The horizontal or the intragenerational interactions which give rise to the mixing angles govern a wide range of phenomena. Their specification is absolutely necessary for a quantitative understanding of the "New Physics". Their manifestations in the low energy region are just as important as baryon number violating processes such as proton decay in constructing grand unified theories.

In spite of a great deal of work in recent years, the theoretical situation regarding horizontal interactions is far from satisfactory. Conventional symmetries and classification schemes are not adequate enough to provide detailed answers. If we take the viewpoint that the phenomenon of replication is due to an underlying compositeness of quarks and leptons, we need new dynamical ideas to explain the observed features. Most of the models studied in the literature contain too much arbitrariness to provide quantitative information regarding such longstanding problems as CP-violation, precise magnitudes of the rare decay modes of old and new particles.

In this review, we have discussed in some detail two models, one within the framework of purely electroweak interactions and the other within that of a grand unified theory. These models are by no means unique. But they have gone far enough to provide numerical values for the mixing angles, CP-violation parameter, and other quantities that can be subjected to experimental verification. The determination of B-lifetime has narrowed down the structure of Kobayashi-Maskawa matrix. Experimental information concerning ε'/ε, flavorchanging neutral currents, top quark mass will help considerably in restricting further the models which at present have too much arbitrariness as they stand.

ACKNOWLEDGEMENTS

This work was supported by the U.S. Department of Energy under contract number DE-AC02-76ER03533. I would also like to thank Professor R. Mohapatra for many valuable discussions.

BIBLIOGRAPHY

This is not intended to be an exhaustive list of all the papers on the subject. I shall refer to other reviews for the most part and a few papers of specific interst to the discussion in the text.

1. For a comprehensive review of both the theoretical and phenomenological aspects of quark mixing angles, see Ling-Lie Chau, Phys. Reports, 95, Number 1, 1983.

2. The list of references to papers that consider $SU(2) \times U(1) \times G_H$ where G_H is discrete or continuous is too long to reproduce here. For a fairly complete list of references, see references 3 and 4 of K.C. Wali, AIP Conf. Proc. No. 72, Particles and Fields subseries No. 23; Weak Interactions as Probes of Unification, VPI 1980, eds. G.B. Collins, L.N.Chang and

J.R. Ficenec. See also R.D. Peccei, Proc. of XXth International Conference on High Energy Physics, Madison, Wisc. (1980), H. Fritzsch and P. Minkowski, Physics Reports, $\underline{73}$ (1981) 69; C. Jarlskog, Proceedings of the International Europhysics Conference on High Energy Physics, Brighton (UK), 1983.

3. Multigenerational grand unified models: H. Georgi, Nucl. Phys. $\underline{B156}$, 126 (1979); P. Frampton, Phys. Lett. $\underline{89B}$, 352 (1980); A. Davidson and K.C. Wali, Phys. Rev. $\underline{D23}$, 477 (1981); A. Davidson, P. Mannheim and K.C. Wali, Phys. Rev. $\underline{D26}$, 1133 (1982). For a more complete list of references see V.P. Nair and K.C. Wali, AIP Conference Proceedings No. 98, Particles and Fields subseries No. 29.

4. Permutation symmetry model: S. Pakvasa and H. Sugawara, Phys. Lett. $\underline{82B}$ (1979) 105; Y. Yamanaka, H. Sugawara and S. Pakvasa, Phys. Rev. $\underline{D25}$ (1982) 1895; T. Brown, N. Deshpande, S. Pakvasa, H. Sugawara, Univ. of Hawaii Preprint, CP Nonconservation and rare processes in S_4 model of permutation symmetry. UH-511-522-84. See this last paper for other related references. Calculability of mixing parameters in a more general framework has been discussed by K. Kang, Particles and Fields. Edited by A.Z. Capri and A.N. Kamal, (Plenum Publishing Corporation, 1983). See also H. Fritzsch, Max-Planck Inst. Preprint, Flavor Mixing and the Masses of leptons and quarks, MPI-PAE/PTH 93/83.

5. Model based on $SO(10) \times U(1)_{PQ}$: A. Davidson, V.P. Nair, and K.C. Wali, Phys. Rev. $\underline{D29}$, 1504 and 1513 (1984); V.P. Nair, L. Michel and K.C. Wali, Phys. Lett. $\underline{138B}$, 128 (1984). References to Peccei-Quinn symmetry, its implementation within the framework of grand unified theory are contained in the above papers.

DISCUSSION

LITTENBERG:

In your SO(10) × U(1) model, what do you get for lepton flavor mixing; like $K_L^0 \to \mu e$ etc.?

WALI:

We have yet to calculate those. We need to understand symmetry breaking very well to calculate such decays. We need to study the Higgs potential.

RAJASKARAN:

Do the mass matrix and mixing matrix of your model have any predictions in the neutrino sector (assuming neutrinos have masses)?

WALI:

Yes, it has. They have to be worked out.

STECH:

Is there a definite prediction such that experiments could support or rule out the model?

WALI:

Once the top quark mass is known, the entire K-M matrix (both real and imaginary parts) is known. I think the model predicts top decay characteristics, and U_{ub} matrix element, also flavor changing neutral currents -- when they are calculated.

Chau (BNL) - Could you give some guiding principle why certain restrictions are put on the mass matrices other than wanting to express mixing matrix in terms of physical quantities like quark masses?

WALI:

Historically, the successful relation $\theta_c = \sqrt{m_d/m_s}$ led to the study of such restrictions. As far as I know there are no guiding principles. In the model SO(10) × U(1)$_{PQ}$, restrictions arise automatically because of U(1)$_{PQ}$, which is needed to avoid the strong CP-problem.

FRITZSCH:

The main reason to look for relations between quark masses and mixing angles has been in part the phenomenological success of the relation $\theta_c \sim \sqrt{m_d/m_s}$. Furthermore in grand unified theories, additional symmetries in the Higgs sector (e.g. discrete symmetries) are easily incorporated and are often naturally embedded in larger symmetries. It is interesting to note that the mass matrices of the type Wali has used and which were introduced around 1979 lead to the nearest neighbor type mixing as observed nowadays in the weak decays of heavy quarks.

SEGRE:
The K-M matrix has only one phase for 3 families, but of course there can be many phases in Higgs couplings, which in turn influence mixing angles.

CHAU:
By general counting of parametrization there should be only one phase, so could you explain the two phases in your formulation?

WALI:
There is only one phase in K-M matrix. But in calculating it in our model, it is a function of the phase differences of the vacuum expectation values as well as the Yukawa couplings. Some of the parameters are eliminated in terms of quark masses. Two phase differences, however, remain. Mixing angles are also functions of these phase differences.

LEE-FRANZINI:
How stringent do you need the flavor-changing neutral current when already in the B sector the experimental limit is 5 times smaller than the theoretical bound?

WALI:
An observed number would be helpful.

WOLFENSTEIN:
When m_t is discovered, your model will predict all K-M angles in terms of two phase factors so that it can be tested eventually?

WALI:
Yes.

THE EXPERIMENTAL DETERMINATION OF WEAK MIXING ANGLES IN THE SIX-QUARK SCHEME

K.Kleinknecht

Institut für Physik der Universität Dortmund

Dortmund, Fed.Rep.Germany

ABSTRACT

The mixing of quarks through the weak interaction is described by a unitary matrix which can be parametrized in terms of three angles and one phase. Experimental results on weak decays of hyperons and B mesons, on neutrino production of charm quarks, and on the B meson lifetime are used to obtain, in a combined fit, values for the three mixing angles in the Kobayashi-Maskawa scheme: $\sin\theta_1 = 0.231 \pm 0.003$, $0.025 < \sin\theta_2 < 0.06$ and $\sin\theta_3 < 0.02$. In the Maiani parametrization, the angles obtained are: $\sin\theta = 0.231 \pm 0.003$, $|\sin\gamma| = 0.044 {\ +\ 0.007 \atop \ -\ 0.005}$ and $\sin\beta < 0.004$.

1. INTRODUCTION

If the mixing of quarks by the weak interaction is described phenomenologically, the six-quark mixing scheme proposed by Kobayashi and Maskawa [1] serves as a useful parametrization of the connection between generation of quarks. The elements U_{ik} of the quark mixing matrix (i = u,c,t; k = d,s,b) are parametrized in terms of three angles θ_1, θ_2 and θ_3, and one phase δ, possibly related to CP violation (Table 1a). If CP violation is due to quark mixing, then this phase δ is related to the parameter ε, describing the admixture of wrong CP parity in the long- and short-lived neutral K-meson states, measured to be $\varepsilon = (2.28 \pm 0.05) \times 10^{-3} \times \exp(i\pi/4)$ [2]. An approximate relation derived by Pakvasa and Sugawara [3] is $|\varepsilon| = |(m_t - m_c)/m_c| \sin2\theta_2 \tan\theta_3 \sin\delta/(2\sqrt{2} \cos\theta_1)$ where m_t and m_c are the top- and charm-quark masses. An alternative parametrization of the matrix U has been given by Maiani [4] in terms of angles θ, γ and β and a phase δ' (see Table 1b).

A more detailed calculation of the CP parameter in the K meson system gives [5-8]

$$\sqrt{2}|\varepsilon| = B\frac{\sin\beta \, \sin\gamma \, \sin\delta'}{\sin\theta}\{-1+ \frac{\eta_3}{\eta_1} \ln\frac{m_t^2}{m_c^2} + \frac{\eta_2}{\eta_1} \frac{m_t^2}{m_c^2} [\sin^2\gamma - \frac{\sin\beta \, \sin\gamma}{\sin\theta}\cos\delta']\} \tag{1}$$

where B is the K^o-$\bar{K}^o$ transition matrix element, normalized to its value for a specific model ("vacuum insertion value"), m_t and m_c are the top and charm quark masses and η_1, η_2, η_3 represent QCD corrections. CP violating amplitudes, in this model, are proportional to the product of three, presumably small, angles.

Experimentally, information on the weak mixing angles comes from measurements of weak decays of light and heavy quarks and from neutrino production of charm quarks as observed in dimuon events, as summarized in previous papers [9-13]. New results on the B-meson lifetime [14,15] and on hyperon semileptonic decays [16] give new stringent constraints. In a recent paper [17], the impact of all these constraints on the weak mixing angles has been analyzed. I give here an updated version of this analysis using the most recently available data on the B lifetime [15,18] and on B-meson semileptonic inclusive decays [19,20]. I first go through constraints on the coupling parameters U_{ik}, and then proceed to derive bounds on the mixing angles.

2. CONSTRAINTS ON MATRIX ELEMENTS

2.1 Light-quark couplings

2.1.1 Coupling U_{ud}

This coupling parameter has been determined from a comparison of measured rates of nuclear beta decays with that of muon decay. Two different evaluations of this quantity have been made and their results are U_{ud} = 0.9730 ± 0.0024 [10] and U_{ud} = 0.9737 ± 0.0025 [21]. Combining these two, one obtains

$$U_{ud} = 0.9733 \pm 0.0024 \tag{2}$$

2.1.2 Coupling U_{us}

In a series of experiments, the WA2 Collaboration has studies five different hyperon semileptonic decays, i.e. the leptonic weak decays $\Sigma^- \rightarrow n e \bar{\nu}$, $\Sigma^- \rightarrow \Lambda e^- \bar{\nu}$, $\Xi^- \rightarrow \Lambda e^- \bar{\nu}$, $\Xi^- \rightarrow \Sigma^0 e^- \bar{\nu}$, and $\Lambda \rightarrow p e^- \bar{\nu}$. Including radiative corrections and using in addition the neutron lifetime [22], this experiments gives [16]

$$U_{us} = 0.231 \pm 0.003. \tag{3}$$

This represents a substantial improvement over former analyses [10, 21], although the value for U_{us} is exactly the same as the one obtained ten years ago [23]. It is still debated at which level corrections for SU(3) breaking effects have to be applied.

2.2 Charm-quark couplings

2.2.1 Coupling U_{cd}

This coupling can be determined from measurements of single charm production in neutrino and antineutrino reactions.

The differential cross-sections for neutrino charm production on isoscalar targets are:

$$\frac{d\sigma^{\nu}}{dxdy} = \frac{G^2 M E_{\nu} x}{\pi} \; [U^2_{cd}(u(x) + d(x)) + |U_{cs}|^2 \, 2s(x)] \qquad (4)$$

$$\frac{d\sigma^{\bar{\nu}}}{dxdy} = \frac{G^2 M E_{\bar{\nu}} x}{\pi} \; [U^2_{cd}(\bar{u}(x) + \bar{d}(x)) + |U_{cs}|^2 \, 2\bar{s}(x)] \qquad (5)$$

where $u(x)$, $d(x)$ and $s(x)$ are the quark density distributions in the proton, G is the Fermi coupling constant, M the nucleon mass, E_{ν} the neutrino laboratory energy, and x and y the Bjorken scaling variables.

Experimentally, the observation of charm production has been done mainly by three methods: 1. direct observation of the short-lived decay of charmed hadrons in emulsions, 2. observation of semileptonic charm decay $c \to s + \mu^{+} + \nu_{\mu}$ in neutrino induced dimuon events, $\nu N \to \mu^{-} \mu^{+} X$, 3. observation of semileptonic charm decay $c \to s + e^{+} + \nu_e$ in dilepton events, $\nu N \to \mu^{-} e^{+} X$. By far the largest event samples have been collected using the second method.

In order to obtain the coupling parameter U_{cs}, the contribution of charm production from the strange sea s and $\bar{s}$ quarks has to be eliminated. According to the cross-section given in (4) and (5), this can be done [24] by using the weighted difference of neutrino and antineutrino cross-sections:

$$\beta U^2_{cd} = \frac{(\sigma^{\nu}_{\mu^{-}\mu^{+}}/\sigma^{\nu}_{\mu^{-}}) - (R\sigma^{\bar{\nu}}_{\mu^{+}\mu^{-}}/\sigma^{\bar{\nu}}_{\mu^{+}})}{1 - R} \; \frac{2}{3} \qquad (6)$$

where R is the ratio of antineutrino to neutrino total cross-sections, $R = \sigma^{\bar{\nu}}_{\mu^{-}}/\sigma^{\nu}_{\mu^{-}} = 0.48 \pm 0.02$ [25], $R^{\nu} = \sigma^{\nu}_{\mu^{-}\mu^{+}}/\sigma^{\nu}_{\mu^{-}}$ is the dimuon to singlemuon cross-section ratio in neutrino induced reactions corrected for the threshold effects due to the charm mass (slow rescaling) [26], and β denotes the semileptonic branching ratio of that mixture if charmed particles which is produced in the neutrino reactions.

In an analysis of their large sample of neutrino- and antineutrino-induced dimuon events, the CDHS collaboration [24] obtains $\beta U_{cd}^2 = (0.41 \pm 0.07)10^{-2}$. With a value of $\beta = (7.1 \pm 1.3)$ % based on e^+e^- results and on the composition of charmed particles in neutrino reactions from emulsion epxeriments, a value of

$$|U_{cd}| = 0.24 \pm 0.03 \tag{7}$$

is obtained.

2.2.2 Coupling U_{cs}
- - - - - - - - - -

In charged current reactions this coupling appears always together with the strange-sea structure function $xs(x)$ or its integral $S = \int xs(x)dx$. The quantity measured is $|U_{cs}|^2 \cdot 2S$. In the absence of an independent determination of S, only the upper limit for 2S given by SU(3) symmetry, $2S \leq \bar{U} + \bar{D}$, and a corresponding lower limit on $|U_{cs}|$ can be obtained. The product $|U_{cs}|^2 \cdot 2S$ can be extracted in three ways from the neutrino and antineutrino dimuon production data [12]. We use here the results from the x distribution of neutrino dimuons [12,24]

$$\frac{|U_{cs}|^2}{U_{cd}^2} = (6.26 \pm 0.73) \frac{1 + \alpha^*}{\alpha} \tag{8}$$

and the one from the cross-sections of neutrino- and antineutrino-induced dimuon production using the semileptonic branching ratio of D mesons:

$$|U_{cs}|^2 = (0.41 + 0.09)\frac{1 + \alpha^*}{\alpha}, \tag{9}$$

where $\alpha = 2S/(\bar{U} + \bar{D})$ is the ratio of momentum fractions carried by strange and non-strange sea quarks in the nucleon, and $\alpha^* = 2S(U_{us}^2 + U_{cs}^2/r_s)/(\bar{U} + \bar{D})$ is the same ratio modified by the threshold suppression factor r_s for the charm-quark msss, which is $r_s = 1.5$ for the experiment considered [24,25].

2.3 Bottom-quark couplings

2.3.1 Ratio $|U_{ub}|/|U_{cb}|$

At electron-positron storage rings, the reaction $e^+e^- \rightarrow \Upsilon(4S) \rightarrow B\bar{B}$ can be used as a B meson source. The most sensitive search for decays of b quarks into u quarks can be done by measuring the inclusive lepton momentum spectrum of B meson decays. Data correponding to an integrated luminosity of 50 pb^{-1} have been collected by the CLEO [20] and CUSB [19] Collaborations at the Cornell Electron Storage Ring. The CLEO group measured the momentum spectra of 3650 electron events and 2115 muon events, while the CUSB group reports about the momentum spectrum of 900 electron events. The method of analysis is discussed in detail in the talk of J.Lee-Franzini [19]. In principle, the two possible semileptonic decays $B \rightarrow \ell\nu X_u$ and $B \rightarrow \ell\nu X_c$ can be distinguished by measuring the end point of the lepton momentum spectrum. In practice, the analysis is model-dependent because of the theoretical uncertainty about which X_u state with which mass is populated in the decay. Altarelli et al. [27] have calculated the expected lepton momentum spectrum using as an input the observed electron momentum spectrum in decays of charmed D mesons. Based on this model, the limits on the $b \rightarrow u$ decay width are

$$\Gamma(b \rightarrow u)/\Gamma(b \rightarrow c) < 0.04 \quad \text{at 90 \% C.L.} \quad \text{(CLEO [20])}$$

$$\Gamma(b \rightarrow u)/\Gamma(b \rightarrow c) < 0.045 \text{ at 90 \% C.L.} \quad \text{(CUSB [19])}$$

By taking into account the ratio of phase space available, one obtains

$$|U_{ub}|/|U_{cb}| < 0.12 \text{ at 90 \% C.L.} \tag{10}$$

2.3.2 B lifetime

The lifetime τ_B of b flavoured hadrons, apart from phase-space factors, depends on the magnitude of the couplings U_{cb} and U_{ub}. In fact [28]

$$\tau_B = 10^{-14} \text{ s}/(3.68 \ |U_{cb}|^2 + 7.8 \ |U_{ub}|^2).$$

The previous upper limit obtained by the JADE Collaboration [29], $\tau_B < 1.4 \times 10^{-12}$ s at 95 % C.L., is significantly improved by the recent measurements of the Mark II and MAC Collaborations [15,18]. In these experiments, semileptonic decays of b-flavoured hadrons are tagged by identifying an electron in an electromagnetic calorimeter or a muon from its penetration through a layer of iron and by requiring that this lepton has a high transverse momentum p_T relative to the thrust axis of a jet. The transverse momenta required are $p_T > 1$ GeV/c for the Mark II experiment and $p_T > 1.5$ GeV/c for the MAC experiment. In addition, the lepton total momentum is required to exceed 2 GeV/c. The experiments are discussed in detail in the talks of W.Ford [18] and G.H.Trilling [15] at this conference. The results are:

$$\tau_B = (12.0 \ ^{+\ 4.5}_{-\ 3.6} \pm 3.0) \times 10^{-13} \text{ s} \qquad \text{Mark II Collab. [15]}$$

$$\tau_B = (16 \pm 4 \pm 3) \times 10^{-13} \text{ s} \qquad \text{MAC Collab. \qquad [18]}$$

(11)

2.4 Combined fit

Using the constraints of eqs.(2,3) and (7) to (11), we obtain for a minimum $\chi = 1.1/3$ D.F. the values $\sin\theta_1 = 0.231 \pm 0.003$, $\alpha = 2S(\bar{U} + \bar{D}) = 0.49 \pm 0.07$, and values of $\sin\theta_2$ and $\sin\theta_3$ with the error contours in the ($\sin\theta_2$, $\sin\theta_3$) plane given in figure 1. These contours vary slightly with the value of the phase δ. From the resulting error contours, we obtain a finite value of $0.025 < \sin\theta_2 < 0.06$, and an upper limit for $\sin\theta_3 < 0.02$.

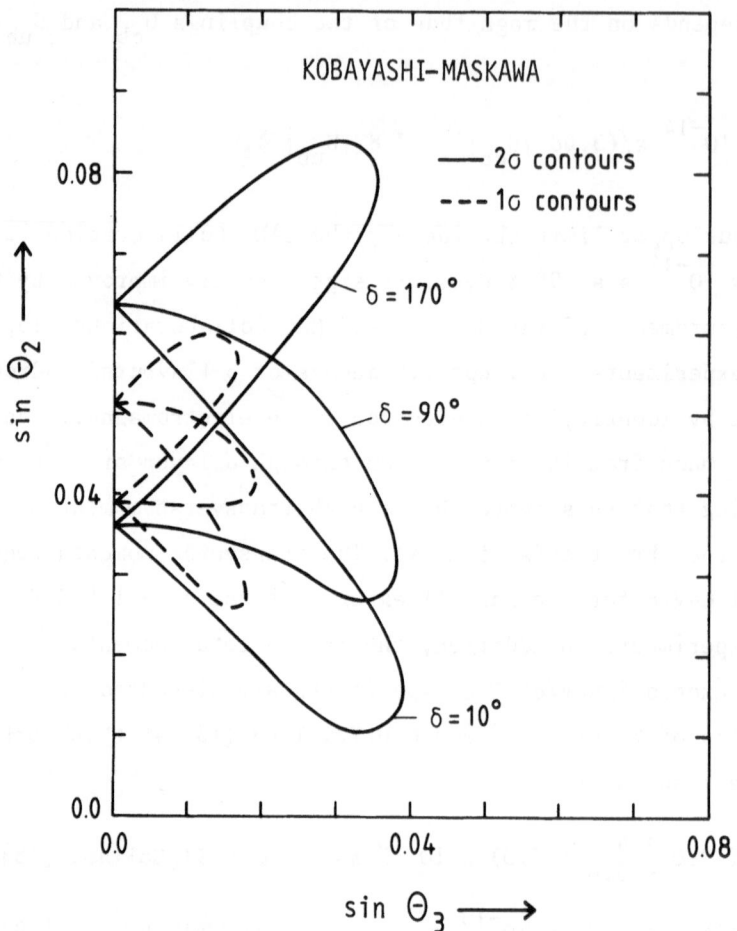

Fig.1: Error contours in the $(\sin\theta_2, \sin\theta_3)$ plane for three values of the phase $\delta(10^\circ, 90^\circ$ and $170^\circ)$. One standard deviation contours (dashed line) and two standard deviation contours (solid lines) are shown.

We conclude from this analysis that the second mixing angle θ_2 is smaller than the first one, θ_1, i.e. $\sin\theta_2/\sin\theta_1 < 0.26$. The third angle, θ_3, is still compatible with zero, with the upper limit $\sin\theta_3 < 0.02$ at the 67 % C.L. This pattern of decreasing mixing angles means that weak transitions between memebers of different quark families are suppressed more for heavy quarks than for light ones.

Analogously for the Maiani parametrization [4], the error contours
in the plane of the parameters sinγ (corresponding approximately to
sinθ$_2$) and sinβ/tanθ (corresponding to sinθ$_3$) are given in figure 2.
Here error contours at the one (1σ) and two (2σ) standard deviation
level are given. These contours are nearly independent of the phase
angle δ' in the Maiani parametrization. The values for the angles
are sinθ = 0.231 ± 0.003, |sinγ| = 0.044 $^{+\ 0.007}_{-\ 0.005}$ and sinβ < 0.004.

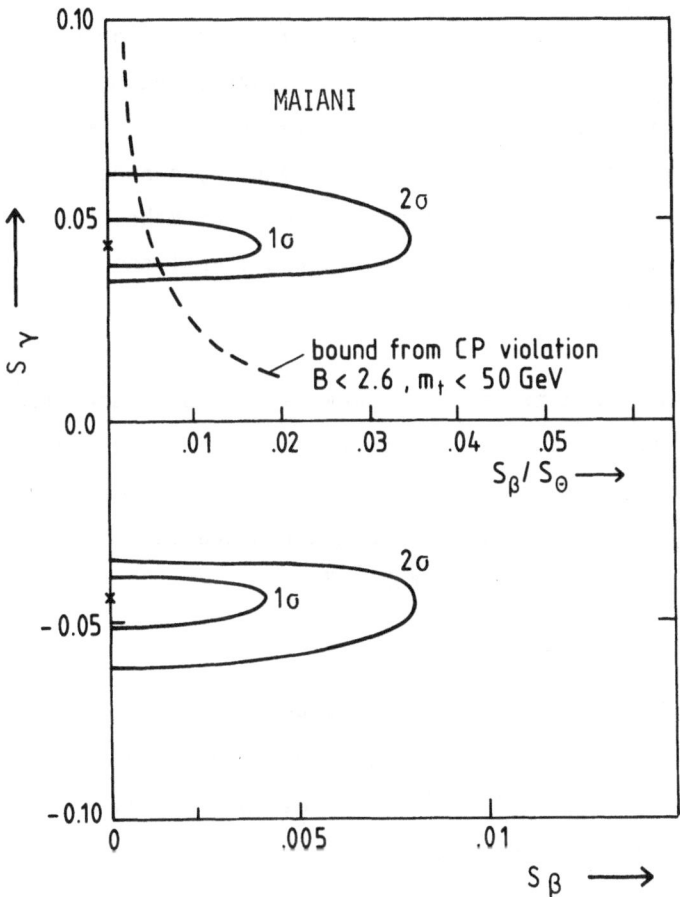

Fig.2: Error contours in the plane of mixing parameters sinγ and
sinβ/sinθ in the Maiani parametrization. One standard deviation (1σ) and two standard deviation (2σ) contours are given.
Also shown is a limit on the range of angles from the
measured value of the CP parameter |ε|.

From the range of values of the mixing angles in either parametrization, the values of the Kobayashi-Maskawa matrix elements can be obtained. These are given in table 2. It is evident that the error on these matrix elements from the common fit is, for most of them, much smaller than the one obtained from individual experimental bounds on one matrix elements.

It also appears from the values in table 2, that, apart from the diagonal elements, and the three off-diagonal elements observed directly up to now, $(U_{us}, U_{cs}$ and $U_{cb})$, there is only one other non-diagonal element (U_{ts}) whose magnitude is such as to allow direct observation, possibly in the reaction $\nu + s \to \mu^- + t$ or in the decay $t \to s + X$. The other two elements, connecting the first and third families, are of a magnitude which makes their detection very difficult, and therefore represent a challenge to future experimentation.

This analysis is done in the framework of a six-quark model. If the number of quark flavours is larger than six, the unitarity condition on the elements of the 3 x 3 matrix U_{ik} is replaced by $\Sigma_{i=1}^{3} |U_{ik}|^2 \leq 1$ and $\Sigma_{k=1}^{3} |U_{ik}|^2 \leq 1$. In this generalized case, the ranges of values for U_{ik} are given in table 3.

The pattern of decreasing mixing angles can be parametrized in still another way as suggested by Wolfenstein [30]. Realizing that experimentally $|\sin\gamma| \sim (\sin\theta)^2$, the parametrization of table 4 is proposed. The experimental limits on these parameters are then: $\lambda = 0.231 \pm 0.003$, $A = 0.82 \begin{smallmatrix} + 0.13 \\ - 0.10 \end{smallmatrix}$ and $\rho^2 + \eta^2 < 0.2$.

The underlying theoretical reason for this pattern of mixing angles is not yet known. A model in which such a pattern emerges has been given by B.Stech at this conference [31].

A description of CP violation in terms of quark mixing requires all three angles to be finite, as shown in eq.(1) and as discussed by K.C.Chou [32]. This means that $\sin\theta_3$ (or $\sin\beta$), although experimentally compatible with zero, has a lower limit depending on the value of the $K^0-\bar{K}^0$ transition matrix element B. If an upper limit of B < 2.6 from theoretical arguments of Guberina et al. [33] is used, the allowed values for the angles $\sin\beta$ and $\sin\gamma$ in figure 2 have to lie above the hyperbolic line drawn from eq.(1). If the vacuum insertion value B = 1 [5] is assumed, the allowed region becomes smaller, and even more so if the PCAC-value B = 0.33 is taken [7].

The question whether CP violation is indeed due to a phaseshift δ in the quark-mixing matrix can be studied further by searching for the second kind of CP violation not due to the mass matrix but to CP violation in the weak transition from the eigenstate $K_2 = (K^0 - K^0)/\sqrt{2}$ with negative CP eigenvalue to a 2π final state. While the first kind of CP violation is described by the parameter ε, the corresponding amplitude for this second kind is called ε'. From present experimental knowledge the amplitude ε' is still compatible with zero, and an upper limit on the amplitude ratio is $|\varepsilon'/\varepsilon| < 0.02$ at 90 % C.L. [2], consistent with superweak models of CP violation [34]. If, however, CP violation is due to the quark mixing matrix, a finite value of ε' is epxected. From the range of mixing matrix angles allowed a lower limit on the ratio of these amplitudes can be derived: $|\varepsilon'/\varepsilon| > 2 \times 10^{-3}$ [35]. Future experiments on the ratio of decay rates of $K_L \rightarrow \pi^0\pi^0$ and $K_L \rightarrow \pi^+\pi^-$ undertaken now at Fermilab, Brookhaven [36] and CERN will show whether this model will emerge as the true picture of CP violation.

ACKNOWLEDGMENT

I thank B. Renk for a valuble discussion and Ling-Lie Chau for organizing a fruitful conference.

Table 1

Parametrizations of the quark mixing matrix

a) Kobayashi-Maskawa parametrization [1]

$$U = \begin{bmatrix} c_1 & s_1 c_3 & s_1 s_3 \\ -s_1 c_2 & c_1 c_2 c_3 - e^{i\delta} s_2 s_3 & c_1 c_2 c_3 + e^{i\delta} s_2 c_3 \\ s_1 s_2 & -c_1 s_2 c_3 - e^{i\delta} c_2 s_3 & -c_1 s_2 s_3 + e^{i\delta} c_2 c_3 \end{bmatrix}$$

b) Maiani parametrization [4]

$$U \begin{bmatrix} c_\beta c_\theta & c_\beta s_\theta & s_\beta \\ -s_\gamma c_\theta s_\beta e^{i\delta'} - s_\theta c_\gamma & c_\gamma c_\theta - s_\gamma s_\beta s_\theta e^{i\delta'} & s_\gamma c_\beta e^{i\delta'} \\ -s_\beta c_\gamma c_\theta + s_\gamma s_\theta e^{-i\delta'} & -c_\gamma s_\beta s_\theta - s_\gamma c_\theta e^{-i\delta'} & c_\gamma c_\theta \end{bmatrix}$$

Table 2

Elements of quark mixing matrix U_{ik} from fit of experimental constraints (1 standard deviation range)

	d	s	b
u	0.9723 - 0.9737	0.228 - 0.234	0.000 - 0.004
c	0.228 - 0.234	0.9711 - 0.9727	0.039 - 0.051
t	0.005 - 0.015	0.038 - 0.050	0.9987 - 0.9993

Table 3

Elements of quark mixing matrix $|U_{ik}|$ from experimental constraints if number of quark flavours is larger than 6

	d	s	b
u	0.9709 - 0.9737	0.228 - 0.234	0.000 - 0.006
c	0.21 - 0.27	0.78 - 0.98	0.039 - 0.051
t	0.00 - 0.12	0.00 - 0.58	0.000 - 0.999

Table 4

Parametrization of the quark mixing matrix according to Wolfenstein [30]

	d	s	b
u	$1 - \lambda^2/2$	λ	$\lambda^3 A(\rho - i\eta)$
c	$-\lambda$	$1 - \lambda^2/2$	$\lambda^2 A$
t	$\lambda^3 A(1 - \rho - i\eta)$	$-\lambda^2 A$	1

REFERENCES

[1]M.Kobayashi and K.Maskawa, Progr.Theor.Phys.49(1973)652

[2]K.Kleinknecht, Ann.Rev.Nucl.Sci.26(1976)1

[3]S.Pakvasa and H.Sugawara, Phys.Rev.D 14(1976)305

[4]L.Maiani, In.Symp.on Lepton and Photon Interactions at High Energies, Hamburg 1977 (DESY, Hamburg 1977), p.877

[5]M.K.Gaillard and B.W.Lee, Phys.Rev.D 10(1974)897

[6]F.J.Gilman and M.B.Wise, Phys.Rev.D 27(1983)1128

[7]J.F.Donoghue et al., Phys.Lett.119 B(1982)412

[8]E.A.Paschos, B.Stech and U.Türke, Phys.Lett.128 B(1983)240

[9]K.Kleinknecht, Proc.10th Int.Neutrino Conf., Balatonfüred, 1982 (Central Res.Inst.Physics, Budapest, 1982), Vol.1, p.115

[10] E.A.Paschos and U.Türke, Phys.Lett.116 B(1982)360

[11] S.Pakvasa, Proc.21st.Int.Conf.on High Energy Physics, Paris, July 26-31, 1982, J.Phys.43, Suppl.12(1982)C3-234

[12] K.Kleinknecht and B.Renk, Z.Phys.C 16(1982)7; Z.Phys.C 20(1983)67

[13] L.L.Chau et al. Phys.Rev.D 27(1983)2145; L.L.Chau. Phys.Rep.95(1983)3

[14] E.Fernandez et al., MAC Collaboration, Phys.Rev.Lett.51(1983)1022

[15] N.S.Lockyer et al., Mark II Collaboration, Phys.Rev.Lett.51 (1983)1316 and paper given by G.H.Trilling at Europhysics Study Conference on Flavour Mixing in Weak Interactions, Erice, March 4-10, 1984

[16] M.Bourquin et al., WA 2 Collaboration, Z.Phys.C 21(1983)27 and paper given by H.W.Siebert at Europhysics Study Conference on Flavour Mixing in Weak Interactions, Erice, March 4-10, 1984

[17] K.Kleinknecht and B.Renk, Phys.Lett.130 B(1983)459

[18] W.T.Ford, paper given at Europhysics Study Conference on Flavour Mixing in Weak Interactions, Erice, March 4-10, 1984

[19] C.Klopfenstein et al., CUSB Collaboration, Phys.Lett.130 B (1983)444; J.Lee-Franzini, paper given at Europhysics Study Conference on Flavour Mixing in Weak Interactions, Erice, March 4-10, 1984

[20] A.Chen et al., CLEO Collaboration, Phys.Rev.Lett.52(1984)1084; P.Avery, paper given at Europhysics Study Conference on Flavour Mixing in Weak Interactions, Erice, March 4-10 1984

[21] R.E.Shrock and L.L.Wang, Phys.Rev.Lett.41(1978)1692 and 42(1979)1589

[22] C.H.Christensen et al., Phys.Rev.D 5(1972)1628; J.Byrne et al., 92 B(1980)274

[23] M.Roos, as quoted in K.Kleinknecht, Weak decays and CP violation, Plenary report at 17th Int.Conf.on High Energy Physics, London, July 1974, ed. by J.R.Smith, Science Research Council London 1974, p.III-23

[24] H.Abramowicz et al., Z.Phys.C 15(1982)19

[25] H.G.J.de Groot et al., Z.Phys.C 1(1979)143

[26] R.Brock, Phys.Rev.Lett.44(1980)1027

[27] G.Altarelli et al., Nucl.Phys.B 208(1982)365

[28] M.K.Gaillard and L.Maiani, Proc.Summer Institute on Quarks and Leptons, Cargèse, 1979 (Plenum Press, New York, 1980), p.433; the phase space values are taken from a more recent analysis of J.Lee-Franzini (ref.[19]).

[29]W.Bartel et al., Phys.Lett.114 B(1982)71

[30]L.Wolfenstein, Phys.Rev.Lett.51(1983)1945

[31]B.Stech, paper given at Europhysics Study Conference on Flavour Mixing in Weak Interactions, Erice, March 4-10, 1984

[32]K.C.Chou, paper given at Europhysics Study Conference on Flavour Mixing in Weak Interactions, Erice, March 4-10, 1984

[33]B.Guberina et al., Phys.Lett.128 B(1983)269

[34]L.Wolfenstein, Phys.Rev.Lett.13(1964)562

[35]F.J.Gilman and J.S.Hagelin, SLAC-PUB-3226 (Sept.1983)

[36]W.M.Morse, paper given at Europhysics Study Conference on Flavour Mixing in Weak Interactions, Erice, March 4-10, 1984

DISCUSSION

WOLFENSTEIN:
In the form of the K-M matrix, I suggested that you showed is
unitary only to order λ^3. To go beyond λ^3 it is necessary to add
some small imaginary force, but, of course, there are no new
parameters.

If you assume unitarity and only three generations only two
numbers are significant in determining the K-M matrix: (1) U_{ud}
from beta-decay and (2) U_{bc} from the B lifetime. The error on
U_{us} must be increased by at least a factor 3 because of the
theoretical uncertainty in SU(3) corrections. Then U_{us} is
determined by unitarity and U_{ud}. In addition to these two
numbers we have the limit on U_{ub} from the limit on b → u
decays. The other information is only relevant on limiting extra
generation effects.

CHAU:
I have two comments: 1) The point made by Wolfenstein that the CP
violation imaginary part happens at 10^{-3} can be put in exact
unitary form, and we can also show that in first order weak
interaction that the CP violation effects in absolute quantities,
like rate differences, has a universal factor of $s_1^2 s_2 s_3 s_\delta$.
(See contributed paper, Chau and Keung). 2) About the point of
measuring V_{ub} by measuring B → ν_τ, had made the calculation
$Br(B \rightarrow \tau\bar{\nu}_\tau) < 10^{-5}$ from the ratio $\left|V_{ub}/V_{cb}\right| < .12$.

TRILLING:
Can you explain the justification for combining CUSB and CLEO
upper limits on $\left|U_{bu}/U_{bc}\right|$ to obtain a reduced upper limit?

LEE-FRANZINI:
I answered the question in my talk. Namely, that the proper way
to get a combined limit is to combine the likelihood functions.
However from my knowledge that they were approximately Gaussian
and centered about zero, the combined limit for $\Gamma(b \rightarrow u)$:
$\Gamma(b \rightarrow c)$ is < .003 at 90% C.L.

SOME THOUGHTS ON SEARCHES FOR PROTON DECAY*

Maurice Goldhaber

Brookhaven National Laboratory

Upton, N.Y. 11973

(Irvine, Michigan, Brookhaven, Cal Tech, Cleveland State,
Hawaii, University College - London Collaboration)

At the Conference I summarized the results of searches for proton decay as reported by various groups at the XIXth Rencontre de Moriond[1] which took place a week earlier. Since the detailed talks given there will be published in the near future, there seems to be no point in repeating the evidence in this report. Instead, I would like to put down some tentative thoughts which might be worth keeping in mind in further searches for proton decay.

The minimal SU(5) theory appears to be contradicted by the experiment of Bionta et XXVIII al.[2] who find a lifetime longer than predicted by this theory. However, A.S. Goldhaber, T. Goldman, and S. Nussinov have recently pointed out that final state interaction due to strong meson field couplings may suppress proton decay by one or two orders of magnitude, which may yield a lifetime approaching the present experimental limit (see 5th Workshop on Grand Unification, and Physics Letters, to be

---

*Work performed under the auspices of the U.S. Department of Energy.

published). We shall have to see whether the gap between experiment and theory will increase or decrease with time, as both are refined.

The Kolar Goldfields, Nusex, IMB, Kamiokande and HPW experiments all find contained events which include proton decay candidates. Improved detectors are being prepared at the Kolar Goldfields and at Soudan (Minnesota); a fine grained detector being installed at Fréjus should approach ∿ 1 kton by the end of the year. The IMB water Cherenkov detector is being upgraded to collect a bigger fraction of the light in order to improve particle identification and energy determination. But even with the good light collection already achieved by Kamiokande, there are still ambiguities left in the assignment of decay modes to candidates, as reported at the XIXth Rencontre de Moriond[1] by Y. Totsuka.

To establish proton decay, if it exists, we have to demonstrate reproducibility, which implies large detectors and long observation times. This may permit the background events from atmospheric neutrinos to become ultimately "self-calibrated", provided that in the mass region of interest, above and below the proton mass, background events with topologies similar to those expected from proton decay can be assumed to vary continuously in intensity.

It is useful to imagine what we could achieve if we were able to build detectors which approach an "ideal" detector, defined as one which allows one to determine, without ambiguity, the different kinds of decay particle including their charges and energies. We shall make the assumption that for all proton decays in which only one fermion is emitted there exists a selection rule which forces the fermion to be either a lepton or an anti-lepton. The charged-current interaction of atmospheric neutrinos will give an unavoidable, "intrinsic" background when the neutrinos produce leptons of the same electric charge as is found in proton decay. Thus, either neutrinos ($\overset{\leftarrow}{\nu}$) or antineutrinos ($\overset{\rightarrow}{\nu}$), but not both, will be responsible for this intrinsic background. For a proton decay in which an uncharged lepton is emitted, neutral-current interactions of both neutrinos and anti-neutrinos will give an intrinsic background, though with a cross-section smaller than for the charged-current interaction. A special case would be the decay modes $p \rightarrow \overleftarrow{\nu} K^+$ and $n \rightarrow \overleftarrow{\nu} K^0$ for which there is no intrinsic background because of the absence of strangeness changing neutral currents (GIM mechanism). With the new and improved detectors now in preparation, it should be possible to identify characteristic kaon decays, though, at finite resolution, some non-intrinsic neutrino background, which could again be "self-calibrated", will have to be taken into account.

Before we can convince ourselves that the proton actually decays (if it does so in the "window of vulnerability" accessible to experiments in the available detectors), we must demand, besides reproducibility, some of the fairly obvious consistency checks, which I discussed at the XIXth Rencontre de Moriond[1]. Thus, e.g., if candidates interpreted as $p \rightarrow \mu^+ K^0$ with $K^0 \rightarrow 2\pi^0$ are observed, then there should be about twice as many candidates $p \rightarrow \mu^+ K^0$ with $K^0 \rightarrow \pi^+\pi^-$. Also, if proton decay candidates are found with a momentum imbalance ascribed to the Fermi momentum of protons bound in nuclei, then one can predict, after nuclear corrections, how many similar events should originate from the unbound protons in H_2O where momenta should be balanced.

More than a dozen candidates of varying "quality" have by now been reported from the experiments in progress. For no single case can we exclude the possibility that it is a neutrino induced background event. But, if we keep an open mind, and for the sake of argument assume that many of these events are indeed due to proton decays, rather than produced by atmospheric neutrinos, it would seem about time that decays from free hydrogen should be observed, i.e., events in which momenta appear balanced.

The many dedicated experimental searches for proton decay begun in the last few years were catalyzed by theoretical predictions. But a catalyst should not be contained in the final result! Is that indeed the case here? In the interpretation of candidates one usually tacitly assumes that one particle is an anti-lepton (e^+, μ^+, $\bar{\nu}$), as expected from SU(5) and related theories where B-L is conserved, rather than a lepton (e^-, μ^-, $\bar{\nu}$), an alternative which an open-minded, phenomenological approach should at least consider. The anti-lepton is usually assumed to be accompanied by a meson, which may show further decay. But since the existing detectors are rather insensitive to the sign of charged leptons, and mesons can suffer charge exchange in the nucleus, these interpretations are often not unique.

Unless there are special selection rules one would expect from phase space considerations that two-body decays predominate.[3] If leptons rather than anti-leptons were emitted, then two-body events could arise in both proton and neutron decay when either a neutrino or anti-neutrino is emitted, but, because of charge conservation, only in neutron decay when a charged lepton is emitted (e.g., $n \rightarrow e^- \pi^+$). It is intriguing that detectors which contain iron as source material, where the ratio of neutrons to protons (n/p) is nearly 45% larger than in water, appear to yield more candidates per ton-year than do those which contain water, though better statistics are needed before this can be considered significant.[4] If one wants to pursue the uncertain hints of n/p dependence (or of some unforeseen nuclear structure effects) one might alternate layers of materials with large n/p--e.g., in an extreme case, bismuth (126/83)--with materials with small n/p, e.g., polyethylene $(CH_2)_n(6/8)$, thus changing n/p between layers by about a factor of two, while also supplying a source of unbound protons. Such an alternation of high Z and low Z layers would also yield--as a bonus--some information on the charges of muons and pions. In bismuth stopping μ^+ and π^+ decay, while μ^- and π^- are absorbed, whereas in polyethylene essentially only π^- is absorbed.

The next few years should show whether some of the candidates for proton decay will pass reproducibility and consistency criteria, with either lepton or anti-lepton emission.

REFERENCES

1. Talks by M. Goldhaber, E. Shumard, J. van der Velde, Y. Tosuka and S. Ragazzi at the XIXth Rencontre de Moriond, LaPlagne, France, March 1984 (to be published).

2. R.M. Bionta, G. Blewitt, C.B. Bratton, B.G. Coretz, S. Errede, G.W. Foster, W. Gajewski, M. Goldhaber, J. Greenberg, T.J. Haines, T.W. Jones, D. Kielczewska, W.R. Kropp, J.G. Learned, E. Lehmann, J.M. LoSecco, P.V. Ramana Murthy, H.S. Park, E. Reines, J. Schultz, E. Shumard, D. Sinclair, D.W. Smith, H.W. Sobel, J.L. Stone, L.R. Sulak, R. Svoboda, J.C. van der Velde and C. Wuest, Phys. Rev. Lett. $\underline{51}$, 27 (1983), and later work.

3. In the decay of another fermion, the τ, which is nearly twice as heavy as the proton, and thus has a considerably larger phase space available for multi-particle decay, it is found that about half or more of the semileptonic decays are two-body decays (see Particle Data Tables).

4. One should however remember that the background produced by atmospheric neutrinos also depends on the ratio n/p.

DISCUSSION

WALI:
Is there a way in your experiment to tell when the proton decays in your arms? How do you know when something is not due to neutrino background?

GOLDHABER:
Only good statistics, reproducibility, consistancy and better knowledge of neutrino background will allow a definite statement on proton decay.

SOERGEL:
What is the technical limit for the size of a water detector?

GOLDHABER:
I see no reasonable technical limit, except that due to absorption of light in water, photo-tubes will have to be repeated for $>$ 50m diameter detectors.

CHAU:
1) Would you mention future plans for measuring proton decays?
2) Since supersymmetric theories predict that a strange particle in the final state is preferred, which experiment has the highest sensitivity?

GOLDHABER:

1) Our Irvine-Michigan-Brookhaven Collaboration plans to upgrade our detector to achieve more light collection. There are various tentative plans for still larger detectors.
2) Probably fine grained detectors, as the one in process of being installed at Fréjus, and also planned elsewhere, are most sensitive, though water detectors have also a reasonable sensitivity.

THE D0 DETECTOR FOR THE FERMILAB COLLIDER

AND THE FUTURE PHYSICS PROGRAM AT CESR

Paolo Franzini

Columbia University

New York, NY 10027

INTRODUCTION

By late '86 the Fermilab superconducting ring is
expected to be operating as a proton-antiproton collider,
sometimes known as TEV I, at a c.m. energy $\sqrt{s}$ = 2 TeV,
with a luminosity of 10^{30} cm^{-2}s^{-1}. Concurrently a
machine improvement program is under way at the CESR
$e^{+}e^{-}$ collider at Cornell, with the aim to raise the
collider luminosity up to values of ~ 5 pb^{-1} day^{-1}.
Two new detectors have been designed to take advantage
of the promises of the new colliders in several fields
of particle physics: the 'D0' detector at Fermilab and
the 'CLEO II' detector at CESR. The design principles
of the two detectors and some examples of the physics
which they can address are presented.

I. The D0 DETECTOR FOR THE FERMILAB COLLIDER

At the new energies and luminosities available at
the Fermilab collider, it will become feasible to study
properties of the weak interactions at an unprecedented
level of accuracies as well as to uncover possible new
phenomena.[1] A conventional detector, the Collider

481

Detector Facility (CDF), is under construction for
experiments at TEV I. A new detector was approved,
late in '83 to operate at the D0 straight section/
interaction region. The preliminary design[2] of this
new detector was completed in the fall of '83 and is
still undergoing refinements and completion. Figure 1
is an artist's conception of the detector. The most
novel feature of the detector, still in search for a
name and generally known as the D0 detector, is in its
emphasis on calorimetry, energy resolution and
hermeticity.

The experience of the UAl and UA2 detectors[3] at
CERN has clearly shown the desirability to complement
more traditional designs with an instrument which
achieves the ultimate hadronic and electromagnetic
calorimetry resolution, segmentation and 4π coverage.

I.1 The D0 DETECTOR DESIGN

At $\sqrt{s}$ = 2 TeV, detection of individual hadrons
becomes irrelevant for most of the physics of interest

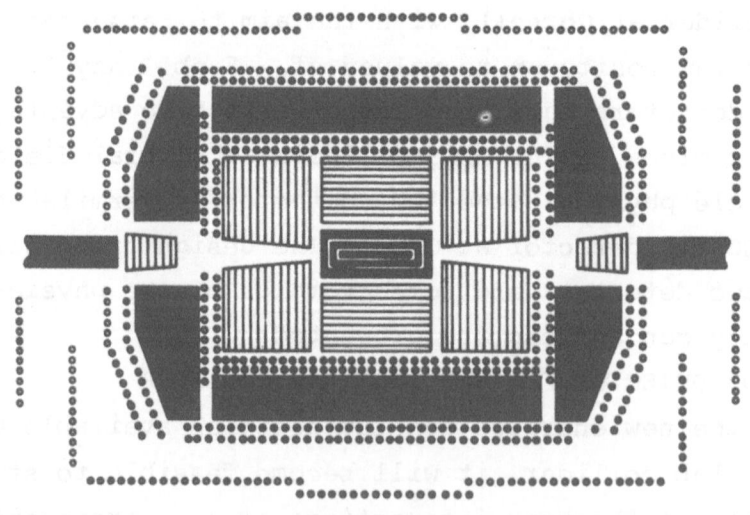

Fig. 1

Table I

"Particle"	Detector
Jet	Calorimeter (Electromagnetic + hadronic)
Electron	Calorimeter (e.m.)
Photon	Calorimeter (e.m.)
Neutrino	Missing Energy → Calorimeter (e.m. + hadr.)
Muon	Magnetized Iron Filter

and the type of "particles" to be detected can be abbreviated to the list given in Table I, where means for their detection are indicated.

With the exception of muons, all particles of interest are best detected by calorimetry. In the case of neutrinos, as well as other more exotic objects such as photinos etc., calorimetry must be hermetic. Thus insensitive regions between detector elements can severely limit the "detectability" of non-interacting particles and coverage down to extremely small angles is essential. In addition jets (and neutrinos balanced by jets) always have an electromagnetic and a hadronic component. Very equal response to e.m. and hadronic energy is therefore required to obtain good energy resolution on an event-by-event basis. Finally good hadronic resolution requires some mechanism for recovering the energy lost in fragmenting nuclei.

All these requirements can be achieved by using uranium[4] as the absorber in which the hadronic and e.m. showering takes place. Frequent sampling of the shower, with significant energy deposit in an active medium with uniform and constant response, is best obtained by interspersing layers of uranium and liquid argon. Measurements of the ion pairs produced in the liquid argon can be reliably performed and related to the local shower particle density and ultimately to the energy of the "particle".

The core of the D0 detector is a uranium-liquid argon calorimeter surrounding the interaction to within 1° of the two beams. Thus 99.98% of the total solid angle is covered. Since e.m. showers develop more rapidly than hadronic ones and have considerably smaller transverse dimensions, the calorimeter is subdivided longitudinally into a fine grained e.m. compartment of approximately 24 radiation lengths (X_o) followed by a coarser grained hadronic compartment of thickness varying from 7 nuclear absorption lengths (λ_o) at 90° to 9 λ_o at small angles to account for the higher energies of jets in the forward/backward region. The calorimeter is subdivided into ~ 5000 e.m. 'towers' and ~ 1000 hadronic towers pointing to the interaction region. This segmentation is chosen to be ~ constant in units of rapidity $(\Delta y \sim 0.05)$ and azimuth. Table II illustrates the quality of the calorimetry for the D0 detector and, in

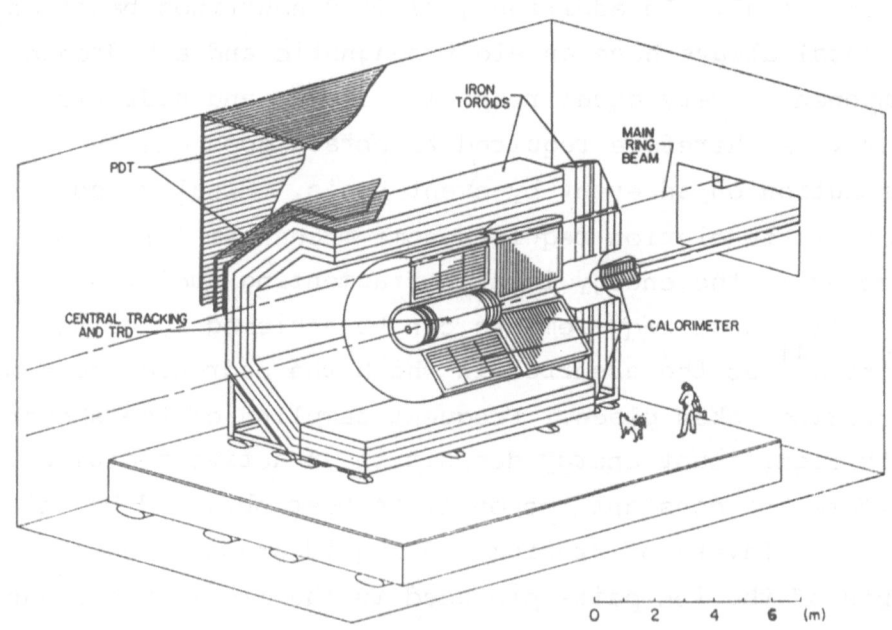

Fig. 2

Table II

Parameter	DØ Detector	CDF Detector
Central Calorimeter		
Coverage	45-135°, 70% of 4π	50-130°, 64% of 4π
Thickness (90°)	7 λ_o	5.6 λ_o
Resolution e.m.	11%/$\sqrt{E}$ + 0.5%	14%/$\sqrt{E}$ + 1.5%
Hadr.	37%/$\sqrt{e}$ + 0.5%	62%/$\sqrt{E}$ + 5.5%
Segmentation e.m.	3360 twrs, 4 smpls	480 twrs, 1 smpl
Hadr.	576 twrs, 4 smpls	384 twrs, 1 smpl
e.m./Hadr. response	1.1	1.35
Forward Calorimeters		
Coverage	2×(5-45°), 29% of 4π	2×(10-50°), 34% of 4π
Thickness (10°)	8.9 λ_o	6.7 λ_o
Resolution e.m.	11%/$\sqrt{E}$ + 0.5%	29%/$\sqrt{E}$ + 3.0%
Hadr.	37%/$\sqrt{E}$ + 0.5%	92%/$\sqrt{E}$ + 6.0%
Segmentation e.m.	1536 twrs, 4 smpls	864 twrs, 2 smpls
Hadr.	672 twrs, 3 smpls	1080 twrs, 1 smpl
Small Angle Calorimeters		
Coverage	2×(1-5°), 0.36% of 4π	2×(2-10°), 1.5% of 4π
Thickness	9.0 λ_o	6.7 λ_o
Resolution e.m.	20%/$\sqrt{E}$ + 0.5%	30%/$\sqrt{E}$ + 3%
Hadr.	50%/$\sqrt{E}$ + 0.5%	105%/$\sqrt{E}$ + 7%
Segmentation e.m.	360 twrs, 4 smpls	1224 twrs, 1 smpl
Hadr.	360 twrs, 3 smpls	1224 twrs, 1 smpl

comparison, for CDF. Tracking around the interaction region is provided by a 67 cm radius, 160 cm long drift chamber, incorporating transition radiation detectors (TRD). Figure 2 shows a partial cutaway and Figure 3 an elevation quarter section of the D0 detector.

The four samplings of the e.m. shower, together with the TRD's, give superior electron vs hadron identification. For isolated electrons the information from the depth segmented e.m. calorimeter and the TRD detectors results in an electron signal contamination of less than 10^{-4} and, for electrons close to a jet, of less than 10^{-2}. In addition the e.m. calorimeter can distinguish γ from π^o and other neutral particles decaying into two or more photons, both by shower shape

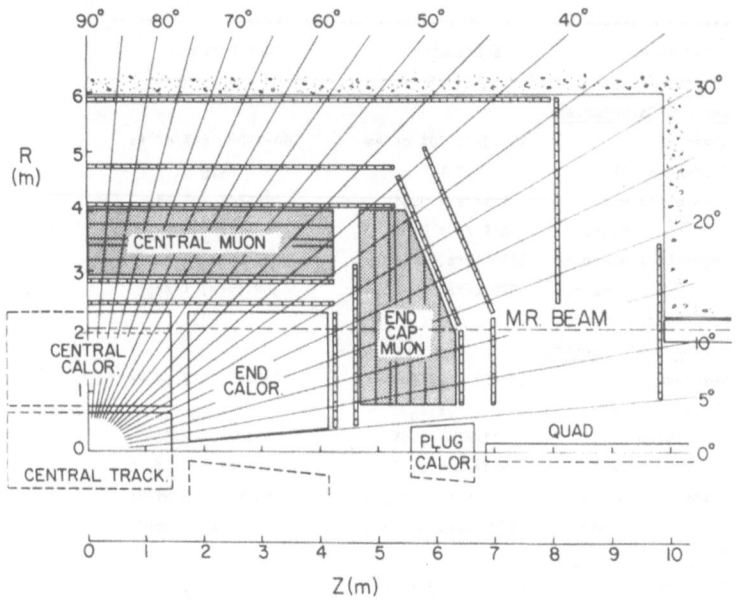

Fig. 3

and conversion depth, thus allowing measurements of direct photon production.

Muons are identified as particles which cross without interaction the entire calorimeter and additional magnetized iron which allows momentum analysis with an accuracy of 17% up to transverse momenta of 300 GeV. The total thickness of material that a muon must cross varies from 13 λ_o at 90° to 18 λ_o at 10°. The incident energy of a single pion, which results on average in one leakage particle, is 600 GeV at 90° and 6000 GeV at 10°. Muons can therefore be identified with negligible background, even in the core of a shower over the whole energy range available. The muon detector system covers 99% of 4π.

I.2 PHYSICS WITH THE D0 DETECTOR

Two specific examples of the physics for which the proposed detector is excellent will now be discussed in more detail, after giving a brief list[1] of some interesting physics for which the D0 detector is uniquely suited.

1. The accurate simultaneous measurement of the $W^{\pm}$ and Z^0 masses.

2. Measurement of W and Z width to 200 MeV accuracy.

3. Measurement of $p\bar{p} \to W\gamma X$, γ/π^0, $\gamma+2$ jets vs. 3 jets.

4. Observe $W \to q\bar{q}$ from di-jet invariant mass.

5. Searches for exotic new objects, resulting in leptons, jets or missing p_T, such as: heavy leptons, SUSY particles, heavy quarks, leptoquarks, etc.

Measurements of W, Z masses and widths and the standard model. With the D0 detectors at Fermilab, it is possible to measure the mass of W and Z by the use of the 'transverse mass' method[5] applied to $W \to e$ and $Z \to e$ in the same way so as to cancel out possible systematic errors in the mass difference or ratio. This is only possible with an excellent missing P_T resolution. We expect in this way to measure $\sin^2\theta_w$ to an accuracy of ± 0.002, a most sensitive test for unification theories. Likewise the ρ parameter can be measured to an accuracy of ± 0.003. Deviation of ρ from unity at the 1% level would be the first sign[6] of a new mass scale below 1 TeV.

Search for gluinos. Figure 4 shows the observed cross section vs. missing P_T in the D0 detector, with the effects of covered angle and resolution.[7] Also shown is the contribution to this cross section from semi-leptonic decays of the (u,d,s), c and b quarks. This very strong signal is in fact a background to searches for gluinos, for example, where one expects the gluino to decay into jets plus an invisible photino, resulting in the dashed line contribution to the cross section in Fig. 4. Neutrinos from heavy flavor decays are however

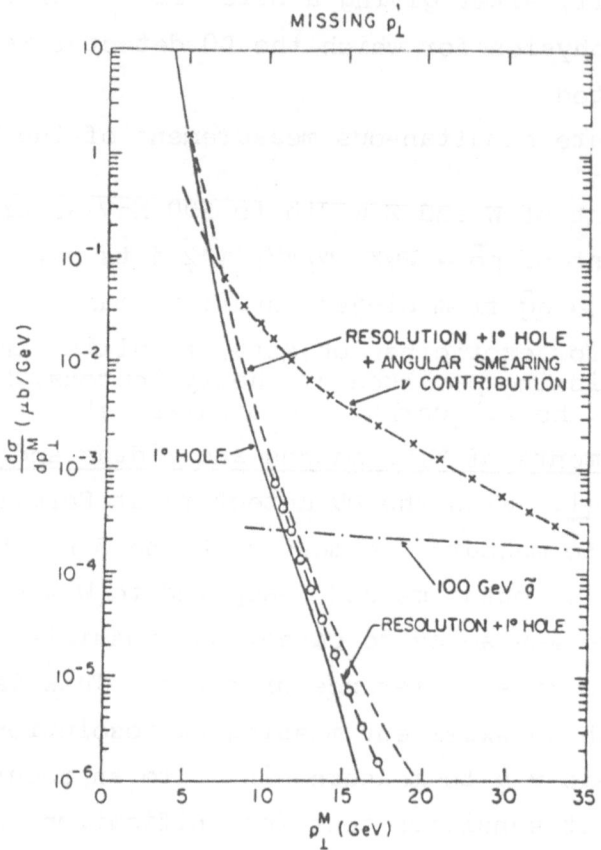

MISSING $p_\perp^{'}$

Fig. 4

accompanied by leptons. Owing to the unique features
of the D0 detector, these leptons can be recognized with
better than 99% efficiency, thus allowing detection of
gluinos up to 150 GeV mass.

II. PHYSICS AT CESR

The physics mission of CESR (and of DORIS) is
unique: the study of b quarks and B mesons. At $\sqrt{s}$ =
10.576 GeV the continuum cross section for e^+e^-
annihilations into hadrons is $\sim$ 2.8 nb. In addition,
at the same energy the fourth upsilon, Υ''', is produced
with a cross section of $\sim$ 1 nb. Since the Υ''' decays

essentially always into a B$\bar{\text{B}}$ pair, one can collect large samples of B mesons almost background free. In addition the B$\bar{\text{B}}$ pair is in a pure $J^{PC} = 1^{--}$ state. There are several questions[8] concerning B mesons and b quarks about which experiment only can provide an answer:

1. Is there B-$\bar{\text{B}}$ mixing?

2. Is there CP violation in the B system?

3. Are there b → u transitions?

4. Are the lifetimes of B$^{\pm}$ and B^{o} equal?

This list is just an example.

II.1   B DECAYS

Most of our present knowledge about the nature of the weak b couplings comes from small samples of leptonic decays of a single B meson, at CESR, PEP and PETRA.[9] Improved measurements of the B lifetime will continue to come from PEP and PETRA and they are fundamental to the understanding of weak interactions. An answer to the first two questions above requires production of large samples of B$\bar{\text{B}}$ pairs and, in the traditional way, the ability to identify events in which both B and $\bar{\text{B}}$ decay leptonically. Comparison of same sign and opposite sign dilepton yields can provide the answers. The use of the lepton sign to tag mesons with b or $\bar{\text{b}}$ contents is however ultimately limited by the fact that the decay chain B → D → ℓ gives the wrong sign leptons. A better, or a least complementary way, is to reconstruct B mesons through their non-leptonic decays. For most reconstructed B^{o} mesons, one has an unambiguous tag of the "b-ness" of the other in the pair. Reconstructed B's result in almost a factor ten improvement in the "signal/noise" ratio for the first two questions above and the only clean way for answering the third and fourth.

II.2 RECONSTRUCTING B MESONS

Reconstructing B mesons is not easy! We know however quite well what is the best that we can ever do. From a knowledge of B and D semileptonic (sl) branching ratios (BR's), we obtain BR(B → no neutrinos) = 0.48; that is, ∼ one B can be reconstructed per produced Υ'''. A very conservative estimate of the efficiency for reconstructing B mesons, using the D^* → Dπ "trick" and final states with charged pions only, has been given by members[10] of the CLEO Collaboration as 0.3%. Hopefully one can gain several factors of two using B^* → B + γ, π^o's detected through two photon decay, K^o's detected from $\pi^+\pi^-$ and $\pi^o\pi^o$ decays. An optimistic estimate for the reconstruction efficiency might therefore be ∼ 10%. Putting together luminosity, cross section and efficiency we predict that in one year of running at CESR one might obtain a sample of 2000 to 150,000 reconstructed B's, a reasonable expectation being ∼ 20,000.

II.3 A NEW DETECTOR FOR CESR

The requirements for a detector capable of achieving the described goals are: i) superior tracking efficiency and momentum resolution over 4π; ii) a matching energy and angular resolution for γ's; iii) excellent e/π/K identification up to ∼ 2 GeV; iv) good muon identification.

While the first requirement is a relatively simple improvement of standard tracking, the second point can only be satisfied, with present technology, by a crystal spectrometer mounted inside the coil providing magnetic field for momentum analysis, forcing the crystal array to have an inner radius of ∼ 1 meter. Finally good muon identification at energies below 2.5 GeV (the end point of B → μνX) is very difficult. Large increases in

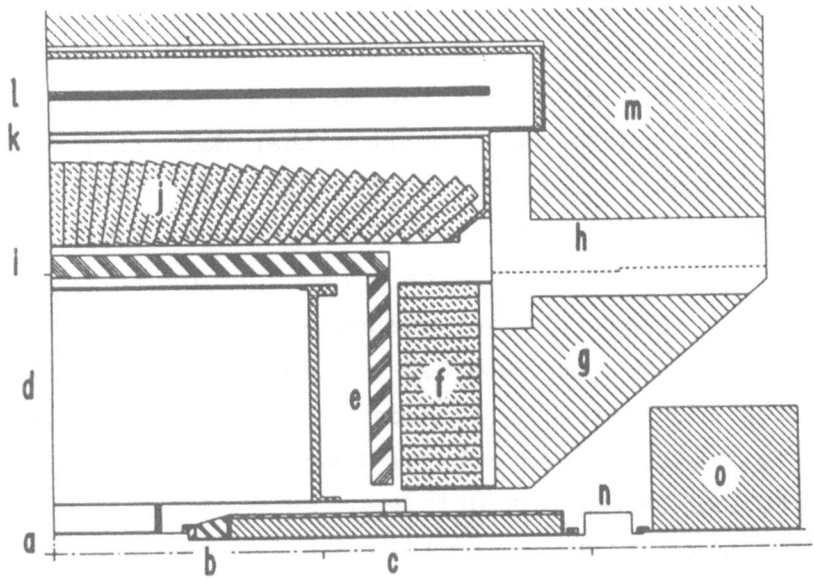

Fig. 5

complexity give only minor improvements. The new
detector proposed at CESR, CLEO II, is a reasonable
compromise, putting most emphasis on best tracking,
very good e.m. calorimetry, good electron identification
and only partial kaon and muon identification.

A cut-away view of CLEO II is shown in Figure 5.
Tracking is provided over 94% of 4π by a vertex chamber
followed by a 51-layer drift chamber up to a radius of
95 cm. Of the 51 layers of sense wires, 40 are used for
dE/dx measurement. Because of the relativistic rise of
specific ionization, its measurement improves electron
identification by a factor ~ 10. π/k identification is
also provided by dE/dx measurements up to ~ 600 MeV/c
and up to about 1 GeV/c by time-of-flight counters.
The proposed e.m. calorimeter is a novel idea. The
calorimeter consists of 8000 CsI(Tl) crystals, whose
scintillation light is detected by four silicon photo-

diodes mounted on the outer face of the crystal. CsI was chosen for several reasons: i) high light output, similar to NaI, better matched to diode response; ii) much less hygroscopic and brittle than NaI; iii) somewhat shorter radiation length than NaI, X_o = 1.6 cm. The approximately 10 times higher light output of CsI with respect to BGO (bismuth germanate), the longer radiation length, the lower cost per volume and the ease of production appear to make CsI an ideal material for a very high accuracy e.m. calorimeter, inside a high magnetic field, around a large tracking chamber.

II.4 FUTURE PHYSICS RESULTS FROM CESR

By late 1984 or early 1985 the luminosity improvement program at CESR should be completed. The CLEO II detector is scheduled for completion by the end of 1987. By 1988 a sample of 10^6 produced Υ''' or 2,000,000 B's should be available. This should yield 30,000 dileptons and 20,000 reconstructed B's of which 5000 are tagged neutral B's. By that time we might expect answers to the four questions above to the following accuracies:

1. B-$\bar{\text{B}}$ mixing. Mixing can be measured to the level of 10% or better, both using dileptons and tagged B's. Comparison of the two methods will give much confidence in a positive answer.

2. CP violation. Since both mixing and CP violation must be large for a positive result, the same limit above essentially applies to the combined effect.

3. b → u transitions. From of the order of 4000 reconstructed B's accompanied by the other B decaying semileptonically, one can positively detect decays like $B \to l + \nu + (\pi, \rho, A_1 \ldots)$ for $\Gamma(B \to l\nu X_u)/\Gamma(B \to l\nu X_c) \approx$ 0.5%, almost ten times lower than the present limit.

4. Ratio of charged to neutral B lifetimes. Using

reconstructed B's, one can measure the equivalent ratio
of the semileptonic BR's to an accuracy around 5%.

ACKNOWLEDGMENTS

The author wishes to thank and acknowledge all
members of the D0 Collaboration and P. Avery, B.
Gittelman, K. Berkelman, S. Stone and N. Mistri of
the CLEO Collaboration for many discussions. He also
wishes to thank J. Lee-Franzini, P. M. Tuts, and R. D.
Schamberger for help with this report. This work
was supported in part by the National Science
Foundation, Washington, DC, USA.

REFERENCES

1. See the many discussions in: Proc. of the 1982 DPF
   Summer Study on Elementary Particle Physics, Eds.
   R. Donaldson, R. Gustafson, F. Paige, APS (1982).

2. D0 Design Report, by physicists from: U of Arizona,
   Brookhaven National Laboratory, Brown U, Columbia U,
   Fermilab, Florida State U, U of Maryland, Michigan
   State U, Northwestern U, U of Pennsylvania, SUNY at
   Stony Brook, VPI. Fermilab Experiment E740 (1983).

3. See for instance the presentations by the UA1 and
   UA2 Collaborations at this Conference.

4. W.J. Willis, V. Radeka, NIM 120 (1974) 221; J.H.
   Cobb et al, NIM 158 (197(93; C.W. Fabjan et al,
   NIM 141 (1977) 61.

5. J. Smith et al, Stony Brook Preprint ITB-SB-83-11
   (1983).

6. See for instance W.J. Marciano, Proc. Int. Symp. on
   Lepton and Photon Interactions, Eds. D.G. Cassel,
   D.L. Kreinick, Cornell University (1983), p. 80.

7. This result was obtained using the ISAJET M.C.
   program of F. E. Paige and S. D. Protopopescu, BNL
   29777 to generate events into the D0 detector.

8. See the talks of J. Lee-Franzini and R. Rückl at
   this Conference.

9. Experimental results on B meson decays were presented
   at this Conference by: P. Avery, W. Ford, J. Lee-
   Franzini and G. Trilling.

10. B. Gittelman et al, CLEO CBX-84-2, Cornell U (1984).

DISCUSSION

SOERGEL:

Could you give more details about the CS-I detector? And what are the chances to get the higher luminosity at CESR?

FRANZINI:

CS-I was chosen because of its large light output and convenient mechanical properties. It will be read out with photodiodes. With regard to the second question I have great faith in the machine experts at CESR.

HANSEN:

What do you plan to do in DØ to get rid of background from converted photons?

FRANZINI:

dE/dx measurements in the tracking chamber can very well distinguish single electrons from pairs.

MORSE:

You mentioned an energy resolution of $1\%/\sqrt{E}$ with a CS-I calorimeter. How well can you calibrate such a calorimeter?

FRANZINI:

CS-I has enough light output to allow source calibration.

FUTURE PLANS AT KEK

Makoto Kobayashi

KEK, National Laboratory for High Energy Physics
Tsukuba
Japan

INTRODUCTION

In this talk, I will briefly discuss the following points: (i) TRISTAN project, (ii) Reinforcement of 12 GeV proton synchrotron, (iii) Post TRISTAN projects, and (iv) Non-accelerator experiments.

Figure 1 shows the layout of KEK site. The site is roughly a rectangle of 1 km × 2 km, located at the north end of Tsukuba Science City. The 12 GeV proton synchrotron (PS) started operation in 1976 and has been a main facility in KEK. The 500 MeV booster of PS is utilized for many purposes. Booster utilization facility includes Neutron scattering experimental facility, Meson science laboratory and Biomedical research facility. Photon factory is a facility utilizing the synchroton orbital radiation of the electron and it consists of the 3 GeV electron linac and the storage ring. Operation of the photon factory was started in 1982 and now it is providing synchrotron radiation to users from various branches of the science.

TRISTAN PROJECT

TRISTAN is an e^--e^+ collider with the energy range of 25 ∿ 35 GeV per beam. The construction started in 1981. The accumulation

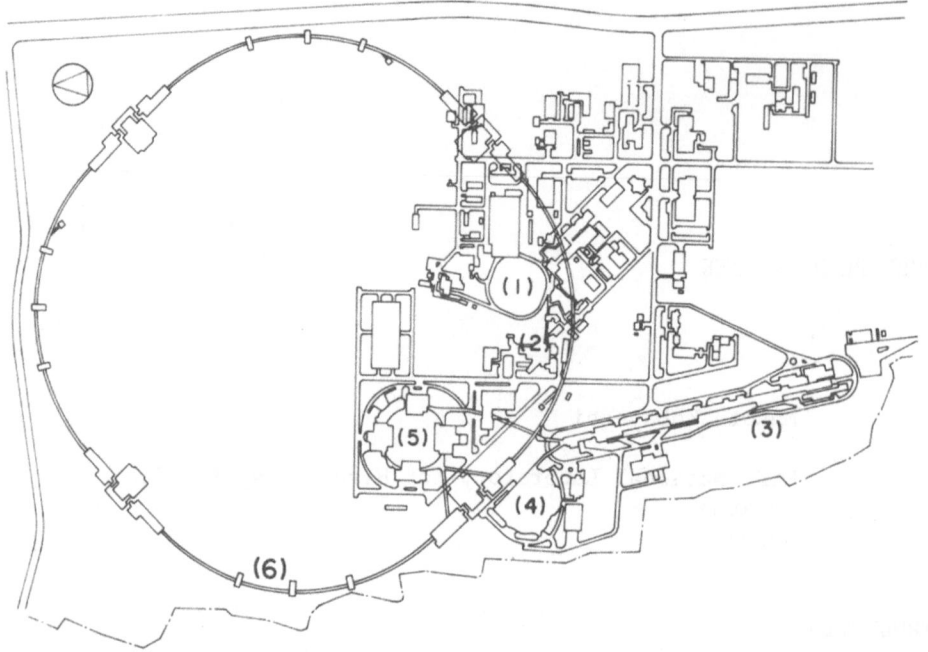

Fig. 1.   Layout of KEK site.   (1) 12 GeV proton synchrotron, (2)
          booster utilization facility, (3) electron linac, (4)
          storage ring for photon factory, (5) TRISTAN accumlation
          ring, (6) TRISTAN main ring.

ring has been already completed and the beam test is now going on.
The main ring and detector construction is in progress.   The physics
experiments will start by the end of 1986.   The general parameters
of TRISTAN accumulation and main rings are shown in Table 1.

     The electron linac of the photon factory is used as an injector
to the accumulation ring.   The positron generator is attached to the
linac and the 250 MeV positron beam is transported to the linac.

     The beam injected from the linac is accelerated up to 8 GeV in
the accumulation ring and transported to the main ring.   The maximum
energy of the main ring depends on the rf cavity system.   Probably
we will take the following two steps: The first step is installation

Table 1. General Parameters of TRISTAN

	Accum. Ring	Main Ring	
Beam energy	6 ᷉ 8	25 ᷉ 35	GeV
Injection energy	2.5 ᷉ 3	6 ᷉ 8	GeV
Circumference	377	3018	m
Average radius of curved section	47.7	346.7	m
Bending radius	23.2	246.5	m
Long straight sections	19.5(×2) +19.1(×2)	194.4(×4)	m
Total length of RF sections	38.1	509.4	m
Total length of acc. RF cavities	29.6	318	m
RF frequency	508.6	508.6	MHz
Revolution frequency	795	99.3	kHz
Max. circul. current	35	22	mA/beam
Radialion loss per turn	4.9(6GeV)	538(35GeV)	MeV
RF peak voltage	10	700	MV
Experimental insertions	5(×2)	6(×4)	m
Max. design luminosity	2×10^{31} (6GeV)	8×10^{31} (27GeV)	$cm^{-2}s^{-1}$
Betatron function at coll. point (β_x^*/β_y^*)	2.0/0.1 (6GeV)	1.12/0.07 (35GeV)	m
Beam size at coll. point(σ_x^*/σ_y^*)	0.71/0.036 (6GeV)	0.507/0.032 (35GeV)	mm

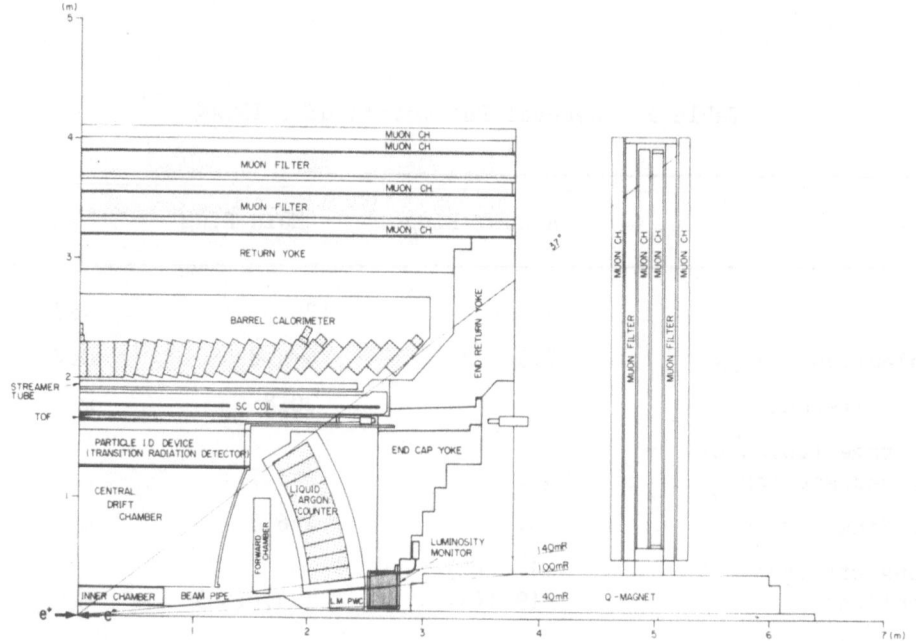

Fig. 2. VENUS detector.

of the normal rf cavities as many as about one-half of the total number of the cavities. At this stage 25 GeV per beam will be achieved. The rest of the rf cavity section will be filled by the superconducting rf cavities as the second step. Then the maximum energy of 35 GeV per beam will be realized.

Research and development on the superconducting rf cavity is in progress. Very recently a three-cell structure of the supercon-ducting cavity was installed to the accumlation ring and the beam test is going to start.

Physics goal of TRISTAN is clear. Top quark search is the most important objective of this project. If there exists the top quark in the TRISTAN energy range, the study of the toponium spectrum and the decays of the top quark will provide valuable informations for understanding the QCD and the electroweak theory. Furthermore search for the Higgs boson in the toponium decay should not be

498

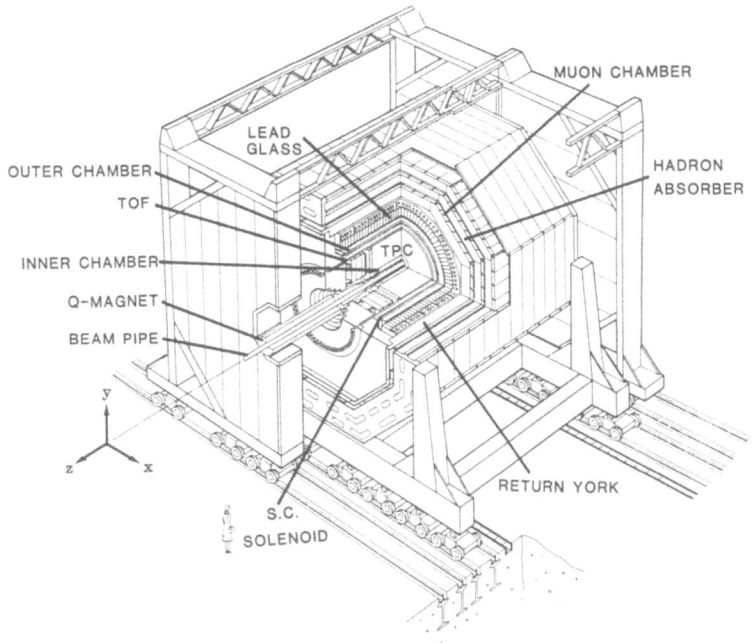

OUTER CHAMBER

TOF

INNER CHAMBER

Q-MAGNET

BEAM PIPE

LEAD GLASS

TPC

MUON CHAMBER

HADRON ABSORBER

RETURN YORK

S.C. SOLENOID

Fig. 3.   TOPAZ detector.

overlooked.   Beside the top quark, searches for all kinds of new
particles such as the heavy leptons, the super symmetric particles,
the techniparticles etc. are also interesting subjects.   As a matter
of course the test of the QCD and the electroweak theory should be
done extensively.   It should be noted that the interference between
the electromagnetic and weak interactions becomes maximum in the
TRISTAN energy range.

TRISTAN main ring has four experimental areas.   So far, the
following three experiments have been approved:

Spokesman

VENUS: N. Nagashima (Osaka)

TOPAZ: T. Kamae (Tokyo)

AMY: S.L. Olsen (Rochester)

Detector designs of these experiments are shown in Fig. 2, 3 and 4.
The fourth experimental area is reserved for the future decision.

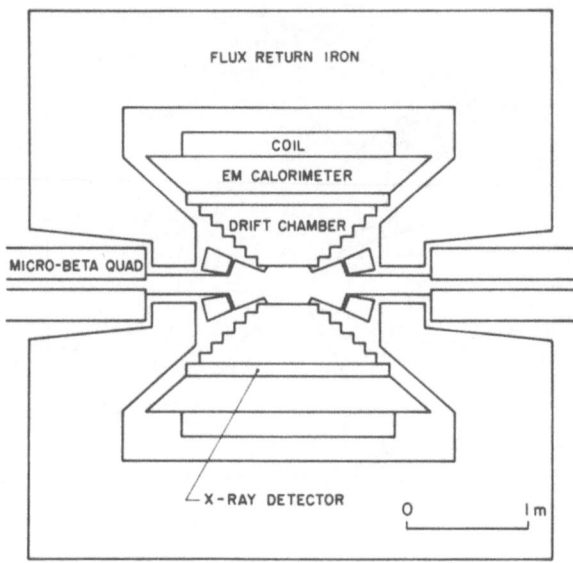

Fig. 4.   AMY detector.

Although the accumulation ring has two experimental areas, we have no plans of experiments in near future.

12 GeV PROTON SYNCHROTRON

The 12 GeV PS has just entered a long shutdown period due to the construction of TRISTAN main ring which passes under the PS facility. The shutdown will continue untill the spring of 1985.

We have some plans of reinforcement of the 12 GeV PS. They are

i)   acceleration of the polarized beam,

ii)   intensity up by a factor ~ 2,

iii)   acceleration of the ion beam.

The acceleration of the polarized beam is already decided and we have some approved experiments as a post shutdown program. The other two are still under discussion.

Table 2. General Parameters of the Proton Synchrotron Complex

	Booster	Main Ring	
Max. energy	0.5	12	GeV
Injection energy	0.02	0.5	GeV
Max. beam intensity	6×10^{11}	4.4×10^{12}	ppp
Repetition rate	20	~ 0.5	Hz
Ring diameter	12	108	m
Forcusing type	AG, Combined	AG, Separated	
Forcusing order	FDFO	FODO	
Betatron frequency	$\nu_H = 2.2$	$\nu_H = 7.1$	
	$\nu_V = 2.3$	$\nu_V = 6.2$	
Bending field	0.197(inj.)	0.15(inj.)	T
	1.1(max.)	1.75(max.)	T
No. of cavities	2	4	
Harmonic number	1	9	
RF range	$1.6 \sim 6.0$	$6.0 \sim 7.9$	MHz

POST TRISTAN PROJECTS

We have already started discussions about post TRISTAN projects
and we are going to have the second workshop on this subject in the
end of this month.

Studies on a hadron collider with the energy of $10 \sim 30$ TeV
range and an $e^- - e^+$ linear collider with the energy of ~ 1 TeV are
under way. There is also a group of people who are considering
a high intensity proton synchrotron.

Although opinions are diverging among the highenergy physicists
and we also need much debate about the international collaboration,
we will try to reach some consensus about the post TRISTAN projects
within one year or so.

NON–ACCELERATOR EXPERIMENTS

There are several important non–accelerator experiments which are going on or planned in Japan. They include nucleon decay experiments, monopole searches, measurements of the neutrino mass, etc.

Among them, KAMIOKANDE is the nucleon decay experiment at Kamioka mine in collaboration of Univ. of Tokyo, Institute for Cosmic Ray Research (Univ. of Tokyo), Niigata Univ., Tsukuba Univ. and KEK. This experiment is of water Cerenkov type with the fiducial volume of $\sim$ 800 tons and use is made of newly developed 20" photo-tubes which enable them to have clean Cerenkov rings and good particle identification. Data taking started in July 1983 and they already have a few interesting events.

Discussion

KANE:
Are there any rare K decay experiments planned for the 12 GeV machine after its improvements?

KOBAYASHI:
Some groups have plans, but they are not yet approved.

RESEARCH POSSIBILITIES FOR HIGH ENERGY

PHYSICS AT DESY - PRESENT AND FUTURE

Volker Soergel

D E S Y

Hamburg

The high energy physics programme at DESY today is based on the two $e^+ e^-$ storage rings DORIS and PETRA. From 1990 onwards we expect the ep-collider HERA to be in operation, and the main activity will go to that machine. PETRA, which is needed as an injector for HERA, will be closed down for high energy experiments by the end of 1986, while DORIS can provide luminosity as long as there is a strong physics interest - there is no date foreseen for a stop of the elementary physics programme at DORIS. DESY has also a synchrotron radiation laboratory, HASYLAB, which uses DORIS as a light source, mainly for X-rays.

DORIS

The DORIS storage ring, completely rebuilt in 1981/82 and now called "DORIS II", can reach 2 x 2.6 GeV. It works with single bunches of about 40 mA at filling and delivers typically about 1000 nb^{-1} per day in the energy range of the Y-resonances. On Febr. 26 this year, an integrated luminosity of 1560 nb^{-1} at the Y' resonance has been achieved.

At the two interaction points, the detectors ARGUS and Crystal Ball are installed. The Crystal Ball programme concentrates on Gamma-transitions in the $b\bar{b}$ system, in continuation of the successful research programme in the $c\bar{c}$ system at SPEAR, from where the detector was transferred to DESY in spring 1982.

ARGUS is a new magnetic detector, which came into operation in autumn 1982. Its central drift chamber has wires with different inclination to allow a good determination of the coordinate along the beam, and gives a good dE/dx measurement. The lead scintillator temperature, the cooling water has 90° C. at the outlet and is used for heating the adjacent large building. In April this year, a vertex chamber with 80 μ resolution and a thin beam pipe will be installed.

DORIS will now operate the major part of 1984 on the Y(4S) resonance for the ARGUS-programme, which concentrates on b-decays, with a look also for B° B̄° mixing.

Once the high energy physics programme comes to an end, DORIS is expected to be used as a dedicated synchrotron radiation facility. At present, one third of the scheduled beam time at DORIS is main user time for synchrotron radiation. In both, the high energy shifts with typically 40 mA per beam at about 5 GeV and the dedicated shifts with typically 80-100 mA at 3.5 GeV, DORIS is a powerful X-ray source, much in demand by a large community of users.

PETRA

After the completion of the energy upgrading programme in which all available space in the four long straight sections was filled with copper cavities, PETRA can reach now 2 x 23.3 GeV. In an energy scan up to 23.15 GeV per beam, no toponium has been seen. The programme for PETRA foresees some more weeks of running close to the highest energy, and to take out later in the year some cavities as to allow for a higher luminosity at somewhat reduced energy - unless something exciting shows up at the high energy running.

High energy physics at PETRA will continue until end of 1986. In 1987 we will begin to transform PETRA into an injector for HERA for 14 GeV electrons and 40 GeV protons, and remove most of the rf-cavities now installed in order to use them for the HERA-electron ring.

At PETRA, all four detectors are presently taking data: TASSO, MARK J, JADE and CELLO. They all have improvement programmes, which are mainly concerned with installations of vertex chambers to improve momentum resolution and to measure short life times of e.g. Tau-leptons and particles with c- or b-quarks. The TASSO vertex chamber is installed since two years already; JADE will install a vertex chamber and a Z-chamber in April this year; MARK J is working on a high precision vertex chamber based on the time-expension principle, planned to be installed in January/February 1985; and CELLO which has at the moment a drift tube chamber around the beam pipe, is building a completely new central drift chamber with dE/dx capability. No other major upgrading is foreseen for the PETRA-detectors for these last 2.5 years of running.

As for the present PETRA physics, I will just shortly present a peculiar event recently observed by CELLO at a cm energy of 43.45 GeV. The event, which is published in Phys. Lett. B141(1984),p.145, shows 2 high energy muons with opposite charge, and in about back to back direction to each muon a high energy jet. The whole event is nearly planar, and practically the total machine energy (apart from 1 GeV) appears in charged particles. Here are some numbers:

The momenta of the muons are 11.0 GeV and 12.6 GeV respectively, and the charged energies in the two jets 10.2 GeV and 9.1 GeV. The invariant masses of the combinations muon - opposite jet turn out to be 20.4 GeV and 22.2 GeV.

Such an event could be produced by a 2 photon-process, but the probability to obtain the cinematical configuration observed is only of the order 10^{-3}. There is no explanation at hand for this event at this moment. We will have to wait for more data obviously. If it is new physics, we should see more of these animals, also in the other detectors.

HERA

The future project of DESY is the electron-proton-collider HERA. These are the characteristics of the planned machine:

proton energy	820 GeV
electron energy	30 GeV
expected luminosity at design energies	$6 \times 10^{-31} cm^{-2} sec^{-1}$

To achieve that, there will be 210 bunches each of protons and electrons in the machine. It is aimed to have the electrons longigudinally polarized at the intersection points with the possibility of both helicities. Electrons and positrons will be available. At the design energies the maximum momentum transfer squared will be $Q^2_{max} \approx 10^5$ GeV^2.

The HERA machine consists of two storage rings, one for electrons and one for protons, mounted in a common underground ring tunnel of 6.3 km circumference. The tunnel is adjacent to the DESY site. It will be 10-20 m below surface, which gives enough shielding to avoid any radiation problem. There are four experimental halls with intersections to allow the installation of four experiments. Only one of the halls is on the DESY site. The existing DESY machines will be used for the injection system, which for the protons will be a Linac, yet to be built, the existing DESY-synchrotron, to be transformed into a proton accelerator, and PETRA; and for the electrons and positrons the existing Linacs, a new injector synchrotron for 9 GeV electrons just under construction, and again PETRA.

A few words about the HERA physics: HERA will be a machine to study electron-quark interactions via spacelike currents, charged and neutral, at momentum transfers squared up to $Q^2 \approx 10^5$ GeV^2. Investigations, where an electron-proton collider is unique, include:

— Study of the properties of the currents up to 500 GeV mass. With polarized electrons one can search for right handed currents up to about this mass, also with heavy neutrinos (for light neutrinos, there exists in decay a good limit from a nice experiment done at TRIUMF).

— Search for flavour changing neutral currents: this relates my talk to the title of the workshop. Flavour changing neutral currents could be very neatly discovered by something strange happening to the lepton side of the event.

— Investigation of the structure of the proton and its con-stitutents to very small dimensions, 10^{-17}cm or smaller.

Other exciting research items include search for supersymme-tric particles, heavy leptons, physics at small Q^2 etc.

The topology of deep inelastic scattering events at HERA energies will be such, that the leptonic part, normally the scattered electrons or a neutrino, and the jet from the struck quark are well separated. This should make experimentation at HERA par-ticularly clean and give good sensitivity to new phenomena.

A brief remark on rates: At $Q^2 > 10^4$ GeV^2, where weak and electromagnetic interactions have the same strength, one expects in the standard model about 10^4 charged current events per year (5000 hours of operation) at the design luminosity. At $Q^2 > 3 \times 10^4$ GeV^2, one expects still about 100 events per year. These Q^2 limits will bring us really into new land, which can be explored with HERA.

The Machine

HERA will consist of two storage rings, intersecting at four places. The proton ring will be built with superconducting magnets in order to reach the required energy. The dipoles have a field of 4.5 Tesla at 820 GeV. We have developed at DESY a dipole magnet which safely reachs this field. The development was based on the Fermilab warm iron magnet, successfully used in the energy saver.

Superconducting quadrupoles for HERA, which reach the required gradients, have been developed at Saclay.

For the electron ring, where only 0.18 Tesla are required at design energy as a bending field, iron magnets will be used with a single aluminum conductor to excite the field. Prototype work is here in preparation. The rf-system for the electron ring will for the initial phase of HERA be the existing rf-system of PETRA, leaving there just enough cavities and clystrons reach the 14 GeV necessary for injection.

In the tunnel the two storage rings will be mounted on top of each other. The lower one will be the electron ring, which will be installed first.

Realization and Schedule

For the realization of HERA some important foreign contributions to the machine are required, which are expected in the form of components. Research organizations and laboratories in five foreign countries have declared their intention, to collaborate with DESY in the construction of HERA by providing various components. The decision of the project is now still pending and subject to some clarifications on governmental level concerning these foreign contributions. We expect however, to obtain the final approval of the project in the beginning of April this year. This would allow us to begin with the civil engineering work in May. If this can be achieved, the expected schedule is as follows:

tunnel and experimental halls finish by end of 1987,
electron ring completed end 1987/beginning 1988,
proton ring installation completed end of 1989,
first ep collisions middle of 1990.

To meet this schedule, we need of course appropriate funding, and we will have to solve our manpower problem, which at the moment is still quite serious.

Experimental Programme

Here is a calendar of events which is foressen to initiate the experimental programme: Letters of intent should be submitted by 30 June 1985. They will be evaluated by the Physics Research Committee, so that first recommendations can be made by Christmas 1985. Technical proposals are then expected for March 1986, and final approval for the first round of experiments is foressen for middle 1986. In this first round not more than three proposals will be approved, so that at least one intersection is kept free for a later decision.

As a final comment let me say that HERA will of course be open to physicists from all over the world.

We hope that the study of ep collisions at high energies with HERA will bring us in the next decade much exciting new insight into the physics of elementary particles.

Question by M. GOLDHABER, BNL: Did you run repeatedly at the energy of the intresting event from CELLO?

Answer: No, this event was obtained during the energy scan. We are scanning at the moment in steps of 40 MeV, and there was in fact a hint for a small peak at a somewhat higher energy, at around 44 GeV. This point was repeated with improved statistics and turned out to be just on the line of constant R. We will ertainly come back to the energy where the CELLO-event was observed.

Will there be any e^+e^- phase at HERA?

At the moment we are not planning an e^+e^- phase before HERA is completed. Such intermediate e^+e^--running would delay ep collisions by several years. We think at this moment that ep physics is the most interesting physics we will be able to do. Maybe, this is wrong, and if there are compelling arguments to do e^+e^- at HERA, a decision could be taken to do an e^+e^- programme. The machine is technically capable for that. With the warm cavities to be installed initially, we could with one or two bunch operation for each beam reach something above 30 GeV per beam. To reach higher energies, a superconducting rf-system would have to be installed. This at the moment is not foreseen in our planning nor in our budget.

STATUS OF RARE DECAY EXPERIMENTS

L.S. Littenberg

Brookhaven National Laboratory
Department of Physics
Upton, New York, 11973

I will discuss a number of current and future rare μ and K
decay experiments. We call these decays 'rare' because we know
they are not common, however, most of them have never been observed
at all. Therefore they are not rare in the sense the white
rhinoceros is, they are rare in the sense that the pink rhinoceros
is rare.

By and large the muon experiments are running now, while the
kaon experiments are still at the proposal stage. I'll give what
results there are for the muon experiments while my discussion of
the kaon work is necessarily of the 'coming attractions' variety.

The physics in both cases is mainly beyond the standard model
($SU(3) \times SU(2) \times U(1)$ with 3 generations). After the triumphs of
this model it may seem ungrateful to look for its warts, but it is
clear that it leaves many questions unanswered. Among others--why
are there multiple generations and how many are there? What is the
relationship between leptons and quarks? Why are the masses what
they are? What is the origin of CP violation? Etc.

In order to answer these questions we have to look beyond the
boundaries of the model. One of the most incisive ways to do this
is to search for processes which are forbidden or at least highly

suppressed in the model. The observation of such a process would clearly signal the presence of new physics, and make possible the elucidation of its properties.

One class of forbidden processes are those which violate lepton flavor conservation. Examples are $\mu \rightarrow e\gamma$, $\mu \rightarrow eee$, $\mu^- A \rightarrow e^- A$, $K_L^0 \rightarrow e\mu$, $K^+ \rightarrow \pi\mu e$, $D^0 \rightarrow \mu e$, $\tau \rightarrow \mu\gamma$, etc. Examples of processes which are highly suppressed by the detailed structure of the standard model are $K_L^0 \rightarrow e^+e^-$ and $K^+ \rightarrow \pi^+ \nu\bar{\nu}$. $K_L^0 \rightarrow e^+e^-$ is suppressed by some 3 or 4 orders of magnitude below $K_L^0 \rightarrow \mu^+\mu^-$ by helicity conservation (and $K_L^0 \rightarrow \mu^+\mu^-$ is already very suppressed by the GIM mechanism). The branching ratio for $K^+ \rightarrow \pi^+ \nu\bar{\nu}$ is suppressed by the GIM mechanism to $\approx 10^{-10}$.

Lepton flavor violation occurs naturally in many types of models, a number of which have been discussed at this meeting. An incomplete list of these models are:

1) Almost standard mode with $m_\nu \neq 0$.
2) Almost standard model with extra generations containing heavy neutrinos.
3) Almost standard model with flavor changing Higgs.
4) GUTs with low mass B & L violating Higgs.
5) Extended technicolor.
6) Some supersymmetric models.
7) Models containing horizontal gauge symmetries.
8) Constituent models.

Figure 1 shows how some of these mechanisms mediate lepton flavor violation. Few or no regularities obtain in all models. Some mechanisms require both quarks and leptons to participate in the LFV, others are diagonal. In some models $\mu \rightarrow eee$ occurs as an electromagnetic correction to $\mu \rightarrow e\gamma$, in others the reverse is true. In most models $K_L^0 \rightarrow \mu e$ occurs with a higher rate than $K^+ \rightarrow \pi^+ e\mu$ because of the phase spce, but in models where the mediating interaction is vector, $K_L^0 \rightarrow \mu e$ does not occur at all. Therefore, one is well advised to look for all accessible processes.

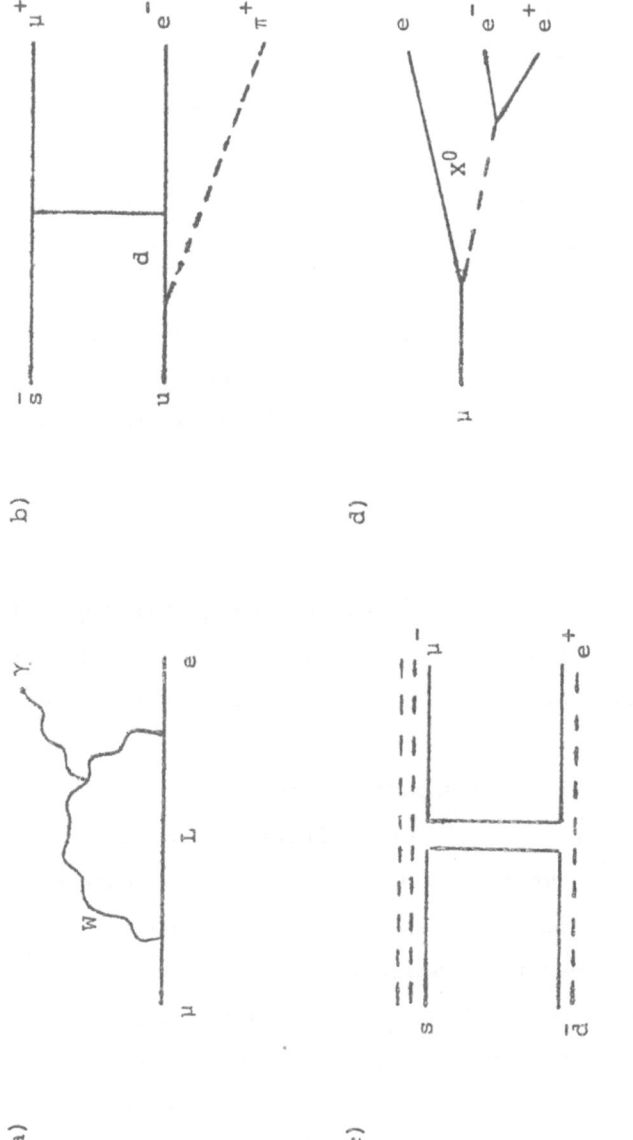

Fig. 1. Flavor violating processes. a) $\mu \to e\gamma$ via a heavy lepton, (b) $K^+ \to \pi e\mu$ via leptoquark exchange, c) $K^0_L \to \mu e$ via constituent exchange, d) $\mu \to eee$ via horizontal gauge boson exchange.

In general the new interaction responsible for LFV is mediated by some heavy object. Typically the rate for the process is $\propto$ $g^2 g'^2 M_H^{-4}$ where M_H is the heavy particle mass and the g's are its coupling constants at the fermion vertices. Table I[1] gives a comparison of the mass probed by different processes in various models for a fixed branching ratio sensitivity of 10^{-10}. The most striking information in the table is the mass scale in some of the models - tens or hundreds of TeV! This is higher than can be accessed even by the proposed SSC. Thus it is fair to say that these are really frontier experiments. The Table also illustrates the variations in relative sensitivity of the different processes as one goes from model to model (or even changes the parameters of a given model).

There have been relatively few rare kaon decay experiments in recent years. New data on lepton flavor violation has been virtually non-existent for at least ten years, and even before that it tended to be a by-product of other, more fashionable physics. Thus the new dedicated experiments anticipate large improvements (100-10000) on the previous branching ratio upper limits of $\approx 10^{-8}$.

By constrast the muon experimentors have been much more constant in their pursuit of lepton flavor violation, and they have pushed the limits down to the $\approx 10^{-10}$ level. The current round of experiments aim to improve these limits by 10-100 fold. I will report on three experiments which are in the process of taking data. They are quite ambitious undertakings aimed at reducing branching ratio upper limits currently around 10^{-9}-10^{-10} down to sensitivities of 10^{-11}-10^{-12}. Although not physically large by modern high energy physics standards (≈ 1 m^3) these are no longer table top experiments. Technologically they are quite sophisticated, employing in one case a large array of sodium iodide crystals, in another a TPC, and in all three, state of the art electronics. Although many of the techniques are familiar to high

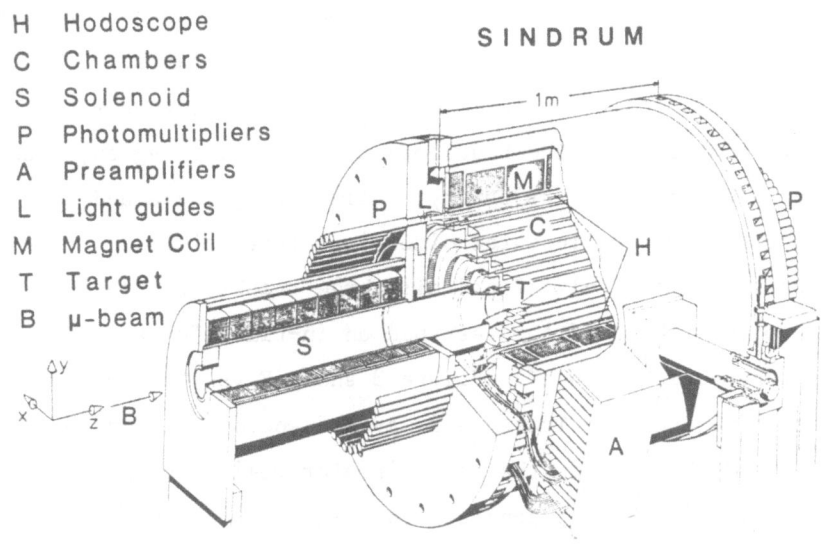

H Hodoscope
C Chambers
S Solenoid
P Photomultipliers
A Preamplifiers
L Light guides
M Magnet Coil
T Target
B μ-beam

Fig. 2.   SINDRUM apparatus

energy experimentalists there are still enough characteristically low energy problems to make this field a separate speciality.

The first experiment I'll discuss is the SINDRUM[2] $\mu^+ \rightarrow e^+e^-e^+$ experiment. A schematic of the apparatus is given in Figure 2. A 'surface' beam[3] of $\approx 10^7$ 28 MeV/c muon/sec impinges on a hollow double-cone target. About a quarter of these stop in the target and charged particles from their decays are analyzed in a solenoidal PWC spectrometer (B = 4kG). There are four low-mass cylindrical chambers with resolutions $\sigma_{r-\phi}$ = 600 μ, σ_z = 300 μ. A triggering hodoscope outside of the chambers required 3 clusters ($p_{min}^{TRIG} \approx$ 20 MeV/c) while on-line use of the chambers required one negative and two positive tracks, with a total p_T < 25 MeV/c.

The main backgrounds to $\mu \rightarrow 3e$ are internally or externally converted radiative μ decay $\mu \rightarrow e\nu\bar{\nu}\gamma$, and normal μ decay in
$\llcorner\rightarrow e^+e^-$

random coincidence with e^+e^- pairs from radiative decay.

About 250k events were written to tape. The on-line cuts were repeated with one-wire precision and it was found that most 'events' seemed to be two superimposed $\mu^+ \to e^+ \nu \bar{\nu}$ events one of which spiraled back from the hodoscope to the target region, thus generating a spurious e^- trajectory. Eliminating these and demanding a good vertex in three dimensions left only 52 candidates. Of these only 17 had all three tracks in time ($\sigma_t \approx 350$ psec). Most of the rest had an in-time e^+e^- pair randomly superposed on an e^+. Figure 3 shows E vs $\left| \Sigma \vec{p} \right|$ for in and out of time events. True $\mu \to 3e$ events would have $\left| \Sigma \vec{p} \right| = 0$ and $E = m_\mu$ and thus lie in the semi-circular region shown in the plots. The in-time events populate the region expected for $\mu^+ \to e^+ \nu_e \bar{\nu}_\mu e^+ e^-$ decays. In fact their observed rate and kinematic distribution agree with that predicted for such decays. Since no event consistent with $\mu \to 3e$ was observed, a 90% confidence level upper limit of $\Gamma(\mu \to e^+ e^- e^+)/\Gamma(\mu^+ \to e^+ \nu_e \bar{\nu}_\mu)$ $< 1.6 \times 10^{-10}$ was extracted. This is to be compared with the previous upper limit[4] of 1.9×10^{-9}. The present data represent only a day or two of running. The ultimate goal of the experiment is a sensitivity of $\approx 10^{-12}$. This certainly seems possible from the statistical point of view. If the reader is tempted to judge from Figure 3(b) whether this will also be true from the point of view of background rejection, keep in mind that the event represented by an open circle is the only survivor of a second, more stringent set of cuts.

In contrast to the SINDRUM apparatus, the Crystal Box experiment[5] at LAMPF is non-magnetic, seeking to exploit the good energy resolution of NaI. In this way one can gain a factor of $\approx$ 2.5 in acceptance for $\mu \to eee$ (8% to 20%), and avoid the complications of a magnetic analysis, at the cost of about a factor 2 in electron energy resolution. In addition, of course, the Crystal Box is sensitive to γ's, and can search for the decays $\mu \to e\gamma$ and $\mu \to e\gamma\gamma$.

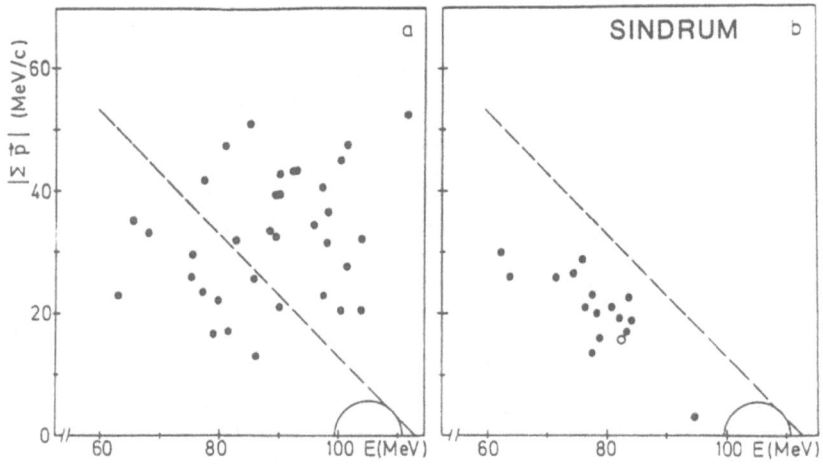

Fig. 3. E vs $|\Sigma \vec{p}|$ for SINDRUM $\mu \rightarrow 3e$ candidates
a) out-of-time events; b) in-time events

Figure 4 shows the apparatus. The Box is composed of 360 6.3
× 6.3 × 30.5 cm and 36 6.3 × 6.3 × 70 cm crystals which subtend
about 60% of 4π for $\mu \rightarrow e\gamma$ events. The resolution of the NaI is
5.8% FWHM at 50 MeV. Charged particles are tracked in a
cylindrical drift chamber ($\sigma \approx 150$ μ). The experiment seeks to
probe the three reactions $\mu \rightarrow e^+e^-e^+$, $\mu \rightarrow e^+\gamma$ and $\mu \rightarrow e^+\gamma\gamma$ to the
10^{-11} level. The present limits on the latter two reactions are
1.9×10^{-9} and 8×10^{-9} respectively[6,7].

At the sensitivities being proposed, the poor duty cycle of
LAMPF ($\approx 6\%$) is becoming a serious impediment. In the Crystal Box
experiment accidentals limit the incident beam rates to 2-5 ×
10^5/sec (average).

The main background to $\mu \rightarrow e\gamma$ stems from random coincidences
between e^+ from $\mu \rightarrow e\nu\bar{\nu}$ and γ from $\mu^+ \rightarrow e^+\nu\bar{\nu}\gamma$. These must be
discriminated against via energy, timing, colinearity and target
position consistency. A preliminary result is expected within a
few months.

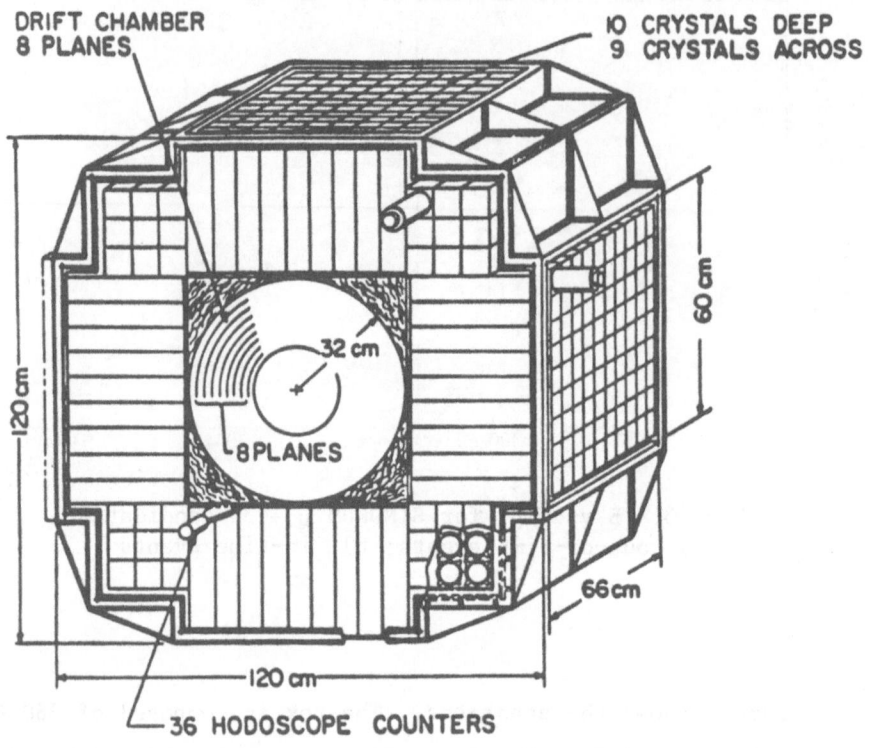

DRIFT CHAMBER
8 PLANES

IO CRYSTALS DEEP
9 CRYSTALS ACROSS

32 cm

8 PLANES

120 cm

60 cm

66 cm

120 cm

36 HODOSCOPE COUNTERS

Fig. 4. The Crystal Box

Other processes that may be studied by the Crystal Box are $\pi^0 \to 3\gamma$, $\pi^0 \to \nu\nu'$, $\pi^0 \to e^+e^-\gamma$, $\pi^+ \to e^+\nu_e\gamma$, and $\pi^0 \to e^+e^-$. There is an approved upgrade[8] of this experiment in which the NaI is reconfigured into two walls, and a magnetic spectrometer is added. In this way a sensitivity of $< 10^{-12}$ can be reached for $\mu \to e^+\gamma$.

The last member of this trilogy is the TRIUMF $\mu - e$ conversion experiment.[9] I remind you that in this case the physics is a little different, basically one has $\mu^- + q \to e^- + q$ wherein q = u or d. Thus unlike μ decay, both leptons and quarks are involved, and unlike K decay, no strangeness change is required. As a result in many models, this process is predicted to occur at the highest branching ratio of any in which lepton flavor is violated. The

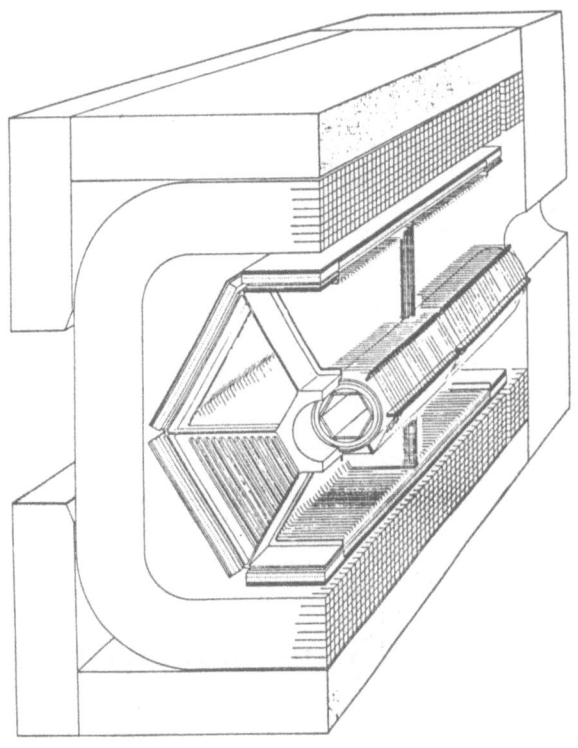

Fig. 5. The TRIUMF TPC

term "branching ratio" for μ − e conversion, means the rate
relative to that of normal μ capture, i.e., $\mu^- + A \rightarrow \nu_\mu + A'$ in
the same material.

The signature for this process is the appearance, in place of
a stopped muon, of an unaccompanied electron of unique energy
m_μ − B where B is a small binding energy. The TRIUMF apparatus
is shown in Figure 5. It consists of a small TPC[10] augmented by
trigger hodoscopes and MWPC's set in a 1T box magnet. The TPC is
divided into 6 segments on each end. One readout plane is shown in
Fig. 6. Each segment has 12 anode wires which measure the Z
position of the tracks to ≈ 1 mm via time-of-flight. These are

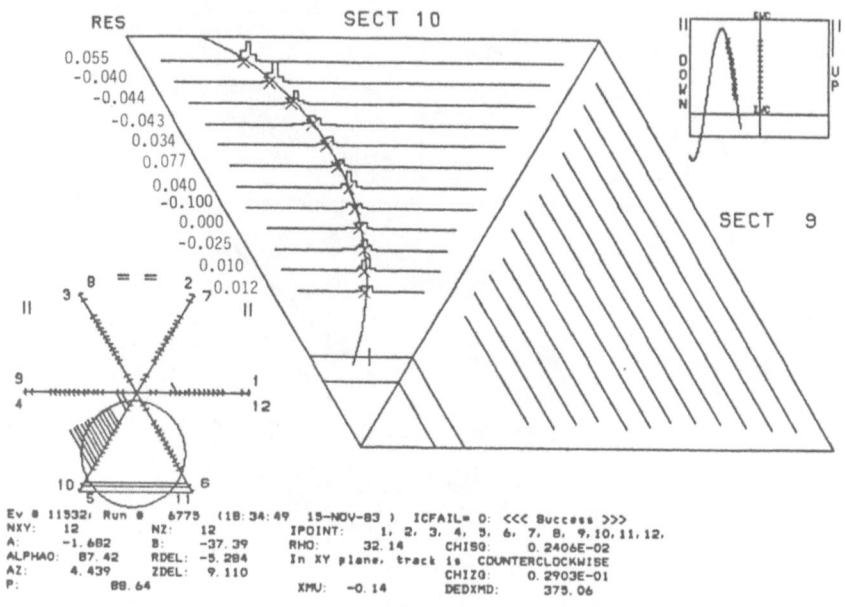

Fig. 6. TPC readout of one track

backed up by 6 mm cathode pads which measure the transverse postion to $\gtrsim 180\ \mu$. This results in a $\Delta p/p$ of $\approx .02$ for 100 MeV/c tracks. The 'cloud' beam provides $10^6\ \mu/\text{sec}$ at 77 MeV/c ($\Delta p/p \approx 10\%$, spot size ≈ 10 cm).

An r.f. separator reduces the π contamination in the beam to 10^{-3}, which is essential for reducing background due to radiative π capture. The target which consists of titanium wool, stops some 80% of the beam muons and of these about 85% are captured before they decay. The minority that decay in orbit constitute the most important background to $\mu \to e$ conversion. This may seem unlikely since electrons from free μ decay never have E > 53 MeV, and boosts given by atomic orbital velocities are generally small.[11] Yet when sensitivities like 10^{-12} are sought, far flung infinitesimally small tails must sometimes be confronted, and in fact, such tails are present in this case. This makes energy resolution crucial.

518

Other possible backgrounds are radiative μ capture wherein the γ converts to a very asymmetric pair, radiative pion capture, and cosmic rays. The TRIUMF experiment has been running for some time, and the experimentors are now confident that the proposed sensitivity of $\approx 10^{-12}$ will be reached. This is to be compared with the previous limit[12] of 7×10^{-11}. A preliminary result is expected in a few weeks.

Kaon decays

It's instructive to compare kaon with muons from the point of view of performing rare decay experiments.

particle	τ	m	possible decay products
μ^+	2197 nsec	105.6 MeV/c^2	e,γ,ν
K^+	12.4 nsec	493.7 MeV/c^2	above + μ + π
K^0	51.8 nsec	497.8 MeV/c^2	above + K^+

The long lifetime of the μ encourages low energy or stopping beams. The fact that all detectable products are electromagnetic can be exploited to simplify the apparatus as in the case of the Crystal Box. Working at low energy and high rates entails a characteristic set of experimental problems.

By contrast the K^+ lifetime is rather short and it is practical to work even at very high energies, although this has never been done. The availability of intense medium energy beams, the possibility of separated beams and the range of applicability of certain effective particle identification techniques (e.g., gas Cerenkovs) has made 3-10 GeV/c a popular interval. It's also possible to stop large numbers of K^+'s and a number of experiments have used this technique. One can get relatively pure K^+ beams (3π:1K or better) in this way, and the stopping geometry facilitates large acceptance and full veto coverage. There are also rather effective characteristically low energy particle i.d. techniques (e.g., observing the π-μ-e decay chain in stopped π^+) which lend themselves to this regime. On the other hand, there are

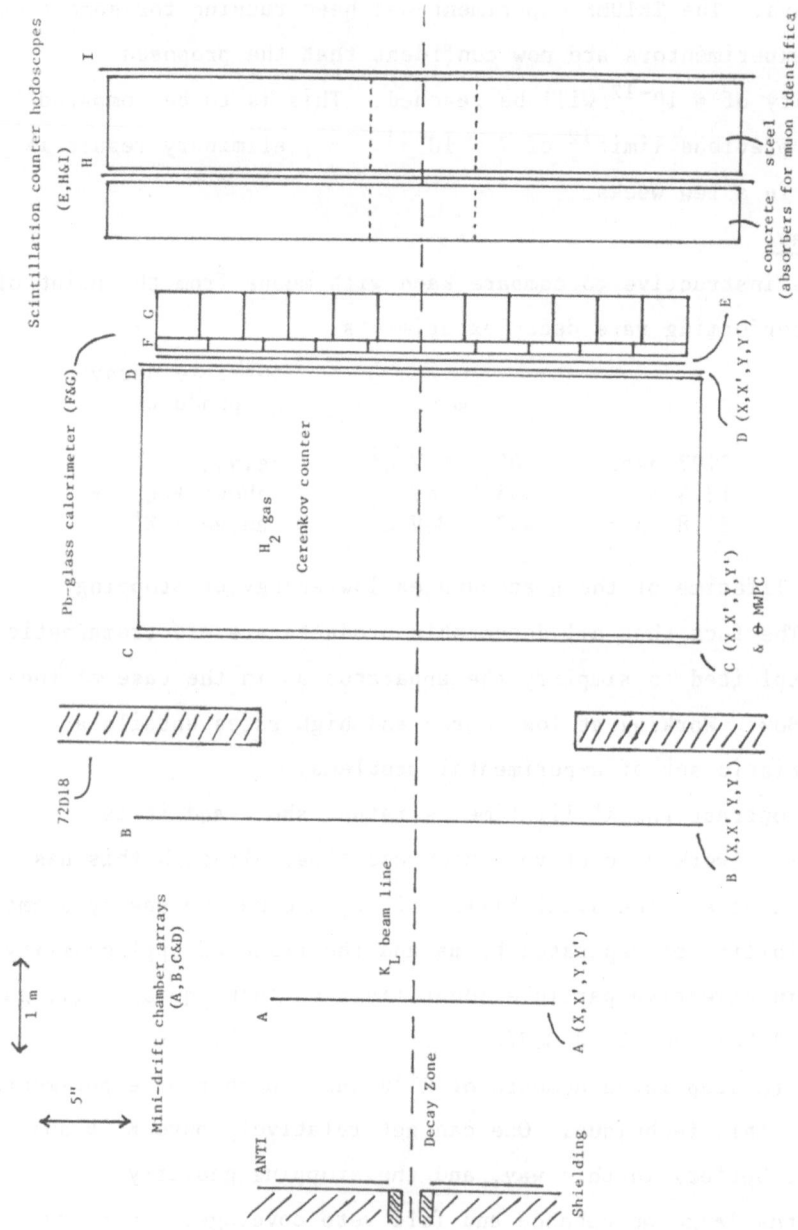

Fig. 7. The Yale-BNL $K_L^0 \rightarrow \mu e$ spectrometer (E780)

also a number of characteristic problems which have to be faced --
for example daughter particle interactions in the stopping target
can serve to compromise resolution and particle identification.

For K^0's there's no stopping option. Rare decay experiments
have been performed at just about every other available energy
however. Attention is currently focussed on intermediate energy
beams. Typically one has a broad bend beam peaked near $\approx$ 8 GeV/c.
To get the highest possible flux one works at 0° production angle
and thus must accept a high contamination of neutrons ($\geq$ 20 n/K).

$K^0_L \rightarrow \mu e$

This is a very appealing process from an experimental point of
view. Two oppositely charged leptons emanate from a common vertex
in a neutral beam. Their maximum p_T is higher than that of any K
decay yet observed. There are no uncharged accompanying particles
so that the effective mass of the e and the μ equals M_k. The
vector sum of the two momenta points back toward the K^0 production
target.

The primary background is the copious decay mode $K^0_L \rightarrow \pi e \nu$.
If the π is misidentified as a μ the event is topologically
equivalent to $K^0_L \rightarrow \mu e$. If, in addition, the ν carries off very
little energy, the event can begin to approach $K^0_L \rightarrow \mu e$
kinematically as well. Since π's decay primarily to μ's,
downstream μ identification devices are of limited usefulness.
Much more important is good spatial and momentum resolution, both
to impose the K $\rightarrow \mu e$ kinematics and to detect possible $\pi \rightarrow \mu \nu$
decays via kinks or other indications. A modern realization of the
classic kaon spectrometer used in this work is shown in Figure 7.
The K^0_L are allowed to decay in vacuum and the charged decay
products are bent parallel to the beam by a dipole magnet. This
permits a relatively large acceptance ($\approx$ 10% in the pictured
spectrometer) and simplifies the trigger. The particle
trajectories are measured by high rate (mini) drift chambers and

the electrons and muons identified respectively by a segmented H_2 Cerenkov counter plus lead glass hodoscope and by a sequence of range-measuring hodoscopes. The electron identification system, which should reject π's or μ's by $> 10^5$:1 is probably overkill for the K $\rightarrow$ μe mode, but it is useful for possible other physics. The range stack can reject pion punch-throughs to a level at which the background is completely dominated by $\pi \rightarrow \mu\nu$ decay. The limiting background is given by the rare K $\rightarrow$ $\pi e\nu$ decays in which the π decays in the middle of the magnet (so the event vertex reconstruction is undisturbed) with its decay plane aligned with the bending plane (so that there is no apparent kink), and with the μ kicked against the sense of the bend (so that the apparent momentum is increased). In addition the neutrino from the original K decay must take off very little cm energy, so that the event initially is kinematically near enough to the K $\rightarrow$ μe for the π momentum mis-measurement to close the gap. One's ability to reject such catastrophic events rises as some power > 2 of the improvement in spatial resolution of the chambers, since the background is distributed over the space of kinematic variables, whereas the target region for true events can be made smaller and smaller. Multiple Coulomb scattering in the chambers finally limits this improvement. In a spectrometer of the dimensions shown here, the chamber resolution corresponding to this limit is about 200μ, which is more or less routinely attainable in modern drift chambers. The residual background level is then reduced to $\approx 10^{-11}$, making practical an experimental sensitivity well beyond 10^{-10}. This is in fact the goal of the Yale-BNL experiment[13] pictured in Figure 7. To achieve this they will work in a 0° neutral beam containing some 2×10^7 K^0_L/pulse (and ≈ 30 times more neutrons). Combining a K^0_L decay probability of $\approx .012$, and a $K^0_L \rightarrow \mu$e acceptance of $\approx .06$, one finds a statistical sensitivity of 0.7×10^{-10}/evt. If the observed chamber rates allow it, the beam will be increased in order to push the sensitivity toward 10^{-11}. This is to be compared with the present limit[14] of $\approx 10^{-8}$. This experiment will also probe the decay $K^0_L \rightarrow e^+e^-$ to the same level of sensitivity. Since the standard model expectation for this branching ratio is a few $\times 10^{-12}$, there is a large window, extending from the present upper limit[14] of $\approx 10^{-8}$ down to the ultimate sensitivity of this experiment, in which to look for new physics. Since the large relative suppression of $K^0_L \rightarrow e^+e^-/K^0_L \rightarrow \mu^+\mu^-$ in the standard model is due to helicity constraints, this decay mode is particularly sensitive to scalar interactions.

The above experiment is approved and in the process of being built. A second K^0_L experiment has recently been proposed.[15] It is noteworthy in that its proponents intend to request a primary beam intensity equal to the entire slow extracted beam of the AGS as it is presently operated (10^{13} protons/pulse). It will be used to create a 0° neutral beam. Like the beam of the previous experiment its momentum will peak at about 8 GeV/c and contain a useful K flux from ≈ 4 to 20 GeV/c. In addition the beam aperture will be opened to ≈ 100 µsr. This combination of primary intensity and beam aperture will result in the truly impressive flux of 6 $\times$ 10^8 K^0_L/pulse. Accompanying them will be the truly daunting flux of $\approx 2 \times 10^{10}$ neutrons/pulse. The apparatus is shown schematically in Figure 8. About 2.2% of the K^0_L decay in the 6 m vacuum decay region. This is followed by low mass drift chamber spectrometer with two magnets. The acceptance of this device for $K^0_L \rightarrow \mu e$ is $\approx 6\%$. Electrons are identified by a Ne-Ar Cerenkov counter and by a lead glass wall whereas µ's are identified in a hadron filter. This combination of decays/pulse and acceptance allows a statistical sensitivity of $\approx 10^{-12}$ (6 $\times$ 10^8 K^0_L/pulse $\times$ .022 decays/pulse $\times$ 1.2 $\times$ 10^6 pulses $\times$ .06 acceptance/decay = 1.5 $\times$ 10^{12}). The anticipated chamber spatial resolution of 100 µ will result in a K mass resolution of ≈ 1.6 MeV/c^2. Although the resolutions here are comparable to those of the Yale-BNL experiment, the opportunity for a second momentum measurement is

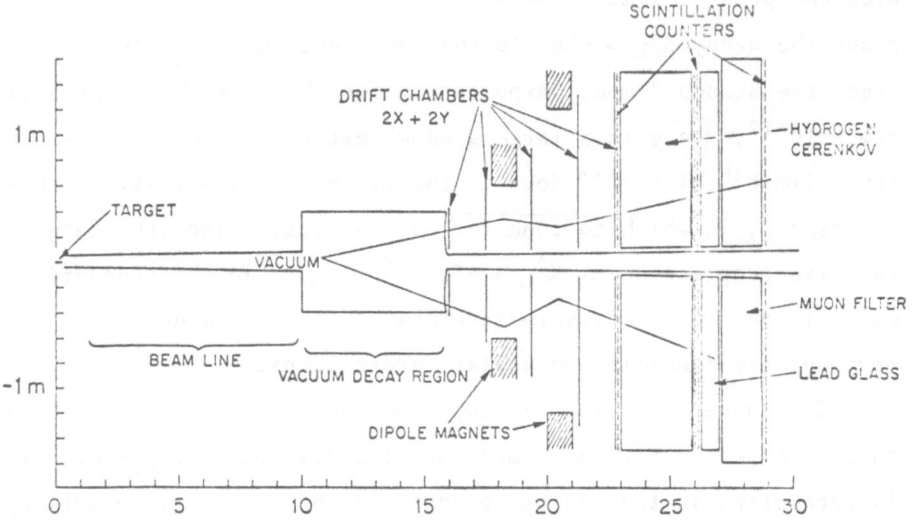

Fig. 8.   LANL, Penn, Princeton, Stanford, Temple, UCLA, et al.
K^0_L spectrometer (E791)

asserted to allow the $K^0_L \to \pi e \nu$ background to be reduced by a
further factor of 10 or more. This is because μ's stemming from
the 'catastrophic' decay discussed above, emerge from the first
section of the spectrometer with a falsely measured momentum.
Thus, if no unexpected background processes crop up, the experiment
can hope to reach its target sensitivity of 10^{-12}.

The problems connected with such an audacious undertaking
should not be underestimated. These are largely consequences of
the extremely high rates. The chamber wire spacing is 1 cm with a
maximum drift time of 100 nsec. This implies that there will
generally be products of at least one extra K^0_L decay in the
chambers in every accepted event. In addition there are the rates
due to the 10^{10} neutrons. These are minimized by transporting the
neutrons through the center of the apparatus in vacuum, but are
still liable to be formidable. Such calculations as have been done

indicate that the rates will be tolerable but in an experiment of
this extreme sensitivity one must worry about higher order effects
such as rate-induced degradation of various resolutions. Achieving
a manageable trigger rate is no mean task under these conditions.
Even after a 10:1 on-line rejection of the principal source of
triggers ($K^0_L \to \pi e \nu$) the fast trigger rate will be

$$\hookrightarrow \mu\nu$$

~ 10^4/pulse. This will have to be beaten down another hundred
fold via on-line reconstruction of the decay kinematics.

Although the difficulties are great, so are the possible
rewards. In addition to improving the sensitivity on $K^0_L \to \mu e$ to
10^{-12}, a very interesting level, the decay $K^0_L \to e^+e^-$ should be
observable at the standard model rate. In addition the
experimentors intend to use their lead glass array to detect γ's,
and thus search for $K^0_L \to \pi^0 e \mu$, $\gamma e \mu$, $e^+e^-\gamma$, $\mu^+\mu^-\gamma$, $\pi^0 e^+e^-$ etc.
This last process which has a standard model prediction[16] of
$\approx 10^{-11}$ is particularly interesting from the point of view of CP
violation. Finally, there is the intention to install a μ
polarimeter and measure the μ^+ polarization in $K^0_L \to \mu^+\mu^-$ of
which there should be 10^4 examples. It is anticipated that the
polarization can be measured to $\approx 14\%$. A signal at this level is
equivalent to the detection of a direct CP violating[17] $K^0_L \to \mu^+\mu^-$
decay at the 7×10^{-11} level, a measurement which is otherwise
impossible because of the 'large' CP-conserving $K^0_L \to \mu^+\mu^-$
signal.

I'll next discuss the BNL-Washington-Yale $K^+ \to \pi^+ e^- \mu^+$
experiment.[18] For the same strength interaction $K^+ \to \pi^+ e \mu$ is some
50 or 100 times less sensitive than $K^0_L \to \mu e$ (partly from
phase space and partly due to the higher K^+ total decay rate).
However, if the lepton flavor violating interaction has no axial
vector or pseudoscalar part, the latter reaction is forbidden.
Thus both processes must be sought. Another reason for studying
$K^+ \to \pi^+ e^\mp \mu^\pm$ is that one can test the strength of the LFV inter-

action with respect to generation number, an important concept in some models.[19]

From an experimental point of view $K^+ \to \pi^+ e^- \mu^+$ offers a number of advantages. 1) K^+ beams are more tractable than K^0_L, they can be focussed, bent, etc. In principle, one can make separated K^+ beams (although in practice most often one can't spare the intensity). 2) Because of the relative charged/neutral decay rate one can get four times more K^+ to decay in a given volume than K^0_L. 3) The K^+ signature beats the K^0_L, four charged particles to two. 4) There's a 4c fit rather than a 3c fit (this is particularly effective if one can measure the beam momentum). 5) Trigger rate problems for $K^+ \to \pi^+ e^- \mu^+$ are much less than those of $K^0_L \to \mu e$. This can be seen from the following comparison

$$\text{leading } K^0 \text{ trigger/decay} = b(K \to \pi e \nu) \times (\pi \to \mu \text{ on line}$$
$$\text{misidentification)}$$
$$= .39 \times .01 = 4 \times 10^{-3}$$

$$K^+ \text{ trigger/decay} = b(\pi^+ \pi^0) \times b(\pi^0 \to \gamma e^+ e^-) \times e \to \mu \text{ on line}$$
$$\text{misidentification)}$$
$$= .22 \times .012 \times .01 = 2.5 \times 10^{-5}.$$

Moreover, this assumes a pessimistic $e \to \mu$ misidentification. 6) One can gain large factors in background rejection through improvements in particle identification. To be fair I should mention a couple of experimental disadvantages: 1) One has to eat the charged beam (usually mostly π^+ and p) and 2) Particle misidentification can hurt one kinematically in a way which is not possible in $K^0_L \to \mu e$. In the latter case if a π from $K^0_L \to \pi e \nu$ is misidentified as μ the reconstructed 2-body effective mass will be reduced by the misattribution of the μ mass to the π and the fit to the K^0_L will be worsened. In the K^+ case there are backgrounds in which this works the opposite way, such as $K^+ \to \pi^+ \pi^0 \to \pi^+ e^+ e^- \gamma$ wherein the e^+ is misidentified as a μ^+. This can raise the reconstructed mass, making up for energy carried off by the photon.

The apparatus of E777 is shown in Figure 9. A 6 GeV beam containing 2×10^7 K$^+$, $2.5 \times 10^8 \pi^+$, and 1.2×10^8 protons/pulse with $\Delta p/p \approx .1$ impinges on a .11 Λ_{K^+} decay tank. Decays are detected in a two-magnet PWC spectrometer (deadened in the beam region). The first magnet kicks the decay products out of the beam and separates them by sign. The second magnet is used for momentum analysis. This arrangement results in a 9% acceptance for K$^+ \to \pi^+ e^- \mu^+$, and, with a 2 mm PWC pitch, a mass resolution of $\approx$ 1%. The statistical sensitivity is then [2 $\times$ 10^7 K$^+$/pulse $\times$ 0.11 decays/K$^+$ $\times$ 10^6 pulses $\times$ .09 acceptances/decay]$^{-1}$ = 0.5 $\times$ 10^{-11}.

Each side of the apparatus is specialized to identify particles of its particular sign. On the negative side two H$_2$ Cerenkov counters in series yield a π or μ rejection of $>$ 10^5:1. A shower counter provides a further factor of $>$ 10^2 or so. On the positive side two gas Cerenkovs provide a rejection of e$^+$ by $>$ 10^4:1. A somewhat less elaborate shower counter provides a further

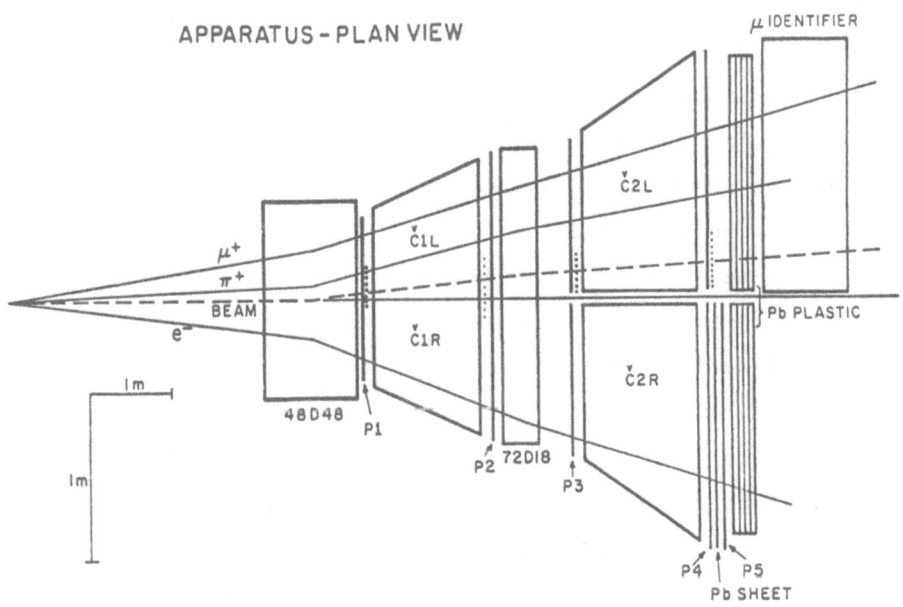

Fig. 9. The E777 K$^+ \to \pi^+ e^- \mu^+$ apparatus

large factor in e^+ rejection. Finally a 7-layer steel-scintillator muon filter serves to reject π^+ punch-throughs by a factor 100.

Unlike the case of $K^0_L \to \mu e$, the backgrounds to $K^+ \to \pi^+ e^- \mu^+$ are quite diverse. I've already mentioned $K^+ \to \pi^+ \pi^0$ followed by Dalitz decay of the π^0 and subsequent e^+ misidentification as a μ^+. The decay $K^+ \to \pi^+ \pi^+ \pi^-$ can produce false $K^+ \to \pi^+ e^- \mu^+$ candidates in a number of different ways involving various combinations of π decays and misidentifications. There are also potential backgrounds stemming from less common K decays. Table 2, taken from the E777 proposal, lists the most important backgrounds and the (calculated) effects of various background factors. The summed number of residual background events is $< 10^{-11}$, permitting a measurement at this level of sensitivity.

A measurement at the 10^{-11} level represents a 500-fold improvement on the present upper limit[20] for the branching ratio for $K^+ \to \pi^+ e^- \mu^+$. In addition the experiment should be able to collect several thousand examples of the rare decay $K^+ \to \pi^+ e^+ e^-$ for which the present world supply is some 30 events.[21] This process is interesting both because of the light it sheds on higher order electroweak processes in the standard model and because of the opportunity it affords to search for decidedly non-standard states decaying into $e^+ e^-$. Another possible by-product of this experiment is a clean high-statistics measurement of the decay $\pi^0 \to e^+ e^-$. $K^+ \to \pi^+ \pi^0$ provides a 'tagged' π^0 for this purpose. This process, which has been measured in two experiments[22], appears at a rate significantly in excess of the standard model prediction.[23]

The third kaon decay experiment I'll discuss is $K^+ \to \pi^+ + X^0$ where X^0 denotes an undetectible neutral particle or particles. In the case that the X^0 is a neutrino-antineutrino pair, this process has a significant standard model rate. However, this branching ratio is small enough, $O(10^{-10})$, in relation to the present upper limit[24] of 1.4×10^{-7}, to allow a very large window for the observation of possible new physics. The lack of significant long distance contributions to $K^+ \to \pi^+ \nu \bar{\nu}$ makes this process, unlike $K^0_L \to \mu^+ \mu^-$ or $K^0 \leftrightarrow \bar{K}^0$ mixing, a clean test of higher order corrections in the standard model. Figure 10 shows the diagrams involved. The prediction depends only on K-M matrix parameters and on quark masses, the principle unknowns at the moment being the CP violating phase δ, and the top quark mass. Recent determinations of the b-quark lifetime[25] have narrowed the range of prediction considerably, as shown by Fig. 11 which is taken from a recent paper of Gilman and Hagelin.[26] For a top quark mass of 100 GeV/c, the branching ratio is predicted to be between 6×10^{-11} and 6×10^{-10}, assuming 3 generations of light neutrinos. Future progress in determining the parameters of the 6 quark model can only help in making the prediction more precise. On the other hand, if this progress is stalled a measurement of the $K^+ \to \pi^+ \nu \bar{\nu}$ rate can provide a valuable constraint on these parameters, as well as a check on the overall consistency of the standard model.

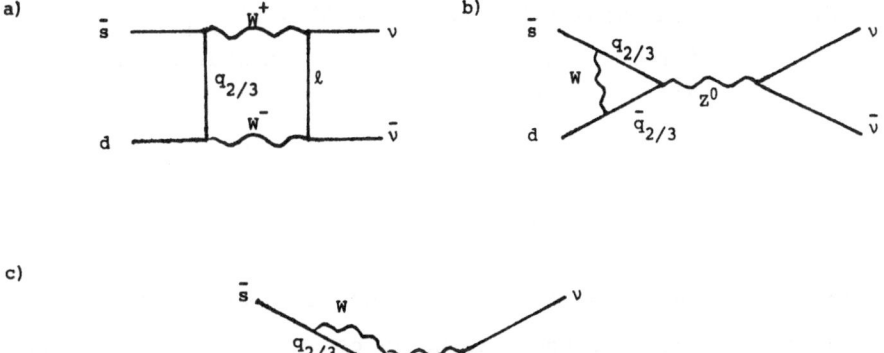

Fig. 10. Standard model contributions to $K^+ \to \pi^+ \nu \bar{\nu}$

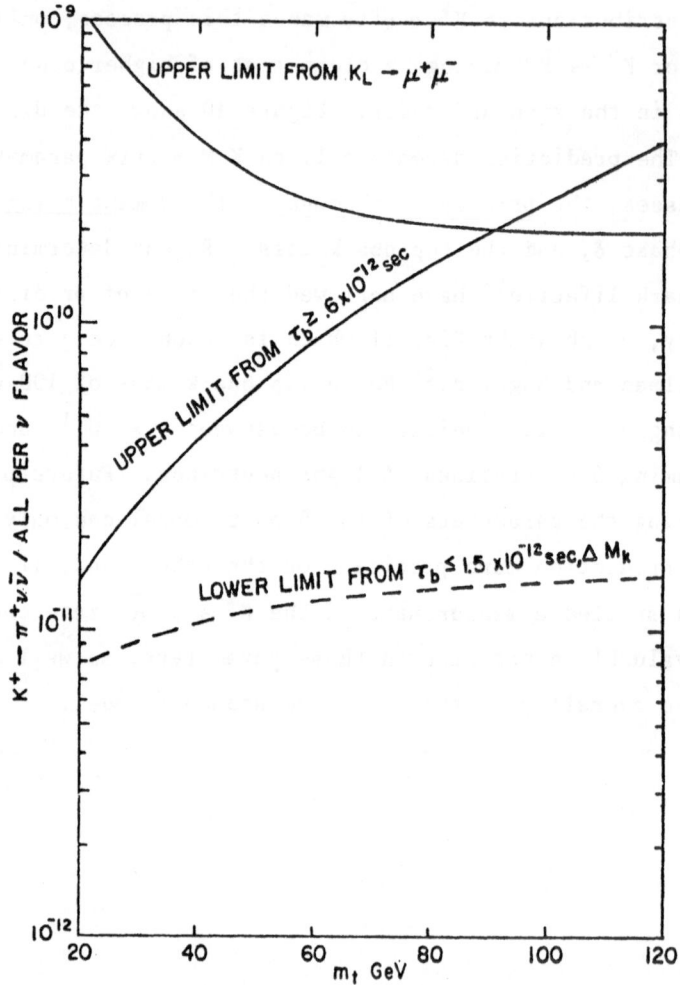

Fig. 11.   Branching ratio limits on $K^+ \to \pi^+ \nu_i \bar{\nu}_i$ vs M_t.
(from Gilman and Hagelin[26]). Measured rate
will be 3 times this.

If on the other hand the measurement falls significantly outside the 6 quark model bounds, it would constitute a clear signal
of new physics. The possibilities for this are numerous and
diverse. As discussed by Ellis and Hagelin[27] and Kane and
Shrock[28] a fourth generation of quarks and leptons could either

subtract or add to the branching ratio. It's possible that lepton
flavor-violating interactions could contribute, giving decays like
$K^+ \rightarrow \pi^+ \nu_e \bar{\nu}_\mu$ etc., although clearly this process is not the best
probe of such physics. There are a number of possible supersym-
metric contributions, some of which are shown in Figure 12. Again
there are possibilities for either constructive or destructive
interference depending on the various masses and mixings. One
interesting possibility is the tree level contribution[29] shown in
Figure 12(h) which could give a contribution almost as large as the
present experimental upper limit. Finally, there is also the
possibility of contributions as yet undreamt of in our philosophy.

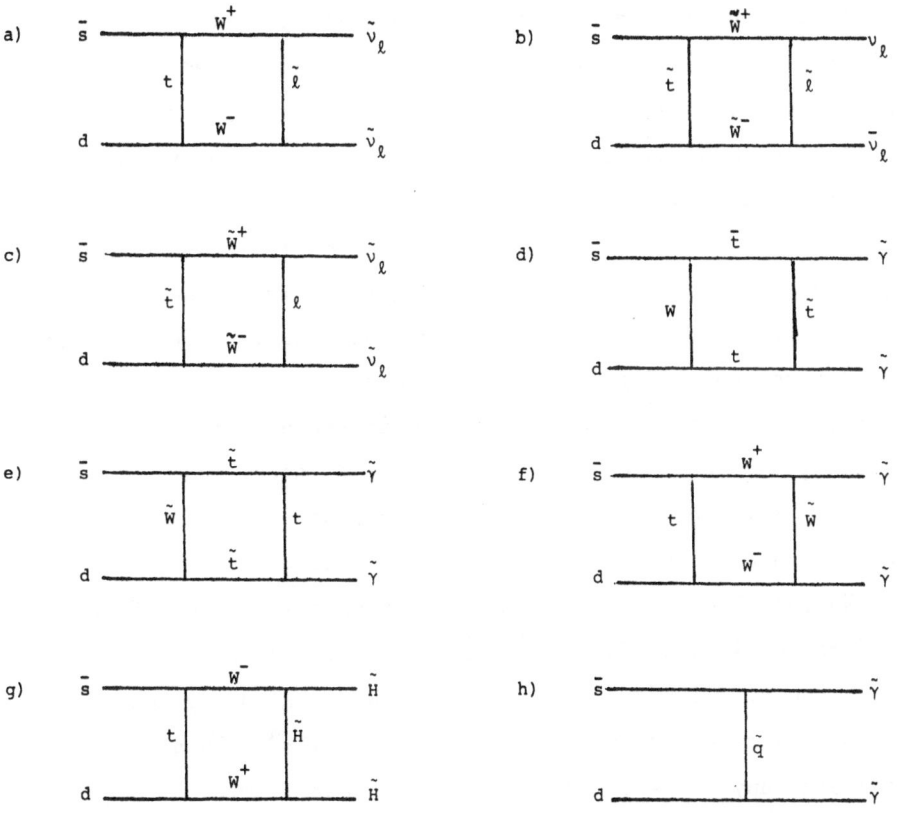

Fig. 12. A selection of possible supersymmetric contributions
to $K^+ \rightarrow \pi^+ X^0$.

The decay $K^+ \to \pi^+ X^0$ when X^0 is a single unobserved particle is also extremely interesting. Although the conventional axion is dead, Wilczek has introduced a successor called the familon.[30] This is the Goldstone boson of a hypothesized family symmetry breaking whose scale F is in the 10^{12} GeV region. The predicted branching ratio is $b(K^+ \to \pi^+ f) = 2.75 \times 10^{13}$ GeV^2/F^2. Cosomological constraints limit F to between 10^9 and 10^{12} GeV, and thus $b(K^+ \to \pi^+ f^0)$ to 2.75×10^{-5} to 2.75×10^{-11}. Other possibilities also exist for X^0.[31]

Although the motivation to search for $K \to \pi^+ + X^0$ is strong, the experimental difficulties are considerable, particularly in the case when X^0 is a two particle system. Unlike the other K decays discussed, there is no strong kinematic signature. It's necessary to veto any charged particles or γ's accompanying the final state pion, and excellent particle identification is needed to establish that it is indeed a pion. Yet the target branching ratio, $\approx 10^{-10}$, is not so much higher than that of the lepton flavor violation experiments discussed above.

All searches for these decays past and present have used a stopping K^+ beam. One is then working directly in the center of mass system and can use kinematics at least to veto certain background reactions. Figure 13 illustrates the kinematical situation. It gives the distribution in range[32] in scintillator of the charged products of various relevant K^+ decay modes. The π^+ from $K^+ \to \pi^+ \nu\bar{\nu}$ has a typical vector spectrum peaking near its maximum range. The two largest K^+ branching ratios, $K^+ \to \mu^+ \nu$ (63.5%) and $K^+ \to \pi^+ \pi^0$ (21.2%) appear as spikes at about 58 and 32 g/cm^2 respectively. The spikes are shown broadened by a range resolution of ≈ 1 g/cm^2. Note that no known K^+ decay mode produces a π^+ with a range larger than that of $K^+ \to \pi^+ \pi^0$. Thus, in principle, any K^+ decay producing a daughter π^+ with a range exceeding that of $\pi^+ \pi^0$ must[33] be a $K^+ \to \pi^+ \nu\bar{\nu}$. This has led most previous experiments to concentrate on the region of phase space above the $\pi^+\pi^0$, at a cost of about a factor 5 in acceptance. In the past the primary residual background was due to $K^+ \rightarrow \mu^+\nu\gamma$ which has a branching ratio of $\approx 6 \times 10^{-3}$. The γ is difficult to veto since it has little energy and tends to follow the μ^+. Since the μ^+ range distribution from this reaction straddles the region of $\pi^+\nu\bar{\nu}$ acceptance, one needs excellent μ^+ rejection to reduce this background to an acceptable level.

The apparatus of AGS experiment 787[34] is shown in Figure 14. An 800 MeV/c separated K^+ beam of $\approx 350,000$/pulse is degraded to $\approx$ 350 MeV/c in a barium fluoride scintillator and brought to rest in a live stopping target. This will consist of a bundle of scintillating fibers which provide a detailed picture of the K^+ stop and subsequent decay. Surrounding the target is a cyndricial drift chamber which covers about 50% of 4π. A large iron bound solenoid enclosing the apparatus provides a 1T field, allowing $\approx 2\%$ momentum measurement on the π^+. Outside the drift chamber is a cylindrical array of range counters about 40 cm thick. These counters serve to distinguish charged pions from muons or electrons in a number of different ways. They make possible measurements of both the range and the energy of the stopping pions. Comparison of these quantities with the measured momentum yields a relative μ rejection of $> 2 \times 10^4$:1. Observation of the $\pi \rightarrow \mu \rightarrow e$ decay chain in the stopping counter yields a further factor $\geq 10^5$:1 against muons. This technique works so well, in spite of the large ambient rates, because of the high effective segmentation of the stopping counters. This is achieved by reading out the range counters on two ends, making possible a determination of the various signal sources to a few cm by means of differential timing and pulse height measurements. Thus one can require that all three signals emanate from the same area of the stopping counter.

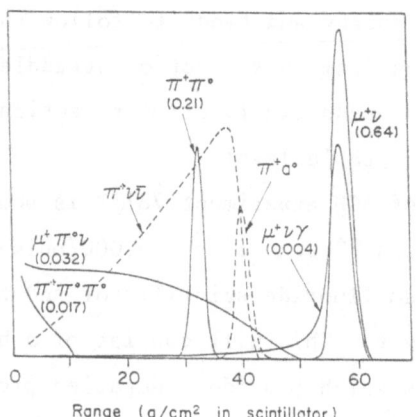

Fig. 13. Range distribution of final state charged particle
from various K^+ decay modes.

Surrounding the range stack is an array of lead-scintillator
"barrel" counters. The use of very thin ($<$ 1 mm) Pb plates allows
a sufficiently low threshold ($\approx$ 20 MeV) for efficient vetoing of
background γ's. This veto coverage is completed by inserting 1 mm
lead sheets between the range counters (outside the π acceptance
region) to make the end regions of the range array into shower
counters and by the use of barium fluoride scintillators in the
foward and backward regions. BaF_2 (radiation length $\approx$ 2 cm), has a
fast scintillation component with decay time $\approx$ 0.6 nsec so that it
can cope with the high rates of the beam region. It will be read
out via low pressure PWC's doped with TMAE.[35]

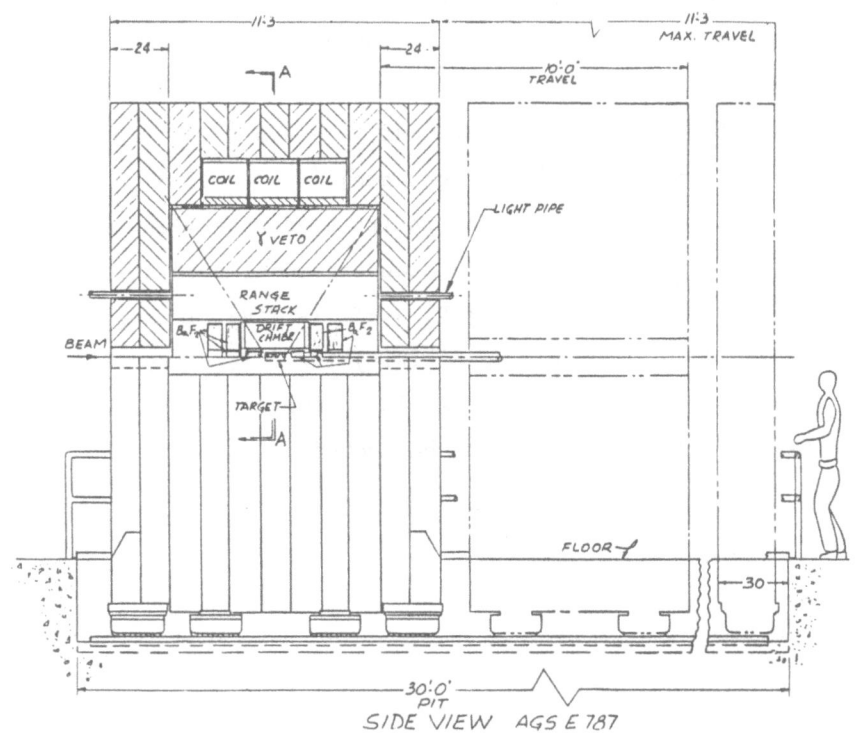

Fig. 14(a). E787 Apparatus (Side View)

The combination of μ rejection and γ vetoing reduces the K⁺ → μ⁺νγ background to a level below that of other processes. In fact the leading residual background is likely to be one of two processes neither of which is a K⁺ decay. One is simply a large angle scattering of a beam π^+. In order to defeat the trigger this must occur within about 50 nsec of a stopping K⁺ which happens to decay very late. The segmented target will afford some protection against this, but the primary line of defense is the introduction of veto Cerenkov counters and scintillator hodoscopes into the beam. The second potential background is due to K⁺ charge exchange in the stopping target. If this occurs at low enough energy the resulting K^0_L may be soft enough to remain in the target beyond the end of the stopping gate delay. It can then decay semi-leptonically--occassionally giving the lepton so little energy that it is lost in the target. This is calculated to give a residual background of $\leq 2\times 10^{-11}$.

The sensitivity of the experiment is $\approx 2.3\ [3.5 \times 10^5$ stops/ pulse $\times\ 2 \times 10^6$ pulses $\times\ .015$ acceptance$]^{-1} = 2 \times 10^{-10}$. This is low enough to be sensitive to the entire window between the current upper limit of 1.4×10^{-7} and the standard model prediction of $\approx 10^{-10}$. Since the acceptance for $K^+ \to \pi^+ f$ is five times greater than that for $K^+ \to \pi^+ \nu\bar{\nu}$, the sensitivity for this decay will be at the 4×10^{-11} level. This permits almost the entire allowed range of the family symmetry scale F to be probed.

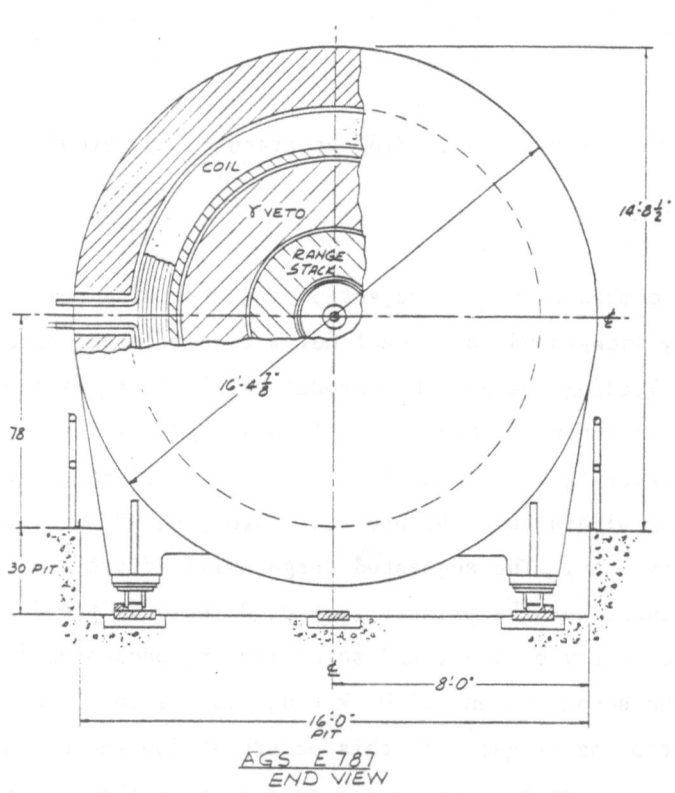

Fig. 14(b). E787 Apparatus (End View)

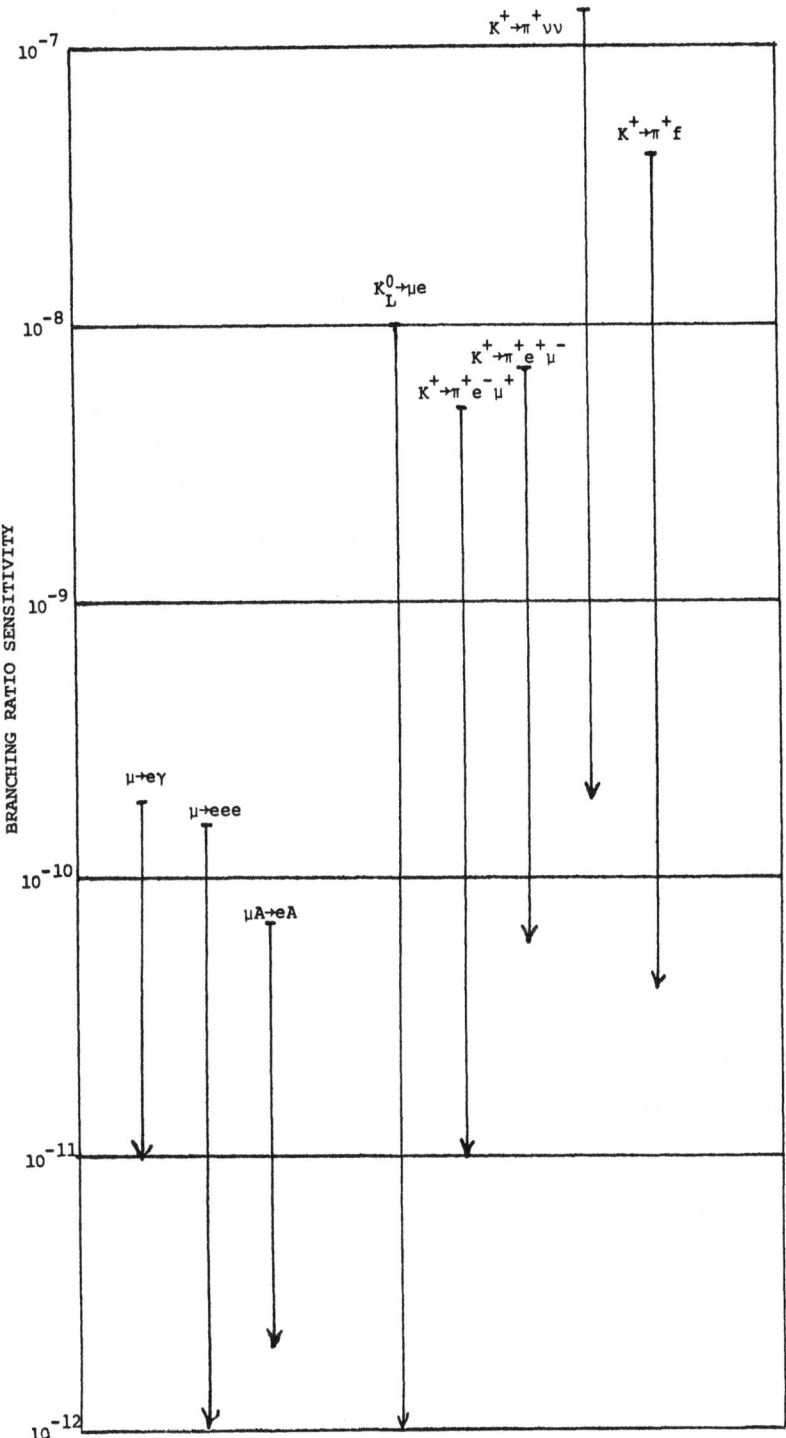

Fig. 15, Current and anticipate branching
ratio limits for rare decays

Table 1. Relative Sensitivities of LFV Processes.

Processes	Multiple Higgs	PS LQ	V LQ	U(1) HGB	SU(2) HGB	Exp BR × 10^{10}
$B(\mu \to e\gamma)$	.011	--	--	*	13θ	1.9
$B(\mu \to e\bar{e}e)$	.040	--	--	89	27θ	1.6
$B(\mu Z \to eZ)$	.35	.14θ	55θ	251	94θ	0.7
$B(K_L^0 \to e\mu)$	.90	.38	197	237	76(0)	100
$B(K^+ \to \pi^+\mu e)$	.12	.03	10	69	47(94)	50

Mass limits in TeV

For branching ratio upper limit of 10^{-10}

$$M_{heavy} \propto (BR)^{-1/4}$$

Table 2

Mode	Evts. Gen.	Total Decays	Signal Events	Bkg Level	x Misident as y	Misident Prob	Final Bkg Level ($\times 10^{-12}$)	Comment
$\pi^+\pi^+\pi^-$	2.5×10^4	4×10^5	0	2.7 10^{-5}	π^- as e^- π^+ as μ^+	2×10^{-7}	<5.4	No events with M_K greater than 445 MeV
$\pi^+\pi^+\pi^-$ $\mu^-\nu$	10^5	1.6 $\times10^7$	37	2.5 $\times10^{-5}$	μ^- as e^- π^+ as μ^+	2×10^{-8}	0.5	
$\pi^+\pi^+\pi^-$ $\mu^+\nu$	10^4	0.9 $\times10^6$	3	3×10^{-5}	π^- as e^-	10^{-7}	3	
$\pi^+\pi^+\pi^-$ $e^-\nu$	1.4×10^5	10^{12}	1	1×10^{-11}	π^+ as μ^+	10	$<.13$	Forshortened decay length of π by 50
$\pi^+\pi^+\pi^-$ $\mu^+\nu$	10^4	8×10^6	0	<1.3 $\times10^{-6}$	μ^- as e^-	10^{-7}	<0.1	
$\pi^+\pi^+\pi^-$ $e^-\nu$ $\mu^+\nu$	10^5	1.8 $\times10^{13}$	1	0.6 $\times10^{-12}$	π^- as e^-	$<10^{-2}$	<0.1	Only good event π^- decayed beyond C1R.
$\pi^0\mu^-\nu$ $e^+e^-\gamma$	2.5×10^4	0.6 $\times10^8$	0	<1.8 $\times10^{-7}$	e^+ as π^+	10^{-6}	<0.18	
$\pi^0\pi^+$ $e^+e^-\gamma$	2.5×10^4	10^7	1	1.1 $\times10^{-6}$	e^+ as μ^+	10^{-6}	1.1	

In addition there are many other interesting decay modes which can be studied with this apparatus. The complimentary decay to that of E777, $K^+ \rightarrow \pi^+ e^+ \mu^-$ can be probed down to $\approx 7 \times 10^{-11}$. The neutral current decay $K^+ \rightarrow \pi^+ \gamma\gamma$ ought to be observed (the standard model prediction is $\approx 10^{-7}$). One can tag π^0's via the decay $K^+ \rightarrow \pi^+ \pi^0$ and search for $\pi^0 \rightarrow$ 'nothing'. Other less exotic but still interesting K^+ decays such as $K^+ \rightarrow e^+ \nu\gamma$ and $\mu^+ \nu\gamma$ can also be studied.

Figure 15 shows the current and anticipated future experimental situation for the processes discussed above.

In the next few years rare decay experiments should provide stringent tests of the standard model. Limits on lepton flavor violating proceses will be pushed down to levels of 10^{-11}–10^{-12} which correspond to new physics scales in the tens or hundreds of TeV. A number of K decays like $K_L^0 \rightarrow e^+ e^-$, $\pi^0 e^+ e^-$, $K^+ \rightarrow \pi^+ \nu\bar{\nu}$, etc., will be probed down to or near the level of the highly suppressed standard model contribution, affording large windows for new physics.

Useful discussions with D. Bryman, L-L. Chau, R. Cousins, F. Gilman, J. Hagelin, C.M. Hoffman, G. Kane, W-Y. Keung, W. Marciano, W. Molzon, W. Morse, M. Schmidt, R. Shrock, H.K. Walter, S. Wojcicki, and M. Zeller are gratefully acknowledged.

This work was supported in part by the U.S. Department of Energy under Contract No. DE-AC02-76CH00016.

REFERENCES

1. This table is a descendent of a similar table by John Ellis, in Intense Medium Energy Sources of Strangeness, Ed., T. Goldman, et al., AIP (1983) p. 191, which is itself a descendent of a table due to O. Shanker, Nucl. Phys. B206, 253 (1982).
2. W. Bertl, et al., (SINDRUM collab). PR-84-01.
3. A.E. Pifer, et al., NIM 135, 39 (1976). H.-W. Reist et al., ibid 153, 61 (1978).

4. S.M. Korenchenko, et al., JETP $\underline{43}$ (1976) 1.
5. LAMPF experiments 400/445. J.D. Bowman, C.M. Hoffman, and H.S. Matis, Spokesmen, Los Alamos National Laboratory Report LA-7444-SR, Rev. 3 (1982).
6. W.W. Kinnison, et al., Phys. Rev. $\underline{D25}$, 2846 (1982).
7. G. Azuelos, et al., Phys. Rev. Lett $\underline{51}$, 164 (1983).
8. LAMPF experiment 444. J.D. Bowman and R.Hofstadter, Spokesmen, Los Alamos National Laboratory Report LA-7444-SR, Rev. 3 (1981), p. 249.
9. D. Bryman, C. Hargrove, et al., TRIUMF proposal 104.
10. C.K. Hargrove, et al., April '83, submitted to NIM.
11. P. Hänggi, et al., Phys. Lett. $\underline{51B}$, 119 (1974).
12. A. Badertscher, et al., Lett. al Nuovo Cimento $\underline{28}$, 401 (1980).
13. AGS experiment E780. R.C. Larsen, et al.
14. The current limits on $K^0_L \to \mu e$ and $K^0_L \to ee$ are based on the experiment of A.R. Clark, et al., Phys. Rev. Lett. $\underline{26}$, 1667 (1971). However, since this experiment failed to detect $K^0_L \to \mu^+\mu^-$ whose branching ratio was subsequently shown to be $\approx 10^{-8}$, it seems prudent to raise the $e\mu$ and ee limits from the claimed 2×10^{-9} level to $\approx 10^{-8}$.
15. AGS Proposal 791, R. Cousins, et al. This experiment is currently at the proposal stage.
16. F.J. Gilman and M.B. Wise, Phys. Rev. $\underline{D21}$, 3150 (1980).
17. Not all direct CP violating mechanisms show up in the polarization. See P. Herceg. Phys. Rev. $\underline{D27}$, 1512 (1983).
18. AGS Experiment E777, D. Lazarus et al.
19. See for example R.N. Cahn and H. Harari, Nuc. Phys. $\underline{B176}$, 135 (1980). $K^+ \to \pi^+e^-\mu^+$ conserves generation number (+ 1 for u, d, e^-, ν_e, + 2 for c, s, μ^-, ν_μ, etc., - for antiparticles), $K^+ \to \pi^+e^+\mu^-$ violates it by 2 units. $K^0 \to \mu^+e^-$ conserves it, $K^0 \to \mu^-e^+$ violates it, but since in a K^0_L beam one doesn't know whether the decay is from the K^0 or $\bar{K}^0$ component, it's not possible to get any information on generation number conservation from neutral decays.
20. A.M. Diamant-Berger, et al., Phys. Lett. $\underline{62B}$, 485 (1976).
21. P. Bloch, et al., Phys. Lett. $\underline{56B}$, 201 (1975).
22. J. Fischer, et al., Phys. Lett. $\underline{73B}$, 364 (1978), erratum, ibid $\underline{76B}$, 663 (1978); J.S. Frank, et al., Phys. Rev. $\underline{D28}$, 423 (1983).
23. P. Herczeg, Phys. Rev. $\underline{D16}$, 712 (1977) and references cited therein.
24. Y. Asano, et al., Phys. Lett. $\underline{107B}$, 159 (1981).
25. E. Fernandez, et al., Phys. Rev. Lett. $\underline{51}$, 1022 (1983). N.S. Lockyer, et al., Phys. Rev. Lett. $\underline{51}$, 1316 (1983).
26. F.J. Gilman and J.S. Hagelin, Phys. Lett. 133B, 443 (1983).
27. J. Ellis and J.S. Hagelin, Nucl. Phys. B217, 189 (1983).
28. G.L. Kane and R.E. Shrock in Intense Medium Energy Sources of Strangeness, Ed. T. Goldman, et al., AIP (1983) p. 123.

29. R. Schrock in the Proceedings of the DPF Summer Study on Elementary Particle Physics and Future Facilities, Snowmass, June-July 1982, R. Donaldson, et al., Eds. p. 291. M. Suzuki, UC-Berkeley-LBL preprint UCB-PTH-82/8.

30. F. Wilczek, Phys. Rev. Lett. $\underline{49}$, 1549 (1982).

31. R. Shrock and M. Suzuki, Phys. Lett. $\underline{110B}$, 250 (1982). M.B. Wise, Phys. Lett. $\underline{103B}$, 121 (1981).

32. These distributions are, of course, just transforms of the cm momentum distributions. However, the various peaks and other kinematical features pull away from each other more in range than in momentum, so that measurements of the former are more valuable.

33. One possible exception is the decay $K^+ \rightarrow \pi^+\gamma\gamma$. This as yet unobserved at the level of 8.4×10^{-6}. If the branching ratio were this large it would be necessary to veto the $\gamma\gamma$ state recoiling against the π^+ with an inefficiency $\leq 10^{-5}$. The γ-γ inefficiency of experiment 787 is calculated to be $< 10^{-5}$, so this decay mode should not be a problem.

34. AGS Experiment E787. I-H. Chiang, et al.

35. D.F. Anderson, et al., in Proc. Wire Chamber Conference, Vienna, 1983 [NIM $\underline{217}$ (1983) 217].

DISCUSSION

SEGRE:
With all the μ's, K's, etc., can you look for time reversal effects such as in $\mu \to e\nu\gamma$ or $(\vec{P}e_\mu + \vec{P}_\gamma)$ effect?

LITTENBERG:
I don't know about the muons but there is a new $K^+ \to \pi^0\mu^+\nu$ polarization experiment being thought about for the AGS.

SIEBERT:
Will you do other "bread and butter" measurements, like K_{e2}, radiative K-decays and such-like?

LITTENBERG:
The planning for "other physics" is still at an early stage. It will be vigorously pursued to the extent that it doesn't compromise the main purpose of the experiment. It seems clear that we can do rather well on $K^+ \to \pi^+\gamma\gamma$, and we ought to be able to increase $K^+ \to e\gamma\gamma$, $\mu\nu\gamma$, $\pi^+e^+e^-$, $\pi^+\mu\mu$, $\pi^0 \to e^+e^-$, etc., but how well we can do remains to be seen.

MOHAPATRA:
Could you tell anything about any proposed or on-going experiments on any limits on muonium to antimuonium conversion?

LITTENBERG:
No.

KLEINKNECHT:
Is the reason for the higher K^+ flux at the AGS compared to KEK mainly the higher proton energy; or are there other factors?

LITTENBERG:
There are a number of factors. We use 2-3 times more protons on target than the KEK experiment. The production of K^+ is higher at 28 GeV/c than at 12. The Japanese group used a 550 MeV/c K^+ beam. We will use 800 or 850 MeV/c beam. In addition, these beam lines are ~ 15m long and there is a large loss of low-momentum K's through decay which is mitigated by raising the momentum. What you pay is an increased K loss through interactions in the degrader and an increased beam size.

GOLDHABER:
Since such intense beams are being built, has someone suggested to look for unknown long-lived particles at a large distance, which might be lost in absorbers in the usual experiments?

LITTENBERG:

The Yale-BNL K_L^0 beam line may be a little short for this work. The beam line for the other K_L^0 is not yet defined; it may be suitable. One possible problem is that people sometimes put absorber in these beams to differentially absorb neutrons.

TRILLING:

What is the time scale for your experiment ($K^+ \to \pi^+ \nu \bar{\nu}$)? What is spatial resolution of target scintillating fibers?

LITTENBERG:

The time scale is anticipated to be 2 years to build and a year or two to run it. The fibers are about 4mm in diameter. In the precise region of the incoming K^+ the resolution may be a little worse than 1 fiber due to occupation of several (3-4) fibers by the K.

THE PERMANENT ELECTRIC DIPOLE MOMENT OF THE NEUTRON AND ATOMIC ^{129}Xe

Blayne Heckel

Physics Department
University of Washington
Seattle, Washington

INTRODUCTION

The most sensitive tests for time reversal symmetry (T)
violation outside of the K°-$\overline{K^{\circ}}$ system come from measurements of
the permanent electric dipole moments (edm) of neutral particles.
It is easy to show that for a spin $\vec{J}$ quantum system having m_J
degeneracy only, the existence of a permanent electric dipole
moment requires a violation of both T and parity (P) symmetry.[1]
The greatest sensitivity in edm measurements is provided by experiments on neutral particles because the edm is observed by its
interaction with an applied electric field, and a zero electric
charge permits an application of a strong electric field.

The neutron edm has received the greatest share of theoretical
and experimental attention. Experimental limits set on the neutron
edm have long provided the most stringent test for the majority of
theories of T-violation. Two experiments, at Grenoble and Gatchina,
are currently collecting data to measure more precisely the neutron
edm. Both experiments measure the magnetic resonance frequency of
bottled ultra-cold neutrons in the presence of an applied electric
field. The published results: $d_n = (0.3 \pm 4.8) \times 10^{-25}$ cm
(Grenoble)[2] and $d_n < 4 \times 10^{-25}$ cm (Gatchina)[3] place the neutron

edm below the 4×10^{-25} e-cm level. The first half of this
presentation will review the current theoretical predictions for
the neutron edm and then describe the Grenoble experiment in some
detail.

Recently the Seattle group[4] has measured the permanent edm of
atomic ^{129}Xe to be d_{xe} = $(-0.3 \pm 1.1) \times 10^{-26}$ cm. This measurement provides an improvement by a factor of 10^4 over the best
previous measurement of an atomic edm. The edm of Xe is most
sensitive to a T-violating electron-nucleon interaction, but can
also be generated by an intrinsic electron edm or an edm of the Xe
nucleus. In the second half of this presentation, each of these
mechanisms will be briefly discussed and a description of the
experiment will be given.

THE NEUTRON ELECTRIC DIPOLE MOMENT

Most of the theories that account for the observed T-violation
in the decay of the K_L° meson also predict a non-zero value for the
neutron edm in the range from 10^{-19} e-cm to 10^{-32} e-cm. Because it
is not yet established as to which of the fundamental interactions
is responsible for the observed T-violation, it is not surprising
that the range of theoretical predictions is so large. The present
experimental upper limit on the neutron edm leaves roughly four
categories of theories that are still within the experimental bounds:
phenomenological milli-weak and super-weak theories, and gauge
theories that attribute the T-violation to the mixing of 6 or more
quarks, or to the exchange of multiple Higgs bosons.

Super-weak theories hypothesize a new super-weak force that is
10^9 times weaker than the weak interaction, and that require a
change in strangeness $\Delta S = \pm 2$. In this case predictions for the
neutron edm are at the 10^{-29} to 10^{-30} e-cm level,[5,6] which will be
difficult to verify experimentally. However, if lepton-quark
interactions are included in the milli-weak or super-weak theories,
then stronger limits on their strengths can possibly be deduced from
the atomic edm measurements.

More recent predictions for the neutron edm come from the attempt to incorporate CP violation into grand unified gauge theories. Kobayashi and Maskawa[7] have shown that a quantum chromo-dynamics gauge theory based on only 4 quarks and a single Higgs scalar multiplet can not account for the known CP violation in K_L^o decay. In models with 6 or more quarks, the CP violation is generated by a phase in the generalized Cabbibo matrix, and gives rise to a neutron edm at the 10^{-28} to 10^{-30} e-cm level.[8] If one instead assumes that additional Higgs multiplets generate the CP violation, then depending upon the masses of the Higgs particles, the predicted neutron edm is in the range from 10^{-24} to 10^{-27} e-cm.[9,10]

An interesting lower limit to the neutron edm comes from the interplay between cosmology and particle physics. If the asymmetry between matter and anti-matter in our universe is due to a CP-violating force in the early life of the universe, then using the observed photon to baryon ratio in the universe as a constraint, Ellis et al.[11,12] have deduced a lower bound for the neutron at the level of 10^{-25} to 10^{-28} e-cm, depending upon the gauge theory used. Finally, Nanopoulos has reported at this conference that super-symmetric gauge theories require a neutron edm of order 10^{-26} e-cm.

The experimental precision of measurements of the neutron edm is approaching the level at which it should be possible to distinguish between several of the theoretical predictions given above. In the past, neutron edm measurements were performed using neutron beam magnetic resonance spectrometers, in which the neutron spin precession frequency would be measured in parallel and anti-parallel electric and magnetic fields. A non-zero edm would cause the neutron spins to precess about the electric field at a frequency ν_E which would either add to or subtract from the Larmor frequency ν_B of the spins precessing about the magnetic field. By reversing the electric field with respect to the magnetic field, a frequency shift of $2\nu_E$ would be sought. The most precise neutron beam measurement was performed by Dress, et al.[13] who found that $d_n < 3 \times 10^{-24}$ cm. The neutron beam experiments have a serious limitation, however: a particle moving with velocity $\vec{V}$ in an electric field $\vec{E}$ sees in its own rest frame a magnetic field given by $\vec{E} \times \vec{V}/C$. This motional magnetic field changes sign when $\vec{E}$ is reversed and can give rise to a frequency shift that is indistinguishable from the shift due to an edm. For this reason, current attempts to measure the neutron edm use ultra-cold neutrons (ucn) that are stored in a bottle, for which the time averaged velocity is nearly 0.

The idea that neutrons with velocity less than roughly 6 m/sec (ucn) would be total internally reflected from the walls of a material vessel was first introduced by Zeldovitch[14] (for a more recent review of ucn see ref. 15). Neutrons in this velocity range are produced in the tail of the Maxwellian neutron spectrum in the core of the thermal nuclear reactor, and can be extracted from the core through a vertical neutron guide tube. When incident upon a magnetized foil of Fe/Co alloy, ucn of one spin state will be reflected by the magnetic potential barrier of the foil, while the opposite spin state neutrons will pass through the foil.[15] In this way, a beam of polarized ucn can be produced.

The experimental apparatus for the Grenoble measurement of the neutron edm is shown in Figure 1. The scientists associated with this experiment are listed under ref. 2. After passing through the Fe/Co polarizing foil, the beam of polarized ucn enters a 5 ℓ neutron storage volume located in the center of a 5 layer μ-metal magnetic shield. After filling the storage volume with neutrons, a shutter is closed allowing the neutrons to be stored for 60 – 100 seconds. The storage volume consists of two 25 cm diameter Be discs separated 10 cm by an insulating BeO cylinder. The Be discs serve as electric field plates and an electric field of 10 to 14 KV/cm is applied as shown in Fig. 1. Inside of the innermost μ-metal shield is a

548

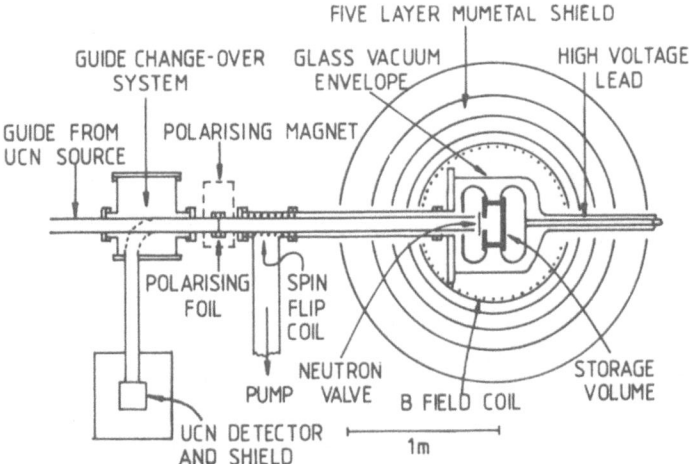

Fig. 1. Apparatus used to observe magnetic resonance in polarized ultra-cold neutrons stored for times up to 100 seconds in magnetic fields of 3 to 40 milligauss.

solenoid which produces a uniform 10 mG magnetic field, $\vec{B}$, parallel to the electric field.

Once the neutron shutter is closed, the neutron spins are rotated into the plane perpendicular to $\vec{B}$ by applying a resonant 30 Hz oscillating magnetic field perpendicular to $\vec{B}$ for 2 seconds. Approximately 80 seconds later, a second oscillating field pulse, coherent with the initial pulse, is applied which leaves the neutrons in a spin state which depends upon the phase difference between the spin precession frequency and the oscillating field frequency. After the second pulse the neutron shutter is opened to allow the neutrons to pass again through the polarizing foil (acting now as an analyzer) and into a ^{3}He proportional counter ucn detector. A curve of the neutron count rate as a function of the frequency of the oscillating field for a neutron storage time of 40 seconds is shown in Figure 2. The curve has a Ramsey separated oscillatory field shape with a linewidth given by approximately 1/2T where T is the neutron storage time. A neutron edm is sought by setting the frequency of the oscillating field to the steepest slope on the

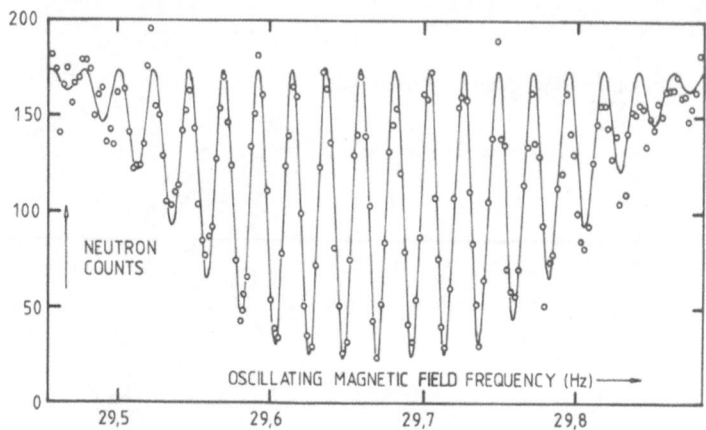

Fig. 2. A neutron magnetic resonance curve obtained with polarized ultra-cold neutrons stored for 40 seconds in a magnetic field of 10 milligauss when an oscillating magnetic field is applied for 4 seconds at the beginning and end of the storage period. The solid line is a theoretical curve fitted using the Ramsey theory which predicts a linewidth at the centre of the pattern of 0.011 Hz.

resonance pattern, and measuring a change in the neutron count rate as the electric field is reversed, which corresponds to a shift in the precession frequency.

The typical neutron count rate in the Grenoble experiment is 50 neutrons per a fill-store-empty cycle, which corresponds to a neutron density in the storage volume of $\sim$ .02 neutrons/cm^3. The cycle is repeated for approximately 45 minutes at which time the electric field polarity is reversed and the cycling begun again. To monitor magnetic field drifts, three Rb magnetometers are mounted near to the neutron storage volume. The sensitivity of the magneto- meters is roughly 5×10^{-8} Gauss per 100 second integration time, while the total magnetic field drift in a day is $\sim 2 \times 10^{-7}$ Gauss. The most serious systematic effect to control is the magnetic field caused by leakage currents across the BeO insulator, which can change in sign and magnitude as the electric field is reversed. The leakage currents are monitored at the 1 nA level electronically, and the electric field strength is chosen so that the total leakage current is less than $\sim$ 20 nA.

The published results of the Grenoble experiment: $d_n = (0.3 \pm 4.8) \times 10^{-25}$ cm is limited solely by neutron counting statistics. Construction is underway for two new ucn sources that are envisioned to increase the neutron count rate by two orders of magnitude. With a more intense ucn source, the prospects are good that a neutron edm measurement at the 10^{-26} e-cm level can be achieved in the next few years.

THE ELECTRIC DIPOLE MOMENT OF ATOMIC ^{129}Xe

There are several mechanisms by which a T-violating force can generate an atomic edm, but the size of the atomic edm to be expected from current theories of T-violation has not been studied in great detail. This is an area where further theoretical work would be very valuable since the experimental sensitivity of the atomic edm measurements has recently surpassed that of the neutron. Because there are no hard predictions for an atomic edm to review, only a description of several sources for an atomic edm will be given in the following paragraphs.

For an atom with a closed electronic shell, such as Xenon, the most direct way to generate an atomic edm is through a short range T-violating force that acts between the electrons and nucleons. The most general non-derivative short range neutral current inter-action that violates both P and T can be written as:[16]

$$H_T - iC_S \frac{G_F}{\sqrt{2}} \sum_{n,e} (\bar{n}n)(\bar{e}\gamma_5 e) + iG_T \frac{G_F}{\sqrt{2}} \sum_{n,e} (\bar{n}\gamma_5\sigma_{\mu\nu}n)(\bar{e}\sigma^{\mu\nu}e) \quad (1)$$

where n,e denote the nucleon, electron wave function, γ_5 and $\sigma_{\mu\nu}$ are Dirac matrices, and where possible isospin dependence in the nucleon current has been neglected. The dimensionless quantities C_S and C_T measure the strength of the scalar (first term) and tensor (second term) interactions relative to the weak coupling constant G_F.

The non-relativistic form of the tensor term in Eq. 1 has terms that look like:

$$V_T = C_T \frac{G_F}{\sqrt{2}} \left[\frac{\delta^3(\vec{r})}{m_e c} \cdot \frac{i\vec{P}_e \cdot \vec{I}}{|I|} + h.c. \right] \qquad (2)$$

where $\vec{I}$ is the nuclear spin. Because there is no explicit dependence upon the electron spin in Eq. 2, V_T can induce an atomic edm in Xe in first order perturbation theory. Preliminary calculations[17,18] indicate that for the case of the ^{129}Xe atom, $d(^{129}Xe) \sim 10^{-20} C_T$ cm.

The non-relativistic form, V_S, of the scalar term in Eq. 1 is linear in the electron spin and hence vanishes to first order when a sum is made over the electronic shell. In third order perturbation theory, however, the hyperfine interaction, $\vec{I} \cdot \vec{J}$, can break the symmetry of the paired electron spins, allowing V_S to induce an atomic edm. For example, consider one of the third order terms for the valence 5p electron in ground state Xenon:

$$d\,(atom) \propto \vec{\nabla}_{\vec{E}} \sum_{n,n'} \frac{<5p|V_S|nS><nS|\vec{I}\cdot\vec{J}|n'S><n'S|e\vec{r}\cdot\vec{E}|5p>}{(E_{5p} - E_{nS})(E_{5p} - E_{n'S})} \qquad (3)$$

where $\vec{E}$ denotes the applied electric field modified by the shielding due to the remaining electrons. An estimate for the size of terms like that given in Eq. 3 gives $d(^{129}Xe) = (10^{-24}$ to $10^{-26})C_S$ cm,[17] where the large uncertainty comes from contributions due to multi-electron effects and relativity.

An intrinsic edm of the electron can also generate an edm on the atom. For the case of Xenon, one must again go to third order and consider terms like that given in Eq. 3, where V_S will now be replaced by $\vec{d}_e \cdot \vec{\nabla} V$ (V being the atomic central potential). Sandars[19] has shown that due to relativistic spin orbit effects, the effective electric field that the electron feels can be orders of magnitude

larger than the applied electric field in heavy atoms. Using
Sandars' enhancement factor, an estimate for Xenon gives
$d(^{129}Xe) \approx 10^{-3} d_e$.[17]

Finally, an edm of the atomic nucleus, due to either an
intrinsic nucleon edm or a T-violating nucleon-nucleon force can
induce an edm on the atom. In this case the situation is even more
unclear because the total electric field averaged over the nuclear
charge distribution is zero. However, due to non-electrostatic
forces, the nuclear charge density and electric dipole density need
not coincide, resulting in a net electric field averaged over the
edm distribution in the nucleus. If one assumes that in the ^{129}Xe
nucleus the rms charge and dipole radii differ by 10%, then a rough
estimate is that $d(^{129}Xe \text{ atom}) \approx 10^{-5} d(^{129}Xe \text{ nucleus})$.[17] In
nuclei possessing nearly degenerate opposite parity levels, large
enhancements of the edm of the nucleus may be expected.[20]

The search for an atomic edm is a rich testing ground for a
variety of T-violating mechanisms. However, before the experimental
limits can be used as constraints on theories of T-violation,
detailed atomic calculations must be performed. Considerable effort
has gone into the calculations of parity violating effects in heavy
atoms, with good agreement between the results of various groups
and with experiment.[21] The calculations for P violation and for
T violation in atoms are very similar in that they both rely upon
a knowledge of the electron wavefunction at the nucleus. Hence
there is good reason to believe that reliable calculations for
atomic edm's may be made.

To measure an atomic edm, the same principle is used as with
the neutron measurement: a difference in atomic precession
frequency is sought as an electric field is reversed with respect
to a magnetic field. A diagram of the Seattle atomic ^{129}Xe edm
apparatus is shown in Figure 3.[4] In the center of a 3 layered
μ-metal shield, three identical gas cells are mounted. Each cell
is made from a 2.5 cm diameter pyrex tube that is vacuum sealed to

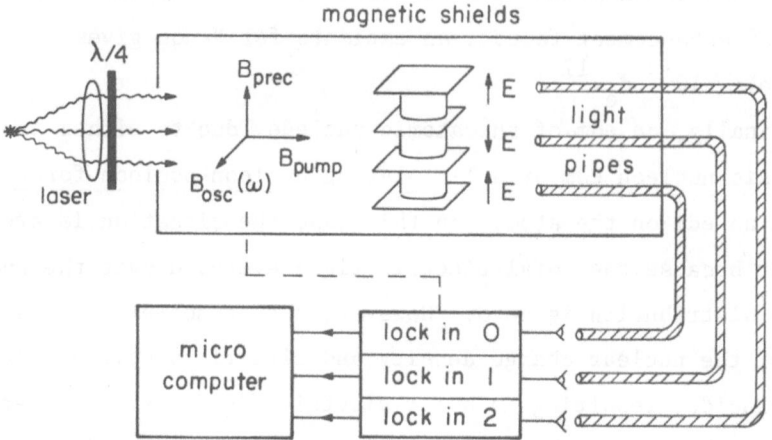

Fig. 3. A schematic view of the atomic edm experimental apparatus.

Pt coated pyrex end-caps, which serve as electric field plates.
The cell walls are coated with a siliconizing material (Surfa Sil)
to increase the Xenon polarization lifetime (~ 500 sec), and to
prevent a conducting monolayer of Rb from forming on the cell walls.
The cells are filled with 2 torr of natural Xenon (26% ^{129}Xe),
200 torr of N_2, and approximately 1 mm^3 of Rb metal. The Rb is
present to polarize the Xe nuclei and to measure the Xe precession
frequency. The N_2 is added to increase the electric field breakdown potential and to aid in the formation of Rb - Xe van der Waals
molecules.

    Circularly polarized light from a commercial diode laser is
tuned to the D_1 line of Rb and illuminates the cells along the
B_{pump} axis. To polarize the ^{129}Xe nuclei, a 10 mG pump field is
applied along the B_{pump} axis which polarizes the Rb atoms through
optical pumping. The Rb polarization is transferred to the Xe
through spin exchange which is enhanced by the formation of Rb - Xe
van der Waals molecules.[22] After the Xe polarization is established
(~ 500 sec) the pump field is turned off suddenly, leaving only a
100 μG precession field directed along B_{prec}. At this time the
^{129}Xe atoms precess about B_{prec} at a frequency of ~ 0.1 Hz. The Rb
atoms precess ~ 0.6 rad. during their polarization lifetime of 2 msec.

To measure the Xe precession frequency, the Rb atoms are used as a magnetometer. A 300 Hz, ν, oscillating field is applied along B_{osc} to precess the Rb spins in the plane perpendicular to B_{osc}. In the absence of any dc magnetic field along B_{osc}, the transmitted light then shows a modulation in intensity at 2ν due to the projection of the Rb polarization along the B_{pump} axis. The Rb atoms also sense the Xe polarization through spin-spin interactions during Rb - Xe collisions. As the Xe atoms precess, they create a quasi-static additional field along B_{osc} (that oscillates at 0.1 Hz) which appears in the transmitted light as a modulation in intensity at frequency ν. The transmitted light is extracted from the cells via plastic optical fibers, and is detected by PIN photodiodes external to the magnetic shielding. The photocell signals are analyzed with lock-in amplifiers referenced to ν. The lock-in outputs, proportional to the amplitude of the Xe polarization along along B_{osc}, are digitized and recorded by a computer. The digitized record for each of the three cells from an early data run is shown in Figure 4. Each signal is numerically fit to an exponentially decaying sine wave function from which the precession frequency can be extracted with an uncertainty of $\sim 5 \times 10^{-7}$ Hz.

Three cells are used to cancel uniform and first derivative drifts in the magnetic field within the shields. As shown in Fig. 3, the electric field direction is alternated between the cells (E $\approx$ 4 KV/cm) and a pseudo edm is calculated by taking the difference between the average precession frequencies of the outer two cells and the center cell. The electric field in each cell is then reversed and a new pseudo edm is measured. A shift between the two pseudo edm values then signals a true edm, independent of uniform drifts and drifts in the linear gradient of the precession field.

After 12 days of data collection, the Seattle group has reported a null result: $d(^{129}Xe) = (-0.3 \pm 1.1) \times 10^{-26}$ cm. The data collection was stopped to construct new quartz cells (to reduce leakage currents to below the 1 nA level) and to install

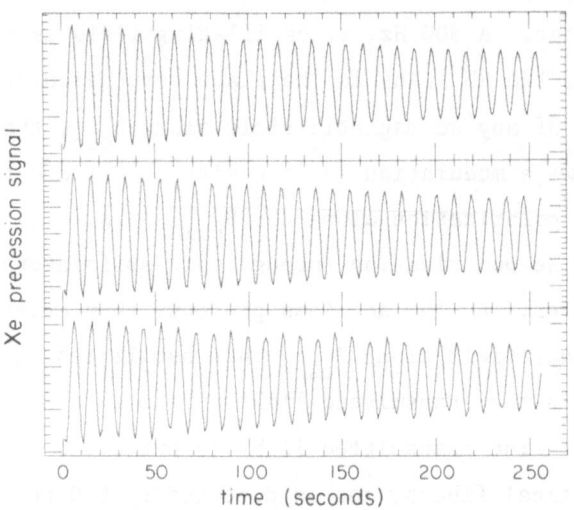

Fig. 4. The xenon magnetization along B_{osc} as a function of time for the three cells, showing the precession and transverse relaxation of the ^{129}Xe nuclear spins. Solid straight lines connect the raw data points, which are digitized output voltages (in arbitary units) from the lock-in amplifiers.

better leakage current monitoring devices. As it is believed that P and T violating effects in atoms scale faster than Z^2 (Z = atomic number),[23] work is proceeding to extend the measuring technique to heavier atoms (notably ^{199}Hg), and to investigate the possibility of examing two atomic species simultaneously in the same cell. It is envisioned that an increase in sensitivity by a factor of 100 should be possible within the next few years.

The searches for the edm's of the neutron and atoms continue to be important tests of theories of T-violation. Improved techniques have recently lowered the experimental upper bounds of

556

the edm's in both cases. The next set of measurements should
reach the level of sensitivity required to test some of the current
theories of T-violation, although better calculations are necessary
in the atomic case before strict theoretical limits can be deduced
from the measurements. It is hoped that the day is nearing when
something other than an upper bound will be reported by these
experiments. The work described in this paper is supported by
NSF Grants PHY 78-08561 and PHY8203512A01.

REFERENCES

1. N. F. Ramsey, Rep. Prog. Phys. $\underline{45}$, 95 (1982).
2. J. M. Pendlebury, K. F. Smith, R. Golub, J. Byrne, T. J.
   Mc Comb, T. J. Sumner, S. M. Burnett, A. R. Taylor, B. Heckel,
   N. F. Ramsey, K. Green, J. Morse, A. I. Kilvington, C. A. Baker,
   S. A. Clark, W. Mampe, P. Ageron, and P. C. Miranda, Phys.
   Lett. $\underline{136\ B}$, 327 (1984).
3. I. S. Alterev et al., Phys. Lett. $\underline{102\ B}$, 13 (1981).
4. T. J. Vold, F. J. Raab, B. Heckel, and E. N. Fortson, Phys.
   Rev. Lett., in press (1984).
5. L. B. Okun, Comm. Nucl. Particle Phys. $\underline{3}$, 135 (1969).
6. L. Wolfenstein, Nucl. Phys. $\underline{B\ 77}$, 375 (1974).
7. M. Kobayashi and K. Maskawa, Prog. Theor. Phys. $\underline{49}$, 652 (1973).
8. D. V. Nanopoulos and A. Yildiz, Phys. Lett. $\underline{87\ B}$, 53 (1979).
9. S. Weinberg, Phys. Rev. Lett. $\underline{37}$, 657 (1976).
10. R. N. Mohapatra and J. C. Pati, Phys. Rev. $\underline{D\ 11}$, 569 (1975).
11. J. Ellis, et al., LAPP-TH-24 (Annecy), (1980).
12. J. Ellis, et al., Nature $\underline{293}$, 41 (1981).
13. W. B. Dress, et al., Phys. Rev. $\underline{D\ 15}$, 9 (1977).
14. Yu. B. Zeldovich, Sov. Phys. JETP $\underline{9}$, 1389 (1959).
15. R. Golub and J. M. Pendlebury, Rep. Prog. Phys. $\underline{42}$, 439 (1979).
16. P. G. H. Sandars, Phys. Lett. $\underline{62\ B}$, 97 (1976).
17. T. G. Vold, Ph.D. Thesis, University of Washington (1984).

18. A. M. Pendrill, private communication (1984).

19. P. G. H. Sandars, Phys. Lett. 14, 194 (1965).

20. W. C. Haxton and E. M. Henley, Phys. Rev. Lett. 51, 1937 (1983).

21. T. P. Emmons, et al, Phys. Rev. Lett. 51, 2089 (1983).

22. N. D. Bhaskar, et al., Phys. Rev. Lett. 50, 105 (1983).

23. M. A. Bouchiat and C. C. Bouchiat, J. Phys. (Paris) 36, 493 (1975).

DISCUSSION

SIEBERT:
What are the prospects for neutron lifetime measurements in magnetic bottles?

HECKEL:
With what is known about the production of ultra-cold neutrons in liquid ^{4}He, a lifetime measurement in a magnetic bottle is technically possible. I am not sure how far along the Bonn group is, however.

An easier method may be to use monochromatic ultra-cold neutrons (ucn), if the new ucn sources are successful. One could then measure the neutron lifetime in a material bottle, as a function of velocity, to possibly correct for the wall losses.

GOLDHABER:
(1) Have you considered Doppler shifted neutron sources?
(2) Do odd mass Pt isotopes depolarize ^{129}Xe?

HECKEL:
(1) Yes, a Doppler shifted ucn source, the "Steyerl Turbine" will be installed at Grenoble in 1986. We expect more than a factor of 100 increase in ucn density with the turbine.
(2) Very weakly, but the Xe doesn't see the Pt in our case -- the Pt is covered with silane and Rb.

STRANGE DECAYS AND NEW PHYSICS

John S. Hagelin

Maharishi International University
Fairfield, IA 52556

ABSTRACT

We consider the rare decays $K_L \to \mu e$, $K^+ \to \pi^+ \mu e$ and $K_L \to e^+ e^-$ in the context of extended technicolor and multiple Higgs theories. Longitudinal polarization in $K_L \to \mu\mu$ is presented as a sensitive test for flavor-changing neutral scalars with CP-violating couplings. The measured b-quark lifetime constrains the KM angles s_2 and s_3, leading to improved bounds on the decay $K^+ \to \pi^+ \nu\nu$ in the standard model. A measurement of this rare decay would test the radiative neutral current structure of the standard model, and would also provide a sensitive probe of new physics.

INTRODUCTION

Strange decays have provided an invaluable laboratory for exploring a wide variety of important physical questions. This article considers the significance of selected rare decays and the importance of improving the experimental bounds on their branching ratios. We first consider the lepton number-violating decays $K_L \to \mu e$ and $K^+ \to \pi^+ \mu e$. Such decays could occur in a supersymmetric scenario in which the supersymmetric scalar partner of the neutrino (the "sneutrino") gets a lepton number-violating vacuum expectation value,[1] however the viability of such a scenario has not yet been fully explored. In contrast, lepton number-violating decays are mainstream features of extended technicolor theories and other models with horizontal gauge interactions. Composite models of quarks and leptons can also give rise to decays in which lepton number is violated.

Secondly, we consider polarization in $K_L \to \mu\mu$ decay and conclude that this is an excellent place to look for technipions and Higgs bosons with flavor-changing couplings.

Finally, upper and lower bounds on the branching ratio for $K^+ \to \pi^+ \nu\nu$ are derived in the standard model for KM angles consistent with the experimentally measured b-quark lifetime. A measurement of $K^+ \to \pi^+$ + "unobserved" outside this allowed range would be evidence for new physics, which could modify the canonical short-distance $sd\nu\nu$ vertex and/or introduce entirely new decay modes.

$\Delta L \neq 0$ DECAYS

Many people[2] regard Higgs fields as an unattractive wart on the face of gauge theory which they would prefer to burn out. The reason is that scalar masses ordinarily acquire huge masses of order $\alpha^n \Lambda^2$ through radiative corrections, where α is a gauge coupling and Λ constitutes a natural cutoff in the theory. In the absence of a natural cutoff $\Lambda \ll M_{P1}$ or M_{GUT}, it is difficult to avoid large radiative masses without repugnant fine-tuning order by order in perturbation theory. One highly successful approach to this problem has been through the development of supersymmetric models,[3] in which radiative corrections to scalar masses automatically cancel order by order in perturbation theory, apart from small corrections of order $\alpha^n \Lambda_{SB}^2$, where $\Lambda_{SB} \lesssim 1$ TeV is the splitting between particles and their supersymmetric partners which acts as an effective cutoff.

A second approach to the problem of light scalars is to replace fundamental Higgs bosons with fermion condensates,[4] which break SU(2)xU(1) symmetry through vacuum expectation values of the form

560

$$\langle 0| \ \overline{\Psi}_Q \Psi_Q \ |0\rangle \propto \Lambda_{TC}^3 \qquad\qquad (1)$$

where Ψ_Q are fermions called "techniquarks" which interact through a new "technicolor" force, and Λ_{TC} is the scale at which this technicolor force confines.

It is unfortunately neccesary to complicate this simple idea in order to give masses to fermions. One solution[5] has been to introduce additional "extended" technicolor (ETC) interactions coupling ordinary fermions to technifermions. This gives masses to ordinary fermions which are of order $M_f \approx \Lambda_{TC}^3/M_{ETC}^2$. These ETC bosons are generally[6] accompanied by phenomenologically interesting "horizontal" extended technicolor (HETC) interactions between ordinary fermions, including lepto-quark gauge bosons of the Pati-Salam[7] (PS) type. In addition to these vector bosons, there should be numerous "technipions"--pseudoscalar bound states of techniquarks. These probably include color triplet lepto-quark bosons (P_{LQ}) with masses of order[6] 150 GeV, charged technipions ($P_\pm$) with masses[8] $O(5\text{-}14)$ GeV, and neutral technipions (P_0) whose masses are given by[9,10,11]

$$M_{P_0} = O(1\tfrac{1}{2} \ \text{GeV}) \ \text{x} \ (300 \ \text{TeV}/M_{PS}) \ \text{x} \ O(2^{0\pm1}). \qquad (2)$$

The non-observation of the decay $K^+ \to \pi^+ P_0$ tells us[11] that $M_{P_0} \gtrsim$ 350 MeV, which in turn implies[2]

$$M_{PS} \lesssim 3000 \ \text{TeV}. \qquad\qquad (3)$$

Given this set of theoretical expectations and rather loose constraints on techniparticle masses, we now turn to $K_L \to \mu e$ and $K^+ \to \pi^+ \mu e$ decays to see what more we can learn about technicolor models. There are four generic mechanisms which can lead to $K_L \to \mu e$. These are illustrated in Fig. 1. Figs. 1a and c involve the exchange of HETC gauge bosons. The cross-channel diagram (Fig. 1a) requires an HETC boson of the PS type. The current experimental bound[12] on the branching ratio for $K_L \to \mu e$

$$BR(K_L \to \mu e)/BR(K^+ \to \mu \nu) < .63 \times 10^{-9} \qquad (4)$$

results[10] in a lower bound on M_{PS}:

$$M_{PS} \gtrsim 300 \text{ TeV.} \qquad (5)$$

In light of the previous upper bound (3) on M_{PS}, the experimental prospects for observing $K_L \to \mu e$ decay based on this mechanism are not discouraging. However, one should be aware that the rate scales as $1/M_{PS}^4$, so that the branching ratio due to this mechanism could still be four orders of magnitude below the current experimental bound.

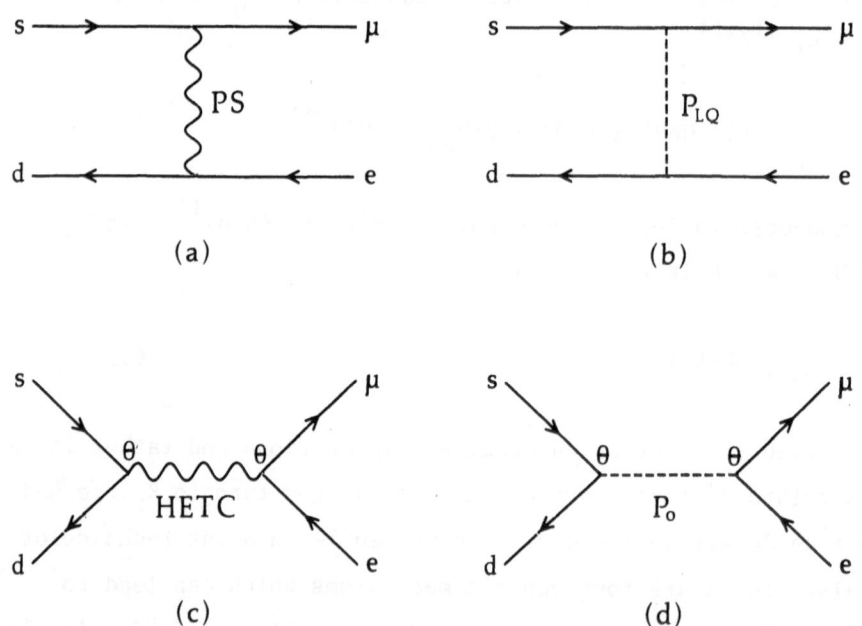

Fig. 1. Diagrams responsible for lepton number-violating decays.

The direct-channel diagram (Fig. 1c) is similar to the cross-channel diagram and gives[10] roughly comparable bounds on the HETC masses. The primary difference is that the direct-channel bound is weakened by the presence of potentially small mixing angles coupling quarks or leptons belonging to different generations.

Even more interesting than these HETC exchange diagrams are graphs involving the exchange of various technipions (Figs. 1b,d). Fig. 1b, which involves the exchange of a lepto-quark boson P_{LQ}, gives[10] a lower bound on the mass of P_{LQ} of 150 GeV, which coincides with the best theoretical estimates[6,9] for the mass of P_{LQ}. Thus ETC theories predict that $K_L \to \mu e$ should[2] occur very close to the current experimental bound.

Finally, $K_L \to \mu e$ can occur through the direct-channel exchange of a neutral technipion P_0 (Fig. 1d). This would place a lower bound on M_{P_0} of order 400 GeV, except for the presence of potentially small angles as in Fig. 1c. In fact, Eq. (2) together with the lower bound (5) on M_{PS} implies that $M_p <3$ GeV! Consistency with Fig. 1d therefore requires that these off-diagonal mixings are indeed very small. We shall return to this point later in our discussion of polarization in $K_L \to \mu\mu$.

The decay $K^+ \to \pi^+ \mu e$ occurs through an analogous set of diagrams and can be used to obtain similar bounds. However the present[13] experimental limit on $K^+ \to \pi^+ \mu e$

$$BR(K^+ \to \pi^+ \mu e)/BR(K^+ \to \pi^0 \mu\nu) < 1.5 \times 10^{-7} \qquad (6)$$

is considerably less constraining. Nevertheless, it should be noted that $K^+ \to \pi^+ \mu e$ and $K_L \to \mu e$ are logically independent: the former results from axial-vector and pseudo-scalar couplings to the s- and d-quarks, while the latter results from vector and scalar couplings. In practice,[14] however, models tend to predict comparable numbers for the ratios (4) and (6).

The decay $K_L \rightarrow e^+e^-$ in ETC models is rather analogous to $K_L \rightarrow \mu e$. In the cross-channel it is suppresed relative to $K_L \rightarrow \mu e$ by the presence of an off-diagonal mixing between the electron and the s-quark. In the direct channel it is <u>enhanced</u> relative to $K_L \rightarrow \mu e$ by the absence of such an angle. However, direct-channel HETC exchange is "helicity suppressed," and the P_0 coupling to the electron is probably also small. Dispite this fact, $K_L \rightarrow e^+e^-$ is competitive with $K_L \rightarrow \mu e$ as a probe for direct-channel P_0 exchange (or any other neutral scalar with flavor-changing couplings), and should be looked for. An observation of $K_L \rightarrow e^+e^-$ with a branching ratio above 5×10^{-12} would signal new and exciting physics.

POLARIZATION IN $K_L \rightarrow \mu\mu$

The measured branching ratio[15]

$$BR(K_L \rightarrow \mu\mu) = (9.1 \pm 1.9) \times 10^{-9} \tag{7}$$

is above the unitarity lower bound but within the range expected from a dispersive 2γ contribution.[16] An improvement in the accuracy of (7) would therefore be of limited theoretical interest. In contrast, however, an observation of longitudinal polarization in $K_L \rightarrow \mu\mu$ would be extremely interesting. Because polarization in $K_L \rightarrow \mu\mu$ is explicitly CP-violating and unobservably small in the standard KM model[17]

$$P \equiv (\mu_R^- - \mu_L^-)/(\mu_R^- + \mu_L^-) \approx 7 \times 10^{-4} \quad \text{(KM)}, \tag{8}$$

an observation of P larger than $\approx 10^{-3}$ would be clear evidence for new physics. Following Herczeg[17] and ignoring the small contribution to P from the CP impurity of the K_L, one can write the amplitude for $K_L \rightarrow \mu\mu$ as the sum of two terms:

$$A(K_L \rightarrow \mu^+\mu^-) = a \ \bar{u}(p_-)\gamma_5 v(p_+) + ib \ \bar{u}(p_-)v(p_+) \tag{9}$$

where the first term, which represents a singlet S-wave, is CP-conserving and the second term, which represents a triplet P-wave, is CP-violating. The amplitude b can arise[17] from an effective Hamiltonian of the form

$$H_{eff} = \frac{G_F}{\sqrt{2}} \, g_{SP} \, si\gamma_5 d \, \bar{\mu}\mu + h.c. \tag{10}$$

in terms of which the polarization P is simply[17]

$$P = (1.8 \times 10^6) \, Re \, g_{SP}. \tag{11}$$

Note that the effective Hamiltonian (10) cannot result from the direct-channel exchange of a gauge boson such as an HETC (Fig. 2c). It also follows that there is no significant contribution to P in the standard KM model or in left-right symmetric models without additional Higgs fields (apart from a possible, small Dirac neutrino

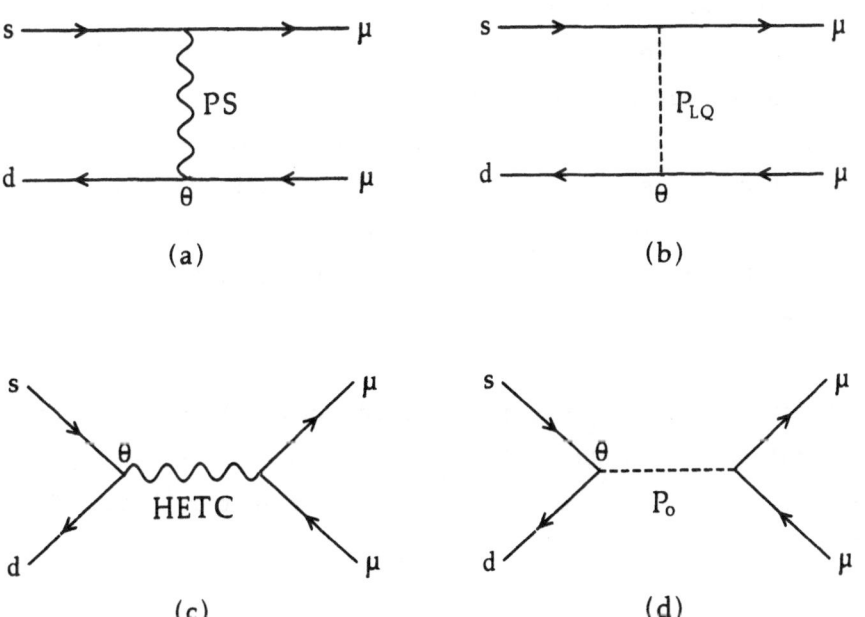

Fig. 2. Diagrams contributing to polarization in $K_L \rightarrow \mu\mu$.

mass term in the "box" contribution to $K_L \to \mu\mu$). It is possible,
however, that cross-channel PS gauge bosons (Fig. 2a) or lepto-quark
salars P_{LQ} (Fig. 2b) could contribute to (10). However, Shankar[18]
and Ellis[2] have pointed out that $K_L \to \mu\mu$ places comparable bounds on
the masses of these particles to those derived from $K_L \to \mu e$ decay,
except for a potentially small off-diagonal mixing θ which <u>weakens</u>
the $K_L \to \mu\mu$ bound. It would therefore seem that $K_L \to \mu e$ is a better
place to look for cross-channel PS bosons and lepto-quark scalars.
[This probably holds true despite the fact that polarization in
$K_L \to \mu\mu$ is an interference effect, and an upper bound on P of 15%
(for example) would correspond to a $(.15)^2 \approx .02$ improvement in the
experimental branching ratio for $K_L \to \mu e$.] The reasons for this
assertion are:

1) A factor of 50 improvement in the experimental bound on
   $K_L \to \mu e$ is not unexpected;

2) The contribution of PS and lepto-quark bosons to polarization in $K_L \to \mu\mu$ is suppressed by a potentially small off-
   diagonal mixing angle θ;

3) Polarization in $K_L \to \mu\mu$ requires CP-violating couplings which
   may or may not be present;

4) Polarization in $K_L \to \mu\mu$ is not lepton number-violating, and
   thus a measurement of P would not constitute clear evidence
   for PS or lepto-quark bosons (e.g., it could signal the
   exchange of direct-channel Higgs bosons with flavor-changing
   couplings).

Polarization in $K_L \to \mu\mu$ is, however, an excellent place to look
for technipions or neutral Higgs bosons with flavor-changing
couplings. Direct channel Higgs exchanges result in[17] an effective
Hamiltonian

$$H_{eff}^{SP} = i \frac{f_{\mu\mu}^S \, f_{sd}^P}{M_{P_0}^2} \, \bar{s}\gamma_5 d \, \bar{\mu}\mu \qquad (12)$$

where $f^S_{\mu\mu}$ (f^P_{sd}) is the scalar (pseudoscalar) coupling to the muons (quarks) necessary to produce polarization. One should note, however, that a related diagram would also generate an effective $\Delta S=2$ interaction which would contribute to $K^0-\bar{K}^0$ mixing:

$$H^{\Delta S=2}_{eff} = \frac{(f^P_{sd})^2}{M^2_{P_0}} \bar{s}\gamma_5 d \; \bar{s}\gamma_5 d \qquad (13)$$

which leads to[17] a lower bound on M_{P_0}:

$$M_{P_0} > 6 \times 10^{-6} \; |f^P_{sd}| \; \text{GeV.} \qquad (14)$$

Combining equations (12) and (14) one observes[17] that $K^0-\bar{K}^0$ mixing places an upper bound on polarization in $K_L \to \mu\mu$:

$$P < 6 \times 10^{-3} \; |f^S_{\mu\mu}/f^P_{sd}| \; . \qquad (15)$$

Thus an experiment which can measure P 15% would be sensitive to $f^S_{\mu\mu}/f^P_{sd} > 20$. This is not an unexpectedly large value for $f^S_{\mu\mu}/f^P_{sd}$ given that f^P_{sd} is off-diagonal in generation space.

In fact, $K^0-\bar{K}^0$ mixing provides an even more stringent constraint on the direct-channel scalar exchange contribution to $K_L \to \mu e$ decay (Fig. 1d). A contribution from Fig. 1d to $K_L \to \mu e$ near the present experimental bound would require the analogous $f_{\mu e}/f_{sd}$ to be >100, where these is no reason to expect this ratio to be greater than one! Since the $K_L \to \mu e$ rate is proportional to $(f_{\mu e})^2$, one does not expect a contribution to $K_L \to \mu e$ from Fig. 1d larger than 10^{-4} times the current experimental bound. We therefore conclude that polarization in $K_L \to \mu\mu$ is a better place to look for flavor-changing technipions and neutral Higgs bosons than $K_L \to \mu e$ decay--provided they have just* the right CP properties.

---

*One must be careful to arrange couplings which do not give too much CP-violation in $K^0-\bar{K}^0$ mixing, for which the constraints are significantly more stringent than those discussed here.

$K^+ \to \pi^+ +$ "UNOBSERVED"

In recent years, the standard model prediction for the branching ratio for $K^+ \to \pi^+ \nu\nu$ has considerably improved,[19,20] making this decay a more precise test of radiatively induced neutral currents as well as a more sensitive probe of new physics. In the standard model, $K^+ \to \pi^+ \nu\nu$ occurs through one-loop diagrams[21,22,23] involving $W^\pm$ and/or Z^0 exchange, giving rise to an effective $sd\nu\nu$ vertex of the form[23]

$$H_{eff} = \frac{G_F}{\sqrt{2}} \frac{\alpha}{\pi \sin^2\theta_W} \bar{s}_L \gamma^\mu d_L \, \bar{\nu}_L \gamma_\mu \nu_L \sum_q U^*_{qs} U_{qd} D(x_q) \qquad (16)$$

where

$$D(x_q) = \frac{1}{8}\left[1 + \frac{3}{(1-x)^2} - \frac{(4-x)^2}{(1-x)^2}\right] x \ln x + \frac{x}{4} - \frac{3}{4}\frac{x}{1-x}, \quad x \equiv \frac{M_q^2}{M_W^2}$$

and the sum over "q" includes all charge 2/3 quarks excluding the up-quark.

The branching for $K^+ \to \pi^+ \nu\nu$ per lepton flavor can be normalized to that for $K^+ \to \pi^0 e^+ \nu$ with the result[19,23]

$$BR(K^+ \to \pi^+ \nu\nu) = \frac{0.61 \times 10^{-6}}{|U_{us}|^2} \left| \sum U^*_{qs} U_{qd} D(x_q) \right|^2 \qquad (17)$$

$$= 0.61 \times 10^{-6} \left| D(x_c) + s_2(s_2+s_3 e^{i\delta}) D(x_t) \right|^2$$

in the approximation of small KM mixing angles (a good approximation in light of recent constraints on the mixing angles from the measured b-quark lifetime). The QCD corrections to (16) have been computed[19] and amount to a very small correction. It has also been assumed that the amplitude for $K^+ \to \pi^+ \nu\nu$ is short-distance dominated, which is believed[19] to be a good approximation.

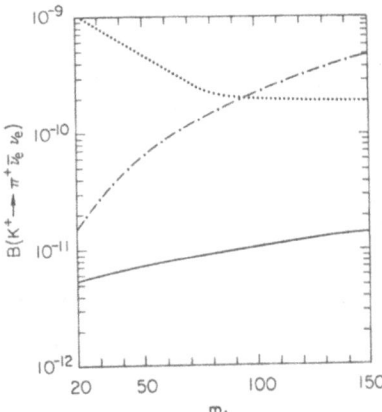

Fig. 3. Lower and upper bounds on BR($K^+ \rightarrow \pi^+ \nu_e \bar{\nu}_e$): solid line--lower bound corresponding to $\tau_b < 1.5 \times 10^{-12}$ sec; dash-dotted line--upper bound corresponding to $\tau_b > 0.6 \mathrm{c} 10^{-12}$ sec; dotted line--previous upper bound from analysis of $K_L \rightarrow \mu\mu$.

A measurement of BR($K^+ \rightarrow \pi^+ \nu\nu$) above or below the range shown in Fig. 3 could be interpreted as evidence for additional generations, which could either increase the branching ratio by providing more decay modes, or even decrease the braching ratio through destructive interference from heavy charge 2/3 quark loops. Alternatively, such a measurement could constitute evidence for interesting new physics, which can come in several varieties.

Since we know[24] that Re $s_2(s_2+s_3e^{i\delta})$ is positive, the two terms in equation (17) arising from the c- and t-quarks interfere constructively, and the c-quark contribution by itself therefore provides a lower bound on the branching ratio 0.5×10^{-11} per neutrino flavor. But one can do better[20] in terms of both a lower and upper bound by including the constructive interference from the t-quark contribution and using the measured b-quark lifetime to constrain the mixing angles s_2 and s_3. Note that

$$\left| s_2+s_3e^{i\delta} \right| = \left| s_3+s_2e^{i\delta} \right| = \left| U_{bc} \right| = 0.059(10^{-12}\text{sec}/\tau_b)^{\frac{1}{2}} \qquad (18)$$

where the number 0.059 is derived[20] from the semi-leptonic branching ratio BR(b→ceν) = 13%, taking the b-quark mass to be 4.7 GeV and the c-quark mass to be 1.5 GeV. [It is trivial to adjust (18) according to fashion.] The mixing angle s_3 is bound directly by the experimental bound on the ratio R ≡ BR(b →ueν)/BR(b →ceν):

$$s_3 < 0.040(10^{-12}\text{sec}/\tau_b)^{\frac{1}{2}} \qquad (19)$$

for R<0.05. This leads through (18) to an allowed range for s_2:

$$0.019(10^{-12}\text{sec}/\tau_b)^{\frac{1}{2}} < s_2 < 0.099(10^{-12}\text{sec}/\tau_b)^{\frac{1}{2}} \qquad (20)$$

which becomes far more restrictive as R decreases.

The constraints (18) and (20) lead[20] to the upper and lower bounds on $K^+ \to \pi^+\nu\nu$ shown in Fig. 3 computed, respectively, for $\tau_b=0.6\times10^{-12}$ sec and 1.5×10^{-12} sec. These upper and lower bounds improve considerably as the b-quark lifetime is further restricted or as the experimental bound on the ratio R decreases below 0.05 (as suggested by evidence presented[25] at this conference). The dotted line in Fig. 3 shows a previous[19] upper bound on $K^+ \to \pi^+\nu\nu$ derived from a short-distance analysis of $K_L \to \mu\mu$ decay.

Supersymmetry can change the predicted rate for $K^+ \rightarrow \pi^+ \nu\nu$ by adding new loop contributions to the effective sd vertex, as well as by adding entirely new modes for $K^+ \rightarrow \pi^+ +$ "nothing". "Nothing" could be a pair of light photinos,[19,26] Higgsinos,[19] or conceivably sneutrinos.[27] The rate for $K^+ \rightarrow \pi \tilde{\gamma} \tilde{\gamma}$ depends on the unknown spectrum of supersymmetric particles, but generally[19,26]

$$BR(K^+ \rightarrow \pi^+ \tilde{\gamma}\tilde{\gamma}) \lesssim 1/10 \; BR(K^+ \rightarrow \pi^+ \nu_e \nu_e). \qquad (21)$$

It is somewhat more likely in the context of today's supersymmetric models that a supersymmetric particle with mass < 0(100) MeV would be a neutral Higgsino. The branching ratio for $K^+ \rightarrow \pi^+ \tilde{H}\tilde{H}$ is likely[19] to be comparable to that for an ordinary neutrino pair, and is therefore more likely to bump up the rate for $K^+ \rightarrow \pi^+ +$ "nothing" than is the photino. Finally, it is conceivable, though not likely in the context of currently fashionable supersymmetric models, that the sneutrino could be the lightest supersymmetric particle with a mass below 0(100) MeV. In this case $K^+ \rightarrow \pi^+ +$ "nothing" could be enhanced significantly.[27]

It is consceivable that $K^+ \rightarrow \pi^+ +$ "nothing" could appear with a two-body rather than a three-body spectrum, for which the present experimental bound is[28]

$$BR(K^+ \rightarrow \pi^+ + \text{"nothing"})_{2-BODY} < 3.8 \times 10^{-8}. \qquad (22)$$

Here it is unlikely that "nothing" would be a conventional axion, since such an axion should[29] already have been observed in this decay if one existed. "Nothing" could, however, be a "familon"[30]-- a massless Goldstone boson of a conjectured family symmetry. In this case one would expect[2,30]

$$BR(K^+ \rightarrow \pi^+ + \text{familon}) \lesssim 10^{-10} \qquad (23)$$

which means that the familon will be observed or ruled out in the coming generation of $K^+ \to \pi^+$ + "nothing" experiments.

Finally, "nothing" could be a π^0, which then decays into unobserved particles. A $K^+ \to \pi^+$ + "nothing" experiment is certainly the best place to "look for" an unobserved π^0, which can decay into massive neutrinos or, under certain circumstances,[26] photinos.

$K^+ \to \pi^+$ + "nothing" is one of the last open frontiers among flavor-changing neutral interactions that has not yet been fully explored. A measurement of $K^+ \to \pi^+ \nu\nu$ would provide a quantitative test of the radiatively induced neutral current structure of the standard model. A measurement of $K^+ \to \pi^+$ + "nothing" outside the standard model prediction would constitute exciting evidence for new physics, which could alter the effective sd$\nu\nu$ vertex or provide entirely new modes such as $K^+ \to \pi^+$ + familon, photinos, Higgsinos, etc.

I would like to thank G. L. Kane for useful discussions related to this work.

REFERENCES

1. L. Hall and M. Suzuki, Nucl. Phys. B231 (1984) 419.
2. J. Ellis, Proc. of the Conf. on Intense Medium Energy Sources of Strangeness, eds. T. Goldman,, H.E. Haber, and H.F.-W. Sadrozinski (UC-Santa Cruz, 1983), p. 191.
3. For reviews and references, see P. Fayet and S. Ferrara, Phys. Rep. 32C (1977) 249;
   P. Fayet, Ecole Normale Superieure preprint LPTENS 82/28 (1982), talk presented at the Int. Conf. on High Energy Physics, Paris 1982.
4. S. Weinberg, Phys. Rev. D13 (1976) 974 and Phys. Rev. D19 (1979) 1277;
   L. Susskind, Phys. Rev. D20 (1079) 2619.
5. E. Eichten and K. D. Lane, Phys. Lett. 90B (1980) 125;
   S. Dimopoulos and L. Susskind, Nucl. Phys. B155 (1979) 237.
6. S. Dimopoulos, Nucl. Phys. B168 (1980) 69;
   M. E. Peskin, Nucl. Phys. B175 (1980) 197;
   J. P. Preskill, Nucl. Phys. B177 (1981) 21.
7. J. C. Pati and A. Salam, Phys. Rev. Lett. 31 (1973) 661; Phys. Rev. D8 (1973) 1240; Phys. Rev. D10 (1974) 275.

8. S. Dimopoulos, Nucl. Phys. B168 (1980) 69;
    S. Chadha and M. E. Peskin, Nucl. Phys. B185 and Nucl. Phys.
    B187 (1981) 541;
    V. Baluni, Univ. of Michigan preprint UM HE 82-27 (1983).
9. R. Binetruy, S. Chadha and M. E. Peskin, Phys. Lett. 107B
    (1981) 425 and Nucl. Phys. B207 (1982) 505.
10. S. Dimopoulos, S. Raby and G. Kane, Nucl. Phys. B182 (1981)
    77; S. Dimopoulos, S. Raby and P. Sikivie, Nucl. Phys. B182
    (1981) 449.
11. J. Ellis, M. K. Gaillard, D. V. Nanopoulos and P. Sikivie,
    Nucl. Phys. B182 (1982) 529;
    J. Ellis, D. V. Nanopoulos and P. Sikivie, Phys. Lett. 101B
    (1981) 387.
12. A. Clark et al., Phys. Rev. Lett. 26 (1971) 1667.
13. A. M. Diamant-Berger et al., Phys. Lett. 62B (1976) 485.
14. J. Ellis and P. Sikivie, Phys. Lett. 104B (1981) 141.
15. Particle Data Group, Phys. Lett. 75B (1978) 1.
16. M. V. Voloshin and E. P. Shabalin, ZHETF Pis'ma 23 (1976) 123
    [JETP Lett. 23 (1976) 107].
17. P. Herczeg, Phys. Rev. D27 (1983) 1512.
18. O. Shankar, Nucl. Phys. B206 (1982) 253.
19. J. Ellis and J. S. Hagelin, Nucl. Phys. B217 (1983) 189.
20. F. J. Gilman and J. S. Hagelin, SLAC-PUB-3226 (1983) to appear
    in Phys. Lett. B.
21. M. K. Gaillard and B. W. Lee, Phys. Rev. D10 (1974) 897.
22. E. Ma and J. Okada, Phys. Rev. D18 (1978) 4219.
23. T. Inami and C. S. Lim, Prog. Theor. Phys. 65 (1981) 297.
24. F. J. Gilman and J. S. Hagelin, Phys. Lett. 126B (1983) 161.
25. J. Lee-Franzini, these proceedings.
26. M. K. Gaillard, Y.-C. Kao, I. -H. Lee and M. Suzuki, Phys.
    Lett. 123B (1983) 241.
27. J. S. Hagelin, G. L. Kane and S. Raby, LANL preprint no.
    LA-UR-83-3711 (1983) to appear in Nucl. Phys. B.
28. Y. Asano et al., Phys. Lett. 107B (1981) 159.
29. T. Goldman and C. M. Hoffman, Phys. Rev. Lett. 40 (1978) 220;
    J.-M. Frere, M. B. Gavela and J. Vermaseren, Phys. Lett. 103B
    (1981) 129.
30. F. Wilczek, Phys. Rev. Lett. 49 (1982) 1549.

DISCUSSION

MOHAPATRA:

I disagree with your statement that the box graph contribution to $K_L \to \mu\bar{\mu}$ in left-right models is negligibly small because the Dirac mass of ν_μ which enters these graphs can be pretty big ($\sim$ 100 MeV) even though the physical m_ν is small.

HAGELIN:

This is precisely the point. $(100 \text{ MeV})^4$ is tiny even compared to the charm-quark contribution which is $(1.5 \text{ GeV})^4$, let alone the top-quark contribution.

KANE:

By the time the experiment for $K^+ \to \pi^+\nu\bar{\nu}$ is completed, probably we will know m_t, τ_b and the BR($b \to u$). Will the standard model prediction then be precise? What will the uncertainty from a knowledge of m_c or other sources be?

HAGELIN:

The standard model prediction is very precise once the quark masses and mixings are known. The major uncertainty for the time being is $|V_{td}|$ or, equivalently, s_2. Our knowledge of s_2 improves as the upper bound on $\Gamma(b \to u)/\Gamma(b \to c)$ improves, but for the foreseeable future, this will remain the major uncertainty.

The uncertainty due to the charm-quark mass is probably small when you consider the fact that the top quark dominates the amplitude for the presently preferred values of the mixing angles.

RÜCKL:

Doesn't the absence of the decay $\tau \to e\tilde{\nu}\tilde{\nu}$ make the decay $K \to \pi\tilde{\nu}\tilde{\nu}$, proposed by you, very unlikely?

HAGELIN:

$\tau \to e\tilde{\nu}_\tau\bar{\nu}_e$, $\tau \to \mu\tilde{\nu}_\tau\bar{\nu}_\mu$, $\mu \to e\tilde{\nu}_\mu\bar{\nu}_e$ all provide useful constraints on $K^+ \to \pi^+\nu\nu$. But for example you could have

$m_{\tilde{\nu}_\tau} \gtrsim m_\tau$ and $m_{\tilde{\nu}_e} + m_{\tilde{\nu}_\mu} \gtrsim m_\mu$ (which would satisfy all the

constraints from τ and μ decays) and still have $K \to \pi\tilde{\nu}_e\bar{\nu}_e$ and

even $K \to \pi\tilde{\nu}_\mu\bar{\nu}_\mu$.

HITLIN:

Is there any theoretical motivation to improve on the "classical" K decay measurements such as $\text{Re}\epsilon$ (it is possible to measure $\text{Re}\epsilon$ as a function of q^2 in $K_{\mu 3}$ decay), the form factors or $\Delta S = - \Delta Q$ amplitudes?

HAGELIN:

This is a theorist- and experimentalist-dependent question. I would personally be interested in a more precise comparison of $|\epsilon|$ from $K_L \rightarrow 2\pi$, δ or $\text{Re}\epsilon$ from $K_L \rightarrow \pi^{\pm} \ell^{\mp} \nu$, and ϕ_{+-}. I understand, however, that these experiments are very hard to improve upon.

In general I think we could say that measurements of ϵ'/ϵ, $K^+ \rightarrow \pi^+ \bar{\nu}_i \nu_i$, $K_L^0 \rightarrow \mu e$, and possibly polarization in $K_L^0 \rightarrow \bar{\mu}\mu$ constitute the most important K experiments.

PARTIAL RATE DIFFERENCES FROM CP VIOLATION AT LEAR

P. Pavlopoulos

CERN, Geneva, Switzerland

1.   INTRODUCTION

The discovery of the quark families and the development of the
standard model, with the recent discovery of the intermediate
bosons, have truly enlightened our understanding of the electroweak
and the strong interactions. Therefore, it is important to ask
what new physics can be learned by studying phenomena which are not
foreseen by and have no direct relation to the standard model.
Besides the decay of the proton, the neutrino masses, and the
neutrino oscillations, the breakdown of the CP invariance in nature
has an impact on problems beyond the standard model, such as the
baryon asymmetry and the grand unification.

From our present knowledge of the production and decay of
c and b quark states in known and planned accelerators and storage
rings, it will be extremely difficult to determine the source of
CP violation in heavy-quark decays. Consequently, the neutral-
kaon system and the hyperon decays still represent the most promising means of investigating the breakdown of the CP invariance.

In a properly chosen convention there are two parameters
describing the CP non-conservation in the neutral-kaon system: a
small complex number ε specifying the CP impurity of the observed

eigenstates K_S^0 and K_L^0, which is a measure of the CP violation in the mass matrix, and the amplitude ε' measuring the violation of the CP symmetry in the decay matrix. Although the ratio $|\varepsilon'/\varepsilon|$ is required in order to quantify the source of CP violation, its magnitude is suppressed, mainly because of the $\Delta I = \frac{1}{2}$ rule. Therefore, any attempts to look for other CP-violating phenomena are very welcome. In this context two main arguments have been discussed:

i)   The partial rate difference between particles and antiparticles introduced and worked out by Chau[1-3].

ii) The measurement of CP violation in decay channels other than into two pions, namely the 3π, $\pi^+\pi^-\gamma$, and 2γ mode.

In order to study experimentally both of these new attempts, it is necessary to have a high-yield clean source of particles and antiparticles, which can be achieved at the Low-Energy Antiproton Ring (LEAR) at CERN[4]. This machine is particularly suited for precise measurements, as required by experiments looking for CP violation, owing to its high performances, i.e. 10^6 antiprotons per second in the momentum range of 50 MeV/c to 2 GeV/c with a momentum spread of better than 10^{-3}. In the middle of 1987 the intensity of LEAR will increase by a factor of 10, reaching 10^7 antiprotons per second, owing to the construction of the Antiproton Collector (ACOL). By using the high-intensity antiproton source at LEAR, it is possible to produce a well-defined source of K^0 and $\bar{K}^0$ mesons through the reactions

$$\bar{p} + p \text{ (at rest)} \begin{cases} K^+\pi^-K^0 \ (2 \times 10^{-3}) \\ K^-\pi^+\bar{K}^0 \ (2 \times 10^{-3}) \end{cases} \tag{1}$$

In these reactions K^0 and $\bar{K}^0$ can be uniquely identified through the detection of the companion K^+ and K^-, respectively. Hence the total rate of tagged K^0's and $\bar{K}^0$'s is $\sim 2 \times 10^8$ per day, assuming a stop rate of 10^6 antiprotons per second ($\sim 10^{11}$ antiprotons per day). The possibility of tagging K^0's and $\bar{K}^0$'s separately, with a flux such as the one provided by LEAR, is unique and is almost inconceivable at external beams.

2. DIFFERENCE IN THE PARTIAL DECAY RATES BETWEEN PARTICLES AND ANTIPARTICLES

Recent theoretical ideas[1-3] have given rise to the possibility of a new effect in the difference of the partial decay rates between particles and antiparticles. It has been argued that besides contributing CP-violation effects in the mass matrix, the complexity in the mixing matrix can also give rise to CP-violation effects in the partial decay rates due to interference between the weak interaction amplitudes and the strong interaction amplitudes. These possible new CP-violating effects are measured by the asymmetry

$$D_K = \frac{|A(K^0 \to 2\pi)|^2 - |A(\bar{K}^0 \to 2\pi)|^2}{|A(K^0 \to 2\pi)|^2 + |A(\bar{K}^0 \to 2\pi)|^2} = 4\left[\text{Re } \Delta_K + \text{Re } \varepsilon'\right] \qquad (2a)$$

or

$$D_\Lambda = \frac{|A(\Lambda \to N\pi)|^2 - |A(\bar{\Lambda} \to \bar{N}\pi)|^2}{|A(\Lambda \to N\pi)|^2 + |A(\bar{\Lambda} \to \bar{N}\pi)|^2} = \text{Re } \Delta_\Lambda + \text{Re } \varepsilon'_\Lambda , \qquad (2b)$$

where the quantity Δ_K (Δ_Λ) appears because of the simultaneous complexity of the coupling constants and the complexity of the amplitudes arising from a possible absorptive part. It is important to notice that Δ_K (Δ_Λ) arises from possible interference in the dominant isospin amplitude and thus it is not suppressed by the $\Delta I = \frac{1}{2}$ rule, whereas the quantity Re ε' arises from the interference of the different isospin amplitudes and thus it is suppressed by the $\Delta I = \frac{1}{2}$ rule. However, as is shown elsewhere[5], the quantity Δ_K (Δ_Λ) is a CPT-violating parameter and thus CPT invariance implies striking limitations to Δ_K (Δ_Λ), resulting in a difference in the partial decay rates of the order of ε':

$$D_K = 4 \text{ Re } \varepsilon' , \qquad D_\Lambda = \text{Re } \varepsilon'$$

with

$$\frac{\varepsilon'}{10} < \varepsilon_\Lambda \leq \varepsilon' . \qquad (3)$$

It is interesting that comparison of the measurements of the asymmetry D_K (D_Λ) with the measurements of $|\varepsilon'/\varepsilon|$ from other methods will provide a very sensitive test for the CPT invariance in the $\Delta S = \pm 1$ current[6,7].

2.1 The neutral-kaon decays

In order to study CP-violation effects in the neutral-kaon system with this novel experimental method with initially pure K^0 and $\bar{K}^0$ states, we consider a detector[7] consisting of a magnetic spectrometer and a gamma-ray calorimeter, suitable for measuring the asymmetry of intensities in different K^0 and $\bar{K}^0$ decay channels (Fig. 1). The magnetic field has to be parallel to the beam in order to stop the $\bar{p}$'s in a "point-like" hydrogen target at the centre of the spectrometer. The initial atomic $p\bar{p}$ state has a precisely defined energy; hence the neutral-kaon direction and momentum are defined through the $K^\pm \pi^\mp$ kinematics, whilst the strangeness of the K^0 and $\bar{K}^0$ is determined by the sign of the charged kaon, which is determined in the magnetic spectrometer. This procedure results in the absolute and independent determination of the flux of K^0's and $\bar{K}^0$'s. The identification of the charged kaons in the relevant momentum range (≤ 750 MeV/c) can be done using time-of-flight (TOF) and conventional Cherenkov detectors, e.g. FC72 or H_2O, at the outer radius of the detector. In order to achieve a high Cherenkov efficiency it is important to extend the photomultiplier sensitivity in the ultraviolet region. The main source of background arises from pions being misidentified as kaons from the reactions $\pi^+\pi^-\pi^0\pi^0$ [10% of all annihilations] and $\pi^+\pi^-\pi^0\pi^0\pi^0$ [23% of all annihilations][8]. In addition to the Cherenkov detector, kinematical constraints will help to isolate the $K^\pm \pi^\mp K^0$ channel. Kaonic channels[9] involving more particles, such as $K^\pm \pi K^0 \pi^0$, can be suppressed by a factor of more than 100 by these kinematical selections. However, pion events, which will be misidentified as kaons because of inefficiencies, will introduce a continuous background in the $K^\pm \pi^\mp$ missing-mass spectrum.

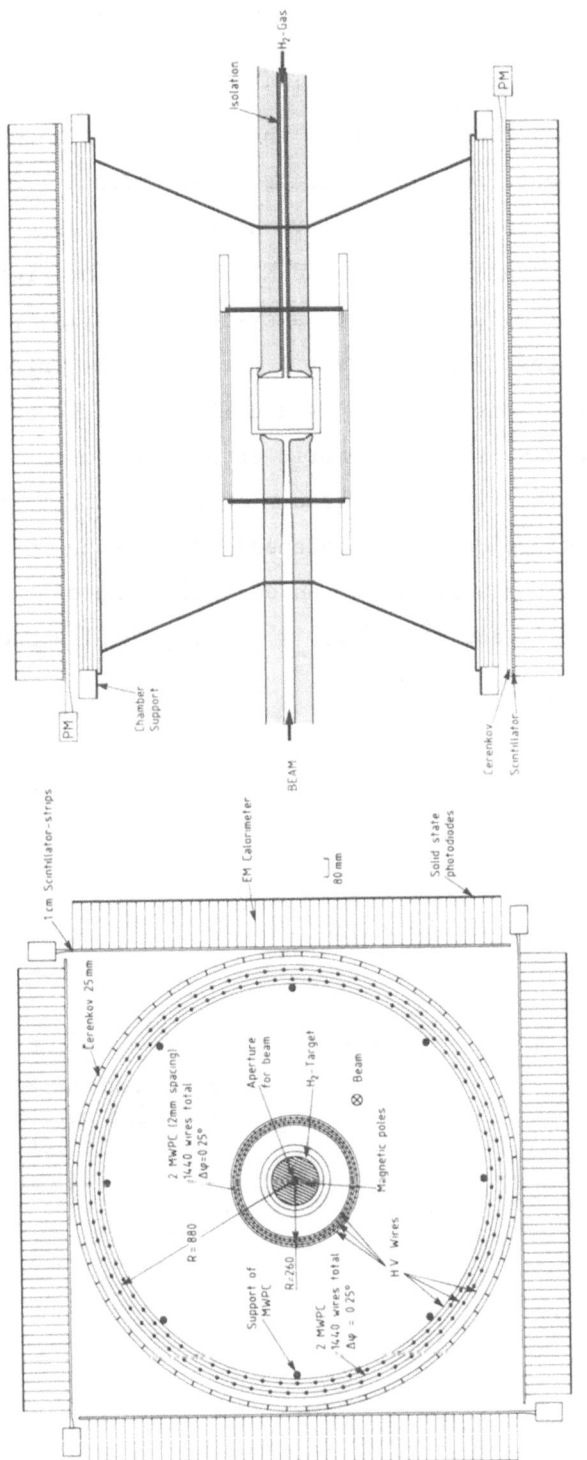

Fig. 1. A schematic view of the experimental set-up.

This small background contribution can by no means create any systematic asymmetry and can be used in monitoring the overall trigger efficiency.

We are aware that in the physical apparatus $K^+\pi^-$ and $K^-\pi^+$ behave differently; total cross-sections, absorption, small-angle scattering, etc., are different, and the detector itself with $\vec{E} \times \vec{B}$ effects cannot be completely charge-symmetric. However, all observables are normalized independently for K^0 and $\bar{K}^0$, and any inefficiency in the K^0 and $\bar{K}^0$ trigger will not create a systematic asymmetry in the partial decay rates. But systematic errors can be introduced by the different interactions of the K^0's and $\bar{K}^0$'s with the target surroundings. Nevertheless, such systematic errors can be eliminated if interactions in the first four decay lengths (~ 25 cm) around the target are minimized, because, independently of the initial state, the neutral kaon contains, after an eigenlifetime of $\sim 4\ \tau_S$, the same amount of K^0's and $\bar{K}^0$'s (Fig. 2). Target self-absorption and

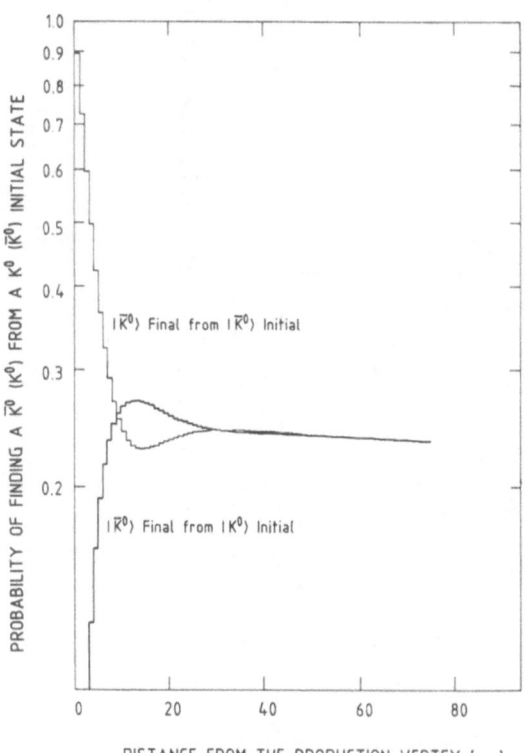

Fig. 2. The $\bar{K}^0$ contents of a neutral kaon as a function of the distance from the production vertex.

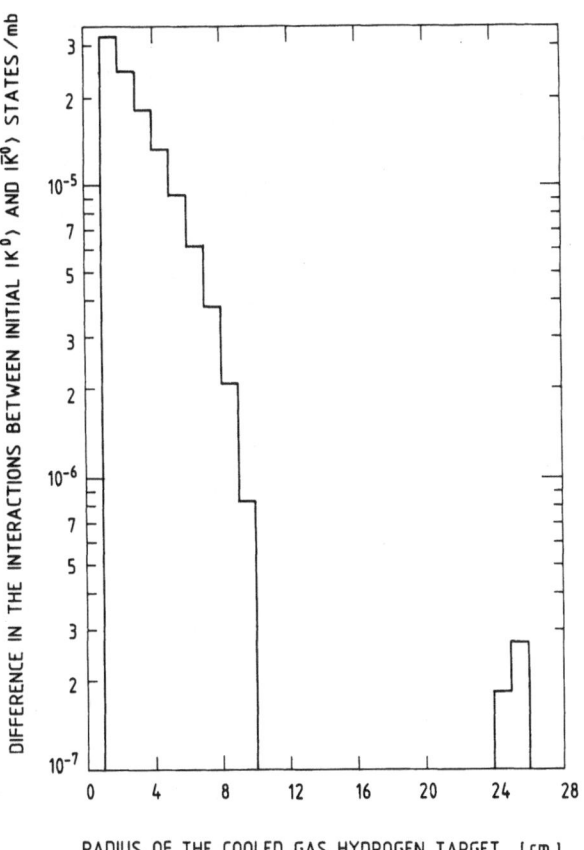

Fig. 3. The difference in the interactions between K^0 and $\bar{K}^0$ as a function of the target radius for a cooled hydrogen gas (27 K, 1 atm) target. The main difference in the interactions comes from the 1 mm thick mylar target window.

interactions with materials around the target -- which are unavoidable -- result in a systematic error of less than 10^{-7} per millibarn in the difference between the tagged K^0's and $\bar{K}^0$'s for the case of a cooled hydrogen-gas target at 1 atm (Fig. 3).

It can be seen that any measurement of the neutral-kaon decay rates should cover the time interval out to $\sim 20\ \tau_S$ (80 cm), in order to cover the region of maximum interference at $14\ \tau_S$. The magnetic spectrometer, with a momentum resolution of a few per cent ($\sigma \sim 4\%$), will enable us to define the neutral-kaon four-momentum and detect the decays over a distance of ~ 80 cm path length. The knowledge of the neutral-kaon momentum vector allows, by a 2C-fit, the determination of the vertex with a resolution of better than 1 cm, without the magnetic analysis of the charged decay pions.

Starting from K^0 and $\bar{K}^0$ states it is possible to define a time-dependent asymmetry factor between K^0 and $\bar{K}^0$ rates given by

$$A(t) = \frac{R[K^0 \rightarrow 2\pi^{\pm}](t) - R[\bar{K}^0 \rightarrow 2\pi^{\pm}](t)}{R[K^0 \rightarrow 2\pi^{\pm}](t) + R[\bar{K}^0 \rightarrow 2\pi^{\pm}](t)}$$

$$= 2\left[\frac{|\eta_{+-}| e^{(\gamma_S/2)t} \cos(\Delta mt - \theta_{+-})}{1 + |\eta_{+-}|^2 e^{\gamma_S t}} - \text{Re }\varepsilon \right] \tag{4}$$

In first order, systematic errors, such as detector efficiencies, solid angles, resolutions, and small contributions from other neutral-kaon decays cancel out, and thus the asymmetry $A(t)$ is free from systematic errors. The time-dependent rate $A(t)$ is illustrated in Fig. 4 from which it is possible to extract $|\eta_{+-}|$ and θ_{+-} since they respectively determine the amplitude and position of the interference effect. Assuming[7] a geometrical acceptance of 60% and an overall trigger efficiency of better than 50%, a resolution of less than 1 cm in defining the decay vertex results in a definition of $|\eta_{+-}|$ with an accuracy of 5×10^{-3} for a 10 day run at LEAR (2×10^9 K^0's and $\bar{K}^0$'s). The measurement of the asymmetry $A(t)$, correcting for the time dependence, is sensitive to the difference between $|\eta_{+-}|$ and $|\varepsilon|$, which is the parameter $|\varepsilon'|$.

The asymmetry $A(t=0)$ is related to the parameter D_K [Eqs. (2a) and (4)] by the expression

$$A(t=0) = D_K = 4(\text{Re }\Delta_K + \text{Re }\varepsilon') . \tag{5}$$

A limit on the asymmetry $A(t=0)$ will give a limit on the magnitude of the CPT-violating parameter[5] Δ_K. The asymmetry $A(t=0)$ can be measured to an accuracy of 7×10^5 at LEAR and would provide a very sensitive test for the CPT invariance in the $\Delta S = \pm 1$ current[5]. This

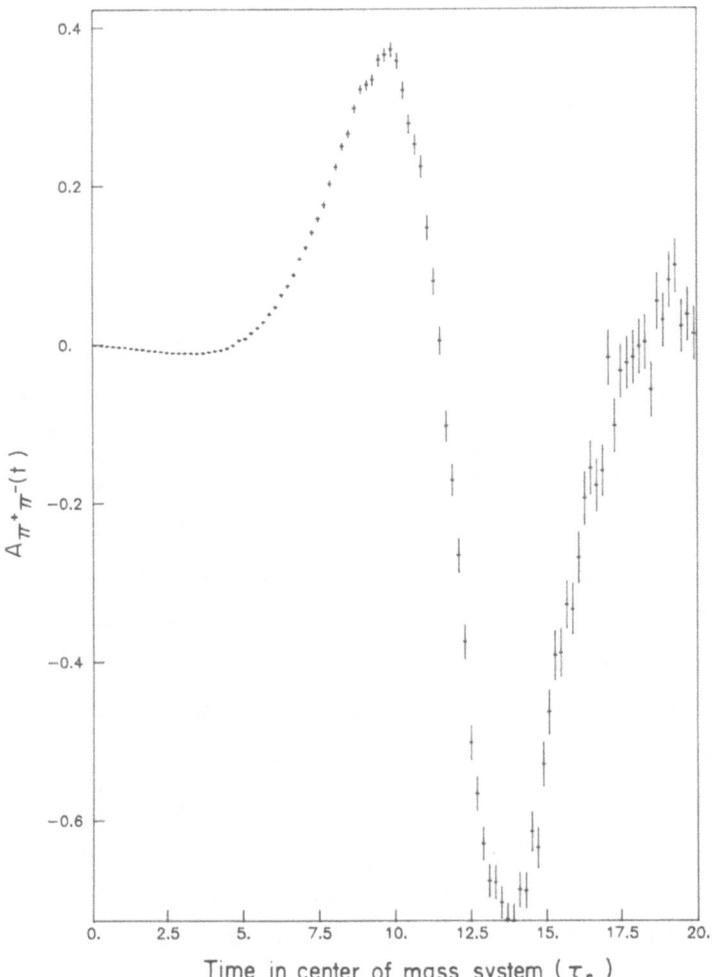

Fig. 4. The asymmetry factor $A(t)$ [Eq. (4)]. The statistical errors correspond to 2×10^9 K^0's and $\bar{K}^0$'s.

measurement will improve the actual limit by roughly one order of magnitude.

In order to determine the partial decay difference in the two neutral-pion decays, we abandon the time-dependent measurement of the neutral decays -- hence we have the cumbersome vertex reconstruction from the gamma-rays -- and use only the integral rates in a time interval, i.e. t=0 to $t_0 \approx 20$ τ_S, given by

$$I^{00} = \frac{\int_0^{t_0} R(K^0 \to \pi^0 \pi^0) dt - \int_0^{t_0} R(\bar{K}^0 \to \pi^0 \pi^0) dt}{\int_0^{t_0} R(K^0 \to \pi^0 \pi^0) dt + \int_0^{t_0} R(\bar{K}^0 \to \pi^0 \pi^0) dt} \approx 4 \operatorname{Re} \eta_{00} - 2 \operatorname{Re} \varepsilon. \tag{6}$$

The detection of the neutral-decay products over a distance of 80 cm will require a photon detector of large solid angle (70-80%), and of relatively good energy resolution and modularity, which can detect photons down to low energies ($\sim$ 10 MeV). The decays into three neutral pions result only from the K_L component; thus these decays represent only $\sim$ 5% of the decays into two neutral pions inside the decay volume considered. We would like to emphasize that, to the order of the achievable statistical accuracy, the integral asymmetry I^{00}, similar to the asymmetry A(t), is free from systematic errors. In a run of 10 days (2×10^9 K^0's and $\bar{K}^0$'s) the neutral asymmetry I^{00} can be defined to an accuracy of $\sim$ 2%.

By measuring the particle (K^0) - antiparticle ($\bar{K}^0$) difference in the partial decays into two charged pions and into two neutral pions we can extract the ratio $|\varepsilon'/\varepsilon|$ [10]. The ratio $|\varepsilon'/\varepsilon|$ as a function of $|\eta_{+-}|$ and I^{00} is given by the expression

$$10|\varepsilon'/\varepsilon| = \left[2 - \frac{I^{00}}{\operatorname{Re} \eta_{+-}} \right]. \tag{7}$$

Consequently, after ten days of running at LEAR we will be able to measure $|\varepsilon'/\varepsilon|$ to an accuracy of 2×10^{-3} (Fig. 5).

To summarize, we can say that this approach with initially pure K^0 and $\bar{K}^0$ states represents a novel experimental method of looking at the CP-violating phenomena. The advantage, as opposed to the standard experimental configuration with K_S and K_L, is mainly the extremely reduced systematic errors. This approach is limited mainly by the statistics and not by systematics. In conclusion, we would like to emphasize that irrespective of the theoretical estimates on

586

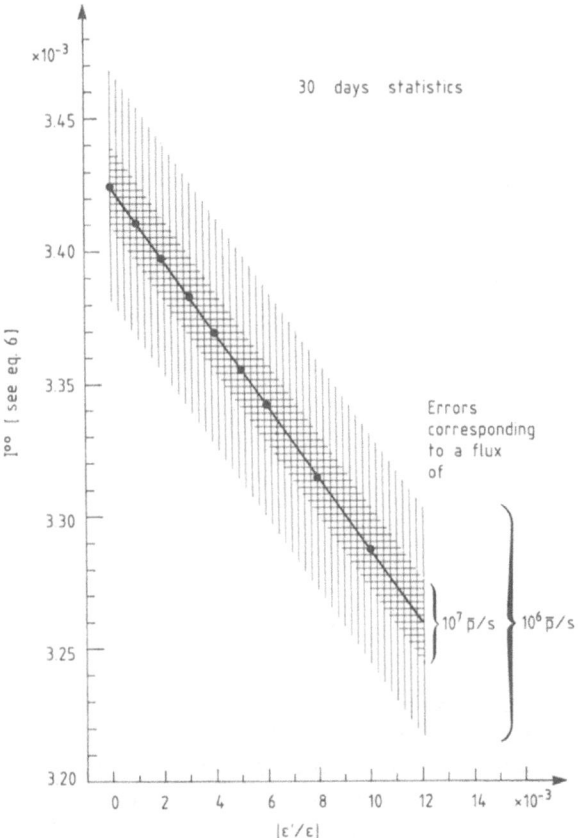

Fig. 5. The experimental sensitivity in defining the CP-violation parameter $|\epsilon'/\epsilon|$ for the two-pion decays.

the magnitude of $|\epsilon'/\epsilon|$, measuring $|\epsilon'/\epsilon|$ from the partial-decay difference between particles and antiparticles gives an independent measurement[1-3] from those existing experiments measuring ϵ'.

2.2 The hyperon decays

Now we turn to the difference in the partial hadronic decays between hyperons and antihyperons. Above 1.43 GeV/c there are the thresholds for strangeness 1 hyperon-antihyperon pairs $\bar{p}p \to \bar{Y}Y$. Up to the highest LEAR momentum $\bar{Y}$ and Y are kinematically confined in a narrow forward cone. Therefore a rather small forward detector, as used in experiment PS185[11], is sufficient to study these reactions

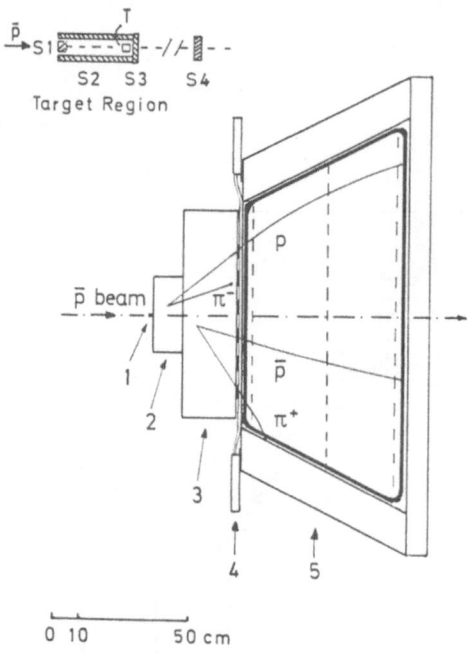

Fig. 6. A schematic view of the set-up of the experiment PS185.
1: target, 2: proportional wire chambers, 3: drift chambers, 4: hodoscope, 5: baryon number identifier. An example of a "perfect track
event" is indicated. The target region is given in a magnified view
with T: target, S1-4: scintillation counters.

with full efficiency (Fig. 6). The experiment PS185 seems to be a
good approach to measure the parameter D_Λ of Eqs. (2b) and (3). A
difference in the rates for $p\bar{p} \to \Lambda\bar{\Lambda} \to \bar{p}n\pi^+\pi^0$ and $p\bar{p} \to \Lambda\bar{\Lambda} \to \bar{n}p\pi^0\pi^-$
would be the signal for CP violation.

The normalized difference

$$\Delta = \frac{\{\bar{p}n\pi^-\pi^0\} - \{\bar{n}p\pi^0\pi^-\}}{\{\bar{p}n\pi^+\pi^0\} + \{\bar{n}p\pi^0\pi^-\}}$$

can be determined to $\sim 10^{-3}$ statistical precision in P5185 within 5
to 10 days with 10^6 $\bar{p}$/s [12].

The problem will be to control *systematic errors*. They occur
mainly (besides "wrong" triggers) owing to unsymmetries in the behaviour of $\bar{\Lambda}$ and Λ and their decay products. There are two principal

experimental differences between the neutral kaons and the hyperons in measuring partial-decay rates. Firstly, the hyperons cannot be tagged and therefore it is not possible to have independent normalization for Λ's and $\bar{\Lambda}$'s. Secondly, the decay products of the hyperons have very different behaviour as compared to the neutral kaons, where the decay products are identical for K^0's and $\bar{K}^0$'s.

a) In the laboratory system $\bar{\Lambda}$ particles are produced preferentially with larger momenta and smaller angles than the Λ particles (forward peaked differential cross-section), except very close to threshold where however cross-sections are unknown and small. The different $\bar{\Lambda}$ and Λ momentum distributions lead to different distributions for the decays in space. The differential production cross-sections will be determined in PS185.

b) Rescattering and absorption of the produced $\bar{\Lambda}$ and Λ in the target and detectors will differ. The short target (~ 2.5 mm polyethylene) and the thin detectors will help to reduce such effects.

c) Also the rescattering and absorption of the decay products ($\bar{p}$ and p, π^+ and π^-) is unsymmetric and needs corrections.

One can control the knowledge of systematic errors by performing the experiment at different $\bar{p}$ beam momenta. With some modifications a similar check of CP violation can be done for $\bar{p}p \rightarrow \bar{\Sigma}^+ \Sigma^+$.

Another good test for CP-violation effects should be a comparison of the decay asymmetries $\bar{\alpha}$ and α for $\bar{\Lambda}$ and Λ. Here $\bar{\alpha} + \alpha = 0$ has to be checked and one can reach a statistical precision of $\sim 5 \times 10^{-3}$ within 5 to 10 d with 10^6 antiprotons per second in experiment PS185. (If $\bar{\alpha} + \alpha \neq 0$ this would simulate an apparent difference in $\bar{\Lambda}$ and Λ polarization[12].)

3. CP VIOLATION IN OTHER THAN THE TWO-PION DECAY MODE

The decay of the long-lived K^0 meson, K_L, into two pions is so far the only clear evidence of CP violation; no other process has yet been uncovered. When studying CP violation in decays of neutral kaons other than into two pions, we cannot look for a simple violation of a selection rule, since both K_L and K_S are allowed to decay into such final states as $\pi^+\pi^-\gamma$ and $\gamma\gamma$, and the branching ratio of $K_S \rightarrow 3\pi$ is extremely small ($\lesssim 10^{-9}$). The following theorem[13] of Sehgal and Wolfenstein forms the basis for identifying CP violation in these channels: "For any possible non-leptonic decay mode of the neutral kaon the observation of an interference effect between K_L and K_S decays in the partial decay rate in this mode is evidence of CP violation". The most feasible way of detecting interference effects seems to be to measure the asymmetry of the decay intensities using a beam of K^0 particles as well as a beam of $\bar{K}^0$ particles. Consequently, our experimental philosophy of starting with pure K^0 and $\bar{K}^0$ states is ideally suited for measuring CP violation in these channels.

3.1 The quantity $|\eta_{000}|$

The actual limit of $|\eta_{000}|^2$ is 0.28, i.e. $|\eta_{000}| < 0.53$, whereas the parameter $|\eta_{+-0}|^2$ is less than 0.12, i.e. $|\eta_{+-0}| < 0.35$. There is a general argument about the major importance of $|\eta_{000}|$ relative to $|\eta_{+-0}|$. If CP were an exact symmetry, $K_S \rightarrow 3\pi^0$ would be forbidden (π^0 are C eigenstates), but $K_S \rightarrow \pi^+\pi^-\pi^0$ would be allowed. However, the $K_S \rightarrow \pi^+\pi^-\pi^0$ would be inhibited by an angular momentum barrier factor of the order of $(Q/m_K)^2 \simeq 1/200$, not so much different from the expected CP-violation effects. Thus the cleanest test for CP violation in the 3π decays is the $3\pi^0$ mode.

We intend to measure[7] the asymmetry of the partial decay rates between $K^0 \rightarrow 3\pi^0$ and $\bar{K}^0 \rightarrow 3\pi^0$

$$I^{000} = \frac{\int_0^{t_0} R(K^0 \rightarrow 3\pi^0)dt - \int_0^{t_0} R(\bar{K}^0 \rightarrow 3\pi^0)dt}{\int_0^{t_0} R(K^0 \rightarrow 3\pi^0)dt + \int_0^{t_0} R(\bar{K}^0 \rightarrow 3\pi^0)dt} . \tag{8}$$

The partial width of K^0's into $3\pi^0$ is

$$R(K^0 \rightarrow 3\pi^0) = \frac{R(K_S \rightarrow 3\pi^0)}{4|p|^2} e^{-t/\tau_S} + \frac{R(K_L \rightarrow 3\pi^0)}{4|p|^2} e^{-t/\tau_L} +$$

$$+ \frac{2|\eta_{000}|}{4|p|^2} R(K_L \rightarrow 3\pi^0) \exp\left[-t(\tau_S + \tau_L)/2\tau_S\tau_L\right] \times$$

$$\times \cos(\Delta mt - \phi_{000}) \tag{9}$$

with $|p|^2 = (1 + 2\ \mathrm{Re}\ \varepsilon)/2$, and the partial width of $\bar{K}^0$'s into $3\pi^0$ is

$$R(\bar{K}^0 \rightarrow 3\pi^0) = \frac{R(K_S \rightarrow 3\pi^0)}{4|q|^2} e^{-t/\tau_S} + \frac{R(K_L \rightarrow 3\pi^0)}{4|q|^2} e^{-t/\tau_L} -$$

$$- \frac{2|\eta_{000}|}{4|q|^2} R(K_L \rightarrow 3\pi^0) \exp\left[-t(\tau_S + \tau_L)/2\tau_S\tau_L\right] \times$$

$$\times \cos(\Delta mt - \phi_{000}) \tag{10}$$

with $|q|^2 = (1 - 2\ \mathrm{Re}\ \varepsilon)/2$.

Similarly to the $2\pi^0$ case the integral asymmetry I^{000} is given by

$$I^{000} = \left[\left(\frac{\tau_S}{\tau_L}\right) \frac{4\ \mathrm{Re}\ |\eta_{000}|}{(1 - e^{-\gamma_L t_0})} - 2\ \mathrm{Re}\ \varepsilon\right] \Big/ \left[1 - \frac{8\ \mathrm{Re}\ \varepsilon\ \mathrm{Re}\ \eta_{000}}{(1 - e^{-\gamma_L t_0})}\left(\frac{\tau_S}{\tau_L}\right)\right]$$

or for $t_0 \ll \tau_L$

$$I^{000} = \left(\frac{\tau_S}{t_0}\right) 4\ \mathrm{Re}\ |\eta_{000}| - 2\ \mathrm{Re}\ \varepsilon . \tag{11}$$

The sensitivity of the measurement of I^{000} is shown in Fig. 7 as a function of the magnitude of $|\eta_{000}|$. For this particular measurement we expect to have an additional systematic error of 10%, coming mainly from the uncertainty in the definition of the integration interval $t_0 \approx 20\ \tau_S$.

To conclude, we would like to emphasize that the only possible place to measure in such a direct way and with such an accuracy the quantity $|\eta_{000}|$ is at LEAR, where one exploits features not accessible from the standard configuration with K_S and K_L beams.

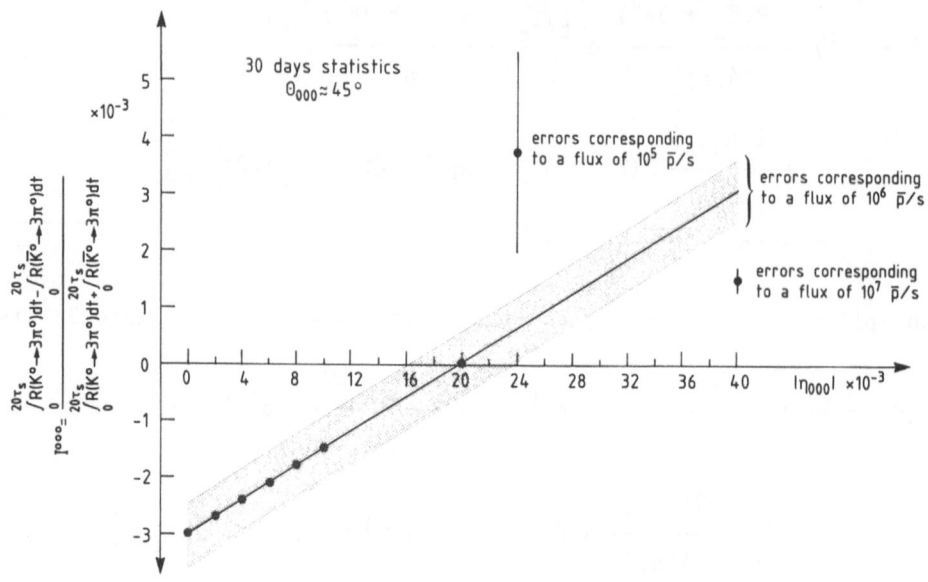

Fig. 7. The experimental sensitivity in defining the CP-violation parameter $|\eta_{000}|$ for the three-pion decays. The parameter ε is here assumed to be independent of the η_{000} magnitude and is equal to 2.3×10^{-3}.

3.2 The two-photon decay

The two-photon decay of the neutral kaon may be of particular interest because, firstly, by involving electromagnetism it does not obey isospin invariance and, secondly, the measurement of the $K_S \to 2\gamma$ branching ratio, which is presently unknown, will check our understanding on neutral-kaon decays[15,16].

The asymmetry between the K^0 and $\bar{K}^0$ decays has the following time dependence in vacuum[17,18]:

$$A(t) = \frac{Y(t) - 2 \operatorname{Re} \varepsilon \, X(t)}{X(t) - 2 \operatorname{Re} \varepsilon \, Y(t)} , \qquad (12)$$

where

$$X(t) = \left\{ R(K_L \to \gamma\gamma) \, e^{-t/\tau_L} + R(K_S \to \gamma\gamma) \, e^{-t/\tau_S} \right\}$$

and

$$Y(t) = 2 \{ |\epsilon_1| \ R(K_S \to \gamma\gamma) \cos(\theta_1 - \Delta m t) +$$
$$+ |\epsilon_2| \ R(K_L \to \gamma\gamma) \cos(\theta_2 + \Delta m t) \} \times \exp \left[\frac{t(\tau_S + \tau_L)}{2 \ \tau_S \tau_L} \right]$$

where t is the proper time and θ_1 and θ_2 are the phases of the complex amplitudes[17]

$$\epsilon_1 = \frac{r_1 - q/p}{r_1 + q/p} \quad \text{and} \quad \epsilon = \frac{r_2 - q/p}{r_2 + q/p}$$

with

$$|r_1| \ e^{i\phi_1} = -A\left[K^0 \to \gamma\gamma \ (CP = +1) \right] \Big/ A\left[\bar{K}^0 \to \gamma\gamma \ (CP = +1) \right]$$

$$|r_2| \ e^{i\phi_2} = A\left[K^0 \to \gamma\gamma \ (CP = -1) \right] \Big/ A\left[\bar{K}^0 \to \gamma\gamma \ (CP = -1) \right]$$

$$p = \frac{1 + \epsilon}{\sqrt{2(1 + |\epsilon|^2)}} \quad \text{and} \quad q = \frac{1 - \epsilon}{\sqrt{2(1 + |\epsilon|^2)}} \ .$$

The sensitivity in measuring the integral asymmetry

$$I^{\gamma\gamma} = \frac{\int_0^{t_0} R(K_0 \to \gamma\gamma) dt - \int_0^{t_0} R(\bar{K}^0 \to \gamma\gamma) dt}{\int_0^{t_0} R(K^0 \to \gamma\gamma) dt + \int_0^{t_0} R(\bar{K}^0 \to \gamma\gamma) dt} \tag{13}$$

at LEAR is presented in Fig. 8 as a function of the phases ϕ_1 and ϕ_2. As was shown a long time ago[17], the asymmetry $I^{\gamma\gamma}$ is always non-vanishing and of order ϵ because of CP violation in the K^0-$\bar{K}^0$ mass matrix. As is shown in Ref. 18, the effects of the CP violation in the decay matrix modify the superweak predictions, as follows:

$$|r_1| = |r_2| \quad \text{and} \quad \phi_1 = 0 \ ; \ \phi_2 = \xi \simeq O(10^{-3}) \ .$$

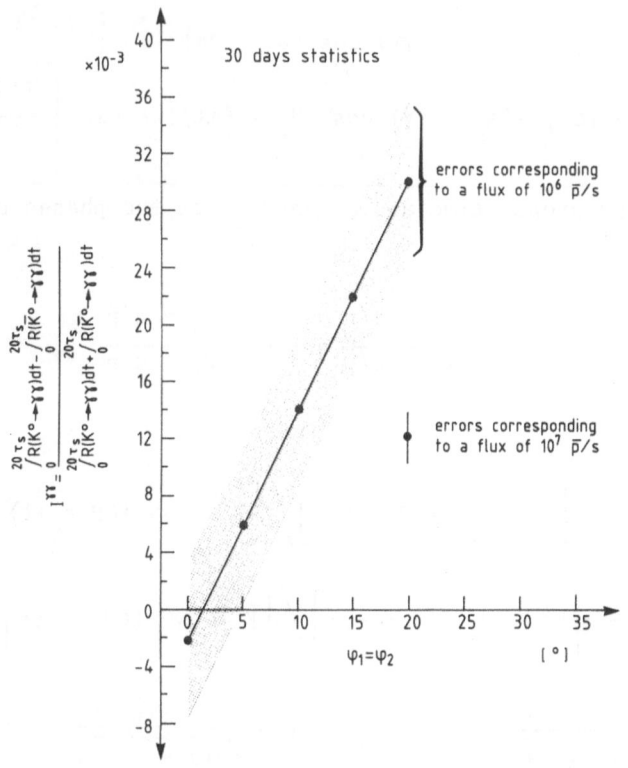

Fig. 8. The dependence of the asymmetry [Eq. (13)] on the relative phase of the amplitudes $A(K^0 \rightarrow \gamma\gamma)$ and $A(\bar{K}^0 \rightarrow \gamma\gamma)$. The magnitudes of these amplitudes were taken to be equal.

Although in the two-photon decay mode we have, for the first time, a milliweak prediction of the order of ε, which is not suppressed by the $\Delta I = \frac{1}{2}$ rule, our experiment is completely insensitive to this effect. Only if the 3π intermediate state has a contribution comparable to the short distance and if, in contradiction to the theoretical predictions[14], η_{000} is very different from ε, then a measurable effect could be seen in $I^{\gamma\gamma}$.

By measuring K^0 $(\bar{K}^0) \rightarrow 2\gamma$ one will verify that CP is indeed violated in this channel and one will measure the branching ratio of the $K_S \rightarrow 2\gamma$.

4. SUMMARY

It is our contention that by using LEAR it is possible to study CP-violating phenomena through the asymmetry of intensities in the partial-decay rates of K^0's and $\bar{K}^0$'s. This experiment is sensitive to the interference terms, as opposed to the standard experiments with K_L beams, which are sensitive to the CP impurities of the observed K_S and K_L states. The detection of such asymmetries will allow the determination, firstly, of the parameter $|\varepsilon'/\varepsilon|$ from a measurement independent from those existing experiments measuring ε' and, secondly, of the violation of the CP invariance in decays other than into two pions. The intensity of stopped $\bar{p}$'s at LEAR and the symmetry of the annihilation at rest provide us with a unique way of studying such processes with reduced systematic errors, using a modest-sized detector with a magnetic-field volume and an electromagnetic calorimeter.

ACKNOWLEDGEMENTS

It is a pleasure to thank E. Gabathuler for his initial ideas, stimulating discussions, and contributions in the early stage of this work. We are indebted to L.L. Chau, A.B. Lahanas and C. Kounnas for useful conversations and encouragement.

REFERENCES

1. L. L. Chau Wang, Phenomenology of CP violation from the Kobayashi-Maskawa Model, *in* "AIP Conf. Proc. No. 72, Particles and Fields", Subseries No. 23, G. B. Collins, L. N. Chang and J. R. Ficenec, eds, AIP, New York (1980)
2. L. L. Chau, Status of electroweak gauge theories, *in* "Proc. 6th European Symposium on NN̄, QQ̄ Interactions", Santiago de Compostela, Spain, 1982, <u>Anales de Fisica</u>, Series A, 79:297 (1983)
3. L. L. Chau, Quark mixing in weak interactions, <u>Phys. Rep.</u> 5:1 (1983)
4. P. Lefèvre, D. Möhl and G. Plass, *in* "Proc. 11th Int. Conf. on High-Energy Accelerators", Geneva, 1980 (Birkhäuser, Basle, 1980), p. 819
5. C. Kounnas, A. B. Lahanas and P. Pavlopoulos, <u>Phys. Lett.</u> 127B:381 (1983)
6. L. Wolfenstein, <u>Nuovo Cimento</u> 63:269 (1969)
7. Basle-Stockholm-Thessaloniki Collaboration, Letter of Intent, Tests of symmetries with K^0 and $\bar{K}^0$ beams, CERN/PSCC/83-28/ PSCC/I65
8. Table XI of R. Armenteros and B. French, *in* "High-Energy Physics", E. H. S. Burhop, ed., Academic Press, New York (1969), Vol. 4
9. Table XIV of R. Armenteros and B. French, *in* "High Energy Physics", E. H. S. Burhop, ed., Academic Press, New York (1969), Vol. 4
10. E. Gabathuler and P. Pavlopoulos, Strong and weak CP violation at LEAR, to be published *in* "Proc. Workshop on Physics at LEAR with Cooled Low Energy Antiprotons", Erice, 1982
11. CERN-Erlangen-Freiburg-Los Alamos- Carnegie Mellon-Saclay-Uppsala Collaboration, Proposal of the experiment PS185, Study of threshold production of $\bar{p}p \to Y\bar{Y}$ at LEAR. CERN/PSCC/ 81-69/PSCC/P49 and extension CERN/PSCC/81-6/PSCC M152
12. K. Kilian, private communication
13. L. M. Sehgal and L. Wolfenstein, <u>Phys. Rev.</u> 162:1362 (1967)
14. L. F. Li and L. Wolfenstein, <u>Phys. Rev.</u> D 21:178 (1980)
15. M. K. Gaillard and B. W. Lee, <u>Phys. Rev.</u> D 10:897 (1974)
16. J. Ellis, M. K. Gaillard and D. V. Nanopoulos, <u>Nucl. Phys.</u> B 109:213 (1976)
17. B. R. Martin and E. de Rafael, <u>Nucl. Phys.</u> B 8:131 (1968)
18. R. Decker, P. Pavlopoulos and G. Zoupanos, CP violation in $K^0(\bar{K}^0) \to 2\gamma$ decays, submitted to <u>Phys. Lett. B</u>

DISCUSSION

KLEINKNECHT:
What about the $K_L \rightarrow 3\pi^0$ background in the $K_L \rightarrow 2\pi^0$
asymmetry? How does an asymmetry in the K^+/K^- tagging effeciency
affect the result?

PAVLOPOULOS:
The K_L-component represents roughly 50% for both K^0 and $\bar{K}^0$. On
the other hand only 3.4% of all K_L willdecay inside the 80 cm
($\sim 20 \ t_S$). Thus the $K_L \rightarrow 3\pi^0$ (BR = 0.215) background
contribute only to $4 \cdot 10^{-3}$ per neutral kaon resulting to a
systematic error in the asymmetry far below the requested
statistical accuracy.

The idea of the experiment is to compare the "partial width" of
the $K^0 \rightarrow 2\pi^0$ with the "partial width" of the $\bar{K}^0 \rightarrow 2\pi^0$, where the
"partial width" is defined for the given time interval. Thus we
have to measure out of a given number of K^0's or $\bar{K}^0$'s how many
they have decayed into two pions. Consequently, we have an
independent normalisation for K^0 and $\bar{K}^0$ decays. The efficiency
of identifying a K^0 should not be equal to the efficiency of
identifying a $\bar{K}^0$ or the number of K^0's has not to be identical to
the number of $\bar{K}^0$'s.

LITTENBERG:
In the cases where you measure only integral asymmetries, how
much would having the vertices buy you? In the case of $K \rightarrow \pi^0\pi^0$,
you could use Dalitz decay of one of the π^0's, at the cost of a
factor of 40 in statistics.

PAVLOPOULOS:
The advantage, as opposed to the standard experimental configuration with K_S and K_L, is mainly the reduced systematic errors.
This approach is limited mainly by the statistics and not by
systematics and therefore we would not like to sacrifice a large
factor in statistics. But there are ideas for a fast reconstruction of the vertex from the gammas rising the knowledge of the K^0
four-momentum (constrained fit), although it is not yet clear how
time consuming they may be.

CHAU:
When will you have results and how much it costs, as compared to
Prof. Kleinknecht's experiment?

PAVLOPOULOS:
The experiment will take a great advantage of the increased
intensity (factor ten) provided by ACOL. Taking into account

that ACOL will be operational for LEAR by the middle of 1987 and
that the experimental data will be evaluated quasi-online, one
can expect results at the beginning of 1988. On your second
question I would prefer to give the estimated costs of the LEAR
experiment and maybe Prof. Kleinknecht can answer about the costs
of his experiment. The LEAR experiment will cost roughly 3 to
3.5M SFr including the calorimeter.

MEASUREMENT OF ε' AT THE AGS

William M. Morse

Brookhaven National Laboratory
Department of Physics
Upton, New York 11973

Since the discovery twenty years ago that CP is violated in the K^0 system, all experimental observations have been consistent with the superweak interpretation that the CP violating decays occur through the K_1 impurity in the K_L^0 wavefunction. The standard $SU(2) \times U(1)$ model, however, had no mechanism to break CP. Now with the discovery of the third quark generation, CP can be violated through an imaginary phase in the quark mixing matrix. Gilman and Hagelin[1] have evaluated CP violation in this model and find that there must be a direct CP violation from K_2 decay. They predict ϵ'/ϵ must be greater than .015, assuming an average B lifetime of 10^{-12} sec, the top mass of 50 GeV, and B = 0.33. I will now discuss AGS experiment 749, a precision measurement of ϵ'/ϵ.

The principle of the measurement is to observe the $\pi^+\pi^-$ and $\pi^0\pi^0$ decay modes of both the K_L^0 and K_S^0 in one experiment. We form the ratio of ratios of the numbers of $\pi^+\pi^-$ and $\pi^0\pi^0$ events from the K_L^0 and K_S^0 runs:

$$ R = \frac{N(K_L^0 \rightarrow \pi^0\pi^0)}{N(K_L^0 \rightarrow \pi^+\pi^-)} \frac{N(K_S^0 \rightarrow \pi^+\pi^-)}{N(K_S^0 \rightarrow \pi^0\pi^0)} . $$

Since we observe the neutral and charged decay modes simultaneously, the normalization cancels in the ratios. The acceptance of the apparatus also cancels when the ratios are computed over K^0 energy and decay position intervals such that the differential intensity variations are small, so that the sum of the ratio of ratios over all intervals is then simply related to ε':

$$\left|\frac{N_{00}}{N_{+-}}\right|^2 = \sum_{ij} W_{ij}R_{ij} = 1 - 6\left|\frac{\varepsilon'}{\varepsilon}\right|$$

where the individual R values are multiplied by their correct statistical weight functions.

A schematic representation of the 0° neutral beam line is shown in Fig. 1. The 28 GeV proton beam from the AGS strikes a 20 cm copper target and is then swept into the beam stop in the first sweeping magnet. The neutral beam solid angle acceptance of 3.5 μ steradians is defined by collimators placed in the first two of the three 18D72 dipole magnets. A one inch lead sheet upstream of the first collimator converts gamma rays. The K_S^0 beam is generated by moving an eighty centimeter carbon regenerator into the neutral beam in the final sweeping magnet, as shown in Fig. 1.

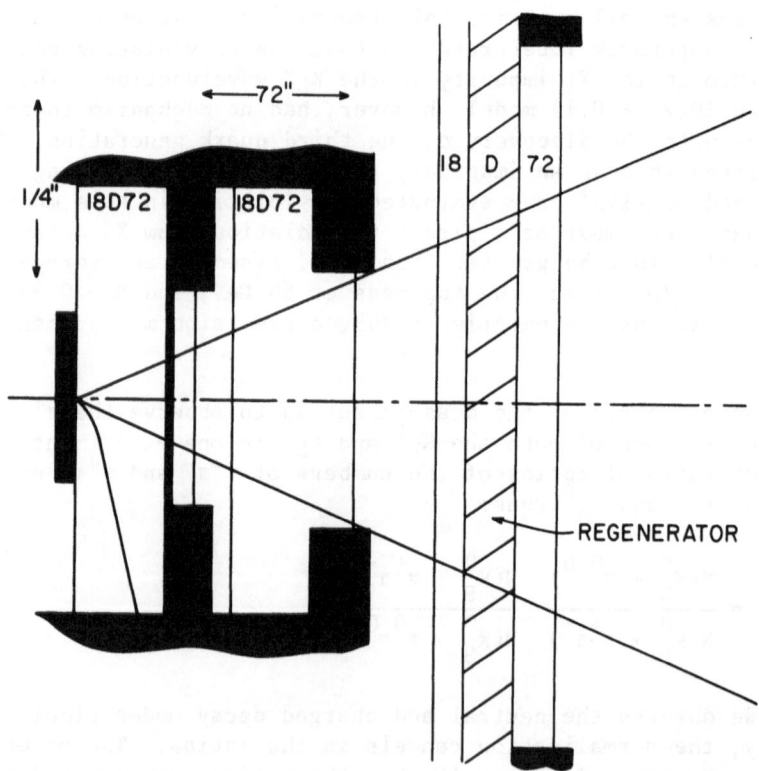

Fig. 1.   Schematic representation of the zero degree neutral beam.

It is important to have a small neutral beam cross section in as much as the decay point of the $\pi^0\pi^0$ decay is determined by converting one of the gammas to an e^+e^- pair after the decay region, reconstructing the trajectories, and then finding the intersection point of the trajectory and the neutral beam. The measured profile of the neutral beam is shown in Fig. 2. This measurement was made by traversing the beam with a small piece of brass, to convert the neutral beam (mostly neutrons), followed by two scintillation counters. A scintillation counter upstream of the brass vetoed charged particles. The FWHM of the beam is about 2 cm.

We ran, in the K_L^0 mode, with typically 10^{12} protons/pulse which generated 3×10^8 neutrons and 10^7 K_L^0. This gives about 4×10^6 counts/sec in the large PWC upstream of the spectrometer magnet. We write about three events to tape per pulse and expect to accumulate about thirty tapes over the course of the experiment.

A schematic representation of the apparatus is shown in Fig. 3. The decay space begins downstream of the third sweeping magnet. Proportional wire chamber arrays (a total of 2560 wires)

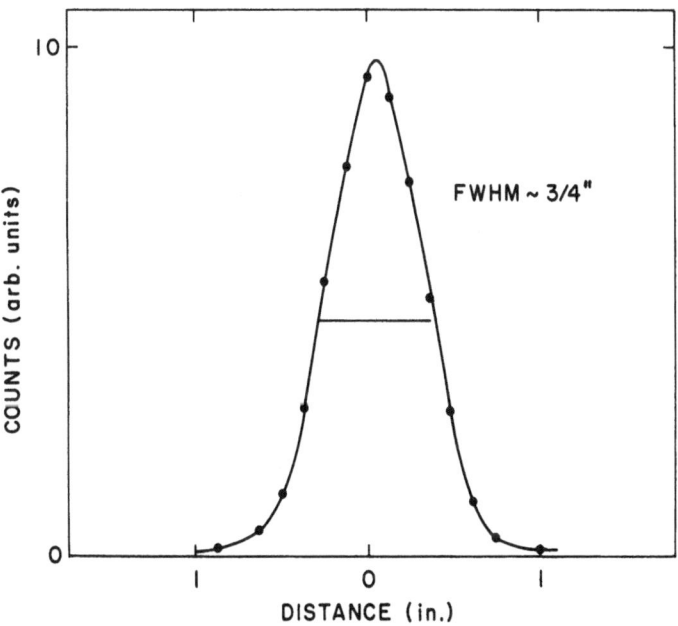

Fig. 2.   Profile of neutral beam measured in the decay space.

are located at locations A, B, C, and D. The 72D18 spectrometer
magnet with a transverse kick of 150 MeV is located between wire
chamber arrays B and C. The lead glass array (208 pieces) is
located downstream of the magnetic spectrometer. The array is
about 1 m × 1 m with a 12 cm × 12 cm hole for the neutral beam.
This is followed by a steel wall for muon identification and a 30
cm × 30 cm lead glass array (8 pieces) with a 10 cm × 10 cm hole
for the neutral beam. The solid angle subtended by this hole from
the decay region is only 10^{-4} steradians.

Charged ($\pi^+\pi^-$) triggers and neutral ($\pi^0\pi^0$) triggers are
collected simultaneously. The charged trigger requires two
particles on opposite sides of the neutral beam, roughly coplanar
with the beam. The neutral trigger requires one and only one gamma
conversion in a 1 mm thick lead sheet downstream of the decay

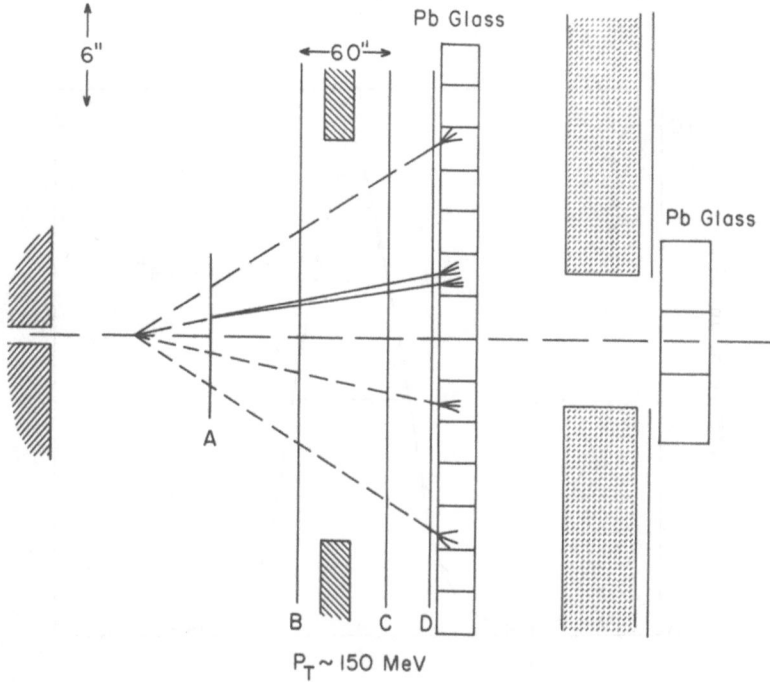

Fig. 3.   Schematic representation of the apparatus illustrating a
$K^0_L \rightarrow \pi^0\pi^0$ decay in the nonbending plane.

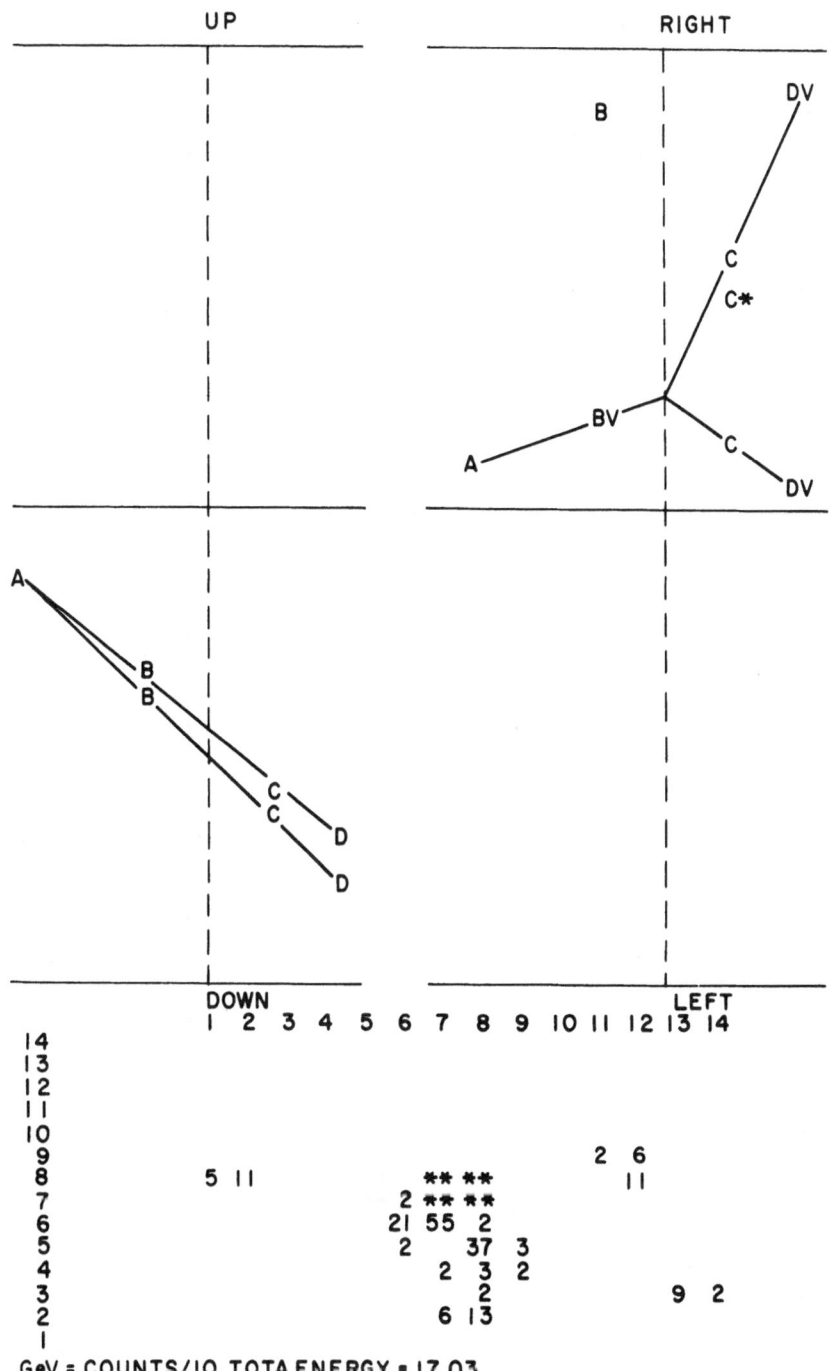

Fig. 4. Typical neutral trigger printout from the ON-LINE program.

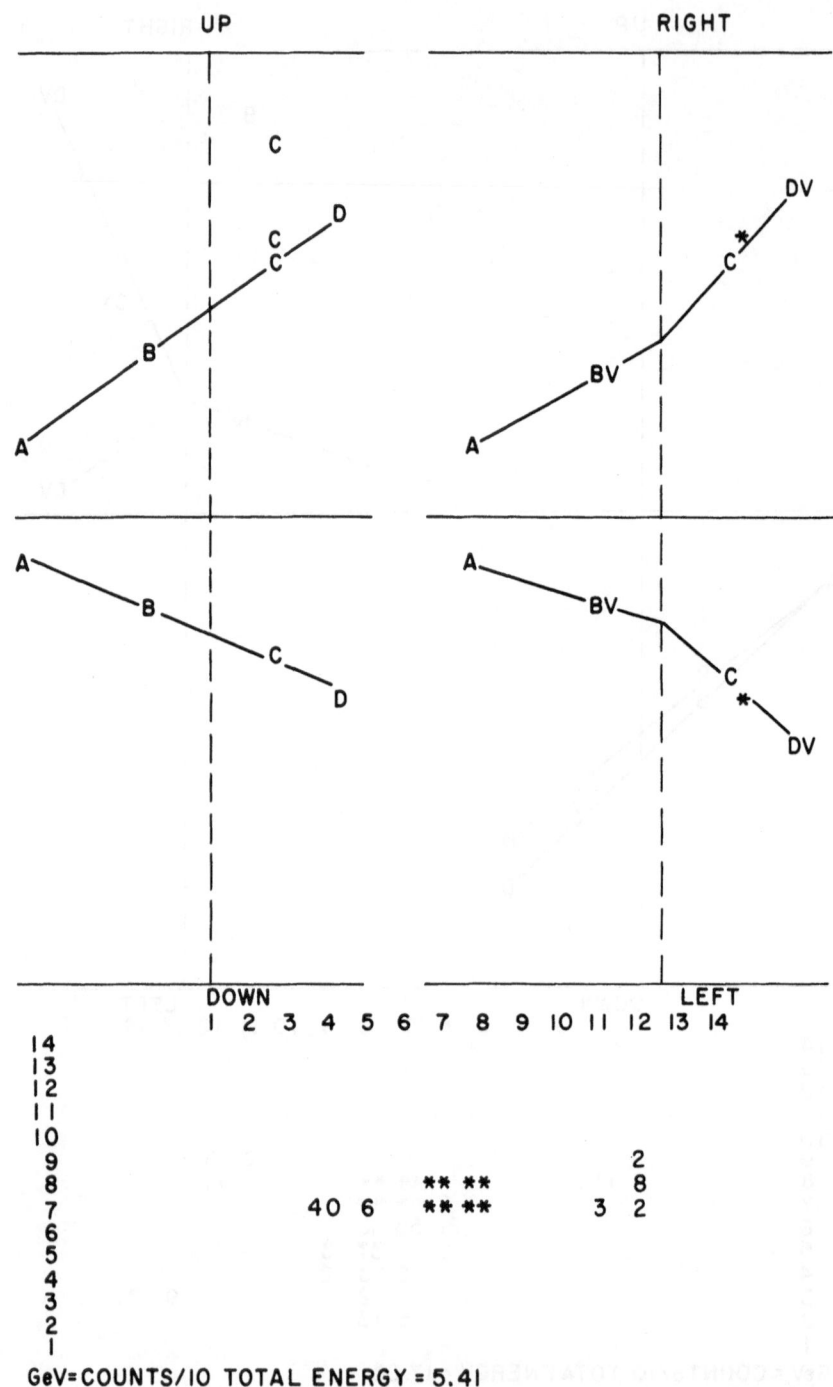

Fig. 5.   Typical charged trigger printout from the ON-LINE
          program.

region, greater than 5 GeV in the lead glass array with at least 1 GeV in both the upper and lower halfs. An example of a neutral decay is shown in Fig. 3. The data for this experiment was collected via our Brookhaven Fastbus system. This system utilizes three Fastbus crates and two Fastbus cable segments. Further details are given in Ref. 2.

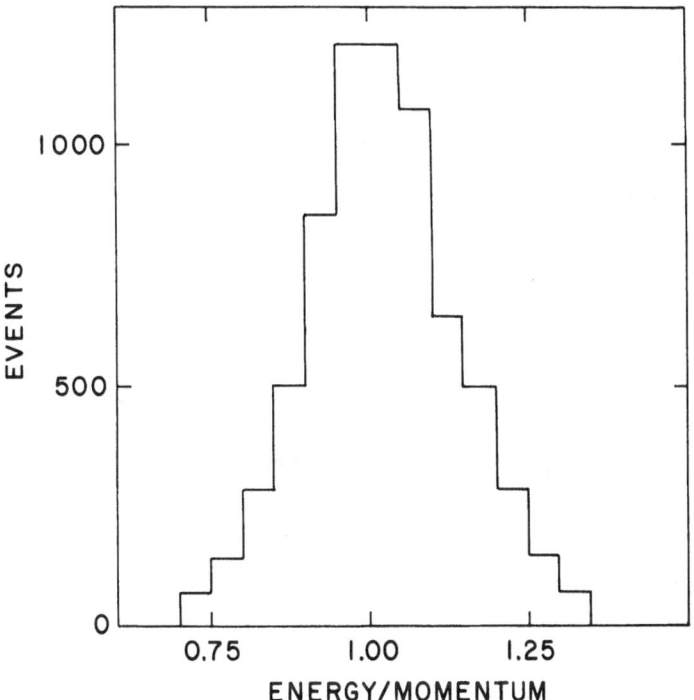

Fig. 6.   Histogram of electron energy from lead glass measurement
          divided by electron momentum from magnetic spectrometer
          measurement.

A computer representation by the ON-LINE program of a typical neutral trigger is shown in Fig. 4. The trajectories of the e^+e^- pair are measured through the magnetic spectrometer and the PWC hits are represented in the upper half of Fig. 4. All chambers have 3 mm wire spacing except the "A" chambers which have 2 mm spacing. The asteriks represent hits in the $C\theta$ chamber which is rotated 10° with respect to the CV chamber. The energy deposited in the 14 × 14 element lead glass array is shown in the lower half of Fig. 4. It is clear that this event had more than four gammas. Furthermore, the rear lead glass array recorded more than 1.5 GeV of energy for this event. This is then a $K_L^0 \to \pi^0\pi^0\pi^0$ decay. A similar representation of a charged trigger is shown in Fig. 5. This event is, however, not a good $\pi^+\pi^-$ event. The energy deposited in the lead glass array equals the momentum of the negative particle as measured in the magnetic spectrometer. This is then a $K_L^0 \to \pi^+e^-\nu$ event. These events are, however, useful as a means of calibrating the lead glass array. A histogram of the electron energy as measured by the pulse height in the lead glass divided by the electron momentum as measured in the magnetic spectrometer is shown in Fig. 6. This distribution has a width compatible with our expected momentum and energy resolution.

The $K_S^0 \to \pi^+\pi^-$ decay is so free of background that we are able to monitor the di-pion effective mass distributions with the ON-LINE program. A histogram of this distribution for a two hour K_S^0 run is shown in Fig. 7. We expect to reduce the background levels and improve the mass resolution in the off-line analysis.

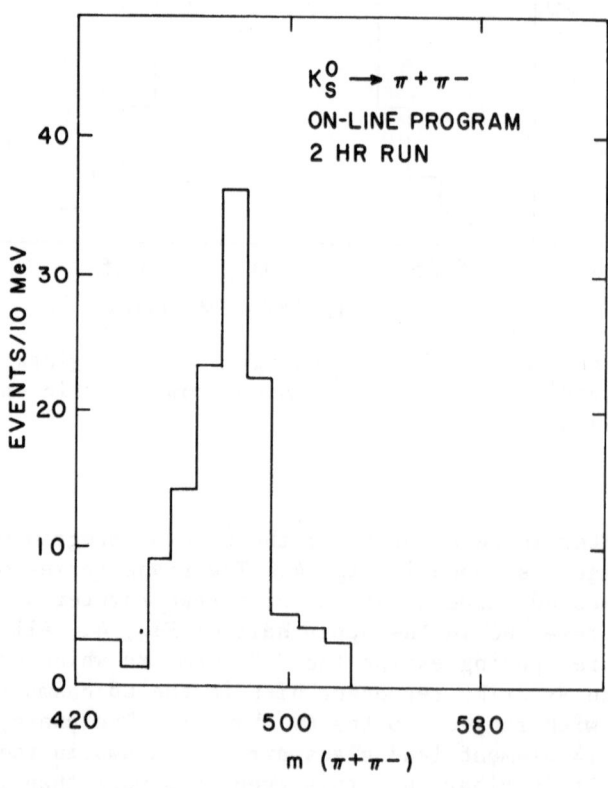

Fig. 7.   Histogram of dipion effective mass from ON-LINE program.

We are currently taking data at the AGS. We expect to accumulate approximately five thousand $K_L^0 \to \pi^0 \pi^0$ events in six hundred hours of data taking. The error in ε'/ε will then be about ± 0.005. We accumulated about five hundred $K_L^0 \to \pi^0 \pi^0$ events from a short test run in May 1983. Our analysis of that data has yielded the result

$$\left| \frac{\varepsilon'}{\varepsilon} \right| = 0.012 \pm 0.015$$

consistent with both the superweak theory and the standard model calculation of Gilman and Hagelin.

This work is supported in part by the U.S. Department of Energy under Contract No. DE-AC02-76CH00016.

REFERENCES

1. F.J. Gilman and J.S. Hagelin, Phys. Lett. <u>133B</u>, 443 (1983).
2. L.B. Leipuner, et al., IEEE Trans. on Nucl. Sci. <u>NS-30</u>, 3773 (1983); G. Benenson, et al., ibid, 3958; and W.M. Morse, et al., ibid, 3964; M.P. Schmidt, et al., IEEE Trans. on Nucl. Sci. <u>NS-31</u>, 203 (1984); J.K. Black, et al., IEEE Trans. on Nucl. Sci. <u>NS-31</u>, 201 (1984).

CP VIOLATION FROM THE STANDARD MODEL

Kuang-chao Chou

Institute of Theoretical Physics
Academia Sinica
Beijing, China

INTRODUCTION

Two decades have passed since the first observation of CP
violation in kaon decay[1]. The subject is still not well understood and the progress is rather slow compared with what has been
achieved in the other branches of weak interactions.

As we all know now, nature has chosen the standard
SU(2) × U(1) gauge model to describe physics at an energy scale
below 100 GeV. Both W and Z^0 bosons have already been seen
within the error predicted by the theory[2]. It is therefore of
great interest to accomodate CP violation in gauge theories which
seem to be the most promising ways from a theoretical point of
view.

As Kobayashi and Maskawa[3] (K-M) first pointed out, CP
violation can occur in the standard model through complex phases
in mass matrix with more than two generations. For three
generations favored by the present experiment there is only one
phase causing CP violation. Could this single phase be sufficient to explain all the CP violation effects? It is certainly
welcome if it could. However, it can not be answered a priori.
The origin of CP violation is closely related to that of the

masses and the number of generations, which in turn are described by physics at much higher energy scales. It would not be a surprise if some new ingredients had to be added to solve the CP problem. We shall wait and see.

Since there were excellent review papers not long ago[4], it is unnecessary for me to repeat all the known results to you. What I would like to report is a recent analysis of CP violation in the K-M model after the measurement of the unexpected long lifetime of the b-quarks.

The outline of this talk is as follows:

I.   Parametrization of the K-M matrix;
II. Physics of ε and ε' in kaon systems;
III. Neutral particle-antiparticle mixing and CP violation in $B^0-\bar{B}^0$ system;
IV. CP violation in partial decay rates of particles and antiparticles;
V.   Concluding remarks.

I.   PARAMETRIZATION OF THE K-M MATRIX

For three generations of quark the K-M matrix containing three angles and one phase is usually expressed in the following form

$$V = \begin{pmatrix} V_{ud} & V_{us} & V_{ub} \\ V_{cd} & V_{cs} & V_{cb} \\ V_{td} & V_{ts} & V_{tb} \end{pmatrix} ,$$

$$= \begin{pmatrix} c_1 & s_1 c_3 & s_1 s_3 \\ -s_1 c_2 & c_1 c_2 c_3 - s_2 s_3 e^{i\delta} & c_1 c_2 s_3 + s_2 c_3 e^{i\delta} \\ -s_1 s_2 & c_1 s_2 c_3 + c_2 s_3 e^{i\delta} & c_1 s_2 s_3 - c_2 c_3 e^{i\delta} \end{pmatrix} . \tag{1.1}$$

where $c_i(s_i)$, $i = 1,2,3$, are the cosine (sine) of the angle θ_i. The Cabbibo angle θ_1 is determined to be[4]

$$s_1 = .227^{+.0104}_{-.0110} \; . \tag{1.2}$$

Recent measurements on b quark lifetime and the branching ratio $\Gamma_{b \to u}/\Gamma_{b \to c}$ have put stringent bounds on the matrix elements $\left| V_{cb}/V_{ub} \right|$. Their values can be found in the talks given by Lee-Franzini and Kleinknecht in this conference.

$$\left| V_{cb} \right| = 0.0435 \pm 0.0047 \; , \tag{1.3}$$

$$\left| V_{ub}/V_{cb} \right| \leq 0.119 \; . \tag{1.4}$$

Both $\left| V_{cb} \right|$ and $\left| V_{ub}/V_{cb} \right|$ have been reduced from the 1983 values[5]

$$\left| V_{cb} \right| = .059^{+.016}_{-.009} \quad \text{and} \quad \left| V_{ub}/V_{cb} \right| \leq .14 \; .$$

The fact that $\left| V_{cb} \right|$ is small and of the order of s_1^2 can be used to simplify the K-M matrix. In a first order approximation where $\mathrm{Re}V_{ij}$ are correct to order s_1^3 and $\mathrm{Im}V_{ij}$ to order s_1^5 we have

$$V = \begin{pmatrix} c_1 & s_1 & s_1 s_3 \\ -s_1 & c_1 - s_2 s_3 e^{i\delta} & s_3 + s_2 e^{i} \\ -s_1 s_2 & s_2 + s_3 e^{i\delta} & -e^{i\delta} \end{pmatrix} \tag{1.5}$$

Writing V_{ts} and V_{cb} in the following form

$$V_{ts} = s_2 + s_3 e^{i\delta} = \left| V_{ts} \right| e^{i\delta_{ts}} \; , \tag{1.6}$$

and

$$V_{cb} = s_3 + s_2 e^{i\delta} = \left| V_{cb} \right| e^{i\delta_{cb}} \; , \tag{1.7}$$

it is easily proved that

$$V_{ts} = e^{i\delta} V_{cb}^* \quad . \tag{1.8}$$

Hence we obtain

$$\left| V_{ts} \right| = \left| V_{cb} \right| = (s_2^2 + s_3^2 + 2s_2 s_3 \cos\delta)^{1/2} \tag{1.9}$$

and

$$\delta = \delta_{ts} + \delta_{cb} \quad . \tag{1.10}$$

One can now redefine the phases of the b and t quarks by a transformation

$$q_b \;\rightarrow\; e^{-i\delta_{cb}} q_b \;, \tag{1.11}$$

$$q_t \;\rightarrow\; e^{i\delta_{ts}} q_t \;, \tag{1.12}$$

and get from Eqs. (1.5)-(1.12) a form first suggested by Wolfenstein[6]

$$V \;=\; \begin{pmatrix} c_1 & s_1 & s_1 s_3 e^{-i\delta_{cb}} \\ -s_1 & c_1 - s_2 s_3 e^{i\delta} & \left| V_{cb} \right| \\ -s_1 s_2 e^{-i\delta_{ts}} & -\left| V_{cb} \right| & 1 \end{pmatrix} \tag{1.13}$$

The phases δ_{ts}, δ_{cb} are related to δ by the following relations:

$$s_2 \sin\delta = \left| V_{cb} \right| \sin\delta_{cb} \;, \tag{1.14}$$

$$s_3 \sin\delta = \left| V_{cb} \right| \sin\delta_{ts} \quad . \tag{1.15}$$

The advantage of the present form for the K-M matrix is that $\text{Im}V_{ij}$ is always proportional to a common factor

$$X_{cp} = s_2 s_3 \sin\delta \quad , \tag{1.16}$$

which is the appropriate parameter measuring CP violation effects in various processes. A similar but rigorous representation of the K-M matrix was obtained recently by Chau and Keung[7].

Since both s_2 and s_3 are proportional to $|V_{cb}|$, it is more convenient in numerical calculations to scale it out. We write

$$s_3 \equiv \alpha \left|V_{cb}\right| = \frac{1}{s_1} \left|V_{ub}\right| \quad , \tag{1.17}$$

$$s_3 \sin\delta \equiv \beta \left|V_{cb}\right| \quad . \tag{1.18}$$

From the experimental bound Eq. (1.4) and the value of s_1 we find

$$\alpha \leq .524 \quad , \tag{1.19}$$

$$\left|\beta\right| = \left|\sin\delta_{ts}\right| \leq \alpha \quad . \tag{1.20}$$

With given α and β, s_2 can be solved from Eq. (1.9) to be

$$s_2 = \left(\sqrt{1-\beta^2} \pm \sqrt{\alpha^2-\beta^2}\right) \left|V_{cb}\right| \quad , \tag{1.21}$$

where we have adopted the convention of positive s_2 and s_3. The solution s_2 with positive (negative) sign in the bracket in Eq. (1.21) corresponds to $\cos\delta < 0$ ($\cos\delta > 0$) in Eq. (1.9).

II. PHYSICS OF ε AND ε' IN KAON SYSTEMS

So far CP violation has been observed only in neutral kaon systems. The ratio of the amplitudes for $K_L \rightarrow 2\pi$ and $K_S \rightarrow 2\pi$

is given in the standard notation as

$$\eta_{+-} = \varepsilon + \varepsilon'/(1 + \omega/\sqrt{2}) \ , \tag{2.1}$$

$$\eta_{00} = \varepsilon - 2\varepsilon'/(1 - \sqrt{2}\omega) \ , \tag{2.2}$$

where $\omega = \text{Re}A_2/\text{Re}A_0$ is known to be approximately .05. In terms of the amplitudes A_I for $K^0 \to 2\pi(I)$, with I the final state isospin, the mass matrix element M_{12}, ε and ε' can be expressed in the following forms

$$\varepsilon = \frac{1}{\sqrt{2}} e^{i\pi/4}\left(\frac{\text{Im}M_{12}}{\Delta M} + \xi_0\right) \ , \tag{2.3}$$

and

$$\varepsilon' = \frac{1}{\sqrt{2}} \omega(\xi_2 - \xi_0) e^{i(\delta_2 - \delta_0)} \ , \tag{2.4}$$

where δ_I are final state interaction phases and

$$\xi_I = \frac{\text{Im}A_I}{\text{Re}A_I} \ . \tag{2.5}$$

The experimental value of ε is well established to be $\text{Re}\varepsilon = 0.00162 \pm .000088$ while that of ε' is still uncertain. As will be discussed later, accurate measurement of $\left|\varepsilon'/\varepsilon\right|$ is extremely important in our understanding of the origin of CP violation.

The mass matrix element for $K^0-\bar{K}^0$ transition consists of a short distance part usually identified to be the contribution from the box diagram and a long distance part determined by the low energy intermediate states

$$M_{12} = M_{12}^{sd} + M_{12}^{soft} \ . \tag{2.6}$$

The mass difference of K_L and K_S is related to $\text{Re}M_{12}$ by the relation

614

$$\Delta M_K = M_S - M_L = 2 \, \text{Re} M_{12}$$

$$= 2(\text{Re} M_{12}^{sd} + \text{Re} M_{12}^{soft}) \ . \tag{2.7}$$

The real part of M_{12}^{soft} has been estimated long ago[8]. Its value
is very sensitive to the small parameter that breaks the SU(3)
symmetry. Even the sign of $\text{Re} M_{12}^{soft}$ can not be determined
reliably. The box diagram contribution to $2 \, \text{Re} M_{12}^{sd}$ is dominated
by charm quark exchange and consists only $1/4 \sim 3/4$ of
ΔM_K.[5] Since $\text{Re} M_{12}$ has nothing to do with CP violation, I shall
use in the following the experimental value of ΔM_K in
evaluating the parameter ε. However, I would like to remark that
if one finds eventually that

$$\Delta M_K \neq 2 \left(\text{Re} M_{12}^{box} + \text{Re} M_{12}^{soft} \right),$$

it will indicate the existence of new $\Delta S = 2$ short distance
interaction besides the box diagram, and thus, possible new
sources of CP violation.

The imaginary part of M_{12}^{soft} can be estimated by current
algebra and Penguin diagram dominance. In this approximation

$$\frac{\text{Im} M_{12}^{soft}}{\text{Re} M_{12}^{soft}} = - 2 \xi_0 \ . \tag{2.8}$$

Using Eqs. (2.7) and (2.8) it is possible to eliminate the soft
part of M_{12} and rewrite Eq. (2.3) in the form[9]

$$\varepsilon = \frac{1}{\sqrt{2}} \, e^{\frac{i\pi}{4}} \left(\frac{\text{Im} M_{12}^{sd}}{\Delta M} + 2 \xi_0 \, \frac{\text{Re} M_{12}^{sd}}{\Delta M} \right) \ , \tag{2.9}$$

where M_{12}^{sd} is calculated by the box diagram[10]

$$M_{12}^{box} = \frac{G_F^2 B_K f_K^2 m_K}{12\pi^2} \{ \eta_{tt} \lambda_t^2 m_t^2 f(\frac{m_t^2}{m_W^2})$$

$$+ \eta_{cc} \lambda_c^2 m_c^2 + 2\eta_{ct} \lambda_c \lambda_t m_c^2 \ln \frac{m_t^2}{m_c^2} \} \quad , \qquad (2.10)$$

where

$$\lambda_i = V_{is} V_{id}^* , \qquad (2.11)$$

$$f(x) = \frac{1}{(1-x)^2} (1 - \frac{11}{4} x + \frac{1}{4} x^2) - \frac{3 x^2 \ln x}{2(1-x)^3} , \qquad (2.12)$$

and $\eta_{tt} = 0.6$, $\eta_{cc} = 0.7$, $\eta_{ct} = 0.4$ are the QCD correction factors. The factor B_K accounts for the uncertainty in determining the matrix element

$$\langle K^0 | [\bar{s}\gamma_\mu (1+\gamma_5)d]^2 | \bar{K}^0 \rangle = -\frac{4}{3} B_K f_K^2 m_K \quad . \qquad (2.13)$$

Current algebraic estimation tells us that B_K is around 1/3, while some lattice calculations[11] give the value about 1, close to the vacuum-insertion value. We shall keep B_K to be a parameter in the following calculations.

Eqs. (2.9)-(2.10) have been used to predict the minimum top quark mass when the K-M angles are given, or the other way around, to set lower bound on the CP violation parameter $X_{cp} = s_2 s_3 \sin\delta$ when the top quark mass is assumed[5,12-15]. In these calculations the second term in Eq. (2.9) proportional to ξ_0 was neglected and the 1983 experimental values of $|V_{cb}|$ and $|V_{ub}/V_{cb}|$ were used.

The parameter ξ_0 could be estimated by using experimental value of $ReA_0 \approx A_0$ and Penguin diagram value of ImA_0. It is found[13-14] that

$$\xi_0 = s_2 s_3 \sin\delta \; c_6 \frac{G_F}{\sqrt{2}} s_1 \frac{\langle 2\pi(I{=}0) | Q_6 | K^0 \rangle}{A_0} \qquad (2.14)$$

where Q_6 is a (V–A)×(V+A) Penguin operator with Wilson coefficient c_6 in the effective Hamiltonian for Penguin diagram. $c_6 = \mathrm{Im} c_6 / s_2 s_2 \sin\delta$ is estimated in the leading logarithmic approximation to all orders in the strong interaction to be −0.1 and is quite stable against the choice of parameters.

For the matrix element $\langle 2\pi(I{=}0) | Q6 | K^0 \rangle$ shall follow the analysis of Gilman and Hagelin[13] to use the bag model value for a conservative estimation. One finds finally that

$$\xi_0 = -\,.54 \; s_2 s_3 \sin\delta \; \left| \frac{c_6}{0.1} \right| \left| \frac{\langle 2\pi(I{=}0) | Q_6 | K^0 \rangle}{1.4 \; \mathrm{GeV}^3} \right| . \qquad (2.15)$$

A similar estimation gives $\xi_2 \approx 0$. The parameter $|\varepsilon'/\varepsilon|$ can then be determined to be

$$|\varepsilon'/\varepsilon| = 8.4 \; s_2 s_3 \sin\delta \; \left| \frac{c_6}{0.1} \right| \left| \frac{\langle 2\pi(I{=}0) | Q_6 | K^0 \rangle}{1.4 \; \mathrm{GeV}^3} \right| \qquad (2.16)$$

Now a combined analysis of $|\varepsilon|$ and $|\varepsilon'/\varepsilon|$ can be made with the help of Eqs. (2.9), (2.10), (2.15) and (2.16) when $s_3(\alpha)$, $s_2 \sin\delta(\beta)$, $|V_{cb}|$ and B_K are given. The results are given in Tables 1–7 and Figs. 1–3, where both m_t and $|\varepsilon'/\varepsilon|$ are shown to be functions of α and β. The minimum top quark mass occurs at the point where $s_3(\alpha)$ saturates its upper bound and δ in the second quadrant where s_2 is larger. Its value increases as B_K and $|V_{cb}|$ decrease. For $B_K = 0.33$ and $|V_{cb}| < .059$ the minimum top quark mass is already over 60 GeV. It is also noted that $|\varepsilon'/\varepsilon|$ is large at the point where m_t is minimum. The present experimental value of $|\varepsilon'/\varepsilon| = -\,.003 \pm .015$ is barely consistent with the one calculated at the point of minimum top quark mass.

TABLE 1. Values of M_t, $|\epsilon'/\epsilon|$, X_{B_d} and X_{B_s} of Eq. (3.10) as functions of $s_2/|V_{cb}|$ for $B_K = 0.33$, $|V_{cb}| = 0.0388$ and $|V_{ub}/V_{cb}| = .119$

$s_2/|V_{cb}|$	.852	1.031	1.110	1.171	1.222	1.266	1.305	1.339	1.395	1.440
M_t (GeV)	325.3	254.3	235.8	226.0	221.0	219.3	220.2	223.4	236.9	162.4
$|\epsilon'/\epsilon|$	.0057	.0065	.0066	.0066	.0065	.0063	.0061	.0058	.0051	.0043
X_{B_d}	2.10	2.10	2.17	2.26	2.38	2.52	2.70	2.90	3.45	4.28
X_{B_s}	56.1	38.3	34.2	32.0	30.96	30.6	30.8	31.5	34.4	40.07

TABLE 2. Same as Table 1 except $|V_{cb}| = .0435$

$s_2/|V_{cb}|$	.852	1.031	1.110	1.771	1.222	1.266	1.305	1.339	1.395	1.440
M_t (GeV)	233.6	180.9	167.6	160.7	157.3	156.3	157.3	159.9	170.3	188.9
$|\epsilon'/\epsilon|$	.0071	.0082	.0083	.0083	.0082	.0079	.0076	.0073	.0064	.0054
X_{B_d}	1.58	1.57	1.62	1.69	1.79	1.90	2.03	2.19	2.63	3.27
X_{B_s}	42.4	28.7	25.5	24.0	23.2	22.96	23.2	23.8	26.2	30.6

TABLE 3. Values of M_t, $|\epsilon'/\epsilon|$, X_{B_d} and X_{B_s} as functions of $s_2/|V_{cb}|$ for $B_K = 0.33$, $|V_{cb}| = 0.0482$ and $|V_{ub}/V_{cb}| = .119$

$s_2/|V_{cb}|$	.852	1.031	1.110	1.171	1.222	1.266	1.305	1.369	1.419	1.458
M_t (GeV)	171.3	132.1	122.4	117.6	115.3	114.8	115.7	121.4	132.5	150.9
$|\epsilon'/\epsilon|$	.0087	.0100	.0103	.0102	.0100	.0097	.0094	.0084	.0073	.0060
X_{B_d}	1.21	1.19	1.22	1.28	1.35	1.44	1.55	1.83	2.26	2.92
X_{B_s}	32.4	21.7	19.2	18.1	17.5	17.4	17.6	19.0	21.8	42607

TABLE 4. Same as Table 3 except $|V_{cb}| = .0588$

$s_2/|V_{cb}|$	.852	1.031	1.110	1.171	1.222	1.266	1.339	1.395	1.440	1.474
M_t (GeV)	91.9	70.4	65.3	63.0	62.1	62.3	64.8	70.3	79.1	93.4
$|\epsilon'/\epsilon|$	.0130	.0150	.0153	.0152	.0149	.0145	.0133	.0117	.0099	.0079
X_{B_d}	.68	.64	.65	.68	.72	.78	.93	1.16	1.52	2.08
X_{B_s}	18.1	11.6	10.2	9.62	9.40	9.43	10.09	11.6	14.2	18.6

TABLE 5. Values of M_t, $|\epsilon'/\epsilon|$, X_{B_d} and X_{B_s} as functions of $s_2/|V_{cb}|$ for $B_K = 1$, $|V_{cb}| = 0.0388$ and $|V_{ub}/V_{cb}| = .119$

$s_2/|V_{cb}|$	.852	1.031	1.110	1.171	1.222	1.305	1.369	1.419	1.458	1.499
M_t (GeV)	113.4	81.9	76.1	73.6	72.9	75.7	82.0	84.2	111.8	161.3
$|\epsilon'/\epsilon|$	.0057	.0065	.0066	.0066	.0065	.0061	.0055	.0047	.0039	.0025
X_{B_d}	.41	.34	.37	.39	.42	.50	.61	.85	1.18	2.21
X_{B_s}	11.0	6.3	5.8	5.5	5.43	5.72	6.30	8.20	10.80	24.9

TABLE 6. Same as Table 5 except $|V_{cb}| = .0435$

$s_2/|V_{cb}|$	.852	1.031	1.110	1.171	1.222	1.305	1.395	1.440	1.474	1.499
M_t (GeV)	71.5	49.5	45.1	43.7	44.0	47.5	58.4	69.8	86.8	114.7
$|\epsilon'/\epsilon|$	.0071	.0082	.0083	.0083	.0082	.0076	.0064	.0054	.0043	.0031
X_{B_d}	.24	.190	.188	.197	.217	.28	.46	.67	1.01	1.64
X_{B_s}	6.5	3.47	2.95	2.79	2.81	3.24	4.66	6.25	9.02	14.14

TABLE 7. Values of M_t, $|\epsilon'/\epsilon|$, X_{B_d} and X_{B_s} as functions of $s_2/|V_{cb}|$ for $B_K = 1$, $|V_{cb}| = 0.0482$ and $|V_{ub}/V_{cb}| = .119$

$s_2/|V_{cb}|$	.852	1.031	1.110	1.171	1.222	1.305	1.369	1.419	1.458	1.487
M_t (GeV)	42.9	24.22	20.8	20.5	21.8	26.9	33.9	42.8	54.6	71.7
$|\epsilon'/\epsilon|$	.0087	.0100	.0103	.0102	.0100	.0094	.0084	.0073	.0060	.0046
X_{B_d}	.12	.062	.054	.058	.071	.12	.21	.34	.55	.92
X_{B_s}	3.30	1.14	0.85	.82	.93 6	1.38	2.14	3.29	5.02	8.06

TABLE 8. Same as Table 7 except $|V_{cb}| = .0435$ and Re $M_{12}^{sd}/\Delta M = 0.36$

$s_2/|V_{cb}|$	.852	1.031	1.110	1.171	1.222	1.266	1.305	1.338	1.395	1.440
M_t (GeV)	82.8	60.4	55.7	53.8	53.5	54.3	55.9	58.3	65.2	75.7
$|\epsilon'/\epsilon|$	.0071	.0082	.0083	.0083	.0082	.0079	.0076	.0073	.0064	.0054
X_{B_d}	.316	.267	.272	.285	.308	.340	.378	.429	.559	.767
X_{B_s}	8.44	4.88	4.29	4.04	4.00	4.11	4.31	4.65	5.58	7.18

TABLE 9. Values of M_t, $|\epsilon'/\epsilon|$, X_{B_d} and X_{B_s} as functions of $s_2/|V_{cb}|$ for $B_K = 1$, $|V_{cb}| = 0.0482$, $|V_{ub}/V_{cb}| = .119$ and $Re\ M_{12}^{sd}/\Delta M = .36$.

$s_2/|V_{cb}|$	.852	1.031	1.110	1.171	1.222	1.266	1.305	1.339	1.395	1.440
M_t (GeV)	54.7	37.7	34.3	33.3	33.6	34.8	36.6	38.9	45.3	54.2
$|\epsilon'/\epsilon|$	.0087	.0100	.0103	.0102	.0100	.0097	.0094	.0089	.0079	.0066
X_{B_d}	.189	.143	.139	.147	.163	.186	.217	.256	.365	.530
X_{B_s}	5.05	2.61	2.19	2.08	2.111	2.25	2.47	2.77	3.64	4.96

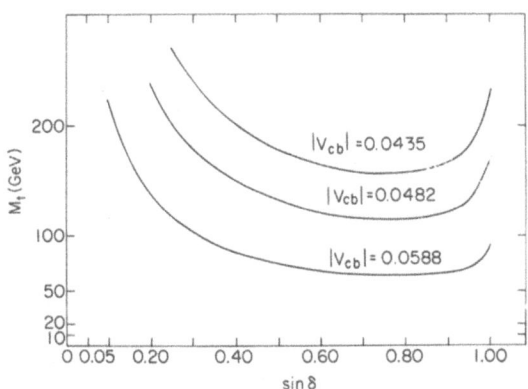

Fig. 1   Top quark mass for B_K = .33, $\left|V_{ub}/V_{cb}\right|$ = .119.

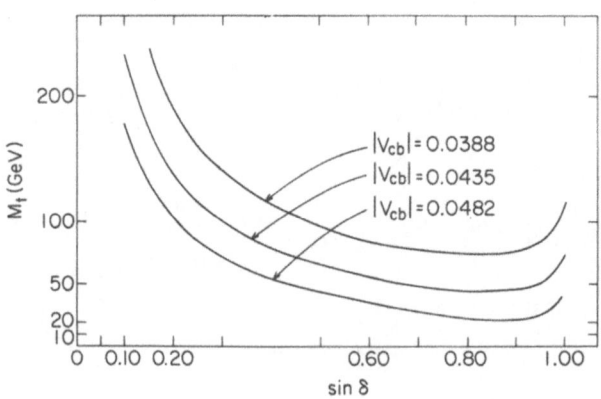

Fig. 2   Top quark mass for $B_K = 1$ and $\left|V_{ub}/V_{cb}\right| = .119.$

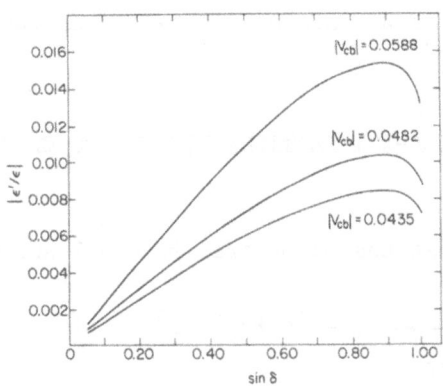

Fig. 3 $\left|\epsilon'/\epsilon\right|$ as function of $\sin\delta$.

To get a definite conclusion we need more accurate measurement on $|\varepsilon'/\varepsilon|$, lower bound on $|V_{ub}/V_{cb}|$ and upper bound on m_t. More reliable theoretical evaluations of B_K, $c_6 \langle 2\pi(I=0)|Q_6|K^0\rangle$ and M_{12}^{soft} are also required. What we could say at the present is that new sources of CP violation besides the K-M phase δ might exist if m_t is found to be around 40 GeV, $|\varepsilon'/\varepsilon| \leq .005$ and $B_K \leq 0.6$.

The second term proportional to ξ_0 in ε is also estimated and the result is given in Tables 8-9. For $B_K = 1$, $2\ \mathrm{Re}M_{12}^{box} \simeq 3/4\ \Delta M_K$, the effect of this term is appreciable at the point where m_t is a minimum. It raises m_t roughly by 30% owing to the negative sign of ξ_0.[16]

III. NEUTRAL PARTICLE-ANTIPARTICLE MIXING AND CP VIOLATION IN $B^0-\bar{B}^0$ SYSTEM.

The mass eigenstates of neutral bosons P and $\bar{P}$ are

$$\left|{P_1 \atop 2}\right\rangle = \frac{1}{(2(1+|\varepsilon_p|^2))^{1/2}} \left[(1+\varepsilon_p)\left|P^0\right\rangle \pm (1-\varepsilon_p)\left|\bar{P}^0\right\rangle\right] , \quad (3.1)$$

where ε_p is determined by the mass matrix elements relating P^0 and $\bar{P}^0$.

$$\frac{1-\varepsilon_P}{1+\varepsilon_P} = \frac{H_{21}}{H_{12}} , \quad (3.2)$$

where $H_{ij} = M_{ij} - i\Gamma_{ij}$ are the mass matrix elements of the neutral $P^0-\bar{P}^0$ system.

For a state being P^0 at $t = 0$, $|\phi(t=0)\rangle = |P^0\rangle$ and later at time t

$$\left|\phi(t)\right\rangle = f_+(t) \left|P^0\right\rangle + \frac{1-\epsilon_P}{1+\epsilon_P} f_-(t) \left|\bar{P}^0\right\rangle , \qquad (3.3)$$

where

$$f_\pm(t) = \frac{1}{2} \left[e^{-im_1 t - \Gamma_1 t/2} \pm e^{-im_2 t - \Gamma_2 t/2} \right] . \qquad (3.4)$$

Mixing of P^0 and $\bar{P}^0$ is necessary for the observation of CP violation effects. There are two cases of maximal mixing. In the kaon case $\delta\Gamma/\Gamma \approx 1$, either with the K^0 or $\bar{K}^0$ to begin with, it will quickly end up as K_L, which is almost an equal mixture of K^0 and $\bar{K}^0$. The second possibility occurs when $\delta m/\Gamma \simeq 1 \gg \delta\Gamma/\Gamma$ In this case, before decaying, the state oscillates quickly between P^0 and $\bar{P}^0$ and appears as an equal mixture of P^0 and $\bar{P}^0$.

Due to the simple fact that the decay width Γ for the D and T particles are K-M angle nonsuppressed, yet δm and $\delta\Gamma$ are always K-M angle suppressed; the values of $\delta m/\Gamma$ and $\delta\Gamma/\Gamma$ are both small and the observation of CP violation in neutral D and T systems is extremely difficult.

The situation is different in B systems. There one expects large mixing effects and possibly large CP violation effects[5],[17-18]. One special feature of the B systems is that the complex parameter ϵ_B is almost imaginary and of the order 1. The imaginary nature follows from the fact that the phase of M_{12} is the same as that of Γ_{12} and the condition $M_{12} \gg \Gamma_{12}$.[18] Therefore the observable effects, depending on the $Re\epsilon_B$ such as the fractional difference of same-sign dilepton production and the asymmetry in semileptonic decay, are very small.

As has been emphasized by Bigi, Carter and Sanda[17] that nonleptonic on shell transition in the bottom sector might produce CP asymmetries of the order 10^{-1}-10^{-2}, whereas the effects due to CP impurities in the mixing is less than 10^{-3}. The effects fall into two categories. The asymmetry for initially pure B^0 and $\bar{B}^0$ states to decay into the same final state f is found to be

$$A = \frac{-xa \; \sin 2\phi}{1+y \; \cos 2\phi} \quad , \qquad\qquad\qquad (3.5)$$

where

$$x = \frac{\Delta M}{\Gamma} \; , \quad y = \frac{\Delta \Gamma}{2\Gamma} \; , \qquad\qquad\qquad (3.6)$$

$$a = \frac{1-y^2}{1+x^2} \; . \qquad\qquad\qquad (3.7)$$

The angle ϕ is defined as

$$\lambda = \left(\frac{1 - \epsilon_B}{1 + \epsilon_B}\right) \frac{M}{\bar{M}} = - \; e^{-2i\phi} \qquad\qquad\qquad (3.8)$$

with M and $\bar{M}$ the amplitudes

$$M = \langle f | H_W | B^0 \rangle \quad , \quad \bar{M} = \langle f | H_W | \bar{B}^0 \rangle \; . \qquad\qquad (3.9)$$

The second type of experiment produces a $B^0\bar{B}^0$ pair with charge parity C and measures the decays of one B meson into a lepton and the other to a hadronic state f. The CP asymmetry was computed to be

$$A = \frac{-2xa^2 \; \sin^2\phi}{1+y^2+y\cos 2\phi} \qquad \text{for C = even} \quad ;$$

$$= 0 \qquad\qquad \text{for C = odd} \quad . \qquad\qquad (3.10)$$

For both B_s and B_d systems the parameter y is small compared to unity and can be neglected in Eqs. (3.5) and (3.10). The parameters x_s and x_d for the B_s and B_d systems are calculated using the box diagram value for ReM_{12} and the experimental value for the B meson lifetime[7,18]. Results of the calculated x are shown in Tables 1-9, which depend very much on the parameters used.

For B_d system the angle ϕ depends on the decay channel

$$tg\phi = \frac{s_3 \sin\delta}{s_2 + s_3 \cos\delta} \quad , \qquad \text{for } b \rightarrow c \text{ decays ;}$$

$$tg\phi = tg\delta \quad , \qquad \text{for } b \rightarrow u \text{ decays .} \qquad (3.11)$$

For B_s system the angle ϕ is very small for the $b \rightarrow c$ channel and

$$tg\phi = \frac{s_2 \sin\delta}{s_3 + s_2 \cos\delta} \quad , \qquad \text{for } b \rightarrow u \text{ decays .} \qquad (3.12)$$

The asymmetries for B_d system can reach 20-40% in a certain range of parameters. However, if $\left| \epsilon'/\epsilon \right|$ is found to be small in future experiments, $\sin\delta$ will become small and the CP violation effects in $B^0 - \bar{B}^0$ system will also be small.

IV. CP VIOLATION IN PARTIAL DECAY RATES OF PARTICLES AND ANTIPARTICLES

To begin with I shall recall some general remarks made by Pais and Treiman[19] concerning the constraints imposed by CPT invariance. Let f_i be a set of final states connected by strong or electromagnetic interactions into which the particle p can decay and $\bar{f}_i$ the corresponding conjugate states. CPT invariance requires the partial width of p and $\bar{p}$ to be equal

$$\sum_i \Gamma(p \rightarrow f_i) = \sum_i \Gamma(\bar{p} \rightarrow \bar{f}_i) \quad , \qquad (3.13)$$

provided this set of final states has no strong or electromagnettic interactions with the other decaying channels. If the set consists of only a single state no information about CP violation can be found. For two final states the decaying amplitudes of the particle have the form

$$a = f M_1 + g M_2 \quad , \qquad (3.14)$$

where f and g are complex coupling constants in the effective weak interaction while M_i, $i = 1,2$ are different amplitudes leading to the same final state. The corresponding amplitudes for the antiparticles are

$$\bar{a} = f\, M_1^* + g\, M_2^* .$$

(3.15)

we can easily calculate the asymmetry to be

$$A = \frac{2\, \text{Im}(f^* g)\, \text{Im}(M_1^* M_2)}{\left|fm_1\right|^2 + \left|gM_2\right|^2 + 2\, \text{Re}(f^* g)\, \text{Re}(M_1^* M_2)}$$

(3.16)

In the case of strange particle decay the asymmetry is found to be very small[20]. In this case

$$\frac{\text{Im}(M_1^* M_2)}{\left|M_1\right|^2 + \left|M_2\right|^2} \propto \sin(\delta_1 - \delta_2) \text{ where } \delta_i \ i = 1,2$$

are final state interaction phases and $\text{Im}(f^* g) \propto s_1^2 s_2 s_3 \sin\delta$, both of which are very small. As the real parts of the corresponding coupling constants f and g in kaon decays are of the order 1, the asymmetry of partial rates for strange decays is very small of the order $10^{-5} - 10^{-6}$.

In general the number of the final hadronic channels increases rapidly as one progresses from strange to bottom decays. The final state interactions are much stronger and $2\, \text{Im}(M_1^* M_2)/(\left|M_1\right|^2 + \left|M_2\right|^2)$ might reach several percent in B-decays. For those decaying channels where $\left|f\right|^2$ and $\left|g\right|^2$ are comparable with $\text{Im}(f^* g) \propto s_1^2 s_2 s_3 \sin\delta$, the asymmetry can be quite big. Examples[21] are given in Table 10. Chau and Cheng proved a theorem using quark diagram techniques to show that the CP violation effects in the asymmetry of partial decay rates of particles and antiparticles are always proportional to $s_1^2 s_2 s_3 \sin\delta$ which is a very small number of the order of

$10^{-5} - 10^{-6}$. The apparent large asymmetry in the B-decays is due to the smallness of the denominator in Eq. (3.16). Therefore for those decay channels where the asymmetry is large the branching ratio is always small. The number of events required to observe the CP violation effects in B-decay is still large, of the order of 10^5-10^6. It is not an easier task than the observation of CP violation in kaon decays.

V. CONCLUDING REMARKS

1) Measurements of $\left|\varepsilon'/\varepsilon\right|$ and the lower limit of $\left|V_{ub}/V_{cb}\right|$ as well as the discovery of the top quark are crucial in finding a consistent picture of CP violation within the standard model. More careful calculation of B_K, $C_6 \langle 2\pi/2=0 | Q_6 | K^0 \rangle$ and M_{12}^{soft} are also necessary. A low value of m_t less than 40 GeV together with B_K around 0.33 might rule out the K-M model with three generations of quarks. In this case the K-M phases might still be the only source of CP violation if more than three generations are found. A reduction of the upper limit of $\left|\varepsilon'/\varepsilon\right|$ not only pushes the theoretical m_t up but also reduces the value of $\sin\delta$, thus making the observation of other CP violation effects more difficult.

2) CP impurities in B^0-$\bar{B}^0$ mass eigenstates are very small. Therefore observations on the fractional difference of same-sign dilepton production and the asymmetries in the semileptonic decays are very difficult.

3) The mixing for B_s^0 systems is large, but the asymmetries in the B_s^0(b→c) channels are small. The mixing for B_d^0 systems and the asymmetries in the decay channels B_d^0(b→c) may reach several percent in a certain range of parameters.

4) CP violation effects in partial decay rates of particles and antiparticles are small for strange particle decays. It is much larger for charged B decays. However the number of events

TABLE 10 Asymmetry and number of events needed to observe CP violation in D and B decays. Taken from Chau and Cheng[21].

Reaction	Amplitudes	A_{tree}	$A_{Penguin}$	Br	No. of Events Needed
$B_u^- \to \pi^0 K^-$	$\frac{1}{\sqrt{2}}[V_{cb}V_{cs}^\star(e)$	0	-6×10^{-2}	3.8×10^{-5}	7.3×10^6
	$+\, V_{ub}V_{us}^\star(a+b+d+e)]$	(0)	(-1.6×10^{-2})	1.4×10^{-4}	(7.3×10^6)
$\to D^- D^{0}\star$	$V_{cb}V_{cd}^\star(a+e)$	-1.6×10^{-2}	1×10^{-3}	3×10^{-3}	1.3×10^6
	$+\, V_{ub}V_{ud}^\star(d+e)$	(-0.86)	(7.3×10^{-4})	(4.1×10^{-3})	(3.3×10^2)
$\to K^- J/\psi$	$V_{cb}V_{cs}^\star(b+e) + V_{ub}V_{us}^\star(\bar{d}+e)$	-3×10^{-4}	$-$	5×10^{-3}	2.2×10^9
	$+\, \sum_{i=u,c,t} V_{ib}V_{is}^\star(e)$				
$B_c^- \to K^- \bar{D}^0$	$V_{cb}V_{cd}^\star(d+e)$	0.39	-7×10^{-2}	2×10^{-5}	3.5×10^5
	$+\, V_{ub}V_{us}^\star(a+e)$	(6.2×10^{-2})	(7.2×10^{-5})	(1.8×10^{-2})	(1.4×10^4)
$\to \pi^- \bar{D}^0$	$V_{cb}V_{cd}(d+e)$	-4×10^{-2}	3.7×10^{-3}	2.3×10^{-4}	2.7×10^6
	$+\, V_{ub}V_{ud}^\star(a+e)$	(-0.86)	(1×10^{-3})	(8.5×10^{-4})	(1.6×10^3)
$\to \pi^0 D^-$	$\frac{1}{\sqrt{2}}[V_{cb}V_{cd}^\star(d+e)$	-0.17	-2.2×10^{-2}	1.3×10^{-5}	2.7×10^6
	$+\, V_{ub}V_{ud}^\star(b+e)]$	(-0.86)	(-5.7×10^{-3})	(4.8×10^{-5})	2.8×10^4
$F^+ \to K^0 \pi^+$	$V_{ud}V_{cd}^\star(a+e)$	-3.6×10^{-4}	-3.3×10^{-5}	1.8×10^{-3}	4.5×10^9
	$+\, V_{us}V_{cs}^\star(d+e)$	(-2×10^{-3})	(-1.7×10^{-5})	(3.4×10^{-3})	(7.4×10^7)

needed to observe the CP violation effects is large, of the order 10^5-10^6.

ACKNOWLEDGMENT

This talk is the result of a collaboration with Wu Yue-liang and Xie Yan-bo. Discussions with Profs. Li Xiao-yuan, Chu Chen-yuan and L.-L. Chau have helped enormously in improving my understanding of the problem. I would like to thank Mrs. Isabell for her kindness and support in typing the manuscript.

REFERENCES

1. J.H. Christenson, J.W. Cronin, V.L. Fitch and R. Turlay, Phys. Rev. Lett. 13 (1964) 138.

2. G. Arnison et al., Phys. Lett. 126B (1983) 398; ibid 129B (1983) 273; P. Bagnaia et al., Phys. Lett. 129B (1983) 130.

3. M. Kobayashi and T. Maskawa, Prog. Theor. Phys. 49 (1973) 652.

4. For a recent review, see L.-L. Chau, "Quark Mixing in Weak Interactions", Phys. Rep. 95, No. 1 (1983).

5. L.-L. Chau and W.-Y. Keung, preprint BNL-23811 (1983).

6. L. Wolfenstein, Phys. Rev. Lett. 51 (1984) 1945.

7. L.-L. Chau and W.-Y. Keung, BNL preprint (1984).

8. C. Itzykson, M. Jacob and G. Mahoux, Nouvo Cim. Suppl. 5 (1967) 978.

9. J.S. Hagelin, Phys. Lett. 117B (1982) 441.

10. T. Inami and C.S. Lim, Prog. Theor. Phys. 65 (1981) 297.

11. N. Cabibbo and C. Martinelli, TH-3774 CERN (1983); R.C. Brower, G. Maturana, M.B. Gavela and R. Gupta, HUTP-84/A004 NUB # 2625 (1983).

12. P.H. Ginsparg, S.L. Glashow and M.B. Wise, Phys. Rev. Lett. 50 (1983) 1415.

13. F.J. Gilman and J.S. Hagelin, preprint SLAC-PUB-3226 (1983).

14. P.H. Ginsparg and M.B. Wise, Phys. Lett. 127B (1983) 265.

15. Pham Xuan-Yen and Vu Xuan-Chi, preprint, PAR LPTHE 82/28 (1983).

16. K.C. Chou, Y.L. Wu and Y.B. Xie, preprint ASITP-84-005 (1984).

17. J.S. Hagelin, Nucl. Phys. B193 (1981) 123;
    B. Carter and A.I. Sanda, Phys. Rev. Lett. 45 (1980) 952;
    Phys. Rev. D23 (1981) 1567;
    L.I. Bigi and A.I. Sanda, Nucl. Phys. B193 (1981) 85;
    Ya. I. Azimov and A.A. Iogansen, Yad. Fix. 33 (1981) 383,
    [Sov. J. of Nucl. Phys. 33 (1981) 205].

18. L.I. Bigi and A.I. Sanda, preprint NSF-ITP 83-168 (1983);
    L. Wolfenstein, preprint CMU-HEG 83-9; NSF-ITP-83-146 (1983);
    E.A. Paschos and U. Türke, preprint NSF-ITP-83-168 (1983).

19. A. Pais and S.B. Treiman, Phys. Rev. D12 (1975) 2744.

20. L.L. Chau and W.Y. Keung, Phys. Rev. D29 (1984) 592.

21. J. Bernabeu and C. Jarlskog, Z. Phys. C8 (1981) 233;
    L.L. Chau and H.Y. Cheng, BNL preprint (1984).

DISCUSSION

PAVLOPOULOS:
Does a t-quark mass of 60-70 GeV lead to a hopelessly small ε'/ε?

CHOU:
No, if B_K is around 0.33, ε'/ε can reach .01 for $m_t \simeq$ 60-70 GeV.

HITLIN:
If $|\varepsilon'/\varepsilon|$ is small and $\sin\delta$ is of the appropriate value, then the t quark mass could be greater than the W or Z mass. Could enough tt pairs be produced at the SPS collider to make the decay $t \rightarrow W + b$ a plausible explanation of the recently reported events at the SPS Collider?

CHOU:
We better keep in mind such possibilities.

CP VIOLATION FROM NON STANDARD MODELS[*]

Gino Segrè

Department of Physics, University of Pennsylvania

Philadelphia, Pennsylvania 19104

1) The Standard Model

The title of my talk allows for a great deal of freedom. I could say that SUSY models are standard etc. and turn to wild speculations, but instead I will be conservative.

I will define the standard model as one in which the invariance group is $SU(3)_c \times SU(2)_L \times U(1)$ broken spontaneously to $SU(3)_c \times U(1)_{e.m.}$. Quarks and leptons are in the usual left handed doublets and right handed singlets. The spontaneous symmetry breaking is due to either one or two Higgs doublets; if there are two, one of them couples only to charge $-1/3$ right handed quarks and the other only to charge $+2/3$ right handed quarks. This ensures that the neutral Higgs couplings are flavor preserving[1]; if we call n and p the quarks which couple to the weak currents and $\psi_L = \begin{pmatrix} p_L \\ n_L \end{pmatrix}$ the left handed doublets, these Yukawa couplings are

$$L_{yuk,quarks} = \sum_{i,j=1}^{F} \{\bar{\psi}_{iL} \Gamma_{ij} n_{jR} H_1 + \bar{\psi}_{iL} \Gamma_{ij} p_{jR} \tilde{H}_2\} + h.c. \qquad (1)$$

where $\tilde{H}_2 = i \tau_2 H_2^\dagger$ and F is the number of families. The vacuum

---

[*]Talk delivered at Europhysics Topical Conference on "Flavor mixing in Weak Interactions" at Ettore Majorana Centre for Scientific Culture, March 1984.

expectation value of $H_{1,2}$ are $v_{1,2}$

$$H_1 = \begin{pmatrix} H_1^+ \\ v_1+\eta_1+i\xi_1 \end{pmatrix} \qquad H_2 = \begin{pmatrix} H_2^+ \\ v_2+\eta_2+i\xi_2 \end{pmatrix} \tag{2}$$

The mass matrix is diagonalized by the usual bi-unitary transformations[2]

$$U_L^\dagger \Gamma v_1 U_R = M^{(-1/3)}$$

$$U_L'^\dagger \Gamma v_2 U_R' = M^{(2/3)} \tag{3}$$

The mass eigenstates are related to the current quark states by (dropping family indices)

$$d_L = U_L^\dagger n_L \qquad\qquad d_R = U_R^\dagger n_R$$

$$u_L = U_L'^\dagger p_L \qquad\qquad u_R = U_R'^\dagger p_R \tag{4}$$

so of course the neutral Higgs boson couplings are flavor diagonal (e.g. η_1 couplings are $\frac{1}{v_1}$ times the mass matrix). The charged Higgs boson couplings are not flavor diagonal however. Dropping the i,j flavor indices again

$$L_{y,q} = \bar{p}_L \Gamma v_1 n_R H_1^+/v_1 + \bar{n}_L \Gamma' v_2 p_R \left(\frac{-H_2^-}{v_2}\right) + \text{h.c.}$$

$$= \bar{u}_L U_L'^\dagger U_L U_L^\dagger \Gamma v_1 \, U_R\left(\frac{H_1^+}{v_1}\right) + \bar{d}_L U_L^\dagger U_L' U_L'^\dagger \Gamma_2 v_2 \, U_R\left(\frac{-H_2^-}{v_2}\right) + \text{h.c.}$$

$$= \bar{u}_L \, A^\dagger M^{(-1/3)} d_R\left(\frac{H_1^+}{v_1}\right) + \bar{d}_L \, A \, M^{(2/3)} \, p_R\left(\frac{-H_2^-}{v_2}\right) + \text{h.c.} \tag{5}$$

where $A = U^\dagger U'$ is just the Kobayashi Maskawa[3] matrix as we see from the weak charged current

$$n_L \gamma^\mu p_L = d_L \, U_L^\dagger U_L' \gamma^\mu \, u_L \quad . \tag{6}$$

A linear combination, $(v_1 H_1^+ + v_2 H_2^+)/\sqrt{v_1^2 + v_2^2}$ is eaten by the W^+ and the orthogonal combination remains as a real Higgs boson. This model, or alternatively the even simpler one in which there is only a single Higgs doublet is what we take as the standard model. In it, we expect[2] $K^0 - \bar{K}^0$ mixing due to the second order weak box diagram with W^+-W^- mixing, with both real and imaginary parts of the mass matrix, CP violation in $K^o \to 2\pi$ predominantly due to penguin diagrams and an ε'/ε of the order of 10^{-2}.

2) The Strong CP Problem

The Q.C.D. Lagrangian admits a term[4] which violates both P and T invariance

$$\mathcal{L}_\theta = \frac{\theta}{32\pi^2} \; F_{\mu\nu} \; \tilde{F}_{\mu\nu} \tag{7}$$

By a chiral U(1) transformation this term can be rotated onto the quark mass matrix[5] or, conversely, making the quark mass matrix real, we have an effective

$$\tilde{\theta} = (\theta + \arg \det M) \tag{8}$$

which cannot be rotated away. By measurements of the electric dipole moment of the neutron[6], and theoretical calculations, we know that[7]

$$\tilde{\theta} \lesssim 10^{-9} \tag{9}$$

The strong CP problem is to find out why $\tilde{\theta}$ is so small. I believe this is the outstanding problem in CP violation; I also believe it does not have a satisfactory answer.

3) The Axion

The cleverest solution to this strong CP problem is the axion. The idea was originally proposed by Peccei and Quinn,[8] in the form of an additional global symmetry $U(1)_{p.q.}$ under which quark and Higgs fields transform non-trivially

$$q_i \to e^{-i\beta\gamma_5/2} \; q_i$$
$$H_1 \to e^{i\beta} \; H_1$$
$$H_2 \to e^{-i\beta} \; H_2 \tag{10}$$

The idea is to rotate the phase θ onto the quark mass matrix, which now has a total phase $\tilde{\theta}$. After QCD condensates are formed, which preserve CP, we have $<0|\bar{q}_L q_R|0< \neq 0$. We now look for the solution of the minimum to the classical potential. It turns out to be at

$\tilde{\theta} = 0$. The price we pay is that the spontaneous breaking of the $U(1)_{P.Q.}$ global symmetry means there is a pseudo-Goldstone boson, the axion.[9] This is analogous to the appearance of the pions as pseudo-Goldstone bosons when the chiral $SU(2) \times SU(2)$ symmetry of the quark mass matrix is broken down to $SU(2)$. In fact the axion mass $m_a \sim \dfrac{m_\pi f_\pi}{v}$, where v is the scale of the breaking of $U(1)_{P.Q.}$ and f_π is the pion decay constant $f_\pi = 95$ MeV. The axion field is

$$a = \frac{v_1 \, \text{Im} \, H_2^O - v_2 \, \text{Im} \, H_1^O}{\sqrt{v_1^2 + v_2^2}} \tag{11}$$

and its coupling to matter is of order $\dfrac{1}{v_{1,2}}$. To see this last statement qualitatively, let a Higgs field $\phi = (v + \rho + ia)$ couple to quarks.

$$\bar{q}_L \, \Gamma \, q_R \phi + \text{h.c.} \Big|_{a \text{ component}} = \bar{q}_L \, \Gamma \, v \, q_R \, \frac{ia}{v} + \text{h.c.}$$

$$= \frac{ia}{v} \, m_q \, \bar{q} \, \gamma_5 q \tag{12}$$

i.e. the coupling is pseudoscalar and proportional to $1/v_{1,2}$. Searches for the axion did not lead to its discovery.[10] A more weakly produced axion is ruled out by astrophysics arguments; once produced in stars they stream out freely, thereby generating an efficient mechanism for energy loss. The solution is to not have them produced copiously, by having them couple very weakly; this argument essentially rules out a range of couplings corresponding to a v of $250 \text{ Gev} \lesssim 10^9 \text{ Gev.}$ [11]

At this point the axion scenario was revived by the idea of the invisible axion.[12] This says, in its simplest form that we include in our model a complex isoscalar spin zero field Φ which transforms non-trivially under $U(1)_{P.Q.}$. In particular the Higgs potential has a term of the form

$$\lambda \Phi^2 \, H_2^\dagger \, H_1 + \text{h.c.} \tag{13}$$

so that under $U(1)_{P.Q.}$ $\Phi \rightarrow e^{-i\beta}\Phi$. The idea is now that Φ has a large v.e.v., $v \gtrsim 10^9$ Gev. Instead of the first order formalism in which $\Phi = (v + \Phi_1 + i\Phi_2)$, it is more convenient to treat the imaginary part as a periodic variable

$$\Phi = (v + \rho)e^{i\frac{\alpha}{v}} \tag{14}$$

We see that the pseudo-Goldstone boson of the theory, the axion, is almost entirely α. The admixtures of Im $H^o_{1,2}$ are of order $v_{1,2}/v \lesssim 10^{-7}$. At this point, we might be tempted to put $v \sim 10^{15}$ Gev, the G.U.T. scale and have Φ be part of a G.U.T. theory, e.g. a complex 24 in SU(5).[13] This was shown to be impossible on cosmological grounds.

The argument, in its most schematic form goes as follows. We put $\alpha = v\theta$; we expect θ at some early time in the universe to be of order one, i.e. α of order v. When Q.C.D. is broken spontaneously and the quarks acquire masses, the instantons introduce a potential in which the axion field begins to oscillate with some amplitude A and a frequency determined by the axion mass m_a

$$\alpha = A \, v \cos m_a t \quad . \tag{15}$$

The energy density in the field is

$$\varepsilon = m_a^2 v^2 A^2 \sim m_\pi^2 f_\pi^2 A^2 \tag{16}$$

The axions are non-relativistic and essentially non-interacting; their number density is therefore $\varepsilon/m_a = m_a v^3 A^2$. As the universe cools this scales like $\frac{1}{R^3}$, i.e. like T^3; the scaling with T of $m_a(T)$ is also known. Putting these together we find that the energy density in axions is greater than the critical density for $v \gtrsim 10^{12}$ Gev.

We have now two scales of symmetry breaking, $v_{1,2} \sim 10^2$ Gev and 10^9 Gev $\lesssim v \lesssim 10^{12}$ Gev. This means that the Higgs potential must be fine tuned. In general the Higgs doublets would have v.e.v.'s of order v once the singlet is introduced. In order to keep the scale of SU(2) breaking at 10^2Gev, we would have to tune dimensionless coupling constants to an accuracy of $v^2/v_{1,2}^2$.

To summarize, we started with the problem of why $\theta \lesssim 10^{-9}$.

We introduced a new global symmetry, a second Higgs doublet, a new Higgs singlet, a new scale v and then fine tuned the Higgs potential parameters. It seems like too much of a good thing.[15]

4) $\tilde{\theta}$ Equal Zero at Tree Level

Having almost ruled out the axion, let us pursue other alternatives. We could just set $\tilde{\theta} = 0$ at tree level, but in general, we expect radiative corrections to give infinite correction terms to $\tilde{\theta}$ since CP is violated. t'Hooft[16] has convincingly argued that a small parameter should exist in a theory only if by setting it equal to zero the symmetry of the theory increases. The radiative corrections will be proportional to the parameter itself and hence are naturally small as well. As an example, consider the electron mass in Q.E.D.: $m_e \to 0$ leads to an additional chiral symmetry. Clearly setting $\tilde{\theta} \to 0$ is not in this category.

One possibility would be to set the up quark mass to zero, since we then have an additional unbroken chiral symmetry which can be used to set $\tilde{\theta} = 0$, but m_u is not equal to zero.[17]

A second possibility is to have CP broken spontaneously or softly so that the corrections to $\tilde{\theta}$ are calculable. In this type of scheme one must arrange that $\tilde{\theta}$ is zero at the tree level and that the previously mentioned corrections are small.[18] We will illustrate this procedure in the next section within the context of left right symmetric models.[19]

Before starting however, we will restate the general argument in schematic form. The pieces of the Lagrangian that need to be considered from the point of view of CP violation are

$$\theta \, F_{\mu\nu} \tilde{F}^{\mu\nu} + \overline{\psi}_{L,i} \, \Gamma^a_{ij} \, \phi_a \, \psi_{R,j} + V(\phi) \tag{17}$$

with generic left and right handed fermion fields $\psi_{L,i}$ and $\psi_{R,i}$ belonging to families labelled by indices i,j and Hermitian Higgs fields ϕ_a. Imposing CP invariance on $\mathcal{L}$ sets $\theta = 0$, makes $V(\phi)$ CP invariant and sets $\Gamma^a_{ij} = \Gamma^a_{i,j}{}^*$. We know nevertheless that CP is violated in nature so we have it broken spontaneously by fixing the

Higgs potential so that one or more of the Higgs fields acquire complex v.e.v.'s (there obviously must be more than one Higgs field in the theory).[20] At this point the fermion mass matrix

$$\bar{\psi}_{L,i} \, M_{ij} \, \psi_{R,j} = \bar{\psi}_{L,i} \, \Gamma_{ij}^{a} \, v_a \, \psi_{Rj} \qquad (18)$$

has in general acquired a phase, since the v_a are complex. We must have an additional symmetry present such that $\Gamma_{ij}^{a} v_a$, has overall phase zero, i.e. so that $\tilde{\theta} = 0$ at tree level. We then proceed to calculate the radiative corrections to Arg det M and show that they are in fact less than 10^{-9}.[7] At that point we're finished with the strong CP problem.

The weak CP problem now comes to the front. The bi-unitary transformation which diagonalizes e.g. the $-1/3$ quarks is

$$U_L^{\dagger} \, M(-\tfrac{1}{3}) U_R = M(-\tfrac{1}{3})_{diag} = \begin{pmatrix} m_d & & 0 \\ & m_s & \\ 0 & & m_b \end{pmatrix} \qquad (19)$$

and similarly the charge $2/3$ quarks

$$U_L'^{\dagger} M(\tfrac{2}{3}) U_R' = M(\tfrac{2}{3})_{diag} = \begin{pmatrix} m_u & & 0 \\ & m_c & \\ 0 & & m_t \end{pmatrix} \qquad (20)$$

The charge gauge boson couplings, when written in terms of the mass eigenstates, have the usual K.-M.[3] matrices $A_{L,R} = U_{L,R}^{\dagger} \, U_{L,R}'$

$$g \, \bar{p}_L \gamma_\alpha \, n_L \, W_L^\alpha + g' \bar{p}_R \gamma_\alpha n_R \, W_R^\alpha$$

$$= g \, \bar{u}_L \, U_L'^{\dagger} U_L \gamma_\alpha \, d_L W_L^\alpha + g' \bar{u}_R U_R'^{\dagger} U_R \gamma_\alpha d_R W_R^\alpha \ . \qquad (21)$$

In addition, we will in general have complex couplings of neutral Higgs bosons which are not flavor diagonal. The matrices which diagonalize $\Gamma_{ij}^{a} v_a$ do not diagonalize the couplings of neutral Higgs bosons to quark mass eigenstates except for the trivial case in which only one Higgs doublet couples to $p_{R,i}$ and one to $n_{R,i}$.[1] We therefore do not expect to have Arg det M = 0 at tree level.

The discrete symmetry imposed to require this is in general incompatible with the neutral Higgs couplings being flavor diagonal.[21] K^O-$\bar{K}^O$ transitions occur therefore not only via W_L-W_L, W_L-W_R and W_R-W_R box diagrams (of course only in the case where there are W_R's present) but also by way of effective terms in the Hamiltonian

$$c_1 (d_L s_R)^2 + c_2 (d_L s_R)(d_R s_L) + c_3 (d_R s_L)^2 \qquad (22)$$

with $c_{1,2,3}$ complex, caused by the exchange of neutral Higgs bosons. The constants $c_{1,2,3}$ depend on the details of the model, but in general the flavor changing neutral Higgs bosons must have masses in the Tev range in order for (22) to be compatible with experimental limits.

There is one model which looks like it provides a solution to all these problems. Assume there are three Higgs doublets, one of which, H_1, couples only to n_R quarks, a second of which, H_2, couples only to p_R quarks, and the third, H_3, which doesn't couple to quarks at all. We clearly have no flavor changing Higgs mediated neutral currents. We next impose CP as a discrete symmetry of $\mathcal{L}$ so $\theta = 0$. CP is now broken spontaneously so

$$\langle H_{1,2} \rangle_o = v_{1,2} \, e^{i\omega_{1,2}} \qquad (23)$$

but the phases $\omega_{1,2}$ can be absorbed by n_R and p_R so the mass matrixes are real. The latter can therefore be diagonalized by orthogonal transformations so the K.-M. matrices are also real. Now for the subtlety: one linear combination of $H_1^{\pm}$ and $H_2^{\pm}$ is eaten by the $W^{\pm}$, but the orthogonal linear combination, which we call, $\phi^{\pm}$, survives as a physical particle.[22] Furthermore $\phi^{\pm}$ is in general a complex linear combination of $H_1^{\pm}$ and $H_2^{\pm}$ (remember also that these mesons' couplings to quarks are not flavor diagonal). The $W_L^{\pm}$ - $W_L^{\mp}$ box diagram is real so it contributes only to Re $M_{K^O-\bar{K}^O}$, but the $W_L^{\pm}$ - $\phi^{\mp}$ box diagram contributes to Im $M_{K^O-\bar{K}^O}$, as well as to Re $M_{K^O-\bar{K}^O}$ so all our problems seem solved. The difficulty is that the model also gives a penguin diagram with a $\phi^{\pm}$ loop rather than the $W^{\pm}$ one and the resulting estimates of ε'/ε in this model are of order 5×10^{-2}, i.e. too big.[23] The electric dipole moment of the neutron may also be too big in this model.[24]

5) <u>Left-Right Symmetric Models</u>

We treat these models as a specific example of the phenomena we discussed in section 4). There are many variants of these models, but the basic ingredients[19] are: i) the electroweak gauge group is extended to $SU(2)_L \times SU(2)_R \times U(1)$ so we have additional gauge bosons $W_R^{\pm}$, Z_R; ii) a left right symmetry is imposed so that the L and R gauge couplings are equal and every particle multiplet has its L or R image; iii) the L↔R symmetry is spontaneously broken so that e.g., $M_{W_L} \neq M_{W_R}$, i.e. L↔R symmetry is not observed in nature. The L↔R symmetry says, for instance, that we have, labelling states by their $SU(2)_L \times SU(2)_R \times U(1)$ quantum mumbers L and R quark doublets

$$\psi_L^i = \begin{pmatrix} p^i \\ n^i \end{pmatrix}_L \sim \left(\frac{1}{2}, 0, \frac{1}{3}\right) \qquad \psi_R^i = \begin{pmatrix} p^i \\ n^i \end{pmatrix}_R \sim \left(0, \frac{1}{2}, \frac{1}{3}\right) \tag{23}$$

with electric charge $Q_{elec.} = T_{3L} + T_{3R} + \frac{Y}{2}$. They are coupled together by Higgs bosons ϕ_a which belong to the $(\frac{1}{2}, \frac{1}{2}, 0)$ representation so that the Yukawa couplings are

$$L_{yuk,quark} = \bar{\psi}_{L,i} \, (\Gamma_{ij}^a \, \Phi_a + \Gamma_{ij}^{a'} \, \tilde{\Phi}_a) \psi_{R,j} + h.c. \tag{24}$$

with $\tilde{\phi} = i\tau_2 \, \Phi*(i\tau_2)^{\dagger}$. [The indices i,j are family indices, as usual. N demotes which of the N Higgs fields we are dealing with; generally we will take N = 1.] The v.e.v. of Φ is

$$<\Phi> = \begin{pmatrix} k & o \\ o & k' \end{pmatrix} \tag{25}$$

where k or k' can be chosen as real by performing an overall rotation. We also have fields which do not couple to fermions with large v.e.v. designed to break the L↔R symmetry.[25] Models with X_L and X_R in the $(\frac{1}{2},0,1)$ and $(0,\frac{1}{2},1)$ representations or Δ_L and Δ_R in the (1,0,2) and (0,1,2) representations have been used to play this role, but we need not concern ourselves with them at this moment.

Our worry is CP violation. First we see that θ is equal to zero by the L↔R symmetry under which $F_{\mu\nu} \tilde{F}^{\mu\nu} \to - F_{\mu\nu} \tilde{F}^{\mu\nu}$. Equation (24) is then written out as

$$L_{\text{yuk.,quarks}} = \bar{\psi}_{Li}(\Gamma^a_{ij} \Phi_a + \Gamma'^a_{ij} \tilde{\Phi}_a) \psi_{Rj}$$

$$+ \bar{\psi}_{Rj} (\Gamma^{a*}_{ij}\Phi^\dagger_a + \Gamma'^{a*}_{ij} \tilde{\Phi}^\dagger_a) \psi_{Li} \tag{26}$$

Under L↔R symmetry $\psi_{Li} ↔ \psi_{Ri}$, $\Phi_a ↔ \Phi^\dagger_a$ so L↔R symmetry implies

$$\Gamma = \Gamma^\dagger \qquad \Gamma' = \Gamma'^\dagger \tag{27}$$

and the mass matrices are

$$M(-1/3)_{ij} = \Gamma^a_{ij} k'_a + \Gamma'^a_{ij} k^*_a$$

$$M(2/3)_{ij} = \Gamma^a_{ij} k_a + \Gamma'^a_{ij} k'^*_a \tag{28}$$

As we see, even though θ is zero, $\tilde{\theta}$ is not. The strategy of calculability suggests we let CP be broken spontaneously.

Even if CP is broken spontaneously, we do not in general have $\tilde{\theta} = 0$. Imposing CP as a discrete symmetry says Γ and Γ' are real matrices. By an overall rotation we can make either k or k' real, but not both so the mass matrix is not real. Of course it may happen that our Higgs potential is such that $\langle\Phi_a\rangle$ is real even though CP is violated, or we may impose additional discrete symmetries to insure that $\langle\Phi_a\rangle$ is real.

CP invariance of a L↔R symmetric theory says

$$\Gamma = \Gamma^* \qquad \Gamma' = \Gamma'^* \qquad . \tag{29}$$

Combining this with (27), we have that Γ and Γ' are symmetric matrices. If we now allow for spontaneous breaking of CP, i.e. $\langle\Phi_a\rangle \neq \langle\Phi^\dagger_a\rangle$, we see that the mass matrices are symmetric, but of course not Hermitian. This is enough to show that[26]

$$U_R = U^*_L J \tag{30}$$

where J is a unitary diagonal matrix, i.e. just phases as diagonal entries. Branco et al.[27] have taken this type of theory a step further by considering models in which U_L is an orthogonal matrix O_L. The left K.-M. matrix is therefore also an orthogonal matrix and ε'/ε is very small (W_R contributions to $K \to 2\pi$ are in the nature of an overall phase so that $I = 2$ and $I = 0$ amplitudes for $K \to 2\pi$ are relatively real).

If we take $L \leftrightarrow R$ symmetric models with CP as a discrete symmetry, we can, as we saw earlier, break CP and $L \leftrightarrow R$ symmetry spontaneously by Φ and $\Delta_{L,R}$ fields. By overall U(1) rotations we take $\langle \Delta_R \rangle$ to be real and $\langle \Phi_a \rangle$ to be of the form (25) with k real. If we now choose the Higgs potential so that $\langle \Delta_L \rangle = 0$ and $|k'/k| \ll 1$, we see that CP violation is small and depends perturbatively on the parameter

$$\frac{k'}{k} = \left| \frac{k'}{k} \right| e^{i\alpha} = re^{i\alpha} \tag{31}$$

D. Chang[28] has recently considered such a model and showed that it almost reproduces the results of the superweak model[29] $\eta_{+-} = \eta_{oo}$ and $\eta_{+-0} = \eta_{ooo}$.

This is all very interesting, but doesn't solve the strong CP problem. To do this, as we stated earlier, you must do something extra such as imposing additional symmetries or enlarging the group.

As a final caveat. we should mention that there are also models in which $\tilde{\theta}$ is not zero at tree level, but naturally (in some sense) small, i.e.[30] $\tilde{\theta} \lesssim 10^{-9}$.

6) Flavor Group Models

It is tempting to speculate that the flavor symmetry is gauged or global. We then may have so-called horizontal interactions[31]. Rather than a general discussion of such models, we will show one particular one, due to A. Nelson[32], in which $\tilde{\theta}$ is zero at tree level and small and calculable at the loop level.

The model is a G.U.T. one in which the gauge group is SU(5). In addition, we have a global SO(3) family group and a global chiral $U(1)_G$ symmetry. At the GUT scale all the symmetries are broken except $SU(3)_c \times SU(2)_L \times U(1)$; we also assume that at this stage all

fermions and Higgs bosons get masses except for the three light
families of fermions and the light Higgs doublet. Labelling states
by their $SU(5) \times SO(3) \times U(1)_G$ quantum numbers we have

Fermions: $[10,3,0]$, $[5,3,0]$, $[10,1,1]$, $[10,1,-1]$

$\qquad\qquad\qquad [5,1,1]$, $[1,3,1]$ ...   (32)

Higgs: $(5,1,0)$, $(24,3,1)$ .....

and a Yukawa coupling

$$L_{Yuk.} = \lambda_u [10,3,0][10,3,0](5,1,0)$$

$$+ \lambda_d [10,3,0][\bar{5},3,0](5,1,0)^*$$

$$+ h [10,3,0][\bar{10},1,-1](24,3,1)$$

$$+ f [\bar{5},3,0][5,1,-1](24,3,1)$$

$$+ m_1 [10,1,1][\bar{10},1,-1] + m_2 [\bar{5},1,1][5,1,-1] \qquad (33)$$

In the above $m_{1,2}$ are bare mass terms allowed by the symmetry. We
naturally would expect them to be of the order of the G.U.T. scale.
Because CP is a discrete symmetry of $\mathcal{L}$, $m_{1,2}$, h, f, λ_u and λ_d are
all real. We have oversimplified the model; we expect there are
other $SO(3)$ triplets than $(24,3,1)$ such as $(1,3,1)$ etc, necessary
in any case to break $SU(5)$, but we wish to present the results as
quickly as possible. In this model, flavor symmetry, CP and $SU(5)$
all get broken simultaneously at the scale of v_i, the v.e.v. of
$(24,3,1)$. The flavor group fermion triplets mix with the flavor
group singlets so the three light quark families are in general
complex linear combinations of these four families. The θ problem
is solved at tree level because complexity is introduced only by
the v.e.v.'s of the $(24,3,1)$; this is enough to ensure Arg det M = 0
(this is not meant to be obvious; a calculation is needed to verify
it).[33] At the loop level the limits on $\tilde{\theta}$ are satisfied if $m_{1,2}$
is sufficiently small $(10^{-3} - 10^{-4})$ compared to $v_{1,2,3}$ since loop
contributions to $\tilde{\theta}$ have $m_{1,2}^2/v_i^2$ suppression factors.

This model is an amusing one, but suffers from some of the same
aesthetic flaws of the invisible axion, to wit it needs a new scale
$(m_{1,2})$, new particles etc.   It is promising though.

648

CONCLUSIONS

We see that the CP problem, or rather problems, remain unsolved. There are many models with suggested solutions, but none of them is compelling. Better measurements, in particular, a new value of ϵ'/ϵ will help a great deal, but the $\tilde{\theta}$ problem is probably still not understood satisfactorily.

REFERENCES

1. S. L. Glashow and S. Weinberg, Phys. Rev. D15, 1958 (1977).
   E. Paschos, ibid. D15, 1966 (1977).
   L. T. Trueman, F. Paige and E. Paschos, ibid D15, 3416 (1977).
   K. Kang and J. E. Kim, Phys. Lett. 64B, 93 (1976).

2. See L. L. Chau, Physics Reports 35, 1 (1983) for discussion and references.

3. M. Kobayashi and T. Maskawa, Prog. Theor. Phys. 49, 652 (1973).

4. G. t'Hooft, Phys. Rev. D14, 3432 (1976).
   C. G. Callan, R. F. Dashen and D. J. Gross, Phys. Lett. 63B, 334 (1976); R. Jackiw and C. Rebbi, Phys. Rev. Lett. 37, 172 (1976).

5. S. L. Adler, Phys. Rev. 177, 2426 (1969); J. S. Bell and R. Jackiw, Nuovo Cimento 60A, 47 (1969).

6. N. F. Ramsey, Ann. Rev. of Nuclear and Particle Science, 1982.

7. V. Baluni, Phys. Rev. D19, 2227 (1979); R. Crewther, P. di Vecchia, G. Veneziano and E. Witten, Phys. Lett. 89B, 123 (1979).

8. R. Peccei and H. Quinn, Phys. Rev. Lett. 38, 1440 (1977), Phya. Rev. D16, 1791 (1977).

9. S. Weinberg, Phys. Rev. Lett. 40, 223 (1978); F. Wilczek, Phys. Rev. Lett. 40, 279 (1978).

10. For a recent review see A. Zehnder, SIN report No. PR-83-03 (1983).

11. D. Dicus, E. W. Kolb and R. Wagoner, Phys. Rev. D18, 1829 (1978), ibid. D22, 839 (1980); K Sato and H. Sato, Prog. Theor. Phys. 54, 1564 (1975); M. Fugugita, S. Watamura and M. Yoshimura, Phys. Rev. Lett. 48, 1522 (1982).

12. J. Kim, Phys. Rev. Lett. $\underline{43}$, 103 (1979); M. Shifman, A. Vainshtein, and V. Zakharov, Nucl. Phys. $\underline{B166}$, 433 (1980); M. Dine, W. Fischler, and M. Srednicki, Phys. Lett. $\underline{104B}$, 199 (1981); H. P. Nilles and S. Raby, Stanford Linear Accelerator Center Report No. SLAC-PhB-2743, 1981 (unpublished).

13. M. B. Wise, H. Georgi and S. L. Glashow, Phys. Rev. Lett. $\underline{47}$, 402 (1981).

14. P. Sikivie, Phys. Rev. Lett. $\underline{48}$, 1556 (1982); L. Abbott and P. Sikivie, Phys. Lett. 120B, $\underline{133}$ (1983); J. Preskill, M. Wise, and F. Wilczek, Phys. Lett. $\underline{120B}$, 27 (1983); M. Dine and W. Fischler, Phys. Lett. $\underline{120B}$, 137 (1983).

15. For some recent attempts to surmount these problems, see e.g., G. Lazarides and Q. Shafi, Phys. Lett. $\underline{115B}$, 21 (1982); S. Barr, D. Reiss and A. Zee, Phys. Lett. $\underline{116B}$, 227 (1982); S. Barr, X. C. Cao and D. Reiss, Phys. Rev. $\underline{D26}$, 2176 (1982).

16. G. t'Hooft, in Recent Developments in Gauge Theories, proceedings of the NATO Advanced Study Institute, Cargese, 1979 (Plenum, New York, 1980).

17. P. Langacker and H. Pagels, Phys. Rev. $\underline{D19}$, 2070 (1979), and references therein.

18. See for example, H. Georgi, Hadronic Journal $\underline{1}$, 155 (1978); M. A. B. Beg and H.-S. Tsao, Phys. Rev. Lett. $\underline{41}$, 378 (1978); R. N. Mohapatra and G. Senjanovic, Phys. Lett. $\underline{79B}$, 283 (1978); S. M. Barr and P. Langacker, Phys. Rev. Lett. $\underline{42}$, 1654 (1979); G. Segrè and H. A. Weldon, Phys. Rev. Lett. $\underline{42}$, 1191 (1979); V. Goffin, G. Segrè and H. A. Weldon, Phys. Rev. $\underline{D21}$, 1410 (1980); S. Barr, Phys. Rev. $\underline{D23}$, 2343 (1981).

19. For an overall view of left right symmetric models see R. N. Mohapatra Lectures at Nato Summer School on Particle Physics, Munich, Germany 1983, to be published by Plenum Press.

20. T. D. Lee, Phys. Rev. $\underline{D8}$, 1226 (1973), Physics Reports $\underline{C9}$, 148 (1979); P. Sikivie, Phys. Lett. $\underline{65B}$, 141 (1976); S. Weinberg, Phys. Rev. Lett. $\underline{37}$, 657 (1976); A. B. Lahanas and

C. E. Vayonakis, Phys. Rev. D19, 2158 (1979).

21. See R. Gatto, G. Morchio and F. Strocchi, Phys. Lett. 80B, 265
    (1979); ibid. 83B, 348 (1979); R. Gatto, G. Morchio, G. Sartori
    and F. Stocchi, Nucl. Phys. B163, 221 (1980); G. Segrè and
    H. A. Weldon, Ann. of Phys. (NY) 124, 37 (1980), Phys. Lett.
    86B, 291 (1979); A. C. Rothman and K. Kang, Phys. Rev. D23,
    2657 (1981), ibid. D24, 167 (1981).

22. For details see, e.g., S. Weinberg, reference 20; G. C. Branco,
    Phys. Rev. Lett. 44, 504 (1980), Phys. Rev. D22, 2905 (1980).

23. This problem has a rather complicated history. The model was
    ruled out by the analysis of N. Deshpande, Phys. Rev. D23,
    2654 (1981) and A. Sanda, Phys. Rev. D23, 2647 (1981). Their
    assumptions of penguin dominance and short distance box explana-
    tion of the K_L-K_S mass difference were criticized by
    J. F. Donogue, E. Golowich, W. Ponce and B. Holstein, Phys. Rev.
    D21, 186 (1980: C. T. Hill and G. G. Ross, Phys. Lett. 94B, 234
    (1980); C. T. Hill, Phys. Lett. 97B, 275 (1980); L. Wolfenstein,
    Nucl. Phys. B160, 501 (1979). These were answered by e.g.
    J. F. Donogue, J. S. Hagelin and B. Holstein, Phys. Rev. D25,
    195 (1982), who again ruled out the model. For a recent dis-
    cussion see N. Deshpande, "CP Violation through Higgs Exchange"
    Univ. of Oregon preprint OITS 217 (1983), who shows $|\varepsilon'/\varepsilon|$ may
    be as small as .016. It is also predicted to be negative. Much
    of the uncertainty hinges on the dispersion contributions to
    K^0-$\bar{K}^0$ mixing. For a recent reference of this question see
    J. F. Donogue and B. Holstein, Phys. Rev. D29, 2088 (1984).

24. G. Beall and N. Deshpande, Phys. Lett. 132B, 427 (1983).

25. For some recent limits on the scale of the breaking of $SU(2)_R$
    imposed by the K_S-K_L mass difference and the neutron electric
    dipole moment see G. Beall, M. Bander and A. Soni, Phys. Rev.
    Lett. 48, 848 (1982); P. L. de Forcrand, Ph.D. Thesis, Univ. of
    Calif. Berkeley (1982) and LBL report 1469 (1982); R. N.
    Mohapatra, G. Senjanovic and M. Tran, Phys. Rev. D28, 546 (1983);

G. Ecker, W. Grimus and H. Neufeld, Phys. Lett. 127B, 365 (1983); ibid, 132B, 467 (1983), Nucl. Phys. B229, 421 (1983); G. Beall and A. Soni, Phys. Rev. Lett. 47, 552 (1981); F. J. Gilman and M. H. Reno, Phys. Lett. 127B, 426 (1983) and SLAC-Pub 3238, Oct. 1983, M. Hwang and R. J. Oakes, Phys. Rev. D28, 546 (1983); J. Trampetic, Phys. Rev. D27, 1565 (1983); A. Datta and A. Raychaudhuri, Phys. Rev. D28, 1170 (1983).

26. R. N. Mohapatra, F. E. Paige and D. Sidhu, Phys. Rev. D17, 2642 (1978).

27. G. C. Branco, J. M. Frere and J. M. Gerard, Nucl. Phys. B221, 317 (1983).

28. D. Chang, Nucl. Phys. B214, 435 (1983).

29. L. Wolfenstein, Phys. Rev. Lett. 13, 1980 (1964) for a discussion of these relations in L↔R theories, see ref. 19, and R. N. Mohapatra and J. C. Pati, Phys. Rev. D11, 566 (1975); L. Wolfenstein in Neutrino 1979, Bergen Conference (A. Hastuft and C. Jarlskog, eds.)

30. A. Masiero, R. N. Mohapatra and R. D. Peccei, Nucl. Phys. B192, 66 (1981).

31. For recent references on CP violation caused by horizontal interactions see R. Decker, J. M. Gerard and G. Zonpanos, CERN preprint Th. 3755 (1983). See also K. C. Wali's contribution to this conference.

32. A. Nelson, Phys. Lett. 136B, 387 (1984), ibid 134B, 422 (1984).

33. S. Barr, Univ. of Washington preprint 40049-9 (1984) has very recently shown what the criteria are for having $\bar{\theta} = 0$ at tree level in a Nelson type model, thus showing how it may be generalized.

DISCUSSION

FRITZSCH:
One might consider the limit $m_u, m_d \to 0$, in which case no
θ-problem exists. On the other hand θ depends on the argument of
det. (mass matrix), which is either zero (if one of the quark
masses vanishes) or non-zero. However the CP-violating physics
effects must be continuously vanishing if one approaches the
limit $m_u, m_d \to 0$. What is the relevant scale in this limit?

SEGRE:
The easiest way is to rotate θ $f \cdot \tilde{f}$ onto the quark mass
matrix by a U(1) chiral transformation so the total phase of the
mass matrix is now $\tilde{\theta} = \theta + \text{Arg det } M$, where Arg det M is found at
tree level. This can be shown to give an effective T-violating
interaction (for $\tilde{\theta}$ small) of the form

$$\mathscr{L} = i\,\tilde{\theta}\,\bar{q}\gamma_5\theta \,\frac{m_u m_d m_s}{m_u m_d + m_d m_s + m_u m_d}$$

so as $m_u \to 0$, $\mathscr{L} \to i\tilde{\theta}m_u\bar{q}\gamma_5 q$. The contribution to the e.d.m/
neutron can then be calculated by current algebra. For e.g. the
diagram

This means calculate

$$\langle N | \,\tilde{\theta}\,\bar{q}\gamma_5 q m_u \,|N\pi\rangle \sim \frac{i\tilde{\theta}m_u}{f_\pi} \langle N | \left[Q_5^\pi, \,\bar{q}\gamma_5 q \right] |N\rangle$$

and use it as an effective $\pi N\bar{N}$ T-violating coupling (see paper by
Crewther, di Vecchia, Veneziano and Witten or a different version
by Baluni). Limit is $\tilde{\theta} < 10^{-9}$ for $m_u \sim 5$ Mev.

GOLDHABER:
If u mass is zero, or close to zero, will there still be
dipole moments for particles containing massive quarks only?

SEGRE:
No. The T violating interaction (for small θ) is

$$i\tilde{\theta}\bar{q}\gamma_5 q \left\{ \frac{m_u m_d m_s}{m_u m_d + m_u m_s + m_d m_s} \right\}$$

so if $m_u \to 0$ (or $m_d \to 0$ or $m_s \to 0$) there is no "strong" T
violation.

COMPOSITE MODELS OF QUARKS, LEPTONS AND GAUGE BOSONS

Hidezumi Terazawa

Institute for Nuclear Study, University of Tokyo
Midori-cho, Tanashi, Tokyo 188, Japan

INTRODUCTION

In this review, I shall discuss some of the current topics in composite models in which quarks and leptons are made of subquarks, the more fundamental particles.[1] They include the following subjects:

1. Minimal Model
2. Nucleon Decays
3. Mass Spectrum of Quarks and Leptons
4. Mass Scale for the Sub-Structure
5. Quarks and Leptons as Nambu-Goldstone Fermions
6. Flavor Mixing
7. Excited Quarks, Leptons and Gauge Bosons

In view of the main theme of this Conference, I shall try to concentrate on the sixth subject, Flavor Mixing, by skipping the details of the other subjects.

1. MINIMAL MODEL

Let us start with introducing the minimal composite model of quarks and leptons as a standard of reference for discussions in this talk. It consists of an isodoublet of subquarks $w = (w_1, w_2)^2$ (called wakems abbreviating _weak_ and _electromagnetic_) and a color quartet of scalar subquarks $C = (C_0, C_1, C_2, C_3)^3$ (called chroms meaning color). The charges of these subquarks are $Q = +1/2$ for w_1, $Q = -1/2$ for w_2 and C_0, and $Q = +1/6$ for C_i ($i = 1,2,3$), satisfying the Nishijima–Gell-Mann rule of $Q = I_3 + \frac{B-L}{2}$ where I_3 is the third component of isospin ($I_3 = +1/2$ for w_1 and $I_3 = -1/2$ for w_2), and B and L are the baryon number and the lepton number, respectively ($B = +1/3$ for C_i and $L = +1$ for C_0). The quarks and leptons of the first generation can be taken as composite states of these subquarks as

$$\nu_e = w_1 C_0 \qquad u_i = w_1 C_i$$
$$e = w_2 C_0 \qquad d_i = w_2 C_i .$$

The gauge bosons as well as the Higgs scalars can also be taken as composite states of subquark–antisubquark pairs as

$$W_\mu^+ = \overline{w}_{2L} \gamma_\mu w_{1L} \qquad W_\mu^- = \overline{w}_{1L} \gamma_\mu w_{2L}$$
$$A_\mu = \frac{\sqrt{3}}{2} (\frac{1}{2} \overline{w}_1 \gamma_\mu w_1 - \frac{1}{2} \overline{w}_2 \gamma_\mu w_2 - \frac{1}{2} i C_0^{\dagger} \overleftrightarrow{\partial}_\mu C_0 + \frac{1}{6} i C_i^{\dagger} \overleftrightarrow{\partial}_\mu C_i)$$
$$Z_\mu = \frac{\sqrt{5}}{2} (\frac{1}{2} \overline{w}_{1L} \gamma_\mu w_{1L} - \frac{1}{2} \overline{w}_{2L} \gamma_\mu w_{2L})$$
$$\qquad - \frac{3\sqrt{5}}{10} (\frac{1}{2} \overline{w}_{1R} \gamma_\mu w_{1R} - \frac{1}{2} \overline{w}_{2R} \gamma_\mu w_{2R} - \frac{1}{2} i C_0^{\dagger} \overleftrightarrow{\partial}_\mu C_0 + \frac{1}{6} i C_i^{\dagger} \overleftrightarrow{\partial}_\mu C_i)$$
$$G_\mu^a = \sqrt{2} i C^{\dagger} \overleftrightarrow{\partial}_\mu \frac{\lambda^a}{2} C \qquad \text{for } a = 1\text{-}8$$
$$\phi = \overline{w}_R w_L \qquad \text{etc.}$$

This minimal composite model may reproduce QFD (the unified gauge theory of Glashow–Salam–Weinberg SU(2)×U(1) for the electroweak interactions of quarks and leptons and QCD (the gauge theory of color SU(3)) for the strong interaction of quarks as effective theories at low energies. It may also reproduce some results of the grand unified SU(5) gauge theory of Georgi and Glashow,[4] including

$$\sin^2\theta_w = tr(I_3)^2/trQ^2 = 3/8$$

and

$$f^2/g^2 = tr(I_3)^2/tr(\lambda^a/2)^2 = 1$$

where θ_w is the weak mixing angle and f and g are the SU(3) gauge coupling constant and the SU(2) one, respectively.

Although the subquark dynamics is totally unknown, it can be assumed to be "subchromodynamics" (the Yang-Mills gauge theory) of subcolor SU(N) in which w and C are N-plet and anti-N-plet of sub-color SU(N), respectively. Then, the model has the global $SU(2)_w$ $\times U(1)_w \times SU(4)_C \times U(1)_C$ symmetry as well as the local subcolor SU(N) symmetry. It naturally produces subcolor singlet states including the composite fermions of (wC) which behave as $(\underline{2},\underline{4})$ in $SU(2)_w$ $\times SU(4)_C$ and the composite bosons of $(\bar{w}w)$ and (C^+C) which behave as $(\underline{3},\underline{1})+(\underline{1},\underline{1})$ and $(\underline{1},\underline{15})+(\underline{1},\underline{1})$, respectively. The best candidate for subcolor symmetry is SU(4). If this is the case, it produces additional subcolor singlet states including the composite fermions of $(www C^+)$ and $(wC^+C^+C^+)$ which behave as $(\underline{4},\underline{4}^*)+2(\underline{2},\underline{4}^*)$ and $(\underline{2},\underline{4})$, respectively, and the composite bosons of (wwww), $(ww C^+C^+)$ and (CCCC) which behave as $(\underline{5},\underline{1})+3(\underline{3},\underline{1})+ 2(\underline{1},\underline{1})$, $(\underline{3},\underline{6}^*)+ (\underline{1},\underline{6}^*)$ and $(\underline{1},\underline{1})$, respectively. These additional composite fermions which behave as $(\underline{2},\underline{4})$ can be taken as quarks and leptons of a higher generation, which is one of the possibilities for the origin of generations (I shall discuss this and other possibilities in more details later). The model can, therefore, correctly accommodate not only three or four generations of composite quarks and leptons, (wC), $(\overline{www}C)$ and $(wC^+C^+C^+)$, depending on whether $2(\underline{2},\underline{4})$ of $(\overline{www}C)$ are degenerate or not, but also composite gauge bosons, $(\bar{w}w)$ and (C^+C), including the electroweak vector bosons, γ, $W^\pm$ and Z, and the gluons, G^a (a = 1-8).

2. NUCLEON DECAYS

It is well known that the prediction of proton decay with the life time of

$$\tau(p \to e^+ \pi^0) \cong 2 \times 10^{29} y \left(\frac{m_X}{4 \times 10^{14} \text{GeV}} \right)^4$$

in the original Georgi-Glashow grand unification model of SU(5)[4] has been jeopardized by the recent Irvine-Michigan-Brookhaven data for proton decays.[5]

It is also one of the serious difficulties in composite models of the Harari-Shupe type[6] that baryon-number changing nucleon decays occur too fast due to a simple exchange of subquarks. In composite models of our type, on the other hand, baryon-number seems to be conserved since the chroms have definite baryon or lepton numbers, i.e., B = 0 and L = 1 for C_0 and B = +1/3 and L = 0 for C_i (i = 1,2,3). This is, however, not true. Due to the possible condensation of subcolor-singlet $(C_0 C_1 C_2 C_3)$, which may happen in the minimal model, baryon-number conservation can possibly break down, causing baryon-number changing nucleon decays.

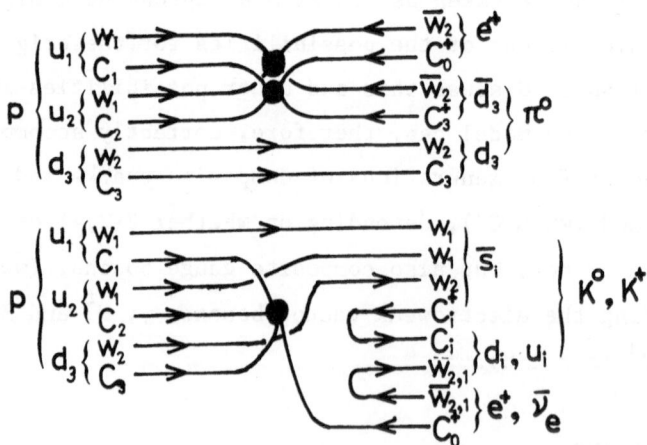

FIG.1. The subquark diagrams for proton decay.

In this picture of baryon-number violation, B-L conserves as in the grand unification gauge model. Also, it is remarkable that generation-changing nucleon decay modes such as $p \to K^0 e^+$, $K^+ \bar{\nu}_e$, $\mu^+ \gamma$, $\mu^+ \gamma\gamma$ etc. may be enhanced compared to generation-conserving ones such as $p \to e^+ \pi^0$, which is the most enhanced in the grand unification gauge model. This is because the former modes are due to the condensation of $(C_0 C_1 C_2 C_3)$ while the latter are due to that of $(C_0 C_1 C_2 C_3)$ and $(w_1 w_1 w_2 w_2)$ (See the details in Fig. 1). Whether this picture of nucleon decays is working or not can be checked by the KAMIOKANDE experiment[7] which is under way.

3. MASS SPECTRUM OF QUARKS AND LEPTONS

One of the most difficult problems to be solved in composite models is to explain the existing mass spectrum of quarks and leptons:

$$
m \begin{pmatrix} \nu & \nu_\mu & \nu_\tau & \cdots \\ e & \mu & \tau & \cdots \\ u & c & t & \cdots \\ d & s & b & \cdots \end{pmatrix} = \begin{pmatrix} \leq 10\text{-}20eV & <.52MeV & <.164GeV & \cdots \\ .5110034(14)MeV & 105.65943(18)MeV & 1784.2(3.2)MeV & \cdots \\ \sim 4\text{-}5MeV & \sim 1.2GeV & \sim 20GeV & \cdots \\ \sim 8\text{-}9MeV & \sim .15GeV & \sim 5GeV & \cdots \end{pmatrix} .
$$

The gross feature of the mass spectrum can be taken as either one of

$$
m_\mu \begin{pmatrix} 0 & 0 & 0 & \cdots \\ 0 & 1 & \sim 16.9 & \cdots \\ 0 & \sim 11 & \geq 190 & \cdots \\ 0 & \sim 1 & \sim 47 & \cdots \end{pmatrix}, \quad m_\tau \begin{pmatrix} 0 & 0 & 0 & \cdots \\ 0 & 0 & 1 & \cdots \\ 0 & \sim 1 & \geq 11 & \cdots \\ 0 & 0 & \sim 2.8 & \cdots \end{pmatrix} \quad \text{or} \quad m_L \begin{pmatrix} 0 & 0 & 0 & 0 & \cdots \\ 0 & 0 & 0 & 1 & \cdots \\ 0 & 0 & \sim 1? & \sim 11? & \cdots \\ 0 & 0 & 0 & \sim 3? & \cdots \end{pmatrix} .
$$

There seems to be no simple empirical mass formula except for[8]

$$
m_{q,\ell} = m_\mu \frac{Q^2}{|B\text{-}L|^3}(n-1)^4 = m_\mu \begin{pmatrix} 0 & 0 & 0 & 0 & \cdots \\ 0 & 1 & 16 & 81 & \cdots \\ 0 & 12 & 192 & 972 & \cdots \\ 0 & 3 & 48 & 243 & \cdots \end{pmatrix}
$$

where Q, B, L and n are charge, baryon number, lepton number and generation number of quarks and leptons, respectively. Since the possibility of the fourth charged lepton with the mass of order $81m_\mu$ has already been excluded by the e^+e^- colliding beam experiments,[9] the $(n-1)^4$ dependence should be modified if there exists more than three generations of quarks and leptons. The exponential dependence of $m_\mu a^{n-1}$ would be better in that case although it does not work well for lower generations. In any case, how to justify this type of mass formulas is a problem to be solved. I shall present an answer to this question later. It is often emphasized that the masses of the fundamental fermions (m_f), quarks and leptons, are much smaller than their size inverses (Λ_f) although the definition of the size of quarks and leptons is not clear. What seems to be more definite is that the masses of the fundamental fermions are much smaller than the masses of the weak bosons $W^\pm$ and Z:

$$m_f \ll m_{W^\pm}, \quad m_Z < \Lambda_f^{-1}?$$

However, this may not be the case either if the top-quark mass is much larger than usually expected. In fact, our sum rule[2]

$$\sqrt{<m_f^2>} = m_{W^\pm}/\sqrt{3} \simeq 46\text{GeV}$$

indicates[10] $m_t \simeq 131$ GeV if there exist only three generations of quarks and leptons. It also indicates that the average mass of quarks and leptons of the fourth generation is of order $m_{W^\pm}$ and m_Z , i.e.,

$$\sqrt{<m_f^2>_{n=4}} \simeq 2m_{W^\pm}/\sqrt{3} \simeq 92\text{GeV} \ ,$$

if there exist four generations of quarks and leptons. Our sum rule which indicates that the average mass of quarks and leptons is of the same order as the masses of the weak bosons seems to be natural in composite models where both the fundamental fermions and the gauge bosons are composites of subquarks although it has originally been derived in a particular model. In any case, the most precise statement is that the masses of quarks and leptons of

lower (at least the first and second) generations are much smaller
than the masses of the weak bosons. But, why? I shall give an
answer to this question later.

4. MASS SCALE FOR THE SUB-STRUCTURE

What is the mass scale for the sub-structure? Where does it
come from? These are also difficult questions to be answered at
this stage. As it stands now, the mass scale can lie between
several 10GeV and 10^{18} GeV.

The smallest possible mass scale comes from the weak boson
mass. The most naive expectation is that the mass of the subquark
w is roughly a half of those of the weak bosons $W^{\pm}$ and Z since the
latter are made of a subquark and an antisubquark, provided that
the subquarks are loosely bound. More precisely, suppose that the
effective Yukawa coupling constant (G_Y) of the composite Higgs
scalar (ϕ) and the subquark (w) is of the same order as the
effective gauge coupling constant (g) of the composite gauge bosons
($W^{\pm}$ and Z), i.e. $G_Y \sim g$. Then, it is natural to expect that the
subquark mass is of the same order as the weak boson mass, i.e. $m_w \sim$
$m_{W^{\pm}}$, m_Z, provided that both of these masses are generated by the
vacuum expectation value of the Higgs scalar, i.e. $m_w \sim G_Y \langle\phi\rangle_0$
and $m_{W^{\pm}}$, $m_Z \sim g\langle\phi\rangle_0$. In fact, our sum rule in a dynamical model[11]
states

$$\sqrt{\langle m_w^2 \rangle} = m_{W^{\pm}}/\sqrt{3} \cong 46\text{GeV} .$$

If indeed the mass scale for the sub-structure is so small, it can
be seen not only in future experiments at LEP, but possible in
experiments at SPS $\bar{p}p$ Collider. The best way to find such sub-
structure is to search for the excited gauge bosons ($W^{\pm'}$ and Z')[11]
at their masses of order 100-200GeV and for the excited quarks and
leptons (u^*, d^*, e^*, μ^*, etc.)[12] at their masses of order 50-100GeV.

The lower bounds on the mass scale for the sub-structure of
order 100GeV-1TeV have been obtained by e^+e^- colliding beam

experiments.[9] The conventional parametrization of the sub-structure by modifying the photon propagator by a factor of $1\pm(g^2/\Lambda^2)$ and the data for $e^+e^- \to e^+e^-$, $\mu^+\mu^-$, $\tau^+\tau^-$ have given the result $\Lambda \gtrsim 100\text{-}200$ GeV. A new parametrization is given by Eichten, Lane and Peskin[13] with the residual interaction of subquark dynamics of the type

$$L = (g^2/2\Lambda^2)[\eta_{LL}\bar{\psi}_L\gamma_\mu\psi_L\bar{\psi}_L\gamma^\mu\psi_L+\cdots] ,$$

where the constants $g^2/4\pi$ and η_{LL} are assumed to be of order unity, and the same experimental data have given the result of $\Lambda \gtrsim 750\text{-}1500$ GeV. It is clear that the magnitude of the lower bound strongly depends on the parametrization of the mass scale for the sub-structure.

A more stringent lower bound of order $10^3\text{-}10^4$ TeV has been obtained from comparison between experimental and theoretical values for the electron and muon anomalous magnetic moments. In most of the model calculations,[14] the effect of the sub-structure on the lepton anomalous moment turns out to be

$$\Delta a_\ell \sim m_\ell/\Lambda$$

where Λ is the size inverse of the lepton or the mass scale of the subquarks of which the lepton is made. On the other hand, the most precise comparison between experimental and theoretical values[15] reports

$$a_\ell^{exp.}-a_\ell^{theor.} = \begin{cases} -(2.51\pm1.54)\times10^{-10} & \text{for } e \\ (0.38\pm1.1)\times10^{-8} & \text{for } \mu . \end{cases}$$

By comparing this with the sub-structure effect, the lower bound of $\Lambda \gtrsim 10^3\text{-}10^4$ TeV can be obtained. However, this bound is also very much model-dependent. In fact, in some model calculations, the sub-structure effect turns out not to be of order m_ℓ/Λ but to be of order $(m_\ell/\Lambda)^2$, in which case only the much weaker lower bound of order $100\text{GeV-}1\text{TeV}$ can be obtained.

An even much larger lower bound of $\Lambda \gtrsim M_{GU} \sim 10^{14}-10^{16}$ GeV, where M_{GU} is the mass scale of grand unification in grand unification gauge theories, may be obtained from a wishful thinking that all the features of grand unification gauge theories be preserved.

An expectation of the largest mass scale for the sub-structure comes from a (pregauge[16] and pregeometric[17]) supergrand unified subquark model[11] in which not only strong and electroweak interactions but also gravity appears as an effective interaction. Suppose that the fundamental length inverse (Λ which is presumably of order of the Planck mass) and the subquark mass (M) are both extremely large and of the same order. Then, it can be shown in a model[18] that the gauge coupling constant (e) and the Newtonian gravitational constant (G_N) are given by these large mass scales as

$$\frac{1}{4e^2} \sim \frac{\Lambda^4}{M^4} \qquad \text{and} \qquad \frac{1}{16\pi G_N} \sim \frac{\Lambda^4}{M^2}$$

so that they are related as

$$e^2 = 16\pi G_N M^2 .$$

This relation suggests that the subquark mass may be of order $(\alpha/G_N)^{1/2}$, which is as large as 10^{18} GeV. Notice that a similar relation has been derived in Kaluza-Klein theories.[19]

From these discussions it has become clear that there seems no definite criterion yet to determine the mass scale for the substructure. I just hope that it is small enough (10^2GeV-10^2 TeV) to be found in experiments by accelerators such as LEP and SSC which will be constructed in the near future.

5. QUARKS AND LEPTONS AS NAMBU-GOLDSTONE FERMIONS

In order to explain the gross feature of the mass spectrum that the quark and lepton masses are much smaller than their size inverses, the following three possibilities have been proposed so far: quarks and leptons are 1) chiral fermions,[20] 2) Nambu-

Goldstone fermions[8] or 3) "quasi-Nambu-Goldstone fermions"[21] although the second and third possibilities may not be independent. In this section, I shall discuss the second possibility that quarks and leptons are composite Nambu-Goldstone fermions of spontaneously broken supersymmetry in some detail in the minimal subquark model.[22]

Although the subquark dynamics is unknown, it may respect supersymmetry with the supercurrent[23]

$$s_\mu = M_w \gamma_\mu wC - i\gamma_\nu \gamma_\mu w \partial^\nu C .$$

If the supersymmetry is spontaneously broken, there will appear massless Nambu-Goldstone fermions. I shall take these massless Nambu-Goldstone fermions as the idealized quarks and leptons of the first generation (ν_e,e,u,d). The non-vanishing but small masses of e, u and d can be caused by the small breaking of supersymmetry. Such small breaking is parametrized by the non-vanishing $M_w - M_C$. Notice that for $M_w - M_C \neq 0$,

$$\partial^\mu s_\mu = -i(M_w^2 - M_C^2)wC .$$

It seems natural to assume that the supersymmetric current is partially conserved as

$$\partial^\mu s_\mu \cong -iF_f^2 m_f f \quad \text{or} \quad s_\mu \cong F_f^2 \gamma_\mu f$$

where f is the quark or lepton field, m_f is the mass and F_f is the "decay constant". This partially conserved supercurrent (PCSC) hypothesis leads to the following formula for the quark or lepton mass

$$m_f = F_f^{-4} <0|\{\bar{s},[S,H(0)]\}|0>$$

$$= (M_w^2 - M_C^2)F_f^{-4} <\bar{w}w + M_w C^\dagger C>_0 \quad \text{for} \quad H = M_w \bar{w}w + M_C^2 C^\dagger C$$

where S is the supercharge. This result indicates that the quark and lepton masses may have the following properties: 1) They are proportional to the small parameter of supersymmetry breaking,

$M_W^2 - M_C^2$. 2) For fixed $M_W^2 - M_C^2$, they may becomes smaller as Λ^{-1} as the mass scale of subquark dynamics or the size inverse of quarks and leptons (Λ) increases since it may be that $F_f \sim O(\Lambda)$ and $\langle \overline{ww} + M_W C^\dagger C \rangle_0 \sim O(\Lambda^3)$. This is on the contrary to a naive expectation. And 3) if the dynamical quantities F_f and $\langle \overline{ww} + M_W C^\dagger C \rangle_0$ are rather universal in a generation of quarks and leptons, the quark and lepton masses should satisfy the relation

$$m_{\nu_e} - m_e \cong m_u - m_d \; .$$

This relation, however, is not well satisfied by the experimental values for the lepton masses ($m_{\nu_e} \cong 0$ and $m_e \cong 0.5$ MeV) and the estimates for the current quark masses ($m_u \cong 4$–5MeV and $m_d \cong 8$–9 MeV)[24] although the signs of both hand sides coincide. There are two ways out from this unsatisfactory feature of this relation. The first way is to discard the universality on which the relation is based and to proceed to difficult calculations of the dynamical quantities of F_f and $\langle \overline{ww} + M_W C^\dagger C \rangle_0$. The second way is to suppose that the subquark masses are all equal, i.e. $m_W = m_C$. Then, the quarks and leptons of the first generation are massless to the lowest order of supersymmetry breaking. Their masses may appear as higher order corrections of supersymmetry breaking which lies in the subquark dynamics. Although the subquark interactions are totally unknown, their residual interactions at the quark and lepton level are known as ordinary strong and electroweak interactions which break supersymmetry. Therefore, the quark and lepton masses can be estimated by calculating the self-masses of quarks and leptons due to the residual interactions at the quark and lepton level. The electromagnetic self-mass of the electron, for example, is given approximately by

$$m_e^{(\gamma)} \cong \frac{\alpha}{\pi} \int dm^2 \rho_2^{(e)}(m^2) \ln \frac{\Lambda_e^2 + m^2}{m^2}$$

where $\rho_2^{(e)}$ is the electron spectral function and Λ_e is the size inverse of the electron. Assume for simplicity that the $\rho_2^{(e)}$ is dominated by the electron and a possible excited electron, i.e., $\rho_2^{(e)}(m^2) \cong \kappa^2 m_e \delta(m^2-m_e^2)+\lambda^2 m_{e*}\delta(m^2-m_{e*}^2)$ where $\kappa^2 \cong 1$ and $\lambda^2 \ll 1$. Then, the only physically interesting case is that there exists an excited electron whose mass and coupling are large enough to satisfy

$$m_e \cong \frac{\alpha\lambda^2}{\pi} m_{e*} \ell n \frac{\Lambda_e^2+m_{e*}^2}{m_{e*}^2} .$$

Such an excited electron may contribute to the anomalous magnetic moment of the electron. Suppose that the effective interaction between the electron and the excited electron is described by[12]

$$L_{e*} = \frac{e\lambda}{2m_{e*}} \bar{e}\sigma_{\mu\nu} e* F^{\mu\nu} + h.c.$$

The contribution of the excited electron to the electron g-2 is estimated to be

$$\Delta a_{e*} \cong - \frac{9\alpha\lambda^2 m_e}{2\pi m_{e*}} \ell n \frac{\Lambda_e^2+m_{e*}^2}{m_{e*}^2}$$

By using the previous relation, this can be reduced to

$$\Delta a_{e*} \cong - \frac{9}{2}\left(\frac{m_e}{m_{e*}}\right)^2 .$$

By comparing this with the most precise comparison between the experimental and theoretical values for the electron g-2,[15] the mass of the excited electron can be estimated to be $m_{e*} \cong (81\pm27)$ GeV. If the logarithmic factor $\ell n(\Lambda_e^2+m_{e*}^2)/m_{e*}^2$ is of order unity, which seems reasonable, the coupling constant can also be estimated to be $\lambda^2 \cong (1.8\pm0.6)\times10^{-3}$. This picture of the composite electron as a Nambu-Goldstone fermion suggests that there may exist an excited electron whose mass lies between 50GeV and 110GeV and whose coupling constant (λ^2) lies between 10^{-3} and 10^{-2}. The possible existence of such an excited electron is not only consistent with the existing e^+e^- colliding beam experimental data[9] but suitable for explaining the anomalous "Z" events lately reported by the UA1 and UA2 experiments at SPS $\bar{p}p$ Collider for production of Z bosons[25] although the strongly interacting, composite or excited weak boson is another promising explanation.[26] Notice that the decay width of the excited electron is large enough, i.e. $\Gamma(e^* \to e\gamma) \cong \alpha\lambda^2 m_{e^*}/2 \cong$ (0.15±0.09)MeV although the coupling constant λ is small.

In the above considerations, possible contributions to the electron mass from the weak interactions are ignored since they are small. It is, however, clear that the quark mass receives a dominant contribution from the gluon since the gluon coupling constant is much larger than the electromagnetic one. Therefore, it is natural to expect that the ratio of the up or down quark mass to the electron mass becomes of order α_s/α if excited quarks have their masses and couplings similar to the excited electron's. This expectation seems satisfactory since it provides an explanation of the fact that the up and down quark masses are roughly by ten times larger than the electron mass. A similar conclusion has recently been reached by Yasuè from different consideration in a different composite model of quarks and leptons.[27]

6. FLAVOR MIXING

There are three ways to understand why the Cabibbo-GIM-KM quark mixing[28] occurs for the weak charged current. The first one is the "hakam mixing", which means that quark mixing is caused by the intrinsic mixing of "hakams" h_i (i = 1,2,$\cdots$,N), the subquarks which have horizontal gauge quantum numbers or generation numbers. If this is the case, since a hakam is shared by quarks and leptons of the same generation, the mixing angles for quarks and leptons would become of the same order of magnitude. This possibility would become relevant if such large lepton mixing is found in neutrino oscillation experiments. However, no definite evidence for neutrino oscillation has yet been reported, as discussed in this Conference by Professor Boehm.[29]

The second possibility is "level mixing",[30] which seems more natural. The weak charged current which has been written in terms of hadrons and in terms of quarks can be most fundamentally written in terms of subquarks as[11]

$$
\begin{aligned}
J_\mu &= \frac{G^\beta}{G^\mu} \bar{p}\gamma_\mu (1 - \frac{g_A^\beta}{g_V^\beta} \gamma_5)n + \frac{G^\Lambda}{G^\mu} \bar{p}\gamma_\mu (1 - \frac{g_A^\Lambda}{g_V^\Lambda} \gamma_5)\Lambda + \cdots \\
&= V_{ud}\bar{u}\gamma_\mu(1-\gamma_5)d + V_{us}\bar{u}\gamma_\mu(1-\gamma_5)s + \cdots \\
&= \bar{w}_1\gamma_\mu(1-\gamma_5)w_2 .
\end{aligned}
$$

Let us also suppose that the origin of quark (and lepton) generations is dynamical. In order words, a quark (or lepton) of a higher generation (u_n or d_n for $n>1$) is considered to be an excited state of its corresponding one of the lowest generation (u or d). Then, the mixing matrix of quarks V_{mn} can be defined by the matrix element of the subquark current between the m-th up-like quark and the n-th down-like quark as[29]

$$
<u_m|\bar{w}_1\gamma_\mu w_2|d_n> = V_{mn}\bar{u}_m\gamma_\mu d_n .
$$

An immediate consequence of this picture is that the mixing matrix elements may vary as functions of momentum transfer between quarks. The algebra of subquark current includes the familiar commutation relation of isospin current,[8]

$$
\delta(x_0-y_0) [V_0^+(x),V_0^-(y)] = 2\delta^4(x-y)V_0^3(x) .
$$

This relation sandwiched between quark states leads to the unitary of the quark mixing matrix, i.e.[31]

$$
VV^\dagger = V^\dagger V = 1
$$

if the intermediate states form a complete set as

$$|u_\ell><u_\ell| = 1 \quad \text{and} \quad |d_\ell><d_\ell| = 1 \, .$$

Furthermore, if the isospin breaking is perturbative, the mixing matrix elements are given by

$$V_{mn} = \frac{<u_m|H_I|u_n>}{m_{u_m}-m_{u_n}} + \frac{<d_m|H_I|d_n>}{m_{d_n}-m_{d_m}} \quad \text{for} \quad m \neq n \, .$$

This indicates that the off-diagonal mixing matrix element between different generations may decrease as fast as or faster than the inverse of mass difference between the relevant quarks. It also indicates that the off-diagonal mixing matrix elements have the antisymmetric property of

$$V_{mn} = V^*_{nm} \, .$$

This property is in excellent agreement with the most recent analysis of the experimental data made by Professor Chau and also discussed in detail by Professor Lee-Franzini,[32] which concludes $V_{us} \cong -V_{cd} \cong 0.23$. These properties of the quark mixing matrix as well as the above mentioned automatic unitarity of it seem to be very natural and can be taken as a successful consequence of the subquark model.

Furthermore, the quark mixing matrix and its momentum dependence have been explicitly calculated by Akama and myself and, independently, by Tomozawa in naive potential models for the subquark-binding force.[33] The results for the quark mixing matrix in case of the square-well potential, for example, shown by

$$V_{mn} \cong \sqrt{\rho} \; \frac{2m \; \sin(n\rho-m)\pi}{(n^2\rho^2-m^2)\pi}$$

$$= \begin{bmatrix} 0.9737\pm0.0025(\text{input}) & 0.20\pm0.01 & -0.092\pm0.003\cdots \\ & (0.217\!\sim\!0.221) & (0\!\sim\!0.098) \\ -0.14\pm0.01 & 0.903\pm0.009 & 0.37\pm0.01 \quad \cdots \\ (0.17\!\sim\!0.23) & (0.66\!\sim\!1) & (0.06\!\sim\!0.73) \\ 0.081\pm0.003 & -0.21\pm0.01 & 0.79\pm0.02 \quad \cdots \\ (0\!\sim\!0.17) & (0\!\sim\!0.72) & (0.67\!\sim\!1) \end{bmatrix},$$

where ρ is a single parameter indicating the order of isospin breaking ($\rho = 0.878\pm0.006$), were roughly consistent with the experimental data[33] (whose absolute values are given in the parentheses). However, as discussed in detail by Professor Chau[32] and in this Conference by Professor Trilling, Professor Ford[34], and Professor Lee-Franzini,[32] the latest data for the lifetime of B-meson show $|V_{cb}| = 0.0435\pm0.0047$, which strongly disagrees with the above results. A possible way out is to discard the naive potential model and to go back to the perturbative picture where

$$\frac{|V_{cb}|}{|V_{us}|}\left(\stackrel{\sim}{=}\frac{|V_{ts}|}{|V_{cd}|}\right) \stackrel{\sim}{=} \frac{m_s}{m_b}\left|\frac{<s|H_I|b>}{<d|H_I|s>}\right| \qquad \text{for} \quad m_s << m_c << m_b << m_t \ .$$

$$\stackrel{\sim}{=} \frac{m_s}{m_b} \qquad \text{if} \quad \left|<s|H_I|b>/<d|H_I|s>\right| \stackrel{\sim}{=} 1 \ .$$

This relation indicates that V_{cb} can be enough (possibly too much) suppressed by the factor of m_s/m_b ($= 0.03 \sim 0.1$) compared to V_{us}. Furthermore, if the matrix elements of the perturbative Hamiltonian (H_I) between quark states whose generation difference is larger than one vanish, which may likely happen due to some quantum number conservation, the quark mixing matrix elements V_{ub} and V_{td} can appear as the second order perturbative effects and can be related to the other elements as

$$|V_{ub}| \stackrel{\sim}{=} \frac{m_s}{m_c} |V_{us}| \cdot |V_{cb}| \qquad \text{if} \quad \left|<u|H_I|c>/<d|H_I|s>\right| \stackrel{\sim}{=} 1$$

and

$$|V_{td}| \stackrel{\sim}{=} |V_{us}| \cdot |V_{cb}|$$

The first relation indicating $|V_{ub}|/|V_{cb}| = 0.03 \sim 0.08$ is perfectly consistent with the experimental upper bound of $|V_{ub}|/|V_{cb}| \leq 0.119$.[32,34] The second relation predicts $|V_{td}| \stackrel{\sim}{=} 0.01$, which can be checked by future experimental results.

Although we have lost some reliability on the naive potential model for the quark mixing matrix, it is still instructive to present the results for the momentum dependence of the individual

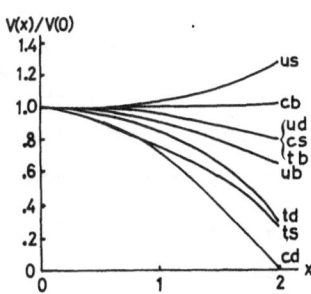

FIG.2. The momentum dependence of the
quark mixing matrix elements.

matrix elements which seems less model-dependent. The results in
case of the square-well potential, for example, are illustrated in
Fig. 2, where x is the product of the momentum transfer and the
quark size. These results show that the quark mixing matrix
elements drastically change when the momentum transfer between
quarks grows up to an order of the size inverse of quarks. We have
suggested that this phenomenon of varying quark mixing matrix
elements may be found in high energy neutrino reactions by measuring the ratio of strangeness changing to non-changing events or
that of di-muon to single-muon events, etc.. Suppose that the
semi-leptonic decays of topped mesons will be analyzed in detail in
future e^+e^- colliding-beam experiments. It will then be possible
to measure the mixing matrix elements $V_{ti}(x)$ (i=d,s,b) for fairly
large momentum transfers between quarks and to find the momentum-
transfer dependence of them. In any case, such possible discovery
of the momentum dependence of the quark mixing matrix elements

would provide one of the most eminent signs for the sub-structure of quarks.

The third possibility is "combinations".[35] If in a composite model there exist more than one composite states which have the same quantum numbers as a quark or lepton, due to different combinations of subquarks, they can be assigned to quark or lepton states of different generations. For example, as mentioned earlier, in the minimal composite model there exist three (or four) generations of composite quarks and leptons corresponding to (wC), ($\overline{www}$C) and (wC†C†C†) states. The quark mixing in this picture of generations is caused by the condensation of (wwww) or (CCCC) (See Fig. 3). If the mixings due to $<wwww>_0$ and $<CCCC>_0$ are of order ε and η ($\varepsilon,\eta<<1$) respectively, the mixing matrix has the structure of either one of

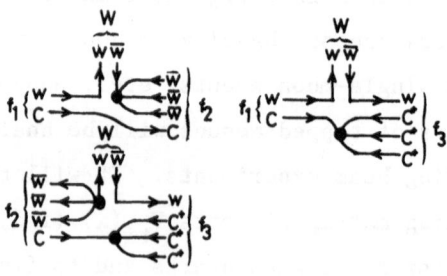

FIG.3. The subquark diagrams for quark mixing.

672

$$|V_{mn}| = \begin{pmatrix} 1 & \varepsilon & \eta \\ \varepsilon & 1 & \varepsilon\eta \\ \eta & \varepsilon\eta & 1 \end{pmatrix}, \quad \begin{pmatrix} 1 & \varepsilon & \varepsilon\eta \\ \varepsilon & 1 & \eta \\ \varepsilon\eta & \eta & 1 \end{pmatrix}, \quad \begin{pmatrix} 1 & \varepsilon\eta & \varepsilon \\ \varepsilon\eta & 1 & \eta \\ \varepsilon & \eta & 1 \end{pmatrix} \quad \text{or} \quad (\varepsilon \leftrightarrow \eta) \ ,$$

depending on which one of (wC), ($\overline{\text{www}}$C) or (wC$^+$C$^+$C$^+$) corresponds to the first, second and third generation of quarks and leptons. Since the experimental data[32,34] indicates that $|V_{cb}|^2 \gg |V_{ub}|^2$, only the second or fifth case where

$$|V_{ub}| \cong |V_{us}| \cdot |V_{cb}| \cong 0.01$$

and

$$|V_{td}| \cong |V_{ts}| \cdot |V_{cd}|$$

can survive. However, this is a rather uncomfortable case where the simplest (wC) states correspond not to the first generation but to the second. There have been proposed more sophisticated models which produce more satisfactory structures of the quark mixing matrix.[35]

7. EXCITED QUARKS, LEPTONS AND GAUGE BOSONS

Recently, it has been emphasized and, in this conference, discussed by Professor Hansen[25] that the SPS $\bar{p}p$ Collider data from the UA1 and UA2 groups for Z production[31] contain the following anomalous events: one of the four UA1 "e^+e^-" events, one of the two UA1 "$\mu^+\mu^-$" events and one of the four UA2 "e^+e^-" events are actually an $e^+e^-\gamma$ event, a $\mu^+\mu^-\gamma$ event and an $e^+e^-\gamma$ event, respectively, in which a very hard γ (the energy $\gtrsim$ 10GeV) is associated with a lepton-antilepton pair. These anomalous events cannot be explained by either one on 1) (internal or external) hard bremsstrahlung, or 2) Higgs production followed by its decay into 2γ, 3) toponium production followed by its decay into $\ell^+\ell^-\gamma$ or 4) heavy lepton production followed by its decay into $\ell\pi^0$ (See Fig. 4). The much more promising explanations of these are either 5) excited-lepton production followed by its decay into $\ell\gamma$ or 6) $\gamma Z'$ production followed by Z' decay into $\ell^+\ell^-$ (the invariant mass of order 50GeV)

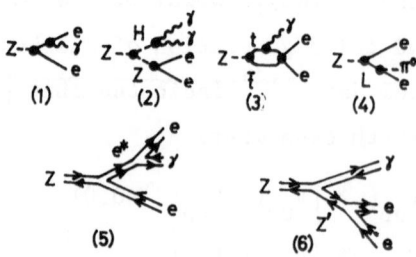

FIG.4. Various processes for the anomalous "e^+e^-" events.

where Z' is another (real or virtual) (pseudoscalar or excited vector) state of Z (See Fig. 4).

If the anomalous events are due to the cascade decay of Z into an excited lepton and a lepton followed by the decay of the excited lepton into a lepton and a photon, the mass of the excited lepton is estimated to be either about 10GeV or about 75GeV, depending on which lepton is associated with the photon. As emphasized in Section 5, the existence of such light excited leptons is yet consistent with the other existing data[9] if the photon coupling λ is sufficiently small. It even agrees with the estimation of m_{e*} $\cong$ (81±27)GeV in the picture of quarks and leptons as composite Nambu-Goldstone fermions discussed in Section 5.

As has been emphasized many times,[36] the phenomenological success of Glashow-Weinberg-Salam theory of electroweak interactions for describing low energy phenomena does not necessarily mean that the gauge bosons $W^{\pm}$ and Z are elementary. There may be many other states and many excited gauge bosons in the $W^{\pm}$ and Z channels. All that are necessary for preserving the success of the G-W-S theory are the sum rules, for example, for "Z" production by e^+e^- colliding beams

$$\int ds \, \frac{\sigma(e^+e^- \to "Z")}{s} = \frac{\pi G_F}{2\sqrt{2}} \, [1+(1-4\sin^2\theta_w)^2]$$

and

$$\int ds \, \sigma(e^+e^- \to "Z") = \frac{\pi^2\alpha}{4\sin^2\theta_w\cos^2\theta_w} [1+(1-4\sin^2\theta_w)^2] \ .$$

Therefore, it is perfectly possible that the existence of relatively light composite bosonic states and excited gauge bosons explains not only these anomalous $e^+e^-\gamma$ (or $\mu^+\mu^-\gamma$) events but also the anomalous $W^{\pm}$ jet events with the invariant masses of "heavy

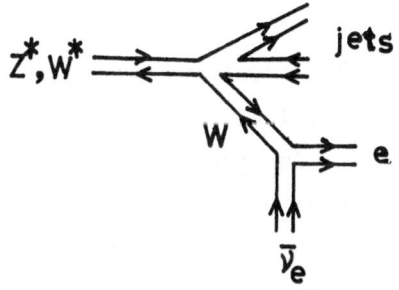

FIG.5. The subquark diagram for the anomalous $W^{\pm}$ jet events.

objects" of order 160–186GeV reported in this Conference by Professor Hansen.[25] Their data suggest that there may exist a composite bosonic state of subquark-antisubquark pair with the mass of order 50GeV and an excited gauge bosonic state of a subquark-antisubquark pair with the mass of order 160–190GeV, which remarkably fits our expectation (See Fig. 5).

In either case 5) or 6), it is highly desirable to observe direct production of excited quarks, leptons, excited gauge bosons and composite bosonic states in future e^+e^- colliding beam experiments at TRISTAN and LEP. For a moment, however, it is worth watching whether the rather large (about 1/4) ratio of the anomalous $e^+e^-\gamma$ (or $\mu^+\mu^-\gamma$) events for Z production and the rather large rate of the anomalous $W^\pm$ jet events for "heavy object" production will remain with higher statistics. If they remain, it may indicate a great discovery of the sub-structure which is even more exciting than that of $W^\pm$ and Z.

CONCLUSION

In concluding this review, I would like to emphasize that the idea of composite models of quarks and leptons (and also gauge bosons as well as Higgs scalars) which was proposed by us in the middle of seventies have just become a subject of experimental relevance in the middle of eighties.

REFERENCES

1. For reviews, see H. Terazawa, in Proc. 1981 INS Symp. on Quark and Lepton Physics, Tokyo, June 25–27, 1981, K. Fujikawa et al., ed., INS, Univ. of Tokyo, Tokyo (1981) p. 296; M.E. Peskin, in Proc. Int. Symp. on Lepton and Photon Interactions at High Energies, Bonn, Aug. 24–29, 1981, W. Pfeil, ed., Univ. Bonn, Bonn (1981) p. 880.

2. H. Terazawa, Y. Chikashige and K. Akama, Phys. Rev. D15:480 (1977).

3. J.C. Pati and A. Salam, Phys. Rev. D10:275 (1974).

4. H. Georgi and S.L. Glashow, Phys. Rev. Lett. 32:438 (1974).

5. For a review, E. Fiorini, in Proc. 1983 Int. Symp. on Lepton
    and Photon Interactions at High Energies, Cornell, Aug.
    4-9, 1983, p. 405.

6. H. Harari, Phys. Lett. 86B:83 (1979); M.A. Shupe, Phys. Lett.
    86B:87 (1979).

7. For a review, M. Koshiba, in Proc. 1983 Int. Symp. on Lepton
    and Photon Interactions at High Energies, Cornell, Aug.
    4-9, 1983. See contributed paper #103.

8. H. Terazawa, Prog. Theor. Phys. 64:1763 (1980).

9. For a review, S. Yamada in Proc. 1983 Int. Symp. on Lepton and
    Photon Interactions at High Energies, Cornell, Aug. 4-9,
    1983, p. 525.

10. H. Terazawa, Phys. Rev. D22:2921 (1980).

11. H. Terazawa, Phys. Rev. D22:184 (1980).

12. F.E. Low, Phys. Rev. Lett. 14: 238 (1965); H. Terazawa, Prog.
    Theor. Phys. 37:204 (1967); H. Terazawa, M. Yasuè, K. Akama
    and M. Hayashi, Phys. Lett. 112B:387 (1982).

13. E.J. Eichten, K.D. Lane and M.E. Peskin, Phys. Rev. Lett.
    50:811 (1983).

14. For example, see H. Terazawa et al. in Ref. 12 and references
    therein.

15. T. Kinoshita and W.B. Linquist, Phys. Rev. Lett. 47:1573
    (1981); T. Kinoshita, B. Nizic and Y. Okamoto, CLNS-83/592
    (Cornell) Dec., 1983.

16. J.D. Bjorken, Ann. Phys. 24:174 (1963); H. Terazawa et al.,
    in Ref. 2.

17. A.D. Sakharov, Doklady Akad, Naud SSSR 177:70 (1967) [Sov.
    Phys. Doklady 12:1040] (1968); K. Akama, Y. Chikashige,
    T. Matsuki and H. Terazawa, Prog. Theor. Phys. 60:868
    (1978).

18. H. Terazawa, Phys. Lett. 133B:57 (1983).

19. T. Kaluza, Sitzungsber. Preuss. Akad. Wiss. K1:966 (1921);
    O. Klein, Z. Phys. 37:895 (1926).

20. G.'t Hooft, in Recent Developments in Gauge Theories,
    G.'t Hooft et al., ed. Plenum Press, New York (1980) p.135.

21. W. Buchmüller, R.D. Peccei and T. Yanagida, Phys. Lett. 124B:
    67 (1983); R. Barbieri, A. Masiero and G. Veneziano, Phys.
    Lett. 128B:179 (1983); O.W. Greenberg, R.N. Mohapatra and
    M. Yasuè, Phys. Lett. 128B:65 (1983).

22. For more details, see H. Terazawa, INS-Report-485 (INS, Univ.
    of Tokyo) Dec., 1983. A similar discussion has been
    presented in W. Bardeen and V. Visnjić, Nucl. Phys.
    B194:422 (1982) and W.A. Bardeen, T.R. Taylor and
    C.K. Zachos, Nucl. Phys. B231:235 (1984).

23. H. Miyazawa, Prog. Theor. Phys. 36:1266 (1966); J. Wess and
    B. Zumino, Nucl. Phys. B70:39 (1974).

24. See for example, S. Weinberg, Transactions of the New York
    Academy of Sciences, Vol. 38 (1977).

25. G. Arnison et. al., Phys. Lett. 126B:398 (1983); P. Bagnaia
    et al., Phys. Lett. 129B:130 (1983); Phys. Today Vol. 36,
    No.11:17 (1983). See also P. Hansen, in this Proceedings
    and private communications.

26. H. Terazawa, Phys. Rev. D7:3663 (1973) and in Ref. 11.

27. M. Yasuè, Physics Publication MdDP-TR-84-66- (Univ. of Maryland)
    Oct. 1983.

28. N. Cabibbo, Phys. Rev. Lett. 10:531 (1963); S.L. Glashow,
    J. Iliopoulos and L. Maiani, Phys. Rev. D2:1285 (1970);
    M. Kobayashi and T. Maskawa, Prog. Theor. Phys. 49:652
    (1973).

29. For a review, see M.H. Shaevitz, in Proc, 1983 Int. Symp. on
    Lepton and Photon Interactions at High Energies, Cornell,
    Aug. 4-9, 1983, p. 132; and F. H. Boehm, in this
    Proceedings.

30. H. Terazawa, Prog. Theor. Phys. 58:1276 (1977).

31. V. Visnjic-Triantafillow, Phys. Rev. $\underline{D25}$:248 (1982); H. Terazawa, in Ref. 8; O.W. Greenberg and J. Sucher, Phys. Lett. $\underline{99B}$:339 (1981).

32. For a review, see L.-L. Chau, Phys. Report Vol. $\underline{95}$, No. 1:1 (1983). For a latest analysis, see L.-L. Chau, and W.-Y. Keung, Phys. Rev. $\underline{D29}$:592 (1984), and J. Lee-Franzini, in this Proceedings.

33. H. Terazawa and K. Akama, Phys. Lett. $\underline{101B}$:190 (1981); Y. Tomozawa, UM HE 81-3 (Univ. of Michigan) 1981, in $\underline{Proc.}$ $\underline{1981\ INS\ Symp.\ on\ Quarks\ and\ Leptons}$, Tokyo, June 25-27, 1981, K. Fujikawa et al., ed., INS, Univ. of Tokyo, Tokyo (1981) p. 319 and Phys. Lett. $\underline{104B}$:136 (1981).

34. For a review, see N.W. Reay (p. 244) and S. Stone (p. 203) in $\underline{PROC.\ 1983\ Int.\ Symp.\ on\ Lepton\ and\ Photon\ Interactions\ at}$ $\underline{High\ Energies}$, Cornell, Aug. 4-9, 1983; and G.H. Trilling and W.T. Ford, in this Proceedings. Also, for the new upper bound on the mass of ν_τ quoted in Section 3, see G.H. Trilling in this Proceedings.

35. R. Casalbuoni, G. Domokos and S. Kövesi-Domokos, Phys. Rev. $\underline{D23}$:462 (1981); K. Matumoto and K. Kakazu, Prog. Theor. Phys. $\underline{65}$:390 (1981); M. Yasué, Prog. Theor. Phys. $\underline{65}$:1995 (1981); S. Weinberg, in $\underline{Proc.\ Workshop\ on\ Weak\ Interactions}$ $\underline{as\ Probes\ of\ Unification}$, VPI, Dec. 4-6, 1980, G.B. Collins et al., ed., AIP, New York (1981) p.521.

36. H. Terazawa, in Ref. 16 and in $\underline{Proc.\ XXI\ Int.\ Conf.\ on\ High}$ $\underline{Energy\ Physics}$, Paris, July 26-31, 1982, P. Petiau and M. Porneuf, ed., J. de Phys. $\underline{C3}$ (1982) p.289.

DISCUSSION

HANSEN:

In the eγ events, you seem to favor the high mass (eγ) as an excited lepton. Then is it not strange that the γ in both the UA1 and UA2 events is emitted with about 20° to the nearest electron?

TERAZAWA:

Either light (~ 10 GeV) or heavy (~ 75 GeV) excited leptons are possible for explaining the anomalous $e^+e^-\gamma$ (and $\mu^+\mu^-\gamma$) events. The light excited lepton whose mass is of order 10 GeV is still consistent with all the available existing experimental data including the e^+e^- colliding beam data if the effective coupling is small.

GOLDHABER:

Would low energy (~ 10 GeV) excited states of the electron have been seen in e^+e^- colliders?

TERAZAWA:

It depends on the effective coupling λ. If λ is smaller than one hundredth, the possible existence of such low mass excited electron is consistent with the e^+e^- colliding beam data since the production cross section is small.

KANE:

There is a problem with interpreting the $e^+e^-\gamma$ events as due to any mechanism where the fermions come from a single current. If e^+e^- come from a Z^0, then the production of jet-jet-γ events must be 20 times larger than $e^+e^-\gamma$. If it is a Z', one can still show that the minimum possible number of jet-jet-γ events is 3 for every $\ell^+\ell^-\gamma$ events, as long as the current coupled to Z' is any combination of SU(2) and U(1) current. There is a general treatment of this in a paper of Duncan and Veltman who give all decays under the assumption of a general current. Probably any interpretation of $\gamma+Z' \rightarrow \gamma\ell^+\ell^-$ is excluded.

TERAZAWA:

The Z' does not necessarily couple to the current. As I said in the talk, the Z' is perhaps a pseudoscalar state of subquark-antisubquark pairs. I think that the excited lepton and the composite gauge boson are both the easiest possibility for explaining the anomalous $e^+e^-\gamma$ (and $\mu^+\mu^-\gamma$) events and the anomalous $W^{\pm}$ jet events. So, why not take the easiest possibility?

680

THEORETICAL UNDERSTANDING OF WEAK DECAYS OF HEAVY QUARKS

R. Rückl

CERN

Geneva

ABSTRACT

The weak decays of hadrons with charm and bottom flavour are reviewed in the framework of the standard model. Particular attention is payed to the impact of strong interactions: short-distance modifications of the weak hamiltonian as well as non-asymptotic effects due to the presence of light quarks and soft gluons.

INTRODUCTION

In the framework of the standard $SU(2)_L \times U(1)$ electroweak theory[1] all heavy fermions decay in a universal manner, with their relative lifetimes solely determined by their masses and weak mixing. Accordingly, the lifetime of the τ-lepton or "free" charm quark can be inferred from the muon lifetime.

$$\tau_\mu \simeq \frac{192\pi^3}{G_F^2 m_\mu^5} \simeq 2.18 \cdot 10^{-6} \text{ sec,} \tag{1}$$

simply by rescaling the mass and accounting for the number of open decay channels:

$$\tau_\tau \simeq \frac{1}{5}\left(\frac{m_\mu}{m_\tau}\right)^5 \tau_\mu \simeq 3.3 \cdot 10^{-13} \text{ sec} \tag{2}$$

and ($m_c \simeq 1.5$ GeV)

$$\tau_c \simeq \frac{1}{5}\left(\frac{m_\mu}{m_c}\right)^5 \tau_\mu \simeq 7 \cdot 10^{-13} \text{ sec .} \tag{3}$$

These free field predictions are in nice agreement with the measured τ and (average) charmed particle lifetime.[2]

Consider the fact that, because of colour confinement, quarks only exist within strongly bound hadronic systems and, hence, any weak process involving quarks is necessarily accompanied by strong interactions, free quark estimates may appear somewhat premature. Yet, the respectable results quoted above have a good reason, namely the asymptotic freedom property of $SU(3)_c$-colour interactions. On the other hand, the typical energy scales of bottom and, in particular, charm decays are relatively moderate: 5 GeV and 2 GeV, respectively. It is, therefore, not too surprising that the more detailed decay properties indeed indicate an appreciable impact of the environment of light quarks and soft gluons, in addition to short-distance gluon effects.

The interplay of strong and weak forces is not only an important issue[3] in the theoretical understanding of weak decays of heavy quarks, but plays also a crucial note[4] in the $\Delta I = 1/2$ rule, in CP-violation, in the determination of weak couplings and mixing angles and numerous other trials to which the standard model is put. Unfortunately, due to the lack of sufficiently general quantitative methods to deal with confinement aspects of QCD, it is not yet possible to analyse the effects of strong interactions solely from first principles. Instead, one usually assumes (with more or less good justification) that the dynamics of a given process can be separated in long and short distance aspects and treats these aspects individually. To be more specific, the short-distance dynamics of weak decays is described by an effective weak hamiltonian H_W^{eff}, which can rigorously be calculated using the well-established short-distance techniques of QCD. The long-range effects of strong interactions, on the other hand, are lumped together in the hadronic matrix elements of H_W^{eff}, which determine the decay amplitudes, $T_{if} = <f|H_W^{eff}|i>$. The latter can only be estimated in certain, sometimes rather crude approximations or in explicit bound state models. Clearly, this is the weak point in the present understanding of heavy quark decays and, therefore, deserves special attention.

My talk is organized as follows. I first discuss in some detail the modifications of the bare weak hamiltonian due to hard gluon interactions, which constitutes the basis of all further considerations. The second part deals with the phenomenology of inclusive charm and bottom decays. Here, the focus is on the short-distance approximation as described by the spectator model, and the non-spectator effects indicated by the charm decay data. In the third part, I rather briefly tackle the more difficult problem of exclusive (2-body) decays, pointing out the main gaps in the present understanding.

THE EFFECTIVE WEAK HAMILTONIAN

The minimal $SU(2)_L \times U(1)$ model, flavour-changing neutral currents[5,6] do not exist. Hence, heavy flavour decays proceed exclusively via charged current interactions mediated by the $W^{\pm}$ bosons. The fundamental couplings of the latter to fermions is given by the lagrangian,

$$\mathcal{L}_{cc} = \frac{g}{2\sqrt{2}} (W^+_\mu J^\mu_- + W^-_\mu J^\mu_+),$$
(4)

where

$$J^\mu_- = \left(J^\mu_+\right)^t = (\overline{u}\,\overline{c}\,\overline{t})\gamma^\mu(1-\gamma^5)\, V \begin{pmatrix} d \\ s \\ b \end{pmatrix} + (\overline{\nu}_e\overline{\nu}_\mu\overline{\nu}_t)\, \gamma^\mu(1-\gamma^5) \begin{pmatrix} e^- \\ \mu^- \\ t \end{pmatrix}$$
(5)

is the standard V-A charged current in the 6-flavour version.[6] For simplicity and since J^μ is, evidently, a colour singlet, the colour indices are suppressed in eq. (5). The symbol V denotes the Kobayashi-Maskawa matrix.[6]

$$V = \begin{pmatrix} V_{ud} & V_{us} & V_{ub} \\ V_{cd} & V_{cs} & V_{cb} \\ V_{td} & V_{ts} & V_{tb} \end{pmatrix} \;,$$
(6)

which relates the weak and mass eigenstates in the quark sector. The present knowledge about quark mixing is reviewed in detail in the talks by Lee-Franzini, Kleinknecht and Siebert at this conference.[7] A crude but, as far as the basic structure is concerned, very transparent approximation for V reads

$$|V| \sim \begin{pmatrix} 1 & s & s^3 \\ -s & 1 & s^2 \\ s^3 & -s^2 & 1 \end{pmatrix},$$
(7)

where $s \sim \sin\theta_c \simeq 0.23$.

From eq. (4) one readily derives the effective hamiltonian to second order in the weak coupling g:

$$H_W^{eff} = \frac{g^2}{8i} \int d^4x\, D_{\mu\nu}\, (x,m_W^2)\, T(J^\mu_+(x)J^\nu_-(0) + h.c.)$$
(8)

Here, $D_{\mu\nu}(x,m_W^2)$ is the W propagator and T denotes time ordering. In the free field approximation and taking the limit $m_W \to \infty$, which is appropriate for charm and bottom decays since $m_c, m_b \ll m_W$, eq. (8) yields the familiar effective hamiltonian,

$$H_w^{(o)} = \frac{G_F}{\sqrt{2}} (J_+^\mu (0) J_{-\mu}(0) + h.c.),\qquad (9)$$

describing local current-current interactions.

The above bare hamiltonian is, or course, affected by strong interactions. The lowest order gluon exchange contributions are illustrated in Fig. 1. As can be directly seen from these diagrams, strong interactions not only renormalize the weak couplings and quark masses, etc. (Fig. 1a), but also induce new effective 4-fermion interactions with different colour (Fig. 1b) and/or Lorentz structure (Fig. 1c). Obviously, semileptonic operators are only affected by corrections to the quark currents of the kind shown in Fig. 1a. In what follows, I shall, therefore, concentrate on the non-leptonic part of the weak hamiltonian given in eq. (8). The local free field approximation of the latter reads:

$$H_{NL}^{(o)} = \frac{G_F}{\sqrt{2}} (\bar{U}VD)_L \, (\bar{D}V^t U)_L,\qquad (10)$$

where

$$U = \begin{pmatrix} u \\ c \\ t \end{pmatrix}, \ D = \begin{pmatrix} d \\ s \\ b \end{pmatrix}\qquad (11)$$

and

$$(\bar{\psi}\psi)_L \equiv \bar{\psi}\gamma^\mu(1-\gamma^5)\psi.\qquad (12)$$

(a)

(b)

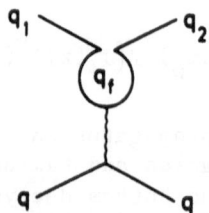

Fig. 1.   o(α_s)-gluon corrections to the bare weak hamiltonian.

The QCD corrected weak hamiltonian H_{NL}^{eff} can rigorously be derived using short distance expansion[8] and renormalization group techniques.[9] In short, the T-product of quark currents in eq. (8) is expanded in terms of local operators with, in general, divergent coefficient functions:

$$T\left(J^\mu(x)J^\nu(o)\right) = \sum_k C_k(x;g_s, m_w, m_q, \ldots, \mu)O_k^{\mu\nu}(0;\mu). \quad (13)$$

Here g_s and m_q are the renormalized QCD coupling constant and quark masses, respectively, and μ is the normalization scale. Note that the operators on the right-hand-side of eq. (13) have perfectly regular matrix elements $<f|O_k^{\mu\nu}(0;\mu)|>$. Because of the heavy mass of the W boson, contributions to the integral in eq. (8) from distances $|x| > 1/m_w$ are strongly suppressed. Hence, the current-current product, eq. (13), is dominated by the local operators with the most singular coefficients. These are[3] dimension six, 4-quark operators $O_k^{(6)}$. Substituting eq. (13) in eq. (8) and taking the limit $m_w \to \infty$, one gets, schematically,

$$H_{NL}^{eff} = \frac{G_F}{\sqrt{2}} \sum_k c_k(\alpha_s, m_w, m_q, \ldots, \mu)O_k^{(6)}(\mu). \quad (14)$$

Obviously, the weak amplitude $T_{if} = <f|H_{NL}^{eff}|i>$, being a measurable quantity, should not depend on the arbitary scale μ chosen to normalize the theory. In other words, the μ-dependence of the coefficients c_k and the operators $O_k^{(6)}$ in eq. (14) must compensate each other. This requirement entails renormalization group equations[3] for the coefficients c_k:

$$(\mu\frac{\partial}{\partial\mu} + \beta(g_s)\frac{\partial}{\partial g_s} + \sum_q \delta_q(g_s)m_q\frac{\partial}{\partial m_q} - \gamma_k(g_s) + \ldots)c_k = 0. \quad (15)$$

Here, $\beta(g_s)$ and $\delta_q(g_s)$ define the running coupling constant of QCD and running quark masses, respectively, and $\gamma_k(g_s)$ are the anomalous dimensions of the operators $O_k^{(6)}$. The coefficients c_k of H_{NL}^{eff} can thus be obtained by solving eq. (15).

Let us first study the limit of zero quark masses and consider the case of realistic quark masses later. In the massless theory with the full SU(6)-flavour symmetry, the expansion eq. (14) only consists of two operators:

$$H_{NL}^{eff} = \frac{G_F}{\sqrt{2}}(C_+(\alpha_s,\frac{m_w}{\mu})O_+ + C_-(\alpha_s,\frac{m_w}{\mu})O_-), \quad (16)$$

with

$$O_\pm = \frac{1}{2}\left((\bar{U}VD)_L(\bar{D}V^+U)_L \pm (\bar{U}VU)_L(\bar{D}V^+D)_L\right), \quad (17)$$

where U and D are defined in eq. (11). 0_+ belongs to different SU(6)-flavour representations, to wit $0_+ \backsim \underline{405}$ and $0_- \backsim \underline{189}$, and hence do not mix under renormalization. The solution of eq. (15) in 1-loop approximation for the β-function and anomalous dimensions $\gamma_\pm(g_s)$,

$$\beta(g_s) = -b \frac{g_s^2}{16\pi^2} \; ; \; b = 11 - \frac{2}{3} f; \; f = 6$$

$$\gamma_\pm(g_s) = -d_\pm \frac{g_s^2}{16\pi^2} \; ; \; d_- = -2d_+ = 8, \tag{18}$$

then gives the well known leading logarithmic (LL) results[10]

$$C_\pm(\alpha_s, \frac{m_w}{\mu}) = \left(\frac{\alpha_s(\mu^2)}{\alpha_s(m_w^2)} \right)^{\frac{d_\pm}{2b}} \tag{19}$$

and

$$\alpha_s(Q^2) = \frac{4\pi}{b \ln \frac{Q^2}{\Lambda_{QCD}^2}} . \tag{20}$$

In next-to-leading log (NLL) approximation,[11]

$$C_\pm(\alpha_s, \frac{m_w}{\mu}) = \left(\frac{\alpha_s(\mu^2)}{\alpha_s(m_w^2)} \right)^{\frac{d_\pm}{2b}} (1 + \frac{\alpha_s(\mu^2) - \alpha_s(m_w^2)}{\pi} \rho_\pm) \tag{21}$$

where

$$\alpha_s(Q^2) = \frac{4\pi}{b \ln \frac{Q^2}{\Lambda_{QCD}^2}} (1 - (102 - \frac{38}{3} f) \frac{\ln\ln \frac{Q^2}{\Lambda_{QCD}^2}}{b^2 \ln \frac{Q^2}{\Lambda_{QCD}^2}} \tag{22}$$

is the running coupling constant in the so-called $\overline{MS}$-scheme ($\Lambda_{QCD} = \Lambda_{\overline{MS}}$). The renormalization scheme independent coefficients $\rho_\pm$ in eq. (21) are given by:

$$\rho_+ = (-\frac{221}{24} + \frac{5}{9} f) \frac{1}{b} + (51 - \frac{19}{3} f) \frac{1}{b^2}$$

$$\rho_- = (\frac{263}{12} - \frac{10}{9} f) \frac{1}{b} + (-102 + \frac{38}{3} f) \frac{1}{b^2} . \tag{23}$$

The numerical significance of these corrections for c, b and t-decays can be read off from the table below $\left(f = 6, \Lambda_{\overline{MS}} = 250\right.$ MeV and $\left.\alpha_s^{LL}(\mu^2) = \alpha_s^{NLL}(\mu^2)\right)$:

$$
\mu(\text{GeV}) =
\begin{array}{c}
2 \\
5; \\
40
\end{array}
\quad C_+ =
\begin{array}{cc}
\text{LL} & \text{NLL} \\
0.77 & \to 0.73 \\
0.85 & \to 0.82; \\
0.97 & \to 0.96
\end{array}
\quad C_- =
\begin{array}{cc}
\text{LL} & \text{NLL} \\
1.69 & \to 1.90 \\
1.39 & \to 1.49 \\
1.07 & \to 1.08
\end{array}
\quad . \qquad (24)
$$

A few remarks may suffice:

(i) The modifications of the bare coefficients $C_\pm(0, m_W/\mu) = 1$ (see eq. (10)) are sizeable, but decrease with increasing scale μ as expected from asymptotic freedom.

(ii) The next-to-leading corrections change the LL-results only moderately and, moreover, reinforce the inequality $C_- > C_+$.

(iii) Referring to the SU(3)-flavour classification of the operators $0_\pm$, C_- being larger than C_+ is known as 8-enhancement in strange particle decays[12] and 6-enhancement in charm decays.[10]

(iv) One net effect of strong interactions are flavour-changing neutral current interactions described by the operator $(C_+ - C_-)/2 \; (\bar{U}VU)_L (\bar{D}V^+D)_L$ in eq. (17).

It is further interesting to note that H_{NL}^{eff}, eq. (16) and (17), can be rewritten in terms of charged current operators only. Using the relation

$$
(\bar{\psi}_1 \lambda^a \psi_2)_L \; (\bar{\psi}_3 \lambda^a \psi_4)_L = -\frac{2}{3}(\bar{\psi}_1 \psi_2)_L (\bar{\psi}_3 \psi_4)_L
$$

$$
+ 2(\bar{\psi}_1 \psi_4)_L (\bar{\psi}_3 \psi_2)_L \qquad (25)
$$

which is a consequence of Fierz identities and $SU(3)_c$-colour algebra, one finds:

$$
H_{NL}^{eff} = \frac{G_F}{\sqrt{2}} \left(\frac{2C_+ + C_-}{3} \; (\bar{U}VD)_L (\bar{D}V^+U)_L \right.
$$

$$
\left. + \frac{C_+ - C_-}{4} \; (\bar{U}\lambda^a VD)_L (\bar{D}\lambda^a V^+U)_L \right) . \qquad (26)
$$

This form very clearly exhibits the effects of hard gluon exchanges, anticipated from the lowest order diagrams of Fig. 1: renormalization of the bare 4-quark interactions (see eq. (10)) and induction of local interactions of colour-octet currents. One also sees that penguin type operators (Fig. 1c) are absent in the flavour symmetry limit. The reason is that, in this limit the contributions from the various quark flavour in the loop of Fig. 1c exactly cancel due to the generalized GIM-mechanism[5,6] or, equivalently, due to the unitarity of the Kobayashi-Maskawa-matrix:

$$\sum_{q_f} V_{q_2 q_f} V^*_{q_1 q_f} = 0. \tag{27}$$

This cancellation is upset by realistic quark masses as explained next.

Generally, when one evolves H^{eff}_{NL} from some asymptotic scale to the physical scale μ of a given process, one crosses several quark thresholds corresponding to the quark masses m_q. In the region $\mu > m_q$, q and all lighter quarks can be considered massless, whence, $H^{eff}_{NL}(\mu)$ has an approximate $SU(n_q)$-flavour symmetry. At $\mu < m_q$, however, the quark flavour q decouples from the effective theory. As a result, the number of flavours f diminishes and the GIM mechanism is partly put out of operation. These effects can approximately be taken into account by renormalizing H^{eff}_{NL} region by region, changing f appropriately. For illustration, at $m_b > \mu > m_c$ one thus gets

$$C_\pm \simeq \left(\frac{\alpha_s(\mu^2)}{\alpha_s(m_b^2)}\right)^{\frac{d_\pm}{2b_4}} \left(\frac{\alpha_s(m_b^2)}{\alpha_s(m_t^2)}\right)^{\frac{d_\pm}{2b_5}} \left(\frac{\alpha_s(m_t^2)}{\alpha_s(m_w^2)}\right)^{\frac{d_\pm}{2b_6}} , \tag{28}$$

where

$$b_f = 11 - \frac{2}{3}f.$$

Numerically, eq. (28) differs very little from the cofficients in the massless limit (see eqs. (19) and (24)). Furthermore, the relevance of penguin operators can roughly be estimated by evaluating the lowest order diagram shown in Fig. 1c. At $m_t < \mu < m_b$, the t-quark flavour is frozen in, generating the effective interaction ($m_{c,s,d,u} \simeq 0$)

$$\frac{G_F}{\sqrt{2}} V_{tb}V^*_{tq}(-\frac{\alpha_s(\mu^2)}{12\pi}\ln\frac{m_t^2}{\mu^2}) (q\lambda^a b)^-_L \sum_f (q_f \lambda_f q_v)^a . \tag{29}$$

Here, q = s or d and the sum runs over all excited quark flavours. All other loop contributions cancel by GIM. Similiarly, at $m_b < \mu < m_c$ the decoupling of the b-quark induces the operator ($m_{s,d,u} \simeq 0$)

$$\frac{G_F}{\sqrt{2}} V_{ub} V_{cb}^* \ (-\frac{\alpha_s(\mu^2)}{12\pi} \ln\frac{m_b^2}{\mu^2}) \ (\bar{u}\lambda^a c)_L \ \sum_f (\bar{q}_f \lambda^a q_f)_v. \tag{30}$$

The above estimates indicate that in heavy flavour decays penguins are not important for several reasons:

(i) small coefficients $(\alpha_s(\mu^2)/12\pi \ln m_t^2, b/\mu^2 \backsim 0.03$ compared with $C_\pm \backsim O\ (1))$,

(ii) small quark mixing $(V_{tb}V_{ts}^* \backsim s^2, V_{tb}V_{td}^* \backsim s^3$ and $V_{ub}V_{cb}^*$ $\backsim s^5$ with respect to the main bottom and charm decays proportional to S^2 and 1, respectively),

(iii) no appreciable enhancement[13] of penguin matrix elements with respect the $<f|0_\pm|i>$.

If anything, penguins may affect the pattern of multiple Cabibbo suppressed B-decays.[14] This is in contrast to strange particle decays where the penguins seem to be essential for the $\Delta I = 1/2$ rule.[15]

The quark mass effects, discussed above somewhat incoherently, can also rigorously be treated in the operator product expansion and renormalization group approach. Technically, however, the renormalization procedure becomes rather involved due to a profileration of operators, which have to be included in the short-distance expansion (in accordance with the effective flavour symmetry), and by their mixing under renormalization. Also the renormalization group equation itself becomes more complicated due to the scale dependence of quark masses and non-zero anomalous dimensions of currents. The results, however, essentially confirm the conclusions drawn above.

To summarize, the QCD corrected hamiltonian for heavy flavour decays is, to a good approximation, given by eq. (9) with the non-leptonic part, eq. (10); modified as detailed in eqs. (16-24). The leading log results for the coefficients $C_\pm(\alpha_s, m_w/\mu)$ are nicely stabilized by the next-to-leading corrections. Numerically, one finds

$$C_+ \simeq 0.74, \ C_- \simeq 1.8 \tag{31}$$

for charm decays, and

$$C_+ \simeq 0.85, \ C_- \simeq 1.4 \tag{32}$$

for bottom decays.

INCLUSIVE DECAYS: THE SPECTATOR MODEL

In order to link theory with experiment one must eventually find ways to calculate hadronic matrix element of the effective weak hamiltonian. This proves not too difficult for inclusive decays of hadrons which contain sufficiently heavy quark Q. Since the typical momentum transfer in the decay of such a system is of order m_Q, one may neglect soft hadronic interactions and bound state effects, once the heavy quark masses is much larger than the ordinary hadronic scales represented by light constituent masses, confinement radius, bound state wavefunctions, etc. Furthermore, the inclusive sum of the hadronic final state may be associated with free quark states carrying the large energy m_Q, similarly as $e^+e^- \to$ hadrons is dual to $e^+e^- \to q\bar{q}$. The resulting parton description is illustrated in Fig. 2. Quite obviously, the decay mechanisms involving a light quark constituent (Fig. 2b) are suppressed by bound state wavefunctions with respect to the dissociation of the heavy quark (Fig. 2a) and can be neglected in the asymptotic limit considered. For illustration, the relative contributions of the processes shown in Fig. 2 to pseudoscalar meson decay rates are given by[3]

$$\frac{\Gamma_{b_1}}{\Gamma_a} = \frac{1}{g} \frac{\Gamma_{b_2}}{\Gamma_a} \simeq (24\pi^2) \left(\frac{f_p}{m_Q}\right)^2 \left(\frac{m_{q_1}^2 + m_{q_2}^2}{m_Q^2}\right) \left(\frac{1}{9}\right) , \qquad (33)$$

where only the leading terms in Λ/m_Q are kept, with Λ being a typical hadronic scale. The parameter f_p is the meson decay constant and characterizes the overlap of the constituent quarks. For p = π,K,D,F,B data or theoretical estimates[16] indicate that

$$f_p \sim 100\text{--}300 \text{ MeV}. \qquad (34)$$

As can be seen from eq. (33), W-exchange and weak annihilation are suppressed by the small constituent overlap (2nd factor), by

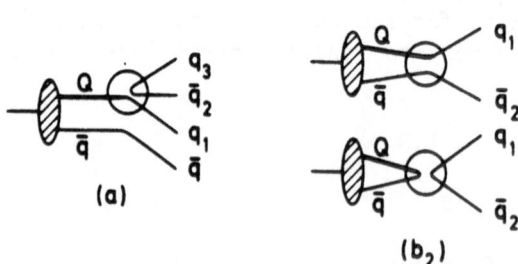

Fig. 2. Parton description of inclusive heavy meson decays via (a) heavy quark dissociation, (b$_1$) W-exchange, (b$_2$) weak annihilation. Circles indicate H_W^{eff} including hard gluon corrections.

helicity conservation (3rd factor) and, this applies to W-exchange only, the the colour degrees of freedom (4th factor). Thus, despite the appreciable phase space enhancement (1st factor), these mechanisms are already subdominant in charm decays and totally negligible in bottom decays. Also baryon decays will ultimately be dominated by heavy quark dissociation. However, since helicity suppression only affects V-A interactions between the heavy quark and a light quark in a spin 1 configuration, W-exchange should play a more important role in baryon decays than in meson decays.

The above arguments show that the inclusive decay properties of sufficiently heavy quark states are essentially determined by the weak decays of the "free" heavy quark corrected for hard gluon interactions. The light quark constituents are merely passive spectators. This fact constitutes the basis of the spectator model.[17] In addition, one usually takes into account gluon bremsstruhlung[18] as well as some non-asymptotic effects such as phase space distortions[19] due to finite quark masses and the Fermi motion[20] inside heavy bound states. The resulting semi- and non-leptonic decay widths are summarized below:

$$\Gamma_{LS} = \sum_{1,q} |V_{Qq}|^2 \; I(\frac{m_q}{m_Q}, \frac{m_e}{m_Q}, 0) \; (1 - \frac{2}{3} \frac{\alpha_s(m_Q^2)}{\pi} f) \; \frac{G_F^2 m_Q^5}{192\pi^3} \tag{35}$$

$$\Gamma_{NL} = \sum_{gi} |V_{Qq1}|^2 |V_{q2q3}|^2 I \left(\frac{m_{q1}}{m_Q}, \frac{m_{q2}}{m_Q}, \frac{m_{q3}}{m_Q}\right) \; \frac{2c_+^2 + c_-^2}{3}$$

$$(1 + \frac{2}{3} \frac{\alpha_s(m_Q^2)}{\pi} h) \frac{G_F^2 m_Q^5}{64\pi^3} \; . \tag{36}$$

Here, $I(x,y,z)$ is the 3-body phase space factor[19] normalized such that $I(o,o,o) = 1$. The subsequent factors arise from radiative gluon corrections to the decay widths and, in the non-leptonic case, in addition from the short-distance modifications of H_{NL}^{eff} specified in eqs. (16) and (17). In the massless limit,[11]

$$f = \pi^2 - \frac{25}{4} = 3.6 \tag{37}$$

and

$$h \simeq \begin{cases} 3.4 \; (m_c = 1.5 \text{ GeV}, \; f = 4) \\ 0.6 \; (m_b = 5 \text{ GeV}, \; f = 5) \; . \end{cases} \tag{38}$$

The last factors in eqs. (35) and (36) are the uncorrected decay widths directly inferable from the μ-decay formula (see eqs. (1-3)). One can further parameterize the heavy quark mass m_Q in terms of the light constituent masses and the relative bound state momenta.[20] For a pseudoscalar meson bound state with mass m_p, energy-momentum conservation requires:

$$m_Q^2 = m_p^2 + m_q^2 - 2m_p \sqrt{\vec{q}^2 + m_q^2}. \tag{39}$$

Effective decay spectra and widths are then obtained from the results for a fixed value of m_Q (such as eqs. (35) and (36)) by averaging the latter over a gaussian distribution of the Fermi momentum $|\vec{q}|$.

Although the spectator model itself is well defined and makes unambiguous predictions for asymptotically heavy quarks, in the mass range of charm and bottom one is still bothered by uncertainties. These mainly concern the question which effective quark masses one ought to use. In particular, the three-body phase space is rather sensitive to masses. In accordance with the spirit of the spectator model, one would argue that current quark masses are the correct choice. On the other hand, if the average hadron multiplicity is as low as in charm decays, it may be more appropriate to use somewhat larger effective masses, in the extreme, constituent quark masses. Obviously, bottom decays are more asymptotic and, hence, this problem is less serious. Moreover, a careful analysis of the semi-leptonic decays reduces this uncertainty by a remarkable amount. Semi-leptonic decays are conceptually simpler and considerably less subject to non-asymptotic effects, disregarded in the spectator model, than non-leptonic decays. I will not go into details (these can be found in Lee-Franzini's talk[21]), but only quote the resulting constraints on the quark masses. The usual procedure is to fit the electron spectra calculated[20] from the $c \rightarrow s e \nu_e$ and $b \rightarrow c e \nu_e$ to the ones measured in $D \rightarrow e \nu X$[22] and $B \rightarrow e \nu X$[23] respectively, letting the mass of the spectator quark m_{sp} and the width p_F of the Fermi momentum distribution $\sim \exp(-|\vec{q}|^2/p_F^2)$ vary between 0 and 300 MeV with $m_{sp} + p_F = 300$ MeV. Good agreement between theory and experiment is obtained with[21]

$$m_c \sim 1.5\text{--}1.65 \text{ GeV},$$
$$m_c\text{--}m_s \sim 1.1\text{--}1.3 \text{ GeV}, \tag{40}$$

and

$$m_b \sim 4.9\text{--}5.05 \text{ GeV},$$
$$m_b\text{--}m_c \sim 3.3\text{--}3.4 \text{ GeV}, \tag{41}$$

respectively. These constraints are taken into account in the following brief examination of the spectator model for non-leptonic decays.

B-Decays

So far, the spectator model is consistent with all known features of inclusive B-decays.[24] However, as a word of warning, the present data refer to an average of $B^{\pm}(60\%)$ and $(\bar{B})^{\circ}$ (40%) decays and may, therefore, hide non-spectator effects similarly as it was the case in the early data on average D-decays. Furthermore, B-decays involve two new weak mixing parameters V_{cb} and V_{ub}, which are constraint[7] by the observed mixing in the u,d,s and c-sector and the unitarity of the Kobayashi-Mashawa matrix, but only with large uncertainties. From a comparison of the experimental semi-leptonic electron spectrum with the two hypotheses, $b \to c e \nu_e$ and $b \to u e \nu_e$, one obtains the important bound[21]

$$|\frac{V_{ub}}{V_{cb}}| < 0.116 \ (90\% \ \text{c.l.}) \ . \tag{42}$$

For definiteness, $|V_{ub}/V_{cb}| = 0.1$ is used throughout the following discussion. Finally, in order to explore the range of inclusive decay properties which can be accommodated in the spectator model, two extreme sets of quark masses are considered:

$$
\begin{array}{lll}
 & \text{(I)} & \text{(II)} \\
m_{u,d} \simeq & 0 & 0.35 \ \text{GeV} \\
m_s \simeq & 0.15 & 0.5 \ \text{GeV} \\
m_c \simeq & 1.4 & 1.8 \ \text{GeV} \\
m_b \simeq & 4.8 & 5.2 \ \text{GeV}
\end{array}
\tag{43}
$$

For both sets, $m_b - m_c \simeq 3.4$ in agreement with eq. (41).

With the above specifications, eqs. (35) and (36) yield[3] the following semi-leptonic branching ratio and lifetime:

$$BR_{e,\mu} \simeq \begin{cases} 12\%, \ \text{(I)} \\ 14.7\%, \ \text{(II)} \end{cases} \tag{44}$$

and

$$\tau_b \cdot |V_{cb}|^2 \simeq \begin{cases} 2.86 \cdot 10^{-15} \ \text{sec, (I)} \\ 3.14 \cdot 10^{-15} \ \text{sec, (II).} \end{cases} \tag{45}$$

Comparison of eq. (44) with the world average[24]

$$BR_{e,\mu} = (11.6 \pm 0.5)\% \tag{46}$$

shows that:

(i) current type quark masses are favoured by the data as they are by theory and

(ii) QCD corrections are absolutely needed in order to reconcile the spectator model with the data (putting $\alpha_s = 0$ would increase $B_{e,\mu}$ to (15.3-18.4)%) substituting the average B-lifetime[21]

$$\tau_B = (1.4 \pm 0.6) \cdot 10^{-12} \text{ sec,} \tag{47}$$

measured by Mac[25] and Mark II,[26] into eq. (45) (I), one finds

$$|V_{cb}| \simeq 0.045, \tag{48}$$

a result which triggered a lot of theoretical speculation. It is interesting that the prediction of the spectator model on the lifetime is not very sensitive to the absolute values of the quark masses (see eq. (45)), as long as one consistently uses either curent or constituent masses. To conclude, the presently quoted semi-leptonic branching ratio of √12% is perfectly consistent with the spectator model. However, a markedly smaller value would be at variance.

Further support for the spectator model comes from the semi-inclusive decays: $B \to (K,D^\circ,J/\psi) + X$. From now on I use the current quark masses given in eq. (43) (I). To the extent that the average number of strange quarks per $(B + \bar{B})$ decay is a measure for the average number of kaons per $B\bar{B}$ event, the prediction[3]

$$<n_s> \simeq 2.86 \tag{49}$$

is nicely confirmed by the experimental result[27]

$$<n_k> \simeq 2.82. \tag{50}$$

Secondly, the inclusive D°-momentum distribution is found[24] to be remarkedly similar to the charm quark ($m_c = 1.86$ GeV) distribution expected from the semi-leptonic decay $b \to ce\nu_c$. This indicates that the D°'s originating in non-leptonic B-decays are also dominantly produced via b-quark dissociation, $b \to c\bar{u}d$, and that there is little communication between the $\bar{u}d$ and the $c\bar{q}$ spectator system (see Fig. 2a). This picture is further corroborated by the branching ratio observed[24] for $B \to J/\psi + X$:

$$BR(B \to J/\psi\, X) = (1^{+0.5}_{-0.4})\% \text{ or } <1.6\% \text{ (90\% c.l.).} \tag{51}$$

The relevant diagram is shown in Fig. 3. Applying the transformation eq. (25) to eqs. (16) and (17) one readily derives the appropriate effective hamiltonian:

$$H_{NL}^{c\bar{c}} = \frac{G_F}{\sqrt{2}} V_{cb} V_{cs}^* \left(\frac{2C_+ - C_-}{3} (c\bar{c})_L (\bar{s}b)_L + \frac{C_+ + C_-}{4} (\bar{c}\lambda^a c)_L (\bar{s}\lambda^a b)_L \right).$$

(52)

If there is indeed no interaction between the $c\bar{c}$-current and the rest of the diagram of Fig. 3, the matrix element $\langle X\ J/\psi | H_{NL}^{c\bar{c}} | B \rangle$ factorizes and one obtains[28,29]

$$\langle X\ J/\psi | H_{NL}^{c\bar{c}} | B \rangle \simeq \frac{G_F}{\sqrt{2}} V_{cb} V_{cs}^* \frac{2C_+ + C_-}{3} \langle J/\psi | \bar{c}\ \gamma^\mu c | o \rangle\ \bar{u}_s \gamma^\mu u_b.$$

(53)

The matrix element $\langle J/\psi | \bar{c}\gamma^\mu c | o \rangle$ is directly measured[30] in $e^+ e^-$ annihilation. From eqs. (53) and (45) (I) one then predicts[28] the branching ratio

$$BR(B \rightarrow J/\psi X) \simeq \left(\frac{2C_+ - C_-}{3} \right)^2 \cdot 12\%.$$

(54)

The above number also includes contributions from the cascades $B \rightarrow X, \psi' \rightarrow J/\psi$. The first factor in eq. (54) is due to the colour mismatch in Fig. 3. Neglecting short-distance effects ($C_+ = C_- = 1$), the $\bar{c}s$-system is produced in a pure colour singlet configuration or, equivalently, the $s\bar{q}$ and $c\bar{c}$-systems from colour singlets only 1/3 of the time. Hence, the decay rate $B \rightarrow J/\psi X$ is suppressed by a factor 1/9. In that case, eq. (54) gives $BR(B \rightarrow J/\psi X) \simeq 1.3\%$ in agreement with the experimental result eq. (51). QCD corrections further decrease the branching ratio. Unfortunately, the leading logarithmic corrections,

$$1 \rightarrow (2C_+ - C_-)^2$$

(55)

are not reliable because of an accidental cancellation of the above factor at $C_- = C_+^{-2} = 2C_+ = 1.59$, a value which is not very different from C_- (5 GeV) $\simeq 1.4$ (see eq. 32). This uncertainty, however, does not vitiate the strong experimental evidence that $B \rightarrow J/\psi + X$ is colour suppressed in accordance with the parton picture of Fig. 3.

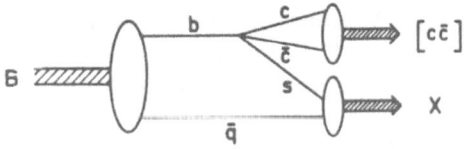

Fig. 3. Quark diagram of the decay $B \rightarrow I/\psi + X$.

I conclude by repeating the initial statement: the spectator model provides a consistent description of inclusive B-decays, at least at the level of the present experimental knowledge.

D,F-Decays

Tempted by the early onset of asymptotic freedom, one originally believed the spectator model to represent a good approximation of charm decays.[10,31] This prejudice was, in particular, supported by the progress made in understanding the $\Delta I = 1/2$ rule in kaon and hyperon decays[12,15] in terms of short-distance corrections to te weak hamiltonian and a simple quark parton description of the amplitudes similar to Fig. 2. Meanwhile, it has become clear that charmed particle decays are considerably more complex than anticipated.

According to the spectator model all weakly decaying charmed hadrons should have almost equal lifetimes and semi-leptonic branching ratios. For the following two sets of masses,

$$
\begin{array}{ll}
\text{(I)} & \text{(II)}
\end{array}
$$

$$
\begin{array}{lll}
m_{u,d} \simeq 0.15 & 0.3 \text{ GeV} & \\
m_s \simeq 0.3 & 0.4 \text{ GeV} & \qquad (56)\\
m_c \simeq 1.6 & 1.7 \text{ GeV} &
\end{array}
$$

both consistent with the constraints, eq. (40), from the semi-leptonic electron spectrum, eqs. (35) and (36) yield[3]

$$
BR_{e,\mu} \simeq \begin{cases} 13\%, \text{ (I)} \\ 19\%, \text{ (II)} \end{cases} \qquad (57)
$$

and

$$
t_c \simeq \begin{cases} 6 \cdot 10^{-13} \text{ sec, (I)} \\ 7.5 \cdot 10^{-13} \text{ sec, (II)} \end{cases} \qquad (58)
$$

Here, $|V_{cs}|^2 \simeq |V_{ud}|^2 \simeq 0.95$ and $|V_{cd}|^2 \simeq |V_{us}|^2 \simeq 0.05$ has been used. Because of the considerable sensitivity to a particular choice of quark masses, the above results are only indicative for what could be considered consistent with the spectator model.

However, the uncertainty of the spectator model predictions is not the main problem. The discovery of substantially differing lifetimes of D and F mesons,[2]

$$\tau(D^{\pm}) = (9.2 \begin{smallmatrix} +1.7 \\ -1.2 \end{smallmatrix}) \cdot 10^{-13} \text{ sec,}$$

$$\tau(\overset{(-)}{D}{}^{\circ}) = (4.4 \begin{smallmatrix} +0.8 \\ -0.6 \end{smallmatrix}) \cdot 10^{-13} \text{ sec,} \tag{59}$$

$$\tau(F^{\pm}) = (1.9 \begin{smallmatrix} +1.3 \\ -0.7 \end{smallmatrix}) \cdot 10^{-13} \text{ sec,}$$

shows that the spectator model itself is inappropriate for charm decays. Non-asymptotic effects involving the light constituent quarks must still be important. Considering the relatively small charm quark mass, this is not too surprising. Similar differences as in the lifetimes are seen in the semi-leptonic branching ratios[2] of $D^{\pm}$ and $\overset{(-)}{D}{}^{\circ}$:

$$BR_e(D^{\pm}) = (19 \begin{smallmatrix} +4 \\ -3 \end{smallmatrix})\% , \tag{60}$$

$$BR_e(\overset{(-)}{D}{}^{\circ}) = (5.3 \begin{smallmatrix} +2.9 \\ -1.3 \end{smallmatrix})\% .$$

Since the Cabibbo allowed ($\Delta I = 0$) semi-leptonic decay rates must be identical for $D^{\pm}$ and $\overset{(-)}{D}{}^{\circ}$, eq. (60) implies

$$\frac{BR_e(D^+)}{BR_e(D^{\circ})} \simeq \frac{\tau(D^+)}{\tau(D^{\circ})} \simeq 3.6 \begin{smallmatrix} +2.1 \\ -1.1 \end{smallmatrix} . \tag{61}$$

The last argument, together with the fact[21] that the semi-leptonic electron spectrum is perfectly consistent with $c \to se^+\nu_e$, identifies the non-leptonic decays as responsible for the lifetime differences.

An interesting question is which, if any, of the charmed mesons is "normal" from the point of view of the spectator model. Because of the uncertainties in both the predictions of eqs. (57) and (58) and the data, one cannot give a totally clear answer. Whereas the $D^{\pm}$ lifetime and semi-leptonic branching ratio can still be accommodated in the spectator model (although they appear somewhat on the large side), it is virtually impossible to obtain a semi-leptonic branching ratio smaller than 10%, as it is observed for the D°. In summary, one must conclude that in inclusive charm decays the spectator model fails by at least a factor 2 to 3.

NON-ASYMPTOTIC EFFECTS

The presence of light constituents in heavy quark bound states can, in various ways, give rise to lifetime differences and other modifications of the asymptotic decay pattern. One possibility are non-spectator interactions of the kind illustrated in Fig. 2b and commonly referred to as "annihilation" processes.[33] It may well be that the quark model estimate, eq. (33) which led to the neglect of these mechanisms, is oversimplified. Another pre-asympototic effect is the interference of identical quarks[34] among the light constituents and the decay products of the heavy quark (see Fig. 2a) due to Pauli's exclusion principle. Both of the above effects depend strongly on QCD bound state properties and are, therefore, difficult to quantify. However, one can certainly gain some qualitative insight in the essential physics points.

Interference Effects in Charmed Meson Decays

On the Cabibbo allowed level, the non-leptonic decay of the charm quark transmutes D and F mesons into four-quark systems with the flavour composition indicated below:

$$(c\bar{q}) \rightarrow \overline{sdu\bar{q}}. \tag{62}$$

In contrast to the D° and F^+ final states, which do not contain identical quarks, the D^+ final state contains two $\bar{d}$-quarks. Consequently, when computing the decay rate for the D^+, one must antisymmetrize the amplitude with respect to the identical quarks in accordance with Fermi statistics.

It is instructive, although maybe not completely reliable in a quantitative sense, to perform this calculation in the non-relativistic quark model. With H_{NL}^{eff} as given in eqs. (16) and (17), and in the limit $p_F \leq m_q \ll m_c$, where p_F is the "mean" Fermi momentum, one obtains[34]

$$\Gamma_{NL}(d^+ \rightarrow s\bar{d}u\bar{d}) \simeq (2C_+^2 + C_-^2) \frac{G_F^2 m_c^5}{192\pi^3} +$$
$$(2C_+^2 - C_-^2) \frac{G_F^2 m_c^2}{\pi} |\phi(0)|^2. \tag{63}$$

The first term is the usual spectator result in leading log approximation (see eq. (36)), whereas the second term arises from Pauli interference. Several comments are in order:

(i) The interference is destructive as a consequence of the QCD short-distance corrections ($2C_+^2 - C_-^2 \simeq -2,3$).

(ii) The amount of interference is determined by the wavefunction at the origin, $\phi(0)$. Thus, the ratio $\Gamma_{interference}/$

$\Gamma_{spectator}$ scales like $(f_D/m_c)^2$ as can be seen if one substitutes the non-relativistic relation

$$f_D^2 \simeq 12 \frac{|\phi(0)|^2}{m_c} \qquad (64)$$

in eq. (63).

(iii) For $f_D \backsim f_\pi$ and $m_c \backsim m_D$, eq. (63) gives, numerically,

$$\frac{\Gamma_{int}}{\Gamma_{spect}} \simeq \frac{C_-^2 - 2C_+^2}{2C_+^2 + C_-^2} \; 16\pi^2 \left(\frac{f_D}{m_c}\right)^2 \simeq 0.5. \qquad (65)$$

Using instead gaussian,[35] coulombic[36] or bag model[36,37] wavefunctions one obtains ratio in the range

$$\frac{\Gamma_{int}}{\Gamma_{spect}} \backsim 0.05-0.2, \; 0.1-0.2 \text{ and } 0.15-0.4, \qquad (66)$$

respectively.

The above results suggest that Pauli interference effects increase the lifetime and semi-leptonic branching ratio of the D^+ by about 20% with respect to the spectator model predictions. Although the latter and also the data are not yet precise enough to provide clear evidence for this effect, one can see a slight indication if one compares eqs. (57-60). Interferences also occur in Cabibbo suppressed D^+ and F^+ decays.[3] However, the D^0 and main F^+ decays are unaffected. The short D^0 and F^+ lifetimes must, therefore, result from other non-spectator interactions.

Annihilation Processes in Charmed Meson Decays

Earlier in the discussion it has been argued that the annihilation processes shown in Fig. 2b may be dismissed because of wavefunction and helicity suppression. The damping by bound state wavefunction is unavoidable. In contrast, helicity suppression is a combined effect of the V-A structure of weak interactions and the quark model used to estimate the annihilation amplitudes. This approximation may indeed be misleading.[33] Since hadronic bound states contain gluons and since gluons carry spin and colour, the $c\bar{q}$-system inside a pseudoscalar meson P may be in a colour singlet or octet, spin 1 state a considerable fraction of the time. Weak annihilation from these states, however, is not inhibited by helicity conservation and, hence, would occur with rates of the order of $(f_p/m_p)^2$ times a large phase space factor (see eq. (33)) relative to the rates of the spectator process, Fig. 2a. Clearly, if this is the case, annihilation is not negligible in charm meson decays.

From Fig. 4, which indicates all possible annihilation processes, it is easy to deduce the qualitative effects annihilation would have on D and F mesons decays:

(1) enhancement of all non-leptonic $(\bar{D})^\circ$ decay modes,

(2) enhancement of Cabibbo suppressed $D^\pm$ decay modes, and

(3) enhancement of both non-leptonic and semi-leptonic $F^\pm$ modes.

We see that the D° and F^+ lifetimes are shortened and the semi-leptonic branching ratio of the D° is lowered with respect to the spectator model expectations. Exactly this is needed as shown by eqs. (57-60).

Whether or not annihilation really explains the data is, at this stage, mainly a quantitative question. A firm answer requires a rather detailed understanding of QCD bound states. One can try a perturbative approach:[38] a gluon is radiated from the initial colour singlet, spin 0 $c\bar{q}$-state turning it into a colour octet, spin 1 system which then annihilates. This gives a rather small effect unless $f_{D,F} \gg f_\pi$:

$$\frac{\Gamma_{annihilation}}{\Gamma_{spectator}} \simeq \frac{2\pi\alpha_s}{27(2C_+^2 + C_-^2)} \begin{cases} (C_+ + C_-)^2 \left(\dfrac{f_D}{m_u}\right)^2 \simeq 0.2 \text{ for } D^\circ \\ \\ (C_+ - C_-)^2 \left(\dfrac{f_F}{m_s}\right)^2 \simeq 0.03 \text{ for } F^+. \end{cases} \qquad (67)$$

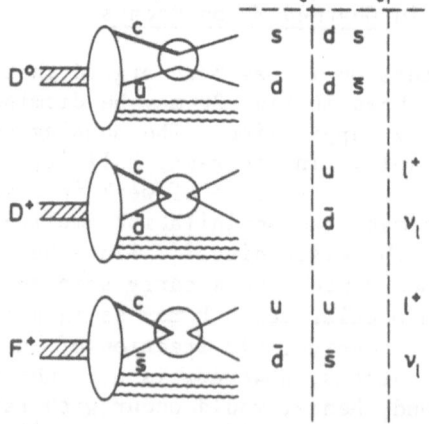

Fig. 4. Gluon enhanced annihilation processes in D and F meson decays. The wavy lines represent a gluonic component carrying spin 1.

The above numbers correspond to the rather optimistic guesses: $\alpha_s \simeq 0.5$ and $(f_D/m_u) \sim (f_F/m_s) \sim 1$. Moreover, the perturbative picture predicts $\tau(F^+) > \tau(D^\circ)$ at variance with experiment (see eq. (59)). Of course, perturbation theory may be inadequate. One has, therefore, also attempted various non-perturbative estimates.[33,39] For example, performing a QCD multipole expansion of the gluon component[39] one can relate the annihilation probability to the gluon condensate $\langle 0|\alpha_s/\pi\, G^a_{\mu\nu}G^{a\mu\nu}|0\rangle \simeq 0.012$ GeV,[4] which is known from the QCD sum rule analysis of the charmonium system.[40] The result,

$$\frac{\Gamma_{annihilation}}{\Gamma_{spectator}} \simeq \begin{cases} 1.5 \text{ for } D^\circ , \\ 0.7 \text{ for } F^+ , \end{cases} \tag{68}$$

suggests that the three decay mechanisms illustrated in Fig. 2 contribute roughly equally. The same conclusion emerges from a straightforward phenomenological analysis[3] in which the annihilation amplitudes are directly determined from the data.

To summarize, the observed deviations of inclusive D° and F^+ decays from the decay pattern predicted by the spectator model may be a preasymptotic effect originating in weak annihilation processes. However, one has so far no clear-cut theoretical proof. It is, therefore, very important to test further qualitative predictions of the annihilation hypotheses such as

(a) $BR_{e,\mu}(F^+) \sim BR_{e,\mu}(D^+)$,

(b) large $BR(F \to n\pi)$,

(c) deviations of the electron spectrum in semi-leptonic F decays from the spectrum produced via $c \to s e \nu_e$, and

(d) unusually frequent Cabibbo-suppressed D^+ decay modes.

Inclusive Charmed Baryon Decays

For sufficiently heavy quarks, the spectator model applies equally to baryons and mesons. Charmed meson decays, on the other hand, show that the asymptotic regime is not yet reached at the charm scale. This leads one to expect pre-asymptotic effects also in charm baryon decays. Let us consider the $\Delta_c^+ = (cdu)$ as an example. The decay of the c-quark produces a five-quark final state

$$(cdu) \to sdudu , \tag{69}$$

which contains two u-quarks. These interfer according to the Pauli principle.[41] Furthermore, the Δ_c^+ can decay via W-exchange between the c and d-constituent,[41,42]

701

$$(dcu) \rightarrow \bar{l}\,suu , \tag{70}$$

a process which is similar to the annihilation mechanisms illustrated in Fig. 2b for mesons. However, in contrast to the meson case, the decay indicated in eq. (70) is not helicity suppressed, even without invoking gluons. The reason is that, in the Λ_c^+, the cd-subsystem is 1/4 of the time in a spin 0 state. While the weak transition of a spin 0 fermion-antifermion system is helicity suppressed, it is perfectly allowed for a spin 0 fermion-fermion system.

A non-relativistic quark model calculation, similarly as in eq. (63), gives[41]

$$\Gamma(\Lambda_c^+ \rightarrow s\bar{d}udu) \simeq (2c_+^2 + c_-^2)\, \frac{G_F^2 m_c^5}{192\pi^3} -$$

$$c_+(2c_- - c_+)\, \frac{G_F^2 m_c^2}{4\pi}\, |\phi_{cu}(o)|^2 , \tag{71}$$

where the first term is the spectator model result and the second term is due to Pauli interference. The W-exchange contribution is given by[42]

$$\Gamma(\Lambda_c^+ \rightarrow suu) \simeq c_-^2\, \frac{G_F^2 m_c^2}{2\pi}\, |\phi_{cd}(o)|^2 . \tag{72}$$

In the above, $|\phi_{cq}(o)|^2$ characterizes the probability for the c-quark and a light constituent to be at the same point. This probabilty can be estimated from $\Sigma_c^+ - \Lambda_c^+$ hyperfine splitting:[43]

$$|\phi_{cq}(o)|^2 \simeq \frac{Gm_u^2 m_c}{16\pi\alpha_s(m_c - m_u)}\, (m_{\Sigma_c^+} - m_{\Lambda_c^+}) \simeq 0.01 \text{ GeV}^3 . \tag{73}$$

Using this estimate and taking into account phase space corrections, one finds[41]

$$\Gamma_{spect} : \Gamma_{int} : \Gamma_{w-ex} \sim \Lambda: \ 0.7: \ 2.7 \tag{74}$$

From that and the spectator model results for the semi-leptonic widths one obtains

$$BR_{e,\mu}(\Lambda_c^+) \simeq (4-7)\%; \ t(\Lambda_c^+) \simeq (2.5-3) \cdot 10^{-13} \text{ sec} \tag{75}$$

in agreement with experiment:[2]

$$BR_e(\Lambda_c^+) = (4.5 \pm 1.7)\%; \ T\Lambda_c^+) = (2.3\, ^{+1.0}_{-0.6}) \cdot 10^{-13} \text{ sec.} \tag{76}$$

A similar analysis is possible for Cabibbo suppressed decays[41] and other weakly decaying baryons.[3] Qualitatively one expects:

$$\tau(A^\circ = (csd)) \leq \tau(\Lambda_c^+ = (cdu)) < \tau(A_s^\circ = (css)) \leq \tau(A^+ = (csu)). \tag{77}$$

The recently reported[44] lifetime of the A^+,

$$\tau(A^+) + (4.8 \, {}^{+4.5}_{-2.0}) \cdot 10^{-13} \text{ sec}, \tag{78}$$

fits nicely into the predicted pattern. As a final remark, if weak annihilation of charmed mesons is indeed enhanced by gluons, a mechanism similar to Fig. 4 could also operate in charm baryon decays.[45] This makes it even more important to test experimentally whether or not the valence quark model predictions described above are correct.

Inclusive Bottom Decays
_____

Before concluding the discussion of non-asymptotic effects, I should make a few comments on the relevance of Pauli interference and weak annihilation to bottom decays. In the valence quark approximation, the ratios of the interference terms (eqs. (63) and (71)) and the W-exchange rate in baryon decays (eq. (72)) to the spectator model width scale like $|\phi_Q(o)|^2/m^3_Q$, where $\phi_Q(o)$ denotes the $Q\bar{q}$ and Qq wavefunctions at the origin. Thus, when going from charm to bottom decays, these effects are expected to decrease by roughly the factor

$$\left|\frac{\phi_b(o)}{\phi_c(o)}\right|^2 \left(\frac{m_c}{m_b}\right)^3 \sim \left(\frac{m_c}{m_b}\right)^{2 \text{ to } 3} \sim 0(10^{-1}). \tag{79}$$

As far as the gluon enhanced annihilation process in meson decays are concerned, extrapolations from charm to bottom are more uncertain. The perturbative approach,[38] eq. (67), suggests a rather slow decrease. On the other hand, it predicts small annihilation contributions anyway. In contrast, the QCD multipole estimate,[39] eq. (68), as well as an analysis[45] based on evolution equations indicates

$$\frac{\Gamma_{ann}(B)}{\Gamma_{spect}(B)} : \frac{\Gamma_{ann}(D)}{\Gamma_{spect}(D)} \sim 1:10. \tag{80}$$

Thus, even if one believes that $\Gamma_{ann}(D)/\Gamma_{spect}(D) \sim 0(1)$ as required in order to explain the observed deviations from the spectator model in D-decays, one would not expect annihilation effects in B-decays larger than of order 10%. The ratio, eq. (80), is also consistent with the scale dependence indicated in eq. (79), which should essentially apply to all processes clamped by bound state wavefunctions.

Although one has good reason to believe that pre-asymptotic effects play a minor role in bottom decays, only experiment can decide. Similarly as in D^o and D^+, weak annihilation and Pauli interference would expose themselves in the inequalities.[3]

$$\tau(B^+) > \tau(\overline{(\overline{B})^o}),$$

$$BR_\ell(B^+) > BR_\ell(\overline{(\overline{B})^o}),$$

$$BR(B^- \rightarrow J/\psi\ X) > BR(B^o \rightarrow J/\psi\ X).$$

TWO-BODY DECAYS

It is quite obvious that bound state properties and other aspects of strong interactions of typical hadronic scales play a more important role in exclusive decays than inclusively. As concerns two-body charm decays, one is at present still quite far from a clear qualitative picture, not to speak about a quantitative description. The original suggestion[17,31] to calculate the relevant matrix elements of the effective weak hamiltonian using a valence quark approximation has turned out completely inadequate in certain cases. In this approximation, the matrix element of a 4-quark operater facterizes in a product of matrix elements of quark currents, similarly as indicated in eq. (53) for the semi-inclusive decay $B \rightarrow J/\psi\ X$. Let us derive the amplitude of $D^o \rightarrow K^-\pi^+$ as an explicit example. The relevant weak hamiltonian can be read off from eqs. (16) and (17):

$$H_{NL}^{eff} = \frac{G_F}{\sqrt{2}}\ V_{cs}^* V_{ud}\left(\frac{C_+ + C_-}{2}\ (\bar{u}d)_L (\bar{s}c)_L + \frac{C_+ - C_-}{2}\ (\bar{u}c)(\bar{s}d)\right). \quad (82)$$

In a valence quark description the above operators give rise to the processes depicted in Fig. 5. Whereas in the first diagram the ud and su quark pairs are produced in colour singlet states and can directly form a π^+ and K^- mesan, respectively, the colour degrees of freedom are mismatched in the second diagram. Correspondingly, one must Fierz-transform the second operater in eq. (82) according to eq. (25) which give 1/3 $(\bar{u}d)_L(\bar{s}c)_L$ + 1/2 $(\bar{u}\lambda^a d)_L(\bar{s}\lambda^a c)_L$. Since colour-octet currents have zero matrix elements between colour-singlet states, one gets

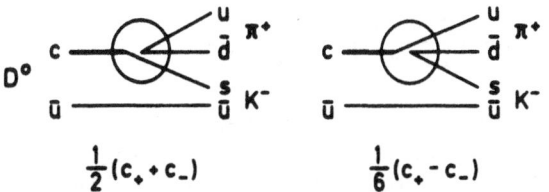

$$\frac{1}{2}(c_+ + c_-) \qquad\qquad \frac{1}{6}(c_+ - c_-)$$

Fig. 5. Valence quark description of the decay $D^\circ \to K^-\pi^+$.

$$\langle K^-\pi^+|H_{NL}^{eff}|D^\circ\rangle = \frac{G_F}{\sqrt{2}} V_{cs}^* V_{ud}\left(\frac{C_+ + C_-}{2} + \frac{C_+ - C_-}{6}\right) \langle\pi^+|(\bar{u}d)_L|0\rangle\langle K^-|(\bar{s}c)_L|D^\circ\rangle$$

$$\tag{83}$$

$$= \frac{G_F}{\sqrt{2}} V_{cs}^* V_{ud} \frac{2C_+ + C_-}{3} f_\pi\left((m_D^2 - m_K^2)f_+(m_\pi^2) + m_\pi^2 f_-(m_\pi^2)\right),$$

where the usual definitions of meson decay constants (f_π) and vector form factors $(f_\pm(q^2))$ have been adopted.

A crucial consequence of the factorization of weak matrix elements and the insertion of the vacuum state is the effective suppression of certain channels due to colour mismatch (see discussion of $B \to J/\psi\, X$ in III.1). For the case at hand, this occurs in the decay $D^\circ \to \bar{K}^\circ\pi^\circ$. According to the quark diagrams of Fig. 6 and the quark presentation of the π°, $1/\sqrt{2}\ (u\bar{u} - d\bar{d})$, the amplitude of $D^\circ \to \bar{K}^\circ\pi^\circ$ is given by

$$\langle\bar{K}^\circ\pi^\circ|H_{NL}^{eff}|D^\circ\rangle = \frac{G_F}{\sqrt{2}} V_{cs}^* V_{ud}\left(\frac{C_+ + C_-}{6} + \frac{C_+ - C_-}{2}\right) \langle\bar{K}^\circ|(\bar{s}d)_L|0\rangle\langle\pi^\circ|(\bar{u}c)_L|D^\circ\rangle$$

$$= \frac{G_F}{\sqrt{2}} V_{cs}^* V_{ud}\left(\frac{2C_+ - C_-}{3}\right) \frac{f_K}{\sqrt{2}}\left((m_D^2 - m_\pi^2)f_+(m_K^2) + m_K^2 f_-(m_K^2)\right).$$

$$\tag{84}$$

One clearly sees that the small factor $2C_+ - C_-/3$ arises from colour mismatch in the contribution (1st diagram of Fig. 6) of the dominant operator of eq. (82). From eqs. (83) and (84) one then predicts a substantial suppression of $D^\circ \to \bar{K}^\circ\pi^\circ$ relative to $D^\circ \to K^-\pi^+$. In the SU(3) limit,

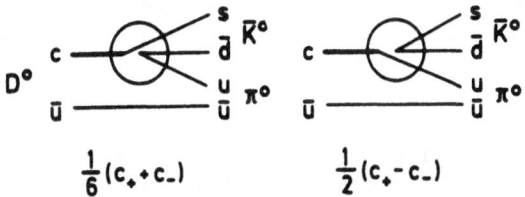

$$\frac{1}{6}(c_+ + c_-) \qquad\qquad \frac{1}{2}(c_+ - c_-)$$

Fig. 6. Valence quark description of the decay $D^\circ \to \bar{K}^\circ\pi^\circ$.

$$\frac{\Gamma(D^\circ \to \overline{K}^\circ \pi^\circ)}{\Gamma(D^\circ_K \pi^+)} \simeq \frac{1}{2} \left(\frac{2C_+ - C_-}{2C_+ + C_-} \right)^2 = \frac{1}{18} \left(\frac{1}{200} \right) , \tag{85}$$

where the first number is the uncorrected ($C_+ = C_- = 1$) result, whereas the second number includes leading log QCD corrections ($C_+ = 0.74$, $C_- = 1.8$). The latter are extremely sensitive to the values of C_+ and C_- as pointed out in III.1, eq. (55). SU(3) breaking effects[46] increase the above result by roughly a factor 2.

At any rate, the valence quark picture outlined above is in striking discrepancy with the measurement[47]

$$\frac{\Gamma(D^\circ \to \overline{K}^\circ \pi^\circ)}{\Gamma(D^\circ \to K^- \pi^+)} = 0.73 \pm 0.35 . \tag{86}$$

Another important consequence of factorization and vacuum saturation is the suppression of annihilation mechanisms such as

$$D^\circ \to s\overline{d} \begin{cases} s(\overline{u}u)d \to K^-\pi^+, \\ s(\overline{d}d)\overline{d} \to \overline{K}^\circ\pi^\circ, \end{cases} \tag{87}$$

where the ($\overline{u}u$) and $\overline{d}d$) quark pairs are picked up from the vacuum. Analoguously to eqs. (83) and (84), one finds

$$<K\pi| H_{NL}^{eff} |D^\circ> \backsim \frac{2C_+ - C_-}{3} <K\pi|(\overline{s}d)_L|o><o|(\overline{u}c)_L|D^\circ> \tag{88}$$

$$\backsim \frac{2C_+ - C_-}{3} f_D <K\pi|P_D^\mu \, \overline{s}\gamma_\mu d|o> ,$$

which vanishes in the SU(3)-limit because of conservation of vector currents:

$$P^\mu \overline{s}\lambda_\mu d = (m_d - m_s)\overline{s}d \underset{m_s = m_d}{\longrightarrow} 0. \tag{89}$$

Thus, in the valence quark approximation annihilation processes contribute only at the level of SU(3)-breaking effects. Moreover, the matrix element eq. (88) is colour suppressed, so that the failure of the prediction eq. (85) on the ratio $\overline{K}^\circ\pi^\circ/K^-\pi^+$ can certainly not be blammed on the neglect of annihilation.

To save the day, a variety of ideas have been put forward. Without going into details, I mention a few of them. The simplest way out is to abandon the QCD values of $C_\pm$ and to assume that, for some reason, $C_- \gg C_+$. This is the sextet-

dominance[48] solution which, in the limit, gives $\bar{K}^\circ\pi^\circ/K^-\pi^+ \backsim 1/2$.
Of course, one must eventually explain why $C_- \gg C_+$. One
possible reason are soft gluon interactions. It has first been
pointed out by Fritzsch[49] that soft gluons may wash out the static
colour structure of the diagrams in Fig. 5 and 6 and, thus,
vibrate colour suppression. Clearly, soft gluon exchanges between
quark lines joining different mesons would spoil factorization and
vacuum saturation of the amplitudes and, therefore, undermine the
basis of colour suppression is the valence quark approximation.
One can show[50] that the net effect amounts to a change of the
ratio C_-/C_+ in eq. (85). Non-factorization also invalidates the
argument for neglecting annihilation processes.[33] The latter, on
the other hand, gives automatically $\bar{K}^\circ\pi^\circ/K^-\pi^+ = 1/2$, since the
final state in $D^\circ \to s\bar{d}$ carries isospin 1/2. Finally, the valence
quark prediction, eq. (85), could also be upset by resonance
effects in the $K\pi$-system as pointed out by Lipkin.[51] Decomposing
the amplitudes according to the isospin of the final states and
allowing for phases subject to final state interactions, one has

$$\frac{\Gamma(D^\circ\to\bar{K}^\circ\pi^\circ)}{\Gamma(D^\circ\to K^-\pi^+)} = \left(\frac{+A_{1/2}R^{i\delta}+\sqrt{2}A_{3/2}R^{i\delta'}}{\sqrt{2}A_{1/2}R^{i\delta}-A_{3/2}R^{i\delta'}}\right)^2 .$$

Since the $I = 1/2$ channel contains many K* resonances, whereas the
$I = 3/2$ channel is exotic, one may get a large phase shift, $\delta-\delta'$,
which could reverse the cancellation, $A_{1/2} \backsim -\sqrt{2}A_{3/2}$, claimed
at the quark level. We see that one can come up with a number of
reasonable explanations for the apparent absence of colour sup-
pression in $D^\circ \to \bar{K}^\circ\pi^\circ$. Unfortunately, one of the above can be
made quantitative at present.

With slight modifications due to SU(6)-breaking,[3] the
valence quark prediction, eq. (85), also applies to the ratios
$\bar{K}^\circ\rho^\circ/K^-\rho^+$ and $\bar{K}^{\circ*}\pi^\circ/K^{-*}\pi^+$. Here, the experimental situation[47]
is unclarified, being consistent with 0-0.1 and 0-1.1 for the
above ratios, respectively.

Turning to the Calibbo-suppressed decay modes, $D^\circ \to K^-K^+$ and
$\pi^-\pi^+$, one encounters some minor problems. The valence quark
approximation in the SU(3)-limit predicts

$$\left.\begin{array}{l}\dfrac{\Gamma(D^\circ\to K^-K^+)}{\Gamma(D^\circ\to K^-\pi^+)} = \left|\dfrac{V_{us}}{V_{ud}}\right|^2 \\[4mm] \dfrac{\Gamma(D^\circ\to\pi^-\pi^+)}{\Gamma(D^\circ\to K^-\pi^+)} = \left|\dfrac{V_{cd}}{V_{cs}}\right|^2\end{array}\right\} \simeq 0.046 - 0.061,^2 \qquad (91)$$

707

whereas, experimentally, one observes[47]

$$\frac{\Gamma(D^{\circ} \rightarrow K^- K^+)}{\Gamma(D^{\circ} \rightarrow K^- \pi^+)} = 0.113 \pm 0.03 \ ,$$

$$\frac{\Gamma(D^{\circ} \rightarrow \pi^- \pi^+)}{\Gamma(D^{\circ} \rightarrow K^- \pi^+)} = 0.033 \pm 0.015 \ . \tag{92}$$

SU(3)-breaking effects[52] increase the results of eq. (91) by roughly 70% and 30%, respectively. This leaves a discrepancy by a factor 1.2-2 between the valence quark approximation and the data. Also, if annihilation processes are really important, as may be indicated by the absence of colour suppression, the predictions eq. (91) would still be valid. The enhancement $K^- K^+/\pi^- \pi^+ \backsim 3$ can then be explained[53] by properly taking into account SU(3)-breaking. I should further point out that the experimental result, eq. (52), is actually consistent with the most general, model-independent SU(3) prediction,[54] but requires a large difference in the magnitude of the reduced matrix elements. Finally, the valence quark approximation yields the interesting relations:[3]

$$\frac{\Gamma(D^+ \rightarrow \bar{K}^{\circ} K^+)}{\Gamma(D^+ \rightarrow \bar{K}^{\circ} \pi^+)} \simeq \left(\frac{2C_+ + C_-}{4C_+}\right)^2 \frac{\Gamma(D^{\circ} \rightarrow K^- K^+)}{\Gamma(D^{\circ} \rightarrow K^- \pi^+)} \simeq (14-18)\% \ , \tag{93}$$

and

$$\frac{\Gamma(D^+ \rightarrow \pi^{\circ} \pi^+)}{\Gamma(D^+ \rightarrow \bar{K}^{\circ} \pi^+)} \simeq \frac{1}{2} \frac{\Gamma(D^{\circ} \rightarrow \pi^- \pi^+)}{\Gamma(D^{\circ} \rightarrow K^- \pi^+)} \simeq 1.5\% \ , \tag{94}$$

which are worth an experimental test. In the above, the data given in eq. (92) and $C_- = (C_+)^{-2} = 1.8-2.1$ have been used.

As exemplified in eqs. (83) and (84), the valence quark approximation does not only predict ratios of partial width, but also the absolute magnitudes.[46] This obviously very important aspect of a critical appraisal of the present understanding can, unfortunately, not be considered further in this talk (see e.g. ref. 55). Also two-body baryon decays[56] must be left out.

To summarize, the valence quark approximation or, more specifically, factorization and vacuum saturation of weak matrix elements, is at variance with the existing data on two-body D decays. The non-observation of colour suppression constitutes the main problem and raises questions about the role of soft hadron physics in exclusive decays. A convincing, even if only qualitative, answer has not yet emerged. In order to unravel the essential dynamical aspects, further and more precise data are needed. In particular, tests of the valence quark predictions[46] on D^+- and F-meson decays would be very desirable. Also decays such as $D^o \rightarrow \bar{K}^o \phi$ and $F^+ \rightarrow \omega \pi^+$, which can only proceed via annihilation, are of great interest. Finally, it is not clear whether or not two-body decays of B-mesons are appreciably simpler. On the one hand, one does certainly not expect D*-resonance effects in the 5 GeV range. Also, annihilation processes are rather unlikely, since the two final quarks move too fast in order to turn into two hadrons by sharing a slow quark pair created fom the vacuum. However, one has no strong arguments against soft gluon effects in typical decays like $B \rightarrow D\pi$. The case $B \rightarrow J/\psi \, X$ (X = K or $K\pi$), shown in III.1 to exhibit colour suppression, is special in the sense that factorization and vacuum saturation should be more reliable due to the small size and the small gluon admixture of the J/ψ.

CONCLUSIONS

The present approach to weak decays of heavy quark states separates the underlying dynamics in short- and long-distance aspects. The former are incorporated in an effective weak hamiltonian, whereas the latter are absorbed in the hadronic matrix elements of this hamiltonian. Distinguishing between these two parts of the whole problem, one can summarize the status of understanding heavy quark decays as follows.

The modifications of the bare weak hamiltonian of the standard $SU(2)_L \times U(1)$ theory arising from hard gluon interactions have been rigorously calculated using operator product expansion and renormalization group techniques. A very important result is that the next-to-leading corrections have "normal" size and reinforce the non-leptonic enhancement found in leading log approximations. Also quark mass effects such as penquin operators are well understood and turn out to play a rather unimportnat role in charm and bottom decays. In short, this part of the problem is in a very trustworthy state.

As concerns the task of calculating weak matrix elements one must further distinguish inclusive and exclusive (2-body) decays. Asymptotically, that is for sufficiently heavy quarks, inclusive decays are dominated by the short-distance properties of the "free" quark decays. Correspondingly, the spectator model should provide a good approximation. Indeed, the existing data on inclusive B-decays, both semi- and non-leptonic modes, are well described by the spectator model. It is further interesting and reassuring that the gluon corrections as predicted by QCD appear necessary to reconcile theory with experiment. Fo not-so-heavy quarks, on the other hand, one expects some influence of the internal structure of the heavy quark bound states. Two prominent pre-asymptotic effects are identical particle interferences between the spectator quarks and the decay products of the heavy quark, and annihilation processes possibly enhanced by the presence of soft gluons. Various estimates suggest that these effects may play a role in charm decays. Actually, the study of pre-asymptotic effects was initiated by the discovery of deviations from the spectator model in non-leptonic charmed meson and baryon decays. The observed differences in lifetimes and semi-leptonic branching ratios are in qualitative agreement with what one expects from Pauli interference and weak annihilation, indicating roughly equal strength of spectator and non-spectator contributions. This conclusion, however, still lacks a quantitiative confirmation by a clear-cut theoretical calculation. In particular, reliable estimates of the annihilation contributions to D- and F-decays, which are negligible unless gluons annul the helicity suppression, require a deeper understanding of QCD bound states. Nevertheless, an upshot have been strong arguments that bottom decays should indeed be short-distance dominated with pre-asymptotic effects not exceeding 10% or so. To summarize, despite some open quantitative questions, the essential physics of inclusive decays appears to be understood.

In comparison, the situation in non-leptonic two-body decays is rather dim. One has not yet succeeded in developing an, at least qualitatively, consistent theoretical framework. The valence quark approximation, in which the matrix elements of the effective weak hamiltonian are factorized in products of matrix elements of quark currents saturated by the vacuum state, seems to fail. The clearest counter-evidence comes from the absence of colour suppression in $D^\circ \to \bar{K}^\circ \pi^\circ$. This failure may be due to soft gluon interactions and/or, unexpectedly, large annihilation contributions. However, also final state resonance effects may considerably confuse the matter. In addition, flavour-symmetry breaking effects have properly to be taken into account as indicated by the Calibbo-suppressed decays $D^\circ \to K^-K^+$ and $\pi^-\pi^+$. It seems that only when further and more accurate data become available one can piece the puzzle together. An understanding of two-body charm decays will not only have direct implications on exclusive B-decays, but also illuminate the origin of the $\Delta I = 1/2$ rule in kaon and hyperon decays. The point is that the present explanation of the $\Delta I = 1/2$ rule as mainly a penquin effect relies on either the soft pion or the valence quark approximation.

As a final remark, the top quark, "predicted" by the standard model and awaiting its definite discovery, is so heavy that the inclusive decays of top flavoured hadrons[57] will clearly reflect the "free" quark dynamics.

I am grateful for discussions with P. Avery, P. Franzini, D. Hitlin, I. Lee-Franzini and B. Stech. Special thanks go to Ling-Lie Chau for inviting me to this stimulating workshop. Without her encourgement and patience this report would not have appeared.

REFERENCES

1. L. B. Okun, Leptons and Quarks (North-Holland Publishing Company, Amsterdam, 1982); H. Fritzsch and P. Minkowski, Phys. Rep. 73 (1981) 67.

2. Particle Data Group, Rev. Mod. Phys. 56 (1984) 1.

3. R. Rückl, Weak Decays of Heavy Flavours, CERN preprint, 1983.

4. See, for example, L.-L. Chau, Phys. Rep. I5C (1983) 1 and these Proceedings.

5. S. L. Glashow, I. Iliopoulos and L. Maiani, Phys. Rev. D2 (1970) 1285.

6. M. Kobayashi and T. Maskawa, Prog. Theor. Phys. 49 (1973) 652.

7. See K. Kleinknecht, I. Lee-Franzini and H. W. Siebert, these Proceedings.

8. K. G. Wilson, Phys. Rev. 179 (1969) 1499 and Phys. Rev. D3 (1971) 1818; W. Zimmerman, Ann. Phys. 77 (1973) 536,570.

9. E. C. G. Stueckelberg and A. Peterman, Helv. Phys. Acta 26 (1953) 499; M. Gell-Mann and F. E. Low, Phys. Rev. 95 (1954) 1300; C. G. Callan, Phys. Rev. D2 (1970) 1541; K. Symanzik, Commun. Math. Phys. 18 (1970) 227.

10. M. K. Gaillard an B. W. Lee, Phys. Rev. Lett. 33 (1974) 108; G. Altarelli and L. Maiani, Phys. Lett. 52B (1974) 351.

11. R. L. Kingsley, S. B. Treiman, F. Wilzek and A. Zee, Phys. Rev. D11 (1975) 1919; I. Ellis, M. K. Gaillard and D. V. Nanopoulos, Nucl. Phys. B100 (1975) 313; G. Altarelli, N. Cabibbo and L. Maiani, Nucl. Phys. B88 (1975) 285.

12. G. Altarelli, G. Curci, G. Martinelli and R. Petrarca; Phys. Lett. 99B (1981) 141 and Nucl. Phys. B187 (1981) 461.

13. A. I. Vainshtein, V. I. Zakharov and M. A. Shifman, Sov. Phys. JETP 45 (1977) 670; M. A. Shifman, A. I. Vainshtein and V. I. Zakharov, Nucl. Phys. B120 (1977) 316;

F. Gilman and M. B. Wise, Phys. Rev. D20 (1979) 2392
and D27 (1983) 1128; B. Guberina, D. Tadic and I.
Trampetic, Nucl. Phys. B 152 (1979) 429; F. Buccella,
M. Lusignoli, L. Maiani and A. Pugliese, Nucl. Phys.
B152 (1979) 461; B. Guberina and R. D. Peccei, Nucl.
Phys. B613 (1980) 289.

14. J. Ellis, M. K. Gaillard, D. Nanopoulos and S. Rudaz,
    Nucl. Phys. B131 (1977) 285.

15. B. Guberina, R. D. Peccei and R. Rückl, Phys. Lett. 90B
    (1980) 169; B. Guberina, Orsay preprint, LPTHE 82/5
    (1982).

16. L. Maiani, in Proc. 21. Int. Conf. on High Energy Physics,
    ed. by P. Petiau and M. Porneuf (Les Editions de
    Physique, Les Ulis, 1982), p. 631.

17. M. K. Gaillard, B. W. Lee and J. L. Rosner, Rev. Mod. Phys.
    47 (1975) 277; J. Ellis, M. K. Gaillard and D. V.
    Nanopoulos, Nucl. Phys. B100 (1975) 313.

18. N. Cabibbo and L. Maiani, Phys. Lett. 79B (1978) 109;
    M. Suzuki, Nucl. Phys. B145 (1978) 420; A. Ali and
    E. Pietarinen, Nucl. Phys. B154 (1979) 519; N. Cabibbo,
    G. Corbo and L. Maiani, Nucl. Phys. B155 (1979) 93;
    G. Corbo, Phys. Lett. 116B (1982) 298 and Nucl. Phys.
    B122 (1983) 99; B. Guberina, R. D. Peccei and R. Rückl,
    Phys. Lett. 91B (1980) 116 and Nucl. Phys. B171 (1980)
    333; Q. Hokim and X. Y. Pham, Orsay preprint, PAR-LPTHE
    83/05 (1983).

19. U. Baur, Diploma Thesis, Universität München, 1982;
    U. Baur and H. Fritzsch, Phys. Lett. 109B (1982) 402;
    J. L. Cortes, X. Y. Pham and A. Tounsi, Phys. Rev. D25
    (1982) 188.

20. A. Ali and E. Pictarinan, Nucl. Phys. B154 (1979) 519;
    G. Altarelli, N. Cabibbo, C. Corbo, L. Maiani and
    G. Martinelli, Nucl. Phys. B208 (1982) 365.

21. I. Lee-Franzini, these Proceedings.

22. W. Bacino et al., Phys. Rev. Lett. 43 (1979) 1073.

23. C. Klopfenstein et al., Phys. Lett. 130B (1983) 444;
    A. Chen et al., Phys. Rev. Lett. 52 (1984) 1084.

24. S. Stone, in Proc. 1983 Intern. Symp. on Lepton and Photon
    Interactions at High Energies, ed. by D. G. Cassel and

D. L. Kreinick (Newman Lab. of Nucl. Studies, Cornell University, Ithaca, New York, 1983), p. 203.

25. E. Fernandez et al., Phys. Rev. Lett. 51 (1983) 1022.

26. N. S. Lockeyer et al., Phys. Rev. Lett. 51 (1983) 1316.

27. P. Avery, these Proceedings.

28. J. H. Kühn, S. Nussinov and R. Rückl, Z. Phys. C5 (1980) 117; J. H. Kühn and R. Rückl, Phys. Lett. 135B (1984) 477.

29. M. B. Wise, Phys. Lett. 89B (1980) 229; T. A. DeGrand and D. Toussaint, Phys. Lett. 89B (1980) 256.

30. B. H. Wiik and G. Wolf, DESY preprint 78/23 (1978).

31. D. Fakirov and B. Stech, Nucl. Phys. B133 (1978) 315; N. Cabibbo and L. Maiani, Phys. Lett. 73B (1978) 418.

32. N. W. Reay, in Proc. 1983 Int. Symp. on Lepton and Photon Interactions at High Energies, ed. by D. G. Cassel and D. L. Kreinick (Newman Lab. of Nucl. Studies, Cornell University, Ithaca, New York, 1983), p. 244.

33. W. Bernreuther, O. Nachtmann and B. Stech, Z. Phys. C4 (1980) 257; H. Fritzsch and P. Minkowski, Phys. Lett. 90B (1980) 455; S. P. Rosen, Phys. Rev. Lett. 44 (1980) 4; M. Bander, D. Silvermann and A. Soni, Phys. Rev. Lett. 44 (1980) 7 and Erratum Phys. Rev. Lett. 44 (1980) 962; V. Barger, J. P. Leveille and P. M. Stevenson, Phys. Rev. D22 (1980) 693.

34. R. D. Peccei and R. Rückl, in Special Topics in Gauge Field Theories, Ahrenshoop Symposium (Akademie der Wissenschaften der DDR, Berlin-Zeuthen, 1981), p. 8. T. Kobayashi and N. Yamazaki, Proc. Theor. Phys. 65 (1981) 775; M. A. Shifman an M. B. Voloshin, ITEP preprint-62 (1984); see also B. Guberina, S. Nussinov, R. D. Peccei and R. Rückl, Phys. Lett. 89B (1979) 111; Y. Koide, Phys. Rev. D20 (1979) 1739; K. Jagannathan and V. S. Mathur, Phys. Rev. D21 (1980) 3165.

35. G. Altarelli and L. Maiani, Phys. Lett. 118B (1982) 414.

36. H. Sawayanagi et al. Phys. Rev. D27 (1983) 2107.

37. N. Bilic, B. Guberina and J. Trampetic, MPI preprint, PAE/PTh 28/84 (1984).

38. M. Bander, D. Silvermann and A. Soni, Phys. Rev. Lett. 44 (1980) 7, 962.

39. K. Shizuya, Phys. Lett. 100B (1981) 79 and 105B (1981) 406.

40. M. Shifman, A. Vainshtein and V. I. Zakharov, Nucl. Phys. B147 (1979) 385, 448; L. J. Reinders, H. R. Rubinstein and S. Yazaki, Phys. Lett. 94B (1980) 203 and 95B (1980) 103; B. Guberina, R. Meckbach, R. D. Peccei and R. Rückl, Nucl. Phys. B184 (1981) 476.

41. R. Rückl, Phys. Lett. 120B (1983) 449.

42. V. Barger, J. P. Leveille and P. M. Stevenson, Phys. Rev. Lett. 44 (1980) 226.

43. D. DeRújula, H. Georgi and S. L. Glashow, Phys. Rev. D12 (1975) 147.

44. S. F. Biagi et al., CERN-EP/84-76 (1984).

45. I. Bigi, Z. Phys. C9 (1981) 197; see also I. Bigi, Nucl. Phys. B177 (1981) 395; M. Suzuki, Nucl. Phys. B177 (1981) 413.

46. D. Fakirov and B. Stech, Nucl. Phys. B133 (1978) 315.

47. G. H. Trilling, Phys. Rep. 75C (1981) 57; R. Schindler, Ph.D. Thesis, SLAC-Report 219 (1979), D. Hitlin, these Proceedings.

48. M. Katuya, Phys. Rev. D18 (1978) 3510; M. Katuya and Y. Koide, Phys. Rev. D19 (1979) 2631; G. Eilam and M. Gronau, Phys. Lett. B96 (1980) 391; see also B. Guberina et al., Y. Koide ref. 34.

49. H. Fritzsch, Phys. Lett. 86B (1979) 343.

50. N. Deshpande, M. Gronau and D. Sutherland, Phys. Lett. 90B (1980) 431; M. Gronau and D. Sutherland, Nucl. Phys. B183 (1981) 367.

51. H. J. Lipkin, Phys. Rev. Lett. 44 (1980) 710.

52. V. Barger and S. Pakvasa, Phys. Rev. Lett. 43 (1979) 812.

53. H. Fritzsch and P. Minkowski, Nucl. Phys. 3171 (1980) 413; I. Bigi, Phys. Lett. 90B (1980) 177.

54. R. L. Kingsley, S. B. Treiman, F. Wilczek and A. Zee, Phys. Rev. D11 (1975) 1919: L.-L. Chau Wang and F. Wilczek, Phys. Rev. Lett. 43 (1979) 816.

55. M. Bonvin and C. Schmid, Nucl. Phys. B194 (1982) 319.

56. J. G. Körner, G. Kramer and J. Willrodt, Phys. Lett. 78B (1978) 492, Erratum 81B (1979) 419 and Z. Phys. C2 (1979) 117; D. Ebert and W. Kallies, Phys. Lett. 131B (1983) 183.

57. N. Cabibbo and L. Maiani, Phys. Lett. 87B (1979) 366.

DISCUSSION

HITLIN:

I would like to point out that there is another possible way to learn about the mechanism of charmed particle decay: through the study of radiative decays. The rate for a particular exclusive channel is comparable to that for a Cabibbo-suppressed decay. The photon energy spectrum may provide information, in particular cases, on whether the $c \rightarrow s$ transition has been accomplished via the W decay or W exchange mechanism.

STECH:

How good or bad is factorization according to present data?

RÜCKL:

In some cases factorization and vacuum insertion gives acceptable results, in other cases it fails. I give you two examples. The $\Delta I = 3/2$ amplitude for $K^+ \rightarrow \pi^+ \pi^0$ is obtained within a factor of 2, whereas the amplitude for $D^+ \rightarrow \bar{K}^0 \pi^+$ disagrees with the data by roughly a factor of 5.

COMPOSITE QUARKS AND THEIR COMPOSITE QUARKS AND THEIR FLAVOR MIXING[+]

Harald Fritzsch

Sektion Physik der Universität München
and
Max-Planck-Institut für Physik und Astrophysik
- Werner Heisenberg Institut für Physik -
Munich (Fed. Rep. Germany)

The standard SU(2) x U(1)-model of the electroweak interactions describes rather well the observed features of the weak and electromagnetic interactions. The scale of the W- and Z- masses is set by the vacuum expectation value of a neutral scalar Higgs field, which is determined by the Fermi constant and is of the order of 300 GeV.

One of the unsolved problems in subnuclear physics is to understand the origin, the observed pattern and the magnitudes of the lepton and quark masses. In the SU(2) x U(1) gauge theory no satisfactory answer is given. The lepton and quark masses arise simply as consequences of the couplings of the leptons and quarks to the neutral scalar field; they reflect the corresponding Yukawa coupling constants. It needs not to be stressed that this cannot be the final

---

[+] Invited Talk given at the Europhysics Conference on Flavor Mixing in Weak Interactions, Erice, Italy, March 1984.
Supported by DFG- contract Fr 412/6-1

answer to the fermion mass problem. Presumably a solution to the mass problem is only possible if one takes into account a substructure of the leptons and quarks[1]. It is difficult to believe that leptons and quarks are pointlike objects, and yet exhibit such a rich spectrum of states as observed.

The leptons and quarks observed thus far seem to come in three families:

I.
$$\left(\begin{array}{c} \nu_e \\ e^- \; [0.5] \end{array} \quad \vdots \quad \begin{array}{c} u \; [5] \\ d \; [9] \end{array} \right)$$

II.
$$\left(\begin{array}{c} \nu_\mu \\ \mu^- \quad 106 \end{array} \quad \vdots \quad \begin{array}{c} c \; [1,200] \\ s \; [180] \end{array} \right)$$

III.
$$\left(\begin{array}{c} \nu_\tau \\ \tau^- \; [1784] \end{array} \quad \vdots \quad \begin{array}{c} t \; [>24,000] \\ b \; [4,800] \end{array} \right)$$

The numbers in the brackets denote the lepton or quark masses in MeV. For the light quarks we have used typical "current algebra masses"[2]. Although no striking regularities of the mass spectrum is observed, some regularities of the lepton- quark mass spectrum are worth mentioning.

a) There exists a definite hierarchical pattern. The ratios of masses of quarks or leptons belonging to different families are large:

$$\frac{m_s}{m_d} \approx 20 \qquad \frac{m_b}{m_s} \approx 27$$

$$\frac{m_c}{m_u} \approx 240 \qquad \frac{m_t}{m_c} > 20$$

$$\frac{m_\mu}{m_e} \approx 207 \qquad \frac{m_\tau}{m_\mu} \approx 17.$$

All charged fermions of the first family (e, u, d) are lighter than the ones of the second family (μ, c, s), and those are lighter than the charged members of the third family (τ, t, b). The neutrinos, which are either massless or have extremely small masses, seem to play a special rôle and will not be discussed further.

b) Inside the various families there is a striking dependence of the masses on the electric charge. The quarks of charge 2/3 are much heavier than the quarks of charge (- 1/3), except in the first family (the u-quark is lighter than the d-quark). The neutrinos, being electrically neutral, have no (or an exceedingly small) mass.

c) The recent results concerning the weak decays of b- flavored particles indicate that the weak decays b → u are strongly suppressed. compared to b → c[3]. Since the weak decays of b-quarks proceed solely via weak interaction mixing, the suppression of b → u supports the hypothesis of the so-called "nearest neighbour mixing" (see e.g. ref. (4)). Note that in the absence of the b → u - decay the weak interaction mixing matrix can be expressed in terms of two angles only (θ_1, θ_2):

$$
\begin{pmatrix} d' \\ s' \\ b' \end{pmatrix} = \begin{pmatrix} c_1 & c_1 & 0 \\ -s_1 c_2 & c_1 c_2 & -s_2 \\ -s_1 s_2 & c_1 s_2 & c_2 \end{pmatrix} \begin{pmatrix} d \\ s \\ b \end{pmatrix} .
$$

($c_i =: \cos\theta_i$, $s_i =: \sin\theta_i$)

The surprisingly long lifetime of the b- flavored particles[5] implies a rather small mixing angle θ_2: $\theta_2 \approx 0.05$. (Compare $\theta_1 \approx 0.22$.) Thus as the masses of the quarks increase, the mixing between the neighbouring families seems to decrease.

Although it is unknown what physical mechanism is responsible for the mass generation, one may speculate that the fermion masses are generated as radiative corrections involving an energy scale of the order of 1 TeV. (Note that the typical fermion mass scale for the fermions of the third family is ~ α · 1 TeV.) This is the same energy scale needed in composite models in which the W- and Z- bosons are composite objects, consisting of two constituents (haplons), bound together by very strong (so-called hyperchromodynamic) forces. The weak interactions are indirect consequences of those forces[6-10].

If the weak interaction turns out to be a remnant of the hyperchromodynamic force, a new interpretation of the relationship between the electromagnetic and weak interaction is required. The W- and Z-bosons cease to be fundamental gauge bosons, but acquire the less prestigious status of bound states of haplons. However the photon remains an elementary object (at least at the scale of the order of 10^{-17} cm, discussed here). As a whole, the SU(2) x U(1)-theory cannot be regarded anymore as a fundamental microscopic theory of the electroweak interactions, but at best can be interpreted as an effective theory, which is useful only at distances larger than the hypercolor confinement scale. It acquires a status comparable to the one of the σ-model in QCD, which correctly describes the chiral dynamics of π-mesons and nucleons at relatively low energies, but fails to be a reasonable description of the strong interaction at high energies.

However I would like to emphasize that at the present time no indication whatsoever comes from the experimental side that leptons, quarks and weak bosons may be bound states of yet smaller constituents. It may well be that the weak force will turn out in the future as a fundamental gauge force, as fundamental as the electromagnetic one and the color force. In fact, interpreting the weak forces as effective forces poses a number of problems which have not been solved in a satisfactory manner. First of all, the weak interactions violate parity, and they do that not in an uncontrolled way, but in a very simple one: only the lefthanded leptons and quarks take part in the charged current interactions. If we interpret the weak interactions as Van der Waals type interactions, the parity violation is a point of worry. How should one interpret the observed parity violation? Does it mean that the lefthanded fermions have a different internal structure than the righthanded ones? Or are we dealing with

720

two or several different hypercolor confinement scales, for example one for the lefthanded fermions, and one for the righthanded fermions, such that the resulting effective theory is similar to the left - right symmetric gauge theory, based on the group $SU(2)_L \times SU(2)_R$?

Another point of concern is the fact that the weak interactions show a number of regularities, e.g. the universality of the weak couplings, which one would not a priori expect if the weak interaction is merely a hypercolor remnant. On the other hand it is well - known that the interaction of pions or ρ-mesons with hadrons shows a number of regularities which can be traced back to current algebra, combined with chiral symmetry or vector meson dominance. Despite the fact that both the ρ-mesons and the pions are quark - antiquark bound states for which one would not a priori expect that their interaction with other hadrons exhibits remarkable simple properties (e.g. the universality of the vector meson couplings), the latter arise as a consequence of the underlying current algebra, which is saturated rather well at low energies by the lowest lying pole (either the pion pole in the case of the divergence of the axial vector current, or the ρ- or A_1-pole in the case of the vector or axial vector current).

In the case of chiral $SU(2) \times SU(2)$ the pole dominance works very well - the predictions of current algebra and PCAC seem to be fulfilled within about 5 %. The universality of the weak interactions is observed to be valid within 1 % in the case of the couplings of the weak currents to electrons, myons, as well as u, d and s- quarks. Much weaker constraints exist for the heavy quarks. It remains to be seen whether the observed universality of the weak interactions will find an explanation along lines similar to the ones used in hadronic physics.

Here I shall concentrate on models in which the W-bosons consist of a haplon and an antihaplon. Since the observed weak interactions exhibit a symmetry $SU(2)_L$, we assume the existence of two haplons, denoted by the doublet $\binom{\alpha}{\beta}$, which carry hypercolor and electric charge. The electric charges are assumed to be: $Q(\alpha) = + 1/2$, $Q(\beta) = -1/2$ (see ref. 7,9).

The spectral functions at energies much above the hypercolor confinement scale are supposed to be described by a continuum of haplon- antihaplon pairs. At low energies the weak amplitudes will be dominated by the lowest lying poles, which are identified with the W-particles. The latter form the triplet:

$$
\begin{pmatrix} W^+ \\ W^3 \\ W^- \end{pmatrix} = \begin{pmatrix} \bar{\beta}\alpha \\ \frac{1}{\sqrt{2}} (\bar{\alpha}\alpha - \bar{\beta}\beta) \\ \bar{\alpha}\beta \end{pmatrix}
$$

The experimental data on the neutral current interaction require a mixing between the photon and the W_3 boson (the neutral, isovector partner of W^+ and W^-), which in the standard SU(2) x U(1) scheme is caused by the spontaneous symmetry breaking. Within our approach this mixing is due to the W_3 - γ transitions, generated dynamically like the ρ-γ transitions in QCD (for an early discussion, based on vector meson dominance, see ref. (11)). The magnitude of $\sin^2\theta_W$ is directly related to the strength of the γ-W_3 transition. The latter is deter- mined by the electric charges of the W-constituents and by the W wave function near the origin. We suppose that in the absence of electromagnetism the weak interactions are mediated by the triplet (W^+, W^-, W^3), where $M(W^+)= M(W^-) = M(W^3) = 0$ $(\wedge_H)$.

After the introduction of the electromagnetic interaction the photon and the W^3- boson mix. We denote the strength of this mixing by a parameter λ, following ref. (12), which is related to g (W- fermion coupling constant) and the effective value of $\sin^2\theta_W$

$$
\sin^2\theta_W = \frac{e}{g} \cdot \lambda
$$

Furthermore one has:

$$
M_W = g \cdot 123 \text{ GeV}
$$

$$
M_Z^2 = \frac{M_W^2}{1-\lambda^2}
$$

The mixing parameter $\sin^2\theta_W$ is determined by the decay constant F_W of the W-boson, which we define in analogy to the decay constants of the ρ_0-meson (F_ρ): $<0|j_\mu^{\ 3}|W^3> = \epsilon_\mu M_W F_W = \epsilon_\mu \cdot M_W^2 / f_W$:

$$\sin^2\theta_W = e^2 / g \cdot F_W / M_W = e^2/gf_W$$

Taking for example $g = 0.65$ and $M_W = 79$ GeV, one obtains $F_W = 123$ GeV, a value which seems not unreasonable for a bound state of the size 10^{-16} cm.

In the SU(2) x U(1) gauge theory the SU(2) coupling constant g is related to e by the relation $g = e/\sin\theta_W$. In bound state models of the weak interactions discussed here this relation need not be true in general. However it has been emphasized[12] that this relation is approximately fulfilled if the lowest lying W-pole dominates the weak spectral function at low energies. This leads to the relation

$$g = M_W / F_W = e/\sin\theta_W \approx 0.65$$

(we have used $\sin^2\theta_W = 0.22$).

It is interesting to note that many aspects of the bound state models can be derived from a local current algebra of the weak currents. We observe that the lefthanded leptons and quarks form doublets of the weak isospin. The weak isospin charges F_i^W (i= 1,2,3) obey the isospin charge algebra

$$[F_i^W, F_j^W] = i\ \epsilon_{ijk}\ F_k^W.$$

Let us assume that these charges can be constructed as integrals over local charge densities $F_{0i}^W(x)$, i.e.,

$$F_i^W(x^0) = \int F_{0i}^W(x)d^3x.$$

Furthermore we suppose that the charge densities obey at equal times the local current algebra

$$[F_{0i}^W(x), F_{0j}^W(y)]_{x^0=y^0} = i\epsilon_{ijk}F_{0k}^W\delta^3(\vec{x}-\vec{y}).$$

The local algebra is trivially fulfilled in a model in which leptons and quarks are pointlike objects and the weak currents are simply bilinear in the lepton and quark fields. However, if leptons and quarks are extended objects, the situation changes. Currents, which are bilinear in the (composite) lepton and quark fields would not obey the local algebra, just like the currents, which are bilinear in nucleon fields, do not obey the local current algebra of QCD. The local algebra becomes a highly non-trivial constraint. It is fulfilled in the haplon models discussed above, in which the currents are bilinear in α and β.

We consider matrix elements of the weak currents between the various fermion fields. In order to do so, we shall assume that the higher families composed of μ, τ, ... etc. are dynamical excitations of the first family (ν_e, e^-, u, d), without specifying in detail the dynamical structure of these states.

Let us look at the form factors of the left-handed weak neutral current $F_{\mu 3}^L(x)$, i.e., the matrix elements of this current between different lefthanded lepton or quark states, e.g., $<e_L^-|F_{\mu 3}(0)|e_L^->$. Denoting these form factors by $F_e(t)$, $F_\mu(t)$, $F_\tau(t)$, etc., the weak isospin algebra requires a universal normalization at $t = 0$, i.e., $F_e = F_{\nu_e} = F_\mu = F_{\nu\mu} = \ldots = 1$. Assuming W dominance to be a reasonably good approximation, we may write for the dependence on the four-momentum transfer t,

$$F_f(t) = \frac{M_W^2}{f_W} \frac{f_W^{ff}}{M_W^2 - t},$$

where f denotes any one of the fermions e^-, ν_e, μ^-, ν_μ, etc. From $F_f(0) = 1$ we obtain the universality relation

$$f_W = f_W^{ff} \equiv g.$$

The neutral W and, because of the weak isospin algebra, also the charged W bosons couple universally to leptons and quarks, $f_W^{ff} \equiv g$. Thus the universality of the weak interactions follows from the W-dominance.

724

The saturation mechanism discussed here makes sense only if the constituents α, β of the W-bosons serve at the same time as constituents of the fermions. Thus both the weak bosons and the fermions should have comparable sizes.

A minimal scheme of haplons would involve only the W-constituents α and β. However it is easy to see that those fields are not enough for building up the quarks and leptons. (Here the situation is unlike QCD. The u and d- quarks as well as the corresponding antiquarks are sufficient to build up the (p,n) doublet and the (ρ^+, ρ^0, ρ^-)- triplet.) Thus further constituents are needed. For example, in some of the schemes discussed in ref. (7,8) new scalar fields of charges 1/2 and 1/6 are added: in this case the leptons and quarks consist of a fermion and a scalar. One may also add further spin 1/2 fields and interpret the leptons and quarks as composed of those objects.

We turn to the problem of fermion masses. A solution of the fermion mass problem is required to give answers to the following two questions:

a) Why are the fermion masses much smaller than 1 TeV ?

b) What is the mechanism responsible for the generation of mass ?

We suppose that the answer to the first question is given by a symmetry. On the scale of the hypercolor interaction, both color and electromagnetism can be viewed as small perturbations. We suppose that in the limit where those interactions are switched off the leptons and quarks are massless; one is dealing with 24 massless states. A chiral symmetry (either a continous or a discrete chiral symmetry) is supposed to provide the reason for the absence of mass[13].

Besides the 24 massless fermions an infinite number of other fermions with mass of order 1 TeV or larger exist. For those states we see no reason why a chiral symmetry should be valid. Such a symmetry if it remained unbroken would require a parity doubling of all massive states. We suppose that the chiral symmetry is strongly broken by the hypercolor dynamics in the heavy fermion sector. The observed SU(2) symmetry of the weak interactions is supposed to be a flavor symmetry of the hypercolor interaction.

If the electromagnetic interaction is introduced, the lepton and quarks acquire a finite electromagnetic self energy (see Fig. (1)), where the heavy fermions (mass ~ 1 TeV $\sim \Lambda_h$) serve as intermediate states. The self energy is finite due to the cut off of Λ_h provided by the finite sizes (of order Λ_h^{-1}) of the leptons and quarks. Note that the leptons and quarks have in particular a finite charge radius of the order of Λ_h^{-1}.

Thus far we have mentioned only the electromagnetic perturbation of hypercolor dynamics. In a similar way one may consider the QCD interaction. However the gluonic self energy of a quark is strongly model dependent. In all bound state models of the type considered here the fermions do have a finite charge radius, but only seldom a finite color radius is implied. In most schemes the color resides on one of the constituents (often a scalar object) of the leptons and quarks. In this case the color self energy can be represented by the renormalized mass of the corresponding constituent, which in general is arbitrary; it may vanish as a result of a symmetry (e.g. as a consequence of supersymmetry and chiral dynamics). If the QCD interaction would be an important interaction for the generation of mass for leptons and quarks, the mass difference inside weak doublets would be small compared to the average mass (e.g. $m_t - m_b \ll \frac{1}{2} (m_t + m_b)$). This is not the case. Instead a very strong dependence of the masses on the electric charges is observed. We conclude: the color force is either excluded from contributing to the lepton or quark masses, or contributes only very little. This has implications for the experimental search for lepton- quark substructure: leptons and quarks are expected to have a charge radius of the order of 10^{-17} cm, but no color radius of this order. Subsequently we shall assume that only the QED interaction is responsible for the fermion mass generation[14].

We analyze the fermion self energy diagram of Fig. 1 somewhat more in detail. In principle, infinitely many heavy (mass > 1 TeV) states will contribute as intermediate states. Since we have no detailed information about the heavy states there is no way to compute the electromagnetic self energy exactly. However, we expect that the lowest contributing intermediate state will dominate. (Something similar is true in hadron physics: the electromagnetic selfenergy of the pion or the proton is dominated by the lowest intermediate state). Under this assumption the resulting fermion mass matrix reads:

$$
M = \frac{\alpha}{\pi} Q^2 K \Lambda_h \begin{pmatrix} |f_1|^2 & f_1 f_2 & f_1 f_3 \\ f_1 f_2 & |f_2|^2 & f_2 f_3 \\ f_1 f_3 & f_2 f_3 & |f_3|^2 \end{pmatrix} + O(\alpha^2).
$$

Q denotes the electric charge of the fermion and K is a parameter of order one, depending on the transition form factors. The elements of the vectors

$$
\vec{f} = \begin{pmatrix} f_1 \\ f_2 \\ f_3 \end{pmatrix}
$$

are measures of the transitions $<i|j_\mu|n>$, where $|i>$ denotes a quark or lepton state, $|n>$ the intermediate heavy state (see Fig. 1), and j_μ the electromagnetic current. It is useful to rewrite M in terms of the matrix

$$
A = \begin{pmatrix} f_1 & f_2 & f_3 \\ 0 & 0 & 0 \\ 0 & 0 & 0 \\ \vdots & \vdots & \vdots \end{pmatrix}
$$

The result is

$$M = \frac{\alpha}{\pi} Q^2 K \wedge_h (A^+ A) + O(\alpha^2).$$

This shows that M is of rank 1, and the diagonalization gives:

$$M = \frac{\alpha}{\pi} Q^2 K \wedge_h |\vec{f}|^2 \cdot \begin{pmatrix} 0 & 0 & 0 \\ 0 & 0 & 0 \\ 0 & 0 & 1 \end{pmatrix} + O(\alpha^2)$$

Thus in the approximation made above (one intermediate state dominates, $O(\alpha^2)$ terms are neglected) only one family of quarks and leptons acquires a mass. The latter is identified with the third family (t, b, τ). The hierarchical pattern of fermion masses starts to emerge.

As soon as we give up the assumption of the dominance of the fermion self energy by one intermediate state, M ceases to be of rank 1. In the case of 2 intermediate states ($|n>$ and $|m>$) the second row of A ceases to be zero and one finds:

$$A = \begin{pmatrix} f_1 & f_2 & f_3 \\ g_1 & g_2 & g_3 \\ 0 & 0 & 0 \\ \vdots & \vdots & \vdots \end{pmatrix}$$

Now M is of rank 2. Still one eigenvalue of M is exactly zero which means that the third and second family are massive. The magnitude of the fermion mass hierarchy is related to the quality of the dominance of the fermion self energy by one intermediate state, i.e. to the magnitude of g_i/f_i.

The situation is as follows: We found that the fermion mass hierarchy is related to the quality of the dominance of the fermion self energy by one or several intermediate states. The neutrinos being neutral remain massless to $O(\alpha)$. The quarks obey the following mass relations:

$$\frac{m_u}{m_d} = \frac{m_c}{m_s} = \frac{m_t}{m_b} = \frac{(\frac{2}{3})^2}{(-\frac{1}{3})^2} = 4$$

Thus the mass matrices for up and down-type quarks are, up to a factor 4, the same. Therefore, no weak interaction mixing exists in this approximation. The factor 4 between u- type and d- type quark masses as well as the absence of weak mixing is a consequence of the SU(2)-invariance of the transition matrix elements $<i|j_\mu|n>$. In order to introduce mixing and departures from this relation one has to allow for a violation of the SU(2)-invariance which is caused by electromagnetic effects of order α^2.

A general ansatz for the $O(\alpha^2)$ mass matrix is:

$$M^{(2)} = (\frac{\alpha}{\pi})^2 \cdot Q^4 \cdot K \cdot \Lambda_h \cdot g$$

where g is a hermitean 3 x 3 matrix.

It is instructive to write down the mass matrices for the quarks of charge (2/3) and (-1/3):

$$M(2/3) = \alpha \cdot 4 \cdot (m^{(1)}) + \alpha^2 \cdot 16 \ (m^{(2)})$$

$$M(-1/3) = \alpha \cdot 1 \cdot (m^{(1)}) + \alpha^2 \cdot 1 \cdot (m^{(2)})$$

$(m^{(1)}, m^{(2)}$ are 3 x 3 matrices). The weak interaction mixing occurs as a consequence of the mismatch between the $O(\alpha)$ and $O(\alpha^2)$ mass matrices[14]. Both $m^{(1)}$ and $m^{(2)}$ are the same for U-type and D-type quarks. However due to the different factors 4 or 16 the U-type and D-type quarks cannot be diagonalized at the same time.

There is no reason for g to factorize. As a result one expects that the heavy quarks are essentially inert against mixing, i.e. they stay almost unmixed, while the light quark spectrum depends strongly on mixing effects:

$$\frac{\theta_2}{\theta_1} \sim 0(\frac{m_s}{m_b}, \frac{m_c}{m_t}) \ll 1.$$

The "light" quarks u and d are indeed very light compared to the quarks of the second and third family. For this reason one may suppose that the light quark masses are entirely due to $0(\alpha^2)$-effects[15]. Lets us consider the first and second generation only. The masses for c and s are predominantly effects of $0(\alpha)$; the mass matrices can be written as follows:

U-type-quarks: $M = \begin{pmatrix} 16\ \widetilde{g}_{11} & 16\ \widetilde{g}_{12} \\ 16\ \widetilde{g}_{12} & 4\ c + 16\ \widetilde{g}_{22} \end{pmatrix}$

D-type-quarks: $M = \begin{pmatrix} \widetilde{g}_{11} & \widetilde{g}_{12} \\ \widetilde{g}_{12} & c + \widetilde{g}_{22} \end{pmatrix}$

(c, $\widetilde{g}_{ij}$: parameters, proportional to $\alpha\wedge_h$ and $\alpha\wedge_h \cdot g_{ij}$).

Four parameters enter in these mass matrices. They determine the four quark masses and the Cabibbo angle, i.e. the Cabibbo angle can be determined as a function of the quark masses.

For the mass values $m_u = 4$ MeV, $m_d = 7.5$ MeV, $m_s = 150$ MeV and $m_c = 1200$ MeV one obtains[15]

$$\sin\theta_c = 0.226,$$

in very good agreement with experiment. For a generalization to six quarks see ref. (15).

This example shows that the pattern of fermion masses and of the weak interaction mixing angles can be understood as a consequence of the hyperchromodynamic forces inside the leptons and quarks and of the electromagnetic interaction, treated perturbatively. Of course, this makes sense only, if the size of the leptons, quarks and weak bosons is of the order of $(1\ TeV)^{-1}$. The effects of the new substructure

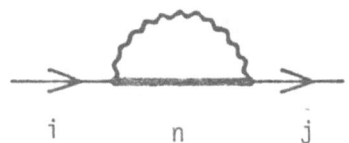

i   n   j

Figure 1. The self energy diagram which is supposed to be responsible for the generation of quark and lepton masses; i,
j denote the quark or lepton states, n the heavy intermediate state (mass $\sim$ 1 TeV). The wavy line represents
the photon.

should soon be discovered in the experiments, e.g. in high energy
proton- antiproton collisions.

ACKNOWLEDGMENTS

My thanks go to Dr. Ling-Lie Chau for the splendid organization
of this Meeting at this Southern tip of Europe, and for inviting
me to give this talk on lepton-quark structure.

REFERENCES

1. For recent review see:

   M.E. Peskin, Proc. 1981 Int. Symp. on Lepton and Photon
   Interactions, Bonn, ed. W. Pfeil.

   H. Harari, Proc. 1982 SLAC Summer Institute, SLAC Rep. n⁰ 259.

   R. Barbieri, Talk at 1983 Int. Symp. on Lepton and Photon Interactions at High Energies, Cornell University, Ithaca, N.Y.

   L. Lyons, Oxford University 52/82 (1982) and 83-0670 (1983).

   H. Fritzsch, Lecture at Int. School on Subnuclear Physics,
   Erice, MPI-PAE/PTh 76 /83 (1983)
   H. Terarawa, these proceedings and references therein.

2. J. Gasser and H. Leutwyler, Phys. Rep. 87 (1982) 77

3. See the talk of J. Lee-Franzini (these proceedings).

4. H. Fritzsch, Nucl. Phys. B 155 (1979) 189;
    Phys. Lett. 85 B (1979) 81.

5. See the talk of W. Ford (these proceedings).

6. L. Abbott and E. Farhi, Nucl. Phys. B 189 (1981) 547

7. H. Fritzsch and G. Mandelbaum, Phys. Lett. 102 B (1981) 319;
    ibid. 109 B (1982) 224.

8. R. Barbieri, R. Mohapatra and A. Masiero,
    Phys. Lett. 105 B (1981) 369.

9. H. Fritzsch, Proceedings of the Int. School on Subnuclear
    Physics, Erice, Sicily (1983).

10. B. Schrempp and F. Schrempp, Nucl. Phys. B 231 (1984) 109

11. P. Hung and J. Sakurai, Nucl. Phys. B 143 (1978) 81.

12. H. Fritzsch, D. Schildknecht, and
    R. Kögerler, Phys. Lett. 114 B (1982) 157;
    R. Kögerler and D. Schildknecht,
    CERN - preprint 3231 (1982).

13. See for example: G t'Hooft, in: Recent Developments in Gauge
    Theories, Plenum, New York (1980), p. 135.

14. U. Baur and H. Fritzsch, Phys. Lett. 134 B (1984) 105.

15. U. Baur and H. Fritzsch, in preparation.

DISCUSSION

RÜCKL:

Your mass matrix for the quarks contains parameters which are the coefficients of the expansion of the electromagnetic masses in powers of α. Thus they should all be of the same order. Is this the case in your fit to the quark mass spectrum?

FRITZSCH:

Yes, this is the case. In lowest order (in α) one obtains the eigenvalues $(0,0,1)$. The (α^2) contributions, which lead to the mixing, are all of the same order. This, by the way, is the reason for the smallness of the $b \to c$ transition.

SEGRE:

Your (spin 1/2) - (spin 1/2) bound states have a mass of $\sim$ 100 GeV. (Spin 1/2) - (spin 0) bound states have mass of order 0. What about the masses of (spin 0) - (spin 0) bound states?

FRITZSCH:

It is true that in essentially all models further exotic particles are expected, for example scalar-scalar bound states. Those states may well have fairly low masses. They would in particular contribute to neutral current processes. At this stage nothing is known about the masses.

GOLDHABER:

1) How many parameters does your theory have?
2) What effect do these ideas have on estimates of proton decay times?

FRITZSCH:

The aim is to construct a theory without any parameters. All dimensional physical quantities can be written as a function of Λ_h, the confinement parameters of hyperchromodynamics. Of course, detailed calculations of masses etc. depend on wave functions, form factors etc., which are not known explicitly, but can be worked out explicitly once the dynamics of the haplons is studied in detail, a stage, which is still far away.
The models I am discussing here have a confinement scale of the order of 1 TeV and have nothing to do with baryon number violating processes.

STECH:

Have you parity violation in your scheme?

FRITZSCH:

The parity violation is introduced by the spectrum of the W-bosons. In composite models it seems natural to have a left-right symmetric theory, i.e. there are light "lefthanded" and relatively heavy "righthanded" weak bosons.

PHAM:

Why the scalar bound state of composite model does not mediate weak interactions, while the vector bound state W does?

FRITZSCH:

The scalar or pseudoscalar particles will also couple to the leptons and quarks, but this coupling constant is unknown. Since the coupling of scalars to fermions is helicity changing, those couplings may be very small. The u-particle which is an iso-scalar may have a fairly large coupling to the fermions - here the couplings are not constrained by decays like   ev.

WALI

What about gluons in this model?

FRITZSCH:

All massive bosons are supposed to be massive, while photons and gluons are considered to be elementary at the scale I am considering here.

THEORETICAL STATUS OF FLAVOUR MIXING MATRIX

Berthold Stech

Institut für Theoretische Physik
der Universität Heidelberg
D-6900 Heidelberg

ABSTRACT

Horizontal symmetries and hints from grand unification can be
used to restrict the fermion mass matrix. Relations between masses
and mixing angles which are independent of the details of the Higgs
part of the Lagrangian are obtained. A scheme is described in which
very massive generations gradually decouple from very light ones
and which shows maximal CP-violation. The corresponding quark mixing matrix is in accord with present experimental information. It
also gives the observed magnitude of the CP-violating ε-parameter
of K-decays within the uncertainties of the B-factor and long-range
contributions. The mass of the top quark is obtained as a function
of the mass of the strange quark. Lepton mixing angles are shortly
discussed.

1. Introduction

The existence of several generations of fermions (quarks and
leptons) constitutes a longstanding unsolved problem. The origin of
the particle spectrum, of the fermion masses and mixing angles are
not understood and one has to orient oneself on the observed regularities. Fortunately, there are some striking features:

i) The spectrum of fermions is not completely irregular: all

735

mass ratios of fermions of the same charge are large and the mixing angles observed in charged current interactions are small.

ii) Neutral current interactions conserve flavour to a high precision.

iii) The gauge anomaly cancels in each generation separately. At the energy scale of the weak vector bosons m_W the forces are described by the fundamental or effective gauge symmetry

$$G = SU_C(3) \times SU_L(2) \times U(1)$$

Because of the existence of several generations this gauge interaction is left invariant under the action of a large global symmetry H_o. This so-called horizontal symmetry consists of the direct product of separate $U(n_G)$ symmetries (n_G denotes the number of generations): The quark and lepton $SU_L(2)$ doublets as well as all $SU_L(2)$ singlets can be transformed independently. Only in case the gauge group G combines all fermions of one generation into an irreducible representation as in the grand unified theories with $G = SO(10)$ or E6, the horizontal symmetry H_o is simply $U(n_G)$.

The combined breaking of the local and global symmetries gives rise to the fermion mass matrix. This fermion mass matrix - in a basis determined by the group G - contains all information on masses and mixing angles[1].

In this report I will concentrate on relations between fermion masses and mixing angles. For a plausible scenario the aims are:

i) To find relations which are independent of specific details such as Higgs potentials, vacuum expectation values and Yukawa couplings, since none of these are reliable.

ii) To find relations which show decoupling properties: If one formally puts the masses of a generation to infinity, the mixing angles to lighter generations should vanish.

iii) These relations should possibly be connected to a symmetry which protects them from suffering large renormalizations.

iv) A connection should exist to a unifying scheme in order to have

particle and antiparticle fields and quarks and leptons on equal
footing.

The restrictions on the fermion mass matrices as suggested by
horizontal symmetries are discussed in section 2. In section 3 properties of mass matrices in grand unified theories are described.
Section 4 contains very definite relations between masses and
mixing angles based on a simple suggestion. In view of the recent
measurements of the B-meson lifetime[2] these relations will be discussed in detail. Some speculative remarks on possible neutrino
mixings are made in section 5. Most of the material presented has
been published or will appear shortly. Two figures showing the connection between the top quark mass and the strange quark mass, and
with the ε-parameter of K^O-decays, are new. Also added is an equation for the neutrino mass matrix. To make the report easy to read,
technical formulas have been avoided.

2. Horizontal Symmetries

The conventional way to obtain the symmetry breaking of the
gauge group G is to add scalar Higgs fields to the Lagrangian. Those
Higgs fields which couple to fermions will also break the horizontal group H_o to a group $H \in H_o$. From the particle spectrum we know
that the horizontal group is severely broken. If H is non-trivial,
several lines can be followed.

A) Local horizontal symmetries. One supposes that H is a local symmetry. The breaking of this symmetry must then occur at a scale
$\geq 10^4$ GeV, in order to suppress neutral flavour-changing processes[3].
So far, models of this type do not make definite predictions about
the fermion mass matrix which is the subject of interest in this
report.

B) Global continuous horizontal symmetries. If H is a global continuous symmetry, the spontaneous breaking induces long-range interactions via Goldstone bosons. To suppress the corresponding flavour-
changing neutral current transitions, the breaking scale must be
very high (order 10^{10} GeV)[4]. Although there are interesting models,

too many ad hoc assumptions have to be made before one arrives at the fermion mass matrix. There exists, however, a notable exception: The identification of a global horizontal axial U(1) symmetry[5] with the Peccei-Quinn (PQ) symmetry[6]. By giving the first, second, and third generations the PQ-charges 5/2, -3/2, and 1/2, respectively, and introducing two scalar fields with PQ charges -1 and +1, the corresponding 3 x 3 mass matrices in generation space take the form

$$M \;=\; \begin{pmatrix} 0 & a & 0 \\ a & 0 & b \\ 0 & b & c \end{pmatrix} \tag{1}$$

M is assumed to be a symmetric matrix having complex elements in general. A third Higgs field, which is an $SU_L(2)$ singlet having PQ charge zero, can be introduced to make the axion "invisible"[7]. Thus, a special solution of the strong CP-problem provides for mass matrices of the form (1), a form which was suggested some time ago[8] from a different point of view. The huge vacuum expectation value required to make the axion invisible can occur in grand unified theories. I refer to the talk by K. C. Wali on a specific $SO(10) \times U_{PQ}(1)$ model for details[9].

C) Discrete horizontal symmetries. At the scale m_W of weak interaction no global continuous symmetry can survive (except for the just mentioned case of an invisible axion). Through spontaneous symmetry breaking Goldstone bosons would appear. Thus, Yukawa interactions which destroy the continuous symmetries must be present. However, a discrete horizontal symmetry can persist, since no Goldstone bosons are connected with it. Such a discrete symmetry can provide constraints on the mass matrices and can protect them from arbitrary renormalization. However, in the case of the standard $SU_L(2) \times U(1)$ model with a single Higgs doublet, horizontal symmetries do not provide for useful constraints on the mass matrix[10]. The gauge group or the number of scalar fields have to be enlarged.

An instructive and interesting example is a left-right symmetric model based on $SU_L(2) \times SU_R(2) \times U(1)$ and 2 generations. A simple discrete symmetry can give rise to the matrix structure[11]

$$M = \begin{pmatrix} o & a \\ a & b \end{pmatrix} \quad a, b \text{ complex} \qquad (2)$$

.

Using this form for the charge 2/3 as well as the charge 1/3 mass matrices, one easily obtains the Cabibbo angle as a function of the quark masses of the 2 generations and of one remaining relative phase. By varying this phase one finds

$$\text{arc tg } \sqrt{|\frac{m_d}{m_s}|} - \text{arc tg } \sqrt{|\frac{m_u}{m_c}|} \leq \Theta_c \leq \text{arc tg } \sqrt{|\frac{m_d}{m_s}|} + \text{arc tg } \sqrt{|\frac{m_u}{m_c}|} \quad (3)$$

A general analysis of discrete symmetries was performed by Ecker and Konetschny[12] for the gauge group $SU_L(2) \times U(1)$, 2 generations and no restriction on the number of scalar fields. It was required that Θ_c should be expressible in terms of mass eigenvalues and relative phases only. They found that only a few forms of mass matrices are obtainable from discrete symmetries. The only realistic one gives

$$|\frac{m_d}{m_s}| - |\frac{m_u}{m_c}| \underset{\sim}{\leq} \Theta_c^2 \underset{\sim}{\leq} |\frac{m_d}{m_s}| + |\frac{m_u}{m_c}| \qquad (4)$$

.

This result is more restrictive than (3). Taking quark mass ratios as given by Leutwyler and Gasser[13], eq. (4) is in accord with the experimental value for θ_c. The corresponding discrete symmetry groups which leave the Lagrangian invariant and give the result (4) are the generalized permutation symmetries $P^N_{n_G=2}$ [14]. $P^N_{n_G}$ is a semi-direct product of the symmetric permutation group S_{n_G} with $n_G - 1$ factors of the cyclic group Z_N.

$$P^N_{n_G} = Z_N \times Z_N \times .. \text{ⓢ} S_{n_G} \qquad (5)$$

.

In a very recent article Ecker[14] applied this symmetry also to the 3 generation problem. Using 3-dimensional representations of P^N_3

739

for the quarks as well as for 3 scalar fields, he obtained a connection between the up-quark mass matrix M_U and the down-quark mass
matrix M_D. If one takes M_U diagonal and adjusts the phases of right-
handed up-quark fields appropriately, the result can be written in
the form

$$M_D = \alpha \, M_u^{(Diagonal)} + h \begin{pmatrix} 0 & W_1 & \varepsilon W_3 \\ \varepsilon W_1 & 0 & W_2 \\ W_3 & \varepsilon W_2 & 0 \end{pmatrix} \qquad (6)$$

α, h real, W_i complex, $\varepsilon = \pm 1$.

Before symmetry breaking, the corresponding Lagrangian is CP invariant and invariant under the infinitely many horizontal groups
P_3^N (N = 6, 7, 8, ...).

   Eq. (6) gives in the limit h → 0 the simple mass relations

$$\frac{m_d}{m_u} = \frac{m_s}{m_c} = \frac{m_b}{m_t} \qquad (7)$$

with all generalized Cabibbo angles equal to zero. (7) is of course
not new but was suggested before by several authors[15-18]. In some
models the connection between the <u>deviations</u> from (7) and the mixing angles lead to the prediction of a relatively long lifetime
of B-mesons[16, 18]. In one of these models, namely in the proposal[18]
to be discussed in detail in section 4, eq. (6) is valid with the
specification $\varepsilon = -1$ and $M_D = M_D^+$. Here we state only that (6) is
compatible with our present knowledge about masses and mixings.
Thus, the symmetric group S_{n_G}, if properly generalized, is a good
candidate for a discrete horizontal symmetry at the scale m_W. Too
many parameters are, however, left free.

3. Mass Matrices in Grand Unified Theories

   In grand unified theories[19] such as SO(10) and E(6) all fermions of one generation are in an irreducible representation of the
group. This implies certain restrictions for the Yukawa coupling

constants. For instance, if the Higgs representation is symmetrically (antisymmetrically) coupled with respect to the gauge group indices, the corresponding coupling constants must form a symmetric (antisymmetric) matrix in generation space. Such rules are independent of the large energy scales involved in grand unified theories and could therefore be of a more general significance.

In SO(10) and E(6) models a charge conjugation operator (C) and a parity operator (P) can be defined[17]. They are conveniently chosen not to act on the generation index of the fermion fields. These operations transform the gauge fields but leave the gauge part of the Lagrangian invariant. Consequently, for suitable Higgs couplings, the total Lagrangian respects C and P invariance. With the scalar fields transforming as if composed of fermions, the Yukawa coupling matrices must be real matrices[20]. After spontaneous symmetry breaking C and P are in general no more conserved. Let us consider the Dirac mass terms which contain fermion anti-fermion Weyl spinor fields. These terms consist of several parts corresponding to the various Higgs fields ϕ^R which acquire vacuum expectation values. For a symmetric Higgs representation R the relevant part of the Lagrangian is even under C, and even under P if $\langle\phi^R\rangle$ is real, and odd under P if $\langle\phi^R\rangle$ is imaginary. For an antisymmetric Higgs representation one has a mass term which is formally odd under C, and odd under P if $\langle\phi^R\rangle$ is real, and even under P if $\langle\phi^R\rangle$ is imaginary[20].

The subgroup of SU(10) and E(6) we are used to consider is SU(5). Of importance is also the group $SU_Q(4)$ [21], the smallest unification group which combines the strictly conserved groups, the colour group and the electromagnetic group:

$$SU_Q(4) \supset SU_C(3) \times U_Q(1) \quad . \tag{8}$$

The group $SU_Q(4)$ is vector-like in the fermion representations of SO(10) and E(6), and thus allows for invariant mass terms. The down-quark fields together with the antilepton field form a "4", and the up-quark fields together with the anti-up-quark fields a "6".

In SO(10) and E(6) at most 3 Higgs fields are needed for the Yukawa interaction; in SO(10) the representation "10", "126", and "120". In table 1 I show the SU(5) and $SU_Q(4)$ transformation properties of those Higgs field components which contribute to the fermion masses. For the neutrinos (N) one has to list the contributions to the Dirac masses ($N_{\bar{\nu}\nu}$) and to the Majorana masses ($N_{\nu\nu}$ and $N_{\bar{\nu}\bar{\nu}}$) of $SU_L(2)$ doublet (ν) and singlet ($\bar{\nu}$) neutrinos. S stands for symmetric, AS for antisymmetric couplings.

Table 1   Higgs representations contributing to fermion masses

SO(10)	10_S	126_S	120_{AS}
$(SU(5), SU_Q(4))$			
up quarks	(5, 1)	(5, 1)	(45, 15)
down quarks and charged leptons	$(5^*, 1)$	$(45^*, 15)$	$(5^*, 1)$ $(45^*, 15)$
$N_{\bar{\nu}\nu}$	(5, 1)	(5, 1)	(5, 1)
$N_{\nu\nu}$	–	(15, 1)	–
$N_{\bar{\nu}\bar{\nu}}$	–	(1, 1)	–

It is seen from the table that the lowest Higgs representation, the "10_S" of SO(10) (or the 27_S of E(6)) gives a symmetric mass matrix in generation space. For a single Higgs field the fermion mass matrices in the corresponding column of table 1 are _proportional_ to each other. The phases of $<\phi^{10}>$ can be absorbed into the fermion fields. Thus, this contribution leads to a real C and P conserving mass term in the effective Lagrangian. If there were not

any additional contributions, one would obtain the simple mass formula (7), and all generalized Cabibbo angles would vanish. Table 1
suggests, however, that at least the "5" plets of SU(5) and the
singlet of SU(5) will give contributions. They are all singlets
with respect to $SU_Q(4)$. Assuming that the other components have negligible vacuum expectation values, one can conclude[18, 20]:

i) The up-quark mass matrix is a symmetric (C-conserving) matrix in
   generation space.

ii) The down-quark and charged lepton mass matrices have a symmetric
    and an antisymmetric (C-violating) part.

To the extent that the symmetric matrices from "10_S" and "126_S" are
proportional to each other, as one might expect from dynamically
generated Higgs couplings, further restrictions occur: The symmetric
parts of the down-quark and lepton matrices are proportional to the
up-quark mass matrix. Consequently the symmetric parts of all fermion mass terms are linearly related and conserve C and P.[x] The
phase factors multiplying the antisymmetric parts of the down-quark
and lepton mass matrices can - in general - not be absorbed any
longer. If they are purely imaginary, CP-violation is, in a sense,
<u>maximal</u>. There is then no P-violation in the Yukawa sector; the mass
matrices are hermitian and one deals with a manifest left-right symmetric fermion mass matrix.

4. <u>Suggestive Relations between Masses and Mixing Angles</u>

     The previous considerations suggest specific relations for the
fermion mass matrices. They have been stated (see refs. 18, 20) in
the form of two assumptions:

$$
\text{I} \quad M_U = S \qquad M_D = S + A
$$

(9)

$$
\text{II} \quad M_U = M_U^+ \qquad M_D = M_D^+ \quad .
$$

---

[x]The proportionality of the symmetric parts of the quark mass matrices also holds if the (5, 1) component of the "126" is suppressed
  due to an approximate Pati-Salam type SU(4) symmetry.

M_U and M_D are the mass matrices of up- and down-quarks, respectively. S denotes a symmetric, A an antisymmetric matrix, and α is a proportionality factor. These relations are supposed to hold simultaneously for an appropriate phase choice of the fermion fields. (The discussion of the leptons is deferred to the next section). Assumption I contains the statement that one deals in the quark sector with only two different matrices, a symmetric (S) and an antisymmetric (A) matrix, an assumption which is not implausible even without the context of grand unified theories. In a basis in which M_U = S is diagonal, one obtains

$$M_D = \alpha \, M_U^{\text{Diagonal}} + A \tag{10}$$

where the new matrix A is again antisymmetric.

According to assumption II, the eigenvalues of M_U and M_D are real - but not necessarily positive - and the matrix A consists of purely imaginary elements. This assumption is also suggestive. It implies "maximal CP-violation": The diagonal elements of M_D in (10) conserve CP, while the off-diagonal elements violate (formal) CP. Charge conjugation invariance is spontaneously broken, parity invariance is not broken by the Yukawa interaction. In the limit of equal masses of the right- and left-handed weak-interaction vector bosons, the entire effective Lagrangian will be symmetric with respect to space reflection.

It is remarkable that the recent proposal of a generalized permutation symmetry of the Yukawa interaction[14] is in accord with eq. (10), as can be seen by comparing it with (6).

The diagonalization of M_D

$$M_D = K \cdot M_D^{\text{Diagonal}} \cdot K^+ \tag{11}$$

defines the unitary matrix K, the Kobayashi-Maskawa matrix. A special parametrization of K corresponds to a new choice of fermion phases. Thus, some care is necessary in using (10) together with (11): For an arbitrary parametrization of K the matrix A will always be hermitian, and all diagonal elements must stay equal to zero, but its antisymmetry property is generally lost. The requirements on A, which replace the antisymmetry property and are independent of the phase choice, are stated in ref. 18, 20. For two generations there is no condition and for 3 generations simply one, namely

$$\det A = 0 \quad .\tag{12}$$

The only consequence of assumptions I and II for the case of <u>two</u> generations is a formula for the Cabibbo angle:

$$\text{tg}^2\theta_c = (-\frac{m_d}{m_s} + \frac{m_u}{m_c})/(1 - \frac{m_d}{m_s} \cdot \frac{m_u}{m_c}) \quad .\tag{13}$$

Several features of this formula are worth noting:

i) The two generations <u>decouple</u> in the formal limit m_s, $m_c \rightarrow \infty$.

ii) Decoupling also occurs in the limit in which the simple mass formula $m_d/m_s = m_u/m_c$ holds. It corresponds to $A \rightarrow 0$.

iii) By using the numerical values for the quark mass ratios, the sign choice $m_d/m_s = - |m_d/m_s|$ is necessary. A negative ratio of the eigenvalues of M_D is in fact required from the form of M_D in the limit $m_u \rightarrow 0$:

$$M_D \rightarrow \begin{pmatrix} 0 & A_{12} \\ A^*_{12} & \alpha m_c \end{pmatrix} \quad .\tag{14}$$

The determinant of this matrix is negative. The sign of m_u/m_c is not determined. If we take it positive, one finds numerically $\sin\theta_c \simeq 0.228$, in perfect agreement with the experimental value of the Cabibbo angle.

To state the consequences of assumptions I and II for <u>3 genera-tions</u>, a suitable parametrization of the mixing matrix is needed. The simplest mass–mixing angle relations will be obtained if one uses besides the Cabibbo angle θ one angle ß connecting the 3rd generation to the first, and another angle γ connecting the 3rd to the second generation. This corresponds to the Maiani form[22] of the mixing matrix:

$$
K_{UD} = \begin{pmatrix}
c_\beta c_\theta & s_\theta c_\beta & s_\beta e^{-i\delta'} \\
-s_\theta c_\gamma - s_\beta s_\gamma c_\theta e^{i\delta'} & c_\theta c_\gamma - s_\beta s_\gamma s_\theta e^{i\delta'} & s_\gamma c_\beta \\
-s_\beta c_\theta c_\gamma e^{i\delta'} + s_\gamma s_\theta & -s_\beta s_\theta c_\gamma e^{i\delta'} - s_\gamma c_\theta & c_\beta c_\gamma
\end{pmatrix} \qquad (15)
$$

$(c_\theta = \cos\theta,\ s_\theta = \sin\theta,\ s_\theta,\ s_\beta,\ s_\gamma \geq 0)$.

In (15) I have connected the CP-violating phase factor with the b → u matrix element:

$$
K_{ub} = s_\beta e^{-i\delta'}, \qquad K_{cb} = s_\gamma c_\beta \qquad . \qquad (16)
$$

For $s_\beta = 0$ all elements of K become real and CP-conservation is manifest.

The diagonal elements of (10) using (11) and (15) give two relations between the mixing angles and the quark masses[18, 20]. Taking for θ either its experimental value or the two generation formula (13), s_β turns out to be a very small quantity, which is <u>zero</u> in the limit $m_u = 0$ and $tg^2\theta = |m_d/m_s|$. The formula for γ is for small s_β independent of s_β , and thereby also independent of the CP-violating phase angle δ'. In a very good approximation one finds the relation

$$
tg^2\gamma \simeq -\frac{m_s + m_d}{m_b} + \frac{m_c}{m_t} \qquad . \qquad (17)
$$

This expression is the same as the one obtained from a two genera-tion model, this time for the second and third generation with $m_s + m_d$ instead of m_s. Thus the b → c matrix element s_γ is very

little affected from the existence of the lightest generation!
In (17) there is again a sign ambiguity. For the quark masses involved here we expect no sign change in the limit A → 0, in which
the simple mass formula (7) is valid. Thus

$$tg^2\gamma \simeq \pm \ (\ |\frac{m_s + m_d}{m_b}| - |\frac{m_c}{m_t}| \ ) \qquad . \tag{18}$$

For m_t large, the positive sign has to be taken. In the following
I will choose this high m_t solution, which is preferred as we will
see later on.

Since now the b → c amplitude γ is approximately known[2, 23],
eq. (18) can be used to obtain the mass of the top quark from the
other quark masses. Of course, all masses entering the mass formulas given here must be scaled to a common scale according to the
renormalization group prescription[24]. Unfortunately, the mass of
the strange quark is not well known[13]. In Fig. 1 I have plotted
the top quark mass ("pole mass") as a function of $m_s = m_s$ (1 GeV),
defined at a scale of 1 GeV and for s_γ = 0.053. The QCD-parameter
Λ_3, defined for 3 flavours, was taken to be 150 MeV. The arrows
indicate the variation of the top quark mass with upper values
for s_γ = 0.063, Λ_3 = 150 MeV, and lower values for s_γ = 0.043
and Λ_3 = 200 MeV. The value for m_s given by Gasser and Leutwyler[13]
is 175 MeV, with an error of about 30 %. This corresponds to top
quark masses between 26 and 60 GeV. Recently, Rubinstein[25]
obtained a relatively small value for m_s, namely 120±20 MeV.
According to Fig. 1, this would correspond to relatively high
values of the top quark mass, $m_{top} \simeq$ 50 GeV. In any case, Fig. 1
can be used as a stringent test of the model once the quark masses
are known.

It was mentioned before that the b → u matrix element s_β is
very small and even zero in the limiting case m_u = 0, $tg^2\theta_c = |m_d/m_s|$.
To get a hold on its value, one can use in (10), (11) for θ_c the two

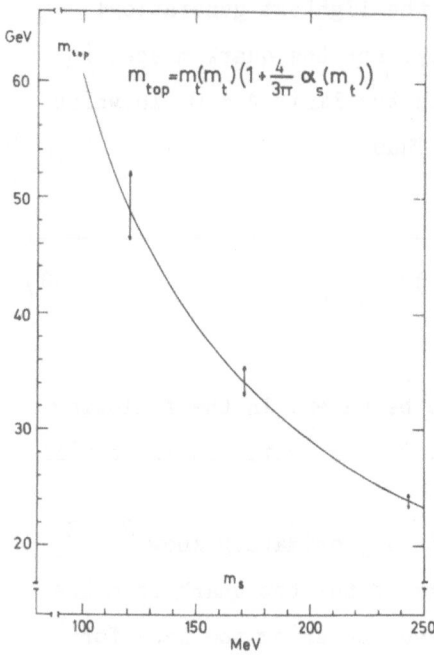

$$m_{top} = m(m_t)\left(1 + \frac{4}{3\pi}\alpha_s(m_t)\right)$$

Fig. 1

The top quark mass (pole mass) as a function of the strange quark mass m_s (1 GeV) for $\gamma = 0.053$, $m_c(m_c) = 1.27$ GeV, $m_b(m_b) = 4.36$ GeV, and $\Lambda_3 = 150$ MeV. The arrows indicate changes due to changes of γ and Λ_3 as described in the text.

generation formula (13). This is only possible if m_u/m_c is a positive number. One then finds

$$\text{tg}\,\beta \approx \sqrt{\frac{m_u}{m_c}}\,\text{tg}\,\gamma \qquad . \tag{19}$$

This formula gives only an estimate for β/γ, since it depends on the precise expression for the Cabibbo angle. It is notable, however, since it is independent of the top quark mass. From (19) I expect roughly[18]

$$|K_{ub}/K_{cb}| \approx 0.06 \quad \text{and} \quad |K_{ub}| \approx 0.003 \qquad . \tag{20}$$

These values lie below the present experimental upper limit[26, 27].

So far, only the diagonal elements of eq. (10) have been used to find mass-mixing angle relations. According to the model, the off-diagonal elements are also restricted. For 3 generations eq. (12) must hold. With masses and angles known so far, this condition can indeed be satisfied. Thereby the CP-violating phase δ' enters in an essential way. For $s_\beta/s_\gamma \gtrsim 0.06$ one obtains

$$\sin \delta' \gtrsim 0.99 \qquad (21)$$

i.e. the phase angle δ' is close to $\pi/2$!

Since the mixing angles θ, β, γ are small, for most purposes the mixing matrix (15) can be presented in the form

$$K = \begin{pmatrix} 1 & \theta & \hat{\beta} \\ -\theta & 1 & \gamma \\ -\hat{\beta}* + \theta\gamma & -\theta\hat{\beta}* - \gamma & 1 \end{pmatrix} \qquad (22)$$

with $\hat{\beta} = e^{-i\delta'}\beta$. The model gives

$$\theta^2 \simeq \left|\frac{m_d}{m_s}\right| + \left|\frac{m_u}{m_c}\right| \qquad\qquad \gamma^2 \simeq \left|\frac{m_s}{m_b}\right| - \left|\frac{m_c}{m_t}\right|$$

$$\beta \approx \sqrt{\frac{m_u}{m_c}}\, \gamma \qquad\qquad |\sin \delta'| \simeq 1 \qquad . \qquad (23)$$

An interesting quantity which can be calculated from these formulas is the ε-parameter in the K^o-$\bar{K}^o$-system. The box diagram formula[28] shows that $|\varepsilon_K|$ is roughly proportional to $\beta\gamma \sin \delta'$. For $\sin \delta' = 1$ and $\beta \simeq 0.06\gamma$, $|\varepsilon_K|$ is then proportional to γ^2. This was used to obtain a rough estimate of β and γ and thus of the B–meson life-time[18]. Today, with some knowledge of γ one can take a more detailed look on $|\varepsilon_K|$. In fig. 2 the box diagram value $|\varepsilon_K^{Box}|$ is plotted versus the top quark mass (pole mass). A range of values for γ is indicated. The ratio β/γ has been kept fixed (= 0.06) in spite of its only approximate validity. A comparison with the experimental value of $|\varepsilon_K|$ involves additional uncertainties. If long-range corrections are neglected, one has

$$|\varepsilon_K|^{Exp} \simeq \frac{B}{1.3} |\varepsilon_K^{Box}| \simeq 2.3 \cdot 10^{-3} \qquad (24)$$

where B is the so-called Bag factor[29]. It is seen that the result for the mixing angles (23) give the correct order of magnitude for $|\varepsilon_K|$. If one disregards the uncertainties mentioned and takes a B–

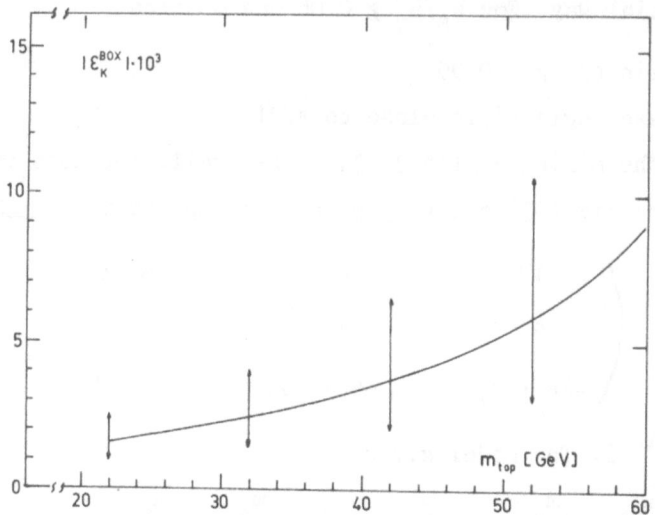

Fig. 2 $|\varepsilon_K^{Box}|$ as a function of m_{top} for $\gamma = 0.053\pm0.01$, $\beta/\gamma = 0.06$, and $\sin \delta' = 1$. The box diagram formula has been evaluated taking $m_{charm} = 1.4$ GeV, $m_W = 83$ GeV, and QCD correction factors corresponding to $\Lambda_{QCD} = 0.1$ GeV.

factor of $\simeq 1/2$, a relatively high mass of the top quark (≈ 50 GeV) fits best[30].

The model gives rather definite and testable predictions for the mixing angle and the CP-phase. Because of the limited knowledge on m_s and of the long-range contributions to ε_K, one obtains, however, for the top quark mass only a range of values.

5. Speculation on Neutrino Mixing

It is tempting to extend the model described in the last section to leptons. Of course, one is here on very speculative grounds. Grand unified theories suggest for the charged lepton matrix M_E the same structure as for M_D (see table 1). In analogy to (9) I will assume that

$$M_E = \alpha_E M_U + A_E , \qquad M_E = M_E^+ \tag{25}$$

where A_E is - with a proper choice of phases - again antisymmetric and hermitian. For 3 generations A_E satisfies, therefore,

$$\det A_E = 0 \quad . \tag{26}$$

The charged-lepton mixing matrix K_E, in the basis in which M_U is diagonal, is defined in analogy to (15) with angles θ_E, β_E, γ_E, and δ'_E. In the pure quark sector the first and 3rd generation almost decouple. This suggests that the charged lepton angles θ_E and γ_E can be obtained from appropriate two-generation models. Hence one has

$$\theta_E^2 \simeq -\frac{m_e}{m_\mu} + \frac{m_u}{m_c} = \left|\frac{m_e}{m_\mu}\right| + \left|\frac{m_u}{m_c}\right|$$

$$\tag{27}$$

$$\gamma_E^2 \simeq -\frac{m_\mu}{m_\tau} + \frac{m_c}{m_t} = \left|\frac{m_\mu}{m_\tau}\right| - \left|\frac{m_c}{m_t}\right|$$

i.e. $\theta_E \simeq 0.09$ and $\gamma_E \simeq 0.18$ for $m_t(m_t) \simeq 40$ GeV.

Here it is evident that sign (m_μ/m_τ) = sign (m_c/m_t) has to be taken <u>negative</u>, since $m_c/m_t \lesssim 0.04$ for $m_t(m_t) \gtrsim 22$ GeV. Thus, the high-mass solution for m_t is the only possible one. Eq. (25-27) now give

$$\beta_E \simeq \sqrt{\frac{m_u}{m_c}} \ \gamma_E \simeq 0.01 \quad \text{and} \quad |\sin \delta'_E| \simeq 0.96 \quad . \tag{28}$$

It is interesting to note that the matrix A_E computed with the help of these mixing angles is roughly equal to the matrix A of eq. (10), with the exception of the smallest element $(A_E)_{12}$.

For the neutrino mass matrix table 1 has to be consulted again. $M_{\nu\bar{\nu}}^{--}$ is symmetric and huge, $M_{\nu\nu}$ is symmetric and very tiny or zero, and $M_{\nu\bar{\nu}}^{-}$, the Dirac-neutrino mass matrix, has symmetric as well as antisymmetric parts and is expected to be of the order of the charged-lepton mass matrix. By transforming the $2n_G$ x $2n_G$ neutrino mass matrix

$$\begin{pmatrix} M_{\nu\nu} & M_{\nu\bar{\nu}}^{-} \\ M_{\nu\nu}^{-} & M_{\nu\nu}^{--} \end{pmatrix}$$

to a block diagonal form, one obtains n_G light neutrinos. Their mass matrix is of the Majorana type and, of course, symmetric:

$$M = M_{\nu\nu} - M_{\nu\bar\nu} \cdot (M_{\bar\nu\bar\nu})^{-1} M_{\bar\nu\nu} \qquad (29)$$

Now, once again the bold assumption is made that only one type of symmetric Yukawa coupling matrix occurs:

$$M_{\nu\nu} = \alpha_{\nu\nu} M_U \qquad\qquad \alpha_{\nu\nu} \ll 1$$

$$M_{\bar\nu\bar\nu} = \alpha_{\bar\nu\bar\nu} M_U \qquad\qquad \alpha_{\bar\nu\bar\nu} \gg 1 \qquad (30)$$

$$M_{\nu\bar\nu} = \alpha_\nu M_U + A_\nu$$

with $A_\nu = A_\nu^+ = - A_\nu^T$, α_ν real. The light neutrino mass matrix (29) is then given by

$$M = (\alpha_{\nu\nu} - \frac{\alpha_\nu^2}{\alpha_{\bar\nu\bar\nu}}) M_U + \frac{1}{\alpha_{\bar\nu\bar\nu}} A_\nu \cdot M_U^{-1} \cdot A_\nu \qquad . \qquad (31)$$

It can be diagonalized by an orthogonal transformation

$$M = K_\nu \cdot M^{diag} \cdot \cdot K_\nu^T \qquad . \qquad (32)$$

The two-component neutrino fields ν_i coupled to the charged leptons are related to the two-component fields $\tilde\nu_i$ composed of mass eigenstates by the equation

$$\nu_i = (K_E^+ K_\nu)_{ik} \tilde\nu_k \qquad . \qquad (33)$$

It is seen that the neutrino mixing matrix $K_E^+ K_\nu$ is in general complex, since K_E has complex elements. Hence CP-violation in the lepton sector can be expected. Since the matrix A_ν should approximately be equal to the matrix A_E, apart from a constant factor, the matrix form of the second term in (31) is known. In this approximation the neutrino mixing angles and the ratio of neutrino masses depend solely on the unknown relative strength of the second to the first term in (31).

A very simple result follows, if the second term in (31) can be neglected. In this case the neutrino masses obey the simple rule

$$m_{\nu_e} : m_{\nu_\mu} : m_{\nu_\tau} = m_u : m_c : m_t$$

and the neutrino mixing angles can be obtained from K_E^+. Numerically one gets

$$\nu_e \leftrightarrow \nu_\mu \qquad \sin^2 2\alpha_{e\mu} \simeq 0.03$$

$$\nu_e \leftrightarrow \nu_\tau \qquad \sin^2 2\alpha_{e\tau} \simeq 0.001$$

$$\nu_\mu \leftrightarrow \nu_\tau \qquad \sin^2 2\alpha_{\mu\tau} \simeq 0.12 \quad .$$

6. Conclusions

Simple relations between fermion masses and mixing angles have been presented. These relations are independent of the details of the Higgs sector of the Lagrangian. Heavy generations decouple from light ones in the high-mass limit. The fermion mass matrix is in accord with grand unification schemes and with the consequences of a set of horizontal symmetries[14]. Unfortunately, only two relations can be derived for the three mixing angles in the 3-generation case. However, by considering the two-generation subsystems separately, all angles can be expressed by mass ratios. The formula for the Cabibbo angle is independent of the top quark mass and in agreement with the experimental value. The B-meson lifetime and the magnitude of ε_K are strongly correlated and gave a first qualitative test of the relations between masses and mixing angles[18]. The $b \to u/b \to c$ amplitude ratio is predicted to be about 0.06, which is a factor 1/2 below the presently known upper limit[26]. If this ratio would turn out to be much smaller, the calculated value of ε_K would become too small. Another important prediction is $\sin \delta' \simeq 1$, which is, however, hard to test. The best tests appear to be measurements of decay channels common to B- and $\bar{B}$-decays[31]. Both predictions, $\beta/\gamma \approx 0.06$ and $\sin \delta' \simeq 1$, are independent of the top quark mass. The definite relation between the top quark mass and the strange quark mass provides another stringent test of the model. The remarks made about the lepton sector are more speculative, since the connection between quark and lepton masses is not yet understood.

If the simple relations between masses and mixing angles are further supported by the data, they might be helpful for attempts to solve the generation problem.

It is a pleasure to thank W. Bernreuther and L. Wolfenstein for valuable discussions.

REFERENCES

1. For a review see: L. L. Chau, Phys. Rep. 95, 2 (1983).
2. E. Fernandez et al., Phys. Rev. Lett. 51, 1022 (1983);
    N. S. Lockyer et al., Phys. Rev. Lett. 51, 1316 (1983).
3. R. Cahn and H. Harari, Nucl. Phys. B176, 135 (1980).
4. F. Wilczek, Phys. Rev. Lett. 49, 1549 (1982).
5. A. Davidson, V. P. Nair, and K. C. Wali,
    Syracuse preprints SU-4217-257, SU-4222-265 (1983).
6. R. Peccei and H. Quinn, Phys. Rev. Lett. 38, 1440 (1979).
7. J. Kim, Phys. Rev. Lett. 43, 103 (1979);
    M. Dine, W. Fishler, and M. Srednicki,
    Phys. Lett. 104B, 199 (1981)
8. H. Fritzsch, Nucl. Phys. B 155, 189 (1979).
9. V. P. Nair, L. Michel, and K. C. Wali,
    Bures-sur-Yvette preprint IHES/P/83/77;
    K. C. Wali, these proceedings.
10. R. Barbieri, R. Gatto, and F. Strocchi,
    Phys. Lett. 74B, 344 (1978).
    D. Wyler, Phys. Rev. D19, 330 (1979).
11. S. Weinberg, N. Y. Academ. Sci. 38 (1977);
    H. Fritzsch, Phys. Lett. 70B, 436 (1977).
12. G. Ecker and W. Konetschny, Phys. Lett. 91B, 225 (1980).
13. J. Gasser and H. Leutwyler, Phys. Rep. 87C, 77 (1982).
14. G. Ecker, Wien preprint UWThPh-1984-3, to be published in
                                                Z. Phys. C.

15. S. Pakvasa and H. Sugawara, Phys. Lett. 82B, 105 (1979);

    D. Grosser, Phys. Lett. 83B, 355 (1979).

16. G. Segre, H. A. Weldon, and J. Weyers,

    Phys. Lett. 83B, 351 (1979);

    A. Davidson, Phys. Lett. 122B, 412 (1983).

17. R. Barbieri and D. V. Nanopoulos, Phys. Lett. 91B, 369 (1980);

    B. Stech, in: "Unification of Fundamental Particle Inter-
    actions", Ed.: S. Ferrara et al., Plenum Press, N.Y. 1980,
    p. 23.

18. B. Stech, Phys. Lett. 130B, 189 (1983).

19. For a review see: J. Ellis, Gauge Theories in High Energy
    Physics, Les Houches 1981, North Holl. Publ. 1983, p. 160.

20. B. Stech, Univ. Heidelberg preprint HD-THEP-83-31, to be
    published in the Proc. Ad. Summer Inst. Munich (1983),
    Plenum Press.

21. M. Abud, H. Ruegg, C. A. Savoy, and F. Bucella,
    Lett. of N. Cim. 19, 494 (1977).

    Y. Achiman and B. Stech, Phys. Lett. 77B, 105 (1979).

22. L. Maiani, Proc. 1977 Sympos. on Lepton and Photon Inter-
    actions, Hamburg.

23. J. Lee-Franzini, these proceedings.

24. O. Nachtmann and W. Wetzel, Nucl. Phys. B187, 333 (1981).

25. H. Rubinstein, private communication.

26. CLEO Collaboration, J. Chauveau, talk at the 7th Intern. Conf.
    on Experimental Meson Spectroscopy, Brookhaven (1983).

    CUSP Collaboration, C. Klopfenstein et al.,

    Phys. Lett. 130B, 444 (1983);

    J. Lee-Franzini, these proceedings.

27. K. Kleinknecht and B. Renk, Phys. Lett. 130B, 459 (1983).

28. T. Inami and C. S. Liu, Progr. Theor. Phys. 65, 297 (1981);

    for recent applications see

    E. A. Paschos, B. Stech, and U. Türke,

    Phys. Lett. 128B, 240 (1983).

A. J. Buras, W. Slominski, and H. Steeger,
Munich preprint MPI-PAE/PTh 77/83.

29. For a recent estimate of B see
J. F. Donoghue, E. Golowich, and R. Holstein,
Phys. Lett. 119B, 412 (1982).

30. P. H. Ginsparg, S. L. Glashow, and M. B. Wise,
Phys. Rev. Lett. 50, 1415 (1983).

31. S. Barshay and J. Geris, Phys. Lett. 84B, 319 (1979);
Ling Lie Chau and Hai-Yang Cheng, BNL preprint (1984).

DISCUSSION

FRITZSCH:

You said, the b → u transition vanishes on m_u → 0. Is this an exact result?

STECH:

If the formula $fg^2\theta_c = |m_d/m_s|$ holds in this case the answer is yes; if it does not hold the answer is no.

SERDAROGLU:

A comment on the 17th neutral weak singlet you mentioned.

STECH:

If SO(1) model is enlarged to E_6 gauge theory, the fundamental representation 27 of E_6 contains 16, 10, of SO(10) and also a weak singlet 1 which is neutral so that the 17th particle you mentioned occurs there naturally. Moreover in an E_6 theory Yukawa coupling in the three graph level may be zero for Majorana ν_R but may be generated by a "Witten diagram" in two loop level.

Note added to

THEORETICAL STATUS OF FLAVOUR MIXING MATRIX

by Berthold Stech, Institut für Theoretische Physik der

Universität Heidelberg, D-6900 Heidelberg

The UA1 collaboration has recently announced[32] the discovery of the top quark. The reported mass (30 to 50 GeV) lies
in the predicted range[18]. According to Fig. 1 the strange quark
mass (defined at the scale of 1 GeV) should then have a value of
about 150 MeV, a very reasonable value. From Fig. 2 it follows
that the effective B-factor as defined by eq. (24) should be
around 1, with large errors, however. For the especially interesting quantity

$$\varepsilon'/\varepsilon \simeq \text{Re } \varepsilon'/\varepsilon \simeq \frac{1}{6} \left(\left| \frac{\eta_{+-}}{\eta_{oo}} \right|^2 - 1 \right)$$

one has the formula

$$\varepsilon'/\varepsilon \simeq - \frac{\text{Im } A_o}{A_2} \frac{1}{\sqrt{2}} \left| \frac{A_2}{A_o} \right|^2 \frac{1}{|\varepsilon|} \tag{34}$$

For $m_t \simeq 40$ GeV the coefficient of the relevant Penguin operator
contained in Im A^o has been estimated[33] to be close to -0.1. The
matrix element of the Penguin operator can be calculated in a
straightforward way in the factorization approximation. Werner
Bernreuther and I repeated this calculation. The result depends
on f_K/f_π and the $\pi\pi$ and πK scalar form factors. (At the relevant
momentum transfers we used the values 1.23, 1.16, and 1.01, respectively.) It also depends on $M = m_\pi^2/(m_u + m_d) \simeq m_K^2/(m_s + m_d)$,
which we took to be 1.4 GeV. We found (using the experimental
values for $|A_o|$, $|A_2|$, and ε, and sign A_2 as obtained by the
factorization approach):

$$\varepsilon'_{\text{fact}}/\varepsilon \simeq 2.0 \frac{s_\beta s_\gamma}{s_\theta} \sin \delta' \tag{35}$$

Unfortunately, the factorization assumption for non-leptonic K-decay amplitudes is hard to justify. It gives a reasonable result for A_2 but not for Re A_0. Thus (35) can be considered only as an order of magnitude estimate of ε'/ε . With $s_\gamma \simeq 0.05$ and $s_\beta/s_\gamma = 0.06$ and $|\sin \delta'| = 1$ as suggested here eq. (35) gives

$$\varepsilon'_{fact.}/\varepsilon \simeq \text{sign} (\sin \delta') \, 1.3 \, 10^{-3} \tag{36}$$

Recent experiments[34] on Re ε'/ε by the Chicago-Saclay collaboration and a Yale-Brookhaven group are very precise, but still cannot confirm or rule out the small value for ε'/ε estimated here.

In a different approach, using the _theoretical_ expression for ε, Gilman and Hagelin[35] obtained a _lower_ bound for $\dot{\varepsilon}'/\varepsilon$, somewhat in conflict with the results of the quoted experiments. However, as pointed out by L.-L. Chau and collaborators[36], this bound depends crucially on the matrix element of the Penguin operator, which was presumably overestimated by previous authors, and on the B-factor. There is certainly not yet any difficulty for the standard model as long as an effective B-factor $B \approx 1$ and Im A_0 as estimated here are tenable.

References

32. C. Rubbia, talk given at the XIth International conference on Neutrino Physics, Nordkirchen, Dortmund, June 1984.

33. F. J. Gilman and M. B. Wise, Phys. Rev. D20, 2392 (1979).

34. K. Nishikawa, talk given at the XXIInd International Conference on High Energy Physics, Leipzig, July 1984.

35. F. J. Gilman and J. S. Hagelin, Phys. Lett. 126B, 111 (1983) J. S. Hagelin, MIU-THP-84/010.

36. L.-L. Chau, talk at APS Washington meeting, April 1984. L.-L. Chau, H.-Y. Cheng, W.-Y. Keung, "The elusive Penguin", BNL preprint July 1984.

CONCLUDING REMARKS AND OUTLOOK

Ling-Lie Chau

Department of Physics
Brookhaven National Laboratory
Upton, New York

The measurements of the b lifetime, $\tau_b \sim 10^{-12}$ sec, have greatly constrained the quark mixing matrix of Kobayashi and Maskawa (K-M). The communication between the second and the third generations of quark states is greatly suppressed compared to that between the first and the second generations, (the suppression is about 5 times in amplitudes more severe than that of Cabibbo). Through unitarity, the rest of the mixing matrix is relatively well determined, including the result that the 2×2 GIM matrix is unitary to a good approximation, (with a deviation less than 10^{-2}). To improve our knowledge of the mixing matrix, we need a more precise value of V_{ub}. What we have now is $10^{-3} < \left|V_{ub}\right| < 10^{-2}$, with the lower bound from the CP-violation parameter ε, and the upper bound from the measurements of the b lifetime and its semileptonic decays. One of the cleanest ways to measure this number is to measure $B_u^{\pm} \rightarrow \tau^{\pm} \overset{\leftarrow}{\nu}_{\tau}$ (here $\overset{\leftarrow}{\nu}$ denotes neutrino, and $\overset{\rightarrow}{\nu}$ denotes antineutrino, a notation invented by M. Goldhaber at this conference). Unfortunately its branching ratio is very small $10^{-6} < Br(B_u \rightarrow \tau \nu_{\tau}) < 10^{-4}$, its precise value depending on the value of $\left|V_{ub}\right|$. I propose this as a challenge to our experimental colleagues.

With this knowledge of the mixing matrix, and the coming precision measurements of multi-channel decays of the charm particles and the beauty particles, we are beginning to be able to analyze the dynamics of nonleptonic decays more systematically. Twenty eight years after the observation of $\Gamma(K^+ \rightarrow \pi^+\pi^0)/\Gamma(K_S^0 \rightarrow \pi^+\pi^-) \approx 1/670$, we are still not sure what is the source for this $\Delta I = 1/2$ rule. The studies of nonleptonic decays in the K and Λ systems are very limited by the decay channels available. Hopefully, by systematically analyzing the many decays of charm and beauty particles we can eventually understand the dynamics of nonleptonic decays.

This is the 20th anniversary of the observation of the CP violation reaction $K_L \rightarrow 2\pi$, which is still the only kind. Observation of CP violation in any other systems will certainly be very interesting. The K-M mechanism does not predict many observable CP violation effects. From our present estimates, the K-M model can produce a $\varepsilon'/\varepsilon \neq 0$, depending upon how important is the so called "Penguin" diagram. As we have heard in this conference, $\left|\varepsilon'/\varepsilon\right|$ may be measured to the 0.002 level by 1986.

There are also CP violation effects in the partial-decay-rate differences of a particle and its antiparticle. CPT dictates that a particle and its antiparticle must have the same decay width; however, their partial decay rates in the CP conjugated states can be different due to CP violation. LEAR is excellently equipped to do such measurements for the K and the hyperon systems. Preliminary estimates, based on our current crude knowledge of how to calculate nonleptonic decays, indicate that percentage partial-decay-rate differences, while extremely small in the K and the hyperon systems, namely $O(10^{-5}-10^{-6})$, can be tens of percent in the B meson decays, though the branching ratios are small: $10^{-4} \sim 10^{-5}$. This indicates that a million B mesons are needed to search for such CP violation effects. The partial decay rate differences in charm meson decays, on the other hand, can be only $10^{-2} - 10^{-3}$ in some optimistic cases, though the branching ratios are relatively larger than for the B exclusive decays. These estimates call for about $10^8 - 10^9$ charm mesons. So, for the next generations of CP-violation search experiments, we need one hundred million charm, one million beauty, (but only one truth!).

Now let us list what is going to happen in the next few years on the topics we discussed at this conference (the following is what we worked out on the blackboard).

1984:

- set bound on $n_\nu \leq 10$, (UA1, UA2, CERN);
- ε'/ε, sensitivity 0.005, (FNAL Exp. #617, BNL Exp. #749);
- high statistics measurements of charm decays, direct measurements of $|V_{cs}|$, $|V_{cd}|$, (MARK III, SLAC);
- g_1/f_1 from $\Sigma^{\mp}_{\text{polarized}} \to n \ e^{\mp} \overset{\leftrightarrow}{\nu}_e$, (FNAL Exp. #715).

1985-86:

- Better measurements of the charm and the beauty lifetimes, (SLAC, CESR, DESY, FNAL, CERN),
- t quark discovered, if $m_t < 70$ GeV, (UA1, UA2, CERN); (if $m_t < 35$ GeV, then Tristan, KEK, will sit on a gold mine of the truth);
- set bound on $B^0 - \bar{B}^0$ mixing $< 50\%$, (CESR);
- measure m_ν to < 5 eV, (ITEP; Tokyo − Oshima);
- measure $\tau_P \to K\nu_\mu > 10^{33}$ years, (IMB);
- Observation of compositeness of W and Z?
- Observation of supersymmetric particles?

1986-87:

- d^e_n to 10^{-26} e-cm., (Grenoble, Gatchina);
- ε'/ε to < 0.002, (FNAL Exp. #631; CERN);
- $\Delta_{K^0 \to 2\pi} \overset{?}{=} 4 \ \text{Re} \ \varepsilon'$, to < 0.002, (LEAR, CERN);
- $K^+ \to \pi^+ \mu e$, sensitivity 10^{-12}, (BNL Exp. #777);

. $K_L \to \mu \bar{e}$, sensitivity 10^{-10}, (BNL Exp. #780);

. $K_L \to \pi$ + "nothing", sensitivity 10^{-10}, (BNL Exp. #787);

. SLC turns on;

. Fermilab $\bar{p}p$ collider turns on;

• Tristan turns on.

1987-88:

. $n_\nu < 4$, (UA1, UA2, SLC, CERN);

. observation of the tri-gauge boson coupling via
$\bar{p}p \to W^+W^-X$, γWX, (Fermilab $\bar{p}p$ collider);

. Higgs boson found, if $m_H \ll 125$ GeV (as bounded by some
elegant theoretical arguments presented at this
conference);

. Beijing e^+e^- collider, BEPC, turns on.

1988-89:

. LEP turns on;

. $B^0-\bar{B}^0$ mixing to 2-10%, (CESR II, Cornell; Argus, DESY);

. $B \to \tau \bar{\nu}_\tau$ observed, $|V_{ub}|$ measured, (CESR II; Argus);

. Δ_{B_u} measured, (CESR II; Argus);

. high precision measurement of charm decays, (BEPC,
Beijing).

. HERA turns on.

We can see that our prospects for the next few years are quite
interesting.

In two years we will observe the 90th anniversary of the
discovery of radioactivity by Becquerel. Our modern history of
studies on weak interactions started with Pauli's proposal of ν
in 1930 and Fermi's construction of Fermi's interaction in 1933.
It has been a very fascinating history. At first the discoveries
of particles were made mainly in cosmic rays, [e^+, 1932; $\mu^\pm$,
1937; $\pi^\pm$, 1947; Λ, Σ, Ξ, $K^\pm$, K^0, 1947 $\cdots$]. Associated
productions were discovered both in cosmic rays and at
accelerators (Cosmotron, BNL, 1955). This marked the the
beginning of the era of doing particle physics using accelerators. Then came the population explosion of the hadrons in the

early sixties. That led to the idea that hadrons are not elementary, and the acceptance of quarks, (層子 particles of layers in Chinese), as constituents of hadrons.

Led by Pauli and Fermi, theorists have made startling advances either directly stimulated by experimental results or based upon the sheer logic and esthetics of reasoning: the questioning of parity conservation, 1956; the unification of electromagnetic and weak interactions; the proposition of the charm and the beauty particles in the seventies. Even more fascinating is that not only do the quark states repeat, but even the history of the discoveries of new quark states, charm and beauty, have been repeating. The recent observation of $W^{\pm}$ and Z at the predicted masses (even with predicted cross sections) is a feat matched only by the observation of the antiproton, as predicted by Dirac's theory.

After these recent milestones, we are again at a point of puzzlement, in this ever on-going process of making discoveries and furthering understandings. Where are the truth particles? Will they be discovered in the same way as the charm and the beauty particles? What is the mass-generating mechanism for the gauge bosons? How many Z^0's and W's are there? How about the W_R's? I tend to believe the "principle of insufficiency" in physics, i.e. if not inhibited by some fundamental principle, a physical phenomenon should happen. The recent discovery of neutral current (1973) is such an example. So I do agree with the composite-model advocates that there should be more W's and Z's, and preons or haplons, etc. But we must make an effort to become capable to calculate and make predictions about where these W's and Z's are, not just "live in the experimental error bars" of the standard model. Do the neutrinos have mass? An important question to ask is if a zero mass fermion exists, what is the fundamental principle for it? The neutrinos were shown not to be supersymmetric fermions [W.A. Bardeen 1975 (unpublish-

ed); D.Z. Freedman and B. de Wit, Phys. Rev. Lett. 35, 827
(1975), W.A. Bardeen, T.R. Taylor and C.K. Zachos, Nucl. Phys.
B231, 235 (1984)]. Unless if we can find some other fundamental
reasons, it is natural for neutrinos to have mass. The proton
probably does decay, (Lee and Yang, could not find a good reason
for it not to decay, Phys. Rev. 98, 1501 (1955)). But how long
does it live? Not only is there no good reason that magnetic
monopoles should not exist, it is beautiful that they do. But
should we repeat ourselves in theory constructions as in constructing the grand unification theories following the same steps
as the electroweak unification? Supersymmetry is surely mathe-
matically and esthetically beautiful. As we have heard, it is
even rescuing some difficulties: the hierarchy problem, the horizon problem, the problem of how to keep the strong CP violation
extremely small (as required by the stringent $d_n^e < 10^{-25}$ e-cm).
However it will take years of theoretical and experimental work
to find if it has a place in physics.

As for doing physics with accelerators - what is beyond the
$\bar{p}p$ colliders, HERA, and LEP? We can not just linearly extrapo-
late in getting higher energies and in the size of experiments.
A technological revolution is called for in building accelerators
and detectors. The many schools and workshops on accelerator
physics which have recently sprung up is an encouraging sign.

I thank you for your interesting talks and enthusiastic
participation. You make me feel life as a physicist is
beautiful. Not only the subject is fascinating, but so are our
fellow physicists. National boundaries disappear before our
common interest in trying to understand the world around us.
This fact is attested to by the many countries represented at
this conference. I thank you all for coming.

Finally, stimulated by the beautiful Chinese poem which
Professor Terazawa presented to us at the end of his talk, I

would like to share with you a few words by Lao Tze (~ 600 BC,
Confucius' contemporary):

生而不有
为而不恃
功成而弗居
是唯弗居，是以不去

giving birth, yet not possessing,
working, yet not taking credit,
work done, then left alone,
in this way, it will last forever.

EDITOR'S NOTES

Our field is very fast moving, and in a very exciting period. Since our conference in March, there have been many new results. I would like to note a few that are particularly relevant to the topics of our conference:

- The discovery of the t quark has been announced by[1] UA1, CERN; 5 events with an electron plus two or more jets and 5 events with a muon plus two or more jets which can be interpreted as $W \to t\bar{b}$, with M_t = 30 to 50 GeV.

- The b lifetime measurements[2]:

$$\tau_b = \{1.6 \pm 0.4 \pm 0.4\} \times 10^{-12} \text{ sec, MAC, SLAC}$$
$$\text{(st.) (sys.)}$$

$$= \{0.85 \pm 0.17 \pm 0.21\} \times 10^{-12} \text{ sec, MARK II, SLAC}$$

$$\{1.25 \pm {}^{0.26}_{0.19} \pm 0.50\} \times 10^{-12} \text{ sec.}$$

$$= \{1.16 {}^{+ 0.37}_{- 0.34} \pm 0.23\} \times 10^{-12} \text{ sec, DELCO, SLAC}$$

$$= \{1.8 {}^{+ 0.5}_{- 0.3} \pm 0.35\} \times 10^{-12} \text{ sec, JADE, PETRA}$$

$$= \{1.79 \pm 0.38 {}^{+ 0.55}_{- 0.40}\} \times 10^{-12} \text{ sec, TASSO, PETRA.}$$

- $\varepsilon'/\varepsilon = -0.0046 \pm 0.0053 \pm 0.0023$ Chicago-Stanford-
                  (St.)   (Sys.),   Saclay Collabora-
                                           tion, FNAL[3]

      $= + 0.0045 \pm 0.0055 \pm 0.008$,   Brookhaven-Yale
                                             Collaboration, BNL

- Charm nonleptonic decays measured by MARK III, SLAC[4]:

$$Br(D^0 \to K^- \pi^+) = (4.9 \pm 0.9 \pm 0.5)\%,$$

$$Br(D^+ \to K^- \pi^+ \pi^+) = (9.1 \pm 1.5 \pm 0.9)\%,$$

	Decay Mode	# Events	Branching ratio (%)
D^0	$\pi^-\pi^+/K^-\pi^+$	33 ± 9	$3.8 \pm 1.0\pm0.5$
	$K^-K^+/K^-\pi^+$	75 ± 10	$12.5\pm 1.8\pm1.0$
	$\bar{K}^0\pi^0/K^-\pi^+$	68 ± 11	$35\pm7\pm7$
D^+	$\pi^0\pi^+/\bar{K}^0\pi^+$		<30 (90% c.ℓ.)
	$\bar{K}^0K^+/\bar{K}^0\pi^+$	28.8 ± 6.4	$29.4\pm7.4\pm5.1$
	$\pi^-\pi^+\pi^+/K^-\pi^+\pi^+$	70.0 ± 18.7	$5.9 \pm 1.6\pm1.0$
	$K^-K^+\pi^+/K^-\pi^+\pi^+$	60.5 ± 13.0	$7.2 \pm 2.4\pm1.5$

For an update of theoretical understandings on quark mixing, heavy quark decays, and CP violations, see Ref. 5.

1. C. Rubbia, talks given at the XIth International Conference on Neutrino Physics, Nordkirchen, Dortmund, June 1984; and DPF Meeting of American Physical Society at Snowmass, July 1984.

2. The τ_b from MAC quoted here is the same as given by W.T. Ford at this conference.
    For other results see review talks by J. Jaros and M. Davier at SLAC Summer Institute, July 23 - August 3, 1984.

3. B.D. Winstein, talk given at the American Physical Society 1984 Spring Meeting, Washington, D.C., April 23-26, 1984.
    K. Nishikawa, talk given at the XXIInd. Int. Conf. on High Energy Physics, Leipzig, July 1984.

4. R. Schindler, talk given at the XXIInd Int. Conf. on High Energy Physics, Leipzig, July 1984, Caltech preprint Caltech-68-1161.
    D. Hitlin, talk given at the SLAC Summer Institute, July 23 - August 3, 1984.

5. L.L. Chau, talks given at the 6th Int. Conf., Vanderbilt Univ., April 5-7, 1984; and the American Physical Society 1984 Spring Meeting, Washinton, D.C., April 23-26, 1984.

PARTICIPANTS

AVERY, Paul R.

Wilson Lababoratory
Cornell University
Ithaca, New York 14853

BARBARO-GALTIERI, Angela

Lawrence Berkeley Lab.
University of California
Berkeley, California 94720

BECKER, Lutz

DESY
Notkestrasse 85
D-2000 Hamburg 52, Germany

BÉG, Mirza A.B.

The Rockefeller University
1230 York Avenue
New York, New York 10021

BELLINI, Gianpaolo

University of Milano
via Celoria 16 - Milano, Italy

BOEHM, Felix H.

California Institute of Technology
Pasadena, California 91125

BOTELLA, Francisco J.

Dept. of Theoretical Physics
Physics Faculty
Burjasot (Valencia) Spain

CABIBBO, Nicola

National Inst. for Nuclear Physics
Piazza del Capressari 70
00187 Rome, Italy

CESTER, Rosanna

CERN-EP Division
CH1211 Geneva 23 Switzerland

CHAU, Ling-Lie

Brookhaven National Laboratory
Upton, New York 11973

CHOU, Kuang-Chao	Inst. for Theoretical Physics Academia Sinica, P.O. Box 2735 Beijing, China
FERRARA, Sergio	CERN-TH Division CH-1211 Geneva 23 Switzerland
FORD, William T.	University of Colorado Boulder, Colorado 80309
FRANCO, Enrico	Univ. Roma, Piazzale Aldo Moro 2 00185 - Rome, Italy
FRANZINI, Paolo	Columbia Univ., N.Y., NY 10027; Cornell University Ithaca, New York 14853
FRITZSCH, Harald	Max-Planck-Institut für Physik Fohringer Ring 6 8 München 40, Germany
GENTILE, Simonetta	Universita degli Studi di Roma Piazzale Aldo Moro 2 00185 - Rome, Italy
GOLDHABER, Maurice	Brookhaven National Laboratory Upton, New York 11973
HAGELIN, John S.	Maharishi International Univ. Fairfield, Iowa 52256
HANSEN, Peter H.	The Niels Bohr Inst. Blegdamsvej 17 DK-2100 Copenhagen, Denmark; and CERN-DP-Division CH-1211 Geneva 23, Switzerland
HECKEL, Blayne	University of Washington Seattle, Washington 98103
HITLIN, David	California Institute of Technology Pasadena, California 91125
HO, Tso-Hsiu	Inst. for Theoretical Physics Academia Sinica, P.O. Box 2735 Beijing, China
KANE, Gordon L.	University of Michigan Ann Arbor, Michigan 48109

·KLEINKNECHT, Konrad

Universität Dortmund
Postf. 500500
46 Dortmund 50, W. Germany;
CERN-EP Division
CH-1211 Geneva 23, Switzerland

KOBAYASHI, Makoto

KEK - Theory Division
Oho-Machi, Tsukuba-Gun
Ibaraki-Ken-305, Japan

LANCERI, Livio

CERN
EP-Division
CH-1211 Geneva 23, Switzerland

LE COMTE, Pierre

ETH Zurich
Laboratorium für Hochenergiephysik
5234 Villigen, Switzerland; and
CERN-EP Division
CH-1211 Geneva 23, Switzerland

LEE-FRANZINI, Juliet

State University of New York at
Stony Brook, New York 11794; and
Wilson Lab. - Cornell University
Ithaca, New York 14853

LITTENBERG, Laurence S.

Brookhaven National Laboratory
Upton, New York 11973

MACHACEK, Marie E.

Northeastern University
360 Huntington Avenue
Boston, Massachusetts 02115

MOHAPATRA, Rabindra N.

University of Maryland
College Park, Maryland 20742

MORSE, William M.

Brookhaven National Laboratory
Upton, New York 11973

NANOPOULOS, Dimetrius V.

CERN-TH Division
CH-1211 Geneva 23, Switzerland

NIU, Kiyoshi

Nagoya University
Nagoya 464, Japan

OLNESS, Fredrick I.

University of Wisconsin
Madison, Wisconsin 53706

PAVLOPOULOS, P.

CERN-EP Division
CH-1211 Geneva 23, Switzerland

PHAM, Yem-Xuan

Université P.M. Curie
Tour 16, 1er étage, 4 Pl. Jussieu
75230 Paris Cedex 05, France

POHL III, Martin

Inst. f. Hockenergiephysik
Eidgen Technische-Hochschule
CH-8093 Zürich, Switzerland

POULARD, Gilbert

CERN-EP Division
CH-1211 Geneva 23, Switzerland

RAJASEKARAN, G.

Madras University, Guindy Campus
Madras - 600 025, India

RÜCKL, Reinhold

CERN-EP Division
CH-1211 Geneva 23, Switzerland

SEGRÉ, Gino C.

University of Pennsylvania
Philadelphia, Pennsylvania 19174

SERDAROGLU, Meral

Bogazici University
P.K. 2, Bebek, Istanbul, Turkey

SIEBERT, Hans-Wolfgang

University of Heidelberg
D-69 Heidelberg, Germany; and
CERN EP Division
CH-1211 Geneva 23, Switzerland

SOERGEL, Volker

DESY
Notkestrasse 85
D-2000 Hamburg 52, Germany

SLIWA, Kryzsztof J.

Fermilab
P.O. Box 500 - MS 122
Batavia, Illinois 60510

STECH, Berthold W.

University of Heidelberg
D-69 Heidelberg, Germany

TADIC, Dubrovko

University of Zagreb
Zagreb, Coratia, Yugoslavia

TERAZAWA, Hidezumi

INS, University of Tokyo
Midori-cho, Tanashi
Tokyo 188, Japan

TRAMPETIC, Josip

Rudjer Boskovic Institute.
Bijenicka 54, P.O. Box 1016
41001 Zagreb, Yugoslavia

TRILLING, George H.   Lawrence Berkeley Laboratory
Berkeley, California, 94720

WALI, Kameshwar C.   Syracuse University
Syracuse, New York 13210

WEBB, James N.   University of Manchester
Schuster Laboratory
Manchester M13 9PL, United Kingdom

WOLFENSTEIN, Lincoln   Carnegie-Mellon Univ.
Pittsburgh, Pennsylvania 15213 and

Yen, Edward   Tsing Hua University
Taipei, China

AUTHOR INDEX

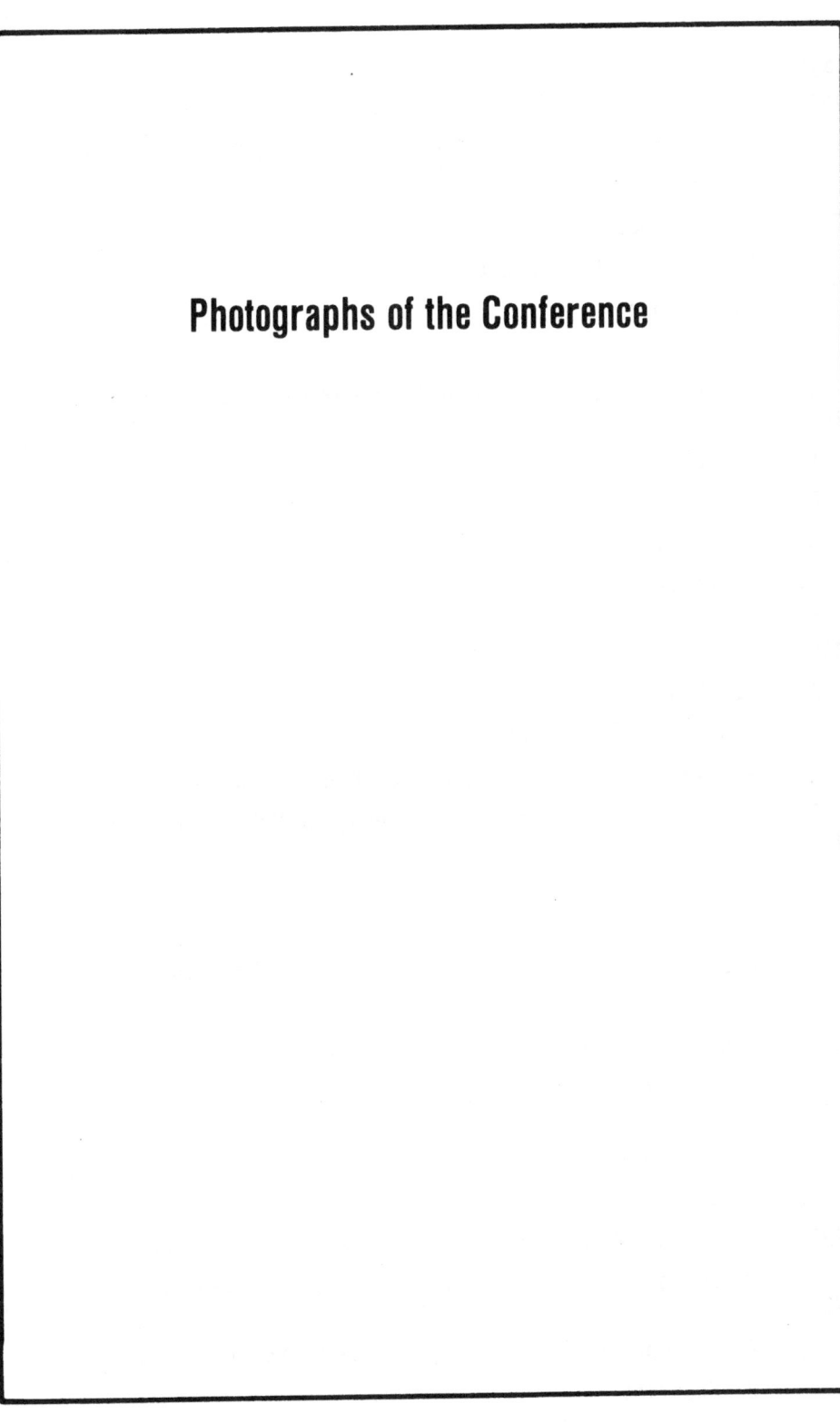

Photographs of the Conference

page

786 Scenery at Erice - (courtesy of Penny and Neil Baggett).

787 Entrance of the Ettore Majorana Centre for Scientific
Culture.

788 In the lecture room:

A. L.-L. Chau, N. Cabibbo

B. N. Cabibbo

C. M. Kobayashi

D. M. Goldhaber, L.-L. Chau, P. Franzini, K.C. Chou
789

A. G. Bellini

B. P. Pavlopoulos

C. G.H. Trilling

D. H. Fritzsch

E. K.C. Chou
790

A. K. Niu

B. H.W. Siebert, M. Pohl, K.C. Wali

C. F. Olness, G. Rajasekaran, G. Kane, K. Sliwa, J. Hagelin

D. W.T. Ford, D. Hitlin, B.W. Stech, R. Mohapatra, Y.X. Pham
791

A. P.H. Hansen

B. J.N. Webb

C. Mrs. Pavlopoulos, P. Pavlopoulos, J.S. Hagelin,
D.V. Nanopoulos

D L.S. Littenberg

E. W.M. Morse
792

A. P. Le Comte

B. A. Barbaro-Galtieri

793 A. Barbaro-Galtieri, J. Lee-Franzini, M.E. Machacek,
M. Serdaroglu, R. Cester, L.-L. Chau, S. Gentile
794

A. P. Pavlopoulos, W.T. Ford, P. Le Comte, H. Terazawa,
P. Franzini

B. Court yard

C. D.V. Nanopoulos

795

A. V. Soergel, B.W. Stech

B. R.N. Mohapatra, V. Soergel, L.-L. Chau, I. Harrity

796 At dinner:

A. K.C. Chou, E. Yen, L.L. Chau

B. S. Ferrara, R.N. Mohapatra

C. L. Wolfenstein, H.W. Siebert, K.C. Chou, H. Terazawa,
    K. Niu, P. Le Comte, P. Franzini, J. Lee-Franzini,
    N. Cabibbo, M.A.B. Bég

797 At the banquet:

A. V. Soergel, K. Kleinknecht, F. Olness, M. Serdaroglu,
    W.T. Ford

B. M. Pohl, R. Rückl

C. D. Tadic, J. Trampetic

798

A. J. Lee-Franzini, R. Cester, K. Kleinknecht, F.J. Olness,
    P. Franzini

B. M. Goldhaber - in the Marsala Room

C. E. Franco

799

A. K.J. Sliwa, H. Terazawa

B. J. Trampetic, R.N. Mohapatra

C. B. Heckel, F.J. Botella

D. F.H. Boehm, G. Poulard

800 The Marsala Folk Dancing Group

801 Dr. A. Gabriele, Signora Pinola

802 At Selinunte:

A. G.H. Trilling, A. Barbaro-Galtieri

B. T.H. Ho, E. Yen

C. M.A.B. Bég

803 Selinunte

788

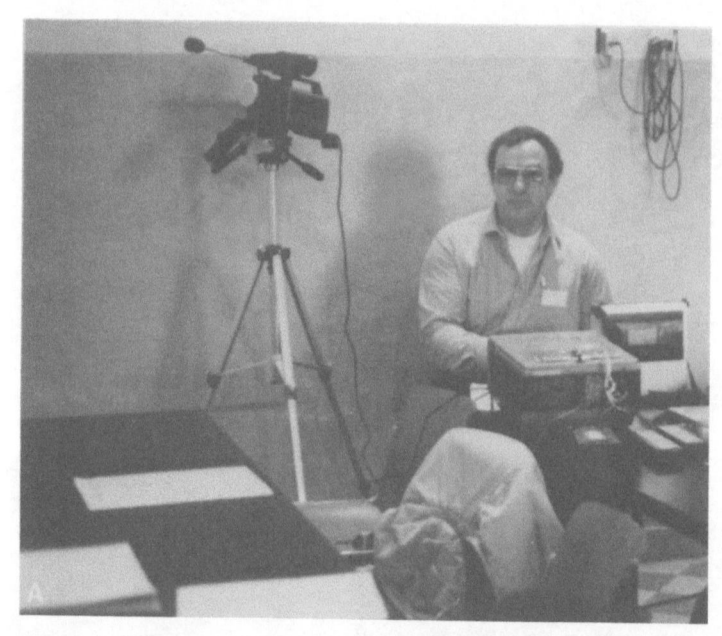

803